EDITION 12

Beginning Algebra

EDITION
12

Beginning Algebra

MARGARET L. LIAL
American River College

JOHN HORNSBY
University of New Orleans

TERRY MCGINNIS

PEARSON

Boston   Columbus   Indianapolis   New York   San Francisco   Hoboken   Amsterdam
Cape Town   Dubai   London   Madrid   Milan   Munich   Paris   Montreal   Toronto   Delhi
Mexico City   São Paulo   Sydney   Hong Kong   Seoul   Singapore   Taipei   Tokyo

Editorial Director: Chris Hoag
Editor in Chief: Michael Hirsch
Editorial Assistant: Chase Hammond
Program Manager: Beth Kaufman
Project Manager: Christine Whitlock
Program Management Team Lead: Karen Wernholm
Project Management Team Lead: Marianne Groth
Media Producer: Shana Siegmund
TestGen Sr. Content Developer: John R. Flanagan
MathXL Executive Content Manager: Rebecca Williams
Marketing Manager: Rachel Ross
Marketing Assistant: Kelly Cross
Senior Author Support/Technology Specialist: Joe Vetere
Rights and Permissions Project Manager: Diahanne Lucas Dowridge
Procurement Specialist: Carol Melville
Associate Director of Design: Andrea Nix
Program Design Lead: Beth Paquin
Text and Cover Design, Production Coordination, Composition, and Illustrations: Cenveo Publisher Services
Cover Image: Boyan Dimitrov/Shutterstock

Library of Congress Cataloging-in-Publication Data
Lial, Margaret L.
 Beginning algebra.—12th edition/Margaret Lial, American River College, John Hornsby, University of New Orleans, Terry McGinnis.
  pages cm
 ISBN 978-0-321-96933-0
1. Algebra. I. Hornsby, John, 1949- II. McGinnis, Terry. III. Title.
 QA152.3.L5 2016
 512.9—dc23

 2013049386

2 3 4 5 6 7 8 9 10—CRK—18 17 16 15

ISBN 13: 978-0-321-96933-0
ISBN 10: 0-321-96933-2

www.pearsonhighered.com

To Margaret L. Lial

On March 16, 2012, the mathematics education community lost one of its most influential members with the passing of our beloved mentor, colleague, and friend Marge Lial. On that day, Marge lost her long battle with ALS. Throughout her illness, Marge showed the remarkable strength and courage that characterized her entire life.

Margaret L. Lial

We would like to share a few comments from among the many messages we received from friends, colleagues, and others whose lives were touched by our beloved Marge:

"What a lady"

"A remarkable person"

"Gracious to everyone"

"One of a kind"

"Truly someone special"

"A loss in the mathematical world"

"A great friend"

"Sorely missed but so fondly remembered"

"Even though our crossed path was narrow, she made an impact and I will never forget her."

"There is talent and there is Greatness. Marge was truly Great."

"Her true impact is almost more than we can imagine."

In the world of college mathematics publishing, Marge Lial was a rock star. People flocked to her, and she had a way of making everyone feel like they truly mattered. And to Marge, they did. She and Chuck Miller began writing for Scott Foresman in 1970. As her illness progressed, she told us that she could no longer continue because "just getting from point A to point B" had become too challenging. That's our Marge—she even gave a geometric interpretation to her illness.

It has truly been an honor and a privilege to work with Marge Lial these past twenty years. While we no longer have her wit, charm, and loving presence to guide us, so much of who we are as mathematics educators has been shaped by her influence. We will continue doing our part to make sure that the Lial name represents excellence in mathematics education. And we remember daily so many of the little ways she impacted us, including her special expressions, "Margisms" as we like to call them. She often ended emails with one of them—the single word "Onward."

We conclude with a poem penned by another of Marge's coauthors, Callie Daniels.

Your courage inspires me
Your strength…impressive
Your wit humors me
Your vision…progressive

Your determination motivates me
Your accomplishments pave my way
Your vision sketches images for me
Your influence will forever stay.

Thank you, dearest Marge.
Knowing you and working with you has been a divine gift.

John Hornsby
Terry McGinnis

CONTENTS

3 Linear Equations and Inequalities in Two Variables; Functions 199

4 Systems of Linear Equations and Inequalities 277

5 Exponents and Polynomials 325

9   Quadratic Equations   591

Additional topics available in MyMathLab®: Using Rational Numbers as Exponents; Complex Numbers; Sets; and Introduction to Calculators

PREFACE

It is with great pleasure that we offer the twelfth edition of *Beginning Algebra*. We have remained true to the original goal that has guided us over the years—to provide the best possible text and supplements package to help students succeed and instructors teach. This edition faithfully continues that process through enhanced explanations of concepts, new and updated examples and exercises, student-oriented features like Vocabulary Lists, Pointers, Cautions, Problem-Solving Hints, and Now Try Exercises, as well as an extensive package of helpful supplements and study aids.

This text is part of a series that includes the following books, all by Lial, Hornsby, and McGinnis:

- *Intermediate Algebra*, Twelfth Edition,
- *Beginning and Intermediate Algebra*, Sixth Edition,
- *Algebra for College Students*, Eighth Edition.

WHAT'S NEW IN THIS EDITION?

We are pleased to offer the following new features:

VOCABULARY LISTS New vocabulary is now given at the beginning of appropriate sections. The list format allows students to preview vocabulary that is introduced in the section and also to review and check-off words they are able to correctly define upon completing a section.

CONCEPT CHECK EXERCISES Each exercise set begins with a set of Concept Check problems that facilitate students' mathematical thinking and conceptual understanding. Problem types include multiple-choice, true/false, matching, completion, and *What Went Wrong?* exercises. Many emphasize new vocabulary.

EXTENDING SKILLS EXERCISES These problems, scattered throughout selected exercise sets, expand on section objectives. Some are challenging in nature.

MIXED REVIEW EXERCISES Each chapter review has been expanded to include a dedicated set of Mixed Review Exercises to help students further synthesize concepts.

MARGIN ANSWERS TO REVIEW COMPONENTS To provide immediate reference and enable students to get the most out of review opportunities, answers are included in the margins for Summary Exercises, Chapter Review Exercises, Mixed Review Exercises, Chapter Tests, and Cumulative Review Exercises.

SPECIFIC CONTENT CHANGES include the following:

- **New Chapter R** provides a thorough review of fractions (Section R.1) as well as decimals and percents (all new Section R.2).

- **Application Sections 2.4, 2.7, 4.4, and 6.6** include new and/or updated problem-solving examples, exercises, and hints.

- **Section 2.5** now covers solving a linear equation in two variables x and y for y as preparation for Chapter 3 on forms of linear equations.

- **The presentation of linear equations in two variables in Chapter 3** includes three new examples of graphing and writing equations of lines, along with several groups of new exercises that make connections between tables, equations, and graphs.

- **Slope-intercept form and point-slope form** are now covered in separate Sections 3.4 and 3.5.

- **Expanded Summary Exercises** in Chapter 2 continue our emphasis on the difference between simplifying expressions and solving equations. A new example in the Chapter 6 Summary Exercises illustrates applying general factoring strategies.

- **Presentations of the following topics have been enhanced and expanded:**

  Applying operations on real numbers (Sections 1.4 and 1.5)
  Solving linear equations in one variable (Section 2.1)
  Solving linear inequalities in one variable with fractional coefficients (Section 2.8)
  Understanding polynomial vocabulary and adding and subtracting polynomials (Section 5.4)
  Factoring trinomials (Section 6.3)
  Factoring perfect square trinomials (Section 6.4)
  Solving equations with rational expressions (Section 7.6)
  Completing the square (Section 9.2)

ACKNOWLEDGMENTS

The comments, criticisms, and suggestions of users, nonusers, instructors, and students have positively shaped this textbook over the years, and we are most grateful for the many responses we have received. The feedback gathered for this edition was particularly helpful.

We especially wish to thank the following individuals who provided invaluable suggestions.

Barbara Aaker, *Community College of Denver*
Kim Bennekin, *Georgia Perimeter College*
Dixie Blackinton, *Weber State University*
Callie Daniels, *St. Charles Community College*
Cheryl Davids, *Central Carolina Technical College*
Robert Diaz, *Fullerton College*
Chris Diorietes, *Fayetteville Technical Community College*
Sylvia Dreyfus, *Meridian Community College*
Sabine Eggleston, *Edison State College*
LaTonya Ellis, *Bishop State Community College*
Beverly Hall, *Fayetteville Technical Community College*
Sandee House, *Georgia Perimeter College*
Joe Howe, *St. Charles Community College*
Lynette King, *Gadsden State Community College*
Linda Kodama, *Windward Community College*
Carlea McAvoy, *South Puget Sound Community College*
James Metz, *Kapi'olani Community College*
Jean Millen, *Georgia Perimeter College*
Molly Misko, *Gadsden State Community College*
Jane Roads, *Moberly Area Community College*
Melanie Smith, *Bishop State Community College*
Erik Stubsten, *Chattanooga State Technical Community College*
Tong Wagner, *Greenville Technical College*
Rick Woodmansee, *Sacramento City College*
Sessia Wyche, *University of Texas at Brownsville*

Over the years, we have come to rely on an extensive team of experienced professionals. Our sincere thanks go to these dedicated individuals at Pearson Arts & Sciences, who worked long and hard to make this revision a success: Chris Hoag, Maureen O'Connor, Michael Hirsch, Rachel Ross, Beth Kaufman, Christine Whitlock, Chase Hammond, Kelly Cross, and Shana Siegmund.

Additionally, Rachel Haskell did a great job helping us update real data applications. We are also grateful to Kathy Diamond and Marilyn Dwyer of Cenveo, Inc., for their excellent production work; Bonnie Boehme for supplying her copyediting expertise; Aptara for their photo research; and Lucie Haskins for producing another accurate, useful index. Callie Daniels, Perian Herring, Jack Hornsby, Paul Lorczak, and Sarah Sponholz did a thorough, timely job accuracy checking manuscript and page proofs and Lisa Collette checked the index.

We particularly thank the many students and instructors who have used this textbook over the years. You are the reason we do what we do. It is our hope that we have positively impacted your mathematics journey. We would welcome any comments or suggestions you might have via email to math@pearson.com.

John Hornsby
Terry McGinnis

STUDENT SUPPLEMENTS

STUDENT'S SOLUTIONS MANUAL

- Provides detailed solutions to the odd-numbered, section-level exercises and to all Now Try Exercises, Relating Concepts, Summary, Chapter Review, Mixed Review, Chapter Test, and Cumulative Review Exercises

ISBNs: 0-321-96981-2, 978-0-321-96981-1

LIAL VIDEO LIBRARY

The **Lial Video Library,** available in MyMathLab, provides students with a wealth of video resources to help them navigate the road to success. All video resources in the library include optional subtitles in English. The **Lial Video Library** includes the following resources:

- **Section Lecture Videos** offer a new navigation menu that allows students to easily focus on the key examples and exercises that they need to review in each section. Optional Spanish subtitles are available.

- **Solutions Clips** show an instructor working through the complete solutions to selected exercises from the text. Exercises with a solution clip are marked in the text and e-book with a Play Button icon ▶.

- **Quick Review Lectures** provide a short summary lecture of each key concept from the Quick Reviews at the end of every chapter in the text.

- **The Chapter Test Prep Videos** provide step-by-step solutions to all exercises from the Chapter Tests. These videos provide guidance and support when students need it the most: the night before an exam. The Chapter Test Prep Videos are also available on YouTube (searchable using author name and book title).

MYWORKBOOK

- Provides Guided Examples and corresponding Now Try Exercises for each text objective

- Refers students to correlated Examples, Lecture Videos, and Exercise Solution Clips

- Includes extra practice exercises for every section of the text with ample space for students to show their work

- Lists the learning objectives and key vocabulary terms for every text section, along with vocabulary practice problems

ISBNs: 0-321-96979-0, 978-0-321-96979-8

NEW MYSLIDENOTES

- Provides a note-taking tool based on the Lecture Slides that accompany the text
- Includes extra examples so that students can apply the concepts and procedures on the slides
- Features guided solutions that break problems into small, manageable steps
- Provides extra vocabulary practice to ensure that students have a firm grasp of new key terms
- Available electronically within MyMathLab

ISBNs: 0-13-393289-3, 978-0-13-393289-8

INSTRUCTOR SUPPLEMENTS

ANNOTATED INSTRUCTOR'S EDITION

- Provides "on-page" answers to all text exercises in an easy-to-read margin format, along with helpful Teaching Tips and extensive Classroom Examples

ISBNs: 0-321-96945-6, 978-0-321-96945-3

INSTRUCTOR'S SOLUTIONS MANUAL (Download only)

- Provides complete answers to all text exercises, including all Classroom Examples and Now Try Exercises

ISBNs: 0-321-96977-4, 978-0-321-96977-4

INSTRUCTOR'S RESOURCE MANUAL WITH TESTS
(Download only)

- Contains two diagnostic pretests, four free-response and two multiple-choice test forms per chapter, and two final exams
- Includes a mini-lecture for each section of the text with objectives, key examples, and teaching tips
- Provides a correlation guide from the eleventh to the twelfth edition

ISBNs: 0-321-96980-4, 978-0-321-96980-4

AVAILABLE FOR STUDENTS AND INSTRUCTORS

MYMATHLAB® ONLINE COURSE (access code required)

MyMathLab from Pearson is the world's leading online resource in mathematics, integrating interactive homework, assessment, and media in a flexible, easy-to-use format. It provides **engaging experiences** that personalize, stimulate, and measure learning for each student. And, it comes from an **experienced partner** with educational expertise and an eye on the future.

To learn more about how MyMathLab combines proven learning applications with powerful assessment, visit www.mymathlab.com or contact your Pearson representative.

MYMATHLAB® READY TO GO COURSE (access code required)

These new "Ready to Go" courses provide students with all the same great MyMathLab features, but make it easier for instructors to get started. Each course includes pre-assigned homework and quizzes to make creating a course even simpler. Ask your Pearson representative about the details for this particular course or to see a copy of this course.

MATHXL® ONLINE COURSE (access code required)

MathXL® is the homework and assessment engine that runs MyMathLab. (MyMathLab is MathXL plus a learning management system.)

With MathXL, instructors can:

- Create, edit, and assign online homework and tests using algorithmically generated exercises correlated at the objective level to the textbook.
- Create and assign their own online exercises and import TestGen tests for added flexibility.
- Maintain records of all student work tracked in MathXL's online gradebook.

With MathXL, students can:

- Take chapter tests in MathXL and receive personalized study plans and/or personalized homework assignments based on their test results.
- Use the study plan and/or the homework to link directly to tutorial exercises for the objectives they need to study.
- Access supplemental animations and video clips directly from selected exercises.

MathXL is available to qualified adopters. For more information, visit our website at www.mathxl.com, or contact your Pearson representative.

TESTGEN®

TestGen® (www.pearsoned.com/testgen) enables instructors to build, edit, print, and administer tests using a computerized bank of questions developed to cover all the objectives of the text. TestGen is algorithmically based, allowing instructors to create multiple but equivalent versions of the same question or test with the click of a button. Instructors can also modify test bank questions or add new questions. The software and testbank are available for download from Pearson Education's online catalog.

POWERPOINT® LECTURE SLIDES

- Present key concepts and definitions from the text
- Available for download at www.pearsonhighered.com

ISBNs: 0-321-96944-8, 978-0-321-96944-6

STUDY SKILLS

Using Your Math Textbook

Your textbook is a valuable resource. You will learn more if you fully make use of the features it offers.

General Features

Locate each general feature and complete any blanks.

- **Table of Contents** Find this at the front of the text. *Mark the chapters and sections you will cover, as noted on your course syllabus.*

- **Answer Section** *Tab this section at the back of the book* so you can refer to it frequently when doing homework. Answers to odd-numbered section exercises are provided.

- **Glossary** Find this feature after the answer section at the back of the text. It provides an alphabetical list of the key terms found in the text, with definitions and section references. *Using the glossary, an equation is a statement that _____.*

- **List of Formulas** Inside the back cover of the text is a helpful list of geometric formulas, along with review information on triangles and angles. Use these for reference throughout the course. *The formula for finding the volume of a cube is _____.*

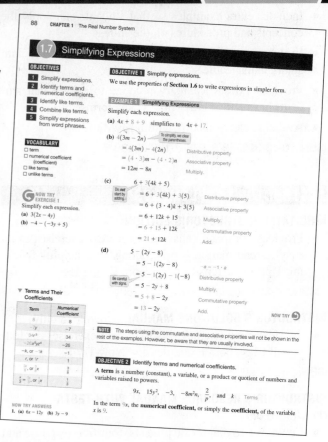

Specific Features

Look through Chapter 2 and give the number of a page that includes an example of each of the following specific features.

- **Objectives** The objectives are listed at the beginning of each section and again within the section as the corresponding material is presented. Once you finish a section, ask yourself if you have accomplished them. *See page _____.*

- **Vocabulary List** Important vocabulary is listed at the beginning of each section. You should be able to define these terms when you finish a section. *See page _____.*

- **Now Try Exercises** These margin exercises allow you to immediately practice the material covered in the examples and prepare you for the exercises. Check your results using the answers at the bottom of the page. *See page _____.*

- **Pointers** These small shaded balloons provide on-the-spot warnings and reminders, point out key steps, and give other helpful tips. *See page _____.*

- **Cautions** These provide warnings about common errors that students often make or trouble spots to avoid. *See page _____.*

- **Notes** These provide additional explanations or emphasize other important ideas. *See page _____.*

- **Problem-Solving Hints** These boxes give helpful tips or strategies to use when you work applications. *See page _____.*

Prealgebra Review

R.1 Fractions

R.2 Decimals and Percents

Study Skills *Reading Your Math Textbook*

R.1 Fractions

OBJECTIVES

1 Learn the definition of *factor*.

2 Write fractions in lowest terms.

3 Convert between improper fractions and mixed numbers.

4 Multiply and divide fractions.

5 Add and subtract fractions.

6 Solve applied problems that involve fractions.

7 Interpret data in a circle graph.

VOCABULARY

☐ natural (counting) numbers
☐ whole numbers
☐ fractions
☐ numerator
☐ denominator
☐ proper fraction
☐ improper fraction
☐ factors
☐ product
☐ prime number
☐ composite number
☐ lowest terms
☐ greatest common factor (GCF)
☐ mixed number
☐ reciprocals
☐ quotient
☐ dividend
☐ divisor
☐ sum
(continued)

In the study of elementary mathematics, the numbers used most often are the **natural (counting) numbers,**

$$1, 2, 3, 4, \ldots,$$

the **whole numbers,**

$$0, 1, 2, 3, 4, \ldots,$$

and **fractions,** such as

$$\frac{1}{2}, \quad \frac{2}{3}, \quad \text{and} \quad \frac{11}{12}.$$

The three dots, or *ellipsis points,* indicate that each list of numbers continues in the same way indefinitely.

The parts of a fraction are named as shown.

$$\text{Fraction bar} \rightarrow \frac{3}{8} \begin{array}{l} \leftarrow \text{Numerator} \\ \leftarrow \text{Denominator} \end{array}$$

NOTE Fractions are a way to represent parts of a whole. In a fraction, the **numerator** gives the number of parts being represented. The **denominator** gives the total number of equal parts in the whole. See **FIGURE 1.**

The shaded region represents $\frac{3}{8}$ of the circle.

FIGURE 1

A fraction is classified as being either a **proper fraction** or an **improper fraction.**

Proper fractions	$\dfrac{1}{5}, \dfrac{2}{7}, \dfrac{9}{10}, \dfrac{23}{25}$	Numerator is **less than** denominator. Value is less than 1.
Improper fractions	$\dfrac{3}{2}, \dfrac{5}{5}, \dfrac{11}{7}, \dfrac{28}{4}$	Numerator is **greater than or equal to** denominator. Value is greater than or equal to 1.

OBJECTIVE 1 Learn the definition of *factor*.

In the statement $3 \times 6 = 18$, the numbers 3 and 6 are **factors** of 18. Other factors of 18 include 1, 2, 9, and 18. The result of the multiplication, 18, is the **product.** We can represent the product of two numbers, such as 3 and 6, in several ways.

$$3 \times 6, \quad 3 \cdot 6, \quad (3)(6), \quad (3)6, \quad 3(6) \qquad \text{Products}$$

We *factor* a number by writing it as the product of two or more numbers.

Multiplication
$$3 \cdot 6 = 18$$
Factors   Product

Factoring
$$18 = 3 \cdot 6$$
Product   Factors

Factoring is the reverse of multiplying two numbers to get the product.

NOTE In algebra, a raised dot · is often used instead of the × symbol to indicate multiplication because × may be confused with the letter *x*.

A natural number greater than 1 is **prime** if it has only itself and 1 as factors. "Factors" are understood here to mean natural number factors.

$$2, 3, 5, 7, 11, 13, 17, 19, 23, 29, 31, 37 \qquad \text{First dozen prime numbers}$$

A natural number greater than 1 that is not prime is a **composite number.**

$$4, 6, 8, 9, 10, 12, 14, 15, 16, 18, 20, 21 \qquad \text{First dozen composite numbers}$$

By agreement, the number 1 is neither prime nor composite.

NOW TRY EXERCISE 1

Write 60 as a product of prime factors.

EXAMPLE 1 Factoring Numbers

Write each number as a product of prime factors.

(a) 35   Write 35 as the product of the prime factors 5 and 7.

$$35 = 5 \cdot 7$$

(b) 24   We show a factor tree on the right, with prime factors circled.

Divide by the least prime factor of 24, which is 2. $24 = 2 \cdot 12$

Divide 12 by 2 to find two factors of 12. $24 = 2 \cdot 2 \cdot 6$

Now factor 6 as $2 \cdot 3$. $24 = \underline{2 \cdot 2 \cdot 2 \cdot 3}$

All factors are prime.   **NOW TRY**

NOTE When factoring, we need not start with the least prime factor. No matter which prime factor we start with, we will *always* obtain the same prime factorization. We verify this in **Example 1(b)** by starting with 3 instead of 2.

Divide 24 by 3. $24 = 3 \cdot 8$

Divide 8 by 2. $24 = 3 \cdot 2 \cdot 4$

Divide 4 by 2. $24 = \underline{3 \cdot 2 \cdot 2 \cdot 2}$

The same prime factors result.

NOW TRY ANSWER
1. $2 \cdot 2 \cdot 3 \cdot 5$

OBJECTIVE 2 Write fractions in lowest terms.

The **basic principle of fractions** is used to write a fraction in *lowest terms*.

Basic Principle of Fractions

If the numerator and denominator of a fraction are multiplied or divided by the same nonzero number, the value of the fraction is not changed.

A fraction is in **lowest terms** when the numerator and denominator have no factors in common (other than 1).

Writing a Fraction in Lowest Terms

Step 1   Write the numerator and the denominator in factored form.

Step 2   Divide the numerator and the denominator by the **greatest common factor (GCF),** the product of all factors common to both.

**NOW TRY
EXERCISE 2**

Write each fraction in lowest terms.

(a) $\dfrac{30}{42}$ (b) $\dfrac{10}{70}$ (c) $\dfrac{72}{120}$

EXAMPLE 2   Writing Fractions in Lowest Terms

Write each fraction in lowest terms.

(a) $\dfrac{10}{15} = \dfrac{2 \cdot 5}{3 \cdot 5} = \dfrac{2 \cdot 1}{3 \cdot 1} = \dfrac{2}{3}$

The factored form shows that 5 is the greatest common factor of 10 and 15. Dividing both numerator and denominator by 5 gives $\dfrac{10}{15}$ in lowest terms as $\dfrac{2}{3}$.

(b) $\dfrac{15}{45}$

By inspection, the greatest common factor of 15 and 45 is 15.

$$\dfrac{15}{45} = \dfrac{15}{3 \cdot 15} = \dfrac{1}{3 \cdot 1} = \dfrac{1}{3}$$ Remember to write 1 in the numerator.

If the GCF is not obvious, factor the numerator and denominator into prime factors.

$$\dfrac{15}{45} = \dfrac{3 \cdot 5}{3 \cdot 3 \cdot 5} = \dfrac{1 \cdot 1}{3 \cdot 1 \cdot 1} = \dfrac{1}{3}$$ The same answer results.

(c) $\dfrac{150}{200} = \dfrac{3 \cdot 50}{4 \cdot 50} = \dfrac{3 \cdot 1}{4 \cdot 1} = \dfrac{3}{4}$ 50 is the greatest common factor of 150 and 200.

Another strategy is to choose *any* common factor and work in stages.

$$\dfrac{150}{200} = \dfrac{15 \cdot 10}{20 \cdot 10} = \dfrac{3 \cdot 5 \cdot 10}{4 \cdot 5 \cdot 10} = \dfrac{3 \cdot 1 \cdot 1}{4 \cdot 1 \cdot 1} = \dfrac{3}{4}$$ The same answer results.

NOW TRY

OBJECTIVE 3 Convert between improper fractions and mixed numbers.

A **mixed number** is a single number that represents the sum of a natural number and a proper fraction.

Mixed number $\longrightarrow 5\dfrac{3}{4} = 5 + \dfrac{3}{4}$

NOW TRY ANSWERS

2. (a) $\dfrac{5}{7}$ (b) $\dfrac{1}{7}$ (c) $\dfrac{3}{5}$

NOW TRY
EXERCISE 3
Write $\frac{92}{5}$ as a mixed number.

EXAMPLE 3   Converting an Improper Fraction to a Mixed Number

Write $\frac{59}{8}$ as a mixed number.

The fraction bar represents division. $\left(\frac{a}{b} \text{ means } a \div b.\right)$ Divide the numerator of the improper fraction by the denominator.

$$\text{Denominator of fraction} \rightarrow 8\overline{)59} \begin{array}{l} 7 \leftarrow \text{Quotient} \\ \leftarrow \text{Numerator of fraction} \\ \underline{56} \\ 3 \leftarrow \text{Remainder} \end{array} \qquad \frac{59}{8} = 7\frac{3}{8}$$

NOW TRY

NOW TRY
EXERCISE 4
Write $11\frac{2}{3}$ as an improper fraction.

EXAMPLE 4   Converting a Mixed Number to an Improper Fraction

Write $6\frac{4}{7}$ as an improper fraction.

Multiply the denominator of the fraction by the natural number and then add the numerator to obtain the numerator of the improper fraction.

$$7 \cdot 6 = 42 \quad \text{and} \quad 42 + 4 = 46$$

The denominator of the improper fraction is the same as the denominator in the mixed number, which is 7 here. Thus, $6\frac{4}{7} = \frac{46}{7}$.

NOW TRY

Multiplying Fractions

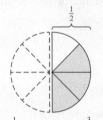

$\frac{3}{4}$ of $\frac{1}{2}$ is equivalent to $\frac{3}{4} \cdot \frac{1}{2}$, which equals $\frac{3}{8}$ of the circle.
FIGURE 2

OBJECTIVE 4   Multiply and divide fractions.

FIGURE 2 illustrates multiplying fractions.

Multiplying Fractions

If $\frac{a}{b}$ and $\frac{c}{d}$ are fractions, then $\quad \dfrac{a}{b} \cdot \dfrac{c}{d} = \dfrac{a \cdot c}{b \cdot d}.$

That is, to multiply two fractions, multiply their numerators and then multiply their denominators.

EXAMPLE 5   Multiplying Fractions

Find each product, and write it in lowest terms.

(a) $\dfrac{3}{8} \cdot \dfrac{4}{9}$

$= \dfrac{3 \cdot 4}{8 \cdot 9}$ Multiply numerators.
Multiply denominators.

$= \dfrac{3 \cdot 4}{2 \cdot 4 \cdot 3 \cdot 3}$ Factor the denominator.

$= \dfrac{1}{2 \cdot 3}$ Divide numerator and denominator by 3 and 4, or by 12.

Remember to write 1 in the numerator.

$= \dfrac{1}{6}$ Make sure the product is in lowest terms.

NOW TRY ANSWERS
3. $18\frac{2}{5}$
4. $\frac{35}{3}$

NOW TRY EXERCISE 5

Find each product, and write it in lowest terms.

(a) $\dfrac{4}{7} \cdot \dfrac{5}{8}$ **(b)** $3\dfrac{2}{5} \cdot 6\dfrac{2}{3}$

(b)

$$2\dfrac{1}{3} \cdot 5\dfrac{1}{4}$$

$$= \dfrac{7}{3} \cdot \dfrac{21}{4} \qquad \text{Write each mixed number as an improper fraction.}$$

$$= \dfrac{7 \cdot 21}{3 \cdot 4} \qquad \begin{array}{l}\text{Multiply numerators.}\\ \text{Multiply denominators.}\end{array}$$

$$= \dfrac{7 \cdot 3 \cdot 7}{3 \cdot 4} \qquad \text{Factor the numerator.}$$

Think: $\frac{49}{4}$ means $49 \div 4$. $= \dfrac{49}{4}, \quad \text{or} \quad 12\dfrac{1}{4} \qquad \begin{array}{l}\text{Write in lowest terms}\\ \text{and as a mixed number.}\end{array}$ **NOW TRY**

NOTE Some students prefer to factor and divide out any common factors *before* multiplying.

$$\dfrac{3}{8} \cdot \dfrac{4}{9} \qquad \text{Example 5(a)}$$

$$= \dfrac{3}{2 \cdot 4} \cdot \dfrac{4}{3 \cdot 3} \qquad \text{Factor.}$$

$$= \dfrac{1}{2 \cdot 3} \qquad \text{Divide out common factors. Multiply.}$$

$$= \dfrac{1}{6} \qquad \text{The same answer results.}$$

▼ Reciprocals

Number	Reciprocal
$\dfrac{3}{4}$	$\dfrac{4}{3}$
$\dfrac{11}{7}$	$\dfrac{7}{11}$
$\dfrac{1}{5}$	5, or $\dfrac{5}{1}$
10, or $\dfrac{10}{1}$	$\dfrac{1}{10}$

A number and its reciprocal have a product of 1. For example,

$$\dfrac{3}{4} \cdot \dfrac{4}{3} = \dfrac{12}{12}, \text{ or } 1.$$

Two numbers are **reciprocals** of each other if their product is 1. Because division is the inverse or opposite of multiplication, we use reciprocals to divide fractions. **FIGURE 3** illustrates dividing fractions.

Dividing Fractions

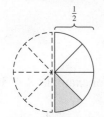

$\dfrac{1}{2} \div 4$ is equivalent to $\dfrac{1}{2} \cdot \dfrac{1}{4}$, which equals $\dfrac{1}{8}$ of the circle.

FIGURE 3

Dividing Fractions

If $\dfrac{a}{b}$ and $\dfrac{c}{d}$ are fractions, then $\dfrac{a}{b} \div \dfrac{c}{d} = \dfrac{a}{b} \cdot \dfrac{d}{c}.$

Multiply by the reciprocal.

That is, to divide by a fraction, multiply by its reciprocal.

As an example of why this procedure works, we know that

$$20 \div 10 = 2 \quad \text{and also that} \quad 20 \cdot \dfrac{1}{10} = 2.$$

The answer to a division problem is a **quotient.** In $\dfrac{a}{b} \div \dfrac{c}{d}$, the first fraction $\dfrac{a}{b}$ is the **dividend,** and the second fraction $\dfrac{c}{d}$ is the **divisor.**

NOW TRY ANSWERS

5. (a) $\dfrac{5}{14}$ **(b)** $\dfrac{68}{3}$, or $22\dfrac{2}{3}$

NOW TRY
EXERCISE 6

Find each quotient, and write it in lowest terms.

(a) $\dfrac{2}{7} \div \dfrac{8}{9}$ **(b)** $3\dfrac{3}{4} \div 4\dfrac{2}{7}$

NOW TRY ANSWERS
6. **(a)** $\dfrac{9}{28}$ **(b)** $\dfrac{7}{8}$

EXAMPLE 6 Dividing Fractions

Find each quotient, and write it in lowest terms.

(a) $\dfrac{3}{4} \div \dfrac{8}{5}$

$= \dfrac{3}{4} \cdot \dfrac{5}{8}$ Multiply by the reciprocal of the divisor.

$= \dfrac{3 \cdot 5}{4 \cdot 8}$ Multiply numerators.
            Multiply denominators.

$= \dfrac{15}{32}$ ◁ Make sure the quotient is in lowest terms.

(b) $\dfrac{3}{4} \div \dfrac{5}{8}$

$= \dfrac{3}{4} \cdot \dfrac{8}{5}$ Multiply by the reciprocal.

$= \dfrac{3 \cdot 4 \cdot 2}{4 \cdot 5}$ Multiply and factor.

$= \dfrac{6}{5}$, or $1\dfrac{1}{5}$

(c) $\dfrac{5}{8} \div 10$ ◁ Think of 10 as $\dfrac{10}{1}$ here.

$= \dfrac{5}{8} \cdot \dfrac{1}{10}$ Multiply by the reciprocal.

$= \dfrac{5 \cdot 1}{8 \cdot 2 \cdot 5}$ Multiply and factor.

$= \dfrac{1}{16}$ ◁ Remember to write 1 in the numerator.

(d) $1\dfrac{2}{3} \div 4\dfrac{1}{2}$

$= \dfrac{5}{3} \div \dfrac{9}{2}$ Write each mixed number as an improper fraction.

$= \dfrac{5}{3} \cdot \dfrac{2}{9}$ Multiply by the reciprocal of the divisor.

$= \dfrac{5 \cdot 2}{3 \cdot 9}$ Multiply numerators.
            Multiply denominators.

$= \dfrac{10}{27}$ The quotient is in lowest terms.   **NOW TRY**

Adding Fractions

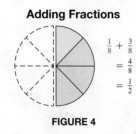

$\dfrac{1}{8} + \dfrac{3}{8}$
$= \dfrac{4}{8}$
$= \dfrac{1}{2}$

FIGURE 4

OBJECTIVE 5 Add and subtract fractions.

The result of adding two numbers is the **sum** of the numbers. For example, $2 + 3 = 5$, so 5 is the sum of 2 and 3.

FIGURE 4 illustrates adding fractions.

Adding Fractions

If $\dfrac{a}{b}$ and $\dfrac{c}{b}$ are fractions, then $\dfrac{a}{b} + \dfrac{c}{b} = \dfrac{a + c}{b}$.

That is, to find the sum of two fractions having the *same* denominator, add the numerators and *keep the same denominator.*

NOW TRY
EXERCISE 7

Find the sum, and write it in lowest terms.

$$\frac{1}{8} + \frac{3}{8}$$

EXAMPLE 7   Adding Fractions (Same Denominator)

Find each sum, and write it in lowest terms.

(a) $\dfrac{3}{7} + \dfrac{2}{7}$

$= \dfrac{3+2}{7}$ Add numerators. Keep the same denominator.

$= \dfrac{5}{7}$

(b) $\dfrac{2}{10} + \dfrac{3}{10}$

$= \dfrac{2+3}{10}$ Add numerators. Keep the same denominator.

$= \dfrac{5}{10}$

$= \dfrac{1}{2}$ Write in lowest terms.

NOW TRY

If the fractions to be added do *not* have the same denominator, we must first rewrite them with a common denominator. For example, to rewrite $\frac{3}{4}$ as an equivalent fraction with denominator 12, think as follows.

$$\frac{3}{4} = \frac{?}{12}$$

We must find the number that can be multiplied by 4 to give 12. Because $4 \cdot 3 = 12$, we multiply the numerator and the denominator by 3.

$$\frac{3}{4} = \frac{3 \cdot 3}{4 \cdot 3} = \frac{9}{12}$$ $\frac{3}{4}$ is equivalent to $\frac{9}{12}$. See **FIGURE 5.**

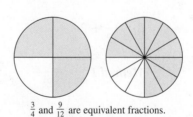

$\frac{3}{4}$ and $\frac{9}{12}$ are equivalent fractions.

FIGURE 5

NOTE   The process of writing an equivalent fraction is the reverse of writing a fraction in lowest terms.

Finding the Least Common Denominator (LCD)

To add or subtract fractions with different denominators, find the **least common denominator (LCD)** as follows.

Step 1   Factor each denominator using prime factors.

Step 2   The LCD is the product of every (different) factor that appears in any of the factored denominators. If a factor is repeated, use the greatest number of repeats as factors of the LCD.

Step 3   Write each fraction with the LCD as the denominator.

NOW TRY
EXERCISE 8

Find each sum, and write it in lowest terms.

(a) $\dfrac{5}{12} + \dfrac{3}{8}$ **(b)** $3\dfrac{1}{4} + 5\dfrac{5}{8}$

EXAMPLE 8   Adding Fractions (Different Denominators)

Find each sum, and write it in lowest terms.

(a) $\dfrac{4}{15} + \dfrac{5}{9}$

Step 1   To find the LCD, factor each denominator using prime factors.

$$15 = 5 \cdot 3 \quad \text{and} \quad 9 = 3 \cdot 3$$

3 is a factor of both denominators.

Step 2 $\text{LCD} = 5 \cdot 3 \cdot 3 = 45$

In this example, the LCD needs one factor of 5 and two factors of 3 because the second denominator has two factors of 3.

Step 3   Write each fraction with 45 as denominator.

$$\frac{4}{15} = \frac{4 \cdot 3}{15 \cdot 3} = \frac{12}{45} \quad \text{and} \quad \frac{5}{9} = \frac{5 \cdot 5}{9 \cdot 5} = \frac{25}{45}$$

At this stage, the fractions are *not* in lowest terms.

$$\frac{4}{15} + \frac{5}{9}$$

$$= \frac{12}{45} + \frac{25}{45} \qquad \text{Use the equivalent fractions with the common denominator.}$$

Make sure the sum is in lowest terms.

$$= \frac{37}{45} \qquad \begin{array}{l}\text{Add numerators.}\\ \text{Keep the same denominator.}\end{array}$$

(b) $3\dfrac{1}{2} + 2\dfrac{3}{4}$

Method 1 $3\dfrac{1}{2} + 2\dfrac{3}{4}$

$$= \frac{7}{2} + \frac{11}{4} \qquad \text{Write each mixed number as an improper fraction.}$$

Think: $\frac{7 \cdot 2}{2 \cdot 2} = \frac{14}{4}$

$$= \frac{14}{4} + \frac{11}{4} \qquad \text{Find a common denominator. The LCD is 4.}$$

$$= \frac{25}{4}, \quad \text{or} \quad 6\frac{1}{4} \qquad \text{Add. Write as a mixed number.}$$

Method 2 $\left.\begin{array}{r} 3\dfrac{1}{2} = 3\dfrac{2}{4} \\[2mm] + 2\dfrac{3}{4} = 2\dfrac{3}{4} \end{array}\right\}$ Write $3\frac{1}{2}$ as $3\frac{2}{4}$. Then add vertically. Add the whole numbers and the fractions separately.

$$5\frac{5}{4} = 5 + 1\frac{1}{4} = 6\frac{1}{4}, \quad \text{or} \quad \frac{25}{4} \qquad \text{The same answer results.}$$

NOW TRY

NOW TRY ANSWERS
8. (a) $\frac{19}{24}$ **(b)** $\frac{71}{8}$, or $8\frac{7}{8}$

The result of subtracting one number from another number is the **difference** of the numbers. For example, $9 - 5 = 4$, so 4 is the difference of 9 and 5.

FIGURE 6 illustrates subtracting fractions.

Subtracting Fractions

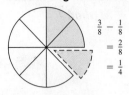

$$\frac{3}{8} - \frac{1}{8}$$
$$= \frac{2}{8}$$
$$= \frac{1}{4}$$

FIGURE 6

Subtracting Fractions

If $\dfrac{a}{b}$ and $\dfrac{c}{b}$ are fractions, then $\qquad \dfrac{a}{b} - \dfrac{c}{b} = \dfrac{a-c}{b}.$

That is, to find the difference of two fractions having the *same* denominator, subtract the numerators and ***keep the same denominator.***

EXAMPLE 9 Subtracting Fractions

Find each difference, and write it in lowest terms.

(a) $\dfrac{15}{8} - \dfrac{3}{8}$

$= \dfrac{15-3}{8}$ Subtract numerators.
Keep the same denominator.

$= \dfrac{12}{8}$

$= \dfrac{3}{2},$ or $1\dfrac{1}{2}$ Write in lowest terms and as a mixed number.

(b) $\dfrac{7}{18} - \dfrac{4}{15}$

$= \dfrac{7 \cdot 5}{2 \cdot 3 \cdot 3 \cdot 5} - \dfrac{4 \cdot 2 \cdot 3}{2 \cdot 3 \cdot 3 \cdot 5}$ $18 = 2 \cdot 3 \cdot 3$ and $15 = 3 \cdot 5$, so the LCD is $2 \cdot 3 \cdot 3 \cdot 5 = 90$.

$= \dfrac{35}{90} - \dfrac{24}{90}$ Write equivalent fractions.

$= \dfrac{11}{90}$ Subtract. The answer is in lowest terms.

(c) $\dfrac{15}{32} - \dfrac{11}{45}$

    Because $32 = 2 \cdot 2 \cdot 2 \cdot 2 \cdot 2$ and $45 = 3 \cdot 3 \cdot 5$, there are no common factors except 1. The LCD is $32 \cdot 45 = 1440$.

$= \dfrac{15 \cdot 45}{32 \cdot 45} - \dfrac{11 \cdot 32}{45 \cdot 32}$

$= \dfrac{675}{1440} - \dfrac{352}{1440}$ Write equivalent fractions.

$= \dfrac{323}{1440}$ Subtract numerators.
Keep the common denominator.

NOW TRY
EXERCISE 9
Find each difference, and write it in lowest terms.

(a) $\dfrac{5}{11} - \dfrac{2}{9}$ **(b)** $4\dfrac{1}{3} - 2\dfrac{5}{6}$

(d) $4\dfrac{1}{2} - 1\dfrac{3}{4}$

Method 1 $4\dfrac{1}{2} - 1\dfrac{3}{4}$

$= \dfrac{9}{2} - \dfrac{7}{4}$ Write each mixed number as an improper fraction.

$= \dfrac{18}{4} - \dfrac{7}{4}$ Find a common denominator. The LCD is 4.

Think: $\dfrac{9 \cdot 2}{2 \cdot 2} = \dfrac{18}{4}$

$= \dfrac{11}{4}$,   or $2\dfrac{3}{4}$ Subtract. Write as a mixed number.

Method 2 $4\dfrac{1}{2} = 4\dfrac{2}{4} = 3\dfrac{6}{4}$ The LCD is 4.
$4\dfrac{2}{4} = 3 + 1 + \dfrac{2}{4} = 3 + \dfrac{4}{4} + \dfrac{2}{4} = 3\dfrac{6}{4}$

$-1\dfrac{3}{4} = 1\dfrac{3}{4} = 1\dfrac{3}{4}$
_____

$2\dfrac{3}{4}$,   or $\dfrac{11}{4}$ The same answer results.

NOW TRY

OBJECTIVE 6 Solve applied problems that involve fractions.

EXAMPLE 10 Adding Fractions to Solve an Applied Problem

NOW TRY
EXERCISE 10
A board is $10\dfrac{1}{2}$ ft long. If it must be divided into four pieces of equal length for shelves, how long must each piece be?

The diagram in **FIGURE 7** appears in the book *Woodworker's 39 Sure-Fire Projects*. Find the height of the desk to the top of the writing surface.

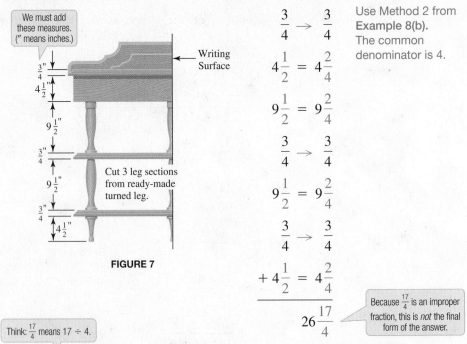

FIGURE 7

$\dfrac{3}{4} \rightarrow \dfrac{3}{4}$

$4\dfrac{1}{2} = 4\dfrac{2}{4}$

$9\dfrac{1}{2} = 9\dfrac{2}{4}$

$\dfrac{3}{4} \rightarrow \dfrac{3}{4}$

$9\dfrac{1}{2} = 9\dfrac{2}{4}$

$\dfrac{3}{4} \rightarrow \dfrac{3}{4}$

$+ 4\dfrac{1}{2} = 4\dfrac{2}{4}$
_____
$26\dfrac{17}{4}$

Use Method 2 from **Example 8(b)**. The common denominator is 4.

Because $\dfrac{17}{4}$ is an improper fraction, this is *not* the final form of the answer.

Think: $\dfrac{17}{4}$ means $17 \div 4$.

Because $\dfrac{17}{4} = 4\dfrac{1}{4}$, we have $26\dfrac{17}{4} = 26 + 4\dfrac{1}{4} = 30\dfrac{1}{4}$. The height is $30\dfrac{1}{4}$ in.

NOW TRY

NOW TRY ANSWERS
9. (a) $\dfrac{23}{99}$ **(b)** $\dfrac{3}{2}$, or $1\dfrac{1}{2}$
10. $2\dfrac{5}{8}$ ft

OBJECTIVE 7 Interpret data in a circle graph.

In a **circle graph,** or **pie chart,** a circle is used to indicate the total of all the data categories represented. The circle is divided into *sectors,* or wedges, whose sizes show the relative magnitudes of the categories. The sum of all the fractional parts must be 1 (for 1 whole circle).

NOW TRY EXERCISE 11

Refer to the circle graph in **FIGURE 8.**

(a) Which region had the least number of Internet users?

(b) Estimate the number of Internet users in Asia.

(c) How many Internet users were there in Asia?

EXAMPLE 11 Using a Circle Graph to Interpret Information

In a recent year, there were about 2100 million (that is, 2.1 billion) Internet users worldwide. The circle graph in **FIGURE 8** shows the fractions of these users living in various regions of the world.

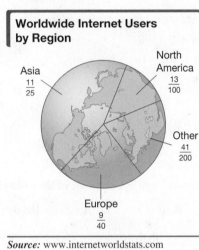

Worldwide Internet Users by Region

North America $\frac{13}{100}$

Asia $\frac{11}{25}$

Other $\frac{41}{200}$

Europe $\frac{9}{40}$

Source: www.internetworldstats.com

FIGURE 8

(a) Which region had the largest share of Internet users? What was that share?

The sector for Asia is the largest. Asia had the largest share of Internet users, $\frac{11}{25}$.

(b) Estimate the number of Internet users in North America.

A share of $\frac{13}{100}$ can be rounded to $\frac{10}{100}$, or $\frac{1}{10}$, and the total number of Internet users, 2100 million, can be rounded to 2000 million (or 2 billion). We multiply $\frac{1}{10}$ by 2000.

$$\frac{1}{10} \cdot 2000 = 200 \text{ million}$$ Approximate number of Internet users in North America

(c) How many Internet users were there in North America?

$$\frac{13}{100} \cdot 2100$$ Multiply the actual fraction from the graph for North America by the number of users.

$$= \frac{13}{100} \cdot \frac{2100}{1}$$ $a = \frac{a}{1}$, for all a.

$$= \frac{27{,}300}{100}$$ Multiply numerators. Multiply denominators.

$$= 273$$ Divide.

Thus, 273 million, or 273,000,000 people in North America used the Internet.

NOW TRY ANSWERS

11. (a) North America

(b) 1000 million, or 1 billion $\left(\frac{11}{25} \text{ is about } \frac{1}{2}.\right)$

(c) 924 million, or 924,000,000

NOW TRY

R.1 Exercises

● *Complete solution available in MyMathLab*

Concept Check   Decide whether each statement is true *or false. If it is false, explain why.*

1. In the fraction $\frac{5}{8}$, 5 is the numerator and 8 is the denominator.

2. The mixed number equivalent of the improper fraction $\frac{31}{5}$ is $6\frac{1}{5}$.

3. The fraction $\frac{7}{7}$ is proper.

4. The number 1 is prime.

5. The fraction $\frac{13}{39}$ is in lowest terms.

6. The reciprocal of $\frac{6}{2}$ is $\frac{3}{1}$.

7. The product of 10 and 2 is 12.

8. The difference of 10 and 2 is 5.

Concept Check   *Choose the letter of the correct response.*

9. Which choice shows the correct way to write $\frac{16}{24}$ in lowest terms?

A. $\frac{16}{24} = \frac{8+8}{8+16} = \frac{8}{16} = \frac{1}{2}$

B. $\frac{16}{24} = \frac{4 \cdot 4}{4 \cdot 6} = \frac{4}{6}$

C. $\frac{16}{24} = \frac{8 \cdot 2}{8 \cdot 3} = \frac{2}{3}$

D. $\frac{16}{24} = \frac{14+2}{21+3} = \frac{2}{3}$

10. Which fraction is *not* equal to $\frac{5}{9}$?

A. $\frac{15}{27}$ **B.** $\frac{30}{54}$ **C.** $\frac{40}{74}$ **D.** $\frac{55}{99}$

11. For the fractions $\frac{p}{q}$ and $\frac{r}{s}$, which one of the following can serve as a common denominator?

A. $q \cdot s$ **B.** $q + s$ **C.** $p \cdot r$ **D.** $p + r$

12. Which fraction with denominator 24 is equivalent to $\frac{5}{8}$?

A. $\frac{21}{24}$ **B.** $\frac{15}{24}$ **C.** $\frac{5}{24}$ **D.** $\frac{10}{24}$

Identify each number as prime, composite, *or* neither. *If the number is composite, write it as a product of prime factors.* **See Example 1.**

13. 19   **14.** 31   **15.** 30   **16.** 50

● **17.** 64   **18.** 81   **19.** 1   **20.** 0

21. 57   **22.** 51   **23.** 79   **24.** 83   **25.** 124

26. 138   **27.** 500   **28.** 700   **29.** 3458   **30.** 1025

Write each fraction in lowest terms. **See Example 2.**

31. $\frac{8}{16}$ **32.** $\frac{4}{12}$ ● **33.** $\frac{15}{18}$ **34.** $\frac{16}{20}$ **35.** $\frac{64}{100}$

36. $\frac{55}{200}$ **37.** $\frac{18}{90}$ **38.** $\frac{16}{64}$ **39.** $\frac{144}{120}$ **40.** $\frac{132}{77}$

Write each improper fraction as a mixed number. **See Example 3.**

41. $\frac{12}{7}$ **42.** $\frac{16}{9}$ **43.** $\frac{77}{12}$ **44.** $\frac{101}{15}$ **45.** $\frac{83}{11}$ **46.** $\frac{67}{13}$

Write each mixed number as an improper fraction. **See Example 4.**

47. $2\frac{3}{5}$ **48.** $5\frac{6}{7}$ **49.** $10\frac{3}{8}$ **50.** $12\frac{2}{3}$ **51.** $10\frac{1}{5}$ **52.** $18\frac{1}{6}$

Find each product or quotient, and write it in lowest terms. **See Examples 5 and 6.**

53. $\dfrac{4}{5} \cdot \dfrac{6}{7}$ **54.** $\dfrac{5}{9} \cdot \dfrac{2}{7}$ **55.** $\dfrac{2}{3} \cdot \dfrac{15}{16}$ **56.** $\dfrac{3}{5} \cdot \dfrac{20}{21}$

▶ **57.** $\dfrac{1}{10} \cdot \dfrac{12}{5}$ **58.** $\dfrac{1}{8} \cdot \dfrac{10}{7}$ **59.** $\dfrac{15}{4} \cdot \dfrac{8}{25}$ **60.** $\dfrac{21}{8} \cdot \dfrac{4}{7}$

61. $21 \cdot \dfrac{3}{7}$ **62.** $36 \cdot \dfrac{4}{9}$ **63.** $3\dfrac{1}{4} \cdot 1\dfrac{2}{3}$ **64.** $2\dfrac{2}{3} \cdot 1\dfrac{3}{5}$

65. $2\dfrac{3}{8} \cdot 3\dfrac{1}{5}$ **66.** $3\dfrac{3}{5} \cdot 7\dfrac{1}{6}$ ▶ **67.** $\dfrac{5}{4} \div \dfrac{3}{8}$ **68.** $\dfrac{7}{5} \div \dfrac{3}{10}$

69. $\dfrac{32}{5} \div \dfrac{8}{15}$ **70.** $\dfrac{24}{7} \div \dfrac{6}{21}$ **71.** $\dfrac{3}{4} \div 12$ **72.** $\dfrac{2}{5} \div 30$

73. $6 \div \dfrac{3}{5}$ **74.** $8 \div \dfrac{4}{9}$ **75.** $6\dfrac{3}{4} \div \dfrac{3}{8}$ **76.** $5\dfrac{3}{5} \div \dfrac{7}{10}$

77. $2\dfrac{1}{2} \div 1\dfrac{5}{7}$ **78.** $2\dfrac{2}{9} \div 1\dfrac{2}{5}$ **79.** $2\dfrac{5}{8} \div 1\dfrac{15}{32}$ **80.** $2\dfrac{3}{10} \div 1\dfrac{4}{5}$

Find each sum or difference, and write it in lowest terms. **See Examples 7–9.**

81. $\dfrac{7}{15} + \dfrac{4}{15}$ **82.** $\dfrac{2}{9} + \dfrac{5}{9}$ ▶ **83.** $\dfrac{7}{12} + \dfrac{1}{12}$ **84.** $\dfrac{3}{16} + \dfrac{5}{16}$

▶ **85.** $\dfrac{5}{9} + \dfrac{1}{3}$ **86.** $\dfrac{4}{15} + \dfrac{1}{5}$ **87.** $\dfrac{3}{8} + \dfrac{5}{6}$ **88.** $\dfrac{5}{6} + \dfrac{2}{9}$

89. $3\dfrac{1}{8} + 2\dfrac{1}{4}$ **90.** $4\dfrac{2}{3} + 2\dfrac{1}{6}$ **91.** $3\dfrac{1}{4} + 1\dfrac{4}{5}$ **92.** $5\dfrac{3}{4} + 1\dfrac{1}{3}$

93. $\dfrac{7}{9} - \dfrac{2}{9}$ **94.** $\dfrac{8}{11} - \dfrac{3}{11}$ ▶ **95.** $\dfrac{13}{15} - \dfrac{3}{15}$ **96.** $\dfrac{11}{12} - \dfrac{3}{12}$

97. $\dfrac{7}{12} - \dfrac{1}{3}$ **98.** $\dfrac{5}{6} - \dfrac{1}{2}$ **99.** $\dfrac{7}{12} - \dfrac{1}{9}$ **100.** $\dfrac{11}{16} - \dfrac{1}{12}$

101. $4\dfrac{3}{4} - 1\dfrac{2}{5}$ **102.** $3\dfrac{4}{5} - 1\dfrac{4}{9}$ **103.** $6\dfrac{1}{4} - 5\dfrac{1}{3}$ **104.** $5\dfrac{1}{3} - 4\dfrac{1}{2}$

Work each problem involving fractions.

105. For each description, write a fraction in lowest terms that represents the region described.

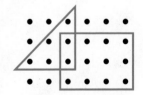

    (a) The dots in the rectangle as a part of the dots in the entire figure

    (b) The dots in the triangle as a part of the dots in the entire figure

    (c) The dots in the overlapping region of the triangle and the rectangle as a part of the dots in the triangle alone

    (d) The dots in the overlapping region of the triangle and the rectangle as a part of the dots in the rectangle alone

106. At the conclusion of the Pearson Learning softball league season, batting statistics for five players were as shown in the table.

Player	At-Bats	Hits	Home Runs
Maureen	36	12	3
Christine	40	9	2
Chase	11	5	1
Joe	16	8	0
Greg	20	10	2

Use this information to answer each question. Estimate as necessary.

(a) Which player got a hit in exactly $\frac{1}{3}$ of his or her at-bats?

(b) Which player got a hit in just less than $\frac{1}{2}$ of his or her at-bats?

(c) Which player got a home run in just less than $\frac{1}{10}$ of his or her at-bats?

(d) Which player got a hit in just less than $\frac{1}{4}$ of his or her at-bats?

(e) Which two players got hits in exactly the same fractional parts of their at-bats? What was the fractional part, expressed in lowest terms?

Use the table to answer Exercises 107 and 108.

107. How many cups of water would be needed for eight microwave servings?

108. How many teaspoons of salt would be needed for five stove-top servings? (*Hint:* 5 servings is halfway between 4 and 6 servings.)

	Microwave	Stove Top		
Servings	1	1	4	6
Water	$\frac{3}{4}$ cup	1 cup	3 cups	4 cups
Grits	3 Tbsp	3 Tbsp	$\frac{3}{4}$ cup	1 cup
Salt (optional)	Dash	Dash	$\frac{1}{4}$ tsp	$\frac{1}{2}$ tsp

Source: Package of Quaker Quick Grits.

The Pride Golf Tee Company, the only U.S. manufacturer of wooden golf tees, has created the Professional Tee System, shown in the figure. Use the information given to work Exercises 109 and 110. (Source: The Gazette.)

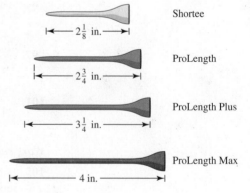

109. Find the difference in length between the ProLength Plus and the once-standard Shortee.

110. The ProLength Max tee is the longest tee allowed by the U.S. Golf Association's *Rules of Golf.* How much longer is the ProLength Max than the Shortee?

Solve each problem. See Example 10.

111. A hardware store sells a 40-piece socket wrench set. The measure of the largest socket is $\frac{3}{4}$ in., while the measure of the smallest is $\frac{3}{16}$ in. What is the difference between these measures?

112. Two sockets in a socket wrench set have measures of $\frac{9}{16}$ in. and $\frac{3}{8}$ in. What is the difference between these two measures?

▶ **113.** A piece of property has an irregular shape, with five sides, as shown in the figure. Find the total distance around the piece of property. (This distance is the **perimeter** of the figure.)

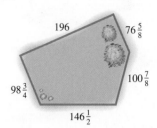

196 $76\frac{5}{8}$

$100\frac{7}{8}$

$98\frac{3}{4}$

$146\frac{1}{2}$

Measurements are in feet.

114. Find the perimeter of the triangle in the figure.

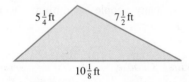

$5\frac{1}{4}$ ft $7\frac{1}{2}$ ft

$10\frac{1}{8}$ ft

115. A board is $15\frac{5}{8}$ in. long. If it must be divided into three pieces of equal length, how long must each piece be?

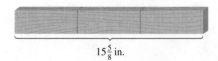

$15\frac{5}{8}$ in.

116. Paul's favorite recipe for barbecue sauce calls for $2\frac{1}{3}$ cups of tomato sauce. The recipe makes enough barbecue sauce to serve seven people. How much tomato sauce is needed for one serving?

117. A cake recipe calls for $1\frac{3}{4}$ cups of sugar. A caterer has $15\frac{1}{2}$ cups of sugar on hand. How many cakes can he make?

118. Kyla needs $2\frac{1}{4}$ yd of fabric to cover a chair. How many chairs can she cover with $23\frac{2}{3}$ yd of fabric?

119. It takes $2\frac{3}{8}$ yd of fabric to make a costume for a school play. How much fabric would be needed for seven costumes?

120. A cookie recipe calls for $2\frac{2}{3}$ cups of sugar. How much sugar would be needed to make four batches of cookies?

121. First published in 1953, the digest-sized *TV Guide* has changed to a full-sized magazine. The full-sized magazine is 3 in. wider than the old guide. What is the difference in their heights? (*Source: TV Guide.*)

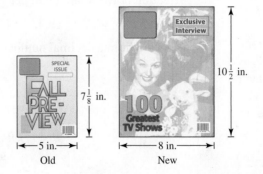

$7\frac{1}{8}$ in.

$10\frac{1}{2}$ in.

5 in.

8 in.

Old

New

122. Under existing standards, most of the holes in Swiss cheese must have diameters between $\frac{11}{16}$ and $\frac{13}{16}$ in. To accommodate new high-speed slicing machines, the U.S. Department of Agriculture wants to reduce the minimum size to $\frac{3}{8}$ in. How much smaller is $\frac{3}{8}$ in. than $\frac{11}{16}$ in.? (*Source:* U.S. Department of Agriculture.)

Approximately 40 *million people living in the United States were born in other countries. The circle graph gives the fractional number from each region of birth for these people. Use the graph to answer each question.* **See Example 11.**

123. What fractional part of the foreign-born population was from other regions?

124. What fractional part of the foreign-born population was from Latin America or Asia?

125. About how many people (in millions) were born in Europe?

126. About how many people (in millions) were born in Latin America?

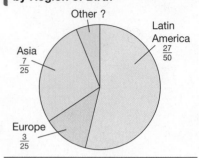

U.S. Foreign-Born Population by Region of Birth

Source: U.S. Census Bureau.

Extending Skills   *Choose the letter of the correct response.*

127. Estimate the best approximation for the sum.

$$\frac{14}{26} + \frac{98}{99} + \frac{100}{51} + \frac{90}{31} + \frac{13}{27}$$

A. 5   **B.** 6   **C.** 7   **D.** 8

128. Estimate the best approximation for the product.

$$\frac{202}{50} \cdot \frac{99}{100} \cdot \frac{21}{40} \cdot \frac{75}{36}$$

A. 3   **B.** 4   **C.** 8   **D.** 16

R.2 Decimals and Percents

OBJECTIVES

1 Write decimals as fractions.

2 Add and subtract decimals.

3 Multiply and divide decimals.

4 Write fractions as decimals.

5 Write percents as decimals and decimals as percents.

6 Write percents as fractions and fractions as percents.

7 Solve applied problems that involve percents.

Fractions are one way to represent parts of a whole. Another way is with a decimal fraction or **decimal,** a number written with a decimal point.

$$9.4, \quad 14.001, \quad 0.25 \qquad \text{Decimal numbers}$$

Each digit in a decimal number has a place value, as shown below.

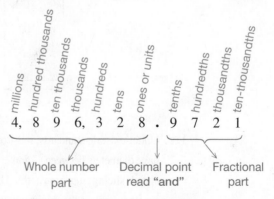

Each successive place value is ten times greater than the place value to its right and one-tenth as great as the place value to its left.

VOCABULARY
☐ decimal
☐ decimal places
☐ divisor
☐ dividend
☐ quotient
☐ terminating decimal
☐ repeating decimal
☐ percent

OBJECTIVE 1 Write decimals as fractions.

Place value is used to write a decimal number as a fraction.

Converting a Decimal to a Fraction

Read the decimal using the correct place value. Write it in fractional form just as it is read.

- The numerator will be the digits to the right of the decimal point.
- The denominator will be a power of 10—that is, 10 for tenths, 100 for hundredths, and so on.

NOW TRY EXERCISE 1

Write each decimal as a fraction. (Do not write in lowest terms.)

(a) 0.8   **(b)** 0.431   **(c)** 2.58

EXAMPLE 1 Writing Decimals as Fractions

Write each decimal as a fraction. (Do not write in lowest terms.)

(a) 0.95
We read 0.95 as "ninety-five hundredths." The equivalent fractional form is

$$0.95 = \frac{95}{100}. \leftarrow \text{For hundredths}$$

(b) 0.056
We read 0.056 as "fifty-six thousandths." The equivalent fractional form is

$$0.056 = \frac{56}{1000}.$$

↗ For thousandths

> Do not confuse 0.056 with 0.56, read "fifty-six *hundredths*," which is the fraction $\frac{56}{100}$.

(c) 4.2095
We read this decimal number, which is greater than 1, as "Four *and* two thousand ninety-five ten-thousandths."

$$4.2095 = 4\frac{2095}{10,000} \quad \text{Write the decimal number as a mixed number.}$$

$$= \frac{42,095}{10,000} \quad \text{Write the mixed number as an improper fraction.}$$

NOW TRY ↩

OBJECTIVE 2 Add and subtract decimals.

EXAMPLE 2 Adding and Subtracting Decimals

Add or subtract as indicated.

(a) 6.92 + 14.8 + 3.217
Place the digits of the decimal numbers in columns by place value, so that tenths are in one column, hundredths in another column, and so on.

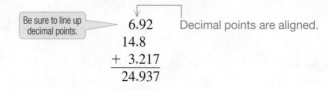

Be sure to line up decimal points.

```
    6.92
   14.8
 +  3.217
 ────────
   24.937
```
Decimal points are aligned.

**NOW TRY
EXERCISE 2**

Add or subtract as indicated.

(a) $68.9 + 42.72 + 8.973$

(b) $351.8 - 2.706$

To avoid errors, attach zeros as placeholders so that all the decimal numbers are the same length.

$$\begin{array}{r} 6.92 \\ 14.8 \\ +\ 3.217 \end{array} \qquad \text{becomes} \qquad \begin{array}{r} 6.920 \\ 14.800 \\ +\ 3.217 \\ \hline 24.937 \end{array}$$

Attach 0s.

> 6.92 is equivalent to 6.920.
> 14.8 is equivalent to 14.800.

(b) $47.6 - 32.509$

$$\begin{array}{r} 47.6 \\ -\ 32.509 \end{array} \qquad \text{becomes} \qquad \begin{array}{r} 47.600 \\ -\ 32.509 \\ \hline 15.091 \end{array}$$

Write the decimal numbers in columns, attaching 0s to 47.6.

(c) $3 - 0.253$

$$\begin{array}{r} 3.000 \\ -\ 0.253 \\ \hline 2.747 \end{array}$$

A whole number is assumed to have the decimal point at the right of the number. Write 3 as 3.000.

NOW TRY

OBJECTIVE 3 Multiply and divide decimals.

Multiplying Decimals

Step 1 Ignore the decimal points, and multiply as if the numbers were whole numbers.

Step 2 Add the number of **decimal places** (digits to the *right* of the decimal point) in each factor. Place the decimal point that many digits from the right in the product.

**NOW TRY
EXERCISE 3**

Multiply.

(a) 9.32×1.4

(b) 0.06×0.004

EXAMPLE 3 Multiplying Decimals

Multiply.

(a) 29.3×4.52

$$\begin{array}{r} 29.3 \\ \times\ 4.52 \\ \hline 586 \\ 1465 \\ 1172 \\ \hline 132.436 \end{array}$$

1 decimal place
2 decimal places
$1 + 2 = 3$

3 decimal places

(b) 31.42×65

$$\begin{array}{r} 31.42 \\ \times\ \ 65 \\ \hline 15710 \\ 18852 \\ \hline 2042.30 \end{array}$$

2 decimal places
0 decimal places
$2 + 0 = 2$

2 decimal places

The final 0 can be dropped, and the product can be written 2042.3.

(c) 0.05×0.3

Here $5 \times 3 = 15$. Be careful placing the decimal point.

2 decimal places   1 decimal place

$$0.05 \quad \times \quad 0.3$$

$$= 0.015$$

Do *not* write 0.150.

$2 + 1 = 3$ decimal places
Attach 0 as a placeholder in the tenths place.

NOW TRY

Divisor → $25\overline{)125}$ $\overset{5}{}$ ← Quotient

↑
Dividend

Remember this terminology for the parts of a division problem.

NOW TRY EXERCISE 4

Divide.

(a) $451.47 \div 14.9$

(b) $7.334 \div 1.3$

(Round the quotient to two decimal places.)

Dividing Decimals

Step 1 Change the **divisor** (the number we are dividing *by*) into a whole number by moving the decimal point as many places as necessary to the right.

Step 2 Move the decimal point in the **dividend** (the number we are dividing *into*) to the right by the same number of places.

Step 3 Move the decimal point straight up, and then divide as with whole numbers to find the **quotient.**

EXAMPLE 4 **Dividing Decimals**

Divide.

(a) $233.45 \div 11.5$

Write the problem as follows. $11.5\overline{)233.45}$

$11.5\overline{)233.4\,5}$ To change the divisor 11.5 into a whole number, move each decimal point one place to the right.

To see why this works, write the division in fractional form and multiply by $\frac{10}{10}$. The result is the same as when we moved the decimal point one place to the right in the divisor and the dividend.

$$\frac{233.45}{11.5} \cdot \frac{10}{10} = \frac{2334.5}{115}$$ Multiplying by $\frac{10}{10}$ is equivalent to multiplying by 1.

Move the decimal point straight up and divide as with whole numbers.

$$
\begin{array}{r}
20.3 \\
115\overline{)2334.5} \\
\underline{230} \\
345 \\
\underline{345} \\
0
\end{array}
$$
Move the decimal point straight up.

In the second step of the division, 115 does not divide into 34, so we used zero as a placeholder in the quotient.

(b) $8.949 \div 1.25$ (Round the quotient to two decimal places.)

$1.25\overline{)8.949}$ Move each decimal point two places to the right.

$$
\begin{array}{r}
7.159 \\
125\overline{)894.900} \\
\underline{875} \\
199 \\
\underline{125} \\
740 \\
\underline{625} \\
1150 \\
\underline{1125} \\
25
\end{array}
$$
Move the decimal point straight up, and divide as with whole numbers. Attach 0s as placeholders.

NOW TRY ANSWERS

4. **(a)** 30.3 **(b)** 5.64

We carried out the division to three decimal places so that we could round to two decimal places, obtaining the quotient 7.16.

NOW TRY

NOTE To round 7.159 in **Example 4(b)** to two decimal places (that is, to the nearest hundredth), we look at the digit to the *right* of the hundredths place. If this digit is 5 or greater, we round up. If it is less than 5, we drop the digit(s) beyond the desired place.

Hundredths place
↓

7.15**9**   9, the digit to the right of the hundredths place, is 5 or greater.

≈ 7.16   Round 5 up to 6. ≈ means "is approximately equal to."

Multiplying or Dividing by Powers of 10 (Shortcuts)

- To *multiply* by a power of 10, *move the decimal point to the right* as many places as the number of zeros.
- To *divide* by a power of 10, *move the decimal point to the left* as many places as the number of zeros.

In both cases, insert 0s as placeholders if necessary.

NOW TRY EXERCISE 5

Multiply or divide as indicated.

(a) 294.72 × 10

(b) 4.793 ÷ 100

EXAMPLE 5 Multiplying and Dividing by Powers of 10

Multiply or divide as indicated.

(a) 48.731 × 100

= 48.73 1   or   4873.1

Move the decimal point two places to the right because 100 has two 0s.

(b) 48.7 ÷ 1000

= 048.7   or   0.0487

Move the decimal point three places to the left because 1000 has three 0s. Insert a 0 in front of the 4 to do this.   NOW TRY

OBJECTIVE 4 Write fractions as decimals.

Converting a Fraction to a Decimal

Because a fraction bar indicates division, write a fraction as a decimal by dividing the numerator by the denominator.

EXAMPLE 6 Writing Fractions as Decimals

Write each fraction as a decimal.

(a) $\dfrac{19}{8}$

$$
\begin{array}{r}
2.375 \\
8)\overline{19.000} \\
16 \\
\overline{30} \\
24 \\
\overline{60} \\
56 \\
\overline{40} \\
40 \\
\overline{0}
\end{array}
$$

Divide 19 by 8. Add a decimal point and as many 0s as necessary.

(b) $\dfrac{2}{3}$

$$
\begin{array}{r}
0.6666\ldots \\
3)\overline{2.0000\ldots} \\
18 \\
\overline{20} \\
18 \\
\overline{20} \\
18 \\
\overline{20}
\end{array}
$$

$\dfrac{2}{3} = 0.6666\ldots$ ← Repeating decimal

$\dfrac{19}{8} = 2.375$ ← Terminating decimal

**NOW TRY
EXERCISE 6**

Write each fraction as a decimal. For repeating decimals, write the answer by first using bar notation and then rounding to the nearest thousandth.

(a) $\dfrac{17}{20}$ **(b)** $\dfrac{2}{9}$

Because the remainder in the division in part (a) is 0, this quotient is a **terminating decimal.** The remainder in the division in part (b) is never 0. Because a number, in this case 2, is always left after the subtraction, this quotient is a **repeating decimal.** A convenient notation for a repeating decimal is a bar over the digit (or digits) that repeats.

$$\frac{2}{3} = 0.6666\ldots, \quad \text{or} \quad 0.\overline{6}$$

We often round repeating decimals to as many places as needed.

$$\frac{2}{3} \approx 0.667 \qquad \text{An approximation to the nearest thousandth}$$

NOW TRY

OBJECTIVE 5 Write percents as decimals and decimals as percents.

The word **percent** means **"per 100."** Percent is written with the symbol **%.** *One percent means "one per one hundred," or "one one-hundredth."*

Percent, Decimal, and Fraction Equivalents

$$1\% = 0.01, \quad \text{or} \quad 1\% = \frac{1}{100}$$

**NOW TRY
EXERCISE 7**

Write each percent as a decimal.

(a) 23% **(b)** 350%

EXAMPLE 7 Writing Percents as Decimals

Write each percent as a decimal.

(a) 73%

We use the fact that $1\% = 0.01$ and convert as follows.

$$73\% = 73 \cdot 1\% = 73 \cdot 0.01 = 0.73$$

(b) $125\% = 125 \cdot 1\% = 125 \cdot 0.01 = 1.25$ $1\% = 0.01$

> A percent greater than 100 represents a number greater than 1.

(c) $3\frac{1}{2}\%$

First write the fractional part as a decimal.

$$3\frac{1}{2}\% = (3 + 0.5)\% = 3.5\%$$

Now write the percent in decimal form.

$$3.5\% = 3.5 \cdot 1\% = 3.5 \cdot 0.01 = 0.035 \qquad 1\% = 0.01 \qquad \text{NOW TRY} $$

**NOW TRY
EXERCISE 8**

Write each decimal as a percent.

(a) 0.71 **(b)** 1.32

EXAMPLE 8 Writing Decimals as Percents

Write each decimal as a percent.

(a) 0.32

This conversion is the opposite of what we did in **Example 7** when we wrote percents as decimals. We use $1\% = 0.01$ in reverse.

$$0.32 = 32 \cdot 0.01 = 32 \cdot 1\% = 32\% \qquad 0.01 = 1\%$$

(b) $0.05 = 5 \cdot 0.01 = 5 \cdot 1\% = 5\%$ $0.01 = 1\%$

(c) $2.63 = 263 \cdot 0.01 = 263 \cdot 1\% = 263\%$

> A number greater than 1 is more than 100%.

NOW TRY

NOW TRY ANSWERS
6. **(a)** 0.85 **(b)** $0.\overline{2}$, 0.222
7. **(a)** 0.23 **(b)** 3.50, or 3.5
8. **(a)** 71% **(b)** 132%

Converting Percents and Decimals (Shortcuts)

- To convert a percent to a decimal, move the decimal point two places to the *left* and drop the % symbol.

- To convert a decimal to a percent, move the decimal point two places to the *right* and attach a % symbol.

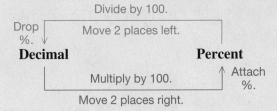

Divide by 100.
Move 2 places left.

Drop %. ↓

Decimal   **Percent**

↑ Attach %.

Multiply by 100.
Move 2 places right.

NOW TRY EXERCISE 9

Convert each percent to a decimal and each decimal to a percent.
(a) 52%   **(b)** 2%
(c) 0.45   **(d)** 3.5

EXAMPLE 9   Converting Percents and Decimals by Moving the Decimal Point

Convert each percent to a decimal and each decimal to a percent.

(a) $45\% = 0.45$ **(b)** $250\% = 2.50$, or 2.5 **(c)** $9\% = 09\% = 0.09$

(d) $0.57 = 57\%$ **(e)** $1.5 = 1.50 = 150\%$ **(f)** $0.327 = 32.7\%$

NOW TRY

OBJECTIVE 6   Write percents as fractions and fractions as percents.

NOW TRY EXERCISE 10

Write each percent as a fraction. Give answers in lowest terms.
(a) 20%   **(b)** 160%

EXAMPLE 10   Writing Percents as Fractions

Write each percent as a fraction. Give answers in lowest terms.

(a) 8%
  We use the fact that $1\% = \frac{1}{100}$, and convert as follows.

$$8\% = 8 \cdot 1\% = 8 \cdot \frac{1}{100} = \frac{8}{100}$$

In lowest terms, $\frac{8}{100} = \frac{2 \cdot 4}{25 \cdot 4} = \frac{2}{25}.$

Thus, $8\% = \frac{2}{25}.$

(b) $175\% = 175 \cdot 1\% = 175 \cdot \frac{1}{100} = \frac{175}{100}$

  In lowest terms,

$$\frac{175}{100} = \frac{7 \cdot 25}{4 \cdot 25} = \frac{7}{4}, \quad \text{or} \quad 1\frac{3}{4}.$$

> A number greater than 1 is more than 100%.

(c) 13.5%

$$= 13\frac{1}{2} \cdot 1\% \qquad \text{Write 13.5 as a mixed number.}$$

$$= \frac{27}{2} \cdot \frac{1}{100} \qquad \begin{array}{l}\text{Write } 13\frac{1}{2} \text{ as an improper fraction.} \\ \text{Use the fact that } 1\% = \frac{1}{100}.\end{array}$$

$$= \frac{27}{200} \qquad \text{Multiply the fractions.}$$

NOW TRY

NOW TRY ANSWERS
9. (a) 0.52   **(b)** 0.02
    (c) 45%   **(d)** 350%
10. (a) $\frac{1}{5}$ **(b)** $\frac{8}{5}$, or $1\frac{3}{5}$

We know that 100% of something is the whole thing. One way to convert a fraction to a percent is to multiply by 100%, which is equivalent to 1.

 NOW TRY EXERCISE 11

Write each fraction as a percent.

(a) $\dfrac{6}{25}$ **(b)** $\dfrac{7}{2}$

EXAMPLE 11   Writing Fractions as Percents

Write each fraction as a percent.

(a) $\dfrac{2}{5}$

$$= \frac{2}{5} \cdot 100\% \qquad \text{Multiply by 1 in the form 100\%.}$$

$$= \frac{2}{5} \cdot \frac{100}{1}\%$$

$$= \frac{2 \cdot 5 \cdot 20}{5 \cdot 1}\% \qquad \text{Multiply and factor.}$$

$$= \frac{2 \cdot 20}{1}\% \qquad \text{Divide out the common factor.}$$

$$= 40\% \qquad \text{Simplify.}$$

(b) $\dfrac{1}{6}$

$$= \frac{1}{6} \cdot 100\%$$

$$= \frac{1}{6} \cdot \frac{100}{1}\%$$

$$= \frac{1 \cdot 2 \cdot 50}{2 \cdot 3 \cdot 1}\%$$

$$= \frac{50}{3}\%$$

$$= 16\frac{2}{3}\%, \quad \text{or} \quad 16.\overline{6}\%$$

NOW TRY

OBJECTIVE 7   Solve applied problems that involve percents.

The decimal form of a percent is generally used in calculations.

 NOW TRY EXERCISE 12

A winter coat is on sale for 60% off. The regular price is $120. Find the amount of the discount and the sale price of the coat.

EXAMPLE 12   Using Percent to Solve an Applied Problem

A DVD with a regular price of $18 is on sale this week at 22% off. Find the amount of the discount and the sale price of the DVD.

The discount is 22% *of* 18. The word *of* here means multiply.

$$
\begin{array}{ccc}
22\% & \text{of} & 18 \\
\downarrow & \downarrow & \downarrow \\
0.22 & \cdot & 18 \qquad \text{Write 22\% as a decimal.} \\
\end{array}
$$

$$= 3.96 \qquad \text{Multiply.}$$

The discount is $3.96. The sale price is found by subtracting.

$$\$18.00 - \$3.96 = \$14.04 \qquad \text{Original price} - \text{discount} = \text{sale price}$$

NOW TRY

NOW TRY ANSWERS
11. **(a)** 24%   **(b)** 350%
12. $72; $48

R.2 Exercises

FOR EXTRA HELP   MyMathLab®

 Complete solution available in MyMathLab

Concept Check   In Exercises 1–4, provide the correct response.

1. In the decimal number 367.9412, name the digit that has each place value.

    (a) tens   **(b)** tenths   **(c)** thousandths   **(d)** ones or units   **(e)** hundredths

2. Write a decimal number that has 5 in the thousands place, 0 in the tenths place, and 4 in the ten-thousandths place.

3. For the decimal number 46.249, round to the place value indicated.

    (a) hundredths   **(b)** tenths   **(c)** ones or units   **(d)** tens

4. Round each decimal to the nearest thousandth.

(a) $0.\overline{8}$ (b) $0.\overline{5}$ (c) 0.9762 (d) 0.8642

Write each decimal as a fraction. (Do not write in lowest terms.) **See Example 1.**

5. 0.4   **6.** 0.6   **7.** 0.64   **8.** 0.82   **9.** 0.138

10. 0.104   **11.** 0.043   **12.** 0.087   **13.** 3.805   **14.** 5.166

Add or subtract as indicated. **See Example 2.**

15. $25.32 + 109.2 + 8.574$ **16.** $90.527 + 32.43 + 589.8$ **17.** $28.73 - 3.12$

18. $46.88 - 13.45$ **19.** $43.5 - 28.17$ **20.** $345.1 - 56.31$

21. $3.87 + 15 + 2.9$ **22.** $8.2 + 1.09 + 12$ **23.** $32.56 + 47.356 + 1.8$

24. $75.2 + 123.96 + 3.897$ ◐ **25.** $18 - 2.789$ **26.** $29 - 8.582$

Multiply or divide as indicated. **See Examples 3–5.**

27. 12.8×9.1 **28.** 34.04×0.56 **29.** 0.2×0.03 **30.** 0.07×0.004

◐ **31.** $78.65 \div 11$ **32.** $73.36 \div 14$ **33.** $19.967 \div 9.74$ **34.** $44.4788 \div 5.27$

35. 57.116×100 **36.** 82.053×100 **37.** 0.094×1000 **38.** 0.025×1000

39. $1.62 \div 10$ **40.** $8.04 \div 10$ **41.** $124.03 \div 100$ **42.** $490.35 \div 100$

Concept Check   *Complete the following table of fraction, decimal, and percent equivalents.*

	Fraction in Lowest Terms (or Whole Number)	Decimal	Percent
43.	$\frac{1}{100}$	0.01	
44.	$\frac{1}{50}$		2%
45.		0.05	5%
46.	$\frac{1}{10}$		
47.	$\frac{1}{8}$	0.125	
48.			20%
49.	$\frac{1}{4}$		
50.	$\frac{1}{3}$		
51.			50%
52.	$\frac{2}{3}$		$66\frac{2}{3}\%$, or $66.\overline{6}\%$
53.		0.75	
54.	1	1.0	

Write each fraction as a decimal. For repeating decimals, write the answer by first using bar notation and then rounding to the nearest thousandth. **See Example 6.**

55. $\frac{3}{8}$ **56.** $\frac{7}{8}$ **57.** $\frac{5}{4}$ **58.** $\frac{9}{5}$

◐ **59.** $\frac{5}{9}$ **60.** $\frac{8}{9}$ **61.** $\frac{1}{6}$ **62.** $\frac{5}{6}$

Write each percent as a decimal. See Examples 7 and 9(a)–9(c).

63. 54%   **64.** 39%   **65.** 7%   **66.** 4%

67. 117%   **68.** 189%   **69.** 2.4%   **70.** 3.1%

71. $6\frac{1}{4}\%$ **72.** $5\frac{1}{2}\%$ **73.** 0.8%   **74.** 0.9%

Write each decimal as a percent. See Examples 8 and 9(d)–9(f).

75. 0.79   **76.** 0.83   **77.** 0.02   **78.** 0.08

79. 0.004   **80.** 0.005   **81.** 1.28   **82.** 2.35

83. 0.4   **84.** 0.6   **85.** 6   **86.** 10

Write each percent as a fraction. Give answers in lowest terms. See Example 10.

87. 51%   **88.** 47%   **89.** 15%   **90.** 35%   **91.** 2%

92. 8%   **93.** 140%   **94.** 180%   **95.** 7.5%   **96.** 2.5%

Write each fraction as a percent. See Example 11.

97. $\frac{4}{5}$ **98.** $\frac{3}{25}$ **99.** $\frac{7}{50}$ **100.** $\frac{9}{20}$ **101.** $\frac{2}{11}$

102. $\frac{4}{9}$ **103.** $\frac{9}{4}$ **104.** $\frac{8}{5}$ **105.** $\frac{13}{6}$ **106.** $\frac{31}{9}$

Solve each problem. See Example 12.

107. What is 50% of 320?   **108.** What is 25% of 120?

109. What is 6% of 80?   **110.** What is 5% of 70?

111. What is 14% of 780?   **112.** What is 26% of 480?

Solve each problem. See Example 12.

113. Elwyn's bill for dinner at a restaurant was $89. He wants to leave a 20% tip. How much should he leave for the tip? What is his total bill for dinner and tip?

114. Gary earns $15 per hour at his job. He recently received a 7% raise. How much per hour was his raise? What is his new hourly rate?

115. Find the discount on a leather recliner with a regular price of $795 if the recliner is on sale at 15% off. What is the sale price of the recliner?

116. A laptop computer with a regular price of $597 is on sale at 20% off. Find the amount of the discount and the sale price of the computer.

In a recent year, approximately 60 million people from other countries visited the United States. The circle graph shows the distribution of these international visitors by country or region. Use the graph to work each problem.

117. About how many travelers visited the United States from Canada?

118. About how many travelers visited the United States from Mexico?

119. What percent of travelers visited the United States from places other than Canada, Mexico, Europe, or Asia? (*Hint:* The sum of the parts of the graph must equal 1 whole, that is, 100%.)

120. Use the answer from **Exercise 119** to find about how many travelers visited the United States from places other than Canada, Mexico, Europe, or Asia.

International Travelers to the United States

Source: U.S. Department of Commerce.

STUDY SKILLS

Reading Your Math Textbook

Take time to read each section and its examples before doing your homework. You will learn more and be better prepared to work the exercises your instructor assigns.

Approaches to Reading Your Math Textbook

Student A learns best by listening to her teacher explain things. She "gets it" when she sees the instructor work problems. She previews the section before the lecture, so she knows generally what to expect. **Student A carefully reads the section in her text *AFTER* she hears the classroom lecture on the topic.**

Student B learns best by reading on his own. He reads the section and works through the examples before coming to class. That way, he knows what the teacher is going to talk about and what questions he wants to ask. **Student B carefully reads the section in his text *BEFORE* he hears the classroom lecture on the topic.**

Which of these reading approaches works best for you—that of Student A or Student B?

Tips for Reading Your Math Textbook

- **Turn off your cell phone.** You will be able to concentrate more fully on what you are reading.

- **Survey the material.** Glance over the assigned material to get an idea of the "big picture." Look at the list of objectives to see what you will be learning.

- **Read slowly.** Read only one section—or even part of a section—at a sitting, with paper and pencil in hand.

- **Pay special attention to important information given in colored boxes or set in bold-face type.**

- **Study the examples carefully.** Pay particular attention to the blue side comments and any pointer balloons.

- **Do the Now Try exercises in the margin on separate paper as you go.** These mirror the examples and prepare you for the exercise set. The answers are given at the bottom of the page.

- **Make study cards as you read.** Make cards for new vocabulary, rules, procedures, formulas, and sample problems.

- **Mark anything you don't understand. *ASK QUESTIONS* in class**—everyone will benefit. Follow up with your instructor, as needed.

Think through and answer each question.

1. Which two or three reading tips will you try this week?

2. Did the tips you selected improve your ability to read and understand the material? Explain.

The Real Number System

Positive and *negative numbers,* shown here to indicate gains and losses, are examples of *real numbers,* the subject of this chapter.

1.1 Exponents, Order of Operations, and Inequality

OBJECTIVE 1 Use exponents.

In **Chapter R,** we factored a number as the product of its prime factors.

> 81   can be written as $3 \cdot 3 \cdot 3 \cdot 3$.   The factor 3 appears four times.

In algebra, repeated factors are written with an *exponent,* so the product $3 \cdot 3 \cdot 3 \cdot 3$ is written as 3^4 and read as "3 to the fourth power."

$$\underbrace{3 \cdot 3 \cdot 3 \cdot 3}_{\text{4 factors of 3}} = 3^{\overset{\text{Exponent}}{4}}$$
Base

The number 4 is the **exponent,** or **power,** and 3 is the **base** in the **exponential expression** 3^4. A natural number exponent, then, tells how many times the base is used as a factor. *A number raised to the first power is simply that number.* For example,

$$5^1 = 5 \quad \text{and} \quad \left(\frac{1}{2}\right)^1 = \frac{1}{2}. \qquad \text{In general, } a^1 = a.$$

EXAMPLE 1   **Evaluating Exponential Expressions**

Find the value of each exponential expression.

(a) 5^2 means $\underbrace{5 \cdot 5}$,   which equals   25.

5 is used as a factor 2 times.

Read 5^2 as "5 to the second power" or, more commonly, "5 squared."

(b) 6^3 means $\underbrace{6 \cdot 6 \cdot 6}$,   which equals   216.

6 is used as a factor 3 times.

Read 6^3 as "6 to the third power" or, more commonly, "6 cubed."

(c) 2^5 means $2 \cdot 2 \cdot 2 \cdot 2 \cdot 2$,   which equals   32.   2 is used as a factor 5 times.

Read 2^5 as "2 to the fifth power."

(d) $\left(\dfrac{2}{3}\right)^3$ means $\dfrac{2}{3} \cdot \dfrac{2}{3} \cdot \dfrac{2}{3}$,   which equals $\dfrac{8}{27}$. $\frac{2}{3}$ is used as a factor 3 times.

(e) $(0.3)^2$ means $0.3(0.3)$,   which equals   0.09.   0.3 is used as a factor 2 times.

 NOW TRY

> **⚠ CAUTION** *Squaring, or raising a number to the second power, is not the same as doubling the number.* For example,
>
> $$3^2 \text{ means } 3 \cdot 3, \text{ not } 2 \cdot 3.$$
>
> Thus $3^2 = 9$, *not* 6. Similarly, cubing, or raising a number to the third power, does *not* mean tripling the number.

OBJECTIVE 2 Use the rules for order of operations.

When an expression involves more than one operation, we often use **grouping symbols,** such as parentheses ( ), to indicate the order in which the operations should be performed.

Consider the following expression.

$$5 + 2 \cdot 3$$

To show that the multiplication should be performed before the addition, we could use parentheses to group $2 \cdot 3$.

$$5 + (2 \cdot 3) \quad \text{equals} \quad 5 + 6, \quad \text{which equals} \quad 11.$$

If the addition is to be performed first, the parentheses should group $5 + 2$.

$$(5 + 2) \cdot 3 \quad \text{equals} \quad 7 \cdot 3, \quad \text{which equals} \quad 21.$$

Other grouping symbols are brackets $[\ \ ]$, braces $\{\ \ \}$, and fraction bars. (For example, in $\frac{8-2}{3}$, the expression $8 - 2$ is "grouped" in the numerator.)

To simplify an expression that involves more than one operation, we use the following rules for **order of operations**. This order is used by most calculators and computers.

Order of Operations

If grouping symbols are present, work within them, innermost first (and above and below fraction bars separately), in the following order.

Step 1   Apply all **exponents.**

Step 2   Do any **multiplications** or **divisions** in order from left to right.

Step 3   Do any **additions** or **subtractions** in order from left to right.

If no grouping symbols are present, start with Step 1.

NOTE   Multiplication is understood in expressions with parentheses such as

$$3(7), \quad (6)2, \quad (-5)(-4), \quad \text{or} \quad 3(4 + 1).$$

EXAMPLE 2   **Using the Rules for Order of Operations**

Find the value of each expression.

(a) $4 + 5 \cdot 6$ A helpful strategy is to label the order in which the
    ② ①   operations should be performed.

    $= 4 + 30$ Multiply.

    $= 34$ Add.

(b) $9(6 + 11)$
    ② ①

    $= 9(17)$ Add inside the parentheses.

    $= 153$ Multiply.

(c) $6 \cdot 8 + 5 \cdot 2$
    ① ③ ②

    $= 48 + 10$ Multiply, working from left to right.

    $= 58$ Add.

NOW TRY
EXERCISE 2
Find the value of each expression.

(a) $15 - 2 \cdot 6$

(b) $8 + 2(5 - 1)$

(c) $6(2 + 4) - 7 \cdot 5$

(d) $8 \cdot 10 \div 4 - 2^3 + 3 \cdot 4^2$

(d) $16 - 3(2 + 3)$ ⟵ Do *not* subtract $16 - 3$ first.

③ ② ①

$= 16 - 3(5)$ Add inside the parentheses.

$= 16 - 15$ Multiply.

$= 1$ Subtract.

(e) $2(5 + 6) + 7 \cdot 3$

② ① ④ ③

$= 2(11) + 7 \cdot 3$ Add inside the parentheses.

$= 22 + 21$ Multiply, working from left to right.

$= 43$ Add.

$2^3 = 2 \cdot 2 \cdot 2$, not $2 \cdot 3$.

(f) $9 - 2^3 + 5$

$= 9 - 2 \cdot 2 \cdot 2 + 5$ Apply the exponent.

$= 9 - 8 + 5$ Multiply.

$= 1 + 5$ Subtract.

$= 6$ Add.

(g) $72 \div 2 \cdot 3 + 4 \cdot 2^3 - 3^3$ ⟵ Think: $3^3 = 3 \cdot 3 \cdot 3$

$= 72 \div 2 \cdot 3 + 4 \cdot 8 - 27$ Apply the exponents.

$= 36 \cdot 3 + 4 \cdot 8 - 27$ Divide.

$= 108 + 32 - 27$ Multiply, working from left to right.

$= 140 - 27$ Add.

$= 113$ Subtract.

Multiplications and divisions are done from left to right as they appear. Then additions and subtractions are done from left to right as they appear.

NOW TRY

OBJECTIVE 3 Use more than one grouping symbol.

In an expression such as $2(8 + 3(6 + 5))$, we often use brackets in place of the outer pair of parentheses.

EXAMPLE 3 Using Brackets and Fraction Bars as Grouping Symbols

Simplify each expression.

(a) $2[8 + 3(6 + 5)]$

$= 2[8 + 3(11)]$ Add inside the parentheses.

$= 2[8 + 33]$ Multiply inside the brackets.

$= 2[41]$ Add inside the brackets.

$= 82$ Multiply.

NOW TRY ANSWERS
2. (a) 3   **(b)** 16
    (c) 1   **(d)** 60

NOW TRY EXERCISE 3

Simplify each expression.

(a) $7[(3^2 - 1) + 4]$

(b) $\dfrac{9(14 - 4) - 2}{4 + 3 \cdot 6}$

(b) $\dfrac{4(5 + 3) + 3}{2(3) - 1}$ Simplify the numerator and denominator separately.

$= \dfrac{4(8) + 3}{2(3) - 1}$ Work inside the parentheses in the numerator.

$= \dfrac{32 + 3}{6 - 1}$ Multiply.

$= \dfrac{35}{5}$ Add and subtract.

$= 7$ Divide.

NOW TRY

NOTE The expression $\frac{4(5 + 3) + 3}{2(3) - 1}$ in **Example 3(b)** can be written as the quotient

$$[4(5 + 3) + 3] \div [2(3) - 1].$$

This shows that the fraction bar "groups" the numerator and denominator separately.

OBJECTIVE 4 Know the meanings of $\neq$, $<$, $>$, $\leq$, and $\geq$.

So far, we have used the equality symbol $=$. The symbols

$$\neq, \quad <, \quad >, \quad \leq, \quad \text{and} \quad \geq \qquad \text{Inequality symbols}$$

are used to express an **inequality,** a statement that two expressions may not be equal. The equality symbol with a slash through it, $\neq$, means "is not equal to."

$$7 \neq 8 \qquad \text{7 is not equal to 8.}$$

If two numbers are not equal, then one of the numbers must be less than the other. The symbol $<$ represents "is less than."

$$7 < 8 \qquad \text{7 is less than 8.}$$

The symbol $>$ means "is greater than."

$$8 > 2 \qquad \text{8 is greater than 2.}$$

To keep the meanings of the symbols $<$ and $>$ clear, remember that the symbol always points to the lesser number.

$$\text{Lesser number} \rightarrow \quad 8 < 15$$

$$15 > 8 \quad \leftarrow \text{Lesser number}$$

The symbol $\leq$ means "is less than or equal to."

$$5 \leq 9 \qquad \text{5 is less than or equal to 9.}$$

If either the $<$ part or the $=$ part is true, then the inequality $\leq$ is true. The statement $5 \leq 9$ is true because $5 < 9$ is true.

The symbol $\geq$ means "is greater than or equal to."

$$9 \geq 5 \qquad \text{9 is greater than or equal to 5.}$$

NOW TRY ANSWERS
3. (a) 84 **(b)** 4

**NOW TRY
EXERCISE 4**

Determine whether each statement is *true* or *false*.

(a) $12 \neq 10 - 2$

(b) $5 > 4 \cdot 2$

(c) $7 \leq 7$

(d) $\dfrac{5}{9} > \dfrac{7}{11}$

EXAMPLE 4 Using Inequality Symbols

Determine whether each statement is *true* or *false*.

(a) $6 \neq 5 + 1$ This statement is false because $6 = 5 + 1$.

(b) $5 + 3 < 19$ The statement $5 + 3 < 19$ is true because $8 < 19$.

(c) $15 \leq 20 \cdot 2$ The statement $15 \leq 20 \cdot 2$ is true because $15 < 40$.

(d) $25 \geq 30$ Both $25 > 30$ and $25 = 30$ are false, so $25 \geq 30$ is false.

(e) $12 \geq 12$ Because $12 = 12$, this statement is true.

(f) $9 < 9$ Because $9 = 9$, this statement is false.

(g) $\dfrac{6}{15} \geq \dfrac{2}{3}$ Find a common denominator.

$\dfrac{6}{15} \geq \dfrac{10}{15}$ Both $\dfrac{6}{15} > \dfrac{10}{15}$ and $\dfrac{6}{15} = \dfrac{10}{15}$ are false, so $\dfrac{6}{15} \geq \dfrac{2}{3}$ is false.   **NOW TRY**

OBJECTIVE 5 Translate word statements to symbols.

EXAMPLE 5 Translating from Words to Symbols

Write each word statement in symbols.

**NOW TRY
EXERCISE 5**

Write each word statement in symbols.

(a) Ten is not equal to eight minus two.

(b) Fifty is greater than fifteen.

(c) Eleven is less than or equal to twenty.

(a) Twelve equals ten plus two.

$$12 = 10 + 2$$

(b) Nine is less than ten.

$$9 < 10$$

(c) Fifteen is not equal to eighteen.

$$15 \neq 18$$

(d) Seven is greater than four.

$$7 > 4$$

(e) Thirteen is less than or equal to forty.

$$13 \leq 40$$

(f) Eleven is greater than or equal to eleven.

$$11 \geq 11$$

NOW TRY

OBJECTIVE 6 Write statements that change the direction of inequality symbols.

Any statement involving $<$ can be converted to one with $>$, and any statement involving $>$ can be converted to one with $<$. *We do this by reversing the order of the numbers and the direction of the symbol.*

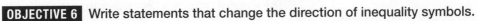

$6 < 10$ becomes $10 > 6$ Interchange numbers.

Reverse symbol.

**NOW TRY
EXERCISE 6**

Write the statement as another true statement with the inequality symbol reversed.

$$8 < 9$$

EXAMPLE 6 Converting between Inequality Symbols

Parts (a)–(c) each show a statement written in two equally correct ways. In each inequality, the inequality symbol points toward the lesser number.

(a) $5 > 2$, $2 < 5$ (b) $\dfrac{1}{2} \leq \dfrac{3}{4}$, $\dfrac{3}{4} \geq \dfrac{1}{2}$ (c) $1.2 \geq 0.5$, $0.5 \leq 1.2$

NOW TRY

NOW TRY ANSWERS

4. (a) true (b) false
   (c) true (d) false
5. (a) $10 \neq 8 - 2$ (b) $50 > 15$
   (c) $11 \leq 20$
6. $9 > 8$

▼ **Summary of Equality and Inequality Symbols**

Symbol	Meaning	Example
=	Is equal to	$0.5 = \frac{1}{2}$ means 0.5 is equal to $\frac{1}{2}$.
≠	Is not equal to	$3 \neq 7$ means 3 is not equal to 7.
<	Is less than	$6 < 10$ means 6 is less than 10.
>	Is greater than	$15 > 14$ means 15 is greater than 14.
≤	Is less than or equal to	$4 \leq 8$ means 4 is less than or equal to 8.
≥	Is greater than or equal to	$1 \geq 0$ means 1 is greater than or equal to 0.

⚠ **CAUTION** Equality and inequality symbols are used to write mathematical **sentences.** Operation symbols $(+, -, \cdot,$ and $\div)$ are used to write mathematical **expressions** that represent a number. Compare the following.

Sentence: $4 < 10$ ← Gives the relationship between 4 and 10

Expression: $4 + 10$ ← Tells how to operate on 4 and 10 to get 14

1.1 Exercises

FOR EXTRA HELP   ▶ MyMathLab®

▶ *Complete solution available in MyMathLab*

Concept Check *Decide whether each statement is* true *or* false. *If it is false, explain why.*

1. $3^2 = 6$

2. $1^3 = 3$

3. $3^1 = 1$

4. The expression 6^2 means that 2 is used as a factor 6 times.

5. When evaluated, $4 + 3(8 - 2)$ is equal to 42.

6. When evaluated, $12 \div 2 \cdot 3$ is equal to 2.

Concept Check *For each expression, label the order in which the operations should be performed. Do not actually perform them.*

7. $18 - 2 + 3$
  ○ ○

8. $28 - 6 \div 2$
  ○ ○

9. $2 \cdot 8 - 6 \div 3$
  ○ ○ ○

10. $40 + 6(3 - 1)$
  ○ ○ ○

11. $3 \cdot 5 - 2(4 + 2)$
  ○ ○ ○ ○

12. $9 - 2^3 + 3 \cdot 4$
  ○○○ ○

Find the value of each exponential expression. **See Example 1.**

13. 3^2

14. 8^2

▶ **15.** 7^2

16. 4^2

17. 12^2

18. 14^2

19. 4^3

20. 5^3

21. 10^3

22. 11^3

23. 3^4

24 6^4

25. 4^5

26. 3^5

27. $\left(\frac{1}{6}\right)^2$

28. $\left(\frac{1}{3}\right)^2$

29. $\left(\frac{2}{3}\right)^4$

30. $\left(\frac{3}{4}\right)^3$

31. $(0.4)^3$

32. $(0.5)^4$

Find the value of each expression. **See Examples 2 and 3.**

33. $64 \div 4 \cdot 2$

34. $250 \div 5 \cdot 2$

▶ **35.** $13 + 9 \cdot 5$

36. $11 + 7 \cdot 6$

37. $25.2 - 12.6 \div 4.2$

38. $12.4 - 9.3 \div 3.1$

39. $\frac{1}{4} \cdot \frac{2}{3} + \frac{2}{5} \cdot \frac{11}{3}$

40. $\frac{9}{4} \cdot \frac{2}{3} + \frac{4}{5} \cdot \frac{5}{3}$

41. $9 \cdot 4 - 8 \cdot 3$

42. $11 \cdot 4 + 10 \cdot 3$ **43.** $20 - 4 \cdot 3 + 5$ **44.** $18 - 7 \cdot 2 + 6$

45. $10 + 40 \div 5 \cdot 2$ **46.** $12 + 64 \div 8 - 4$ **47.** $18 - 2(3 + 4)$

48. $30 - 3(4 + 2)$ **49.** $3(4 + 2) + 8 \cdot 3$ **50.** $9(1 + 7) + 2 \cdot 5$

51. $18 - 4^2 + 3$ **52.** $22 - 2^3 + 9$ **53.** $2 + 3[5 + 4(2)]$

54. $5 + 4[1 + 7(3)]$ **55.** $5[3 + 4(2^2)]$ **56.** $6[2 + 8(3^3)]$

▶ 57. $3^2[(11 + 3) - 4]$ **58.** $4^2[(13 + 4) - 8]$ **59.** $\dfrac{6(3^2 - 1) + 8}{8 - 2^2}$

60. $\dfrac{2(8^2 - 4) + 8}{29 - 3^3}$ **61.** $\dfrac{4(6 + 2) + 8(8 - 3)}{6(4 - 2) - 2^2}$ **62.** $\dfrac{6(5 + 1) - 9(1 + 1)}{5(8 - 6) - 2^3}$

Extending Skills   *Insert one pair of parentheses in each expression so that the given value results when the operations are performed.*

63. $3 \cdot 6 + 4 \cdot 2$ **64.** $2 \cdot 8 - 1 \cdot 3$ **65.** $10 - 7 - 3$

   $= 60$ $= 42$ $= 6$

66. $15 - 10 - 2$ **67.** $8 + 2^2$ **68.** $4 + 2^2$

   $= 7$ $= 100$ $= 36$

First simplify both sides of each inequality. Then determine whether the given statement is true *or* false. ***See Examples 2–4.***

▶ 69. $9 \cdot 3 - 11 \le 16$ **70.** $6 \cdot 5 - 12 \le 18$

71. $5 \cdot 11 + 2 \cdot 3 \le 60$ **72.** $9 \cdot 3 + 4 \cdot 5 \ge 48$

73. $0 \ge 12 \cdot 3 - 6 \cdot 6$ **74.** $10 \le 13 \cdot 2 - 15 \cdot 1$

75. $45 \ge 2[2 + 3(2 + 5)]$ **76.** $55 \ge 3[4 + 3(4 + 1)]$

77. $[3 \cdot 4 + 5(2)] \cdot 3 > 72$ **78.** $2 \cdot [7 \cdot 5 - 3(2)] \le 58$

79. $\dfrac{3 + 5(4 - 1)}{2 \cdot 4 + 1} \ge 3$ **80.** $\dfrac{7(3 + 1) - 2}{3 + 5 \cdot 2} \le 2$

81. $3 \ge \dfrac{2(5 + 1) - 3(1 + 1)}{5(8 - 6) - 4 \cdot 2}$ **82.** $7 \le \dfrac{3(8 - 3) + 2(4 - 1)}{9(6 - 2) - 11(5 - 2)}$

Write each statement in words, and decide whether it is true *or* false. ***See Examples 4 and 5.***

83. $5 < 17$ **84.** $8 < 12$ **85.** $5 \ne 8$ **86.** $6 \ne 9$

87. $7 \ge 14$ **88.** $6 \ge 12$ **89.** $15 \le 15$ **90.** $21 \le 21$

91. $\dfrac{1}{3} = \dfrac{3}{10}$ **92.** $\dfrac{10}{6} = \dfrac{3}{2}$ **93.** $2.5 > 2.50$ **94.** $1.80 > 1.8$

Write each word statement in symbols. ***See Example 5.***

95. Fifteen is equal to five plus ten.   **96.** Twelve is equal to twenty minus eight.

▶ 97. Nine is greater than five minus four.   **98.** Ten is greater than six plus one.

99. Sixteen is not equal to nineteen.   **100.** Three is not equal to four.

101. One-half is less than or equal to two-fourths.

102. One-third is less than or equal to three-ninths.

Write each statement with the inequality symbol reversed while keeping the same meaning.
See Example 6.

103. $5 < 20$ **104.** $30 > 9$ **105.** $\dfrac{4}{5} > \dfrac{3}{4}$

106. $\dfrac{5}{4} < \dfrac{3}{2}$ **107.** $2.5 \geq 1.3$ **108.** $4.1 \leq 5.3$

One way to measure a person's cardiofitness is to calculate how many METs, or metabolic units, he or she can reach at peak exertion. One MET is the amount of energy used when sitting quietly. To calculate ideal METs, we can use the following expressions.

$$14.7 - \text{age} \cdot 0.13 \quad \text{For women}$$
$$14.7 - \text{age} \cdot 0.11 \quad \text{For men}$$

(Source: New England Journal of Medicine.)

109. A 40-yr-old woman wishes to calculate her ideal MET.

    (a) Write the expression, using her age.

    (b) Calculate her ideal MET. (*Hint:* Use the rules for order of operations.)

    (c) Researchers recommend that a person reach approximately 85% of his or her MET when exercising. Calculate 85% of the ideal MET from part (b). Then refer to the following table. What activity listed in the table can the woman do that is approximately this value?

Activity	METs	Activity	METs
Golf (with cart)	2.5	Skiing (water or downhill)	6.8
Walking (3 mph)	3.3	Swimming	7.0
Mowing lawn (power mower)	4.5	Walking (5 mph)	8.0
Ballroom or square dancing	5.5	Jogging	10.2
Cycling	5.7	Skipping rope	12.0

Source: Harvard School of Public Health.

    (d) Repeat parts (a)–(c) for a 55-yr-old man.

110. Repeat parts (a)–(c) of **Exercise 109** using your age and gender.

The table shows the number of pupils per teacher in U.S. public schools in selected states.

111. Which states had a number greater than 13.8?

112. Which states had a number that was at most 14.7?

113. Which states had a number not less than 13.8?

114. Which states had a number less than 13.0?

State	Pupils per Teacher
Alaska	16.2
Texas	14.7
California	24.1
Wyoming	12.5
Maine	12.3
Idaho	17.6
Missouri	13.8

Source: National Center for Education Statistics.

STUDY SKILLS

Taking Lecture Notes

Study the set of sample math notes given here.

- **Include the date and title** of the day's lecture topic.
- **Include definitions,** written here in parentheses—don't trust your memory.
- **Skip lines and write neatly** to make reading easier.
- **Emphasize direction words** (like *simplify*) with their explanations.
- **Mark important concepts with stars, underlining, etc.**
- **Use two columns,** which allows an example and its explanation to be close together.
- **Use brackets and arrows** to clearly show steps, related material, etc.

With a partner or in a small group, compare lecture notes. Answer each question.

1. What are you doing to show main points in your notes (such as boxing, using stars, etc.)?

2. In what ways do you set off explanations from worked problems and subpoints (such as indenting, using arrows, circling, etc.)?

3. What new ideas did you learn by examining your class-mates' notes?

4. What new techniques will you try in your notes?

January 2

Exponents

Exponents used to show repeated multiplication.

$3 \cdot 3 \cdot 3 \cdot 3$ *can be written* 3^4 ← *exponent (how many times it's multiplied)*

base (the number being multiplied)

Read 3^2 *as 3 to the 2nd power or 3 squared*

3^3 *as 3 to the 3rd power or 3 cubed*

3^4 *as 3 to the 4th power*

etc.

Simplifying an expression with exponents

actually do the repeated multiplication

2^3 *means* $2 \cdot 2 \cdot 2$ *and* $2 \cdot 2 \cdot 2 = 8$

★*Careful!* 5^2 *means* $5 \cdot 5$ *NOT* $5 \cdot 2$
so $5^2 = 5 \cdot 5 = 25$ *BUT* $5^2 \neq 10$

Example

Simplify $(2^4) \cdot (3^2)$

$2 \cdot 2 \cdot 2 \cdot 2 \quad 3 \cdot 3$

$16 \quad \cdot \quad 9$

144

Explanation

Exponents mean multiplication.

Use 2 as a factor 4 times.
Use 3 as a factor 2 times.
$2 \cdot 2 \cdot 2 \cdot 2$ *is 16*
$3 \cdot 3$ *is 9* } $16 \cdot 9$ *is 144*

simplified result is 144
(no exponents left)

1.2   Variables, Expressions, and Equations

OBJECTIVES

1. Evaluate algebraic expressions, given values for the variables.
2. Translate word phrases to algebraic expressions.
3. Identify solutions of equations.
4. Identify solutions of equations from a set of numbers.
5. Distinguish between *equations* and *expressions*.

A **constant** is a fixed, unchanging number. A **variable** is a symbol, usually a letter, used to represent an unknown number.

$$5, \quad \frac{3}{4}, \quad 8\frac{1}{2}, \quad 10.8 \quad \text{Constants} \quad \bigg| \quad a, \quad x, \quad y, \quad z \quad \text{Variables}$$

An **algebraic expression** is a sequence of constants, variables, operation symbols, and/or grouping symbols formed according to the rules of algebra.

$$x + 5, \quad 2m - 9, \quad 8p^2 + 6(p - 2) \quad \text{Algebraic expressions}$$

$2m$ means $2 \cdot m$, the product of 2 and m.

$6(p - 2)$ means the product of 6 and $p - 2$.

VOCABULARY
☐ constant
☐ variable
☐ algebraic expression
☐ equation
☐ solution
☐ set
☐ element

NOW TRY EXERCISE 1

Find the value of each expression for $x = 6$.

(a) $9x$ **(b)** $4x^2$

OBJECTIVE 1 Evaluate algebraic expressions, given values for the variables.

To *evaluate* an expression means to find its *value*. An algebraic expression can have different numerical values for different values of the variables.

EXAMPLE 1 Evaluating Algebraic Expressions

Find the value of each expression for $x = 5$.

(a) $8x$ ◁ Multiplication is understood.

$= 8 \cdot x$

$= 8 \cdot 5$ Let $x = 5$.

$= 40$ Multiply.

(b) $3x^2$

$= 3 \cdot x^2$ [$5^2 = 5 \cdot 5$]

$= 3 \cdot 5^2$ Let $x = 5$.

$= 3 \cdot 25$ Square 5.

$= 75$ Multiply.   **NOW TRY**

! CAUTION

$3x^2$ means $3 \cdot x^2$, **not** $3x \cdot 3x$.   See **Example 1(b)**.

Unless parentheses are used, the exponent refers only to the variable or constant just before it. We use parentheses to write $3x \cdot 3x$ with exponents as $(3x)^2$.

NOW TRY EXERCISE 2

Find the value of each expression for $x = 4$ and $y = 7$.

(a) $3x + 4y$ **(b)** $\dfrac{6x - 2y}{2y - 9}$

(c) $4x^2 - y^2$

EXAMPLE 2 Evaluating Algebraic Expressions

Find the value of each expression for $x = 5$ and $y = 3$.

(a) $2x + 7y$ We could use parentheses and write $2(5) + 7(3)$.

[Follow the rules for order of operations.]

$= 2 \cdot 5 + 7 \cdot 3$ Let $x = 5$ and $y = 3$.

$= 10 + 21$ Multiply.

$= 31$ Add.

(b) $\dfrac{9x - 8y}{2x - y}$

$= \dfrac{9 \cdot 5 - 8 \cdot 3}{2 \cdot 5 - 3}$ Let $x = 5$ and $y = 3$.

$= \dfrac{45 - 24}{10 - 3}$ Multiply.

$= \dfrac{21}{7}$ Subtract.

$= 3$ Divide.

(c) $x^2 - 2y^2$ [$3^2 = 3 \cdot 3$]

$= 5^2 - 2 \cdot 3^2$ Let $x = 5$ and $y = 3$.

[$5^2 = 5 \cdot 5$]

$= 25 - 2 \cdot 9$ Apply the exponents.

$= 25 - 18$ Multiply.

$= 7$ Subtract.   **NOW TRY**

NOW TRY ANSWERS
1. (a) 54   **(b)** 144
2. (a) 40   **(b)** 2   **(c)** 15

Write each word phrase as an algebraic expression, using x as the variable.

(a) The sum of a number and 10

(b) A number divided by 7

(c) The product of 3 and the difference of 9 and a number

OBJECTIVE 2   Translate word phrases to algebraic expressions.

EXAMPLE 3   Using Variables to Write Word Phrases as Algebraic Expressions

Write each word phrase as an algebraic expression, using x as the variable.

(a) The sum of a number and 9

$$x + 9, \quad \text{or} \quad 9 + x \qquad \text{"Sum" is the answer to an addition problem.}$$

(b) 7 minus a number

$$7 - x \qquad \text{"Minus" indicates subtraction.}$$

The expression $x - 7$ is incorrect. We cannot subtract in either order and obtain the same result.

(c) A number subtracted from 12

$$12 - x \quad \boxed{\text{Be careful with order.}}$$

Compare this result with "12 subtracted from a number," which is $x - 12$.

(d) The product of 11 and a number

$$11 \cdot x, \quad \text{or} \quad 11x$$

(e) 5 divided by a number

$$5 \div x, \quad \text{or} \quad \frac{5}{x} \quad \boxed{\frac{x}{5} \text{ is } not \text{ correct here.}}$$

(f) The product of 2 and the difference of a number and 8

We are multiplying 2 times "something." This "something" is the difference of a number and 8, written $x - 8$. We use parentheses around this difference.

$$2 \cdot (x - 8), \quad \text{or} \quad 2(x - 8) \quad \boxed{\begin{array}{l} 8 - x, \text{ which means the difference} \\ \text{of 8 and a number, is } not \text{ correct.} \end{array}}$$

OBJECTIVE 3   Identify solutions of equations.

An **equation** is a statement that two algebraic expressions are equal. ***An equation always includes the equality symbol, $=$.***

$$x + 4 = 11, \qquad 2y = 16, \qquad 4p + 1 = 25 - p,$$
$$\frac{3}{4}x + \frac{1}{2} = 0, \qquad z^2 = 4, \qquad 4(m - 0.5) = 2m$$

Equations

To **solve an equation** means to find the value of the variable that makes the equation true. Such a value of the variable is a **solution** of the equation.

EXAMPLE 4   Deciding Whether a Number Is a Solution of an Equation

Decide whether each equation has the given number as a solution.

(a) $5p + 1 = 36; \quad 7$

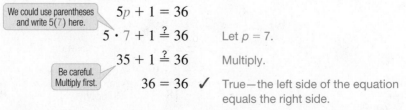

$$5p + 1 = 36$$
$$\boxed{\begin{array}{l} \text{We could use parentheses} \\ \text{and write 5(7) here.} \end{array}} \quad 5 \cdot 7 + 1 \overset{?}{=} 36 \qquad \text{Let } p = 7.$$
$$35 + 1 \overset{?}{=} 36 \qquad \text{Multiply.}$$
$$\boxed{\begin{array}{l} \text{Be careful.} \\ \text{Multiply first.} \end{array}} \quad 36 = 36 \quad \checkmark \quad \begin{array}{l} \text{True—the left side of the equation} \\ \text{equals the right side.} \end{array}$$

The number 7 is a solution of the equation.

**NOW TRY
EXERCISE 4**

Decide whether the equation has the given number as a solution.

$8k + 5 = 61; \quad 7$

(b) $9m - 6 = 32; \quad 4$

$$9m - 6 = 32$$
$$9 \cdot 4 - 6 \overset{?}{=} 32 \qquad \text{Let } m = 4.$$
$$36 - 6 \overset{?}{=} 32 \qquad \text{Multiply.}$$
$$30 = 32 \qquad \text{False—the left side does } not \text{ equal the right side.}$$

The number 4 is not a solution of the equation.

NOW TRY

OBJECTIVE 4 Identify solutions of equations from a set of numbers.

A **set** is a collection of objects. In mathematics, these objects are usually numbers. The objects that belong to a set are its **elements.** They are written between **braces { }.**

$$\{1, 2, 3, 4, 5\} \longleftarrow \text{The set containing the elements 1, 2, 3, 4, and 5}$$

**NOW TRY
EXERCISE 5**

Write the word statement as an equation. Use x as the variable. Then find the solution of the equation from the set $\{0, 2, 4, 6, 8, 10\}$.

The sum of a number and nine is equal to the difference of 25 and the number.

EXAMPLE 5 Finding a Solution from a Given Set

Write each word statement as an equation. Use x as the variable. Then find the solution of the equation from the following set.

$$\{0, 2, 4, 6, 8, 10\}$$

(a) The sum of a number and four is six.

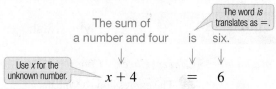

One by one, mentally substitute each number from the given set $\{0, 2, 4, 6, 8, 10\}$ in $x + 4 = 6$. Because $2 + 4 = 6$ is true, 2 is the only solution.

(b) Nine more than five times a number is 49.

Start with $5x$, and then add 9 to it.   The word *is* translates as $=$.

$$5x + 9 = 49 \qquad 5 \cdot x = 5x$$

Substitute each of the given numbers. The solution is 8 because

$$5 \cdot 8 + 9 = 49 \text{ is true.}$$

(c) The sum of a number and 12 is equal to four times the number.

The sum of a number and 12   is equal to   four times the number.

$$x + 12 \qquad = \qquad 4x \qquad 4 \cdot x = 4x$$

Substituting each of the given numbers leads to a true statement only for $x = 4$ because

$$4 + 12 = 4(4) \text{ is true.}$$

NOW TRY

OBJECTIVE 5   Distinguish between *equations* and *expressions*.

To distinguish between equations and expressions, remember the following.

> *An equation is a sentence—it has something on the left side, an = symbol, and something on the right side.*
>
> *An expression is a phrase that represents a number.*

$$\underbrace{4x + 5}_{\text{Left side}} = \underbrace{9}_{\text{Right side}} \qquad 4x + 5$$

Equation
(to solve)

Expression
(to simplify or evaluate)

**NOW TRY
EXERCISE 6**

Decide whether each of the following is an *equation* or an *expression*.

(a) $2x + 5 = 6$

(b) $2x + 5 - 6$

NOW TRY ANSWERS
6. **(a)** equation   **(b)** expression

EXAMPLE 6   Distinguishing between Equations and Expressions

Decide whether each of the following is an *equation* or an *expression*.

(a) $2x - 3$ Ask, "Is there an equality symbol?" The answer is no, so this is an expression.

(b) $2x - 3 = 8$ Because there is an equality symbol with something on either side of it, this is an equation.

(c) $5x^2 + 2y^2$ There is no equality symbol. This is an expression.

NOW TRY

1.2 Exercises

FOR EXTRA HELP   ▶ MyMathLab®

▶ *Complete solution available in MyMathLab*

Concept Check   *Choose the letter(s) of the correct response.*

1. The expression $8x^2$ means _____ .

   A. $8 \cdot x \cdot 2$ **B.** $8 \cdot x \cdot x$ **C.** $8 + x^2$ **D.** $8x \cdot 8x$

2. If $x = 2$ and $y = 1$, then the value of xy is _____ .

   A. $\dfrac{1}{2}$ **B.** 1 **C.** 2 **D.** 3

3. The sum of 15 and a number x is represented by _____ .

   A. $15 + x$ **B.** $15 - x$ **C.** $x - 15$ **D.** $15x$

4. Which of the following are expressions? Which are equations?

   A. $6x = 7$ **B.** $6x + 7$ **C.** $6x - 7$ **D.** $6x - 7 = 0$

Give a short explanation.

5. Explain why $2x^3$ is not the same as $2x \cdot 2x \cdot 2x$.

6. Why are "7 less than a number" and "7 is less than a number" translated differently?

7. When evaluating the expression $5x^2$ for $x = 4$, explain why 4 must be squared *before* multiplying by 5.

8. Suppose that the directions on a test read "*Solve each expression.*" How could we politely correct the person who wrote these directions?

*Find the value of each expression for **(a)** x = 4 and **(b)** x = 6.* **See Example 1.**

9. $x + 7$ **10.** $x - 3$ **11.** $4x$ **12.** $6x$ ▶ **13.** $4x^2$

14. $5x^2$ **15.** $\dfrac{x + 1}{3}$ **16.** $\dfrac{x - 2}{5}$ **17.** $\dfrac{3x - 5}{2x}$ **18.** $\dfrac{4x - 1}{3x}$

19. $3x^2 + x$ **20.** $2x + x^2$ **21.** $6.459x$ **22.** $3.275x$

*Find the value of each expression for **(a)** x = 2 and y = 1 and **(b)** x = 1 and y = 5.* **See Example 2.**

▶ **23.** $8x + 3y + 5$ **24.** $4x + 2y + 7$ **25.** $3(x + 2y)$ **26.** $2(2x + y)$

27. $x + \dfrac{4}{y}$ **28.** $y + \dfrac{8}{x}$ **29.** $\dfrac{x}{2} + \dfrac{y}{3}$ **30.** $\dfrac{x}{5} + \dfrac{y}{4}$

31. $\dfrac{2x + 4y - 6}{5y + 2}$ **32.** $\dfrac{4x + 3y - 1}{x}$ **33.** $2y^2 + 5x$ **34.** $6x^2 + 4y$

35. $\dfrac{3x + y^2}{2x + 3y}$ **36.** $\dfrac{x^2 + 1}{4x + 5y}$ **37.** $0.841x^2 + 0.32y^2$ **38.** $0.941x^2 + 0.25y^2$

Write each word phrase as an algebraic expression, using x as the variable. **See Example 3.**

▶ **39.** Twelve times a number   **40.** Fifteen times a number

41. Nine added to a number   **42.** Six added to a number

43. Four subtracted from a number   **44.** Seven subtracted from a number

45. A number subtracted from seven   **46.** A number subtracted from four

47. The difference of a number and 8   **48.** The difference of 8 and a number

49. 18 divided by a number   **50.** A number divided by 18

51. The product of 6 and four less than a number   **52.** The product of 9 and five more than a number

Decide whether each equation has the given number as a solution. **See Example 4.**

53. $4m + 2 = 6$;   1   **54.** $2r + 6 = 8$;   1

▶ **55.** $2y + 3(y - 2) = 14$;   3   **56.** $6x + 2(x + 3) = 14$;   2

57. $6p + 4p + 9 = 11$; $\dfrac{1}{5}$ **58.** $2x + 3x + 8 = 20$; $\dfrac{12}{5}$

59. $3r^2 - 2 = 46$;   4   **60.** $2x^2 + 1 = 19$;   3

61. $\dfrac{3}{8}x + \dfrac{1}{4} = 1$;   2   **62.** $\dfrac{7}{10}x + \dfrac{1}{2} = 4$;   5

63. $0.5(x - 4) = 80$;   20   **64.** $0.2(x - 5) = 70$;   40

Write each word statement as an equation. Use x as the variable. Then find the solution of the equation from the set $\{2, 4, 6, 8, 10\}$. **See Example 5.**

▶ **65.** The sum of a number and 8 is 18.   **66.** A number minus three equals 1.

67. One more than twice a number is 5.   **68.** The product of a number and 3 is 6.

69. Sixteen minus three-fourths of a number is 13.

70. The sum of six-fifths of a number and 2 is 14.

71. Three times a number is equal to 8 more than twice the number.

72. Twelve divided by a number equals $\dfrac{1}{3}$ times that number.

Decide whether each of the following is an equation or an expression. See Example 6.

▶ **73.** $3x + 2(x - 4)$ **74.** $8y - (3y + 5)$ **75.** $7t + 2(t + 1) = 4$

76. $9r + 3(r - 4) = 2$ **77.** $x + y = 9$ **78.** $x + y - 9$

A ***mathematical model*** is an equation that describes the relationship between two quantities. For example, the life expectancy at birth of Americans can be approximated by the equation

$$y = 0.180x - 283,$$

where *x* is a year between 1960 and 2010 and *y* is age in years. (Source: Centers for Disease Control and Prevention.)

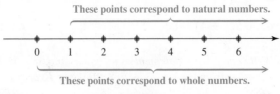

Use this model to approximate life expectancy (to the nearest year) in each of the following years.

79. 1960   **80.** 1975

81. 1995   **82.** 2010

1.3   Real Numbers and the Number Line

OBJECTIVES

1. Classify numbers and graph them on number lines.
2. Tell which of two real numbers is less than the other.
3. Find the additive inverse of a real number.
4. Find the absolute value of a real number.
5. Interpret meanings of real numbers from a table of data.

VOCABULARY

☐ natural (counting) numbers
☐ whole numbers
☐ number line
☐ integers
☐ signed numbers
☐ rational numbers
☐ graph
☐ coordinate
☐ irrational numbers
☐ real numbers
☐ additive inverse (opposite)
☐ absolute value

OBJECTIVE 1 Classify numbers and graph them on number lines.

The set of numbers used for counting is the *natural numbers*. The set of *whole numbers* includes 0 with the natural numbers.

Natural Numbers and Whole Numbers

$\{1, 2, 3, 4, 5, \ldots\}$ is the set of **natural numbers** (or **counting numbers**).

$\{0, 1, 2, 3, 4, 5 \ldots\}$ is the set of **whole numbers.**

We can represent numbers on a **number line** like the one in **FIGURE 1**.

These points correspond to natural numbers.

To draw a number line, choose any point on the line and label it 0. Then choose any point to the right of 0 and label it 1. Use the distance between 0 and 1 as the scale to locate, and then label, other points.

These points correspond to whole numbers.

FIGURE 1

The natural numbers are located to the right of 0 on the number line. For each natural number, we can place a corresponding number to the left of 0, labeling the points -1, -2, -3, and so on, as shown in **FIGURE 2** on the next page. Each is the **opposite,** or **negative,** of a natural number. The natural numbers, their opposites, and 0 form the set of *integers*.

> ### Integers
>
> $\{\ldots, -3, -2, -1, 0, 1, 2, 3, \ldots\}$ is the set of **integers.**

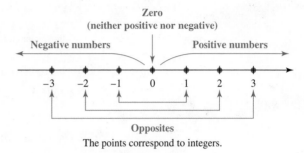

Zero
(neither positive nor negative)

Negative numbers   Positive numbers

Opposites
The points correspond to integers.

FIGURE 2

Positive numbers and *negative numbers* are **signed numbers.**

NOW TRY EXERCISE 1

Use an integer to express the number in boldface italics in the following statement.

At its deepest point, the floor of West Okoboji Lake sits *136* ft below the water's surface. (*Source:* www.watersafetycouncil.org)

> **EXAMPLE 1** Using Negative Numbers

Use an integer to express the number in boldface italics in each statement.

(a) The lowest Fahrenheit temperature ever recorded was *129°* below zero at Vostok, Antarctica, on July 21, 1983. (*Source: World Almanac and Book of Facts.*)

Use -129 because "below zero" indicates a negative number.

(b) General Motors had a loss of about $*399* billion in 2013. (*Source: The Wall Street Journal.*)

A loss indicates a negative "profit," here -399.　　　　NOW TRY

Fractions, reviewed in **Section R.1,** are examples of *rational numbers.*

> ### Rational Numbers
>
> $\{x \mid x$ is a quotient of two integers, with denominator not $0\}$ is the set of **rational numbers.**
>
> 　　(Read the part in the braces as "the set of all numbers x such that x is a quotient of two integers, with denominator not 0.")

NOTE　The set symbolism used in the definition of rational numbers,

$$\{x \mid x \text{ has a certain property}\},$$

is **set-builder notation.** We use this notation when it is not possible to list all the elements of a set.

Because any number that can be written as the quotient of two integers (that is, as a fraction) is a rational number, *all integers, mixed numbers, terminating (or ending) decimals, and repeating decimals are rational.*

▼ Rational Numbers

Rational Number	Equivalent Quotient of Two Integers
-5	$\frac{-5}{1}$ (means $-5 \div 1$)
$1\frac{3}{4}$	$\frac{7}{4}$ (means $7 \div 4$)
0.23 (terminating decimal)	$\frac{23}{100}$ (means $23 \div 100$)
$0.3333\ldots$, or $0.\overline{3}$ (repeating decimal)	$\frac{1}{3}$ (means $1 \div 3$)
4.7	$\frac{47}{10}$ (means $47 \div 10$)

To **graph** a number, we place a dot on a number line at the point that corresponds to the number. The number is the **coordinate** of the point.

**NOW TRY
EXERCISE 2**

Graph each rational number on a number line.

$$-3, \quad \frac{17}{8}, \quad -2.75, \quad 1\frac{1}{2}, \quad -\frac{3}{4}$$

EXAMPLE 2   Graphing Rational Numbers

Graph each rational number on a number line.

$$-\frac{3}{2}, \quad -\frac{2}{3}, \quad 0.5, \quad 1\frac{1}{3}, \quad \frac{23}{8}, \quad 3.25, \quad 4$$

To locate the improper fractions on a number line, write them as mixed numbers or decimals. The graph is shown in **FIGURE 3**.

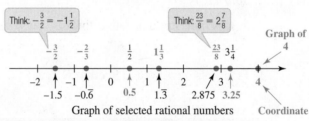

Graph of selected rational numbers

FIGURE 3

Think of the graph of a set of numbers as a picture of the set.

NOW TRY

Not all numbers are rational. For example, the square root of 2, written $\sqrt{2}$, cannot be written as a quotient of two integers. Because of this, $\sqrt{2}$ is an *irrational number*. (See **FIGURE 4**.)

This square has diagonal of length $\sqrt{2}$. The number $\sqrt{2}$ is an irrational number.

FIGURE 4

Irrational Numbers

$\{x \mid x$ is a nonrational number represented by a point on a number line$\}$ is the set of **irrational numbers**.

The decimal form of an irrational number neither terminates nor repeats.

Both rational and irrational numbers can be represented by points on a number line and together form the set of *real numbers*. See **FIGURE 5** on the next page.

NOW TRY ANSWER

2.

$-2.75 \quad -\frac{3}{4} \qquad 1\frac{1}{2} \quad \frac{17}{8}$

$-3 \quad -2 \quad -1 \quad 0 \quad 1 \quad 2 \quad 3$

Real Numbers

$\{x \mid x$ is a rational or an irrational number$\}$ is the set of **real numbers**.*

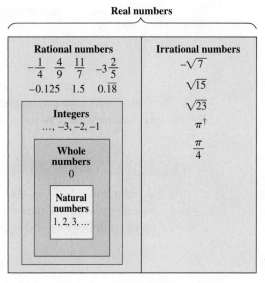

FIGURE 5

NOW TRY
EXERCISE 3

List the numbers in the following set that belong to each set of numbers.

$$\left\{-7, -\tfrac{4}{5}, 0, \sqrt{3}, 2.7, \pi, 13\right\}$$

(a) Whole numbers

(b) Integers

(c) Rational numbers

(d) Irrational numbers

EXAMPLE 3   Determining Whether a Number Belongs to a Set

List the numbers in the following set that belong to each set of numbers.

$$\left\{-5, -\tfrac{2}{3}, 0, 0.\overline{6}, \sqrt{2}, 3\tfrac{1}{4}, 5, 5.8\right\}$$

(a) Natural numbers: 5

(b) Whole numbers: 0 and 5
   The whole numbers consist of the natural (counting) numbers and 0.

(c) Integers: $-5, 0,$ and 5

(d) Rational numbers: $-5, -\tfrac{2}{3}, 0, 0.\overline{6}\left(\text{or } \tfrac{2}{3}\right), 3\tfrac{1}{4}\left(\text{or } \tfrac{13}{4}\right), 5,$ and $5.8\left(\text{or } \tfrac{58}{10}\right)$
   Each of these numbers can be written as the quotient of two integers.

(e) Irrational numbers: $\sqrt{2}$

(f) Real numbers: All the numbers in the set are real numbers.   NOW TRY

OBJECTIVE 2   Tell which of two real numbers is less than the other.

Given any two different positive integers, we can determine which number is less than the other. Positive numbers decrease as the corresponding points on a number line go to the left. For example,

$8 < 12$ because 8 is to the left of 12 on a number line.

This ordering is extended to all real numbers by definition.

NOW TRY ANSWERS
3. (a) 0, 13
   (b) $-7, 0, 13$
   (c) $-7, -\tfrac{4}{5}, 0, 2.7, 13$
   (d) $\sqrt{3}, \pi$

*An example of a number that is not a real number is the square root of a negative number, such as $\sqrt{-5}$.

†The value of the irrational number π (pi) is approximately 3.141592654. The decimal digits continue forever with no repeated pattern.

Ordering of Real Numbers

For any two real numbers a and b, **a is less than b** if a lies to the left of b on a number line. See **FIGURE 6**.

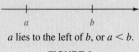

a lies to the left of b, or $a < b$.

FIGURE 6

This means that any negative number is less than 0, and any negative number is less than any positive number. Also, 0 is less than any positive number.

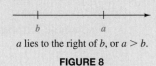

NOW TRY
EXERCISE 4

Determine whether the statement is *true* or *false*.

$$-8 \leq -9$$

EXAMPLE 4   Determining the Order of Real Numbers

Is the statement $-3 < -1$ *true* or *false*?

Locate -3 and -1 on a number line. See **FIGURE 7**. Because -3 lies to the left of -1 on the number line, -3 is less than -1. The statement $-3 < -1$ is true.

-3 lies to the left of -1, so $-3 < -1$.

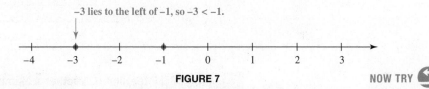

FIGURE 7

NOW TRY

Ordering of Real Numbers

For any two real numbers a and b, **a is greater than b** if a lies to the right of b on a number line. See **FIGURE 8**.

a lies to the right of b, or $a > b$.

FIGURE 8

In **FIGURE 7** above, $-1 > -3$ because -1 lies to the right of -3 on the number line.

OBJECTIVE 3   Find the additive inverse of a real number.

By a property of the real numbers, for any real number x (except 0), there is exactly one number on a number line the same distance from 0 as x, but on the *opposite* side of 0. See **FIGURE 9**. Such pairs of numbers are *additive inverses,* or *opposites,* of each other.

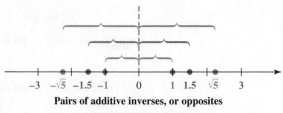

Pairs of additive inverses, or opposites

FIGURE 9

Additive Inverse

The **additive inverse** of a number x is the number that is the same distance from 0 on a number line as x, but on the *opposite* side of 0.

▼ **Additive Inverses**

Number	Additive Inverse
7	−7
−3	−(−3), or 3
0	0
19	−19
$-\frac{2}{3}$	$\frac{2}{3}$
0.52	−0.52

The additive inverse of a nonzero number is found by changing the sign of the number.

A nonzero number and its additive inverse have opposite signs.

We indicate the additive inverse of a number by writing the symbol − in front of the number. For example, the additive inverse of 7 is written −7 (read "negative 7"). We could write the additive inverse of −3 as −(−3), but we know that 3 is the opposite of −3. Because a number can have only one additive inverse, 3 and −(−3) must represent the same number.

$$-(-3) = 3$$

This idea can be generalized.

Double Negative Rule

For any real number x, the following holds.

$$-(-x) = x$$

OBJECTIVE 4 Find the absolute value of a real number.

Because additive inverses are the same distance from 0 on a number line, a number and its additive inverse have the same *absolute value*. The **absolute value** of a real number x, written $|x|$ and read **"the absolute value of x,"** can be defined as the distance between 0 and the number on a number line.

$|2| = 2$ The distance between 2 and 0 on a number line is 2 units.

$|-2| = 2$ The distance between −2 and 0 on a number line is also 2 units.

Distance is a physical measurement, which is never negative. ***Therefore, the absolute value of a number is never negative.***

Absolute Value

For any real number x, the absolute value of x is defined as follows.

$$|x| = \begin{cases} x & \text{if } x \geq 0 \\ -x & \text{if } x < 0 \end{cases}$$

By this definition, if x is a positive number or 0, then its absolute value is x itself. For example, since 8 is a positive number,

$$|8| = 8.$$

If x is a negative number, then its absolute value is the additive inverse of x.

$|-8| = -(-8) = 8$ The additive inverse of −8 is 8.

$$|x| = \begin{cases} x & \text{if } x \geq 0 \\ -x & \text{if } x < 0 \end{cases} \quad \text{Definition of absolute value}$$

The "$-x$" in the second part of the definition of absolute value does *not* represent a negative number. Because x is negative in the second part, it follows that $-x$ represents the *opposite* of a negative number—that is, a positive number. ***The absolute value of a number is never negative.***

NOW TRY
EXERCISE 5
Find each absolute value.
(a) $|4|$ **(b)** $|-4|$
(c) $-|-4|$

EXAMPLE 5 Finding Absolute Values

Find each absolute value.

(a) $|0| = 0$ **(b)** $|5| = 5$ **(c)** $|-5| = -(-5) = 5$

(d) $-|5| = -(5) = -5$ **(e)** $-|-5| = -(5) = -5$

(f) $|8 - 2| = |6| = 6$ **(g)** $-|8 - 2| = -|6| = -6$

Absolute value bars are grouping symbols. In parts (f) and (g), we perform any operations inside the absolute value bars *before* finding the absolute value.

NOW TRY

OBJECTIVE 5 Interpret meanings of real numbers from a table of data.

NOW TRY
EXERCISE 6
In the table for **Example 6,** which category represents a decrease for both years?

EXAMPLE 6 Interpreting Data

The Consumer Price Index (CPI) measures the average change in prices of goods and services purchased by urban consumers in the United States. The table shows the percent change in the CPI for selected categories of goods and services from 2009 to 2010 and from 2010 to 2011. Use the table to answer each question.

Category	Change from 2009 to 2010	Change from 2010 to 2011
Appliances	−4.5	−1.2
Education	4.4	4.2
Gasoline	18.4	26.4
Housing	−0.4	1.3
Medical care	3.4	3.0

Source: U.S. Bureau of Labor Statistics.

(a) Which category in which year represents the greatest percent decrease?
We must find the negative number with the greatest absolute value. The number that satisfies this condition is −4.5, so the greatest percent decrease was shown by appliances from 2009 to 2010.

(b) Which category in which year represents the least change?
We must find the number (either positive, negative, or zero) with the least absolute value. From 2009 to 2010, housing showed the least change, a decrease of 0.4%.

NOW TRY

NOW TRY ANSWERS
5. (a) 4 **(b)** 4 **(c)** −4
6. appliances

1.3 Exercises

FOR EXTRA HELP   MyMathLab®

▶ *Complete solution available in MyMathLab*

Concept Check   *Complete each statement.*

1. The number _____ is a whole number, but not a natural number.

2. The natural numbers, their additive inverses, and 0 form the set of _____ .

3. The additive inverse of every negative number is a (*negative / positive*) number.

4. If x and y are real numbers with $x > y$, then x lies to the (*left / right*) of y on a number line.

5. A rational number is the _____ of two integers, with the _____ not equal to 0.

6. Decimal numbers that neither terminate nor repeat are _____ numbers.

7. *Concept Check*   Match each expression in Column I with its value in Column II. Choices in Column II may be used once, more than once, or not at all.

I	II
(a) $\lvert -9 \rvert$	**A.** 9
(b) $-(-9)$	**B.** -9
(c) $-\lvert -9 \rvert$	**C.** Neither A nor B
(d) $-\lvert -(-9) \rvert$	**D.** Both A and B

8. *Concept Check*   Fill in each blank with the correct values: The opposite of -5 is _____, while the absolute value of -5 is _____. The additive inverse of -5 is _____, while the additive inverse of the absolute value of -5 is _____.

9. Students often say "Absolute value is always positive." Is this true? Explain.

10. *Concept Check*   *True* or *false:* If a is negative, then $\lvert a \rvert = -a$.

Concept Check   *Exercises 11–28 check understanding of the various sets of numbers.*

In Exercises 11–16, give a number that satisfies the given condition.

11. An integer between 3.6 and 4.6

12. A rational number between 2.8 and 2.9

13. A whole number that is not positive and is less than 1

14. A whole number greater than 3.5

15. An irrational number that is between $\sqrt{12}$ and $\sqrt{14}$

16. A real number that is neither negative nor positive

In Exercises 17–22, decide whether each statement is true *or* false.

17. Every natural number is positive.

18. Every whole number is positive.

19. Every integer is a rational number.

20. Every rational number is a real number.

21. Some numbers are both rational and irrational.

22. Every terminating decimal is a rational number.

In Exercises 23–28, give three numbers between -6 *and 6 that satisfy each given condition.*

23. Positive real numbers but not integers

24. Real numbers but not positive numbers

25. Real numbers but not whole numbers

26. Rational numbers but not integers

27. Real numbers but not rational numbers

28. Rational numbers but not negative numbers

In Exercises 29–32, use an integer to express each number in boldface italics representing a change. In Exercises 33 and 34, use a rational number. **See Example 1.**

29. Between July 1, 2011, and July 1, 2012, the population of the United States increased by approximately **2,259,105**. (*Source:* U.S. Census Bureau.)

30. Between 2011 and 2012, the number of movie screens in the United States increased by **207**. (*Source:* Motion Picture Association of America.)

31. From 2011 to 2012, attendance at the first game of the World Series went from 46,406 to 42,982, a decrease of **3424**. (*Source:* Major League Baseball.)

32. In 1935, there were 15,295 banks in the United States. By 2012, the number was 7083, a decrease of **8212** banks. (*Source:* Federal Deposit Insurance Corporation.)

33. On Friday, August 23, 2013, the Dow Jones Industrial Average (DJIA) closed at 15,010.51. On the previous day it had closed at 14,963.74. Thus, on Friday, it closed up **46.77** points. (*Source: The Washington Post.*)

34. On Tuesday, August 27, 2013, the NASDAQ closed at 3578.52. On the previous day, it had closed at 3657.57. Thus, on Tuesday, it closed down **79.05** points. (*Source: The Washington Post.*)

*Graph each number on a number line. **See Example 2.***

35. $0, 3, -5, -6$

36. $2, 6, -2, -1$

37. $-2, -6, -4, 3, 4$

38. $-5, -3, -2, 0, 4$

39. $\dfrac{1}{4}, 2\dfrac{1}{2}, -3.8, -4, -1\dfrac{5}{8}$

40. $5.25, 4\dfrac{5}{9}, -2\dfrac{1}{3}, 0, -3\dfrac{2}{5}$

*For Exercises 41 and 42, **see Example 3**. List all numbers from each set that are the following.*

(a) natural numbers   **(b)** whole numbers   **(c)** integers
(d) rational numbers   **(e)** irrational numbers **(f)** real numbers

41. $\left\{ -9, -\sqrt{7}, -1\dfrac{1}{4}, -\dfrac{3}{5}, 0, 0.\overline{1}, \sqrt{5}, 3, 5.9, 7 \right\}$

42. $\left\{ -5.3, -5, -\sqrt{3}, -1, -\dfrac{1}{9}, 0, 0.\overline{27}, 1.2, 3, \sqrt{11} \right\}$

*For each number, find **(a)** the additive inverse and **(b)** the absolute value. **See Objective 3 and Example 5.***

43. -7

44. -4

45. 8

46. 10

47. $-\dfrac{3}{4}$

48. $-\dfrac{2}{5}$

49. 5.6

50. 8.1

*Find each absolute value. **See Example 5.***

51. $|-6|$

52. $|-14|$

53. $-|12|$

54. $-|19|$

55. $-\left|-\dfrac{2}{3}\right|$

56. $-\left|-\dfrac{4}{5}\right|$

57. $|6-3|$

58. $|9-4|$

59. $-|6-3|$

60. $-|9-4|$

*Select the lesser of the two given numbers. **See Examples 4 and 5.***

61. $-11, -3$

62. $-8, -13$

63. $-7, -6$

64. $-16, -17$

65. $4, |-5|$

66. $4, |-3|$

67. $|-3.5|, |-4.5|$

68. $|-8.9|, |-9.8|$

69. $-|-6|, -|-4|$

70. $-|-2|, -|-3|$

71. $|5-3|, |6-2|$

72. $|7-2|, |8-1|$

Decide whether each statement is true *or false. **See Examples 4 and 5.***

73. $-5 < -2$

74. $-8 > -2$

75. $-4 \le -(-5)$

76. $-6 \le -(-3)$

77. $|-6| < |-9|$

78. $|-12| < |-20|$

79. $-|8| > |-9|$

80. $-|12| > |-15|$

81. $-|-5| \ge -|-9|$

82. $-|-12| \le -|-15|$

83. $|6-5| \ge |6-2|$

84. $|13-8| \le |7-4|$

The table shows the percent change in the Consumer Price Index (CPI) for selected categories of goods and services from 2009 to 2010 and from 2010 to 2011. Use the table to answer each question. ***See Example 6.***

85. Which category in which year represents the greatest percentage increase?

86. Which category in which year represents the greatest percentage decrease?

87. Which category in which year represents the least change?

88. Which categories represent a decrease for both years?

Category	Change from 2009 to 2010	Change from 2010 to 2011
Communication	−0.3	−1.6
Fuel and other utilities	1.7	2.9
Medical care	3.4	3.0
Public transportation	6.3	7.2
Shelter	−0.4	1.3

Source: U.S. Bureau of Labor Statistics.

STUDY SKILLS

Completing Your Homework

You are ready to do your homework **AFTER** you have read the corresponding textbook section and worked through the examples and Now Try exercises.

Homework Tips

- **Survey the exercise set.** Glance over the problems that your instructor has assigned to get a general idea of the types of exercises you will be working. Skim directions, and note any references to section examples.

- **Work problems neatly.** Use pencil and write legibly, so others can read your work. Skip lines between steps. Clearly separate problems from each other.

- **Show all your work.** It is tempting to take shortcuts. Include ALL steps.

- **Check your work frequently to make sure you are on the right track.** It is hard to unlearn a mistake. For all odd-numbered problems, answers are given in the back of the book.

- **If you have trouble with a problem, refer to the corresponding worked example in the section.** The exercise directions will often reference specific examples to review. Pay attention to every line of the worked example to see how to get from step to step.

- **If you are having trouble with an even-numbered problem, work the corresponding odd-numbered problem.** Check your answer in the back of the book, and apply the same steps to work the even-numbered problem.

- **Do some homework problems every day.** This is a good habit, even if your math class does not meet each day.

- **Mark any problems you don't understand.** Ask your instructor about them.

Think through and answer each question.

1. What is your instructor's policy regarding homework?

2. Think about your current approach to doing homework. Be honest in your assessment.

   (a) What are you doing that is working well?

   (b) What improvements could you make?

3. Which one or two homework tips will you try this week? Why?

1.4 Adding and Subtracting Real Numbers

VOCABULARY
☐ sum
☐ addends
☐ difference
☐ minuend
☐ subtrahend

**NOW TRY
EXERCISE 1**

Use a number line to find each sum.

(a) $3 + 5$ **(b)** $-1 + (-3)$

OBJECTIVE 1 Add two numbers with the same sign.

Recall that the answer to an addition problem is a **sum.** The numbers being added are the **addends.**

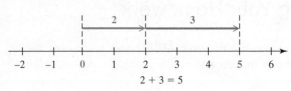

EXAMPLE 1 Adding Numbers (Same Sign) on a Number Line

Use a number line to find each sum.

(a) $2 + 3$

Step 1 Start at 0 and draw an arrow 2 units to the *right*. See **FIGURE 10**.

Step 2 From the right end of that arrow, draw another arrow 3 units to the *right*.

The number below the end of the second arrow is 5, so $2 + 3 = 5$.

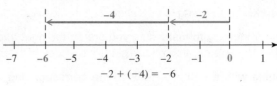

FIGURE 10

(b) $-2 + (-4)$

(We put parentheses around -4 due to the $+$ and $-$ symbols next to each other.)

Step 1 Start at 0 and draw an arrow 2 units to the *left*. See **FIGURE 11**.

Step 2 From the left end of the first arrow, draw a second arrow 4 units to the *left* to represent the addition of a *negative* number.

The number below the end of the second arrow is -6, so $-2 + (-4) = -6$.

FIGURE 11

NOW TRY

In **Example 1(b),** the sum of the two negative numbers -2 and -4 is a negative number whose distance from 0 is the sum of the distance of -2 from 0 and the distance of -4 from 0. *That is, the sum of two negative numbers is the negative of the sum of their absolute values.*

Adding Signed Numbers (Same Sign)

To add two numbers with the *same* sign, add their absolute values. The sum has the same sign as the addends.

Examples: $2 + 4 = 6$ and $-2 + (-4) = -6$

NOW TRY ANSWERS
1. **(a)** 8 **(b)** -4

NOW TRY
EXERCISE 2
Find each sum.

(a) $-6 + (-11)$

(b) $-\dfrac{2}{5} + \left(-\dfrac{1}{2}\right)$

EXAMPLE 2 Adding Two Negative Numbers

Find each sum.

(a) $-9 + (-2)$ Both addends are negative.

$= \underset{\underset{\text{Sign of each addend}}{\uparrow}}{-}(|-9| + |-2|)$ Add the absolute values of the addends.

$= -(9 + 2)$ Take the absolute values.

$= -11$ The sum of two negative numbers is negative.

(b) $-\dfrac{1}{4} + \left(-\dfrac{2}{3}\right)$ LCD 12   Both addends are negative.

Think: $\left|-\dfrac{3}{12}\right| = \dfrac{3}{12}$ and $\left|-\dfrac{8}{12}\right| = \dfrac{8}{12}$ $= -\dfrac{3}{12} + \left(-\dfrac{8}{12}\right)$ Write equivalent fractions using the LCD, 12.

$= -\dfrac{11}{12}$ Add the absolute values of the addends. Use the common negative sign.

(c) $-2.6 + (-4.7)$ Both addends are negative.

$= -7.3$ Add the absolute values of the addends. Use the common negative sign.

NOW TRY

OBJECTIVE 2 Add two numbers with different signs.

NOW TRY
EXERCISE 3
Use a number line to find the sum.

$$4 + (-8)$$

EXAMPLE 3 Adding Numbers (Different Signs) on a Number Line

Use a number line to find the sum $-2 + 5$.

Step 1   Start at 0 and draw an arrow 2 units to the *left*. See **FIGURE 12**.

Step 2   From the left end of this arrow, draw a second arrow 5 units to the *right* to represent the addition of a *positive* number.

The number below the end of the second arrow is 3, so $-2 + 5 = 3$.

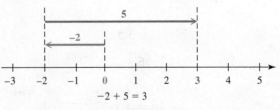

FIGURE 12   NOW TRY

Adding Signed Numbers (Different Signs)

To add two numbers with *different* signs, find their absolute values and subtract the lesser absolute value from the greater. The sum has the same sign as the addend with greater absolute value.

Examples: $-2 + 6 = 4$ and $2 + (-6) = -4$

NOW TRY ANSWERS
2. (a) -17 **(b)** $-\dfrac{9}{10}$
3. -4

**NOW TRY
EXERCISE 4**

Find each sum.

(a) $7 + (-4)$

(b) $\dfrac{2}{3} + \left(-2\dfrac{1}{9}\right)$

(c) $-5.7 + 3.7$

(d) $-10 + 10$

EXAMPLE 4   Adding Signed Numbers (Different Signs)

Find each sum.

(a) $-12 + 5$

$\;\; |{-12}|\;\; |5|$ Find the absolute value of each addend,
$= -(12 - 5)$ and subtract the lesser from the greater.

 Use the sign of the addend
 with greater absolute value.

$= -7$

(b) $-8 + 12$

$= +(12 - 8)$ Find the absolute value of each addend,
 and subtract the lesser from the greater.

 Use the sign of the addend
 with greater absolute value.

$= 4$ The $+$ symbol is understood.

(c) $\dfrac{5}{6} + \left(-1\dfrac{1}{3}\right)$

$= \dfrac{5}{6} + \left(-\dfrac{4}{3}\right)$ Write the mixed number as an improper fraction.

$= \dfrac{5}{6} + \left(-\dfrac{8}{6}\right)$ Find a common denominator.

$= -\left(\dfrac{8}{6} - \dfrac{5}{6}\right)$ $\left|\dfrac{5}{6}\right| = \dfrac{5}{6}$ and $\left|-\dfrac{8}{6}\right| = \dfrac{8}{6}$;
 Subtract the lesser absolute value from the greater.

 Use a $-$ symbol because $\left|-\dfrac{8}{6}\right| > \left|\dfrac{5}{6}\right|$.

$= -\dfrac{3}{6}$ Subtract the fractions.

$= -\dfrac{1}{2}$ Write in lowest terms.

(d) $8.1 + (-4.6)$ $|8.1| = 8.1$ and $|-4.6| = |4.6|$;
 Subtract the lesser absolute value from the greater.

$= +(8.1 - 4.6)$

 $|8.1| > |-4.6|$

$= 3.5$

(e) $-16 + 16$ $|-16| = 16$ and $|16| = 16$;
 The difference of the absolute values is 0,
$= 0$ which is neither positive nor negative.

(f) $42 + (-42)$ ***In general, when additive inverses are added,***
 ***the sum is 0.***
$= 0$

NOW TRY

OBJECTIVE 3   Use the definition of subtraction.

Recall that the answer to a subtraction problem is a **difference.** In the subtraction $x - y$, x is the **minuend** and y is the **subtrahend.**

$$x \quad - \quad y \quad = \quad z$$
$$\uparrow \qquad\quad \uparrow \qquad\quad\; \uparrow$$
Minuend   Subtrahend   Difference

NOW TRY EXERCISE 5

Use a number line to find the difference.

$$6 - 2$$

EXAMPLE 5 Subtracting Numbers on a Number Line

Use a number line to find the difference $7 - 4$.

Step 1 Start at 0 and draw an arrow 7 units to the *right*. See **FIGURE 13**.

Step 2 From the right end of the first arrow, draw a second arrow 4 units to the *left* to represent the *subtraction*.

The number below the end of the second arrow is 3, so $7 - 4 = 3$.

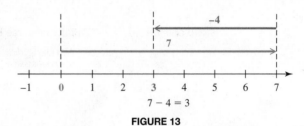

$$7 - 4 = 3$$

FIGURE 13

NOW TRY

The procedure used in **Example 5** to find the difference $7 - 4$ is exactly the same procedure that would be used to find the sum $7 + (-4)$.

$$7 - 4 \quad \text{is equal to} \quad 7 + (-4).$$

This suggests that *subtracting* a positive number from a greater positive number is the same as *adding* the additive inverse of the lesser number to the greater.

Definition of Subtraction

For any real numbers x and y, the following holds.

$$x - y = x + (-y)$$

To subtract y from x, add the additive inverse (or opposite) of y to x. That is, change the subtrahend to its opposite and add.

Example: $4 - 9$
$$= 4 + (-9)$$
$$= -5$$

EXAMPLE 6 Subtracting Signed Numbers

Find each difference.

(a) $12 - 3$

Change − to +.

$$= 12 + (-3)$$

No change ⟋   ⟍ Additive inverse of 3

$$= 9$$

12 has the greater absolute value, so the sum is positive.

(b) $5 - 7$

Change − to +.

$$= 5 + (-7)$$

No change ⟋   ⟍ Additive inverse of 7

$$= -2$$

−7 has the greater absolute value, so the sum is negative.

NOW TRY ANSWER

5. 4

NOW TRY
EXERCISE 6

Find each difference.

(a) $-5 - (-11)$

(b) $4 - 15$

(c) $-\dfrac{5}{7} - \dfrac{1}{3}$

(d) $5.25 - (-3.24)$

(c) $-8 - 15$

Change $-$ to $+$.

$= -8 + (-15)$

No change ⟶ ⟵ Additive inverse of 15

$= -23$ | The sum of two negative numbers is negative.

(d) $-3 - (-5)$

Change $-$ to $+$.

$= -3 + 5$

No change ⟶ ⟵ Additive inverse of -5

$= 2$ | 5 has the greater absolute value, so the sum is positive.

(e) $\dfrac{3}{8} - \left(-\dfrac{4}{5}\right)$

$= \dfrac{15}{40} - \left(-\dfrac{32}{40}\right)$ Write equivalent fractions using the LCD, 40.

$= \dfrac{15}{40} + \dfrac{32}{40}$ Definition of subtraction

$= \dfrac{47}{40}, \quad \text{or} \quad 1\dfrac{7}{40}$ Add the fractions. Write as a mixed number.

(f) $-8.75 - (-2.41)$

$= -8.75 + 2.41$ Definition of subtraction

$= -6.34$ Add the decimals.   **NOW TRY**

Uses of the Symbol $-$

We use the symbol $-$ for three purposes.

1. *It can represent subtraction,* as in $9 - 5 = 4$.

2. *It can represent negative numbers,* such as -10, -2, and -3.

3. *It can represent the additive inverse (or opposite) of a number,* as in "the additive inverse (or opposite) of 8 is -8."

We may see more than one use of $-$ in the same expression, such as $-6 - (-9)$, where -9 is subtracted from -6. The meaning of the symbol $-$ depends on its position in the algebraic expression.

OBJECTIVE 4 Use the rules for order of operations when adding and subtracting signed numbers.

EXAMPLE 7 Using the Rules for Order of Operations

Perform each indicated operation.

(a) $-6 - [2 - (8 + 3)]$ ⟵ Work from the inside out.

$= -6 - [2 - 11]$ Add inside the parentheses.

$= -6 - [2 + (-11)]$ Definition of subtraction

$= -6 - [-9]$ Add inside the brackets.

$= -6 + 9$ Definition of subtraction

$= 3$ Add.

NOW TRY ANSWERS
6. (a) 6 **(b)** -11
  (c) $-\dfrac{22}{21}$, or $-1\dfrac{1}{21}$
  (d) 8.49

**NOW TRY
EXERCISE 7**

Perform each indicated operation.

(a) $8 - [(-3 + 7) - (3 - 9)]$

(b) $3|6 - 9| - |4 - 12|$

(b) $5 + [(-3 - 2) - (4 - 1)]$ Work within each set of parentheses inside the brackets.

$= 5 + [(-3 + (-2)) - 3]$

$= 5 + [(-5) - 3]$

$= 5 + [(-5) + (-3)]$ Show all steps to avoid sign errors.

$= 5 + [-8]$

$= -3$

(c) $\dfrac{2}{3} - \left[\dfrac{1}{12} - \left(-\dfrac{1}{4}\right) \right]$

$= \dfrac{8}{12} - \left[\dfrac{1}{12} - \left(-\dfrac{3}{12}\right) \right]$ Write equivalent fractions using the LCD, 12.

$= \dfrac{8}{12} - \left[\dfrac{1}{12} + \dfrac{3}{12} \right]$ Work inside the brackets.

$= \dfrac{8}{12} - \dfrac{4}{12}$ Add inside the brackets.

$= \dfrac{4}{12}$ Subtract.

$= \dfrac{1}{3}$ Write in lowest terms.

(d) $|4 - 7| + 2|6 - 3|$ $2|6-3|$ means $2 \cdot |6-3|$.

$= |-3| + 2|3|$ Work within the absolute value bars.

$= 3 + 2 \cdot 3$ Evaluate the absolute values.

Be careful. Multiply first. $= 3 + 6$ Multiply.

$= 9$ Add.   **NOW TRY**

OBJECTIVE 5   Translate words and phrases involving addition and subtraction.

▼ **Words and Phrases That Indicate Addition**

Word or Phrase	Example	Numerical Expression and Simplification
Sum of	The *sum of* −3 and 4	−3 + 4, which equals 1
Added to	5 *added to* −8	−8 + 5, which equals −3
More than	12 *more than* −5	−5 + 12, which equals 7
Increased by	−6 *increased by* 13	−6 + 13, which equals 7
Plus	3 *plus* 14	3 + 14, which equals 17

EXAMPLE 8   Translating Words and Phrases (Addition)

Write a numerical expression for each phrase, and simplify the expression.

(a) The sum of −8 and 4 and 6

$$-8 + 4 + 6 \quad \text{simplifies to} \quad -4 + 6, \quad \text{which equals 2.}$$

Add in order from left to right.

**NOW TRY
EXERCISE 8**

Write a numerical expression
for the phrase, and simplify
the expression.

The sum of −3 and 7,
increased by 10

(b) 3 more than −5, increased by 12

$$(-5 + 3) + 12 \quad \text{simplifies to} \quad -2 + 12, \quad \text{which equals} \quad 10.$$

Here we *simplified* each expression by performing the operations.   **NOW TRY**

▼ **Words and Phrases That Indicate Subtraction**

Word, Phrase, or Sentence	Example	Numerical Expression and Simplification
Difference of	The *difference of* −3 and −8	−3 − (−8) simplifies to −3 + 8, which equals 5
Subtracted from*	12 *subtracted from* 18	18 − 12, which equals 6
From . . . , subtract	From 12, subtract 8.	12 − 8 simplifies to 12 + (−8), which equals 4
Less	6 *less* 5	6 − 5, which equals 1
Less than*	6 *less than* 5	5 − 6 simplifies to 5 + (−6), which equals −1
Decreased by	9 *decreased by* −4	9 − (−4) simplifies to 9 + 4, which equals 13
Minus	8 *minus* 5	8 − 5, which equals 3

*Be careful with order when translating.

⊘ CAUTION   When subtracting two numbers, be careful to write them in the correct order. In general,

$$x - y \neq y - x. \qquad \text{For example, } 5 - 3 \neq 3 - 5.$$

Think carefully before interpreting an expression involving subtraction.

**NOW TRY
EXERCISE 9**

Write a numerical expression
for each phrase, and simplify
the expression.

(a) The difference of 5 and
−8, decreased by 4

(b) 7 less than −2

EXAMPLE 9   Translating Words and Phrases (Subtraction)

Write a numerical expression for each phrase, and simplify the expression.

(a) The difference of −8 and 5

When "difference of" is used, write the numbers in the order given.

$$-8 - 5 \quad \text{simplifies to} \quad -8 + (-5), \quad \text{which equals} \quad -13.$$

(b) 4 subtracted from the sum of 8 and −3

Here the operation of addition is also used, as indicated by the words *sum of*. First, add 8 and −3. Next, subtract 4 from this sum.

$$[8 + (-3)] - 4 \quad \text{simplifies to} \quad 5 - 4, \quad \text{which equals 1.}$$

(c) 4 less than −6

Here, 4 must be taken *from* −6, so write −6 first.

Be careful with order. $$-6 - 4 \quad \text{simplifies to} \quad -6 + (-4), \quad \text{which equals} \quad -10.$$

Notice that "4 less than −6" differs from "4 *is less than* −6." The second of these is symbolized $4 < -6$ (which is a false statement).

(d) 8, decreased by 5 less than 12

First, write "5 less than 12" as 12 − 5. Next, subtract 12 − 5 from 8.

$$8 - (12 - 5) \quad \text{simplifies to} \quad 8 - 7, \quad \text{which equals} \quad 1.$$

NOW TRY

NOW TRY ANSWERS
8. $(-3 + 7) + 10$; 14
9. (a) $[5 - (-8)] - 4$; 9
   (b) $-2 - 7$; −9

**NOW TRY
EXERCISE 10**

Find the difference between a gain of 226 yd on the football field by the Chesterfield Bears and a loss of 7 yd by the New London Wildcats.

EXAMPLE 10 Solving an Application Involving Subtraction

The record-high temperature in the United States is 134°F, recorded at Death Valley, California, in 1913. The record low is −80°F, at Prospect Creek, Alaska, in 1971. See **FIGURE 14**. What is the difference between these highest and lowest temperatures? (*Source: National Climatic Data Center.*)

We must subtract the lowest temperature from the highest temperature.

Order of numbers matters in subtraction.

$$134 - (-80)$$
$$= 134 + 80 \quad \text{Definition of subtraction}$$
$$= 214 \quad \text{Add.}$$

The difference between the two temperatures is 214°F.

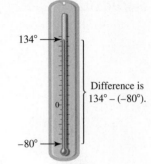

134°

Difference is
134° − (−80°).

0

−80°

FIGURE 14

NOW TRY

OBJECTIVE 6 Use signed numbers to interpret data.

**NOW TRY
EXERCISE 11**

Refer to **FIGURE 15** and use a signed number to represent the change in enrollment from 1985 to 1990.

EXAMPLE 11 Using a Signed Number to Interpret Data

The bar graph in **FIGURE 15** shows public high school (grades 9–12) enrollment in millions of students for selected years from 1980 to 2010.

Public High School Enrollment

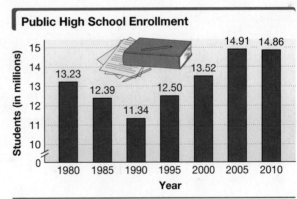

Source: U.S. National Center for Education Statistics.

FIGURE 15

(a) Use a signed number to represent the change in enrollment in millions from 2000 to 2005.
Start with the number for 2005. Subtract from it the number for 2000.

2005   2000
↓   ↓
$$14.91 \ - \ 13.52 = +1.39 \text{ million students}$$ ← A positive number indicates an *increase*. The bar for 2005 is "higher" than the bar for 2000.

(b) Use a signed number to represent the change in enrollment in millions from 2005 to 2010.
Start with the number for 2010. Subtract from it the number for 2005.

2010   2005
↓   ↓
$$14.86 \ - \ 14.91 = -0.05 \text{ million students}$$ ← A negative number indicates a *decrease*. The bar for 2010 is "lower" than the bar for 2005.

NOW TRY ANSWERS
10. 233 yd
11. −1.05 million students

NOW TRY

1.4 Exercises

FOR EXTRA HELP   MyMathLab®

▶ *Complete solution available in MyMathLab*

Concept Check   *Complete each of the following.*

▶ **1.** The sum of two negative numbers will always be a (*positive*/*negative*) number. Give a number-line illustration using the sum $-2 + (-3)$.

2. The sum of a number and its opposite will always be _____ .

▶ **3.** When adding a positive number and a negative number, where the negative number has the greater absolute value, the sum will be a (*positive*/*negative*) number. Give a number-line illustration using the sum $-4 + 2$.

4. To simplify the expression $8 + [-2 + (-3 + 5)]$, one should begin by adding _____ and _____, according to the rules for order of operations.

5. By the definition of subtraction, in order to perform the subtraction $-6 - (-8)$, we must add the opposite of _____ to _____ to obtain _____ .

6. "The difference of 7 and 12" translates as _____ , while "the difference of 12 and 7" translates as _____ .

Concept Check   *Suppose that x represents a positive number and y represents a negative number. Determine whether the given expression must represent a positive number or a negative number.*

7. $x - y$ **8.** $y - x$ **9.** $y - |x|$ **10.** $x + |y|$

Find each sum. See Examples 1–7.

▶ **11.** $-6 + (-2)$ **12.** $-9 + (-2)$ **13.** $-5 + (-7)$

14. $-11 + (-5)$ **15.** $6 + (-4)$ **16.** $11 + (-8)$

17. $4 + (-6)$ **18.** $3 + (-7)$ **19.** $-16 + 7$

20. $-13 + 6$ **21.** $6 + (-6)$ **22.** $-11 + 11$

23. $-\dfrac{1}{3} + \left(-\dfrac{4}{15}\right)$ **24.** $-\dfrac{1}{4} + \left(-\dfrac{5}{12}\right)$ ▶ **25.** $-\dfrac{1}{6} + \dfrac{2}{3}$

26. $-\dfrac{6}{25} + \dfrac{19}{20}$ **27.** $\dfrac{5}{8} + \left(-\dfrac{17}{12}\right)$ **28.** $\dfrac{9}{10} + \left(-\dfrac{11}{8}\right)$

29. $2\dfrac{1}{2} + \left(-3\dfrac{1}{4}\right)$ **30.** $1\dfrac{3}{8} + \left(-2\dfrac{1}{4}\right)$ **31.** $-3.5 + 12.4$

32. $-12.5 + 21.3$ **33.** $-2.34 + (-3.67)$ **34.** $-1.25 + (-6.88)$

35. $4 + [13 + (-5)]$ **36.** $6 + [12 + (-3)]$ **37.** $8 + [-2 + (-1)]$

38. $12 + [-3 + (-4)]$ **39.** $-2 + [5 + (-1)]$ **40.** $-8 + [9 + (-2)]$

41. $-6 + [6 + (-9)]$ **42.** $-3 + [3 + (-8)]$ **43.** $[(-9) + (-3)] + 12$

44. $[(-8) + (-6)] + 14$ **45.** $-6.1 + [3.2 + (-4.8)]$ **46.** $-9.4 + [5.8 + (-7.9)]$

47. $[-3 + (-4)] + [5 + (-6)]$ **48.** $[-8 + (-3)] + [4 + (-6)]$

49. $[-4 + (-3)] + [8 + (-1)]$ **50.** $[-5 + (-9)] + [16 + (-2)]$

51. $[-4 + (-6)] + [-3 + (-8)] + [12 + (-11)]$

52. $[-2 + (-11)] + [-12 + (-2)] + [18 + (-6)]$

Find each difference. See Examples 1–7.

53. $4 - 7$ **54.** $8 - 13$ ▶ **55.** $5 - 9$ **56.** $6 - 11$

57. $-7 - 1$ **58.** $-9 - 4$ **59.** $-8 - 6$ **60.** $-9 - 5$

61. $7 - (-2)$ **62.** $9 - (-2)$ **63.** $-6 - (-2)$ **64.** $-7 - (-5)$

65. $2 - (3 - 5)$ **66.** $-3 - (4 - 11)$ **67.** $\frac{1}{2} - \left(-\frac{1}{4}\right)$

68. $\frac{1}{3} - \left(-\frac{1}{12}\right)$ **69.** $-\frac{3}{4} - \frac{5}{8}$ **70.** $-\frac{5}{6} - \frac{1}{2}$

71. $\frac{5}{8} - \left(-\frac{1}{2} - \frac{3}{4}\right)$ **72.** $\frac{9}{10} - \left(\frac{1}{8} - \frac{3}{10}\right)$ **73.** $3.4 - (-8.2)$

74. $5.7 - (-11.6)$ **75.** $-6.4 - 3.5$ **76.** $-4.4 - 8.6$

Perform each indicated operation. See Examples 1–7.

▶ **77.** $(4 - 6) + 12$ **78.** $(3 - 7) + 4$ **79.** $(8 - 1) - 12$

80. $(9 - 3) - 15$ **81.** $6 - (-8 + 3)$ **82.** $8 - (-9 + 5)$

83. $2 + (-4 - 8)$ **84.** $6 + (-9 - 2)$ **85.** $|-5 - 6| + |9 + 2|$

86. $|-4 + 8| + |6 - 1|$ **87.** $|-8 - 2| - |-9 - 3|$ **88.** $|-4 - 2| - |-8 - 1|$

89. $\left(-\frac{3}{4} - \frac{5}{2}\right) - \left(-\frac{1}{8} - 1\right)$ **90.** $\left(-\frac{3}{8} - \frac{2}{3}\right) - \left(-\frac{9}{8} - 3\right)$

91. $\left(-\frac{1}{2} + 0.25\right) - \left(-\frac{3}{4} + 0.75\right)$ **92.** $\left(-\frac{3}{2} - 0.75\right) - \left(0.5 - \frac{1}{2}\right)$

93. $-9 + [(3 - 2) - (-4 + 2)]$ **94.** $-8 - [(-4 - 1) + (9 - 2)]$

95. $-3 + [(-5 - 8) - (-6 + 2)]$ **96.** $-4 + [(-12 + 1) - (-1 - 9)]$

97. $-9.12 + [(-4.8 - 3.25) + 11.279]$ **98.** $-7.62 - [(-3.99 + 1.427) - (-2.8)]$

Write a numerical expression for each phrase, and simplify the expression. See Examples 8 and 9.

99. The sum of -5 and 12 and 6   **100.** The sum of -3 and 5 and -12

101. 14 added to the sum of -19 and -4 **102.** -2 added to the sum of -18 and 11

▶ **103.** The sum of -4 and -10, increased by 12   **104.** The sum of -7 and -13, increased by 14

105. $\frac{2}{7}$ more than the sum of $\frac{5}{7}$ and $-\frac{9}{7}$ **106.** 1.85 more than the sum of -1.25 and -4.75

▶ **107.** The difference of 4 and -8 **108.** The difference of 7 and -14

109. 8 less than -2 **110.** 9 less than -13

111. The sum of 9 and -4, decreased by 7   **112.** The sum of 12 and -7, decreased by 14

113. 12 less than the difference of 8 and -5 **114.** 19 less than the difference of 9 and -2

*The table gives scores (above or below par—that is, above or below the score "standard")
for selected golfers during the 2013 PGA Tour Championship. Write a signed number that
represents the total score (above or below par) for the four rounds for each golfer.*

	Golfer	Round 1	Round 2	Round 3	Round 4
115.	Steve Stricker	−4	+1*	−2	−5
116.	Phil Mickelson	+1	−3	0	−2
117.	Charl Schwartzel	−2	+9	+7	−4
118.	Kevin Streelman	−1	+2	+4	−3

*Golf scoring commonly includes a + symbol with a score over par.
Source: *The Gazette.*

Solve each problem. **See Example 10.**

119. Based on 2020 population projections, New York will lose
5 seats in the U.S. House of Representatives, Pennsylva-
nia will lose 4 seats, and Ohio will lose 3. Write a signed
number that represents the total number of seats these three
states are projected to lose. (*Source:* Population Reference
Bureau.)

120. Michigan is projected to lose 3 seats in the U.S. House
of Representatives and Illinois 2 in 2020. The states
projected to gain the most seats are California with
9, Texas with 5, Florida with 3, Georgia with 2, and
Arizona with 2. Write a signed number that represents
the algebraic sum of these changes. (*Source:* Population
Reference Bureau.)

121. The surface, or rim, of a canyon is at altitude 0. On a hike down into the canyon, a party
of hikers stops for a rest at 130 m below the surface. The hikers then descend another
54 m. Write the new altitude as a signed number.

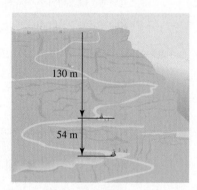

130 m

54 m

122. A pilot announces to the passengers that the current altitude of their plane is 34,000 ft.
Because of turbulence, the pilot is forced to descend 2100 ft. Write the new altitude as a
signed number.

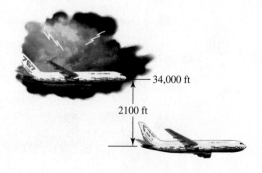

34,000 ft

2100 ft

123. The lowest temperature ever recorded in Arkansas was −29°F. The highest temperature ever recorded there was 149°F more than the lowest. What was this highest temperature? (*Source: National Climatic Data Center.*)

124. On January 23, 1943, the temperature rose 49°F in two minutes in Spearfish, South Dakota. If the starting temperature was −4°F, what was the temperature two minutes later?

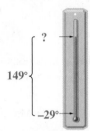

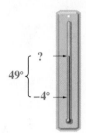

▶ **125.** The lowest temperature ever recorded in Illinois was −36°F on January 5, 1999. The lowest temperature ever recorded in Utah was on February 1, 1985, and was 33°F lower than Illinois's record low. What is the record low temperature for Utah? (*Source: National Climatic Data Center.*)

126. The lowest temperature ever recorded in South Carolina was −19°F. The lowest temperature ever recorded in Wisconsin was 36° lower than South Carolina's record low. What is the record low temperature for Wisconsin? (*Source:* National Climatic Data Center.)

127. Nadine enjoys playing Triominoes every Wednesday night. Last Wednesday, on four successive turns, her scores were

$$-19, \quad 28, \quad -5, \quad \text{and} \quad 13.$$

What was her final score for the four turns?

128. Bruce also enjoys playing Triominoes. On five successive turns, his scores were

$$-13, \quad 15, \quad -12, \quad 24, \quad \text{and} \quad 14.$$

What was his total score for the five turns?

129. In 2005, Americans saved −0.5% of their after-tax incomes. In July 2013, they saved 4.4%. (*Source:* U.S. Bureau of Economic Analysis.)

   (a) Express the difference between these amounts as a positive number.

   (b) How could Americans have a negative personal savings rate in 2005?

130. In 2000, the U.S. federal budget had a surplus of $236 billion. In 2015, the federal budget projected a deficit of $576 billion. Express the difference between these amounts as a positive number. (*Source:* U.S. Treasury Department.)

131. In 2005, undergraduate college students had an average of $4906 in student loans. This average increased $788 by 2010, then dropped $154 by 2012. What was the average amount in student loans in 2012? (*Source:* The College Board.)

132. Among entertainment expenditures, the average annual spending per U.S. household on fees and admissions was $526 in 2001. This amount increased $80 by 2006 and then decreased $12 by 2011. What was the average household expenditure for fees and admissions in 2011? (*Source:* U.S. Bureau of Labor Statistics.)

133. In August, Susan began with a checking account balance of $904.89. Her withdrawals and deposits for August are as follows:

Withdrawals	Deposits
$35.84	$85.00
$26.14	$120.76
$3.12	

Assuming no other transactions, what was her account balance at the end of August?

134. In September, Jeffery began with a checking account balance of $904.89. His withdrawals and deposits for September are as follows:

Withdrawals	Deposits
$41.29	$80.59
$13.66	$276.13
$84.40	

Assuming no other transactions, what was his account balance at the end of September?

135. Linda owes $870.00 on her MasterCard account. She returns two items costing $35.90 and $150.00 and receives credit for these on the account. Next, she makes a purchase of $82.50 and then two more purchases of $10.00 each. She makes a payment of $500.00. She then incurs a finance charge of $37.23. How much does she still owe?

136. Marcial owes $679.00 on his Visa account. He returns three items costing $36.89, $29.40, and $113.55 and receives credit for these on the account. Next, he makes purchases of $135.78 and $412.88 and two purchases of $20.00 each. He makes a payment of $400. He then incurs a finance charge of $24.57. How much does he still owe?

The graph shows annual returns in percent for Class A shares of the Invesco S&P 500 Index Fund from 2009 to 2013. Use a signed number to represent the change in percent return for each period. **See Example 11.**

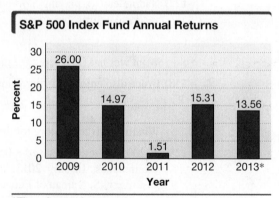

S&P 500 Index Fund Annual Returns

*Through second quarter
Source: Invesco.

137. 2009 to 2010

138. 2010 to 2011

139. 2011 to 2012

140. 2009 to 2013

The two tables show the heights of some selected mountains and the depths of some selected trenches. Use the information given to answer Exercises 141–146 on the next page.

Mountain	Height (in feet)
Foraker	17,400
Wilson	14,246
Pikes Peak	14,110

Trench	Depth (in feet, as a negative number)
Philippine	−32,995
Cayman	−24,721
Java	−23,376

Source: World Almanac and Book of Facts.

141. What is the difference between the height of Mt. Foraker and the depth of the Philippine Trench?

142. What is the difference between the height of Pikes Peak and the depth of the Java Trench?

143. How much deeper is the Cayman Trench than the Java Trench?

144. How much deeper is the Philippine Trench than the Cayman Trench?

145. How much higher is Mt. Wilson than Pikes Peak?

146. If Mt. Wilson and Pikes Peak were stacked one on top of the other, how much higher would they be than Mt. Foraker?

STUDY SKILLS

Using Study Cards

You may have used "flash cards" in other classes. In math, "study cards" can help you remember terms and definitions, procedures, and concepts. Use study cards to

- Help you understand and learn the material;
- Quickly review when you have a few minutes;
- Review before a quiz or test.

One of the advantages of study cards is that you learn while you are making them.

Vocabulary Cards

Put the word and a page reference on the front of the card. On the back, write the definition, an example, any related words, and a sample problem (if appropriate).

Procedure ("Steps") Cards

Write the name of the procedure on the front of the card. Then write each step in words. On the back of the card, put an example showing each step.

Make a vocabulary card and a procedure card for material you are learning now.

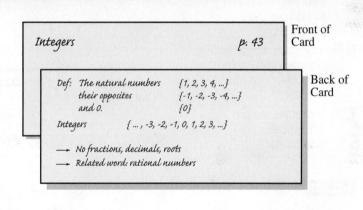

Front of Card

Back of Card

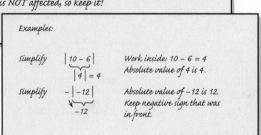

Front of Card

Back of Card

1.5 Multiplying and Dividing Real Numbers

OBJECTIVES

1 Find the product of a positive number and a negative number.

2 Find the product of two negative numbers.

3 Identify factors of integers.

4 Use the reciprocal of a number to apply the definition of division.

5 Use the rules for order of operations when multiplying and dividing signed numbers.

6 Evaluate algebraic expressions given values for the variables.

7 Translate words and phrases involving multiplication and division.

8 Translate simple sentences into equations.

VOCABULARY
☐ product
☐ factor
☐ multiplicative inverse (reciprocal)
☐ quotient
☐ dividend
☐ divisor

The result of multiplication is a **product.** We know that the product of two positive numbers is positive. We also know that the product of 0 and any positive number is 0, so we extend that property to all real numbers.

Multiplication Property of 0

For any real number x, the following hold.

$$x \cdot 0 = 0 \quad \text{and} \quad 0 \cdot x = 0$$

OBJECTIVE 1 Find the product of a positive number and a negative number.

Look at the following pattern.

$$3 \cdot 5 = 15$$
$$3 \cdot 4 = 12$$
$$3 \cdot 3 = 9$$
$$3 \cdot 2 = 6$$
$$3 \cdot 1 = 3$$
$$3 \cdot 0 = 0$$
$$3 \cdot (-1) = ?$$

The products decrease by 3.

What should $3 \cdot (-1)$ equal? The product $3 \cdot (-1)$ represents the sum

$$-1 + (-1) + (-1), \quad \text{which equals} \quad -3,$$

so the product should be -3. Also, $3 \cdot (-2)$ and $3 \cdot (-3)$ represent the sums

$$-2 + (-2) + (-2), \quad \text{which equals} \quad -6$$

and

$$-3 + (-3) + (-3), \quad \text{which equals} \quad -9.$$

These results maintain the pattern in the list.

Multiplying Signed Numbers (Different Signs)

For any positive real numbers x and y, the following hold.

$$x(-y) = -(xy) \quad \text{and} \quad (-x)y = -(xy)$$

That is, the product of two numbers with different signs is negative.

Examples: $6(-3) = -18 \quad \text{and} \quad (-6)3 = -18$

**NOW TRY
EXERCISE 1**

Find each product.

(a) $-11(9)$ **(b)** $3.1(-2.5)$

NOW TRY ANSWERS
1. (a) -99 **(b)** -7.75

EXAMPLE 1 Multiplying Signed Numbers (Different Signs)

Find each product.

(a) $8(-5)$
$= -(8 \cdot 5)$
$= -40$

The product of two numbers with *different* signs is *negative*.

(b) $-9\left(\dfrac{1}{3}\right)$
$= -\left(9 \cdot \dfrac{1}{3}\right)$
$= -3$

(c) $-6.2(4.1)$
$= -(6.2 \cdot 4.1)$
$= -25.42$

NOW TRY

OBJECTIVE 2 Find the product of two negative numbers.

Look at another pattern.

$$-5(4) = -20$$
$$-5(3) = -15$$
$$-5(2) = -10$$
$$-5(1) = -5$$
$$-5(0) = 0$$
$$-5(-1) = ?$$

The products increase by 5.

The numbers in color on the left side of the equality symbols decrease by 1 for each step down the list. The products on the right increase by 5 for each step down the list. To maintain this pattern, $-5(-1)$ should be 5 more than $-5(0)$, or 5 more than 0, so

$$-5(-1) = 5.$$

The pattern continues with

$$-5(-2) = 10$$
$$-5(-3) = 15$$
$$-5(-4) = 20$$
$$-5(-5) = 25, \quad \text{and so on.}$$

These results suggest the next rule.

Multiplying Two Negative Numbers

For any positive real numbers x and y, the following holds.

$$-x(-y) = xy$$

That is, the product of two negative numbers is positive.

Example: $-5(-4) = 20$

NOW TRY
EXERCISE 2

Find each product.

(a) $-8(-11)$ **(b)** $-\dfrac{1}{7}\left(-\dfrac{5}{2}\right)$

EXAMPLE 2 Multiplying Two Negative Numbers

Find each product.

(a) $-9(-2)$
$= 18$

(b) $-\dfrac{2}{3}\left(-\dfrac{3}{2}\right)$
$= 1$

(c) $-0.5(-1.25)$
$= 0.625$

The product of two numbers with the *same* sign is *positive*.   **NOW TRY**

The following box summarizes multiplying signed numbers.

Multiplying Signed Numbers

The product of two numbers with the *same* sign is *positive*.

The product of two numbers with *different* signs is *negative*.

Examples: $7(3) = 21$, $-7(-3) = 21$, $-7(3) = -21$, and $7(-3) = -21$

NOW TRY ANSWERS

2. (a) 88   **(b)** $\frac{5}{14}$

OBJECTIVE 3 Identify factors of integers.

The definition of **factor** from **Section R.1** can be extended to integers. If the product of two integers is a third integer, then each of the two integers is a *factor* of the third.

▼ Integer Factors

Integer	18	20	15	7	1
Pairs of Factors	1, 18	1, 20	1, 15	1, 7	1, 1
	2, 9	2, 10	3, 5	−1, −7	−1, −1
	3, 6	4, 5	−1, −15		
	−1, −18	−1, −20	−3, −5		
	−2, −9	−2, −10			
	−3, −6	−4, −5			

▼ Reciprocals

Number	Reciprocal (Multiplicative Inverse)
4	$\frac{1}{4}$
0.3, or $\frac{3}{10}$	$\frac{10}{3}$
−5	$\frac{1}{-5}$, or $-\frac{1}{5}$
$-\frac{5}{8}$	$-\frac{8}{5}$

A number and its reciprocal have a product of **1.** For example,

$$4 \cdot \frac{1}{4} = \frac{4}{4}, \text{ or } 1.$$

0 has no reciprocal because the product of 0 and any number is 0, not 1.

OBJECTIVE 4 Use the reciprocal of a number to apply the definition of division.

The definition of division depends on the idea of a *reciprocal,* or *multiplicative inverse,* of a number.

> ### Reciprocal, or Multiplicative Inverse
>
> Pairs of numbers whose product is 1 are **reciprocals,** or **multiplicative inverses,** of each other.

Recall that the answer to a division problem is a **quotient.** For example, we can write the quotient of 15 and 3 as $15 \div 3$, which equals 5. We obtain the same answer if we multiply $15 \cdot \frac{1}{3}$, the reciprocal of 3. This suggests the next definition.

> ### Definition of Division
>
> For any real numbers x and y, where $y \neq 0$, the following holds.
>
> $$x \div y = x \cdot \frac{1}{y}$$
>
> That is, to divide two numbers, multiply the first number (the **dividend**) by the reciprocal, or multiplicative inverse, of the second number (the **divisor**).
>
> *Example:* $15 \div 3$
>
> $$= 15 \cdot \frac{1}{3}$$
>
> $$= 5$$

Recall that an equivalent form of $x \div y$ is $\frac{x}{y}$, where the fraction bar represents division. ***In algebra, quotients are usually represented with a fraction bar.*** For example,

$$15 \div 3 \quad \text{is equivalent to} \quad \frac{15}{3}.$$

NOTE The following forms all represent division, where $y \neq 0$.

$$x \div y, \quad \frac{x}{y}, \quad x/y, \quad \text{and} \quad y\overline{)x}$$

Because division is defined in terms of multiplication, the rules for multiplying signed numbers also apply to dividing them.

> ### Dividing Signed Numbers
>
> The quotient of two numbers with the *same* sign is *positive*.
>
> The quotient of two numbers with *different* signs is *negative*.
>
> *Examples:* $\dfrac{15}{3} = 5, \quad \dfrac{-15}{-3} = 5, \quad \dfrac{15}{-3} = -5, \quad \text{and} \quad \dfrac{-15}{3} = -5$

NOW TRY
EXERCISE 3

Find each quotient.

(a) $\dfrac{-10}{5}$ **(b)** $\dfrac{-1.44}{-0.12}$

(c) $-\dfrac{3}{8} \div \dfrac{7}{10}$

EXAMPLE 3 Dividing Signed Numbers

Find each quotient.

(a) $\dfrac{8}{-2} = -4$ **(b)** $\dfrac{-100}{5} = -20$ **(c)** $\dfrac{-4.5}{-0.09} = 50$

(d) $-\dfrac{1}{8} \div \left(-\dfrac{3}{4} \right)$

$= -\dfrac{1}{8} \cdot \left(-\dfrac{4}{3} \right)$ Multiply by the reciprocal of the divisor.

$= \dfrac{1}{6}$ Multiply the fractions. Write in lowest terms.

NOW TRY

Consider the quotient $\dfrac{12}{3}$.

$$\dfrac{12}{3} = 4 \qquad \text{because} \qquad 4 \cdot 3 = 12.$$

> Multiply to check a division problem.

Using this relationship between multiplication and division, we investigate division by 0. Consider the quotient $\dfrac{0}{3}$.

$$\dfrac{0}{3} = 0 \qquad \text{because} \qquad 0 \cdot 3 = 0.$$

Now consider $\dfrac{3}{0}$.

$$\dfrac{3}{0} = ?$$

We need to find a number that when multiplied by 0 will equal 3, that is, $? \cdot 0 = 3$. *No* real number satisfies this equation because the product of any real number and 0 must be 0. Thus,

$\dfrac{x}{0}$ **is not a number, and** *division by 0 is undefined*. **If a division problem involves division by 0, write "undefined."**

NOW TRY ANSWERS

3. (a) -2 **(b)** 12 **(c)** $-\dfrac{15}{28}$

Division Involving 0

For any real number x, where $x \neq 0$, the following hold.

$$\frac{0}{x} = 0 \quad \text{and} \quad \frac{x}{0} \text{ is undefined.}$$

Examples: $\frac{0}{-10} = 0$ and $\frac{-10}{0}$ is undefined.

From the definitions of multiplication and division of real numbers,

$$\frac{-40}{8} = -5 \quad \text{and} \quad \frac{40}{-8} = -5, \quad \text{so} \quad \frac{-40}{8} = \frac{40}{-8}.$$

Based on this example, the quotient of a positive number and a negative number can be expressed in three equivalent forms.

Equivalent Forms

For any positive real numbers x and y, the following are equivalent.

$$\frac{-x}{y}, \quad \frac{x}{-y}, \quad \text{and} \quad -\frac{x}{y}$$

Similarly, the quotient of two negative numbers can be expressed as a quotient of two positive numbers.

Equivalent Forms

For any positive real numbers x and y, the following are equivalent.

$$\frac{-x}{-y} \quad \text{and} \quad \frac{x}{y}$$

NOTE Although we use the forms $\frac{-x}{y}$, $\frac{x}{-y}$, and $\frac{-x}{-y}$ in our work with algebraic expressions, we generally give final answers in the form $-\frac{x}{y}$ or $\frac{x}{y}$.

OBJECTIVE 5 Use the rules for order of operations when multiplying and dividing signed numbers.

EXAMPLE 4 Using the Rules for Order of Operations

Perform each indicated operation.

(a) $-9(2) - (-3)(2)$

$\quad = -18 - (-6)$ Multiply.

$\quad = -18 + 6$ Definition of subtraction

$\quad = -12$ Add.

(b) $-5(-2 - 3)$

$\quad = -5(-5)$ Subtract inside the parentheses.

$\quad = 25$ Multiply.

**NOW TRY
EXERCISE 4**

Perform each indicated operation.

(a) $-4(6) - (-5)(5)$

(b) $\dfrac{12(-4) - 6(-3)}{-4(7 - 16)}$

(c)

$$-6 + 2(3 - 5)$$ ← Begin inside the parentheses.

Do *not* add first.

$$= -6 + 2(-2)$$ Subtract inside the parentheses.

$$= -6 + (-4)$$ Multiply.

$$= -10$$ Add.

(d) $\dfrac{5(-2) - 3(4)}{2(1 - 6)}$ Simplify the numerator and denominator separately.

$$= \dfrac{-10 - 12}{2(-5)}$$ Multiply in the numerator.
Subtract inside the parentheses in the denominator.

$$= \dfrac{-22}{-10}$$ Subtract in the numerator.
Multiply in the denominator.

$$= \dfrac{11}{5}$$ Write in lowest terms.   **NOW TRY** ↻

OBJECTIVE 6 Evaluate algebraic expressions given values for the variables.

EXAMPLE 5 Evaluating Algebraic Expressions

Evaluate each expression for $x = -1$, $y = -2$, and $m = -3$.

**NOW TRY
EXERCISE 5**

Evaluate $\dfrac{3x^2 - 12}{y}$ for $x = -4$ and $y = -3$.

(a) $(3x + 4y)(-2m)$ Use parentheses around substituted negative values to avoid errors.

$$= [3(-1) + 4(-2)][-2(-3)]$$ Substitute the given values for the variables.

$$= [-3 + (-8)][6]$$ Multiply.

$$= [-11]6$$ Add inside the brackets.

$$= -66$$ Multiply.

(b)

$$2x^2 - 3y^2$$ Think: $(-2)^2 = -2(-2) = 4$

$$= 2(-1)^2 - 3(-2)^2$$ Substitute -1 for x and -2 for y.

Think: $(-1)^2 = -1(-1) = 1$

$$= 2(1) - 3(4)$$ Apply the exponents.

$$= 2 - 12$$ Multiply.

$$= -10$$ Subtract.

(c) $\dfrac{4y^2 + x}{m}$

$$= \dfrac{4(-2)^2 + (-1)}{-3}$$ Substitute -2 for y, -1 for x, and -3 for m.

$$= \dfrac{4(4) + (-1)}{-3}$$ Apply the exponent.

$$= \dfrac{16 + (-1)}{-3}$$ Multiply.

$$= \dfrac{15}{-3}$$ Add.

$$= -5$$ Divide.   **NOW TRY** ↻

NOW TRY ANSWERS

4. **(a)** 1   **(b)** $-\frac{5}{6}$

5. -12

OBJECTIVE 7   Translate words and phrases involving multiplication and division.

▼ **Words and Phrases That Indicate Multiplication**

Word or Phrase	Example	Numerical Expression and Simplification
Product of	The *product of* −5 and −2	−5(−2), which equals 10
Times	13 *times* −4	13(−4), which equals −52
Twice (meaning "2 times")	*Twice* 6	2(6), which equals 12
Triple (meaning "3 times")	*Triple* 4	3(4), which equals 12
Of (used with fractions)	$\frac{1}{2}$ *of* 10	$\frac{1}{2}$(10), which equals 5
Percent of	12% *of* −16	0.12(−16), which equals −1.92
As much as	$\frac{2}{3}$ *as much as* 30	$\frac{2}{3}$(30), which equals 20

NOW TRY EXERCISE 6

Write a numerical expression for each phrase, and simplify the expression.

(a) Twice the sum of −10 and 7

(b) 40% of the difference of 45 and 15

EXAMPLE 6   Translating Words and Phrases (Multiplication)

Write a numerical expression for each phrase, and simplify the expression.

(a) The product of 12 and the sum of 3 and −6

$$12[3 + (-6)] \quad \text{simplifies to} \quad 12[-3], \quad \text{which equals} \quad -36.$$

(b) Twice the difference of 8 and −4

$$2[8 - (-4)] \quad \text{simplifies to} \quad 2[12], \quad \text{which equals} \quad 24.$$

> The "difference of *a* and *b*" means *a* − *b*.

(c) Two-thirds of the sum of −5 and −3

$$\frac{2}{3}[-5 + (-3)] \quad \text{simplifies to} \quad \frac{2}{3}[-8], \quad \text{which equals} \quad -\frac{16}{3}.$$

(d) 15% of the difference of 14 and −2

> Remember that 15% = 0.15.

$$0.15[14 - (-2)] \quad \text{simplifies to} \quad 0.15[16], \quad \text{which equals} \quad 2.4.$$

(e) Double the product of 3 and 4

> Double means "2 times."

$$2 \cdot (3 \cdot 4) \quad \text{simplifies to} \quad 2(12), \quad \text{which equals} \quad 24.$$

NOW TRY

▼ **Phrases That Indicate Division**

Phrase	Example	Numerical Expression and Simplification
Quotient of	The *quotient of* −24 and 3	$\frac{-24}{3}$, which equals −8
Divided by	−16 *divided by* −4	$\frac{-16}{-4}$, which equals 4
Ratio of	The *ratio of* 2 to 3	$\frac{2}{3}$

NOW TRY ANSWERS

6. (a) 2(−10 + 7); −6
   (b) 0.40(45 − 15); 12

When translating a phrase involving division into a fraction, we write the first number named as the numerator and the second as the denominator.

**NOW TRY
EXERCISE 7**

Write a numerical expression for the phrase, and simplify the expression.

The quotient of 21 and the sum of 10 and −7

EXAMPLE 7 Interpreting Words and Phrases (Division)

Write a numerical expression for each phrase, and simplify the expression.

(a) The quotient of 14 and the sum of −9 and 2

"Quotient" indicates division. $\dfrac{14}{-9+2}$ simplifies to $\dfrac{14}{-7}$, which equals −2.

(b) The product of 5 and −6, divided by the difference of −7 and 8

$\dfrac{5(-6)}{-7-8}$ simplifies to $\dfrac{-30}{-15}$, which equals 2. **NOW TRY**

OBJECTIVE 8 Translate simple sentences into equations.

EXAMPLE 8 Translating Sentences into Equations

Write each sentence as an equation, using x as the variable. Then find the solution of the equation from the set of integers between −12 and 12, inclusive.

(a) Three times a number is −18.

The word *times* indicates multiplication.   The word *is* translates as =.

$$3 \cdot x = -18, \quad \text{or} \quad 3x = -18 \qquad 3 \cdot x = 3x$$

The integer between −12 and 12, inclusive, that makes this statement true is −6 because $3(-6) = -18$. The solution of the equation is −6.

(b) The sum of a number and 9 is 12.

$$x + 9 = 12$$

Because $3 + 9 = 12$, the solution of this equation is 3.

(c) The difference of a number and 5 is 0.

$$x - 5 = 0$$

Because $5 - 5 = 0$, the solution of this equation is 5.

(d) The quotient of 24 and a number is −2.

$$\frac{24}{x} = -2$$

Here, x must be a negative number because the numerator is positive and the quotient is negative. Because $\dfrac{24}{-12} = -2$, the solution is −12. **NOW TRY**

**NOW TRY
EXERCISE 8**

Write each sentence as an equation, using x as the variable. Then find the solution of the equation from the set of integers between −12 and 12, inclusive.

(a) The sum of a number and −4 is 7.

(b) The difference of −8 and a number is −11.

⚠ **CAUTION** In **Examples 6 and 7** above, the *phrases* translate as *expressions*, while in **Example 8,** the *sentences* translate as *equations*. *An expression is a phrase. An equation is a sentence with something on the left side, an = symbol, and something on the right side.*

$$\underset{\uparrow}{\dfrac{5(-6)}{-7-8}} \qquad \underset{\uparrow}{3x = -18}$$

Expression   Equation

NOW TRY ANSWERS

7. $\dfrac{21}{10 + (-7)}$; 7

8. (a) $x + (-4) = 7$; 11

  (b) $-8 - x = -11$; 3

1.5 Exercises

    MyMathLab®

▶ *Complete solution available in MyMathLab*

Concept Check   *Fill in each blank with one of the following.*

greater than 0   less than 0   equal to 0

1. The product or the quotient of two numbers with the same sign is _____.

2. The product or the quotient of two numbers with different signs is _____.

3. If three negative numbers are multiplied, the product is _____.

4. If two negative numbers are multiplied and then their product is divided by a negative number, the result is _____.

5. If a negative number is squared and the result is added to a positive number, the result is _____.

6. The reciprocal of a negative number is _____.

7. If three positive numbers, five negative numbers, and zero are multiplied, the product is _____.

8. The cube of a negative number is _____.

▶ **9.** *Concept Check*   Complete this statement: The quotient formed by any nonzero number divided by 0 is _____, and the quotient formed by 0 divided by any nonzero number is _____. Give an example of each quotient.

10. *Concept Check*   Which expression is undefined?

A. $\dfrac{4+4}{4+4}$ **B.** $\dfrac{4-4}{4+4}$ **C.** $\dfrac{4-4}{4-4}$ **D.** $\dfrac{4-4}{4}$

Find each product. **See Examples 1 and 2.**

11. $5(-6)$ **12.** $3(-4)$ **13.** $-5(-6)$ **14.** $-3(-4)$ **15.** $-10(-12)$

16. $-9(-5)$ **17.** $3(-11)$ **18.** $3(-15)$ **19.** $-0.5(0)$ **20.** $-0.3(0)$

21. $-6.8(0.35)$ **22.** $-4.6(0.24)$ **23.** $-\dfrac{3}{8} \cdot \left(-\dfrac{10}{9}\right)$ **24.** $-\dfrac{5}{6} \cdot \left(-\dfrac{16}{15}\right)$

25. $\dfrac{2}{15}\left(-1\dfrac{1}{4}\right)$ **26.** $\dfrac{3}{7}\left(-1\dfrac{5}{9}\right)$ **27.** $-8\left(-\dfrac{3}{4}\right)$ **28.** $-6\left(-\dfrac{2}{3}\right)$

Find all integer factors of each number. **See Objective 3.**

29. 32   **30.** 36   ▶ **31.** 40   **32.** 50   **33.** 31   **34.** 17

Find each quotient. **See Example 3** *and the discussion of division involving 0.*

▶ **35.** $\dfrac{15}{5}$ **36.** $\dfrac{35}{5}$ **37.** $\dfrac{-42}{6}$ **38.** $\dfrac{-28}{7}$ **39.** $\dfrac{-32}{-4}$

40. $\dfrac{-35}{-5}$ ▶ **41.** $\dfrac{96}{-16}$ **42.** $\dfrac{38}{-19}$ **43.** $\dfrac{-8.8}{2.2}$ **44.** $\dfrac{-4.6}{0.23}$

45. $-\dfrac{4}{3} \div \left(-\dfrac{1}{8}\right)$ **46.** $-\dfrac{6}{5} \div \left(-\dfrac{1}{3}\right)$ **47.** $-\dfrac{5}{6} \div \dfrac{8}{9}$ **48.** $-\dfrac{7}{10} \div \dfrac{3}{4}$

49. $\dfrac{0}{-5}$ **50.** $\dfrac{0}{-9}$ **51.** $\dfrac{11.5}{0}$ **52.** $\dfrac{15.2}{0}$

Perform each indicated operation. See Example 4.

53. $7 - 3 \cdot 6$

54. $8 - 2 \cdot 5$

55. $-10 - (-4)(2)$

56. $-11 - (-3)(6)$

▶ **57.** $-7(3 - 8)$

58. $-5(4 - 7)$

59. $7 + 2(4 - 1)$

60. $5 + 3(6 - 4)$

61. $-4 + 3(2 - 8)$

62. $-8 + 4(5 - 7)$

63. $(12 - 14)(1 - 4)$

64. $(8 - 9)(4 - 12)$

65. $(7 - 10)(10 - 4)$

66. $(5 - 12)(19 - 4)$

67. $(-2 - 8)(-6) + 7$

68. $(-9 - 4)(-2) + 10$

69. $3(-5) + |3 - 10|$

70. $4(-8) + |4 - 15|$

71. $\dfrac{-5(-6)}{9 - (-1)}$

72. $\dfrac{-12(-5)}{7 - (-5)}$

73. $\dfrac{-21(3)}{-3 - 6}$

74. $\dfrac{-40(3)}{-2 - 3}$

75. $\dfrac{-10(2) + 6(2)}{-3 - (-1)}$

76. $\dfrac{-12(4) + 5(3)}{-14 - (-3)}$

77. $\dfrac{3^2 - 4^2}{7(-8 + 9)}$

78. $\dfrac{5^2 - 7^2}{2(3 + 3)}$

79. $\dfrac{8(-1) - |(-4)(-3)|}{-6 - (-1)}$

80. $\dfrac{-27(-2) - |6 \cdot 4|}{-2(3) - 2(2)}$

81. $\dfrac{-13(-4) - (-8)(-2)}{(-10)(2) - 4(-2)}$

82. $\dfrac{-5(2) + [3(-2) - 4]}{-3 - (-1)}$

A few years ago, the following question and expression appeared on boxes of Swiss Miss Hot Cocoa Mix:

On average, how many mini-marshmallows are in one serving?

$$3 + 2 \times 4 \div 2 - 3 \times 7 - 4 + 47$$

83. The box gave 92 as the answer. What is the *correct* answer?

84. *WHAT WENT WRONG?* Explain the algebraic error that somebody at the company made in calculating the answer.

Evaluate each expression for $x = 6$, $y = -4$, and $a = 3$. See Example 5.

85. $5x - 2y + 3a$

86. $6x - 5y + 4a$

▶ **87.** $(2x + y)(3a)$

88. $(5x - 2y)(-2a)$

89. $\left(\dfrac{1}{3}x - \dfrac{4}{5}y\right)\left(-\dfrac{1}{5}a\right)$

90. $\left(\dfrac{5}{6}x + \dfrac{3}{2}y\right)\left(-\dfrac{1}{3}a\right)$

91. $(-5 + x)(-3 + y)(3 - a)$

92. $(6 - x)(5 + y)(3 + a)$

93. $-2y^2 + 3a$

94. $5x - 4a^2$

95. $\dfrac{2y - x}{a - 3}$

96. $\dfrac{xy + 8a}{x - 6}$

Write a numerical expression for each phrase, and simplify the expression. See Examples 6 and 7.

97. The product of -9 and 2, added to 9

98. The product of 4 and -7, added to -12

▶ **99.** Twice the product of -1 and 6, subtracted from -4

100. Twice the product of -8 and 2, subtracted from -1

101. Nine subtracted from the product of 1.5 and −3.2

102. Three subtracted from the product of 4.2 and −8.5

103. The product of 12 and the difference of 9 and −8

104. The product of −3 and the difference of 3 and −7

▶ **105.** The quotient of −12 and the sum of −5 and −1

106. The quotient of −20 and the sum of −8 and −2

107. The sum of 15 and −3, divided by the product of 4 and −3

108. The sum of −18 and −6, divided by the product of 2 and −4

109. Two-thirds of the difference of 8 and −1

110. Three-fourths of the sum of −8 and 12

111. 20% of the product of −5 and 6

112. 30% of the product of −8 and 5

113. The sum of $\frac{1}{2}$ and $\frac{5}{8}$, times the difference of $\frac{3}{5}$ and $\frac{1}{3}$

114. The sum of $\frac{3}{4}$ and $\frac{1}{2}$, times the difference of $\frac{2}{3}$ and $\frac{1}{6}$

115. The product of $-\frac{1}{2}$ and $\frac{3}{4}$, divided by $-\frac{2}{3}$

116. The product of $-\frac{2}{3}$ and $-\frac{1}{5}$, divided by $\frac{1}{7}$

Write each sentence as an equation, using x as the variable. Then find the solution of the equation from the set of integers between −12 and 12, inclusive. **See Example 8.**

▶ **117.** The quotient of a number and 3 is −3.

118. The quotient of a number and 4 is −1.

119. 6 less than a number is 4.

120. 7 less than a number is 2.

121. When 5 is added to a number, the result is −5.

122. When 6 is added to a number, the result is −3.

*To find the **average (mean)** of a group of numbers, we add the numbers and then divide the sum by the number of values added. For example, we find the average of* 14, 8, 3, 9, *and* 1, *as follows.*

$$\frac{14 + 8 + 3 + 9 + 1}{5} \quad \begin{matrix} \leftarrow \text{Given numbers} \\ \leftarrow \text{Number of values} \end{matrix}$$

$$= \frac{35}{5} \qquad \text{Add.}$$

$$\text{Average} \rightarrow \; = 7 \qquad \text{Divide.}$$

Find the average of each group of numbers.

123. 23, 18, 13, −4, and −8

124. 18, 12, 0, −4, and −10

125. −15, 29, 8, −6

126. −17, 34, 9, −2

127. All integers between −10 and 14, inclusive

128. All even integers between −18 and 4, inclusive

*The operation of division is used in **divisibility tests**. A divisibility test allows us to determine whether a given number is divisible (without remainder) by another number.*

129. An integer is divisible by 2 if its last digit is divisible by 2, and not otherwise. Show that

(a) 3,473,986 is divisible by 2   and   (b) 4,336,879 is not divisible by 2.

130. An integer is divisible by 3 if the sum of its digits is divisible by 3, and not otherwise. Show that

(a) 4,799,232 is divisible by 3   and   (b) 2,443,871 is not divisible by 3.

131. An integer is divisible by 4 if its last two digits form a number divisible by 4, and not otherwise. Show that

(a) 6,221,464 is divisible by 4 and (b) 2,876,335 is not divisible by 4.

132. An integer is divisible by 5 if its last digit is divisible by 5, and not otherwise. Show that

(a) 3,774,595 is divisible by 5 and (b) 9,332,123 is not divisible by 5.

133. An integer is divisible by 6 if it is divisible by both 2 and 3, and not otherwise. Show that

(a) 1,524,822 is divisible by 6 and (b) 2,873,590 is not divisible by 6.

134. An integer is divisible by 8 if its last three digits form a number divisible by 8, and not otherwise. Show that

(a) 2,923,296 is divisible by 8 and (b) 7,291,623 is not divisible by 8.

135. An integer is divisible by 9 if the sum of its digits is divisible by 9, and not otherwise. Show that

(a) 4,114,107 is divisible by 9 and (b) 2,287,321 is not divisible by 9.

136. An integer is divisible by 12 if it is divisible by both 3 and 4, and not otherwise. Show that

(a) 4,253,520 is divisible by 12 and (b) 4,249,474 is not divisible by 12.

SUMMARY EXERCISES Performing Operations with Real Numbers

Operations with Signed Numbers

Addition

Same sign Add the absolute values of the numbers. The sum has the same sign as the addends.

Different signs Find the absolute values of the numbers, and subtract the lesser absolute value from the greater. The sum has the same sign as the addend with greater absolute value.

Subtraction

Add the additive inverse (or opposite) of the subtrahend to the minuend.

Multiplication and Division

Same sign The product or quotient of two numbers with the same sign is positive.

Different signs The product or quotient of two numbers with different signs is negative.

Division by 0 is undefined.

Perform each indicated operation.

1. $14 - 3 \cdot 10$

2. $-3(8) - 4(-7)$

3. $(3 - 8)(-2) - 10$

4. $-6(7 - 3)$

5. $7 + 3(2 - 10)$

6. $-4[(-2)(6) - 7]$

7. $(-4)(7) - (-5)(2)$

8. $-5[-4 - (-2)(-7)]$

9. $40 - (-2)[8 - 9]$

1. -16 **2.** 4
3. 0 **4.** -24
5. -17 **6.** 76
7. -18 **8.** 90
9. 38

10. 4
11. −5   12. 5
13. $-\dfrac{7}{2}$, or $-3\dfrac{1}{2}$
14. 4
15. 13   16. $\dfrac{5}{4}$, or $1\dfrac{1}{4}$
17. 9   18. $\dfrac{37}{10}$, or $3\dfrac{7}{10}$
19. 0   20. 25
21. 14   22. undefined
23. −4   24. $\dfrac{6}{5}$, or $1\dfrac{1}{5}$
25. −1   26. $\dfrac{52}{37}$, or $1\dfrac{15}{37}$
27. $\dfrac{17}{16}$, or $1\dfrac{1}{16}$ 28. $-\dfrac{2}{3}$
29. 3.33   30. 1.02
31. 0   32. 24
33. −7   34. −3
35. −1   36. $\dfrac{1}{2}$
37. $-\dfrac{5}{13}$ 38. 5
39. $-\dfrac{8}{27}$ 40. 4

10. $\dfrac{5(-4)}{-7-(-2)}$

11. $\dfrac{-3-(-9+1)}{-7-(-6)}$

12. $\dfrac{5(-8+3)}{13(-2)+(-7)(-3)}$

13. $\dfrac{6^2-8}{-2(2)+4(-1)}$

14. $\dfrac{16(-8+5)}{15(-3)+(-7-4)(-3)}$

15. $\dfrac{9(-6)-3(8)}{4(-7)+(-2)(-11)}$

16. $\dfrac{2^2+4^2}{5^2-3^2}$

17. $\dfrac{(2+4)^2}{(5-3)^2}$

18. $\dfrac{4^3-3^3}{-5(-4+2)}$

19. $\dfrac{-9(-6)+(-2)(27)}{3(8-9)}$

20. $|-4(9)|-|-11|$

21. $\dfrac{6(-10+3)}{15(-2)-3(-9)}$

22. $\dfrac{3^2-5^2}{(-9)^2-9^2}$

23. $\dfrac{(-10)^2+10^2}{-10(5)}$

24. $-\dfrac{3}{4}\div\left(-\dfrac{5}{8}\right)$

25. $\dfrac{1}{2}\div\left(-\dfrac{1}{2}\right)$

26. $\dfrac{8^2-12}{(-5)^2+2(6)}$

27. $\left[\dfrac{5}{8}-\left(-\dfrac{1}{16}\right)\right]+\dfrac{3}{8}$

28. $\left(\dfrac{1}{2}-\dfrac{1}{3}\right)-\dfrac{5}{6}$

29. $-0.9(-3.7)$

30. $-5.1(-0.2)$

31. $|-2(3)+4|-|-2|$

32. $40+2(-5-3)$

Evaluate each expression for $x=-2$, $y=3$, and $a=4$.

33. $-x+y-3a$

34. $(x-y)-(a-2y)$

35. $\left(\dfrac{1}{2}x+\dfrac{2}{3}y\right)\left(-\dfrac{1}{4}a\right)$

36. $\dfrac{2x+3y}{a-xy}$

37. $\dfrac{x^2-y^2}{x^2+y^2}$

38. $-x^2+3y$

39. $\left(\dfrac{x}{y}\right)^3$

40. $\left(\dfrac{a}{x}\right)^2$

1.6 Properties of Real Numbers

OBJECTIVES

1 Use the commutative properties.
2 Use the associative properties.
3 Use the identity properties.
4 Use the inverse properties.
5 Use the distributive property.

In the basic properties covered in this section, a, b, and c represent real numbers.

OBJECTIVE 1 Use the commutative properties.

The word *commute* means to go back and forth. We might commute to work or to school. If we travel from home to work and follow the same route from work to home, we travel the same distance each time. The **commutative properties** say that if two numbers are added or multiplied in either order, the result is the same.

Commutative Properties

$$a+b=b+a \quad \text{Addition}$$

$$ab=ba \quad \text{Multiplication}$$

VOCABULARY
□ identity element for addition
   (additive identity)
□ identity element for
   multiplication
   (multiplicative identity)

**NOW TRY
EXERCISE 1**

Use a commutative property
to complete each statement.

(a) $7 + (-3) = -3 +$ _____

(b) $(-5)4 = 4 \cdot$ _____

EXAMPLE 1 Using the Commutative Properties

Use a commutative property to complete each statement.

(a) $-8 + 5 = 5 +$ __?__   Notice that the "order" changed.

$-8 + 5 = 5 + (-8)$ Commutative property of addition

(b) $(-2)7 =$ __?__ (-2)

$-2(7) = 7(-2)$ Commutative property of multiplication   NOW TRY

OBJECTIVE 2 Use the associative properties.

When we *associate* one object with another, we think of those objects as being grouped together. The **associative properties** say that when we add or multiply three numbers, we can group the first two together or the last two together and obtain the same answer.

Associative Properties

$$(a + b) + c = a + (b + c) \quad \text{Addition}$$

$$(ab)c = a(bc) \quad \text{Multiplication}$$

**NOW TRY
EXERCISE 2**

Use an associative property to
complete each statement.

(a) $-9 + (3 + 7) =$ _____

(b) $5[(-4) \cdot 9] =$ _____

EXAMPLE 2 Using the Associative Properties

Use an associative property to complete each statement.

(a) $-8 + (1 + 4) = (-8 +$ __?__ $) + 4$ The "order" is the same. The "grouping" changed.

$-8 + (1 + 4) = (-8 + 1) + 4$ Associative property of addition

(b) $[2 \cdot (-7)] \cdot 6 = 2 \cdot$ __?__

$[2 \cdot (-7)] \cdot 6 = 2 \cdot [(-7) \cdot 6]$ Associative property of multiplication

NOW TRY

By the associative property, the sum (or product) of three numbers will be the same no matter how the numbers are "associated" in groups. Parentheses can be left out if a problem contains only addition (or multiplication). For example,

$$(-1 + 2) + 3 \quad \text{and} \quad -1 + (2 + 3) \quad \text{can be written as} \quad -1 + 2 + 3.$$

EXAMPLE 3 Distinguishing between Properties

Decide whether each statement is an example of a commutative property, an associative property, or both.

(a) $(2 + 4) + 5 = 2 + (4 + 5)$

The order of the three numbers is the same on both sides of the equality symbol. The only change is in the *grouping,* or association, of the numbers. This is an example of the associative property.

(b) $6 \cdot (3 \cdot 10) = 6 \cdot (10 \cdot 3)$

The same numbers, 3 and 10, are grouped on each side. On the left, the 3 appears first, but on the right, the 10 appears first. The only change involves the *order* of the numbers, so this is an example of the commutative property.

NOW TRY ANSWERS
1. (a) 7   (b) -5
2. (a) $(-9 + 3) + 7$
   (b) $[5 \cdot (-4)] \cdot 9$

NOW TRY
EXERCISE 3

Is $5 + (7 + 6) = 5 + (6 + 7)$ an example of a commutative property or an associative property?

NOW TRY
EXERCISE 4

Find each sum or product.

(a) $8 + 54 + 7 + 6 + 32$

(b) $5(37)(20)$

(c) $(8 + 1) + 7 = 8 + (7 + 1)$

Both the order and the grouping are changed. On the left, the order of the three numbers is 8, 1, and 7. On the right, it is 8, 7, and 1. On the left, the 8 and 1 are grouped. On the right, the 7 and 1 are grouped. Therefore, *both* properties are used.

NOW TRY

EXAMPLE 4   Using the Commutative and Associative Properties

Find each sum or product.

(a) $23 + 41 + 2 + 9 + 25$

$= (41 + 9) + (23 + 2) + 25$

$= 50 + 25 + 25$

$= 100$

> Use the commutative and associative properties.

(b) $25(69)(4)$

$= 25(4)(69)$

$= 100(69)$

$= 6900$

NOW TRY

OBJECTIVE 3   Use the identity properties.

If a child wears a costume on Halloween, the child's appearance is changed, but his or her *identity* is unchanged. The identity of a real number is unchanged when identity properties are applied.

The **identity properties** say that the sum of 0 and any number equals that number, and the product of 1 and any number equals that number.

Identity Properties			
$a + 0 = a$	and	$0 + a = a$	Addition
$a \cdot 1 = a$	and	$1 \cdot a = a$	Multiplication

The number 0 leaves the identity, or value, of any real number unchanged by addition, so 0 is the **identity element for addition,** or the **additive identity.** Because multiplication by 1 leaves any real number unchanged, 1 is the **identity element for multiplication,** or the **multiplicative identity.**

NOW TRY
EXERCISE 5

Use an identity property to complete each statement.

(a) $\dfrac{2}{5} \cdot \underline{\quad} = \dfrac{2}{5}$

(b) $8 + \underline{\quad} = 8$

EXAMPLE 5   Using the Identity Properties

Use an identity property to complete each statement.

(a) $-3 + \underline{\ ?\ } = -3$

$-3 + 0 = -3$

Identity property of addition

(b) $\underline{\ ?\ } \cdot \dfrac{1}{2} = \dfrac{1}{2}$

$1 \cdot \dfrac{1}{2} = \dfrac{1}{2}$

Identity property of multiplication

NOW TRY

NOW TRY ANSWERS

3. commutative
4. **(a)** 107   **(b)** 3700
5. **(a)** 1   **(b)** 0

**NOW TRY
EXERCISE 6**

Simplify.

(a) $\dfrac{16}{20}$ (b) $\dfrac{2}{5} + \dfrac{3}{20}$

EXAMPLE 6   Using the Identity Property to Simplify Expressions

Simplify. In part (a), write in lowest terms. In part (b), perform the operation.

(a) $\dfrac{49}{35}$

$= \dfrac{7 \cdot 7}{5 \cdot 7}$	Factor.
$= \dfrac{7}{5} \cdot \dfrac{7}{7}$	Write as a product.
$= \dfrac{7}{5} \cdot 1$	Divide.
$= \dfrac{7}{5}$	Identity property

(b) $\dfrac{3}{4} + \dfrac{5}{24}$

$= \dfrac{3}{4} \cdot 1 + \dfrac{5}{24}$	Identity property
$= \dfrac{3}{4} \cdot \dfrac{6}{6} + \dfrac{5}{24}$	Use $1 = \dfrac{6}{6}$ to obtain a common denominator.
$= \dfrac{18}{24} + \dfrac{5}{24}$	Multiply.
$= \dfrac{23}{24}$	Add. **NOW TRY**

OBJECTIVE 4   Use the inverse properties.

Each day before we go to work or school, we probably put on our shoes. Before we go to sleep at night, we probably take them off, and this leads to the same situation that existed before we put them on. These operations from everyday life are examples of *inverse* operations.

The **inverse properties** of addition and multiplication lead to the additive and multiplicative identities, respectively. Recall that $-a$ is the **additive inverse,** or **opposite,** of a and $\dfrac{1}{a}$ is the **multiplicative inverse,** or **reciprocal,** of the nonzero number a.

Inverse Properties

$$a + (-a) = 0 \quad \text{and} \quad -a + a = 0 \qquad \text{Addition}$$

$$a \cdot \dfrac{1}{a} = 1 \quad \text{and} \quad \dfrac{1}{a} \cdot a = 1 \quad (a \neq 0) \qquad \text{Multiplication}$$

**NOW TRY
EXERCISE 7**

Use an inverse property to complete each statement.

(a) $10 + \underline{\hspace{1cm}} = 0$

(b) $-9 \cdot \underline{\hspace{1cm}} = 1$

EXAMPLE 7   Using the Inverse Properties

Use an inverse property to complete each statement.

(a) $\underline{\hspace{0.5cm}?\hspace{0.5cm}} + \dfrac{1}{2} = 0$

$-\dfrac{1}{2} + \dfrac{1}{2} = 0$

(b) $4 + \underline{\hspace{0.5cm}?\hspace{0.5cm}} = 0$

$4 + (-4) = 0$

(c) $-0.75 + \dfrac{3}{4} = \underline{\hspace{0.5cm}?\hspace{0.5cm}}$

$-0.75 + \dfrac{3}{4} = 0$

The inverse property of addition is used in parts (a)–(c).

(d) $\underline{\hspace{0.5cm}?\hspace{0.5cm}} \cdot \dfrac{5}{2} = 1$

$\dfrac{2}{5} \cdot \dfrac{5}{2} = 1$

(e) $-5(\underline{\hspace{0.5cm}?\hspace{0.5cm}}) = 1$

$-5\left(-\dfrac{1}{5}\right) = 1$

(f) $4(0.25) = \underline{\hspace{0.5cm}?\hspace{0.5cm}}$

$4(0.25) = 1$

The inverse property of multiplication is used in parts (d)–(f).   **NOW TRY**

NOW TRY ANSWERS

6. (a) $\dfrac{4}{5}$ (b) $\dfrac{11}{20}$

7. (a) -10 (b) $-\dfrac{1}{9}$

NOW TRY
EXERCISE 8
Simplify.

$$-\frac{1}{3}x + 7 + \frac{1}{3}x$$

EXAMPLE 8   Using Properties to Simplify an Expression

Simplify.

$$
\begin{aligned}
-2x + 10 + 2x & \\
= (-2x + 10) + 2x & \qquad \text{Order of operations} \\
= [10 + (-2x)] + 2x & \qquad \text{Commutative property} \\
= 10 + [(-2x) + 2x] & \qquad \text{Associative property} \\
= 10 + 0 & \qquad \text{Inverse property} \\
= 10 & \qquad \text{Identity property} \qquad \text{NOW TRY} \circlearrowleft
\end{aligned}
$$

For *any* value of *x*, $-2x$ and $2x$ are additive inverses.

NOTE   The steps of **Example 8** may be skipped when we actually do the simplification.

OBJECTIVE 5   Use the distributive property.

The word *distribute* means "to give out from one to several." Consider the following expressions.

$$2(5 + 8) \quad \text{equals} \quad 2(13), \quad \text{which equals} \quad 26.$$
$$2(5) + 2(8) \quad \text{equals} \quad 10 + 16, \quad \text{which equals} \quad 26.$$

Because both expressions equal 26,

$$2(5 + 8) = 2(5) + 2(8).$$

This is an example of the *distributive property of multiplication with respect to addition,* the only property involving *both* addition and multiplication. With this property, a product can be changed to a sum or difference. This idea is illustrated in **FIGURE 16**.

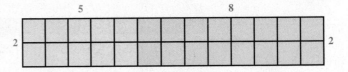

The area of the left part is 2(5) = 10.
The area of the right part is 2(8) = 16.
The total area is 2(5 + 8) = 2(13) = 26,
or the total area is 2(5) + 2(8) = 10 + 16 = 26.
Thus, 2(5 + 8) = 2(5) + 2(8).

FIGURE 16

The **distributive property** says that multiplying a number *a* by a sum of numbers $b + c$ gives the same result as multiplying *a* by *b* and *a* by *c* and then adding the two products.

Distributive Property

$$a(b + c) = ab + ac \qquad \text{and} \qquad (b + c)a = ba + ca$$

As the arrows show, the *a* outside the parentheses is "distributed" over the *b* and *c* inside. The distributive property is also valid for multiplication over subtraction.

$$a(b - c) = ab - ac \qquad \text{and} \qquad (b - c)a = ba - ca$$

The distributive property can be extended to more than two numbers.

$$a(b + c + d) = ab + ac + ad$$

NOW TRY ANSWER
8. 7

NOW TRY EXERCISE 9

Use the distributive property to rewrite each expression.

(a) $2(p + 5)$

(b) $-5(4x + 1)$

(c) $6(2r + t - 5z)$

EXAMPLE 9 **Using the Distributive Property**

Use the distributive property to rewrite each expression.

(a) $5(9 + 6)$ We could write $5(9) + 5(6)$ here.

$= 5 \cdot 9 + 5 \cdot 6$ The factor 5 is "distributed" to the numbers 9 and 6.

Multiply first. $= 45 + 30$ Multiply.

$= 75$ Add.

(b) $4(x + 5 + y)$

$= 4x + 4 \cdot 5 + 4y$ Distributive property

$= 4x + 20 + 4y$ Multiply.

(c) $-\dfrac{1}{2}(4x + 3)$

Think: $-\frac{1}{2}(4x) = \left(-\frac{1}{2} \cdot 4\right)x = \left(-\frac{1}{2} \cdot \frac{4}{1}\right)x$

$= -\dfrac{1}{2}(4x) + \left(-\dfrac{1}{2}\right)(3)$ Distributive property

This step is often omitted. $= -2x + \left(-\dfrac{3}{2}\right)$ Multiply.

$= -2x - \dfrac{3}{2}$ Definition of subtraction

(d) $3(k - 9)$ Be careful here.

$= 3[k + (-9)]$ Definition of subtraction

$= 3k + 3(-9)$ Distributive property

$= 3k - 27$ Multiply; definition of subtraction

(e) $-2(3x - 4)$

$= -2[3x + (-4)]$ Definition of subtraction

$= -2(3x) + (-2)(-4)$ Distributive property

$= (-2 \cdot 3)x + (-2)(-4)$ Associative property

$= -6x + 8$ Multiply.

(f) $8(3r + 11t + 5z)$

$= 8(3r) + 8(11t) + 8(5z)$ Distributive property

This step is often omitted. $= (8 \cdot 3)r + (8 \cdot 11)t + (8 \cdot 5)z$ Associative property

$= 24r + 88t + 40z$ Multiply.   NOW TRY

❗ **CAUTION**   In practice, we often omit the first step in **Examples 9(d) and 9(e),** where we rewrote the subtraction as addition of the additive inverse.

$3(k - 9)$ **Example 9(d)**

$= 3k - 3(9)$ Be careful not to make a sign error.

$= 3k - 27$ Multiply.

NOW TRY ANSWERS

9. **(a)** $2p + 10$ **(b)** $-20x - 5$

    (c) $12r + 6t - 30z$

The expression $-a$ may be interpreted as $-1 \cdot a$. Using this result and the distributive property, we can *remove* (or *clear*) *parentheses* from some expressions.

**NOW TRY
EXERCISE 10**

Write each expression without parentheses.

(a) $-(2 - r)$

(b) $-(2x - 5y - 7)$

EXAMPLE 10   Using the Distributive Property to Remove (Clear) Parentheses

Write each expression without parentheses.

(a) $-(2y + 3)$

The $-$ symbol indicates a factor of -1.

$= -1 \cdot (2y + 3)$ $-a = -1 \cdot a$

$= -1 \cdot 2y + (-1) \cdot 3$ Distributive property

$= -2y - 3$ Multiply; definition of subtraction

(b) $-(-9w - 2)$

$= -1(-9w - 2)$

$= -1(-9w) - 1(-2)$

$= 9w + 2$

We can also interpret the negative sign in front of the parentheses to mean the *opposite* of each of the terms within the parentheses.

$-1(-9w - 2)$

$\downarrow \quad \downarrow$

$= +9w + 2$

(c) $-(-x - 3y + 6z)$

$= -1(-1x - 3y + 6z)$ Be careful with signs.

$= -1(-1x) - 1(-3y) - 1(6z)$ Distributive property

$-1(-1x) = 1x = x$

$= x + 3y - 6z$ Multiply.   **NOW TRY**

Here is a summary of the basic properties of real numbers.

Properties of Addition and Multiplication

For any real numbers a, b, and c, the following properties hold.

Commutative Properties $a + b = b + a$ $ab = ba$

Associative Properties $(a + b) + c = a + (b + c)$

$(ab)c = a(bc)$

Identity Properties   There is a real number 0 such that

$a + 0 = a$ and $0 + a = a.$

There is a real number 1 such that

$a \cdot 1 = a$ and $1 \cdot a = a.$

Inverse Properties   For each real number a, there is a single real number $-a$ such that

$a + (-a) = 0$ and $(-a) + a = 0.$

For each nonzero real number a, there is a single real number $\frac{1}{a}$ such that

$a \cdot \frac{1}{a} = 1$ and $\frac{1}{a} \cdot a = 1.$

Distributive Properties $a(b + c) = ab + ac$ $(b + c)a = ba + ca$

NOW TRY ANSWERS
10. (a) $-2 + r$
  (b) $-2x + 5y + 7$

1.6 Exercises

FOR EXTRA HELP MyMathLab®

 Complete solution available in MyMathLab

1. *Concept Check* Match each item in Column I with the correct choice(s) from Column II. Choices may be used once, more than once, or not at all.

I

(a) Identity element for addition

(b) Identity element for multiplication

(c) Additive inverse of a

(d) Multiplicative inverse, or reciprocal, of the nonzero number a

(e) The number that is its own additive inverse

(f) The two numbers that are their own multiplicative inverses

(g) The only number that has no multiplicative inverse

(h) An example of the associative property

(i) An example of the commutative property

(j) An example of the distributive property

II

A. $(5 \cdot 4) \cdot 3 = 5 \cdot (4 \cdot 3)$

B. 0

C. $-a$

D. -1

E. $5 \cdot 4 \cdot 3 = 60$

F. 1

G. $(5 \cdot 4) \cdot 3 = 3 \cdot (5 \cdot 4)$

H. $5(4 + 3) = 5 \cdot 4 + 5 \cdot 3$

I. $\dfrac{1}{a}$

2. *Concept Check* Complete each statement.

The commutative property allows us to change the (*order/grouping*) of the addends in a sum or the factors in a product.

The associative property allows us to change the (*order/grouping*) of the addends in a sum or the factors in a product.

Concept Check Tell whether or not the following everyday activities are commutative.

3. Washing your face and brushing your teeth

4. Putting on your left sock and putting on your right sock

5. Preparing a meal and eating a meal

6. Starting a car and driving away in a car

7. Putting on your socks and putting on your shoes

8. Getting undressed and taking a shower

Concept Check Work each problem involving the properties of real numbers.

9. Use parentheses to show how the associative property can be used to give two different meanings to the phrase "foreign sales clerk."

10. Use parentheses to show how the associative property can be used to give two different meanings to the phrase "defective merchandise counter."

11. Evaluate the following expressions.

$$25 - (6 - 2) \quad \text{and} \quad (25 - 6) - 2.$$

Do you think subtraction is associative?

12. Evaluate the following expressions.

$$180 \div (15 \div 3) \quad \text{and} \quad (180 \div 15) \div 3.$$

Do you think division is associative?

13. Complete the table and each statement beside it.

Number	Additive Inverse	Multiplicative Inverse
5		
−10		
$-\frac{1}{2}$		
$\frac{3}{8}$		
x		$(x \neq 0)$
−y		$(y \neq 0)$

In general, a number and its additive inverse have (*the same / opposite*) signs.

A number and its multiplicative inverse have (*the same / opposite*) signs.

14. The following conversation took place between one of the authors of this book and his son, Jack, when Jack was 4 years old.

DADDY: "Jack, what is 3 + 0?"
JACK: "3."
DADDY: "Jack, what is 4 + 0?"
JACK: "4. And Daddy, *string* plus zero equals *string*!"

What property of addition did Jack recognize?

Use a commutative or an associative property to complete each statement. State which property is used. **See Examples 1 and 2.**

15. $-15 + 9 = 9 + \underline{\quad}$

16. $6 + (-2) = -2 + \underline{\quad}$

17. $-8 \cdot 3 = \underline{\quad} \cdot (-8)$

18. $-12 \cdot 4 = 4 \cdot \underline{\quad}$

19. $(3 + 6) + 7 = 3 + (\underline{\quad} + 7)$

20. $(-2 + 3) + 6 = -2 + (\underline{\quad} + 6)$

21. $7 \cdot (2 \cdot 5) = (\underline{\quad} \cdot 2) \cdot 5$

22. $8 \cdot (6 \cdot 4) = (8 \cdot \underline{\quad}) \cdot 4$

Decide whether each statement is an example of a commutative, *an* associative, *an* identity, *an* inverse, *or the* distributive property. **See Examples 1, 2, 3, 5, 6, 7, and 9.**

23. $4 + 15 = 15 + 4$

24. $3 + 12 = 12 + 3$

25. $5 \cdot (13 \cdot 7) = (5 \cdot 13) \cdot 7$

26. $-4 \cdot (2 \cdot 6) = (-4 \cdot 2) \cdot 6$

▶ 27. $-6 + (12 + 7) = (-6 + 12) + 7$

28. $(-8 + 13) + 2 = -8 + (13 + 2)$

29. $-9 + 9 = 0$

30. $1 + (-1) = 0$

▶ 31. $\frac{2}{3}\left(\frac{3}{2}\right) = 1$

32. $\frac{5}{8}\left(\frac{8}{5}\right) = 1$

33. $1.75 + 0 = 1.75$

34. $-8.45 + 0 = -8.45$

35. $(4 + 17) + 3 = 3 + (4 + 17)$

36. $(-8 + 4) + 12 = 12 + (-8 + 4)$

37. $2(x + y) = 2x + 2y$

38. $9(t + s) = 9t + 9s$

▶ 39. $-\frac{5}{9} = -\frac{5}{9} \cdot \frac{3}{3} = -\frac{15}{27}$

40. $-\frac{7}{12} = -\frac{7}{12} \cdot \frac{7}{7} = -\frac{49}{84}$

41. $4(2x) + 4(3y) = 4(2x + 3y)$

42. $6(5t) - 6(7r) = 6(5t - 7r)$

Find each sum or product. **See Example 4.**

43. $97 + 13 + 3 + 37$

44. $49 + 199 + 1 + 1$

▶ 45. $1999 + 2 + 1 + 8$

46. $2998 + 3 + 2 + 17$

47. $159 + 12 + 141 + 88$

48. $106 + 8 + (-6) + (-8)$

49. $843 + 627 + (-43) + (-27)$ **50.** $1846 + 1293 + (-46) + (-93)$

51. $5(47)(2)$ **52.** $2(79)5$

53. $-4 \cdot 5 \cdot 93 \cdot 5$ **54.** $2 \cdot 25 \cdot 67 \cdot (-2)$

Simplify each expression. ***See Examples 7 and 8.***

▶ 55. $6t + 8 - 6t + 3$ **56.** $9r + 12 - 9r + 1$ **57.** $\dfrac{2}{3}x - 11 + 11 - \dfrac{2}{3}x$

58. $\dfrac{1}{5}y + 4 - 4 - \dfrac{1}{5}y$ **59.** $\left(\dfrac{9}{7}\right)(-0.38)\left(\dfrac{7}{9}\right)$ **60.** $\left(\dfrac{4}{5}\right)(-0.73)\left(\dfrac{5}{4}\right)$

61. $t + (-t) + \dfrac{1}{2}(2)$ **62.** $w + (-w) + \dfrac{1}{4}(4)$

63. *Concept Check* A student used the distributive property to rewrite the expression $-3(4 - 6)$ as shown.

$$-3(4 - 6)$$
$$= -3(4) - 3(6)$$
$$= -12 - 18$$
$$= -30$$

This answer is incorrect. ***WHAT WENT WRONG?*** Rewrite the given expression correctly.

64. *Concept Check* A student wrote the expression $-(3x + 4)$ without parentheses as shown.

$$-(3x + 4)$$
$$= -1(3x + 4)$$
$$= -1(3x) + 4$$
$$= -3x + 4$$

This answer is incorrect. ***WHAT WENT WRONG?*** Rewrite the given expression correctly.

65. Explain how the procedure for changing $\dfrac{3}{4}$ to $\dfrac{9}{12}$ requires the use of the multiplicative identity element, 1.

66. Explain how the procedure for changing $\dfrac{9}{12}$ to $\dfrac{3}{4}$ requires the use of the multiplicative identity element, 1.

Use the distributive property to rewrite each expression. ***See Example 9.***

67. $5(9 + 8)$ **68.** $6(11 + 8)$ **▶ 69.** $4(t + 3)$

70. $5(w + 4)$ **71.** $7(z - 8)$ **72.** $8(x - 6)$

73. $-8(r + 3)$ **74.** $-11(x + 4)$ **75.** $-\dfrac{1}{4}(8x + 3)$

76. $-\dfrac{1}{3}(9x + 5)$ **77.** $-5(y - 4)$ **78.** $-9(g - 4)$

79. $2(6x + 5)$ **80.** $3(3x + 4)$ **81.** $-3(2x - 5)$

82. $-4(3x - 2)$ **83.** $-6(8x + 1)$ **84.** $-5(4x + 1)$

85. $-\dfrac{4}{3}(12y + 15z)$ **86.** $-\dfrac{2}{5}(10b + 20a)$ **87.** $8(3r + 4s - 5y)$

88. $2(5u - 3v + 7w)$ **89.** $-3(8x + 3y + 4z)$ **90.** $-5(2x - 5y + 6z)$

Write each expression without parentheses. ***See Example 10.***

▶ 91. $-(4t + 3m)$ **92.** $-(9x + 12y)$ **93.** $-(-5c - 4d)$

94. $-(-13x - 15y)$ **95.** $-(-q + 5r - 8s)$ **96.** $-(-z + 5w - 9y)$

1.7 Simplifying Expressions

OBJECTIVES

1 Simplify expressions.
2 Identify terms and numerical coefficients.
3 Identify like terms.
4 Combine like terms.
5 Simplify expressions from word phrases.

VOCABULARY

☐ term
☐ numerical coefficient (coefficient)
☐ like terms
☐ unlike terms

NOW TRY EXERCISE 1

Simplify each expression.

(a) $3(2x - 4y)$

(b) $-4 - (-3y + 5)$

OBJECTIVE 1   Simplify expressions.

We use the properties of **Section 1.6** to write expressions in simpler form.

EXAMPLE 1   Simplifying Expressions

Simplify each expression.

(a) $4x + 8 + 9$ simplifies to $4x + 17$.

(b) $4(3m - 2n)$ — To simplify, we clear the parentheses.

$$= 4(3m) - 4(2n) \qquad \text{Distributive property}$$
$$= (4 \cdot 3)m - (4 \cdot 2)n \qquad \text{Associative property}$$
$$= 12m - 8n \qquad \text{Multiply.}$$

(c) $6 + 3(4k + 5)$ Do **not** start by adding.

$$= 6 + 3(4k) + 3(5) \qquad \text{Distributive property}$$
$$= 6 + (3 \cdot 4)k + 3(5) \qquad \text{Associative property}$$
$$= 6 + 12k + 15 \qquad \text{Multiply.}$$
$$= 6 + 15 + 12k \qquad \text{Commutative property}$$
$$= 21 + 12k \qquad \text{Add.}$$

(d) $5 - (2y - 8)$

$$= 5 - 1(2y - 8) \qquad -a = -1 \cdot a$$
$$= 5 - 1(2y) - 1(-8) \qquad \text{Distributive property}$$
$$= 5 - 2y + 8 \qquad \text{Multiply.}$$ Be careful with signs.
$$= 5 + 8 - 2y \qquad \text{Commutative property}$$
$$= 13 - 2y \qquad \text{Add.} \qquad \text{NOW TRY}$$

NOTE   The steps using the commutative and associative properties will not be shown in the rest of the examples. However, be aware that they are usually involved.

OBJECTIVE 2   Identify terms and numerical coefficients.

A **term** is a number (constant), a variable, or a product or quotient of numbers and variables raised to powers.

$$9x, \quad 15y^2, \quad -3, \quad -8m^2n, \quad \frac{2}{p}, \quad \text{and} \quad k \qquad \text{Terms}$$

In the term $9x$, the **numerical coefficient,** or simply the **coefficient,** of the variable x is 9.

▼ **Terms and Their Coefficients**

Term	Numerical Coefficient
8	8
$-7y$	-7
$34r^3$	34
$-26x^5yz^4$	-26
$-k$, or $-1k$	-1
r, or $1r$	1
$\frac{3x}{8}$, or $\frac{3}{8}x$	$\frac{3}{8}$
$\frac{x}{3} = \frac{1x}{3}$, or $\frac{1}{3}x$	$\frac{1}{3}$

NOW TRY ANSWERS

1. **(a)** $6x - 12y$ **(b)** $3y - 9$

> ❗ **CAUTION** It is important to be able to distinguish between **terms** and **factors**. Consider the following expressions.

$$8x^3 + 12x^2$$ This expression has **two terms**, $8x^3$ and $12x^2$.
Terms are separated by $a +$ or $-$ symbol.

$$(8x^3)(12x^2)$$ This is a **one-term** expression.
The **factors** $8x^3$ and $12x^2$ are multiplied.

OBJECTIVE 3 Identify like terms.

Terms with exactly the same variables that have the same exponents on the variables are **like terms**.

Like Terms	**Unlike Terms**	
$9t$ and $4t$	$4y$ and $7t$	Different variables
$6x^2$ and $-5x^2$	$17x$ and $-8x^2$	Different exponents
$-2pq$ and $11pq$	$4xy^2$ and $4xy$	Different exponents
$3x^2y$ and $5x^2y$	$-7wz^3$ and $2xz^3$	Different variables

OBJECTIVE 4 Combine like terms.

The distributive property

$$a(b + c) = ab + ac \quad \text{can be written "in reverse" as} \quad ab + ac = a(b + c).$$

This last form, which may be used to find the sum or difference of like terms, provides justification for **combining like terms**.

NOW TRY
EXERCISE 2
Simplify each expression.
(a) $4x + 6x - 7x$ **(b)** $z + z$
(c) $4p^2 - 3p^2$

EXAMPLE 2 Combining Like Terms

Simplify each expression.

(a) $-9m + 5m$
$= (-9 + 5)m$ Distributive property in reverse
$= -4m$

(b) $6r + 3r + 2r$
$= (6 + 3 + 2)r$
$= 11r$

(c) $4x + x$
$= 4x + 1x$ $x = 1x$
$= (4 + 1)x$
$= 5x$

(d) $16y^2 - 9y^2$
$= (16 - 9)y^2$
$= 7y^2$

(e) $32y + 10y^2$ These unlike terms cannot be combined.   NOW TRY

> ❗ **CAUTION** Only like terms may be combined.

NOW TRY ANSWERS
2. (a) $3x$ **(b)** $2z$ **(c)** p^2

Simplifying an Expression

An expression has been simplified when the following conditions have been met.

- All grouping symbols have been removed.
- All like terms have been combined.
- Operations have been performed, when possible.

EXAMPLE 3   Simplifying Expressions

Simplify each expression.

(a) $14y + 2(6 + 3y)$ *Start by distributing the 2.*

$$= 14y + 2(6) + 2(3y) \qquad \text{Distributive property}$$

$14y + 6y$
$= (14 + 6)y$
$= 20y$

$$= 14y + 12 + 6y \qquad \text{Multiply.}$$

$$= 20y + 12 \qquad \text{Combine like terms.}$$

(b) $9k - 6 - 3(2 - 5k)$ *Be careful with signs.*

$$= 9k - 6 - 3(2) - 3(-5k) \qquad \text{Distributive property}$$

$$= 9k - 6 - 6 + 15k \qquad \text{Multiply.}$$

$$= 24k - 12 \qquad \text{Combine like terms.}$$

(c) $-(2 - r) + 10r$

$$= -1(2 - r) + 10r \qquad -a = -1 \cdot a, \text{ or } -1(a)$$

$$= -1(2) - 1(-r) + 10r \qquad \text{Distributive property}$$

Be careful with signs. $= -2 + 1r + 10r \qquad \text{Multiply.}$

$$= -2 + 11r \qquad \text{Combine like terms.}$$

Alternatively, $-(2 - r)$ can be thought of as the *opposite* of $(2 - r)$—that is, $-2 + r$—which can then be added to $10r$ to obtain $-2 + 11r$.

(d) $100[0.03(x + 4)]$

$$= [(100)(0.03)](x + 4) \qquad \text{Associative property}$$

$$= 3(x + 4) \qquad \text{Multiply.}$$

$$= 3x + 3(4) \qquad \text{Distributive property}$$

$$= 3x + 12 \qquad \text{Multiply.}$$

(e) $5(2a - 6) - 3(4a - 9)$

$$= 5(2a) + 5(-6) - 3(4a) - 3(-9) \qquad \text{Distributive property twice}$$

$$= 10a - 30 - 12a + 27 \qquad \text{Multiply.}$$

$$= -2a - 3 \qquad \text{Combine like terms.}$$

**NOW TRY
EXERCISE 3**

Simplify each expression.

(a) $5k - 6 - (3 - 4k)$

(b) $\frac{1}{4}x - \frac{2}{3}(x - 9)$

(f) $-\frac{2}{3}(x - 6) - \frac{1}{6}x$

$= -\frac{2}{3}x - \frac{2}{3}(-6) - \frac{1}{6}x$ Distributive property

$= -\frac{2}{3}x + 4 - \frac{1}{6}x$ Multiply.

$= -\frac{4}{6}x + 4 - \frac{1}{6}x$ Find a common denominator.

$= -\frac{5}{6}x + 4$ Combine like terms.   NOW TRY

NOTE   **Examples 2 and 3** suggest that like terms may be combined by adding or subtracting the coefficients of the terms and keeping the same variable factors.

OBJECTIVE 5   Simplify expressions from word phrases.

EXAMPLE 4   Translating Words into a Mathematical Expression

**NOW TRY
EXERCISE 4**

Translate the phrase into a mathematical expression using x as the variable, and simplify.

Twice a number, subtracted from the sum of the number and 5

Translate the phrase into a mathematical expression using x as the variable, and simplify.

The sum of 9, five times a number,
four times the number, and
six times the number

The word "sum" indicates that the terms should be added. Use x for the number.

$9 + 5x + 4x + 6x$ simplifies to $9 + 15x$.   Combine like terms.

This is an expression to be simplified, *not* an equation to be solved.

NOW TRY

NOW TRY ANSWERS
3. **(a)** $9k - 9$ **(b)** $-\frac{5}{12}x + 6$
4. $(x + 5) - 2x; -x + 5$

1.7 Exercises

FOR EXTRA HELP   ▶ MyMathLab®

▶ *Complete solution available in MyMathLab*

Concept Check   *Choose the letter of the correct response.*

1. Which expression is a simplified form of $-(6x - 3)$?

  A. $-6x - 3$ **B.** $-6x + 3$ **C.** $6x - 3$ **D.** $6x + 3$

2. Which is an example of a term with numerical coefficient 5?

  A. $5x^3y^7$ **B.** x^5 **C.** $\frac{x}{5}$ **D.** 5^2xy^3

3. Which is an example of a pair of like terms?

  A. $6t, 6w$ **B.** $-8x^2y, 9xy^2$ **C.** $5ry, 6yr$ **D.** $-5x^2, 2x^3$

4. Which is a correct translation for "six times a number, subtracted from the product of eleven and the number" (if x represents the number)?

  A. $6x - 11x$ **B.** $11x - 6x$ **C.** $(11 + x) - 6x$ **D.** $6x - (11 + x)$

5. *Concept Check*   A student simplified the expression $7x - 2(3 - 2x)$ as shown.

$$7x - 2(3 - 2x)$$
$$= 7x - 2(3) - 2(2x)$$
$$= 7x - 6 - 4x$$
$$= 3x - 6$$

WHAT WENT WRONG? Find the correct simplified answer.

6. *Concept Check*   A student simplified the expression $3 + 2(4x - 5)$ as shown.

$$3 + 2(4x - 5)$$
$$= 5(4x - 5)$$
$$= 5(4x) + 5(-5)$$
$$= 20x - 25$$

WHAT WENT WRONG? Find the correct simplified answer.

Simplify each expression. **See Example 1.**

7. $4r + 19 - 8$ **8.** $7t + 18 - 4$ **9.** $7(3x - 4y)$

10. $8(2p - 9q)$ ▶ **11.** $5 + 2(x - 3y)$ **12.** $8 + 3(s - 6t)$

13. $-2 - (5 - 3p)$ **14.** $-10 - (7 - 14r)$

15. $6 + (4 - 3x) - 8$ **16.** $-12 + (7 - 8x) + 6$

In each term, give the numerical coefficient of the variable(s). **See Objective 2.**

▶ **17.** $-12k$ **18.** $-11y$ **19.** $3m^2$ **20.** $9n^6$

21. xw **22.** pq **23.** $-x$ **24.** $-t$

25. $\dfrac{x}{2}$ **26.** $\dfrac{x}{6}$ **27.** $\dfrac{2x}{5}$ **28.** $\dfrac{8x}{9}$

29. $-0.5x^3$ **30.** $-1.75x^2$ **31.** 10 **32.** 15

Identify each group of terms as like *or* unlike. **See Objective 3.**

▶ **33.** $8r, -13r$ **34.** $-7x, 12x$ **35.** $5z^4, 9z^3$ **36.** $8x^5, -10x^3$

37. $4, 9, -24$ **38.** $7, 17, -83$ **39.** x, y **40.** t, s

Simplify each expression. **See Examples 1–3.**

41. $7y + 6y$ **42.** $5m + 2m$ **43.** $-6x - 3x$

44. $-4z - 8z$ ▶ **45.** $12b + b$ **46.** $19x + x$

47. $3k + 8 + 4k + 7$ **48.** $15z + 1 + 4z + 2$

49. $-5y + 3 - 1 + 5 + y - 7$ **50.** $2k - 7 - 5k + 6 - 1 + 2$

51. $-2x + 3 + 4x - 17 + 20$ **52.** $r - 6 - 12r - 4 + 16$

53. $16 - 5m - 4m - 2 + 2m$ **54.** $6 - 3z - 2z - 5 - 2z$

55. $-10 + x + 4x - 7 - 4x$ **56.** $-p + 10p - 3p - 4 - 5p$

57. $1 + 7x + 11x - 1 + 5x$ **58.** $-r + 2 - 5r - 2 + 4r$

59. $-\dfrac{4}{3} + 2t + \dfrac{1}{3}t - 8 - \dfrac{8}{3}t$ **60.** $-\dfrac{5}{6} + 8x + \dfrac{1}{6}x - 7 - \dfrac{7}{6}$

61. $6y^2 + 11y^2 - 8y^2$ **62.** $-9m^3 + 3m^3 - 7m^3$

63. $2p^2 + 3p^2 - 8p^3 - 6p^3$ **64.** $5y^3 + 6y^3 - 3y^2 - 4y^2$

▶ **65.** $2(4x + 6) + 3$ **66.** $4(6y + 9) + 7$

67. $-6 - 4(y - 7)$ **68.** $-4 - 5(t - 13)$

69. $13p + 4(4 - 8p)$ **70.** $5x + 3(7 - 2x)$

71. $3t - 5 - 2(2t - 4)$ **72.** $8p + 6 - 3(3p - 1)$

73. $100[0.05(x + 3)]$ **74.** $100[0.06(x + 5)]$

75. $10[0.3(5 - 3x)]$ **76.** $10[0.5(8 - 2z)]$

77. $-5(5y - 9) + 3(3y + 6)$ **78.** $-3(2t + 4) + 8(2t - 4)$

79. $2(5r + 3) - 3(2r - 3)$ **80.** $3(2y - 5) - 4(5y - 7)$

81. $8(2k - 1) - (4k - 3)$ **82.** $6(3p - 2) - (5p + 1)$

83. $-\dfrac{4}{3}(y - 12) - \dfrac{1}{6}y$ **84.** $-\dfrac{7}{5}(t - 15) - \dfrac{1}{2}t$

85. $\dfrac{1}{2}(2x + 4) - \dfrac{1}{3}(9x - 6)$ **86.** $\dfrac{1}{4}(8x + 16) - \dfrac{1}{5}(20x - 15)$

87. $-\dfrac{2}{3}(5x + 7) - \dfrac{1}{3}(4x + 8)$ **88.** $-\dfrac{3}{4}(7x + 9) - \dfrac{1}{4}(5x + 7)$

89. $-7.5(2y + 4) - 2.9(3y - 6)$ **90.** $8.4(6t - 6) + 2.4(9 - 3t)$

91. $-2(-3k + 2) - (5k - 6) - 3k - 5$ **92.** $-2(3r - 4) - (6 - r) + 2r - 5$

93. $-4(-3x + 3) - (6x - 4) - 2x + 1$ **94.** $-5(8x + 2) - (5x - 3) - 3x + 17$

Extending Skills   *Write each of the following as a mathematical expression, and simplify.*

95. Add $3x - 2$ to $4x + 8$.   **96.** Add $8t + 5$ to $10t - 8$.

97. Subtract $x - 7$ from $5x + 1$.   **98.** Subtract $3x - 5$ from $2x - 3$.

Translate each phrase into a mathematical expression using x as the variable, and simplify. See Example 4.

▶ **99.** Five times a number, added to the sum of the number and three

100. Six times a number, added to the sum of the number and six

101. A number multiplied by -7, subtracted from the sum of 13 and six times the number

102. A number multiplied by 5, subtracted from the sum of 14 and eight times the number

103. Six times a number added to -4, subtracted from twice the sum of three times the number and 4 (*Hint: Twice* means two times.)

104. Nine times a number added to 6, subtracted from triple the sum of 12 and 8 times the number (*Hint: Triple* means three times.)

RELATING CONCEPTS   For Individual or Group Work (Exercises 105–108)

A manufacturer has fixed costs of $1000 to produce gizmos. Each gizmo costs $5 to make. The fixed cost to produce gadgets is $750, and each gadget costs $3 to make. **Work Exercises 105–108 in order.**

105. Write an expression for the cost to make x gizmos. (*Hint:* The cost will be the sum of the fixed cost and the cost per item times the number of items.)

106. Write an expression for the cost to make y gadgets.

107. Write an expression for the total cost to make x gizmos and y gadgets.

108. Simplify the expression from **Exercise 107.**

STUDY SKILLS

Reviewing a Chapter

Your textbook provides material to help you prepare for quizzes or tests in this course. Refer to the **Chapter 1 Summary** as you read through the following techniques.

Chapter Reviewing Techniques

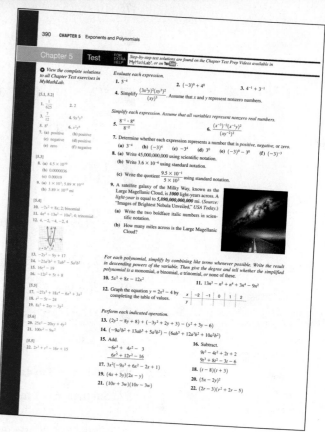

- **Review the Key Terms and any New Symbols.** Make a study card for each. Include a definition, an example, a sketch (if appropriate), and a section or page reference.

- **Take the Test Your Word Power quiz** to check your understanding of new vocabulary. The answers immediately follow.

- **Read the Quick Review.** Pay special attention to the headings. Study the explanations and examples given for each concept. Try to think about the whole chapter.

- **Reread your lecture notes.** Focus on what your instructor has emphasized in class, and review that material in your text.

- **Look over your homework.** Pay special attention to any trouble spots.

- **Work the Review Exercises.** They are grouped by section. Answers are included in the margin for quick reference.

  ▶ Pay attention to direction words, such as *simplify, solve,* and *evaluate.*

  ▶ Are your answers exact and complete? Did you include the correct labels, such as $, cm², ft, etc.?

  ▶ Make study cards for difficult problems.

- **Work the Mixed Review Exercises.** They are in mixed-up order. Check your answers in the margin.

- **Take the Chapter Test under test conditions.**

  ▶ Time yourself.

  ▶ Use a calculator or notes (if your instructor permits them on tests).

  ▶ Take the test in one sitting.

  ▶ Show all your work.

  ▶ Check your answers in the margin. Section references are provided.

Reviewing a chapter will take some time. Avoid rushing through your review in one night. Use the suggestions over a few days or evenings to better understand the material and remember it longer.

Follow these reviewing techniques to prepare for your next test.

1. How much time did you spend reviewing for your test? Was it enough?

2. How did the reviewing techniques work for you?

3. What will you do differently when reviewing for your next test?

Key Terms

1.1
exponent (power)
base
exponential expression
inequality

1.2
constant
variable
algebraic expression
equation
solution
set
element

1.3
natural (counting) numbers
whole numbers
number line
integers
signed numbers
rational numbers
graph
coordinate
irrational numbers
real numbers
additive inverse (opposite)
absolute value

1.4
sum
addends
difference
minuend
subtrahend

1.5
product
factor
multiplicative inverse
(reciprocal)
quotient
dividend
divisor

1.6
identity element for addition
(additive identity)
identity element for
multiplication
(multiplicative identity)

1.7
term
numerical coefficient
(coefficient)
like terms
unlike terms

New Symbols

a^n	n factors of a
$[\ ]$	brackets
$=$	is equal to
$\neq$	is not equal to
$<$	is less than
$>$	is greater than
$\leq$	is less than or equal to

$\geq$	is greater than or equal to
$\{\ \}$	set braces
$\{x \mid x \text{ has a certain property}\}$	set-builder notation

$-x$	additive inverse, or opposite, of x
$\lvert x \rvert$	absolute value of x
$\dfrac{1}{x}$	multiplicative inverse, or reciprocal, of x (where $x \neq 0$)

$a(b),\ (a)b,\ (a)(b),\ a \cdot b,$ or ab	a times b
$a \div b,\ \dfrac{a}{b},\ a/b,$ or $b\overline{)a}$	a divided by b

Test Your Word Power

See how well you have learned the vocabulary in this chapter.

1. An **exponent** is
   A. a symbol that tells how many numbers are being multiplied
   B. a number raised to a power
   C. a number that tells how many times a factor is repeated
   D. one of two or more numbers that are multiplied.

2. A **variable** is
   A. a symbol used to represent an unknown number
   B. a value that makes an equation true
   C. a solution of an equation
   D. the answer in a division problem.

3. An **integer** is
   A. a positive or negative number
   B. a natural number, its opposite, or zero
   C. any number that can be graphed on a number line
   D. the quotient of two numbers.

4. The **absolute value** of a number is
   A. the graph of the number
   B. the reciprocal of the number
   C. the opposite of the number
   D. the distance between 0 and the number on a number line.

5. A **term** is
   A. a numerical factor

   B. a number, a variable, or a product or quotient of numbers and variables raised to powers
   C. one of several variables with the same exponents
   D. a sum of numbers and variables raised to powers.

6. A **numerical coefficient** is
   A. the numerical factor of the variable(s) in a term
   B. the number of terms in an expression
   C. a variable raised to a power
   D. the variable factor in a term.

ANSWERS

1. C; *Example:* In 2^3, the number 3 is the exponent (or power), so 2 is a factor three times, and $2^3 = 2 \cdot 2 \cdot 2 = 8$.   **2.** A; *Examples:* a, b, c
3. B; *Examples:* $-9, 0, 6$ **4.** D; *Examples:* $\lvert 2 \rvert = 2$ and $\lvert -2 \rvert = 2$ **5.** B; *Examples:* $6, \frac{x}{2}, -4ab^2$ **6.** A; *Examples:* The term 3 has numerical
coefficient 3, $8z$ has numerical coefficient 8, and $-10x^4y$ has numerical coefficient -10.

Quick Review

CONCEPTS

EXAMPLES

1.1   Exponents, Order of Operations, and Inequality

Order of Operations
Work within any parentheses or brackets and above and below fraction bars first. Always follow this order.

Step 1   Apply all exponents.

Step 2   Do any multiplications or divisions in order from left to right.

Step 3   Do any additions or subtractions in order from left to right.

Simplify $36 - 4(2^2 + 3)$.

$$36 - 4(2^2 + 3)$$
$$= 36 - 4(4 + 3) \qquad \text{Apply the exponent.}$$
$$= 36 - 4(7) \qquad \text{Add inside the parentheses.}$$
$$= 36 - 28 \qquad \text{Multiply.}$$
$$= 8 \qquad \text{Subtract.}$$

1.2   Variables, Expressions, and Equations

To *evaluate* an expression means to find its *value*. Evaluate an expression with a variable by substituting a given number for the variable.

Find the value of $2x + y^2$ for $x = 3$ and $y = -4$.

$$2x + y^2$$
$$= 2(3) + (-4)^2 \qquad \text{Substitute.}$$
$$= 6 + 16 \qquad \text{Multiply. Apply the exponent.}$$
$$= 22 \qquad \text{Add.}$$

Values of a variable that make an equation true are solutions of the equation.

Is 2 a solution of $5x + 3 = 18$?

$$5(2) + 3 \overset{?}{=} 18 \qquad \text{Let } x = 2.$$
$$13 = 18 \qquad \text{False}$$

2 is not a solution.

1.3   Real Numbers and the Number Line

Ordering Real Numbers
a is less than b if a lies to the left of b on a number line.

a is greater than b if a lies to the right of b on a number line.

The additive inverse, or opposite, of x is $-x$.

The absolute value of x, written $|x|$, is the distance between x and 0 on a number line.

Graph -2, 0, and 3.

$$-2 < 3 \qquad 3 > 0 \qquad 0 < 3$$

$$-(5) = -5 \qquad -(-7) = 7 \qquad -0 = 0$$

$$|13| = 13 \qquad |0| = 0 \qquad |-5| = 5$$

1.4   Adding and Subtracting Real Numbers

Adding Two Signed Numbers
Same sign   Add their absolute values. The sum has the same sign as the addends.

Different signs   Subtract their absolute values. The sum has the sign of the addend with greater absolute value.

Add.
$$9 + 4 = 13$$
$$-8 + (-5) = -13$$
$$7 + (-12) = -5$$
$$-5 + 13 = 8$$

Definition of Subtraction

$$x - y = x + (-y)$$

Subtract.

$5 - (-2)$	$-3 - 4$	$-2 - (-6)$
$= 5 + 2$	$= -3 + (-4)$	$= -2 + 6$
$= 7$	$= -7$	$= 4$

CONCEPTS	**EXAMPLES**

1.5   Multiplying and Dividing Real Numbers

Multiplying and Dividing Two Signed Numbers
Same sign   The product (or quotient) is *positive*.

Multiply or divide.

$$6 \cdot 5 = 30 \qquad -7(-8) = 56 \qquad \frac{-24}{-6} = 4$$

Different signs   The product (or quotient) is *negative*.

$$-6(5) = -30 \qquad \frac{-18}{9} = -2 \qquad \frac{49}{-7} = -7$$

Definition of Division

$$x \div y = x \cdot \frac{1}{y} \quad \text{(where } y \neq 0)$$

$$\frac{10}{2} = 10 \cdot \frac{1}{2} = 5$$

0 divided by a nonzero number equals 0.
Division by 0 is undefined.

$$\frac{0}{5} = 0 \qquad \frac{5}{0} \text{ is undefined.}$$

1.6   Properties of Real Numbers

Commutative Properties

$$a + b = b + a$$
$$ab = ba$$

$$7 + (-1) = -1 + 7$$
$$5(-3) = (-3)5$$

Associative Properties

$$(a + b) + c = a + (b + c)$$
$$(ab)c = a(bc)$$

$$(3 + 4) + 8 = 3 + (4 + 8)$$
$$[-2(6)]4 = -2[(6)4]$$

Identity Properties

$$a + 0 = a \qquad 0 + a = a$$
$$a \cdot 1 = a \qquad 1 \cdot a = a$$

$$-7 + 0 = -7 \qquad 0 + (-7) = -7$$
$$9 \cdot 1 = 9 \qquad 1 \cdot 9 = 9$$

Inverse Properties

$$a + (-a) = 0 \qquad -a + a = 0$$
$$a \cdot \frac{1}{a} = 1 \qquad \frac{1}{a} \cdot a = 1 \quad \text{(where } a \neq 0)$$

$$7 + (-7) = 0 \qquad -7 + 7 = 0$$
$$-2\left(-\frac{1}{2}\right) = 1 \qquad -\frac{1}{2}(-2) = 1$$

Distributive Properties

$$a(b + c) = ab + ac$$
$$(b + c)a = ba + ca$$
$$a(b - c) = ab - ac$$

$$5(4 + 2) = 5(4) + 5(2)$$
$$(4 + 2)5 = 4(5) + 2(5)$$
$$9(5 - 4) = 9(5) - 9(4)$$

1.7   Simplifying Expressions

Only like terms may be combined. We use the distributive property.

Simplify each expression.

$$-3y^2 + 6y^2 + 14y^2$$
$$= (-3 + 6 + 14)y^2$$
$$= 17y^2$$

$$4(3 + 2x) - 6(5 - x)$$
$$= 4(3) + 4(2x) - 6(5) - 6(-x)$$
$$= 12 + 8x - 30 + 6x$$
$$= 14x - 18$$

Chapter 1   Review Exercises

Answers (left column):

1. 625
2. $\dfrac{27}{125}$
3. $\dfrac{1}{64}$
4. 0.001
5. 27
6. 17
7. 4
8. 399
9. 39
10. 5
11. true
12. true
13. false
14. $13 < 17$
15. $5 + 2 \neq 10$
16. $\dfrac{2}{3} \geq \dfrac{4}{6}$
17. 30
18. 60
19. 14
20. 13
21. $x + 6$
22. $8 - x$
23. $6x - 9$
24. $12 + \dfrac{3}{5}x$
25. yes
26. no
27. $2x - 6 = 10;\ 8$
28. $4x = 8;\ 2$

29.

30.

31. rational numbers, real numbers
32. rational numbers, real numbers
33. natural numbers, whole numbers, integers, rational numbers, real numbers
34. irrational numbers, real numbers
35. -10
36. -9
37. $-\dfrac{3}{4}$
38. $-|23|$
39. true
40. true
41. true
42. true

1.1   *Find the value of each exponential expression.*

1. 5^4
2. $\left(\dfrac{3}{5}\right)^3$
3. $\left(\dfrac{1}{8}\right)^2$
4. $(0.1)^3$

Find the value of each expression.

5. $8 \cdot 5 - 13$
6. $16 + 12 \div 4 - 2$
7. $20 - 2(5 + 3)$
8. $7[3 + 6(3^2)]$
9. $\dfrac{9(4^2 - 3)}{4 \cdot 5 - 17}$
10. $\dfrac{6(5 - 4) + 2(4 - 2)}{3^2 - (4 + 3)}$

Decide whether each statement is true *or* false.

11. $12 \cdot 3 - 6 \cdot 6 \leq 0$
12. $3[5(2) - 3] > 20$
13. $9 \leq 4^2 - 8$

Write each word statement in symbols.

14. Thirteen is less than seventeen.
15. Five plus two is not equal to ten.
16. Two-thirds is greater than or equal to four-sixths.

1.2   *Find the value of each expression for* $x = 6$ *and* $y = 3$.

17. $2x + 6y$
18. $4(3x - y)$
19. $\dfrac{x}{3} + 4y$
20. $\dfrac{x^2 + 3}{3y - x}$

Write each word phrase as an algebraic expression, using x as the variable.

21. Six added to a number
22. A number subtracted from eight
23. Nine subtracted from six times a number
24. Three-fifths of a number added to 12

Decide whether each equation has the given number as a solution.

25. $5x + 3(x + 2) = 22;\quad 2$
26. $\dfrac{t + 5}{3t} = 1;\quad 6$

Write each word statement as an equation. Use x as the variable. Then find the solution of the equation from the set $\{0, 2, 4, 6, 8, 10\}$.

27. Six less than twice a number is 10.
28. The product of a number and 4 is 8.

1.3   *Graph each number on a number line.*

29. $-4, -\dfrac{1}{2}, 0, 2.5, 5$
30. $-3, -1\dfrac{1}{2}, \dfrac{2}{3}, 2.25, 3$

Classify each number, using the sets natural numbers, whole numbers, integers, rational numbers, irrational numbers, *and* real numbers.

31. $\dfrac{4}{3}$
32. $0.\overline{63}$
33. 19
34. $\sqrt{6}$

Select the lesser of the two given numbers.

35. $-10, 5$
36. $-8, -9$
37. $-\dfrac{2}{3}, -\dfrac{3}{4}$
38. $0, -|23|$

Decide whether each statement is true *or* false.

39. $12 > -13$
40. $0 > -5$
41. $-9 < -7$
42. $-13 \geq -13$

43. (a) 9   (b) 9
44. (a) 0   (b) 0
45. (a) −6   (b) 6
46. (a) $\dfrac{5}{7}$ (b) $\dfrac{5}{7}$
47. 12   **48.** −3
49. −19   **50.** −7

51. −6   **52.** −4
53. −17   **54.** $-\dfrac{29}{36}$
55. −21.8   **56.** −14
57. −10   **58.** −19
59. −11   **60.** −1
61. 7
62. $-\dfrac{43}{35}$, or $-1\dfrac{8}{35}$
63. 10.31   **64.** −12
65. 2   **66.** −3
67. (−31 + 12) + 19; 0
68. [−4 + (−8)] + 13; 1
69. −4 − (−6); 2
70. [4 + (−8)] − 5; −9
71. $26.25   **72.** −10°F
73. −$29   **74.** −10°
75. 38   **76.** 14,840.95
77. 36   **78.** −105
79. $\dfrac{1}{2}$ **80.** 10.08
81. −20   **82.** −10
83. −24   **84.** −35
85. 4   **86.** −20
87. $-\dfrac{3}{4}$ **88.** 11.3
89. −1

*For each number, find (**a**) the additive inverse and (**b**) the absolute value.*

43. −9   **44.** 0   **45.** 6   **46.** $-\dfrac{5}{7}$

Find each absolute value.

47. $|-12|$ **48.** $-|3|$ **49.** $-|-19|$ **50.** $-|9-2|$

1.4   *Perform each indicated operation.*

51. $-10+4$ **52.** $14+(-18)$ **53.** $-8+(-9)$

54. $\dfrac{4}{9}+\left(-\dfrac{5}{4}\right)$ **55.** $-13.5+(-8.3)$ **56.** $(-10+7)+(-11)$

57. $[-6+(-8)+8]+[9+(-13)]$ **58.** $(-4+7)+(-11+3)+(-15+1)$

59. $-7-4$ **60.** $-12-(-11)$

61. $5-(-2)$ **62.** $-\dfrac{3}{7}-\dfrac{4}{5}$

63. $2.56-(-7.75)$ **64.** $(-10-4)-(-2)$

65. $(-3+4)-(-1)$ **66.** $-(-5+6)-2$

Write a numerical expression for each phrase, and simplify the expression.

67. 19 added to the sum of −31 and 12   **68.** 13 more than the sum of −4 and −8

69. The difference of −4 and −6   **70.** Five less than the sum of 4 and −8

Solve each problem.

71. George found that his checkbook balance was −$23.75, so he deposited $50.00. What is his new balance?

72. The low temperature in Yellowknife, in the Canadian Northwest Territories, one January day was −26°F. It rose 16° that day. What was the high temperature?

73. Reginald owed a friend $28. He repaid $13, but then borrowed another $14. What positive or negative amount represents his present financial status?

74. If the temperature drops 7° below its previous level of −3°, what is the new temperature?

75. A quarterback passed for a gain of 8 yd, was sacked for a loss of 12 yd, and then threw a 42 yd touchdown pass. What positive or negative number represents the total net yardage for the plays?

76. On Friday, August 30, 2013, the Dow Jones Industrial Average closed at 14,810.31, down 30.64 from the previous day. What was the closing value the previous day? (*Source: The Washington Post.*)

1.5   *Perform each indicated operation.*

77. $-12(-3)$ **78.** $15(-7)$ **79.** $-\dfrac{4}{3}\left(-\dfrac{3}{8}\right)$ **80.** $-4.8(-2.1)$

81. $5(8-12)$ **82.** $(5-7)(8-3)$ **83.** $2(-6)-(-4)(-3)$

84. $3(-10)-5$ **85.** $\dfrac{-36}{-9}$ **86.** $\dfrac{220}{-11}$

87. $-\dfrac{1}{2}\div\dfrac{2}{3}$ **88.** $\dfrac{-33.9}{-3}$ **89.** $\dfrac{-5(3)-1}{8-4(-2)}$

90. 2
91. 1 **92.** 0.5
93. −18 **94.** −18
95. 125 **96.** −423
97. −4(5) − 9; −29
98. $\frac{5}{6}[12 + (-6)]$; 5
99. $\frac{12}{8 + (-4)}$; 3
100. $\frac{-20(12)}{15 - (-15)}$; −8
101. 8x = −24; −3
102. $\frac{x}{3} = -2$; −6

103. identity property
104. identity property
105. inverse property
106. inverse property
107. associative property
108. associative property
109. distributive property
110. commutative property
111. 7y + 14 **112.** −48 + 12t
113. 6s + 15y **114.** 4r − 5s

115. 11m **116.** 16p²
117. 16p² + 2p **118.** −4k + 12
119. −2m + 29 **120.** −5k − 1
121. −2(3x) − 7x; −13x
122. (5 + 4x) + 8x; 5 + 12x

90. $\dfrac{5(-2) - 3(4)}{-2[3 - (-2)] - 1}$ **91.** $\dfrac{10^2 - 5^2}{8^2 + 3^2 - (-2)}$ **92.** $\dfrac{(0.6)^2 + (0.8)^2}{(-1.2)^2 - (-0.56)}$

Evaluate each expression for x = −5, y = 4, and z = −3.

93. $6x - 4z$ **94.** $5x + y - z$ **95.** $5x^2$ **96.** $z^2(3x - 8y)$

Write a numerical expression for each phrase, and simplify the expression.

97. Nine less than the product of −4 and 5

98. Five-sixths of the sum of 12 and −6

99. The quotient of 12 and the sum of 8 and −4

100. The product of −20 and 12, divided by the difference of 15 and −15

Write each sentence as an equation, using x as the variable. Then find the solution from the set of integers between −12 and 12, inclusive.

101. 8 times a number is −24. **102.** The quotient of a number and 3 is −2.

1.6 *Decide whether each statement is an example of a* commutative, *an* associative, *an* identity, *an* inverse, *or the* distributive property.

103. $6 + 0 = 6$ **104.** $5 \cdot 1 = 5$

105. $-\dfrac{2}{3}\left(-\dfrac{3}{2}\right) = 1$ **106.** $17 + (-17) = 0$

107. $5 + (-9 + 2) = [5 + (-9)] + 2$ **108.** $w(xy) = (wx)y$

109. $3(x + y) = 3x + 3y$ **110.** $(1 + 2) + 3 = 3 + (1 + 2)$

Use the distributive property to rewrite each expression. Simplify if possible.

111. $7(y + 2)$ **112.** $-12(4 - t)$ **113.** $3(2s + 5y)$ **114.** $-(-4r + 5s)$

1.7 *Simplify each expression.*

115. $2m + 9m$ **116.** $15p^2 - 7p^2 + 8p^2$

117. $5p^2 - 4p + 6p + 11p^2$ **118.** $-2(3k - 5) + 2(k + 1)$

119. $7(2m + 3) - 2(8m - 4)$ **120.** $-(2k + 8) - (3k - 7)$

Translate each phrase into a mathematical expression using x as the variable, and simplify.

121. Seven times a number, subtracted from the product of −2 and three times the number

122. A number multiplied by 8, added to the sum of 5 and four times the number

Chapter 1   Mixed Review Exercises

Complete the table.

1. 3; 3; $-\dfrac{1}{3}$ **2.** 12; −12; $\dfrac{1}{12}$

3. $-\dfrac{2}{3}; \dfrac{2}{3}; \dfrac{2}{3}$ **4.** 0.2; 0.2; 5

	Number	Absolute Value	Additive Inverse	Multiplicative Inverse
1.	−3			
2.	12			
3.				$-\dfrac{3}{2}$
4.			−0.2	

5. rational numbers, real numbers

6. 37

7. $\frac{8}{3}$, or $2\frac{2}{3}$ 8. $-\frac{1}{24}$

9. 2

10. $-\frac{28}{15}$, or $-1\frac{13}{15}$

11. $-\frac{3}{2}$, or $-1\frac{1}{2}$

12. $\frac{25}{36}$

13. 16   14. 77.6

15. 11   16. $16t - 36$

17. $8x^2 - 21y^2$ 18. 24

19. $-47°F$ 20. 14,776 ft

5. To which of the following sets does $0.\overline{6}$ belong: natural numbers, whole numbers, integers, rational numbers, irrational numbers, real numbers?

6. Evaluate $(x + 6)^3 - y^3$ for $x = -2$ and $y = 3$.

Perform each indicated operation.

7. $\dfrac{6(-4) + 2(-12)}{5(-3) + (-3)}$ 8. $\dfrac{3}{8} - \dfrac{5}{12}$ 9. $\dfrac{8^2 + 6^2}{7^2 + 1^2}$

10. $-\dfrac{12}{5} \div \dfrac{9}{7}$ 11. $2\dfrac{5}{6} - 4\dfrac{1}{3}$ 12. $\left(\dfrac{5}{6}\right)^2$

13. $[(-2) + 7 - (-5)] + [-4 - (-10)]$ 14. $-16(-3.5) - 7.2(-3)$

15. $-8 + [(-4 + 17) - (-3 - 3)]$ 16. $-4(2t + 1) - 8(-3t + 4)$

17. $5x^2 - 12y^2 + 3x^2 - 9y^2$ 18. $(-8 - 3) - 5(2 - 9)$

Solve each problem.

19. The highest temperature ever recorded in Iowa was 118°F. The lowest temperature ever recorded in the state was 165° lower than the highest temperature. What is the record low temperature for Iowa? (*Source:* National Climatic Data Center.)

20. The top of Mt. Whitney, visible from Death Valley, has an altitude of 14,494 ft above sea level. The bottom of Death Valley is 282 ft below sea level. Using 0 as sea level, find the difference between these two elevations. (*Source:* World Almanac and Book of Facts.)

Chapter 1   Test

FOR EXTRA HELP Step-by-step test solutions are found on the Chapter Test Prep Videos available in MyMathLab® or on YouTube™.

▶ *View the complete solutions to all Chapter Test exercises in MyMathLab.*

[1.1]
1. true   2. false

[1.3]
3. rational numbers, real numbers
4. $-|-8|$, or -8

[1.5]
5. $\dfrac{-6}{2 + (-8)}$; 1   6. negative

[1.4, 1.5]
7. 4
8. $-\dfrac{17}{6}$, or $-2\dfrac{5}{6}$
9. 2   10. 6
11. 108   12. $\dfrac{30}{7}$, or $4\dfrac{2}{7}$

[1.5]
13. -70 14. 3

Decide whether each statement is true *or* false.

1. $4[-20 + 7(-2)] \le 135$ 2. $\left(\dfrac{1}{2}\right)^2 + \left(\dfrac{2}{3}\right)^2 = \left(\dfrac{1}{2} + \dfrac{2}{3}\right)^2$

3. To which of the following sets does $-\dfrac{2}{3}$ belong: natural numbers, whole numbers, integers, rational numbers, irrational numbers, real numbers?

4. Select the lesser number of 6 and $-|-8|$.

5. Write a numerical expression for the phrase, and simplify the expression.

The quotient of -6 and the sum of 2 and -8

6. If a and b are both negative, is $\dfrac{a + b}{a \cdot b}$ positive or negative?

Perform each indicated operation.

7. $-2 - (5 - 17) + (-6)$ 8. $-5\dfrac{1}{2} + 2\dfrac{2}{3}$

9. $-6.2 - [-7.1 + (2.0 - 3.1)]$ 10. $4^2 + (-8) - (2^3 - 6)$

11. $(-5)(-12) + 4(-4) + (-8)^2$ 12. $\dfrac{30(-1 - 2)}{-9[3 - (-2)] - 12(-2)}$

Evaluate each expression for $x = -2$ and $y = 4$.

13. $3x - 4y^2$ 14. $\dfrac{5x + 7y}{3(x + y)}$

[1.3–1.5]
15. 7000 m   **16.** 15
17. −$1.09 trillion

[1.6]
18. D
19. A   **20.** E
21. B   **22.** C
23. distributive property
24. (a) −18   **(b)** −18
  (c) The distributive property
    assures us that the answers
    must be the same, because
    $a(b + c) = ab + ac$ for all
    a, b, c.

[1.7]
25. $21x$ **26.** $15x − 3$

Solve each problem.

15. The highest elevation in Argentina is Mt. Aconcagua, which is 6960 m above sea level. The lowest point in Argentina is the Valdés Peninsula, 40 m below sea level. Find the difference between the highest and lowest elevations.

16. For a certain system of rating relief pitchers, 3 points are awarded for a save, 3 points are awarded for a win, 2 points are subtracted for a loss, and 2 points are subtracted for a blown save. If Craig Kimbrel of the Atlanta Braves has 4 saves, 3 wins, 2 losses, and 1 blown save, how many points does he have?

17. For 2012, the U.S. federal government collected $2.45 trillion in revenues, but spent $3.54 trillion. Write the federal budget deficit as a signed number. (*Source*: U.S. Department of Management and Budget.)

Match each statement in Column I with the property it illustrates in Column II.

I	**II**
18. $3x + 0 = 3x$	**A.** Commutative property
19. $(5 + 2) + 8 = 8 + (5 + 2)$	**B.** Associative property
20. $-3(x + y) = -3x + (-3y)$	**C.** Inverse property
21. $-5 + (3 + 2) = (-5 + 3) + 2$	**D.** Identity property
22. $-\dfrac{5}{3}\left(-\dfrac{3}{5}\right) = 1$	**E.** Distributive property

23. What property is used to clear parentheses and write $3(x + 1)$ as $3x + 3$?

24. Consider the expression $-6[5 + (-2)]$.

  (a) Evaluate it by first working within the brackets.

  (b) Evaluate it by using the distributive property.

  (c) Why must the answers in parts (a) and (b) be the same?

Simplify each expression.

25. $8x + 4x − 6x + x + 14x$ **26.** $5(2x − 1) − (x − 12) + 2(3x − 5)$

2

Linear Equations and Inequalities in One Variable

Solving *linear equations,* the subject of this chapter, can be thought of in terms of the concept of balance.

2.1 The Addition Property of Equality

An **equation** is a statement asserting that two algebraic expressions are equal.

> ⚠ **CAUTION**   *Remember that an equation includes an equality symbol.*

Equation	Expression
↓	↓
Left side → $x - 5 = 2$ ← Right side	$x - 5$
An equation can be solved.	An expression **cannot** be solved. (It can be *evaluated* for a given value, or *simplified*.)

OBJECTIVE 1   Identify linear equations.

> **Linear Equation in One Variable**
>
> A **linear equation in one variable** (here x) can be written in the form
>
> $$Ax + B = C,$$
>
> where A, B, and C are real numbers and $A \neq 0$.
>
> *Examples:* $4x + 9 = 0$, $2x - 3 = 5$,   and $x = 7$ Linear equations
>
> $x^2 + 2x = 5$, $x^3 = -1$, $\dfrac{1}{x} = 6$,   and $|2x + 6| = 0$ *Non*linear equations

A **solution** of an equation is a number that makes the equation true when it replaces the variable. An equation is solved by finding its **solution set,** the set of all solutions. Equations with exactly the same solution sets are **equivalent equations.**

A linear equation in x is *solved* by using a series of steps to produce a simpler equivalent equation of the form

$$x = \text{a number} \quad \text{or} \quad \text{a number} = x.$$

OBJECTIVE 2   Use the addition property of equality.

In the linear equation $x - 5 = 2$, both $x - 5$ and 2 represent the same number because that is the meaning of the equality symbol. To solve the equation, we change the left side from $x - 5$ to just x, as follows.

$x - 5 = 2$	Given equation
$x - 5 + 5 = 2 + 5$	Add 5 to *each* side to keep them equal.
$x + 0 = 7$	Additive inverse property
$x = 7$	Additive identity property

> Add 5. It is the opposite (additive inverse) of -5, and $-5 + 5 = 0$.

To check that 7 is the solution, we replace x with 7 in the original equation.

CHECK	$x - 5 = 2$	Original equation
	$7 - 5 \overset{?}{=} 2$	Let $x = 7$.
	$2 = 2$ ✓	True

> The left side equals the right side.

> We write a solution set using set braces.

The final equation is true, so 7 is the solution and $\{7\}$ is the solution set.

To solve the equation $x - 5 = 2$, we used the **addition property of equality.**

Addition Property of Equality

If A, B, and C represent real numbers, then the equations

$$A = B \quad \text{and} \quad A + C = B + C \quad \text{are equivalent.}$$

That is, the same number may be added to each side of an equation without changing the solution set.

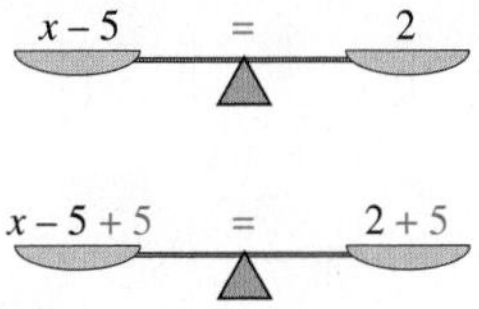

In this property, any quantity that represents a real number C can be added to each side of an equation to obtain an equivalent equation.

NOTE Equations can be thought of in terms of a balance. Thus, adding the *same* quantity to each side does not affect the balance. See **FIGURE 1**.

$$x - 5 = 2$$

$$x - 5 + 5 = 2 + 5$$

FIGURE 1

NOW TRY EXERCISE 1

Solve $x - 13 = 4$.

EXAMPLE 1 Applying the Addition Property of Equality

Solve $x - 16 = 7$.

Our goal is to get an equivalent equation of the form $x = $ a number.

$$x - 16 = 7$$

$$x - 16 + 16 = 7 + 16 \qquad \text{Add 16 to each side.}$$

$$x = 23 \qquad \text{Combine like terms.}$$

CHECK Substitute 23 for x in the *original* equation.

$$x - 16 = 7 \qquad \text{Original equation}$$

$$23 - 16 \overset{?}{=} 7 \qquad \text{Let } x = 23.$$

7 *is not* the solution. $\quad 7 = 7 \; \checkmark \qquad$ True

A true statement results, so 23 is the solution and $\{23\}$ is the solution set.

NOW TRY

> ⊘ **CAUTION** *The final line of the check does* not *give the solution to the problem.* It confirms that the value found is actually a solution.

NOW TRY EXERCISE 2

Solve $t - 5.7 = -7.2$.

EXAMPLE 2 Applying the Addition Property of Equality

Solve $x - 2.9 = -6.4$.

Our goal is to isolate x. $\quad x - 2.9 = -6.4$

$$x - 2.9 + 2.9 = -6.4 + 2.9 \qquad \text{Add 2.9 to each side.}$$

$$x = -3.5$$

CHECK $\qquad\qquad x - 2.9 = -6.4 \qquad \text{Original equation}$

$$-3.5 - 2.9 \overset{?}{=} -6.4 \qquad \text{Let } x = -3.5.$$

$$-6.4 = -6.4 \; \checkmark \qquad \text{True}$$

NOW TRY ANSWERS
1. $\{17\}$
2. $\{-1.5\}$

A true statement results, so the solution set is $\{-3.5\}$.　　　　NOW TRY

In **Section 1.4,** subtraction was defined as addition of the opposite. Thus, we can also use the following when solving an equation.

Addition Property of Equality Extended to Subtraction

The same number may be *subtracted* from each side of an equation without changing the solution set.

**NOW TRY
EXERCISE 3**

Solve $-15 = x + 12$.

EXAMPLE 3   Applying the Addition Property of Equality

Solve $-7 = x + 22$.

   Here, the variable x is on the right side of the equation.

$$-7 = x + 22 \quad \text{The variable can be isolated on *either* side.}$$

$$-7 - 22 = x + 22 - 22 \qquad \text{Subtract 22 from each side.}$$

$$-29 = x, \quad \text{or} \quad x = -29 \qquad \text{Rewrite; a number} = x, \text{ or } x = \text{a number.}$$

CHECK $-7 = x + 22$ Original equation

$$-7 \stackrel{?}{=} -29 + 22 \qquad \text{Let } x = -29.$$

$$-7 = -7 \ \checkmark \qquad \text{True}$$

The check confirms that the solution set is $\{-29\}$.   NOW TRY

NOTE   In **Example 3,** what happens if we subtract $-7 - 22$ incorrectly, obtaining $x = -15$ (instead of $x = -29$) as the last line of the solution? A check should indicate an error.

CHECK $-7 = x + 22$ Original equation from **Example 3**

$$-7 \stackrel{?}{=} -15 + 22 \qquad \text{Let } x = -15.$$

The left side does ***not*** equal the right side. $-7 = 7$ False

The false statement indicates that -15 is *not* a solution of the equation. If this happens, rework the problem.

**NOW TRY
EXERCISE 4**

Solve $x - 5 = 2x$.

EXAMPLE 4   Subtracting a Variable Term

Solve $6x - 8 = 7x$.

$$6x - 8 = 7x$$

$$6x - 8 - 6x = 7x - 6x \qquad \text{Subtract } 6x \text{ from each side.}$$

$$-8 = x \qquad \text{Combine like terms.}$$

CHECK $6x - 8 = 7x$ Original equation

$$6(-8) - 8 \stackrel{?}{=} 7(-8) \qquad \text{Let } x = -8.$$

Use parentheses when substituting to avoid errors. $-48 - 8 \stackrel{?}{=} -56$ Multiply.

$$-56 = -56 \ \checkmark \qquad \text{True}$$

NOW TRY ANSWERS
3. $\{-27\}$
4. $\{-5\}$

A true statement results, so the solution set is $\{-8\}$.   NOW TRY

What happens in **Example 4** if we start by subtracting $7x$ from each side?

$$6x - 8 = 7x \qquad \text{Original equation from \textbf{Example 4}}$$

$$6x - 8 - 7x = 7x - 7x \qquad \text{Subtract } 7x \text{ from each side.}$$

$$-8 - x = 0 \qquad \text{Combine like terms.}$$

$$-8 - x + 8 = 0 + 8 \qquad \text{Add 8 to each side.}$$

$$-x = 8 \qquad \text{Combine like terms.}$$

This result gives the value of $-x$, but not of x itself. However, it does say that the additive inverse of x is 8, which means that x must be -8.

$$x = -8 \qquad \text{Same result as in \textbf{Example 4}}$$

(This result can also be justified by the multiplication property of equality, covered in **Section 2.2.**) We can make the following generalization.

If a is a number and $-x = a$, then $x = -a$.

**NOW TRY
EXERCISE 5**

Solve $\frac{2}{3}x + 4 = \frac{5}{3}x$.

EXAMPLE 5   Subtracting a Variable Term (Fractional Coefficients)

Solve $\frac{3}{5}x + 15 = \frac{8}{5}x$.

$$\frac{3}{5}x + 15 = \frac{8}{5}x \qquad \text{Original equation}$$

$$\frac{3}{5}x + 15 - \frac{3}{5}x = \frac{8}{5}x - \frac{3}{5}x \qquad \text{Subtract } \frac{3}{5}x \text{ from each side.}$$

$\boxed{\text{From now on we will skip this step.}}\!\!\rightarrow 15 = 1x \qquad \frac{3}{5}x - \frac{3}{5}x = 0; \; \frac{8}{5}x - \frac{3}{5}x = \frac{5}{5}x = 1x$

$$15 = x \qquad \text{Multiplicative identity property}$$

Check by replacing x with 15 in the original equation. The solution set is $\{15\}$.

NOW TRY

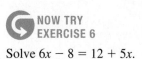
**NOW TRY
EXERCISE 6**

Solve $6x - 8 = 12 + 5x$.

EXAMPLE 6   Applying the Addition Property of Equality Twice

Solve $8 - 6p = -7p + 5$.

$$8 - 6p = -7p + 5$$

$$8 - 6p + 7p = -7p + 5 + 7p \qquad \text{Add } 7p \text{ to each side.}$$

$$8 + p = 5 \qquad \text{Combine like terms.}$$

$$8 + p - 8 = 5 - 8 \qquad \text{Subtract 8 from each side.}$$

$$p = -3 \qquad \text{Combine like terms.}$$

CHECK $8 - 6p = -7p + 5 \qquad \text{Original equation}$

$$8 - 6(-3) \overset{?}{=} -7(-3) + 5 \qquad \text{Let } p = -3.$$

$$8 + 18 \overset{?}{=} 21 + 5 \qquad \text{Multiply.}$$

$$26 = 26 \; \checkmark \qquad \text{True}$$

NOW TRY ANSWERS
5. $\{4\}$
6. $\{20\}$

The check results in a true statement, so the solution set is $\{-3\}$.   **NOW TRY**

NOTE *There are often several correct ways to solve an equation.* In the equation

$$8 - 6p = -7p + 5, \quad \text{See Example 6.}$$

we could begin by adding $6p$ (instead of $7p$) to each side. Combining like terms and subtracting 5 from each side gives $3 = -p$. (Try this.) If $3 = -p$, then $-3 = p$, and the variable has been isolated on the right side of the equation. The same solution results.

OBJECTIVE 3 Simplify, and then use the addition property of equality.

NOW TRY EXERCISE 7

Solve.

$$5x - 10 - 12x$$
$$= 4 - 8x - 9$$

EXAMPLE 7 Combining Like Terms When Solving

Solve $3t - 12 + t + 2 = 5 + 3t + 2$.

$3t - 12 + t + 2 = 5 + 3t + 2$	
$4t - 10 = 7 + 3t$	Combine like terms.
$4t - 10 - 3t = 7 + 3t - 3t$	Subtract $3t$ from each side.
$t - 10 = 7$	Combine like terms.
$t - 10 + 10 = 7 + 10$	Add 10 to each side.
$t = 17$	Combine like terms.

CHECK		
	$3t - 12 + t + 2 = 5 + 3t + 2$	Original equation
	$3(17) - 12 + 17 + 2 \stackrel{?}{=} 5 + 3(17) + 2$	Let $t = 17$.
	$51 - 12 + 17 + 2 \stackrel{?}{=} 5 + 51 + 2$	Multiply.
	$58 = 58 \ \checkmark$	True

The check results in a true statement, so the solution set is $\{17\}$. **NOW TRY**

NOW TRY EXERCISE 8

Solve.

$$4(3x - 2) - (11x - 4) = 3$$

EXAMPLE 8 Using the Distributive Property When Solving

Solve $3(2 + 5x) - (1 + 14x) = 6$.

$3(2 + 5x) - (1 + 14x) = 6$	
Be sure to distribute to all terms within the parentheses. $\ \ 3(2 + 5x) - 1(1 + 14x) = 6$	$-(1 + 14x) = -1(1 + 14x)$
$3(2) + 3(5x) - 1(1) - 1(14x) = 6$	Distributive property
Be careful here. $\ 6 + 15x - 1 - 14x = 6$	Multiply.
$x + 5 = 6$	Combine like terms.
$x + 5 - 5 = 6 - 5$	Subtract 5 from each side.
$x = 1$	Combine like terms.

Check by substituting 1 for x in the original equation. The solution set is $\{1\}$.

NOW TRY

! CAUTION *Be careful to apply the distributive property correctly* in a problem like that in **Example 8,** or a sign error may result.

2.1 Exercises

FOR EXTRA HELP

 MyMathLab®

▶ *Complete solution available in MyMathLab*

Concept Check   Complete each statement with the correct response. The following terms may be used once, more than once, or not at all.

linear	expression	solution set	multiplication
equation	addition	equivalent equations	variable

1. A(n) _____ includes an equality symbol, while a(n) _____ does not.

2. A(n) _____ equation in one _____ (here x) can be written in the form $Ax + B (= / \neq) C$.

3. Equations that have exactly the same solution set are _____.

4. The _____ property of equality states that the same expression may be added to or subtracted from each side of an equation without changing the _____.

5. *Concept Check*   Decide whether each of the following is an *equation* or an *expression*. If it is an equation, solve it. If it is an expression, simplify it.

(a) $5x + 8 - 4x + 7$

(b) $-6y + 12 + 7y - 5$

(c) $5x + 8 - 4x = 7$

(d) $-6y + 12 + 7y = -5$

6. *Concept Check*   Which pairs of equations are equivalent equations?

A. $x + 2 = 6$ and $x = 4$

B. $10 - x = 5$ and $x = -5$

C. $x + 3 = 9$ and $x = 6$

D. $4 + x = 8$ and $x = -4$

7. *Concept Check*   Which of the following are *not* linear equations in one variable?

A. $x^2 - 5x + 6 = 0$

B. $x^3 = x$

C. $3x - 4 = 0$

D. $7x - 6x = 3 + 9x$

8. Explain how to check a solution of an equation.

Solve each equation, and check the solution. **See Examples 1–6.**

9. $x - 3 = 9$

10. $x - 9 = 8$

11. $x - 12 = 19$

12. $x - 18 = 22$

13. $x - 6 = -9$

14. $x - 5 = -7$

15. $r + 8 = 12$

16. $x + 7 = 11$

17. $x + 28 = 19$

18. $x + 47 = 26$

19. $x + \dfrac{1}{4} = -\dfrac{1}{2}$

20. $x + \dfrac{2}{3} = -\dfrac{1}{6}$

21. $7 + r = -3$

22. $8 + k = -4$

▶ **23.** $2 = p + 15$

24. $5 = z + 19$

25. $-4 = x - 14$

26. $-7 = x - 22$

27. $-\dfrac{1}{3} = x - \dfrac{3}{5}$

28. $-\dfrac{1}{4} = x - \dfrac{2}{3}$

▶ **29.** $x - 8.4 = -2.1$

30. $x - 15.5 = -5.1$

31. $t + 12.3 = -4.6$

32. $x + 21.5 = -13.4$

33. $3x = 2x + 7$

34. $5x = 4x + 9$

35. $10x + 4 = 9x$

36. $8t + 5 = 7t$

37. $8x - 3 = 9x$

38. $6x - 4 = 7x$

39. $6t - 2 = 5t$

40. $4z - 6 = 3z$

▶ **41.** $\dfrac{2}{5}w - 6 = \dfrac{7}{5}w$

42. $\dfrac{2}{7}z - 2 = \dfrac{9}{7}z$

43. $\dfrac{1}{2}x + 5 = -\dfrac{1}{2}x$

44. $\dfrac{1}{5}x + 7 = -\dfrac{4}{5}x$

45. $5.6x + 2 = 4.6x$ **46.** $9.1x + 5 = 8.1x$ **47.** $1.4x - 3 = 0.4x$

48. $1.9t - 6 = 0.9t$ **49.** $5p = 4p$ **50.** $8z = 7z$

51. $3x + 7 - 2x = 0$ **52.** $5x + 4 - 4x = 0$ **53.** $3x + 7 = 2x + 4$

54. $9x + 5 = 8x + 4$ **55.** $8t + 6 = 7t + 6$ **56.** $13t + 9 = 12t + 9$

57. $-4x + 7 = -5x + 9$ **58.** $-6x + 3 = -7x + 10$ **59.** $5 - x = -2x - 11$

60. $3 - 8x = -9x - 1$ **61.** $1.2y - 4 = 0.2y - 4$ **62.** $7.7r - 6 = 6.7r - 6$

Solve each equation, and check the solution. ***See Examples 7 and 8.***

63. $3x + 6 - 10 = 2x - 2$ **64.** $8x + 4 - 8 = 7x - 1$

65. $5t + 3 + 2t - 6t = 4 + 12$ **66.** $4x - 6 + 3x - 6x = 3 + 10$

67. $6x + 5 + 7x + 3 = 12x + 4$ **68.** $4x + 3 + 8x + 1 = 11x + 2$

69. $5.2q - 4.6 - 7.1q = -0.9q - 4.6$ **70.** $4.0x + 2.7 - 9.6x = -4.6x + 2.7$

71. $\dfrac{5}{7}x + \dfrac{1}{3} = \dfrac{2}{5} - \dfrac{2}{7}x + \dfrac{2}{5}$ **72.** $\dfrac{6}{7}s - \dfrac{3}{4} = \dfrac{4}{5} - \dfrac{1}{7}s + \dfrac{1}{6}$

73. $(5y + 6) - (3 + 4y) = 10$ **74.** $(8r + 3) - (1 + 7r) = 6$

75. $2(p + 5) - (9 + p) = -3$ **76.** $4(k + 6) - (8 + 3k) = -5$

77. $-6(2b + 1) + (13b - 7) = 0$ **78.** $-5(3w - 3) + (16w + 1) = 0$

79. $10(-2x + 1) = -19(x + 1)$ **80.** $2(-3r + 2) = -5(r - 3)$

Extending Skills   *Solve each equation, and check the solution.* ***See Example 8.***

81. $-2(8p + 2) - 3(2 - 7p) - 2(4 + 2p) = 0$

82. $-5(1 - 2z) + 4(3 - z) - 7(3 + z) = 0$

83. $4(7x - 1) + 3(2 - 5x) - 4(3x + 5) = -6$

84. $9(2m - 3) - 4(5 + 3m) - 5(4 + m) = -3$

Concept Check   *Work each problem.*

85. Write an equation that requires the use of the addition property of equality, where 6 must be added to each side and the solution is a negative number.

86. Write an equation that requires the use of the addition property of equality, where $\frac{1}{2}$ must be subtracted from each side and the solution is a positive number.

Write an equation using the information given in the problem. Use x as the variable. Then solve the equation.

87. Three times a number is 17 more than twice the number. Find the number.

88. One added to three times a number is three less than four times the number. Find the number.

89. If six times a number is subtracted from seven times the number, the result is -9. Find the number.

90. If five times a number is added to three times the number, the result is the sum of seven times the number and 9. Find the number.

STUDY SKILLS

Managing Your Time

Many college students juggle a difficult schedule and multiple responsibilities, including school, work, and family demands.

Time Management Tips

- **Read the syllabus for each class.** Understand class policies, such as attendance, late homework, and make-up tests. Find out how you are graded.

- **Make a semester or quarter calendar.** Put test dates and major due dates for *all* your classes on the *same* calendar. Try using a different color pen for each class.

- **Make a weekly schedule.** After you fill in your classes and other regular responsibilities, block off some study periods. Aim for 2 hours of study for each 1 hour in class.

- **Choose a regular study time and place** (such as the campus library). Routine helps.

- **Keep distractions to a minimum.** Get the most out of the time you have set aside for studying by limiting interruptions. Turn off your cell phone. Take a break from social media. Avoid studying in front of the TV.

- **Make "to-do" lists.** Number tasks in order of importance. Cross off tasks as you complete them.

- **Break big assignments into smaller chunks.** Make deadlines for each smaller chunk so that you stay on schedule.

- **Take breaks when studying.** Do not try to study for hours at a time. Take a 10-minute break each hour or so.

- **Ask for help when you need it.** Talk with your instructor during office hours. Make use of the learning center, tutoring center, counseling office, or other resources available at your school.

Think through and answer each question.

1. Evaluate when and where you are currently studying. Are the places you named quiet and comfortable? Are you studying when you are most alert?

2. How many hours do you have available for studying this week?

3. Which two or three of the above suggestions will you try this week to improve your time management?

4. Once the week is over, evaluate how these suggestions worked. What will you do differently next week?

2.2   The Multiplication Property of Equality

OBJECTIVES

1   Use the multiplication property of equality.

2   Simplify, and then use the multiplication property of equality.

OBJECTIVE 1   Use the multiplication property of equality.

The addition property of equality is not enough to solve some equations. Consider the following.

$$3x + 2 = 17$$

$$3x + 2 - 2 = 17 - 2 \qquad \text{Subtract 2 from each side.}$$

$$3x = 15 \qquad \text{Combine like terms.}$$

The coefficient of x is 3, not 1 as desired. The **multiplication property of equality** is needed to change $3x = 15$ to an equation of the form

$$x = \text{ a number.}$$

Because $3x = 15$, both $3x$ and 15 must represent the same number. Multiplying both $3x$ and 15 by the same number will result in an equivalent equation.

Multiplication Property of Equality

If A, B, and C represent real numbers, where $C \neq 0$, then the equations

$$A = B \quad \text{and} \quad AC = BC \quad \text{are equivalent.}$$

That is, each side of an equation may be multiplied by the same nonzero number without changing the solution set.

In $3x = 15$, we must change $3x$ to $1x$, or x. To do this, we multiply each side of the equation by $\frac{1}{3}$, the *reciprocal* of 3, because $\frac{1}{3} \cdot 3 = \frac{3}{3} = 1$.

$$3x = 15$$

$$\frac{1}{3} \cdot 3x = \frac{1}{3} \cdot 15 \qquad \text{Multiply each side by } \frac{1}{3}, \text{ the reciprocal of 3.}$$

$$\left(\frac{1}{3} \cdot 3 \right) x = \frac{1}{3} \cdot 15 \qquad \text{Associative property}$$

The product of a number and its reciprocal is 1.

$$1x = 5 \qquad \text{Multiplicative inverse property}$$

$$x = 5 \qquad \text{Multiplicative identity property}$$

The solution is 5. We can check this result in the original equation.

Just as the addition property of equality permits *subtracting* the same number from each side of an equation, the multiplication property of equality permits *dividing* each side of an equation by the same nonzero number.

$$3x = 15$$

$$\frac{3x}{3} = \frac{15}{3} \qquad \text{Divide each side by 3.}$$

$$x = 5 \qquad \text{Same result as above}$$

Multiplication Property of Equality Extended to Division

We can divide each side of an equation by the same nonzero number without changing the solution. *Do not, however, divide each side by a variable, because the variable might be equal to 0.*

NOTE It is usually easier to multiply on each side of an equation if the coefficient of the variable is a fraction, and divide on each side if the coefficient is an integer.

To solve $\frac{3}{4}x = 12$, it is easier to multiply by $\frac{4}{3}$ than to divide by $\frac{3}{4}$.

To solve $5x = 20$, it is easier to divide by 5 than to multiply by $\frac{1}{5}$.

NOW TRY EXERCISE 1

Solve $8x = 80$.

EXAMPLE 1 Applying the Multiplication Property of Equality

Solve $5x = 60$.

$$5x = 60$$ *Our goal is to isolate x.*

$$\frac{5x}{5} = \frac{60}{5}$$ Divide each side by 5, the coefficient of x.

Dividing by 5 is the same as multiplying by $\frac{1}{5}$.

$$x = 12$$ $\frac{5x}{5} = \frac{5}{5}x = 1x = x$

CHECK Substitute 12 for x in the original equation.

$$5x = 60$$ Original equation

$$5(12) \stackrel{?}{=} 60$$ Let $x = 12$.

$$60 = 60 \checkmark$$ True

A true statement results, so the solution set is $\{12\}$. **NOW TRY**

NOW TRY EXERCISE 2

Solve $10x = -24$.

EXAMPLE 2 Applying the Multiplication Property of Equality

Solve $25x = -30$.

$$25x = -30$$

To avoid errors later, show the division as a separate step.

$$\frac{25x}{25} = \frac{-30}{25}$$ Divide each side by 25, the coefficient of x.

$$x = -\frac{30}{25}$$ $\frac{-a}{b} = -\frac{a}{b}$

$$x = -\frac{6}{5}$$ Write in lowest terms.

CHECK $$25x = -30$$ Original equation

$$\frac{25}{1}\left(-\frac{6}{5}\right) \stackrel{?}{=} -30$$ Let $x = -\frac{6}{5}$.

$$-30 = -30 \checkmark$$ True

NOW TRY ANSWERS
1. $\{10\}$
2. $\left\{-\frac{12}{5}\right\}$

The check confirms that the solution set is $\left\{-\frac{6}{5}\right\}$. **NOW TRY**

NOW TRY EXERCISE 3

Solve $7.02 = -1.3x$.

EXAMPLE 3   Solving a Linear Equation (Decimal Coefficient)

Solve $6.09 = -2.1x$.

$$6.09 = -2.1x \quad \text{[Isolate } x \text{ on the right.]}$$

$$\frac{6.09}{-2.1} = \frac{-2.1x}{-2.1} \quad \text{Divide each side by } -2.1.$$

$$-2.9 = x, \quad \text{or} \quad x = -2.9$$

Check by replacing x with -2.9 in the original equation. The solution set is $\{-2.9\}$.

NOW TRY

NOW TRY EXERCISE 4

Solve $\frac{x}{5} = -7$.

EXAMPLE 4   Solving a Linear Equation (Fractional Coefficient)

Solve $\frac{x}{4} = 3$.

$$\frac{x}{4} = 3$$

$$\frac{1}{4}x = 3 \qquad \frac{x}{4} = \frac{1x}{4} = \frac{1}{4}x$$

$$4 \cdot \frac{1}{4}x = 4 \cdot 3 \qquad \text{Multiply each side by 4, the reciprocal of } \frac{1}{4}.$$

$$\boxed{4 \cdot \frac{1}{4}x = 1x = x} \quad x = 12 \qquad \text{Multiplicative inverse property; multiplicative identity property}$$

CHECK

$$\frac{x}{4} = 3 \qquad \text{Original equation}$$

$$\frac{12}{4} \overset{?}{=} 3 \qquad \text{Let } x = 12.$$

$$3 = 3 \; \checkmark \qquad \text{True}$$

A true statement results, so the solution set is $\{12\}$.

NOW TRY

NOW TRY EXERCISE 5

Solve $\frac{4}{7}z = -16$.

EXAMPLE 5   Solving a Linear Equation (Fractional Coefficient)

Solve $\frac{3}{4}w = 6$.

$$\frac{3}{4}w = 6$$

$$\frac{4}{3} \cdot \frac{3}{4}w = \frac{4}{3} \cdot 6 \qquad \text{Multiply each side by } \frac{4}{3}, \text{ the reciprocal of } \frac{3}{4}.$$

$$\boxed{\text{Reciprocals have a product of 1.}} \quad 1 \cdot w = \frac{4}{3} \cdot \frac{6}{1} \qquad \text{Multiplicative inverse property}$$

$$w = 8 \qquad \text{Multiplicative identity property; multiply fractions.}$$

Check to confirm that the solution set is $\{8\}$.

NOW TRY

NOW TRY ANSWERS

3. $\{-5.4\}$

4. $\{-35\}$

5. $\{-28\}$

In **Section 2.1,** we obtained $-x = 8$ in our alternative solution to **Example 4.** We reasoned that because the additive inverse (or opposite) of x is 8, then x must equal -8. We can use the multiplication property of equality to obtain the same result.

**NOW TRY
EXERCISE 6**
Solve $-x = -9$.

EXAMPLE 6 Applying the Multiplication Property of Equality

Solve $-x = 8$.

$$-x = 8$$
$$-1x = 8 \qquad \qquad -x = -1x$$
$$-1(-1x) = -1(8) \qquad \text{Multiply each side by } -1.$$
$$[-1(-1)]x = -8 \qquad \text{Associative property; multiply.}$$
These steps are usually omitted.
$$1x = -8 \qquad \text{Multiplicative inverse property}$$
$$x = -8 \qquad \text{Multiplicative identity property}$$

CHECK
$$-x = 8 \qquad \text{Original equation}$$
$$-(-8) \stackrel{?}{=} 8 \qquad \text{Let } x = -8.$$
$$8 = 8 \checkmark \qquad \text{True}$$

A true statement results, so $\{-8\}$ is the solution set. NOW TRY

OBJECTIVE 2 Simplify, and then use the multiplication property of equality.

**NOW TRY
EXERCISE 7**
Solve $9n - 6n = 21$.

EXAMPLE 7 Combining Like Terms When Solving

Solve $5m + 6m = 33$.

$$5m + 6m = 33$$
$$11m = 33 \qquad \text{Combine like terms.}$$
$$\frac{11m}{11} = \frac{33}{11} \qquad \text{Divide by 11.}$$
$$m = 3 \qquad \text{Multiplicative identity property; divide.}$$

CHECK
$$5m + 6m = 33 \qquad \text{Original equation}$$
$$5(3) + 6(3) \stackrel{?}{=} 33 \qquad \text{Let } m = 3.$$
$$15 + 18 \stackrel{?}{=} 33 \qquad \text{Multiply.}$$
$$33 = 33 \checkmark \qquad \text{True}$$

NOW TRY ANSWERS
6. $\{9\}$ **7.** $\{7\}$

A true statement results, so the solution set is $\{3\}$. NOW TRY

2.2 Exercises

FOR EXTRA HELP   ▶ MyMathLab®

▶ *Complete solution available in MyMathLab*

1. *Concept Check* Tell whether to use the addition or multiplication property of equality to solve each equation. *Do not actually solve.*

(a) $3x = 12$ **(b)** $3 + x = 12$ **(c)** $-x = 4$ **(d)** $-12 = 6 + x$

Concept Check Choose the letter of the correct response.

2. Which equation does *not* require the use of the multiplication property of equality?

A. $3x - 5x = 6$ **B.** $-\dfrac{1}{4}x = 12$ **C.** $5x - 4x = 7$ **D.** $\dfrac{x}{3} = -2$

3. In the solution of a linear equation, the next-to-the-last step reads "$-x = -\frac{3}{4}$." Which of the following is the solution of this equation?

   A. $-\frac{3}{4}$ **B.** $\frac{3}{4}$ **C.** -1 **D.** $\frac{4}{3}$

4. Which of the following is the solution of the equation $-x = -24$?

   A. 24   **B.** -24 **C.** 1   **D.** -1

Concept Check *By what number is it necessary to multiply both sides of each equation to isolate x on the left side? Do not actually solve.*

5. $\frac{4}{5}x = 8$ **6.** $\frac{2}{3}x = 6$ **7.** $\frac{x}{10} = 5$ **8.** $\frac{x}{100} = 10$

9. $-\frac{9}{2}x = -4$ **10.** $-\frac{8}{3}x = -11$ **11.** $-x = 0.75$ **12.** $-x = 0.48$

Concept Check *By what number is it necessary to divide both sides of each equation to isolate x on the left side? Do not actually solve.*

13. $6x = 5$ **14.** $7x = 10$ **15.** $-4x = 16$ **16.** $-13x = 26$

17. $0.12x = 48$ **18.** $0.21x = 63$ **19.** $-x = 25$ **20.** $-x = 50$

Solve each equation, and check the solution. **See Examples 1–6.**

21. $6x = 36$ **22.** $8x = 64$ **23.** $2m = 15$ **24.** $3m = 10$

25. $4x = -20$ **26.** $5x = -60$ **27.** $-7x = 28$ **28.** $-9x = 36$

29. $10t = -36$ **30.** $10s = -54$ **31.** $-6x = -72$ **32.** $-4x = -64$

33. $4r = 0$ **34.** $7x = 0$ **35.** $-x = 12$ **36.** $-t = 14$

37. $-x = -\frac{3}{4}$ **38.** $-x = -\frac{1}{2}$ **39.** $0.2t = 8$ **40.** $0.9x = 18$

41. $-0.3x = 9$ **42.** $-0.5x = 20$ **43.** $0.6x = -1.44$ **44.** $0.8x = -2.96$

45. $-9.1 = -2.6x$ **46.** $-7.2 = -4.5x$ **47.** $-2.1m = 25.62$ **48.** $-3.9x = 32.76$

49. $\frac{1}{4}x = -12$ **50.** $\frac{1}{5}p = -3$ **51.** $\frac{z}{6} = 12$ **52.** $\frac{x}{5} = 15$

53. $\frac{x}{7} = -5$ **54.** $\frac{r}{8} = -3$ **55.** $\frac{2}{7}p = 4$ **56.** $\frac{3}{8}x = 9$

57. $-\frac{5}{6}t = -15$ **58.** $-\frac{3}{4}z = -21$ **59.** $-\frac{7}{9}x = \frac{3}{5}$ **60.** $-\frac{5}{6}x = \frac{4}{9}$

Solve each equation, and check the solution. **See Example 7.**

61. $4x + 3x = 21$ **62.** $8x + 3x = 121$ **63.** $6r - 8r = 10$

64. $3p - 7p = 24$ **65.** $\frac{2}{5}x - \frac{3}{10}x = 2$ **66.** $\frac{2}{3}x - \frac{5}{9}x = 4$

67. $7m + 6m - 4m = 63$ **68.** $9r + 2r - 7r = 68$ **69.** $-6x + 4x - 7x = 0$

70. $-5x + 4x - 8x = 0$ **71.** $8w - 4w + w = -3$ **72.** $9x - 3x + x = -4$

73. $\frac{1}{3}x - \frac{1}{4}x + \frac{1}{12}x = 3$ **74.** $\frac{2}{5}x + \frac{1}{10}x - \frac{1}{20}x = 18$

75. $0.9w - 0.5w + 0.1w = -3$ **76.** $0.5x - 0.6x + 0.3x = -1$

Concept Check *Work each problem.*

77. Write an equation that requires the use of the multiplication property of equality, where each side must be multiplied by $\frac{2}{3}$ and the solution is a negative number.

78. Write an equation that requires the use of the multiplication property of equality, where each side must be divided by 100 and the solution is not an integer.

Write an equation using the information given in the problem. Use x as the variable. Then solve the equation.

79. When a number is multiplied by 4, the result is 6. Find the number.

80. When a number is multiplied by −4, the result is 10. Find the number.

81. When a number is divided by −5, the result is 2. Find the number.

82. If twice a number is divided by 5, the result is 4. Find the number.

2.3 More on Solving Linear Equations

OBJECTIVES

1 Learn and use the four steps for solving a linear equation.
2 Solve equations with no solution or infinitely many solutions.
3 Solve equations with fractions or decimals as coefficients.
4 Write expressions for two related unknown quantities.

VOCABULARY
☐ conditional equation
☐ identity
☐ contradiction
☐ empty (null) set

OBJECTIVE 1 Learn and use the four steps for solving a linear equation.

We now apply *both* properties of equality to solve linear equations.

Solving a Linear Equation in One Variable

Step 1 **Simplify each side separately.** Use the distributive property as needed.
- Clear any parentheses.
- Clear any fractions or decimals.
- Combine like terms.

Step 2 **Isolate the variable terms on one side.** Use the addition property of equality so that all terms with variables are on one side of the equation and all constants (numbers) are on the other side.

Step 3 **Isolate the variable.** Use the multiplication property of equality to obtain an equation that has just the variable with coefficient 1 on one side.

Step 4 **Check.** Substitute the value found into the *original* equation. If a true statement results, write the solution set. If not, rework the problem.

Remember that when we solve an equation, our primary goal is to isolate the variable on one side of the equation.

EXAMPLE 1 Solving a Linear Equation

Solve $-6x + 5 = 17$.

Step 1 There are no parentheses, fractions, or decimals in this equation, so this step is not necessary.

Our goal is to isolate x. → $-6x + 5 = 17$

Step 2 $-6x + 5 - 5 = 17 - 5$ Subtract 5 from each side.

$-6x = 12$ Combine like terms.

Step 3 $\dfrac{-6x}{-6} = \dfrac{12}{-6}$ Divide each side by −6.

$x = -2$

NOW TRY EXERCISE 1

Solve $7 + 2m = -3$.

Step 4   Check by substituting -2 for x in the original equation.

CHECK $-6x + 5 = 17$ Original equation

$-6(-2) + 5 \overset{?}{=} 17$ Let $x = -2$.

$12 + 5 \overset{?}{=} 17$ Multiply.

$17 = 17$ ✓   True

The check confirms that -2 is the solution. The solution set is $\{-2\}$.   NOW TRY

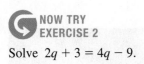

NOW TRY EXERCISE 2

Solve $2q + 3 = 4q - 9$.

EXAMPLE 2   Solving a Linear Equation

Solve $3x + 2 = 5x - 8$.

Step 1   There are no parentheses, fractions, or decimals in the equation.

Our goal is to isolate x. ⟶ $3x + 2 = 5x - 8$

Step 2 $3x + 2 - 5x = 5x - 8 - 5x$ Subtract $5x$ from each side.

$-2x + 2 = -8$ Combine like terms.

$-2x + 2 - 2 = -8 - 2$ Subtract 2 from each side.

$-2x = -10$ Combine like terms.

Step 3 $\dfrac{-2x}{-2} = \dfrac{-10}{-2}$ Divide each side by -2.

$x = 5$

Step 4   Check by substituting 5 for x in the original equation.

CHECK $3x + 2 = 5x - 8$ Original equation

$3(5) + 2 \overset{?}{=} 5(5) - 8$ Let $x = 5$.

$15 + 2 \overset{?}{=} 25 - 8$ Multiply.

$17 = 17$ ✓   True

The check confirms that 5 is the solution. The solution set is $\{5\}$.   NOW TRY

NOTE   *Remember that the variable can be isolated on either side of the equation.* In **Example 2,** x will be isolated on the right if we begin by subtracting $3x$.

$3x + 2 = 5x - 8$ Equation from **Example 2**

$3x + 2 - 3x = 5x - 8 - 3x$ Subtract $3x$ from each side.

$2 = 2x - 8$ Combine like terms.

$2 + 8 = 2x - 8 + 8$ Add 8 to each side.

$10 = 2x$ Combine like terms.

$\dfrac{10}{2} = \dfrac{2x}{2}$ Divide each side by 2.

$5 = x$ The same solution results.

There are often several equally correct ways to solve an equation.

NOW TRY ANSWERS
1. $\{-5\}$
2. $\{6\}$

**NOW TRY
EXERCISE 3**

Solve.

$3(z - 6) - 5z = -7z + 7$

EXAMPLE 3   Solving a Linear Equation

Solve $4(k - 3) - k = k - 6$.

Step 1   Clear parentheses using the distributive property.

$$4(k - 3) - k = k - 6$$

$4(k) + 4(-3) - k = k - 6$	Distributive property
$4k - 12 - k = k - 6$	Multiply.
$3k - 12 = k - 6$	Combine like terms.

Step 2

$3k - 12 - k = k - 6 - k$	Subtract k.
$2k - 12 = -6$	Combine like terms.
$2k - 12 + 12 = -6 + 12$	Add 12.
$2k = 6$	Combine like terms.

Step 3

$\dfrac{2k}{2} = \dfrac{6}{2}$	Divide by 2.
$k = 3$	

Step 4   **CHECK**

$4(k - 3) - k = k - 6$	Original equation
$4(3 - 3) - 3 \stackrel{?}{=} 3 - 6$	Let $k = 3$.
$4(0) - 3 \stackrel{?}{=} 3 - 6$	Work inside the parentheses.
$-3 = -3$ ✓	True

The solution set is $\{3\}$.

NOW TRY

**NOW TRY
EXERCISE 4**

Solve.

$5x - (x + 9) = x - 4$

EXAMPLE 4   Solving a Linear Equation

Solve $8z - (3 + 2z) = 3z + 1$.

Step 1

$8z - (3 + 2z) = 3z + 1$	
$8z - 1(3 + 2z) = 3z + 1$	Multiplicative identity property
$8z - 1(3) - 1(2z) = 3z + 1$	Distributive property
$8z - 3 - 2z = 3z + 1$	Multiply.
$6z - 3 = 3z + 1$	Combine like terms.

> Be careful with signs.

Step 2

$6z - 3 - 3z = 3z + 1 - 3z$	Subtract $3z$.
$3z - 3 = 1$	Combine like terms.
$3z - 3 + 3 = 1 + 3$	Add 3.
$3z = 4$	Combine like terms.

Step 3

$\dfrac{3z}{3} = \dfrac{4}{3}$	Divide by 3.
$z = \dfrac{4}{3}$	

NOW TRY ANSWERS
3. $\{5\}$

4. $\left\{\dfrac{5}{3}\right\}$

Step 4   Check that $\left\{\dfrac{4}{3}\right\}$ is the solution set.

NOW TRY

⚠ CAUTION   In an expression such as $8z - (3 + 2z)$ in **Example 4,** the − sign acts like a factor of −1 and affects the sign of *every* term within the parentheses.

$$8z - (3 + 2z) \leftarrow \text{Left side of the equation in \textbf{Example 4}}$$

$$= 8z - 1(3 + 2z)$$

$$= 8z + (-1)(3 + 2z)$$

$$= 8z - 3 - 2z$$
$$\qquad \uparrow \quad \uparrow$$
Change to − in *both* terms.

NOW TRY
EXERCISE 5

Solve.

$$24 - 4(7 - 2t) = 4(t - 1)$$

EXAMPLE 5   Solving a Linear Equation

Solve $4(4 - 3x) = 32 - 8(x + 2)$.

Step 1
$$4(4 - 3x) = 32 - 8(x + 2) \quad \boxed{\text{Be careful with signs.}}$$

$$16 - 12x = 32 - 8x - 16 \qquad \text{Distributive property}$$

$$16 - 12x = 16 - 8x \qquad \text{Combine like terms.}$$

Step 2
$$16 - 12x + 8x = 16 - 8x + 8x \qquad \text{Add } 8x.$$

$$16 - 4x = 16 \qquad \text{Combine like terms.}$$

$$16 - 4x - 16 = 16 - 16 \qquad \text{Subtract 16.}$$

$$-4x = 0 \qquad \text{Combine like terms.}$$

Step 3
$$\frac{-4x}{-4} = \frac{0}{-4} \qquad \text{Divide by } -4.$$

$$x = 0$$

Step 4   **CHECK**
$$4(4 - 3x) = 32 - 8(x + 2) \qquad \text{Original equation}$$

$$4[4 - 3(0)] \stackrel{?}{=} 32 - 8(0 + 2) \qquad \text{Let } x = 0.$$

$$4(4 - 0) \stackrel{?}{=} 32 - 8(2) \qquad \text{Multiply and add.}$$

$$4(4) \stackrel{?}{=} 32 - 16 \qquad \text{Subtract and multiply.}$$

$$16 = 16 \; \checkmark \qquad \text{True}$$

Because a true statement results, the solution set is $\{0\}$.   **NOW TRY**

NOTE   As the check in **Example 5** confirms, it is perfectly acceptable for an equation to have solution set $\{0\}$.

OBJECTIVE 2   Solve equations with no solution or infinitely many solutions.

Each equation so far has had exactly one solution. An equation with exactly one solution is a **conditional equation** because it is only true under certain conditions. Some equations may have no solution or infinitely many solutions.

NOW TRY ANSWER
5. $\{0\}$

**NOW TRY
EXERCISE 6**

Solve.

$-3(x - 7) = 2x - 5x + 21$

EXAMPLE 6   Solving an Equation That Has Infinitely Many Solutions

Solve $5x - 15 = 5(x - 3)$.

$$5x - 15 = 5(x - 3)$$

$$5x - 15 = 5x - 15 \qquad \text{Distributive property}$$

$$5x - 15 - 5x = 5x - 15 - 5x \qquad \text{Subtract } 5x.$$

Notice that the variable "disappeared." $\qquad -15 = -15 \qquad \text{Combine like terms.}$

$$-15 + 15 = -15 + 15 \qquad \text{Add 15.}$$

$$0 = 0 \qquad \text{True}$$

Solution set: {all real numbers}

Because the last statement $(0 = 0)$ is true, *any* real number is a solution. We could have predicted this from the second line in the solution.

$$5x - 15 = 5x - 15 \leftarrow \text{This is true for } any \text{ value of } x.$$

Try several values for x in the original equation to see that they all satisfy it.

   An equation with both sides exactly the same, like $0 = 0$, is an **identity.** An identity is true for *all* replacements of the variables. As shown above, we write the solution set as

{all real numbers}.   NOW TRY

❗ **CAUTION**   In **Example 6,** do not write {0} as the solution set. While 0 is a solution, there are infinitely many other solutions. *For the solution set to be {0}, the last line must include a variable, such as x, and read x = 0 (as in Example 5), not 0 = 0.*

**NOW TRY
EXERCISE 7**

Solve.

$-4x + 12 = 3 - 4(x - 3)$

EXAMPLE 7   Solving an Equation That Has No Solution

Solve $2x + 3(x + 1) = 5x + 4$.

$$2x + 3(x + 1) = 5x + 4$$

$$2x + 3x + 3 = 5x + 4 \qquad \text{Distributive property}$$

$$5x + 3 = 5x + 4 \qquad \text{Combine like terms.}$$

$$5x + 3 - 5x = 5x + 4 - 5x \qquad \text{Subtract } 5x.$$

Again, the variable "disappeared." $\qquad 3 = 4 \qquad \text{False}$

There is no solution.   Solution set: $\varnothing$

A false statement $(3 = 4)$ results. A **contradiction** is an equation that has no solution. Its solution set is the **empty set,** or **null set,** symbolized $\varnothing$.   NOW TRY

❗ **CAUTION**   **Do not** write {∅} to represent the empty set.

NOW TRY ANSWERS
6. {all real numbers}
7. $\varnothing$

▼ Solution Sets of Equations

Type of Equation	Final Equation in Solution	Number of Solutions	Solution Set
Conditional (See Examples 1–5.)	x = a number	One	{a number}
Identity (See Example 6.)	A true statement with no variable, such as 0 = 0	Infinitely many	{all real numbers}
Contradiction (See Example 7.)	A false statement with no variable, such as 3 = 4	None	∅

OBJECTIVE 3 Solve equations with fractions or decimals as coefficients.

To avoid messy computations, we clear an equation of fractions by multiplying each side by the least common denominator (LCD) of all the fractions in the equation. Doing this will give an equation with only *integer* coefficients.

NOW TRY EXERCISE 8

Solve.

$$\frac{1}{2}x + \frac{5}{8}x = \frac{3}{4}x - 6$$

EXAMPLE 8 Solving a Linear Equation (Fractional Coefficients)

Solve $\frac{2}{3}x - \frac{1}{2}x = -\frac{1}{6}x - 2$.

Step 1 The LCD of all the fractions in the equation is 6.

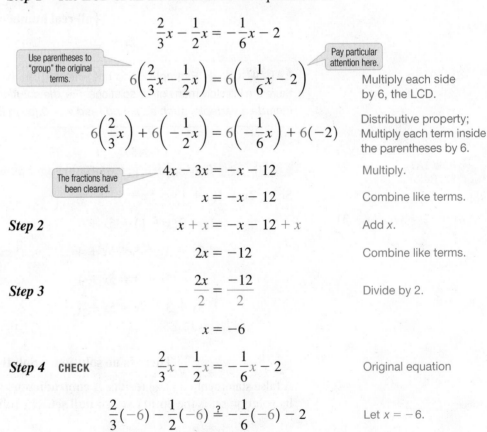

$$\frac{2}{3}x - \frac{1}{2}x = -\frac{1}{6}x - 2$$

Use parentheses to "group" the original terms.

Pay particular attention here.

$$6\left(\frac{2}{3}x - \frac{1}{2}x\right) = 6\left(-\frac{1}{6}x - 2\right)$$ Multiply each side by 6, the LCD.

$$6\left(\frac{2}{3}x\right) + 6\left(-\frac{1}{2}x\right) = 6\left(-\frac{1}{6}x\right) + 6(-2)$$ Distributive property; Multiply each term inside the parentheses by 6.

The fractions have been cleared.

$$4x - 3x = -x - 12$$ Multiply.

$$x = -x - 12$$ Combine like terms.

Step 2 $$x + x = -x - 12 + x$$ Add x.

$$2x = -12$$ Combine like terms.

Step 3 $$\frac{2x}{2} = \frac{-12}{2}$$ Divide by 2.

$$x = -6$$

Step 4 CHECK $$\frac{2}{3}x - \frac{1}{2}x = -\frac{1}{6}x - 2$$ Original equation

$$\frac{2}{3}(-6) - \frac{1}{2}(-6) \overset{?}{=} -\frac{1}{6}(-6) - 2$$ Let $x = -6$.

$$-4 + 3 \overset{?}{=} 1 - 2$$ Multiply.

$$-1 = -1 \ ✓$$ True

The check confirms that the solution set is {−6}.

NOW TRY ↻

! CAUTION *When clearing an equation of fractions, be sure to multiply every term on each side of the equation by the LCD.*

NOW TRY EXERCISE 9

Solve.

$$\frac{2}{3}(x + 2) - \frac{1}{2}(3x + 4) = -4$$

EXAMPLE 9 Solving a Linear Equation (Fractional Coefficients)

Solve $\frac{1}{3}(x + 5) - \frac{3}{5}(x + 2) = 1$.

Step 1 We clear the parentheses first. Then we clear the fractions.

$$\frac{1}{3}(x + 5) - \frac{3}{5}(x + 2) = 1$$ ⟵ Study Step 1 carefully.

$$\frac{1}{3}(x) + \frac{1}{3}(5) - \frac{3}{5}(x) - \frac{3}{5}(2) = 1$$ Distributive property

$$\frac{1}{3}x + \frac{5}{3} - \frac{3}{5}x - \frac{6}{5} = 1$$ Multiply.

$$15\left(\frac{1}{3}x + \frac{5}{3} - \frac{3}{5}x - \frac{6}{5}\right) = 15(1)$$ Multiply each side by 15, the LCD.

$$15\left(\frac{1}{3}x\right) + 15\left(\frac{5}{3}\right) + 15\left(-\frac{3}{5}x\right) + 15\left(-\frac{6}{5}\right) = 15(1)$$ Distributive property

Think: $15\left(\frac{1}{3}x\right)$
$= \left(\frac{15}{1} \cdot \frac{1}{3}\right)x$
$= 5x$

$$5x + 25 - 9x - 18 = 15$$ Multiply.

$$-4x + 7 = 15$$ Combine like terms.

Step 2 $$-4x + 7 - 7 = 15 - 7$$ Subtract 7.

$$-4x = 8$$ Combine like terms.

Step 3 $$\frac{-4x}{-4} = \frac{8}{-4}$$ Divide by -4.

$$x = -2$$

Step 4 Check to confirm that $\{-2\}$ is the solution set. **NOW TRY** ↺

EXAMPLE 10 Solving a Linear Equation (Decimal Coefficients)

Solve $0.1t + 0.05(20 - t) = 0.09(20)$.

Step 1 $$0.1t + 0.05(20 - t) = 0.09(20)$$ ⟵ Clear the parentheses first.

$$0.1t + 0.05(20) + 0.05(-t) = 0.09(20)$$ Distributive property

$$0.1t + 1 - 0.05t = 1.8$$ (*) Multiply.

The decimals here are expressed as tenths (0.1 and 1.8) and hundredths (0.05). We choose the least exponent on 10 to eliminate the decimal points, which will make all coefficients integers. Here, we multiply by 10^2—that is, 100.

Now clear the decimals. ⟶ $$100(0.1t + 1 - 0.05t) = 100(1.8)$$ Multiply by 100.

$$100(0.1t) + 100(1) + 100(-0.05t) = 100(1.8)$$ Distributive property

$$10t + 100 - 5t = 180$$ Multiply.

$$5t + 100 = 180$$ Combine like terms.

NOW TRY ANSWER
9. $\{4\}$

Solve.

$0.05(13 - t) - 0.2t = 0.08(30)$

Step 2

$$5t + 100 - 100 = 180 - 100 \quad \text{Subtract 100.}$$
$$5t = 80 \quad \text{Combine like terms.}$$

Step 3

$$\frac{5t}{5} = \frac{80}{5} \quad \text{Divide by 5.}$$
$$t = 16$$

Step 4   Check to confirm that $\{16\}$ is the solution set.

NOW TRY

NOTE   In **Example 10,** multiplying by 100 is the same as moving the decimal points two places to the right.

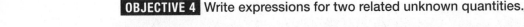

$$0.10t + 1.00 - 0.05t = 1.80 \quad \text{Equation (*) from \textbf{Example 10}}$$
with 0s included as placeholders

$$10t + 100 - 5t = 180 \quad \text{Multiply by 100.}$$

OBJECTIVE 4 Write expressions for two related unknown quantities.

EXAMPLE 11 **Translating a Phrase into an Algebraic Expression**

Perform each translation.

(a) Two numbers have a sum of 23. If one of the numbers is represented by x, find an expression for the other number.

First, suppose that the sum of two numbers is 23, and one of the numbers is 10. To find the other number, we would subtract 10 from 23.

$$23 - 10 \;\leftarrow\; \text{This gives 13 as the other number.}$$

Instead of using 10 as one of the numbers, we use x. The other number would be obtained in the same way—by subtracting x from 23.

$$23 - x. \;\leftarrow\; \begin{array}{l} x - 23 \text{ is not correct.} \\ \text{Subtraction is not} \\ \text{commutative.} \end{array}$$

To check, find the sum of the two numbers.

$$x + (23 - x)$$
$$= 23, \quad \text{as required.}$$

(b) Two numbers have a product of 24. If one of the numbers is represented by x, find an expression for the other number.

Suppose that one of the numbers is 4. To find the other number, we would divide 24 by 4.

$$\frac{24}{4} \;\leftarrow\; \begin{array}{l} \text{This gives 6 as the other number.} \\ \text{The product } 6 \cdot 4 \text{ is 24.} \end{array}$$

In the same way, if x is one of the numbers, then we divide 24 by x to find the other number.

$$\frac{24}{x} \;\leftarrow\; \text{The other number}$$

NOW TRY

NOW TRY
EXERCISE 11

Two numbers have a sum of 18. If one of the numbers is represented by m, find an expression for the other number.

NOW TRY ANSWERS
10. $\{-7\}$
11. $18 - m$

2.3 Exercises

FOR EXTRA HELP   ▶ MyMathLab®

▶ *Complete solution available in MyMathLab*

Concept Check   Using the methods of this section, what should we do first when solving each equation? Do not actually solve.

1. $7x + 8 = 1$

2. $7x - 5x + 15 = 8 + x$

3. $3(2t - 4) = 20 - 2t$

4. $\dfrac{3}{4}z = -15$

5. $\dfrac{2}{3}x - \dfrac{1}{6} = \dfrac{3}{2}x + 1$

6. $0.9x + 0.3(x + 12) = 6$

7. *Concept Check*   Suppose that when solving three linear equations, we obtain the final results shown in parts (a)–(c). Fill in the blanks in parts (a)–(c), and then match each result with the solution set in choices A–C for the *original* equation.

(**a**) $6 = 6$ (The original equation is a(n) _____.)   **A.** $\{0\}$

(**b**) $x = 0$ (The original equation is a(n) _____ equation.)   **B.** $\{$all real numbers$\}$

(**c**) $-5 = 0$ (The original equation is a(n) _____.)   **C.** $\varnothing$

8. *Concept Check*   Which equation does *not* have $\{$all real numbers$\}$ as its solution set?

A. $5x = 4x + x$ **B.** $2(x + 6) = 2x + 12$ **C.** $\dfrac{1}{2}x = 0.5x$ **D.** $3x = 2x$

Solve each equation, and check the solution. See Examples 1–7.

9. $3x + 2 = 14$

10. $4x + 3 = 27$

11. $-5z - 4 = 21$

12. $-7w - 4 = 10$

13. $4p - 5 = 2p$

14. $6q - 2 = 3q$

15. $2x + 9 = 4x + 11$

16. $7p + 8 = 9p - 2$

17. $5m + 8 = 7 + 3m$

18. $4r + 2 = r - 6$

19. $-12x - 5 = 10 - 7x$

20. $-16w - 3 = 13 - 8w$

▶ **21.** $12h - 5 = 11h + 5 - h$

22. $-4x - 1 = -5x + 1 + 3x$

23. $7r - 5r + 2 = 5r + 2 - r$

24. $9p - 4p + 6 = 7p + 6 - 3p$

▶ **25.** $3(4x + 2) + 5x = 30 - x$

26. $5(2m + 3) - 4m = 2m + 25$

▶ **27.** $-2p + 7 = 3 - (5p + 1)$

28. $4x + 9 = 3 - (x - 2)$

▶ **29.** $11x - 5(x + 2) = 6x + 5$

30. $6x - 4(x + 1) = 2x + 4$

▶ **31.** $6(3w + 5) = 2(10w + 10)$

32. $4(2x - 1) = -6(x + 3)$

33. $-(4x + 2) - (-3x - 5) = 3$

34. $-(6k - 5) - (-5k + 8) = -3$

▶ **35.** $3(2x - 4) = 6(x - 2)$

36. $3(6 - 4x) = 2(-6x + 9)$

37. $6(4x - 1) = 12(2x + 3)$

38. $6(2x + 8) = 4(3x - 6)$

Solve each equation, and check the solution. See Examples 8–10.

▶ **39.** $\dfrac{3}{5}t - \dfrac{1}{10}t = t - \dfrac{5}{2}$

40. $-\dfrac{2}{7}r + 2r = \dfrac{1}{2}r + \dfrac{17}{2}$

41. $\dfrac{3}{4}x - \dfrac{1}{3}x + 5 = \dfrac{5}{6}x$

42. $\dfrac{1}{5}x - \dfrac{2}{3}x - 2 = -\dfrac{2}{5}x$

43. $\dfrac{1}{7}(3x + 2) - \dfrac{1}{5}(x + 4) = 2$

44. $\dfrac{1}{4}(3x - 1) + \dfrac{1}{6}(x + 3) = 3$

45. $-\dfrac{1}{4}(x-12)+\dfrac{1}{2}(x+2)=x+4$

46. $\dfrac{1}{9}(p+18)+\dfrac{1}{3}(2p+3)=p+3$

47. $\dfrac{2}{3}k-\left(k-\dfrac{1}{2}\right)=\dfrac{1}{6}(k-51)$

48. $-\dfrac{5}{6}q-(q-1)=\dfrac{1}{4}(-q+80)$

49. $0.75x-3.2=0.55-0.5x$

50. $1.35x-0.6=1.65+2.1x$

51. $0.8t+0.15=2t-1.35$

52. $-0.12p+3.4=0.84+5p$

53. $0.2(60)+0.05x=0.1(60+x)$

54. $0.3(30)+0.15x=0.2(30+x)$

55. $1.00x+0.05(12-x)=0.10(63)$

56. $0.92x+0.98(12-x)=0.96(12)$

57. $0.6(10{,}000)+0.8x=0.72(10{,}000+x)$

58. $0.2(5000)+0.3x=0.25(5000+x)$

Solve each equation, and check the solution. **See Examples 1–10.**

59. $10(2x-1)=8(2x+1)+14$

60. $9(3k-5)=12(3k-1)-51$

61. $24-4(7-2t)=4(t-1)$

62. $8-2(2-x)=4(x+1)$

63. $4(x+8)=2(2x+6)+20$

64. $4(x+3)=2(2x+8)-4$

65. $\dfrac{1}{2}(x+2)+\dfrac{3}{4}(x+4)=x+5$

66. $\dfrac{1}{3}(x+3)+\dfrac{1}{6}(x-6)=x+3$

67. $0.1(x+80)+0.2x=14$

68. $0.3(x+15)+0.4(x+25)=25$

69. $9(v+1)-3v=2(3v+1)-8$

70. $8(t-3)+4t=6(2t+1)-10$

Perform each translation. **See Example 11.**

71. Two numbers have a sum of 11. One of the numbers is q. What expression represents the other number?

72. Two numbers have a sum of 34. One of the numbers is r. What expression represents the other number?

73. The product of two numbers is 9. One of the numbers is x. What expression represents the other number?

74. The product of two numbers is -6. One of the numbers is m. What expression represents the other number?

75. A baseball player got 65 hits one season. He got h of the hits in one game. What expression represents the number of hits he got in the rest of the games?

76. A hockey player scored 42 goals in one season. He scored n goals in one game. What expression represents the number of goals he scored in the rest of the games?

77. Monica is x years old. What expression represents her age 15 yr from now? 5 yr ago?

78. Chandler is y years old. What expression represents his age 4 yr ago? 11 yr from now?

79. Cliff has r quarters. Express the value of the quarters in cents.

80. Claire has y dimes. Express the value of the dimes in cents.

81. A clerk has t dollars, all in \$5 bills. What expression represents the number of \$5 bills?

82. A clerk has v dollars, all in \$10 bills. What expression represents the number of \$10 bills?

83. A plane ticket costs x dollars for an adult and y dollars for a child. Find an expression that represents the total cost for 3 adults and 2 children.

84. A concert ticket costs p dollars for an adult and q dollars for a child. Find an expression that represents the total cost for 4 adults and 6 children.

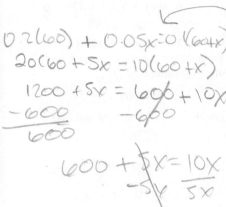

STUDY SKILLS

Using Study Cards Revisited

We introduced study cards previously. Another type of study card follows.

Practice Quiz Cards

Write a problem with direction words (like *solve, simplify*) on the front of the card, and work the problem on the back. Make one for each type of problem you learn.

Make a practice quiz card for material you are learning now.

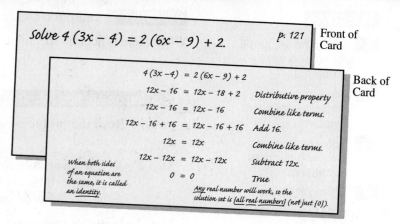

Front of Card

Solve $4(3x - 4) = 2(6x - 9) + 2$.   p. 121

Back of Card

$$4(3x - 4) = 2(6x - 9) + 2$$
$$12x - 16 = 12x - 18 + 2 \quad \text{Distributive property}$$
$$12x - 16 = 12x - 16 \quad \text{Combine like terms.}$$
$$12x - 16 + 16 = 12x - 16 + 16 \quad \text{Add 16.}$$
$$12x = 12x \quad \text{Combine like terms.}$$
$$12x - 12x = 12x - 12x \quad \text{Subtract 12x.}$$
$$0 = 0 \quad \text{True}$$

When both sides of an equation are the same, it is called an identity.

Any real number will work, so the solution set is {all real numbers} (not just {0}).

SUMMARY EXERCISES Applying Methods for Solving Linear Equations

Concept Check Decide whether each of the following is an **equation** *or an* **expression**. *If it is an equation, solve it. If it is an expression, simplify it.*

1. $x + 2 = -3$

2. $4p - 6 + 3p - 8$

3. $-(m - 1) - (3 + 2m)$

4. $6q - 9 = 12 + 3q$

5. $5x - 9 = 3(x - 3)$

6. $\frac{2}{3}x + 8 = \frac{1}{4}x$

7. $2 - 6(z + 1) - 4(z - 2) - 10$

8. $7(p - 2) + p = 2(p + 2)$

9. $\frac{1}{2}(x + 10) - \frac{2}{3}x$

10. $-4(k + 2) + 3(2k - 1)$

Solve each equation, and check the solution.

11. $-6z = -14$

12. $2m + 8 = 16$

13. $12.5x = -63.75$

14. $-x = -12$

15. $\frac{4}{5}x = -20$

16. $7m - 5m = -12$

17. $-x = 6$

18. $\frac{x}{-2} = 8$

19. $4x + 2(3 - 2x) = 6$

20. $x - 16.2 = 7.5$

21. $7m - (2m - 9) = 39$

22. $2 - (m + 4) = 3m - 2$

23. $-3(m - 4) + 2(5 + 2m) = 29$

24. $-0.3x + 2.1(x - 4) = -6.6$

25. $0.08x + 0.06(x + 9) = 1.24$

26. $3(m + 5) - 1 + 2m = 5(m + 2)$

27. $-2t + 5t - 9 = 3(t - 4) - 5$

28. $2.3x + 13.7 = 1.3x + 2.9$

29. $0.2(50) + 0.8r = 0.4(50 + r)$

30. $r + 9 + 7r = 4(3 + 2r) - 3$

31. $2(3 + 7x) - (1 + 15x) = 2$

32. $0.6(100 - x) + 0.4x = 0.5(92)$

33. $\frac{1}{4}x - 4 = \frac{3}{2}x + \frac{3}{4}x$

34. $\frac{3}{4}(z - 2) - \frac{1}{3}(5 - 2z) = -2$

Answers (left column)

1. equation; $\{-5\}$
2. expression; $7p - 14$
3. expression; $-3m - 2$
4. equation; $\{7\}$
5. equation; $\{0\}$
6. equation; $\left\{-\dfrac{96}{5}\right\}$
7. expression; $-10z - 6$
8. equation; $\{3\}$
9. expression; $-\dfrac{1}{6}x + 5$
10. expression; $2k - 11$
11. $\left\{\dfrac{7}{3}\right\}$ 12. $\{4\}$
13. $\{-5.1\}$ 14. $\{12\}$
15. $\{-25\}$ 16. $\{-6\}$
17. $\{-6\}$ 18. $\{-16\}$
19. $\{\text{all real numbers}\}$
20. $\{23.7\}$
21. $\{6\}$ 22. $\{0\}$
23. $\{7\}$ 24. $\{1\}$
25. $\{5\}$ 26. $\varnothing$
27. $\varnothing$ 28. $\{-10.8\}$
29. $\{25\}$
30. $\{\text{all real numbers}\}$
31. $\{3\}$ 32. $\{70\}$
33. $\{-2\}$ 34. $\left\{\dfrac{14}{17}\right\}$

2.4 Applications of Linear Equations

VOCABULARY

☐ consecutive integers
☐ consecutive even (or odd) integers
☐ degree
☐ complementary angles
☐ right angle
☐ supplementary angles
☐ straight angle

OBJECTIVE 1 Learn the six steps for solving applied problems.

While there is not one specific method, we suggest the following.

Solving an Applied Problem

Step 1 **Read** the problem carefully. *What information is given? What is to be found?*

Step 2 **Assign a variable** to represent the unknown value. Make a sketch, diagram, or table, as needed. If necessary, express any other unknown values in terms of the variable.

Step 3 **Write an equation** using the variable expression(s).

Step 4 **Solve** the equation.

Step 5 **State the answer.** Label it appropriately. *Does it seem reasonable?*

Step 6 **Check** the answer in the words of the *original* problem.

OBJECTIVE 2 Solve problems involving unknown numbers.

EXAMPLE 1 Finding the Value of an Unknown Number

The product of 4, and a number decreased by 7, is 100. What is the number?

Step 1 **Read** the problem carefully. We are asked to find a number.

Step 2 **Assign a variable** to represent the unknown quantity.

Let x = the number.

Step 3 **Write an equation.**

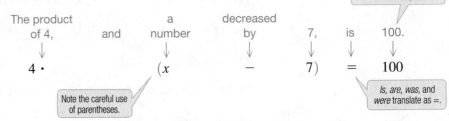

Writing a "word equation" is often helpful.

Note the careful use of parentheses.

Is, are, was, and were translate as =.

Step 4 **Solve.**

$$4(x - 7) = 100 \qquad \text{Equation from Step 3}$$

$$4x - 28 = 100 \qquad \text{Distributive property}$$

$$4x - 28 + 28 = 100 + 28 \qquad \text{Add 28.}$$

$$4x = 128 \qquad \text{Combine like terms.}$$

$$\frac{4x}{4} = \frac{128}{4} \qquad \text{Divide by 4.}$$

$$x = 32$$

**NOW TRY
EXERCISE 1**

The product of 7, and a number increased by 3, is −63. What is the number?

Step 5 **State the answer.** The number is 32.

Step 6 **Check.** When 32 is decreased by 7, we obtain $32 - 7 = 25$. If 4 is multiplied by 25, we obtain 100, as required. The answer, 32, is correct.

NOW TRY

 CAUTION Because of the commas in the problem in **Example 1,** writing the equation as

$$4x - 7 = 100 \text{ is } incorrect.$$

This equation corresponds to the statement "The product of 4 and a number, decreased by 7, is 100."

**NOW TRY
EXERCISE 2**

If 5 is added to a number, the result is 7 less than three times the number. Find the number.

EXAMPLE 2 Finding the Value of an Unknown Number

If 6 is subtracted from five times a number, the result is 9 more than twice the number. Find the number.

Step 1 **Read** the problem. We are asked to find a number.

Step 2 **Assign a variable** to represent the unknown quantity.

Let $x =$ the number.

Step 3 **Write an equation.**

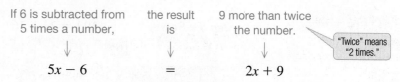

If 6 is subtracted from 5 times a number,	the result is	9 more than twice the number.
↓	↓	↓
$5x - 6$	$=$	$2x + 9$

"Twice" means "2 times."

Step 4 **Solve.**

$$5x - 6 - 2x = 2x + 9 - 2x \quad \text{Subtract } 2x.$$
$$3x - 6 = 9 \quad \text{Combine like terms.}$$
$$3x - 6 + 6 = 9 + 6 \quad \text{Add 6.}$$
$$3x = 15 \quad \text{Combine like terms.}$$
$$\frac{3x}{3} = \frac{15}{3} \quad \text{Divide by 3.}$$
$$x = 5$$

Step 5 **State the answer.** The number is 5.

Step 6 **Check.** If 6 is subtracted from 5 times the number, we have

$$5 \cdot 5 - 6, \quad \text{which equals 19.}$$

Nine more than twice the number would be

$$2 \cdot 5 + 9, \quad \text{which also equals 19.}$$

The answer, 5, checks.

NOW TRY

OBJECTIVE 3 Solve problems involving sums of quantities.

PROBLEM-SOLVING HINT To solve problems involving sums of quantities, choose a variable to represent one of the unknowns. ***Then represent the other quantity in terms of the same variable.*** (See **Example 11** in **Section 2.3**.)

NOW TRY ANSWERS
1. −12
2. 6

**NOW TRY
EXERCISE 3**

In the 2012 Summer Olympics in London, England, Germany won 21 fewer medals than Great Britain. The two countries won a total of 109 medals. How many medals did each country win? (*Source: World Almanac and Book of Facts.*)

EXAMPLE 3   Finding Numbers of Olympic Medals

In the 2012 Summer Olympics in London, England, the United States won 16 more medals than China. The two countries won a total of 192 medals. How many medals did each country win? (*Source: World Almanac and Book of Facts.*)

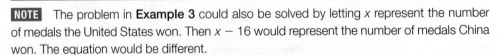

Step 1   **Read** the problem. We are given the total number of medals and asked to find the number each country won.

Step 2   **Assign a variable.**

Let x = the number of
                   medals China won.

Then $x + 16$ = the number of medals the United States won.

Step 3   **Write an equation.**

The total	is	the number of medals China won	plus	the number of medals the United States won.
↓	↓	↓	↓	↓
192	=	x	+	$(x + 16)$

Step 4   **Solve.**

$192 = 2x + 16$	Combine like terms.
$192 - 16 = 2x + 16 - 16$	Subtract 16.
$176 = 2x$	Combine like terms.
$\dfrac{176}{2} = \dfrac{2x}{2}$	Divide by 2.

$$88 = x, \quad \text{or} \quad x = 88 \leftarrow \text{Medals China won}$$

Step 5   **State the answer.** The variable x represents the number of medals China won, so China won 88 medals.

$$x + 16$$
$$= 88 + 16$$
$$= 104 \leftarrow \text{Medals the United States won}$$

Step 6   **Check.** The United States won 104 medals and China won 88, so the total number of medals was $104 + 88 = 192$. Because $104 - 88 = 16$, the United States won 16 more medals than China. This agrees with the information given in the problem. The answer checks.

NOW TRY

NOTE   The problem in **Example 3** could also be solved by letting x represent the number of medals the United States won. Then $x - 16$ would represent the number of medals China won. The equation would be different.

$$192 = x + (x - 16) \qquad \text{Alternative equation for \textbf{Example 3}}$$

The solution of this equation is 104, which is the number of U.S. medals. The number of Chinese medals would be $104 - 16 = 88$. **The answers are the same,** whichever approach is used, even though the equation and its solution are different.

> **! CAUTION** The nature of an applied problem may restrict the set of possible solutions. For instance, an answer such as −33 medals or $25\frac{1}{2}$ medals would be inappropriate in **Example 3.** Be sure that an answer is reasonable given the context of the problem.

NOW TRY EXERCISE 4

In one week, the owner of Carly's Coffeehouse found that the number of orders for bagels was $\frac{2}{3}$ the number of orders for chocolate scones. If the total number of orders for the two items was 525, how many orders were placed for bagels?

EXAMPLE 4 Finding the Number of Orders for Tea

The owner of Terry's Coffeehouse found that on one day the number of orders for tea was $\frac{1}{3}$ the number of orders for coffee. If the total number of orders for the two drinks was 76, how many orders were placed for tea?

Step 1 **Read** the problem. It asks for the number of orders for tea.

Step 2 **Assign a variable.** Because of the way the problem is stated, let the variable represent the number of orders for *coffee*.

Let x = the number of orders for coffee.

Then $\frac{1}{3}x$ = the number of orders for tea.

Step 3 **Write an equation.** Use the fact that the total number of orders was 76.

The total	is	orders for coffee	plus	orders for tea.
↓	↓	↓	↓	↓
76	=	x	+	$\frac{1}{3}x$

Remember the x in $\frac{1}{3}x$.

Step 4 **Solve.**

$$76 = \frac{4}{3}x \qquad x = 1x = \frac{3}{3}x;$$
Combine like terms.

$$\frac{3}{4}(76) = \frac{3}{4}\left(\frac{4}{3}x\right) \qquad \text{Multiply by } \frac{3}{4}, \text{ the reciprocal of } \frac{4}{3}.$$

Be careful. This is *not* the answer.

$$57 = x \leftarrow \text{Number of orders for coffee}$$

Step 5 **State the answer.** In this problem, ***x does not represent the quantity that we must find.*** The number of orders for tea was $\frac{1}{3}x$.

$$\frac{1}{3}(57) = 19 \leftarrow \text{Number of orders for tea}$$

Step 6 **Check.** The number of orders for tea, 19, is one-third the number of orders for coffee, 57, and

$$19 + 57 = 76, \quad \text{as required.}$$

This agrees with the information given in the problem. The answer checks.

NOW TRY

> **PROBLEM-SOLVING HINT** In **Example 4,** it was easier to let the variable represent the quantity that was *not* specified. This required extra work in Step 5 to find the number of orders for tea. In some cases, this approach is easier.

NOW TRY ANSWER
4. 210 bagel orders

**NOW TRY
EXERCISE 5**

At the Sherwood Estates pool party, each resident brought four guests. If a total of 175 people visited the pool that day, how many were residents and how many were guests?

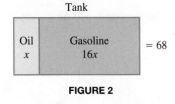

FIGURE 2

EXAMPLE 5   Analyzing a Gasoline-Oil Mixture

A lawn trimmer uses a mixture of gasoline and oil. The mixture contains 16 oz of gasoline for each 1 ounce of oil. If the tank holds 68 oz of the mixture, how many ounces of oil and how many ounces of gasoline does it require when it is full?

Step 1   **Read** the problem. We must find how many ounces of oil and gasoline are needed to fill the tank.

Step 2   **Assign a variable.**

Let x = the number of ounces of oil required.

Then $16x$ = the number of ounces of gasoline required.

A diagram like the one in **FIGURE 2** is helpful.

Step 3   **Write an equation.**

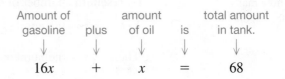

$$16x + x = 68$$

Step 4   **Solve.** $17x = 68$ Combine like terms.

$$\frac{17x}{17} = \frac{68}{17}$$ Divide by 17.

$$x = 4$$

Step 5   **State the answer.** When full, the lawn trimmer requires 4 oz of oil, and

$$16x = 16(4) \text{ oz}$$
$$= 64 \text{ oz of gasoline.}$$

Step 6   **Check.** Because $4 + 64 = 68$, and 64 is 16 times 4, the answer checks.

NOW TRY

PROBLEM-SOLVING HINT   Sometimes we must find three unknown quantities. ***When three unknowns are compared in pairs, let the variable represent the unknown found in both pairs.***

EXAMPLE 6   Dividing a Board into Pieces

A project calls for three pieces of wood. The longest piece must be twice the length of the middle-sized piece. The shortest piece must be 10 in. shorter than the middle-sized piece. If a board 70 in. long is to be used, how long must each piece be?

Step 1   **Read** the problem. There will be three answers.

Step 2   **Assign a variable.** The middle-sized piece appears in both pairs of comparisons, so let x represent the length, in inches, of the middle-sized piece.

Let x = the length of the middle-sized piece.

Then $2x$ = the length of the longest piece,

and $x - 10$ = the length of the shortest piece.

NOW TRY ANSWER
5. 35 residents; 140 guests

**NOW TRY
EXERCISE 6**

Over a 6-hr period, a basketball player spent twice as much time lifting weights as practicing free throws and 2 hr longer watching game films than practicing free throws. How many hours did he spend on each task?

A sketch is helpful. See **FIGURE 3**.

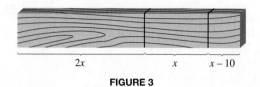

$$2x \qquad x \qquad x-10$$

FIGURE 3

Step 3   Write an equation.

Longest	plus	middle-sized	plus	shortest	is	total length.
↓	↓	↓	↓	↓	↓	↓
$2x$	$+$	x	$+$	$(x-10)$	$=$	70

Step 4   Solve.

$$4x - 10 = 70 \qquad \text{Combine like terms.}$$
$$4x - 10 + 10 = 70 + 10 \qquad \text{Add 10.}$$
$$4x = 80 \qquad \text{Combine like terms.}$$
$$\frac{4x}{4} = \frac{80}{4} \qquad \text{Divide by 4.}$$
$$x = 20$$

Step 5   State the answer. The middle-sized piece is 20 in. long, the longest piece is $2(20) = 40$ in. long, and the shortest piece is $20 - 10 = 10$ in. long.

Step 6   Check. The lengths sum to 70 in. All problem conditions are satisfied.

NOW TRY

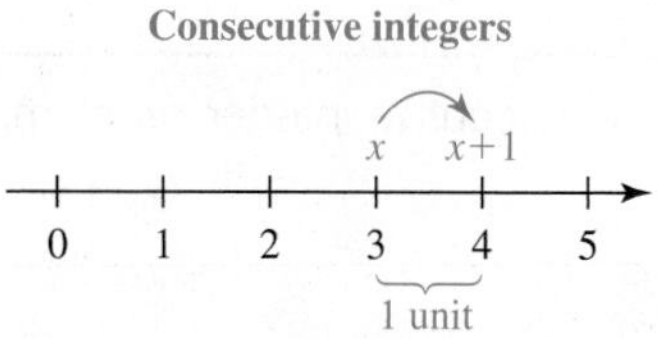

Consecutive integers

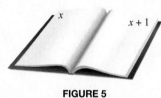

FIGURE 4

OBJECTIVE 4   Solve problems involving consecutive integers.

Two integers that differ by 1 are **consecutive integers.** For example, 3 and 4, 16 and 17, and -2 and -1 are pairs of consecutive integers. See **FIGURE 4**.

In general, if x represents an integer, then x + 1 represents the next greater consecutive integer.

EXAMPLE 7   Finding Consecutive Integers

Two pages that face each other in this book have 269 as the sum of their page numbers. What are the page numbers?

Step 1   Read the problem. Because the two pages face each other, they must have page numbers that are consecutive integers.

FIGURE 5

Step 2   Assign a variable.

Let $x =$ the lesser page number.

Then $x + 1 =$ the greater page number.

FIGURE 5 illustrates the situation.

NOW TRY ANSWER
6. practicing free throws: 1 hr;
lifting weights: 2 hr;
watching game films: 3 hr

Step 3   Write an equation. The sum of the page numbers is 269.

$$x + (x + 1) = 269$$

NOW TRY EXERCISE 7

Two pages that face each other in a book have 593 as the sum of their page numbers. What are the page numbers?

Step 4   **Solve.**

$$x + (x + 1) = 269 \quad \text{Equation from Step 3}$$
$$2x + 1 = 269 \quad \text{Combine like terms.}$$
$$2x + 1 - 1 = 269 - 1 \quad \text{Subtract 1.}$$
$$2x = 268 \quad \text{Combine like terms.}$$
$$\frac{2x}{2} = \frac{268}{2} \quad \text{Divide by 2.}$$
$$x = 134$$

Step 5   **State the answer.** The lesser page number is 134, and the greater page number is $134 + 1 = 135$. (Your book is opened to these two pages.)

Step 6   **Check.** The sum of 134 and 135 is 269. The answer checks.   **NOW TRY**

Consecutive *even* **integers,** such as 2 and 4, and 8 and 10, differ by 2. Similarly, **consecutive *odd* integers,** such as 1 and 3, and 9 and 11, also differ by 2. See **FIGURE 6.**

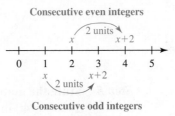

FIGURE 6

In general, if x represents an even (or odd) integer, then x + 2 represents the next greater consecutive even (or odd) integer, respectively.

In this book, we list consecutive integers in increasing order.

PROBLEM-SOLVING HINT   If x = the lesser (least) integer in a consecutive integer problem, then the following apply.

- For two consecutive integers, use $x, \quad x + 1$.
- For two consecutive *even* integers, use $x, \quad x + 2$.
- For two consecutive *odd* integers, use $x, \quad x + 2$.

EXAMPLE 8   **Finding Consecutive Odd Integers**

If the lesser of two consecutive odd integers is doubled, the result is 7 more than the greater of the two integers. Find the two integers.

Step 1   **Read** the problem. We must find two consecutive odd integers.

Step 2   **Assign a variable.**

Let x = the lesser consecutive odd integer.

Then $x + 2$ = the greater consecutive odd integer.

Step 3   **Write an equation.**

If the lesser is doubled,	the result is	7	more than	the greater.
↓	↓	↓	↓	↓
$2x$	$=$	7	$+$	$(x + 2)$

NOW TRY ANSWER
7. 296, 297

**NOW TRY
EXERCISE 8**

Find two consecutive odd integers such that the sum of twice the lesser and three times the greater is 191.

Step 4 **Solve.** $2x = 9 + x$ Combine like terms.

$2x - x = 9 + x - x$ Subtract x.

$x = 9$ Combine like terms.

Step 5 **State the answer.** The lesser integer is 9. The greater is $9 + 2 = 11$.

Step 6 **Check.** When 9 is doubled, we obtain 18, which is 7 more than the greater odd integer, 11. The answer checks.   NOW TRY

OBJECTIVE 5 Solve problems involving supplementary and complementary angles.

An angle can be measured by a unit called the **degree** (°), which is $\frac{1}{360}$ of a complete rotation. See **FIGURE 7**.

- Two angles whose sum is 90° are **complementary**, or *complements* of each other.
- An angle that measures 90° is a **right angle.**
- Two angles whose sum is 180° are **supplementary**, or *supplements* of each other.
- An angle that measures 180° is a **straight angle.**

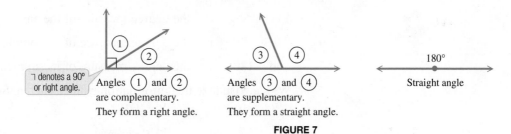

⌐ denotes a 90°
or right angle.

Angles ① and ②
are complementary.
They form a right angle.

Angles ③ and ④
are supplementary.
They form a straight angle.

180°

Straight angle

FIGURE 7

PROBLEM-SOLVING HINT   Let x represent the degree measure of an angle.

90 − x represents the degree measure of its complement.

180 − x represents the degree measure of its supplement.

EXAMPLE 9   Finding the Measure of an Angle

Find the measure of an angle whose complement is five times its measure.

Step 1 **Read** the problem. We must find the measure of an angle, given information about the measure of its complement.

Step 2 **Assign a variable.**

Let x = the degree measure of the angle.

Then $90 - x$ = the degree measure of its complement.

Step 3 **Write an equation.**

Measure of the   5 times the measure
complement   is   of the angle.
↓   ↓   ↓
$90 - x$ $=$ $5x$

NOW TRY
EXERCISE 9

Find the measure of an angle whose complement is twice its measure.

Step 4   **Solve.**

$$90 - x = 5x$$ Equation from Step 3

$$90 - x + x = 5x + x$$ Add x.

$$90 = 6x$$ Combine like terms.

$$\frac{90}{6} = \frac{6x}{6}$$ Divide by 6.

$$15 = x, \quad \text{or} \quad x = 15$$

Step 5   **State the answer.** The measure of the angle is 15°.

Step 6   **Check.** If the angle measures 15°, then $90° - 15° = 75°$ is the measure of its complement. 75° is equal to five times 15°, as required.   **NOW TRY**

NOW TRY
EXERCISE 10

Find the measure of an angle whose supplement is 46° less than three times its complement.

EXAMPLE 10   Finding the Measure of an Angle

Find the measure of an angle whose supplement is 10° more than twice its complement.

Step 1   **Read** the problem. We are to find the measure of an angle, given information about its complement and its supplement.

Step 2   **Assign a variable.**

Let x = the degree measure of the angle.

Then $90 - x$ = the degree measure of its complement,

and $180 - x$ = the degree measure of its supplement.

We can visualize this information using a sketch. See **FIGURE 8**.

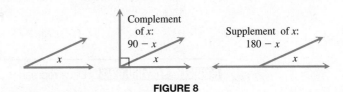

FIGURE 8

Step 3   **Write an equation.**

Supplement	is	10	more than	twice	its complement.
↓	↓	↓	↓	↓	↓
$180 - x$	$=$	10	$+$	$2 \cdot$	$(90 - x)$

Be sure to use parentheses here.

Step 4   **Solve.**

$$180 - x = 10 + 180 - 2x$$ Distributive property

$$180 - x = 190 - 2x$$ Combine like terms.

$$180 - x + 2x = 190 - 2x + 2x$$ Add $2x$.

$$180 + x = 190$$ Combine like terms.

$$180 + x - 180 = 190 - 180$$ Subtract 180.

$$x = 10$$

Step 5   **State the answer.** The measure of the angle is 10°.

Step 6   **Check.** The complement of 10° is 80° and the supplement of 10° is 170°. Also, 170° is equal to 10° more than twice 80° (that is, $170 = 10 + 2(80)$ is true). Therefore, the answer checks.

NOW TRY

2.4 Exercises

  MyMathLab®

▶ *Complete solution available in MyMathLab*

Concept Check *Which choice would **not** be a reasonable answer? Justify your response.*

1. A problem requires finding the number of cars on a dealer's lot.

A. 0 **B.** 45 **C.** 1 **D.** $6\frac{1}{2}$

2. A problem requires finding the number of hours a light bulb is on during a day.

A. 0 **B.** 4.5 **C.** 13 **D.** 25

3. A problem requires finding the distance traveled in miles.

A. −10 **B.** 1.8 **C.** $10\frac{1}{2}$ **D.** 50

4. A problem requires finding the time in minutes.

A. 0 **B.** 10.5 **C.** −5 **D.** 90

Solve each problem. In each case, give an equation using x as the variable and give the answer. ***See Examples 1 and 2.***

5. The sum of a number and 9 is −26. What is the number?

6. The difference of a number and 11 is −31. What is the number?

7. The product of 8, and a number increased by 6, is 104. What is the number?

8. The product of 5, and 3 more than twice a number, is 85. What is the number?

▶ **9.** If 2 is added to five times a number, the result is equal to 5 more than four times the number. Find the number.

10. If four times a number is added to 8, the result is equal to three times the number, added to 5. Find the number.

11. If 2 is subtracted from a number and this difference is tripled, the result is 6 more than the number. Find the number.

12. If 3 is added to a number and this sum is doubled, the result is 2 more than the number. Find the number.

13. When 6 is added to $\frac{3}{4}$ of a number, the result is 4 less than the number. Find the number.

14. When $\frac{2}{3}$ of a number is added to 10, the result is 5 more than the number. Find the number.

15. The sum of three times a number and 7 more than the number is the same as the difference of −11 and twice the number. What is the number?

16. If 4 is added to twice a number and this sum is multiplied by 2, the result is the same as if the number is multiplied by 3 and 4 is added to the product. What is the number?

Complete the six suggested problem-solving steps to solve each problem.

17. The 150-member Iowa legislature includes 4 fewer Democrats than Republicans. (No other parties are represented.) How many Democrats and how many Republicans are there in the legislature? (*Source:* www.legis.iowa.gov) **(See Example 3.)**

Step 1 **Read** the problem carefully.

We must find the number of Democrats and the number of _____.

Step 2 **Assign a variable.**

Let x = the number of Republicans.

Then _____ = the number of _____.

Step 3 **Write an equation.**

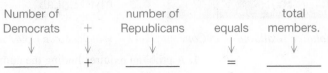

Complete Steps 4–6 to solve the problem.

18. The sum of two consecutive even integers is 254. Find the integers. **(See Example 8.)**

   Step 1 **Read** the problem carefully.

   We must find two consecutive _____ .

   Step 2 **Assign a variable.**

   Let $x =$ the lesser of the two _____ even integers.

   Then _____ = the greater of the two consecutive even integers.

   Step 3 **Write an equation.**

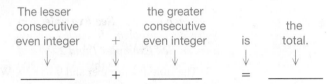

Complete Steps 4–6 to solve the problem.

Solve each problem. See Example 3.

19. New York and Ohio are among the states with the most remaining drive-in movie screens. New York has 2 more screens than Ohio, and there are 56 screens total in the two states. How many drive-in movie screens remain in each state? (*Source:* www.drive-ins.com)

20. Two of the most watched episodes in television were the final episodes of *M*A*S*H* and *Cheers*. The total number of viewers for these two episodes was about 92 million, with 8 million more people watching the *M*A*S*H* episode than the *Cheers* episode. How many people watched each episode? (*Source:* Nielsen Media Research.)

21. During the 113th session, the U.S. Senate had a total of 98 Democrats and Republicans. There were 6 more Democrats than Republicans. How many members of each party were there? (*Source:* www.thegreenpapers.com)

22. During the 113th session, the total number of Democrats and Republicans in the U.S. House of Representatives was 432. There were 32 more Republicans than Democrats. How many members of each party were there? (*Source:* www.thegreenpapers.com)

23. Madonna and Bruce Springsteen had the two top-grossing concert tours for 2012, together generating about $427 million in ticket sales. If Bruce Springsteen took in $29 million less than Madonna, how much did each tour generate? (*Source:* www.billboard.com)

24. The Toyota Camry and the Honda Accord were the top-selling passenger cars in the United States in 2012. Accord sales were 27 thousand less than Camry sales, and 691 thousand of the two cars were sold. How many of each car were sold? (*Source:* www.edmunds.com)

25. In the 2012–2013 NBA regular season, the Miami Heat won two more than four times as many games as they lost. The Heat played 82 games. How many wins and losses did the team have? (*Source:* www.NBA.com)

26. In the 2013 regular baseball season, the Cleveland Indians won 48 fewer than twice as many games as they lost. They played 162 regular-season games. How many wins and losses did the team have? (*Source:* www.MLB.com)

27. A one-cup serving of orange juice contains 3 mg less than four times the amount of vitamin C as a one-cup serving of pineapple juice. Servings of the two juices contain a total of 122 mg of vitamin C. How many milligrams of vitamin C are in a serving of each type of juice? (*Source:* U.S. Agriculture Department.)

28. A one-cup serving of pineapple juice has 9 more than three times as many calories as a one-cup serving of tomato juice. Servings of the two juices contain a total of 173 calories. How many calories are in a serving of each type of juice? (*Source:* U.S. Agriculture Department.)

Solve each problem. ***See Examples 4 and 5.***

29. In one day, a store sold $\frac{2}{3}$ as many DVDs as Blu-ray discs. The total number of DVDs and Blu-ray discs sold that day was 280. How many DVDs were sold?

30. A workout that combines weight training and aerobics burns a total of 371 calories. If weight training burns $\frac{2}{5}$ as many calories as aerobics, how many calories does weight training burn?

31. The world's largest taco contained approximately 1 kg of onion for every 6.6 kg of grilled steak. The total weight of these two ingredients was 617.6 kg. To the nearest tenth of a kilogram, how many kilograms of each ingredient were used to make the taco? (*Source: Guinness World Records.*)

32. As of 2013, the combined population of China and India was estimated at 2.5 billion. If there were about 0.9 as many people living in India as China, what was the population of each country, to the nearest tenth of a billion? (*Source:* U.S. Census Bureau.)

33. The value of a "Mint State-63" (uncirculated) 1950 Jefferson nickel minted at Denver is twice the value of a 1945 nickel in similar condition minted at Philadelphia. Together, the total value of the two coins is $24.00. What is the value of each coin? (*Source:* Yeoman, R., *A Guide Book of United States Coins.*)

34. U.S. five-cent coins are made from a combination of nickel and copper. For every 1 lb of nickel, 3 lb of copper are used. How many pounds of copper would be needed to make 560 lb of five-cent coins? (*Source:* The United States Mint.)

▶ 35. A recipe for whole-grain bread calls for 1 oz of rye flour for every 4 oz of whole-wheat flour. How many ounces of each kind of flour should be used to make a loaf of bread weighing 32 oz?

36. A medication contains 9 mg of active ingredients for every 1 mg of inert ingredients. How much of each kind of ingredient would be contained in a single 250-mg caplet?

Solve each problem. ***See Example 6.***

37. An office manager booked 55 airline tickets. He booked 7 more tickets on American Airlines than United Airlines. On Southwest Airlines, he booked 4 more than twice as many tickets as on United. How many tickets did he book on each airline?

38. A mathematics textbook editor spent 7.5 hr making telephone calls, writing e-mails, and attending meetings. She spent twice as much time attending meetings as making telephone calls and 0.5 hr longer writing e-mails than making telephone calls. How many hours did she spend on each task?

39. A party-length submarine sandwich that is 59 in. long is cut into three pieces. The middle piece is 5 in. longer than the shortest piece, and the shortest piece is 9 in. shorter than the longest piece. How long is each piece?

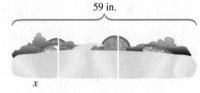
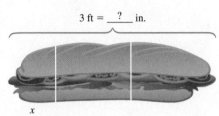

40. A three-foot-long deli sandwich must be split into three pieces so that the middle piece is twice as long as the shortest piece and the shortest piece is 8 in. shorter than the longest piece. How long should the three pieces be?

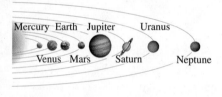

41. The United States earned 104 medals at the 2012 Summer Olympics. The number of silver medals earned was the same as the number of bronze medals. The number of gold medals was 17 more than the number of silver medals. How many of each kind of medal did the United States earn? (*Source: World Almanac and Book of Facts.*)

42. China earned a total of 88 medals at the 2012 Summer Olympics. The number of gold medals earned was 15 more than the number of bronze medals. The number of bronze medals earned was 4 fewer than the number of silver medals. How many of each kind of medal did China earn? (*Source: World Almanac and Book of Facts.*)

43. Venus is 31.2 million mi farther from the sun than Mercury, while Earth is 57 million mi farther from the sun than Mercury. If the total of the distances from these three planets to the sun is 196.2 million mi, how far away from the sun is Mercury? (All distances given here are *mean* (*average*) distances.) (*Source: The New York Times Almanac.*)

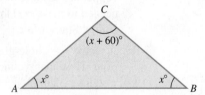

44. Saturn, Jupiter, and Uranus have a total of 156 known satellites (moons). Jupiter has 5 more satellites than Saturn, and Uranus has 35 fewer satellites than Saturn. How many known satellites does Uranus have? (*Source: http://solarsystem.nasa.gov*)

45. The sum of the measures of the angles of any triangle is 180°. In triangle *ABC*, angles *A* and *B* have the same measure, while the measure of angle *C* is 60° greater than each of angles *A* and *B*. What are the measures of the three angles?

46. In triangle *ABC*, the measure of angle *A* is 141° more than the measure of angle *B*. The measure of angle *B* is the same as the measure of angle *C*. Find the measure of each angle. (*Hint:* See Exercise 45.)

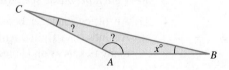

Solve each problem. See Examples 7 and 8.

47. The numbers on two consecutively numbered gym lockers have a sum of 137. What are the locker numbers?

48. The numbers on two consecutive checkbook checks have a sum of 357. What are the numbers?

▶ **49.** Two pages that are back-to-back in this book have 203 as the sum of their page numbers. What are the page numbers?

50. Two apartments have numbers that are consecutive integers. The sum of the numbers is 59. What are the two apartment numbers?

▶ **51.** Find two consecutive even integers such that the lesser added to three times the greater gives a sum of 46.

52. Find two consecutive even integers such that six times the lesser added to the greater gives a sum of 86.

53. Find two consecutive odd integers such that 59 more than the lesser is four times the greater.

54. Find two consecutive odd integers such that twice the greater is 17 more than the lesser.

55. When the lesser of two consecutive integers is added to three times the greater, the result is 43. Find the integers.

56. If five times the lesser of two consecutive integers is added to three times the greater, the result is 59. Find the integers.

Extending Skills   Solve each problem.

57. If the sum of three consecutive even integers is 60, what is the first of the three even integers? (*Hint:* If x and $x + 2$ represent the first two consecutive even integers, how would we represent the third consecutive even integer?)

58. If the sum of three consecutive odd integers is 69, what is the third of the three odd integers?

59. If 6 is subtracted from the third of three consecutive odd integers and the result is multiplied by 2, the answer is 23 less than the sum of the first and twice the second of the integers. Find the integers.

60. If the first and third of three consecutive even integers are added, the result is 22 less than three times the second integer. Find the integers.

Solve each problem. See Examples 9 and 10.

61. Find the measure of an angle whose complement is four times its measure.

62. Find the measure of an angle whose complement is five times its measure.

63. Find the measure of an angle whose supplement is eight times its measure.

64. Find the measure of an angle whose supplement is three times its measure.

▶ **65.** Find the measure of an angle whose supplement measures 39° more than twice its complement.

66. Find the measure of an angle whose supplement measures 38° less than three times its complement.

67. Find the measure of an angle such that the difference between the measures of its supplement and three times its complement is 10°.

68. Find the measure of an angle such that the sum of the measures of its complement and its supplement is 160°.

2.5 Formulas and Additional Applications from Geometry

VOCABULARY

☐ formula
☐ area
☐ perimeter
☐ vertical angles

NOW TRY EXERCISE 1

Find the value of the remaining variable.

$$P = 2a + 2b;$$
$$P = 78, a = 12$$

A **formula** is an equation in which variables are used to describe a relationship. For example, formulas exist for finding perimeters and areas of geometric figures, calculating money earned on bank savings, and converting among measurements.

$$P = 4s, \quad \mathcal{A}^* = \pi r^2, \quad I = prt, \quad F = \frac{9}{5}C + 32 \qquad \text{Formulas}$$

Many of the formulas used in this book are given inside the back cover.

OBJECTIVE 1 Solve a formula for one variable, given values of the other variables.

The **area** of a plane (two-dimensional) geometric figure is a measure of the surface covered by the figure. Area is measured in square units.

EXAMPLE 1 Using Formulas to Evaluate Variables

Find the value of the remaining variable in each formula.

(a) $\mathcal{A} = LW; \quad \mathcal{A} = 64, L = 10$

This formula gives the area of a rectangle. See **FIGURE 9**.

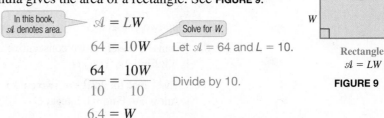

In this book, $\mathcal{A}$ denotes area.

$$\mathcal{A} = LW$$

Solve for W.

$$64 = 10W \qquad \text{Let } \mathcal{A} = 64 \text{ and } L = 10.$$

$$\frac{64}{10} = \frac{10W}{10} \qquad \text{Divide by 10.}$$

$$6.4 = W$$

Rectangle
$\mathcal{A} = LW$

FIGURE 9

The width is 6.4. Because $10(6.4) = 64$, the given area, the answer checks.

(b) $\mathcal{A} = \frac{1}{2}h(b + B); \quad \mathcal{A} = 210, B = 27, h = 10$

This formula gives the area of a trapezoid. See **FIGURE 10**.

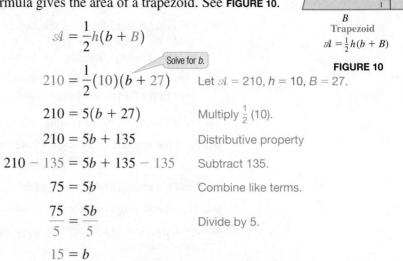

Trapezoid
$\mathcal{A} = \frac{1}{2}h(b + B)$

FIGURE 10

$$\mathcal{A} = \frac{1}{2}h(b + B)$$

Solve for b.

$$210 = \frac{1}{2}(10)(b + 27) \qquad \text{Let } \mathcal{A} = 210, h = 10, B = 27.$$

$$210 = 5(b + 27) \qquad \text{Multiply } \frac{1}{2}(10).$$

$$210 = 5b + 135 \qquad \text{Distributive property}$$

$$210 - 135 = 5b + 135 - 135 \qquad \text{Subtract 135.}$$

$$75 = 5b \qquad \text{Combine like terms.}$$

$$\frac{75}{5} = \frac{5b}{5} \qquad \text{Divide by 5.}$$

$$15 = b$$

The length of the shorter parallel side, b, is 15. This answer checks because

$$\frac{1}{2}(10)(15 + 27) = 210, \quad \text{as required.}$$

NOW TRY

* In this book, we use $\mathcal{A}$ to denote area.

OBJECTIVE 2 Use a formula to solve an applied problem.

The **perimeter** of a plane (two-dimensional) geometric figure is the measure of the outer boundary of the figure. For a polygon (e.g., a rectangle, square, or triangle), it is the sum of the lengths of the sides.

NOW TRY EXERCISE 2

Kurt's garden is in the shape of a rectangle. The length is 10 ft less than twice the width, and the perimeter is 160 ft. Find the dimensions of the garden.

EXAMPLE 2 Finding the Dimensions of a Rectangular Yard

Cathleen's backyard is in the shape of a rectangle. The length is 5 m less than twice the width, and the perimeter is 80 m. Find the dimensions of the yard.

Step 1 **Read** the problem. We must find the dimensions of the yard.

Step 2 **Assign a variable.** Let $W =$ the width of the lot, in meters. The length is 5 meters less than twice the width, so the length is $L = 2W - 5$. See FIGURE 11.

FIGURE 11

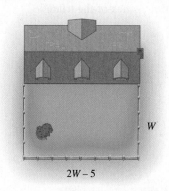

Step 3 **Write an equation.** Use the formula for the perimeter of a rectangle.

$$P = 2L + 2W \quad \text{Perimeter of a rectangle}$$

$$\text{Perimeter} = 2 \cdot \text{Length} + 2 \cdot \text{Width}$$

$$80 = 2(2W - 5) + 2W \quad \text{Substitute 80 for perimeter } P \text{ and } 2W - 5 \text{ for length } L.$$

Step 4 **Solve.**

$$
\begin{aligned}
80 &= 4W - 10 + 2W && \text{Distributive property}\\
80 &= 6W - 10 && \text{Combine like terms.}\\
80 + 10 &= 6W - 10 + 10 && \text{Add 10.}\\
90 &= 6W && \text{Combine like terms.}\\
\frac{90}{6} &= \frac{6W}{6} && \text{Divide by 6.}\\
15 &= W
\end{aligned}
$$

We must also find the length.

Step 5 **State the answer.** The width is 15 m and the length is $2(15) - 5 = 25$ m.

Step 6 **Check.** If the width is 15 m and the length is 25 m, the perimeter is

$$2(25) + 2(15) = 80 \text{ m}, \quad \text{as required.}$$

NOW TRY

EXAMPLE 3 Finding the Dimensions of a Triangle

The longest side of a triangle is 3 ft longer than the shortest side. The medium side is 1 ft longer than the shortest side. If the perimeter of the triangle is 16 ft, what are the lengths of the three sides?

Step 1 **Read** the problem. We must find the lengths of the sides of a triangle.

Step 2 **Assign a variable.**

Let $s =$ the length of the shortest side, in feet,

$s + 1 =$ the length of the medium side, in feet, and,

$s + 3 =$ the length of the longest side in feet.

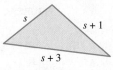

It is a good idea to draw a sketch. See FIGURE 12.

FIGURE 12

NOW TRY ANSWER

2. width: 30 ft; length: 50 ft

**NOW TRY
EXERCISE 3**

The perimeter of a triangle is 30 ft. The longest side is 1 ft longer than the medium side, and the shortest side is 7 ft shorter than the medium side. What are the lengths of the three sides of the triangle?

Step 3   Write an equation. Use the formula for the perimeter of a triangle.

$$P = a + b + c \qquad \text{Perimeter of a triangle}$$
$$16 = s + (s + 1) + (s + 3) \qquad \text{Substitute.}$$

Step 4   Solve.

$$16 = 3s + 4 \qquad \text{Combine like terms.}$$
$$12 = 3s \qquad \text{Subtract 4.}$$
$$4 = s \qquad \text{Divide by 3.}$$

Step 5   State the answer. The shortest side, s, has length 4 ft.

$$s + 1 = 4 + 1 = 5 \text{ ft} \qquad \text{Length of medium side}$$
$$s + 3 = 4 + 3 = 7 \text{ ft} \qquad \text{Length of longest side}$$

Step 6   Check. The medium side, 5 ft, is 1 ft longer than the shortest side, and the longest side, 7 ft, is 3 ft longer than the shortest side. The perimeter is

$$4 + 5 + 7 = 16 \text{ ft,} \quad \text{as required.} \qquad \text{NOW TRY}$$

**NOW TRY
EXERCISE 4**

The area of a triangle is 77 cm². The base is 14 cm. Find the height of the triangle.

EXAMPLE 4   Finding the Height of a Triangular Sail

The area of a triangular sail of a sailboat is 126 ft². (Recall that "ft²" means "square feet.") The base of the sail is 12 ft. Find the height of the sail.

Step 1   Read the problem. We must find the height of the triangular sail.

Step 2   Assign a variable. Let $h =$ the height of the sail, in feet. See **FIGURE 13**.

Step 3   Write an equation. Use the formula for the area of a triangle.

$$\mathscr{A} = \frac{1}{2}bh \qquad \mathscr{A} \text{ is area, } b \text{ is base,} \\ \text{and } h \text{ is height.}$$

$$126 = \frac{1}{2}(12)h \qquad \text{Let } \mathscr{A} = 126, b = 12.$$

Step 4   Solve.

$$126 = 6h \qquad \text{Multiply.}$$
$$21 = h \qquad \text{Divide by 6.}$$

FIGURE 13

Step 5   State the answer. The height of the sail is 21 ft.

Step 6   Check to see that the values $\mathscr{A} = 126$, $b = 12$, and $h = 21$ satisfy the formula for the area of a triangle.

$$126 = \frac{1}{2}(12)(21) \text{ is true.} \qquad \text{NOW TRY}$$

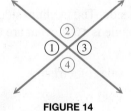

FIGURE 14

OBJECTIVE 3 Solve problems involving vertical angles and straight angles.

FIGURE 14 shows two intersecting lines forming angles that are numbered ①, ②, ③, and ④. Angles ① and ③ lie "opposite" each other. They are **vertical angles.** Another pair of vertical angles is ② and ④. *Vertical angles have equal measures.*

Consider angles ① and ②. When their measures are added, we obtain 180°, the measure of a straight angle. There are three other angle pairs that form straight angles: ② and ③, ③ and ④, and ① and ④.

**NOW TRY
EXERCISE 5**

Find the measure of each marked angle in the figure.

EXAMPLE 5 Finding Angle Measures

Refer to the appropriate figure in each part.

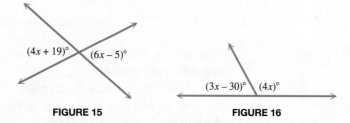

FIGURE 15   **FIGURE 16**

(a) Find the measure of each marked angle in **FIGURE 15**.

The marked angles are vertical angles, so they have equal measures.

$$4x + 19 = 6x - 5 \qquad \text{Set } 4x + 19 \text{ equal to } 6x - 5.$$
$$19 = 2x - 5 \qquad \text{Subtract } 4x.$$
$$24 = 2x \qquad \text{Add 5.}$$

This is *not* the answer. $\quad 12 = x \qquad$ Divide by 2.

Replace x with 12 in the expression for the measure of each angle.

$4x + 19$		$6x - 5$	
$= 4(12) + 19$	Let $x = 12$.	$= 6(12) - 5$	Let $x = 12$.
$= 48 + 19$	Multiply.	$= 72 - 5$	Multiply.
$= 67$	Add.	$= 67$	Subtract.

The angles have equal measures, as required. Each measures 67°.

(b) Find the measure of each marked angle in **FIGURE 16**.

The measures of the marked angles must add to 180° because together they form a straight angle. (They are also *supplements* of each other.)

$$(3x - 30) + 4x = 180 \qquad \text{Supplementary angles sum to 180°.}$$
$$7x - 30 = 180 \qquad \text{Combine like terms.}$$
$$7x = 210 \qquad \text{Add 30.}$$

Don't stop here. $\quad x = 30 \qquad$ Divide by 7.

Replace x with 30 in the expression for the measure of each angle.

$3x - 30$		$4x$	
$= 3(30) - 30$	Let $x = 30$.	$= 4(30)$	Let $x = 30$.
$= 90 - 30$	Multiply.	$= 120$	Multiply.
$= 60$	Subtract.		

The two angles measure 60° and 120°, which add to 180°, as required.   **NOW TRY**

> ⊘ **CAUTION** In **Example 5,** the answer is *not* the value of x. **Remember to substitute the value of the variable into the expression given for each angle.**

NOW TRY ANSWER
5. 32°, 32°

OBJECTIVE 4   Solve a formula for a specified variable.

Sometimes we want to rewrite a formula in terms of a *different* variable in the formula. For example, consider $\mathcal{A} = LW$, the formula for the area of a rectangle.

How can we rewrite $\mathcal{A} = LW$ in terms of W?

The process whereby we do this involves **solving for a specified variable,** or **solving a literal equation.**

To solve a formula for a specified variable, we use the *same* steps that we used to solve an equation with just one variable. Consider the parallel reasoning to solve each of the following for x.

$3x + 4 = 13$		$ax + b = c$	
$3x + 4 - 4 = 13 - 4$	Subtract 4.	$ax + b - b = c - b$	Subtract b.
$3x = 9$		$ax = c - b$	
$\dfrac{3x}{3} = \dfrac{9}{3}$	Divide by 3.	$\dfrac{ax}{a} = \dfrac{c - b}{a}$	Divide by a.
$x = 3$	Equation solved for x	$x = \dfrac{c - b}{a}$	Formula solved for x

When we solve a formula for a specified variable, we treat the specified variable as if it were the ONLY variable in the equation, and we treat the other variables as if they were numbers.

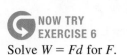
NOW TRY EXERCISE 6
Solve $W = Fd$ for F.

EXAMPLE 6   Solving for a Specified Variable

Solve $\mathcal{A} = LW$ for W.

W is multiplied by L, so we undo the multiplication by dividing each side by L.

$$\mathcal{A} = LW \quad \text{← Our goal is to isolate } W.$$

$$\frac{\mathcal{A}}{L} = \frac{LW}{L} \qquad\qquad \text{Divide by } L.$$

$$\frac{\mathcal{A}}{L} = W, \quad \text{or} \quad W = \frac{\mathcal{A}}{L} \qquad \frac{LW}{L} = \frac{L}{L} \cdot W = 1 \cdot W = W \qquad \textbf{NOW TRY}$$

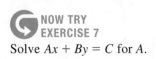
NOW TRY EXERCISE 7
Solve $Ax + By = C$ for A.

EXAMPLE 7   Solving for a Specified Variable

Solve $P = 2L + 2W$ for L.

$$P = 2L + 2W \quad \text{← Our goal is to isolate } L.$$

$$P - 2W = 2L + 2W - 2W \qquad\qquad \text{Subtract } 2W.$$

$$P - 2W = 2L \qquad\qquad \text{Combine like terms.}$$

$$\frac{P - 2W}{2} = \frac{2L}{2} \qquad\qquad \text{Divide by 2.}$$

$$\frac{P - 2W}{2} = L, \quad \text{or} \quad L = \frac{P - 2W}{2} \qquad \frac{2L}{2} = \frac{2}{2} \cdot L = 1 \cdot L = L$$

$$\frac{P - 2W}{2} \neq P - W$$

NOW TRY

NOW TRY ANSWERS

6. $F = \dfrac{W}{d}$

7. $A = \dfrac{C - By}{x}$

NOW TRY
EXERCISE 8
Solve $S = \frac{1}{2}(a + b + c)$
for a.

EXAMPLE 8 Solving for a Specified Variable

Solve $\mathcal{A} = \frac{1}{2}h(b + B)$ for B.

$$\mathcal{A} = \frac{1}{2}h(b + B)$$

Our goal is to isolate B.

$$\mathcal{A} = \frac{1}{2}hb + \frac{1}{2}hB \qquad \text{Clear the parentheses using the distributive property.}$$

$$2 \cdot \mathcal{A} = 2\left(\frac{1}{2}hb + \frac{1}{2}hB\right) \qquad \text{Multiply each side by 2 to clear the fractions.}$$

$$2 \cdot \mathcal{A} = 2 \cdot \frac{1}{2}hb + 2 \cdot \frac{1}{2}hB \qquad \text{Distributive property}$$

$$2\mathcal{A} = hb + hB \qquad \text{Multiply; } 2 \cdot \frac{1}{2} = \frac{2}{2} = 1$$

$$2\mathcal{A} - hb = hb + hB - hb \qquad \text{Subtract } hb.$$

$$2\mathcal{A} - hb = hB \qquad \text{Combine like terms.}$$

$$\frac{2\mathcal{A} - hb}{h} = \frac{hB}{h} \qquad \text{Divide by } h.$$

$$\frac{2\mathcal{A} - hb}{h} = B, \quad \text{or} \quad B = \frac{2\mathcal{A} - hb}{h} \qquad \text{NOW TRY} \; \text{◀}$$

NOW TRY
EXERCISE 9
Solve each equation for y.
(a) $5x + y = 3$
(b) $x - 2y = 8$

EXAMPLE 9 Solving for a Specified Variable

Solve each equation for y.

(a) $$2x - y = 7$$

Our goal is to isolate y.

$$2x - y - 2x = 7 - 2x \qquad \text{Subtract } 2x.$$

$$-y = 7 - 2x \qquad \text{Combine like terms.}$$

$$-1(-y) = -1(7 - 2x) \qquad \text{Multiply by } -1.$$

$$y = -7 + 2x \qquad \text{Multiply; distributive property}$$

$$y = 2x - 7 \qquad -a + b = b - a$$

We could have added y and subtracted 7 from each side of the equation to isolate y on the right, giving $2x - 7 = y$, a different form of the same result. There is often more than one way to solve for a specified variable.

(b) $$-3x + 2y = 6$$

$$-3x + 2y + 3x = 6 + 3x \qquad \text{Add } 3x.$$

$$2y = 3x + 6 \qquad \text{Combine like terms; commutative property}$$

$$\frac{2y}{2} = \frac{3x + 6}{2} \qquad \text{Divide by 2.}$$

Be careful here.

$$y = \frac{3x}{2} + \frac{6}{2} \qquad \frac{a + b}{c} = \frac{a}{c} + \frac{b}{c}$$

$\frac{3x}{2} = \frac{3}{2} \cdot \frac{x}{1} = \frac{3}{2}x$

$$y = \frac{3}{2}x + 3 \qquad \text{Simplify.}$$

NOW TRY ANSWERS
8. $a = 2S - b - c$
9. (a) $y = -5x + 3$
   (b) $y = \frac{1}{2}x - 4$

Although we could have given our answer as $y = \frac{3x + 6}{2}$, we simplified further in preparation for our work in **Chapter 3.**

NOW TRY

2.5 Exercises

   FOR EXTRA HELP ▶ MyMathLab®

▶ *Complete solution available in MyMathLab*

Concept Check Decide whether perimeter or area would be used to solve a problem concerning the measure of the quantity.

1. Carpeting for a bedroom

2. Sod for a lawn

3. Fencing for a yard

4. Baseboards for a living room

5. Tile for a bathroom

6. Fertilizer for a garden

7. Determining the cost of replacing a linoleum floor with a wood floor

8. Determining the cost of planting rye grass in a lawn for the winter

A formula is given along with the values of all but one of the variables. Find the value of the variable that is not given. Use 3.14 as an approximation for π (pi). **See Example 1.**

9. $P = 2L + 2W$ (perimeter of a rectangle); $L = 8, W = 5$

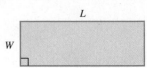

10. $P = 2L + 2W$; $L = 6, W = 4$

11. $A = \dfrac{1}{2}bh$ (area of a triangle); $b = 8, h = 16$

12. $A = \dfrac{1}{2}bh$; $b = 10, h = 14$

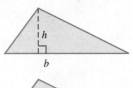

13. $P = a + b + c$ (perimeter of a triangle); $P = 12, a = 3, c = 5$

14. $P = a + b + c$; $P = 15, a = 3, b = 7$

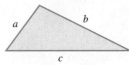

▶ **15.** $d = rt$ (distance formula); $d = 252, r = 45$

16. $d = rt$; $d = 100, t = 2.5$

17. $A = \dfrac{1}{2}h(b + B)$ (area of a trapezoid); $A = 91, h = 7, b = 12$

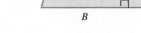

18. $A = \dfrac{1}{2}h(b + B)$; $A = 75, b = 19, B = 31$

19. $C = 2\pi r$ (circumference of a circle); $C = 16.328$

20. $C = 2\pi r$; $C = 8.164$

21. $C = 2\pi r$; $C = 20\pi$

22. $C = 2\pi r$; $C = 100\pi$

23. $A = \pi r^2$ (area of a circle); $r = 4$

24. $A = \pi r^2$; $r = 12$

25. $S = 2\pi rh$; $S = 120\pi, h = 10$

26. $S = 2\pi rh$; $S = 720\pi, h = 30$

*The **volume** of a three-dimensional geometric figure is a measure of the space occupied by the figure. For example, the volume of a gasoline tank determines how many gallons of gasoline it would take to completely fill the tank. Volume is measured in cubic units.*

In Exercises 27–32, a formula for the volume (V) of a three-dimensional object is given, along with values for the other variables. Evaluate V. (Use 3.14 as an approximation for π.) **See Example 1.**

27. $V = LWH$ (volume of a rectangular box); $L = 10, W = 5, H = 3$

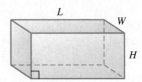

28. $V = LWH$; $L = 12, W = 8, H = 4$

29. $V = \dfrac{1}{3}Bh$ (volume of a pyramid); $B = 12, h = 13$

30. $V = \dfrac{1}{3}Bh$; $B = 36, h = 4$

31. $V = \dfrac{4}{3}\pi r^3$ (volume of a sphere); $r = 12$

32. $V = \dfrac{4}{3}\pi r^3$; $r = 6$

Simple interest I in dollars is calculated using the formula

$$I = prt,$$ Simple interest formula

where p represents principal, or amount, in dollars that is invested or borrowed, r represents annual interest rate, expressed as a decimal, and t represents time, in years.

   *In Exercises 33–38, find the value of the remaining variable in the simple interest formula. **See Example 1.** (Hint: Write percents as decimals. **See Section R.2.**)*

33. $p = \$7500, r = 4\%, t = 2$ yr

34. $p = \$3600, r = 3\%, t = 4$ yr

35. $I = \$33, r = 2\%, t = 3$ yr

36. $I = \$270, r = 5\%, t = 6$ yr

37. $I = \$180, p = \$4800, r = 2.5\%$

38. $I = \$162, p = \$2400, r = 1.5\%$

*Solve each problem. **See Examples 2 and 3.***

39. The length of a rectangle is 9 in. more than the width. The perimeter is 54 in. Find the length and the width of the rectangle.

40. The width of a rectangle is 3 ft less than the length. The perimeter is 62 ft. Find the length and the width of the rectangle.

▶ **41.** The perimeter of a rectangle is 36 m. The length is 2 m more than three times the width. Find the length and the width of the rectangle.

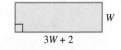

42. The perimeter of a rectangle is 36 yd. The width is 18 yd less than twice the length. Find the length and the width of the rectangle.

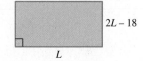

▶ **43.** The longest side of a triangle is 3 in. longer than the shortest side. The medium side is 2 in. longer than the shortest side. If the perimeter of the triangle is 20 in., what are the lengths of the three sides?

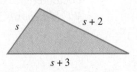

44. The perimeter of a triangle is 28 ft. The medium side is 4 ft longer than the shortest side, while the longest side is twice as long as the shortest side. What are the lengths of the three sides?

45. Two sides of a triangle have the same length. The third side measures 4 m less than twice that length. The perimeter of the triangle is 24 m. Find the lengths of the three sides.

46. A triangle is such that its medium side is twice as long as its shortest side and its longest side is 7 yd less than three times its shortest side. The perimeter of the triangle is 47 yd. What are the lengths of the three sides?

Use a formula to write an equation, and then solve each problem. (Use 3.14 as an approximation for π.) **Formulas are found inside the back cover of this book. See Examples 2–4.**

47. One of the largest fashion catalogues in the world was published in Hamburg, Germany. Each of the 212 pages in the catalogue measured 1.2 m by 1.5 m. What was the perimeter of a page? What was the area? (*Source: Guinness World Records.*)

48. One of the world's largest mandalas (sand paintings) measures 12.24 m by 12.24 m. What is the perimeter of the sand painting? To the nearest hundredth of a square meter, what is the area? (*Source: Guinness World Records.*)

▶ **49.** The area of a triangular road sign is 70 ft². If the base of the sign measures 14 ft, what is the height of the sign?

50. The area of a triangular advertising banner is 96 ft². If the height of the banner measures 12 ft, what is the measure of the base?

51. A prehistoric ceremonial site dating to about 3000 B.C. was discovered in southwestern England. The site is a nearly perfect circle, consisting of nine concentric rings that probably held upright wooden posts. Around this timber temple is a wide, encircling ditch enclosing an area with a diameter of 443 ft. Find this enclosed area to the nearest thousand square feet. (*Hint:* Find the radius. Then use $A = \pi r^2$.) (*Source: Archaeology.*)

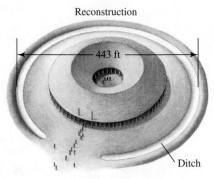

Reconstruction

52. The Rogers Centre in Toronto, Canada, is the first stadium with a hard-shell, retractable roof. The steel dome is 630 ft in diameter. To the nearest foot, what is the circumference of this dome? (*Source:* www.ballparks.com)

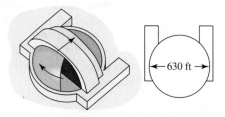

53. One of the largest drums ever constructed was made from Japanese cedar and cowhide, with radius 7.87 ft. What was the area of the circular face of the drum? What was the circumference of the drum? Round answers to the nearest hundredth. (*Source: Guinness World Records.*)

54. A drum played at the Royal Festival Hall in London had radius 6.5 ft. What was the area of the circular face of the drum? What was the circumference of the drum? (*Source: Guinness World Records.*)

55. The survey plat depicted here shows two lots that form a trapezoid. The measures of the parallel sides are 115.80 ft and 171.00 ft. The height of the trapezoid is 165.97 ft. Find the combined area of the two lots. Round the answer to the nearest hundredth of a square foot.

56. Lot A in the survey plat is in the shape of a trapezoid. The parallel sides measure 26.84 ft and 82.05 ft. The height of the trapezoid is 165.97 ft. Find the area of Lot A. Round the answer to the nearest hundredth of a square foot.

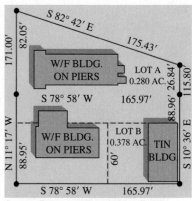

Source: Property survey in New Roads, Louisiana.

57. The U.S. Postal Service requires that any box sent by Priority Mail® have length plus girth (distance around) totaling no more than 108 in. The maximum volume that meets this condition is contained by a box with a square end 18 in. on each side. What is the length of the box? What is the maximum volume? (*Source:* United States Postal Service.)

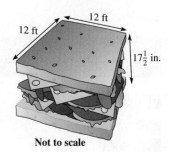

58. One of the world's largest sandwiches, made by Wild Woody's Chill and Grill in Roseville, Michigan, was 12 ft long, 12 ft wide, and $17\frac{1}{2}$ in. $\left(1\frac{11}{24}\text{ ft}\right)$ thick. What was the volume of the sandwich? (*Source:* Guinness World Records.)

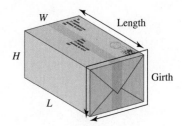

Not to scale

Find the measure of each marked angle. See Example 5.

▶ 59.
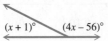
$(x + 1)°$ $(4x - 56)°$

60.
$(10x + 7)°$ $(7x + 3)°$

61.
$(8x - 1)°$
$(5x)°$

62.
$(4x)°$
$(3x + 13)°$

63.
$(5x - 129)°$ $(2x - 21)°$

64.
$(3x + 45)°$ $(7x + 5)°$

65.
$(10x + 15)°$

$(12x - 3)°$

66.
$(11x - 37)°$
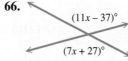
$(7x + 27)°$

Solve each formula for the specified variable. See Examples 6–8.

▶ 67. $d = rt$ for t **68.** $d = rt$ for r **69.** $\mathcal{A} = bh$ for b

70. $\mathcal{A} = LW$ for L **71.** $C = \pi d$ for d **72.** $P = 4s$ for s

73. $V = LWH$ for H **74.** $V = LWH$ for W **75.** $I = prt$ for r

76. $I = prt$ for p **77.** $\mathcal{A} = \dfrac{1}{2}bh$ for h **78.** $\mathcal{A} = \dfrac{1}{2}bh$ for b

79. $V = \dfrac{1}{3}\pi r^2 h$ for h **80.** $V = \pi r^2 h$ for h **81.** $P = a + b + c$ for b

82. $P = a + b + c$ for a ▶ **83.** $P = 2L + 2W$ for W **84.** $A = p + prt$ for r

▶ **85.** $y = mx + b$ for m **86.** $y = mx + b$ for x **87.** $Ax + By = C$ for y

88. $Ax + By = C$ for x ▶ **89.** $M = C(1 + r)$ for r **90.** $A = p(1 + rt)$ for t

91. $P = 2(a + b)$ for a **92.** $P = 2(a + b)$ for b **93.** $S = \dfrac{1}{2}(a + b + c)$ for b

94. $S = \dfrac{1}{2}(a + b + c)$ for c **95.** $C = \dfrac{5}{9}(F - 32)$ for F **96.** $\mathcal{A} = \dfrac{1}{2}h(b + B)$ for b

Solve each equation for y. **See Example 9.**

97. $6x + y = 4$ **98.** $3x + y = 6$ **99.** $5x - y = 2$ **100.** $4x - y = 1$

101. $-3x + 5y = -15$ **102.** $-2x + 3y = -9$ **103.** $x - 3y = 12$ **104.** $x - 5y = 10$

2.6   Ratio, Proportion, and Percent

OBJECTIVES

1. Write ratios.
2. Solve proportions.
3. Solve applied problems using proportions.
4. Find percents and percentages.

VOCABULARY

☐ ratio
☐ proportion
☐ terms of a proportion
☐ extremes
☐ means
☐ cross products of a proportion
☐ percent
☐ percentage
☐ base

OBJECTIVE 1 Write ratios.

A **ratio** is a comparison of two quantities using a quotient.

> **Ratio**
>
> The ratio of a number a to a number b (where $b \neq 0$) is written as follows.
>
> $$a \text{ to } b, \quad a:b, \quad \text{or} \quad \dfrac{a}{b}$$
>
> Writing a ratio as a quotient $\dfrac{a}{b}$ is most common in algebra.
>
> *Examples:*   2 to 3,   2:3, $\dfrac{2}{3}$

EXAMPLE 1   Writing Word Phrases as Ratios

Write a ratio for each word phrase. Express fractions in lowest terms.

(a) 5 hr to 3 hr $\dfrac{5 \text{ hr}}{3 \text{ hr}} = \dfrac{5}{3}$

(b) 6 hr to 3 days

First convert 3 days to hours.

$$3 \text{ days} = 3 \cdot 24 = 72 \text{ hr} \qquad \text{1 day} = 24 \text{ hr}$$

Now write the ratio using the common unit of measure, hours.

$$\dfrac{6 \text{ hr}}{3 \text{ days}} = \dfrac{6 \text{ hr}}{72 \text{ hr}} = \dfrac{6}{72} = \dfrac{1}{12} \qquad \text{Write in lowest terms.} \qquad \textbf{NOW TRY}$$

**NOW TRY
EXERCISE 1**

Write a ratio for each word phrase. Express fractions in lowest terms.

(a) 7 in. to 4 in.

(b) 45 sec to 2 min

 **NOW TRY
EXERCISE 2**

A supermarket charges the following prices for a certain brand of liquid detergent.

Size	Price
75 oz	$ 8.94
100 oz	$13.97
150 oz	$19.97

Which size is the best buy? What is the unit price (to the nearest thousandth) for that size?

Applications of ratios occur regularly in everyday life. For example, automobile manufacturers report "miles per gallon" (abbreviated mpg) for their vehicles. This is a ratio found by dividing number of miles driven by number of gallons of gasoline used.

Another application of ratios is in *unit pricing,* to see which size of an item offered in different sizes produces the best price per unit.

EXAMPLE 2 Finding Price per Unit

A Jewel-Osco supermarket charges the following prices for a jar of extra crunchy peanut butter.

Peanut Butter

Size	Price
18 oz	$3.49
28 oz	$4.99
40 oz	$6.79

Which size is the best buy? That is, which size has the lowest unit price?

To find the best buy, write ratios comparing the price for each size jar to the number of units (ounces) per jar.

Peanut Butter

Size	Unit Price (dollars per ounce)	
18 oz	$\dfrac{\$3.49}{18} = \$0.194$	To find the price per ounce, the number of ounces goes in the denominator.
28 oz	$\dfrac{\$4.99}{28} = \$0.178$	(Results are rounded to the nearest thousandth.)
40 oz	$\dfrac{\$6.79}{40} = \$0.170$	← Best buy

Because the 40-oz size produces the lowest unit price, it is the best buy. Buying the largest size does not always provide the best buy, although it often does, as in this case.

NOW TRY

OBJECTIVE 2 Solve proportions.

A ratio is used to compare two numbers or amounts. A **proportion** says that two ratios are equal. For example, the proportion

$$\frac{3}{4} = \frac{15}{20}$$

A proportion is a special type of equation.

says that the ratios $\frac{3}{4}$ and $\frac{15}{20}$ are equal. In the proportion

$$\frac{a}{b} = \frac{c}{d} \quad (\text{where } b, d \neq 0),$$

$a, b, c,$ and d are the **terms** of the proportion. The terms a and d are the **extremes,** and the terms b and c are the **means.** We read the proportion $\frac{a}{b} = \frac{c}{d}$ as

"a is to b as c is to d."

NOW TRY ANSWERS

1. **(a)** $\frac{7}{4}$ **(b)** $\frac{3}{8}$

2. 75 oz; $0.119 per oz

Multiplying each side of this proportion by the common denominator, bd, gives the following.

$$\frac{a}{b} = \frac{c}{d}$$

$$bd \cdot \frac{a}{b} = bd \cdot \frac{c}{d} \qquad \text{Multiply each side by } bd.$$

$$\frac{b}{b}(d \cdot a) = \frac{d}{d}(b \cdot c) \qquad \text{Associative and commutative properties}$$

$$ad = bc \qquad \text{Commutative and identity properties}$$

We can also find the products ad and bc by multiplying diagonally.

$$ad = bc$$

$$\frac{a}{b} \diagdown \frac{c}{d}$$

For this reason, ad and bc are the **cross products of the proportion.**

Cross Products of a Proportion

If $\frac{a}{b} = \frac{c}{d}$, then the cross products ad and bc are equal—that is, **the product of the extremes equals the product of the means.**

Also, if $ad = bc$, then $\frac{a}{b} = \frac{c}{d}$ (where $b, d \neq 0$).

NOTE If $\frac{a}{c} = \frac{b}{d}$, then $ad = cb$, or $ad = bc$. This means that the two proportions are equivalent, and the proportion

$$\frac{a}{b} = \frac{c}{d} \quad \text{can also be written as} \quad \frac{a}{c} = \frac{b}{d} \quad \text{(where } c, d \neq 0\text{)}.$$

Sometimes one form is more convenient to work with than the other.

NOW TRY
EXERCISE 3

Decide whether each proportion is *true* or *false*.

(a) $\frac{1}{3} = \frac{33}{100}$ **(b)** $\frac{4}{13} = \frac{16}{52}$

EXAMPLE 3 Deciding Whether Proportions Are True

Decide whether each proportion is *true* or *false*.

(a) $\frac{3}{4} = \frac{15}{20}$

Check to see whether the cross products are equal.

$$4 \cdot 15 = 60$$

$$\frac{3}{4} \diagdown \frac{15}{20}$$

$$3 \cdot 20 = 60$$

The cross products are equal, so the proportion is true.

(b) $\frac{6}{7} = \frac{30}{32}$

The cross products, $6 \cdot 32 = 192$ and $7 \cdot 30 = 210$, are not equal, so the proportion is false.

NOW TRY

Four numbers are used in a proportion. If any three of these numbers are known, the fourth can be found.

NOW TRY EXERCISE 4

Solve.

$$\frac{9}{7} = \frac{x}{56}$$

EXAMPLE 4 Finding an Unknown in a Proportion

Solve $\frac{5}{9} = \frac{x}{63}$.

$$\frac{5}{9} = \frac{x}{63}$$ Solve for *x*.

$5 \cdot 63 = 9 \cdot x$ Cross products must be equal.

$315 = 9x$ Multiply.

$35 = x$ Divide by 9.

Check by substituting 35 for *x* in the proportion. The solution set is $\{35\}$.

NOW TRY

NOW TRY EXERCISE 5

Solve.

$$\frac{k - 3}{6} = \frac{3k + 2}{4}$$

EXAMPLE 5 Solving an Equation Using Cross Products

Solve $\frac{m - 2}{5} = \frac{m + 1}{3}$.

$$\frac{m - 2}{5} = \frac{m + 1}{3}$$ Be sure to use parentheses.

$3(m - 2) = 5(m + 1)$ (*) Cross products

$3m - 6 = 5m + 5$ Distributive property

$-2m - 6 = 5$ Subtract 5*m*.

$-2m = 11$ Add 6.

$m = -\dfrac{11}{2}$ Divide by -2.

The solution set is $\left\{-\dfrac{11}{2}\right\}$.

NOW TRY

NOTE When we set cross products equal to each other, we are actually multiplying each ratio in the proportion by a common denominator.

$$\frac{m - 2}{5} = \frac{m + 1}{3}$$ See **Example 5**.

$$15\left(\frac{m - 2}{5}\right) = 15\left(\frac{m + 1}{3}\right)$$ Multiply each ratio by 15, the LCD.

$$15\left(\frac{m-2}{5}\right)$$
$$= 15 \cdot \frac{1}{5}(m - 2)$$
$$= 3(m - 2)$$

$3(m - 2) = 5(m + 1)$ This is equation (*) from **Example 5**.

⚠ **CAUTION** *The cross-product method cannot be used directly if there is more than one term on either side of the equality symbol.*

$$\underbrace{\frac{m - 1}{5} = \frac{m + 1}{3} - 4}_{\text{2 terms}}, \qquad \underbrace{\frac{x}{3} + \frac{5}{4} = \frac{1}{2}}_{\text{2 terms}}$$

Do *not* use the cross-product method to solve equations in this form.

NOW TRY ANSWERS
4. $\{72\}$ 5. $\left\{-\frac{12}{7}\right\}$

OBJECTIVE 3 Solve applied problems using proportions.

EXAMPLE 6   Applying Proportions

NOW TRY
EXERCISE 6

Twenty gallons of gasoline costs $69.80. How much would 27 gal of the same gasoline cost?

After Lee Ann pumped 5.0 gal of gasoline, the display showing the price read $18.10. When she finished pumping the gasoline, the price display read $52.49. How many gallons did she pump?

To solve this problem, set up a proportion, with prices in the numerators and gallons in the denominators. Let x = the number of gallons she pumped.

$$\text{Price} \longrightarrow \frac{\$18.10}{5.0} = \frac{\$52.49}{x} \longleftarrow \text{Price}$$
$$\text{Gallons} \longrightarrow \qquad\qquad \longleftarrow \text{Gallons}$$

> Be sure that numerators represent the *same* quantities and denominators represent the *same* quantities.

$$18.10x = 5.0(52.49) \qquad \text{Cross products}$$
$$18.10x = 262.45 \qquad \text{Multiply.}$$
$$x = 14.5 \qquad \text{Divide by 18.10.}$$

She pumped 14.5 gal. Check this answer. Notice that the way the proportion was set up uses the fact that the unit price is the same, no matter how many gallons are purchased.

NOW TRY

OBJECTIVE 4 Find percents and percentages.

A percent is a ratio where the second number is always 100. For example,

50% represents the ratio of 50 to 100, that is, $\frac{50}{100}$, or, as a decimal, 0.50.

27% represents the ratio of 27 to 100, that is, $\frac{27}{100}$, or, as a decimal, 0.27.

The word **percent** means **"per 100."** One percent means "one per 100."

$$1\% = 0.01, \quad \text{or} \quad 1\% = \frac{1}{100} \qquad \begin{array}{l}\text{Percent, decimal, and fraction}\\ \text{equivalents (Section R.2)}\end{array}$$

We can solve a percent problem involving $x\%$ by writing it as a proportion. The amount, or **percentage,** is compared to the **base** (the whole amount).

$$\frac{amount}{base} = \frac{x}{100}$$

We can also write this proportion as follows.

$$\frac{amount}{base} = \text{percent (as a decimal)} \qquad \begin{array}{l}\frac{x}{100} \text{ or } 0.01x \text{ is equivalent}\\ \text{to } x \text{ percent.}\end{array}$$

$$\textbf{amount} = \textbf{percent (as a decimal)} \cdot \textbf{base} \qquad \text{Basic percent equation}$$

EXAMPLE 7   Solving Percent Equations

Solve each problem.

(a) What is 15% of 600?

Let n = the number. The word *of* indicates multiplication.

$$\text{What} \quad \text{is} \quad 15\% \quad \text{of} \quad 600? \qquad \boxed{\begin{array}{l}\text{Translate each word or}\\ \text{phrase to write the equation.}\end{array}}$$
$$\downarrow \qquad \downarrow \qquad \downarrow \qquad \downarrow \qquad \downarrow$$
$$n \quad = \quad 0.15 \quad \cdot \quad 600 \qquad \text{Write the percent equation.}$$
$$n = 90 \qquad \boxed{\begin{array}{l}\text{Write 15\% as}\\ \text{a decimal.}\end{array}} \qquad \text{Multiply.}$$

Thus, 90 is 15% of 600.

NOW TRY ANSWER
6. $94.23

NOW TRY
EXERCISE 7

Solve each problem.

(a) What is 20% of 70?

(b) 40% of what number is 130?

(c) 121 is what percent of 484?

(b) 32% of what number is 64?

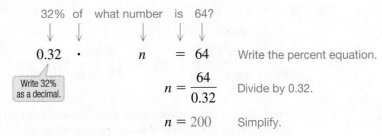

$$0.32 \cdot n = 64 \qquad \text{Write the percent equation.}$$

$$n = \frac{64}{0.32} \qquad \text{Divide by 0.32.}$$

$$n = 200 \qquad \text{Simplify.}$$

32% of 200 is 64.

(c) 90 is what percent of 360?

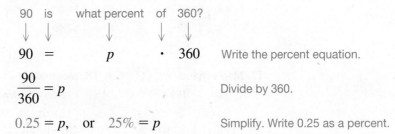

$$90 = p \cdot 360 \qquad \text{Write the percent equation.}$$

$$\frac{90}{360} = p \qquad \text{Divide by 360.}$$

$$0.25 = p, \quad \text{or} \quad 25\% = p \qquad \text{Simplify. Write 0.25 as a percent.}$$

Thus, 90 is 25% of 360.

NOW TRY

NOW TRY
EXERCISE 8

A winter coat is on a clearance sale for $48. The regular price is $120. What percent of the regular price is the savings?

EXAMPLE 8 **Solving an Applied Percent Problem**

A newspaper ad offered a set of tires at a sales price of $258. The regular price was $300. What percent of the regular price was the savings?

The savings on the tires amounted to $300 − $258 = $42. We can now restate the problem: What percent of 300 is 42?

What percent of 300 is 42?

$$p \cdot 300 = 42 \qquad \text{Write the percent equation.}$$

$$p = \frac{42}{300} \qquad \text{Divide by 300.}$$

$$p = 0.14, \quad \text{or} \quad 14\% \qquad \text{Simplify. Write 0.14 as a percent.}$$

The sale price represents a 14% savings.

NOW TRY

NOW TRY ANSWERS
7. (a) 14 **(b)** 325 **(c)** 25%
8. 60%

2.6 Exercises

FOR EXTRA HELP   ▶ MyMathLab®

▶ *Complete solution available in MyMathLab*

1. *Concept Check* Match each ratio in Column I with the ratio equivalent to it in Column II.

I	II
(a) 75 to 100	**A.** 80 to 100
(b) 5 to 4	**B.** 50 to 100
(c) $\frac{1}{2}$	**C.** 3 to 4
(d) 4 to 5	**D.** 15 to 12

2. *Concept Check* Which of the following represent a ratio of 4 days to 2 weeks?

A. $\frac{4}{2}$ **B.** $\frac{4}{7}$ **C.** $\frac{4}{14}$

D. $\frac{2}{1}$ **E.** $\frac{2}{7}$ **F.** $\frac{1}{2}$

G. $\frac{2}{4}$ **H.** $\frac{7}{2}$ **I.** 2

Write a ratio for each word phrase. Express fractions in lowest terms. ***See Example 1.***

▶ **3.** 40 mi to 30 mi
4. 60 ft to 70 ft

5. 120 people to 90 people
6. 72 dollars to 220 dollars

▶ **7.** 20 yd to 8 ft
8. 30 in. to 8 ft
9. 24 min to 2 hr

10. 16 min to 1 hr
11. 60 in. to 2 yd
12. 720 sec to 1 hr

Find the best buy for each item. Give the unit price to the nearest thousandth for that size. ***See Example 2.*** (*Source:* Various grocery stores.)

13. Granulated Sugar

Size	Price
4 lb	$3.29
10 lb	$7.49

14. Applesauce

Size	Price
23 oz	$1.99
48 oz	$3.49

15. Orange Juice

Size	Price
64 oz	$2.99
89 oz	$4.79
128 oz	$6.49

16. Salad Dressing

Size	Price
8 oz	$1.69
16 oz	$1.97
36 oz	$5.99

17. Maple Syrup

Size	Price
8.5 oz	$5.79
12.5 oz	$7.99
32 oz	$16.99

18. Mouthwash

Size	Price
16.9 oz	$3.39
33.8 oz	$3.49
50.7 oz	$5.29

19. Tomato Ketchup

Size	Price
32 oz	$1.79
36 oz	$2.69
40 oz	$2.49
64 oz	$4.38

20. Grape Jelly

Size	Price
12 oz	$1.05
18 oz	$1.73
32 oz	$1.84
48 oz	$2.88

21. Laundry Detergent

Size	Price
87 oz	$7.88
131 oz	$10.98
263 oz	$19.96

22. Spaghetti Sauce

Size	Price
14 oz	$1.79
24 oz	$1.77
48 oz	$3.65

Decide whether each proportion is true *or* false. ***See Example 3.***

▶ **23.** $\dfrac{5}{35} = \dfrac{8}{56}$
24. $\dfrac{4}{12} = \dfrac{7}{21}$
25. $\dfrac{120}{82} = \dfrac{7}{10}$

26. $\dfrac{27}{160} = \dfrac{18}{110}$
27. $\dfrac{\frac{1}{2}}{5} = \dfrac{1}{10}$
28. $\dfrac{\frac{1}{3}}{6} = \dfrac{1}{18}$

Solve each equation. ***See Examples 4 and 5.***

▶ **29.** $\dfrac{k}{4} = \dfrac{175}{20}$
30. $\dfrac{x}{6} = \dfrac{18}{4}$
31. $\dfrac{49}{56} = \dfrac{z}{8}$
32. $\dfrac{20}{100} = \dfrac{z}{80}$

33. $\dfrac{x}{24} = \dfrac{15}{16}$
34. $\dfrac{x}{4} = \dfrac{12}{30}$
35. $\dfrac{z}{2} = \dfrac{z+1}{3}$

36. $\dfrac{m}{5} = \dfrac{m-2}{2}$
▶ **37.** $\dfrac{3y-2}{5} = \dfrac{6y-5}{11}$
38. $\dfrac{2r+8}{4} = \dfrac{3r-9}{3}$

39. $\dfrac{5k+1}{6} = \dfrac{3k-2}{3}$
40. $\dfrac{x+4}{6} = \dfrac{x+10}{8}$
41. $\dfrac{2p+7}{3} = \dfrac{p-1}{4}$

42. $\dfrac{3m-2}{5} = \dfrac{4-m}{3}$
43. $\dfrac{2(x-4)}{3} = \dfrac{4(x-3)}{5}$
44. $\dfrac{9(x-3)}{6} = \dfrac{6(x-2)}{2}$

Solve each problem. (Assume that all items are equally priced.) ***See Example 6.***

45. If 16 candy bars cost $20.00, how much do 24 candy bars cost?

46. If 12 ring tones cost $30.00, how much do 8 ring tones cost?

47. Eight quarts of oil cost $14.00. How much do 5 qt of oil cost?

48. Four tires cost $398.00. How much do 7 tires cost?

49. If 9 pairs of jeans cost $121.50, find the cost of 5 pairs.

50. If 7 shirts cost $87.50, find the cost of 11 shirts.

51. If 6 gal of premium unleaded gasoline costs $22.56, how much would it cost to completely fill a 15-gal tank?

52. If sales tax on a $16.00 DVD is $1.32, find the sales tax on a $120.00 DVD player.

Solve each problem. (In Exercises 57–60, round answers to the nearest tenth.) ***See Example 6.***

53. Biologists tagged 500 fish in North Bay. At a later date, they found 7 tagged fish in a sample of 700. Estimate the total number of fish in North Bay to the nearest hundred.

54. Researchers at West Okoboji Lake tagged 840 fish. A later sample of 1000 fish contained 18 that were tagged. Approximate the fish population in West Okoboji Lake to the nearest hundred.

55. The distance between Kansas City, Missouri, and Denver is 600 mi. On a certain wall map, this is represented by a length of 2.4 ft. On the map, how many feet would there be between Memphis and Philadelphia, two cities that are actually 1000 mi apart?

56. The distance between Singapore and Tokyo is 3300 mi. On a certain wall map, this distance is represented by 11 in. The actual distance between Mexico City and Cairo is 7700 mi. How far apart are they on the same map?

57. A wall map of the United States has a distance of 8.5 in. between Memphis and Denver, two cities that are actually 1040 mi apart. The actual distance between St. Louis and Des Moines is 333 mi. How far apart are St. Louis and Des Moines on the map?

58. A wall map of the United States has a distance of 8.0 in. between New Orleans and Chicago, two cities that are actually 912 mi apart. The actual distance between Milwaukee and Seattle is 1940 mi. How far apart are Milwaukee and Seattle on the map?

59. On a world globe, the distance between Capetown and Bangkok, two cities that are actually 10,080 km apart, is 12.4 in. The actual distance between Moscow and Berlin is 1610 km. How far apart are Moscow and Berlin on this globe?

60. On a world globe, the distance between Rio de Janeiro and Hong Kong, two cities that are actually 17,615 km apart, is 21.5 in. The actual distance between Paris and Stockholm is 1605 km. How far apart are Paris and Stockholm on this globe?

61. According to the directions on a bottle of Armstrong® Concentrated Floor Cleaner, for routine cleaning, $\frac{1}{4}$ cup of cleaner should be mixed with 1 gal of warm water. How much cleaner should be mixed with $10\frac{1}{2}$ gal of water?

62. The directions on the bottle mentioned in **Exercise 61** also specify that, for extra-strength cleaning, $\frac{1}{2}$ cup of cleaner should be used for each gallon of water. How much cleaner should be mixed with $15\frac{1}{2}$ gal of water for extra-strength cleaning?

63. On September 23, 2013, the exchange rate between euros and U.S. dollars was 1 euro to $1.3492. Ashley went to Rome and exchanged her U.S. currency for euros, receiving 300 euros. How much in U.S. dollars did she exchange? (*Source:* www.xe.com/ucc)

64. If 8 U.S. dollars can be exchanged for 103.0 Mexican pesos, how many pesos can be obtained for $65? (Round to the nearest tenth.)

*Two triangles are **similar** if they have the same shape (but not necessarily the same size). Similar triangles have sides that are proportional. The figure shows two similar triangles. Notice that the ratios of the corresponding sides all equal $\frac{3}{2}$.*

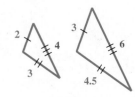

$$\frac{3}{2} = \frac{3}{2}, \quad \frac{4.5}{3} = \frac{3}{2}, \quad \frac{6}{4} = \frac{3}{2}$$

If we know that two triangles are similar, we can set up a proportion to solve for the length of an unknown side.

    Find the lengths x and y as needed in each pair of similar triangles.

65.

66.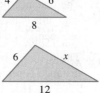

67.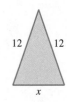

68.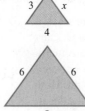

69.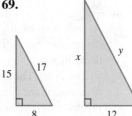

70.

*For Exercises 71 and 72, **(a)** draw a sketch consisting of two right triangles depicting the situation described, and **(b)** solve the problem. (Source: Guinness World Records.)*

71. An enlarged version of the chair used by George Washington at the Constitutional Convention casts a shadow 18 ft long at the same time a vertical pole 12 ft high casts a shadow 4 ft long. How tall is the chair?

72. One of the tallest candles ever constructed was exhibited at the 1897 Stockholm Exhibition. If it cast a shadow 5 ft long at the same time a vertical pole 32 ft high cast a shadow 2 ft long, how tall was the candle?

The Consumer Price Index (CPI) provides a means of determining the purchasing power of the U.S. dollar from one year to the next. Using the period from 1982 to 1984 as a measure of 100.0, the CPI for selected years from 1999 through 2011 is shown in the table. To use the CPI to predict a price in a particular year, we set up a proportion and compare it with a known price in another year.

$$\frac{\text{price in year } A}{\text{index in year } A} = \frac{\text{price in year } B}{\text{index in year } B}$$

Year	Consumer Price Index
1999	166.6
2001	177.1
2003	184.0
2005	195.3
2007	207.3
2009	214.5
2011	224.9

Source: U.S. Bureau of Labor Statistics.

Use the CPI figures in the table to find the amount that would be charged for using the same amount of electricity that cost $225 in 1999. Give answers to the nearest dollar.

73. in 2001   **74.** in 2003   **75.** in 2007   **76.** in 2011

Children are often given antibiotics in liquid form, called an oral suspension. Pharmacists make up these suspensions by mixing medication in powder form with water. They use proportions to calculate the volume of the suspension for the amount of medication that has been prescribed. In Exercises 77 and 78, do each of the following.

(a) *Find the total amount of medication in milligrams to be given over the full course of treatment.*

(b) *Write a proportion that can be solved to find the total volume of the liquid suspension that the pharmacist will prepare. Use x as the variable.*

(c) *Solve the proportion to determine the total volume of the oral suspension.*

77. Logan's pediatric nurse practitioner has prescribed 375 mg of Amoxil a day for 7 days to treat his ear infection. The pharmacist uses 125 mg of Amoxil in each 5 mL of the suspension. (*Source:* www.drugs.com)

78. An Amoxil oral suspension can also be made by using 250 mg for each 5 mL of suspension. Ava's pediatrician prescribed 900 mg a day for 10 days to treat her bronchitis. (*Source:* www.drugs.com)

Solve each problem. ***See Examples 7 and 8.***

79. What is 18% of 780?

80. What is 23% of 480?

81. 42% of what number is 294?

82. 18% of what number is 108?

83. 120% of what number is 510?

84. 140% of what number is 315?

85. 4 is what percent of 50?

86. 8 is what percent of 64?

87. What percent of 30 is 36?

88. What percent of 48 is 96?

89. Clayton earned 48 points on a 60-point geometry project. What percent of the total points did he earn?

90. On a 75-point algebra test, Grady scored 63 points. What percent of the total points did he score?

91. A laptop computer that has a regular price of $700 is on sale for $504. What percent of the regular price is the savings?

92. An all-in-one desktop computer that has a regular price of $980 is on sale for $833. What percent of the regular price is the savings?

93. Tyler has a monthly income of $1500. His rent is $480 per month. What percent of his monthly income is his rent?

94. Lily has a monthly income of $2200. She has budgeted $154 per month for entertainment. What percent of her monthly income did she budget for entertainment?

95. Anna saved $1950, which was 65% of the amount she needed for a used car. What was the total amount she needed for the car?

96. Bryn had $525, which was 70% of the total amount she needed for a deposit on an apartment. What was the total deposit she needed?

2.7 Further Applications of Linear Equations

OBJECTIVES

1. Use percent in solving problems involving rates.
2. Solve problems involving mixtures.
3. Solve problems involving simple interest.
4. Solve problems involving denominations of money.
5. Solve problems involving distance, rate, and time.

OBJECTIVE 1 Use percent in solving problems involving rates.

Recall from **Sections R.2 and 2.6** that the word "percent" means "per 100." One percent means "one per 100."

$$1\% = 0.01, \quad \text{or} \quad 1\% = \frac{1}{100} \quad \text{Percent, decimal, and fraction equivalents}$$

PROBLEM-SOLVING HINT Mixing different concentrations of a substance or different interest rates involves percents. To obtain the amount of pure substance or the interest, we multiply as follows.

Mixture Problems	Interest Problems (annual)
base · rate (%) = percentage	principal · rate (%) = interest
$b \cdot r = p$	$p \cdot r = I$

In an equation, percent is always written as a decimal (or a fraction).

NOW TRY EXERCISE 1

(a) How much pure alcohol is in 70 L of a 20% alcohol solution?

(b) Find the annual simple interest if $3200 is invested at 2%.

EXAMPLE 1 Using Percents to Find Percentages

(a) If a chemist has 40 L of a 35% acid solution, then the amount of pure acid in the solution is found by multiplying.

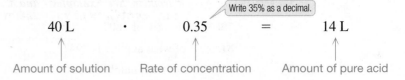

Write 35% as a decimal.

40 L · 0.35 = 14 L

Amount of solution · Rate of concentration = Amount of pure acid

(b) If $1300 is invested for one year at 3% simple interest, the amount of interest earned in the year is calculated as follows.

3% = 0.03, *not* 0.30

$1300 · 0.03 = $39

Principal · Interest rate · Interest earned · **NOW TRY**

PROBLEM-SOLVING HINT In the applications that follow, using a table helps organize the information in a problem and more easily set up an equation, which is usually the most difficult step.

OBJECTIVE 2 Solve problems involving mixtures.

EXAMPLE 2 Solving a Mixture Problem

A chemist mixes 20 L of a 40% acid solution with some 70% acid solution to obtain a mixture that is 50% acid. How many liters of the 70% acid solution should she use?

Step 1   **Read** the problem. Note the percent of each solution and of the mixture.

Step 2   **Assign a variable** to represent the unknown quantity.

Let $x =$ the number of liters of 70% acid solution needed.

As in **Example 1(a),** the amount of pure acid in this solution is the product of the percent of strength and the number of liters of solution.

$$0.70x \qquad \text{Liters of pure acid in } x \text{ liters of 70\% solution}$$

The amount of pure acid in the 20 L of 40% solution is found similarly.

$$0.40(20) \qquad \text{Liters of pure acid in the 40\% solution}$$

The new solution will contain $(x + 20)$ liters of 50% solution. The amount of pure acid in this solution is again found by multiplying.

$$0.50(x + 20) \qquad \text{Liters of pure acid in the 50\% solution}$$

FIGURE 17 illustrates this information, which is organized in the table.

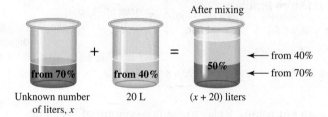

Liters of Solution	Rate (as a decimal)	Liters of Pure Acid	
x	0.70	$0.70x$	← Sum
20	0.40	$0.40(20)$	← must equal
$x + 20$	0.50	$0.50(x + 20)$	←

FIGURE 17

Step 3   **Write an equation.** The number of liters of pure acid in the 70% solution added to the number of liters of pure acid in the 40% solution will equal the number of liters of pure acid in the final mixture.

Pure acid in 70% solution	plus	pure acid in 40% solution	is	pure acid in 50% solution.
↓	↓	↓	↓	↓
$0.70x$	$+$	$0.40(20)$	$=$	$0.50(x + 20)$

Refer to the last column of the table.

Step 4   **Solve.** First clear the parentheses. Then clear the decimals.

$0.70x = 0.7x$ and $0.50x = 0.5x.$

$$0.7x + 8 = 0.5x + 10 \qquad \text{Multiply; distributive property}$$
$$10(0.7x + 8) = 10(0.5x + 10) \qquad \text{Multiply by 10.}$$
$$7x + 80 = 5x + 100 \qquad \text{Distributive property}$$
$$2x + 80 = 100 \qquad \text{Subtract } 5x.$$
$$2x = 20 \qquad \text{Subtract 80.}$$
$$x = 10 \qquad \text{Divide by 2.}$$

Step 5   **State the answer.** The chemist needs to use 10 L of 70% solution.

NOW TRY
EXERCISE 2

A certain seasoning is 70% salt. How many ounces of this seasoning must be mixed with 30 oz of dried herbs containing 10% salt to obtain a seasoning that is 50% salt?

Step 6 **Check.** If 10 L of 70% solution are used, the amounts of pure acid are the same.

$$0.70(10) + 0.40(20) \quad \text{Sum of the two solutions}$$
$$= 7 + 8$$
$$= 15$$

$$0.50(10 + 20) \quad \text{Mixture}$$
$$= 0.50(30)$$
$$= 15 \quad \text{NOW TRY}$$

⚠ **CAUTION** In a mixture problem, the concentration of the final mixture must be *between* the concentrations of the two solutions making up the mixture.

NOW TRY
EXERCISE 3

How many liters of a 25% saline solution must be mixed with a 10% saline solution to obtain 15 L of a 15% solution?

EXAMPLE 3 Solving a Mixture Problem

How many ounces of a seasoning that is 15% pepper must be mixed with a version that is 30% pepper to obtain 9 oz of a seasoning that is 20% pepper?

Step 1 **Read** the problem. We are given the *total* amount of the mixture. We must find the amount of the seasoning that is 15% pepper.

Step 2 **Assign a variable.** Use the fact that the total mixture is 9 oz.

Let x = the number of ounces of seasoning that is 15% pepper.

Then $9 - x$ = the number of ounces of seasoning that is 30% pepper.

Ounces of Seasoning	Rate (as a decimal)	Ounces of Pepper
x	0.15	$0.15x$
$9 - x$	0.30	$0.30(9 - x)$
9	0.20	$0.20(9)$

Use a table to organize the given information.

Step 3 **Write an equation.** Refer to the last column of the table.

Pepper in 15% seasoning	plus	pepper in 30% seasoning	is	pepper in 20% seasoning.
↓	↓	↓	↓	↓
$0.15x$	$+$	$0.30(9 - x)$	$=$	$0.20(9)$

Step 4 **Solve.**

$$0.15x + 2.7 - 0.3x = 1.8 \quad \text{Distributive property; multiply.}$$

To multiply by 100, move the decimal point in each term 2 places to the right.

$$15x + 270 - 30x = 180 \quad \text{Multiply by 100.}$$
$$-15x + 270 = 180 \quad \text{Combine like terms.}$$
$$-15x = -90 \quad \text{Subtract 270.}$$
$$x = 6 \quad \text{Divide by } -15.$$

Step 5 **State the answer.** 6 oz of seasoning that is 15% pepper is needed. (This means that $9 - 6 = 3$ oz of the 30% pepper seasoning is needed, although the problem does not specifically ask for this amount.)

Step 6 **Check.** The ounces of pepper before and after mixing are the same.

$$0.15(6) + 0.30(9 - 6) \quad \text{Sum of the two seasonings}$$
$$= 0.9 + 0.9$$
$$= 1.8$$

$$0.20(9) \quad \text{Mixture}$$
$$= 1.8$$

NOW TRY

NOW TRY ANSWERS
2. 60 oz
3. 5 L

OBJECTIVE 3 Solve problems involving simple interest.

The formula for simple interest

$$I = prt \quad \text{becomes} \quad I = pr \quad \text{when time } t = 1 \text{ (for annual interest),}$$

as shown in the Problem-Solving Hint at the beginning of this section. Multiplying the total amount (principal) by the rate (rate of interest) gives the percentage (amount of interest).

NOW TRY
EXERCISE 4

A financial advisor invests some money in a municipal bond paying 3% annual interest and $5000 more than that amount in a certificate of deposit paying 4% annual interest. To earn $410 per year in interest, how much should he invest at each rate?

EXAMPLE 4 Solving a Simple Interest Problem

Susan plans to invest some money at 2% and $2000 more than this amount at 4%. To earn $380 per year in interest, how much should she invest at each rate?

Step 1 **Read** the problem. There will be two answers.

Step 2 **Assign a variable.**

Let $x =$ the amount invested at 2% (in dollars).

Then $x + 2000 =$ the amount invested at 4% (in dollars).

Amount Invested (in dollars)	Rate (as a decimal)	Interest for One Year (in dollars)
x	0.02	$0.02x$
$x + 2000$	0.04	$0.04(x + 2000)$

Use a table to organize the given information.

Step 3 **Write an equation.** Multiply amount by rate to obtain interest earned. The two amounts of interest must total $380.

Interest at 2%	plus	interest at 4%	is	total interest.
↓	↓	↓	↓	↓
$0.02x$	$+$	$0.04(x + 2000)$	$=$	380

Step 4 **Solve.**

$$0.02x + 0.04x + 80 = 380 \qquad \text{Distributive property}$$
$$2x + 4x + 8000 = 38{,}000 \qquad \text{Multiply by 100.}$$
$$6x + 8000 = 38{,}000 \qquad \text{Combine like terms.}$$
$$6x = 30{,}000 \qquad \text{Subtract 8000.}$$
$$x = 5000 \qquad \text{Divide by 6.}$$

Step 5 **State the answer.** At 2%, she should invest $5000. At 4%, she should invest

$$\$5000 + \$2000 = \$7000.$$

Step 6 **Check.** The sum of the two interest amounts is

$$0.02(\$5000) + 0.04(\$7000)$$
$$= \$100 + \$280$$
$$= \$380, \quad \text{as required.}$$

NOW TRY

OBJECTIVE 4   Solve problems involving denominations of money.

PROBLEM-SOLVING HINT   To obtain the total value in problems that involve different denominations of money or items with different monetary values, we multiply as follows.

Money Denominations Problems

number · value of one item = total value

Examples: 30 dimes have a value of 30($0.10) = $3.
15 five-dollar bills have a value of 15 ($5) = $75.

NOW TRY EXERCISE 5

Clayton has saved $5.65 in dimes and quarters. He has 10 more quarters than dimes. How many of each denomination of coin does he have?

EXAMPLE 5   Solving a Money Denominations Problem

A bank teller has 25 more $5 bills than $10 bills. The total value of the money is $200. How many of each denomination of bill does she have?

Step 1   **Read** the problem. We must find the number of each denomination of bill.

Step 2   **Assign a variable.**

Let x = the number of $10 bills.

Then $x + 25$ = the number of $5 bills.

Number of Bills	Denomination (in dollars)	Total Value (in dollars)
x	10	$10x$
$x + 25$	5	$5(x + 25)$

Step 3   **Write an equation.** Multiplying the number of bills by the denomination gives the monetary value. The value of the tens added to the value of the fives must be $200.

$$\begin{array}{ccccc} \text{Value of} & & \text{value of} & & \\ \text{tens} & \text{plus} & \text{fives} & \text{is} & \$200. \\ \downarrow & \downarrow & \downarrow & \downarrow & \downarrow \\ 10x & + & 5(x + 25) & = & 200 \end{array}$$

Step 4   **Solve.**

$$10x + 5x + 125 = 200 \qquad \text{Distributive property}$$
$$15x + 125 = 200 \qquad \text{Combine like terms.}$$
$$15x = 75 \qquad \text{Subtract 125.}$$
$$x = 5 \qquad \text{Divide by 15.}$$

Step 5   **State the answer.** The teller has 5 tens and $5 + 25 = 30$ fives.

Step 6   **Check.** The teller has $30 - 5 = 25$ more fives than tens. The value is

$$5(\$10) + 30(\$5) = \$200, \quad \text{as required.}$$

NOW TRY

OBJECTIVE 5   Solve problems involving distance, rate, and time.

If a car travels at an average rate of 50 mph for 2 hr, then it travels

$$50 \times 2 = 100 \text{ mi.}$$

This is an example of the basic relationship between distance, rate, and time.

distance = rate · time,   or $d = rt$

By solving, in turn, for r and t in the formula $d = rt$, we obtain two other equivalent forms of the formula.

Forms of the Distance Formula

$$d = rt \qquad r = \frac{d}{t} \qquad t = \frac{d}{r}$$

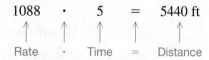

**NOW TRY
EXERCISE 6**

It took a driver 6 hr to travel from St. Louis to Fort Smith, a distance of 400 mi. What was the driver's rate, to the nearest hundredth?

EXAMPLE 6 **Finding Distance, Rate, or Time**

Solve each problem using a form of the distance formula.

(a) The speed (rate) of sound is 1088 ft per sec at sea level at 32°F. Find the distance sound travels in 5 sec under these conditions.

We must find distance, given rate and time, using $d = rt$ (or $rt = d$).

$$1088 \quad \cdot \quad 5 \quad = \quad 5440 \text{ ft}$$

Rate $\cdot$ Time $=$ Distance

(b) The winner of the first Indianapolis 500 race (in 1911) was Ray Harroun, driving a Marmon Wasp at an average rate of 74.59 mph. (*Source: Universal Almanac.*) How long did it take him to complete the 500 mi?

We must find time, given rate and distance, using $t = \frac{d}{r}$ $\left(\text{or } \frac{d}{r} = t\right)$.

Distance $\longrightarrow$ $\dfrac{500}{74.59} = 6.70$ hr (rounded) $\longleftarrow$ Time
Rate $\longrightarrow$

To convert 0.70 hr to minutes, we multiply by 60 to obtain $0.70(60) = 42$. It took Harroun about 6 hr, 42 min, to complete the race.

(c) At the 2012 Olympic Games, American swimmer Missy Franklin won a gold medal with a time of 58.33 sec in the women's 100-m backstroke swimming event. (*Source: World Almanac and Book of Facts.*) Find her rate.

We must find rate, given distance and time, using $r = \frac{d}{t}$ $\left(\text{or } \frac{d}{t} = r\right)$.

Distance $\longrightarrow$ $\dfrac{100}{58.33} = 1.71$ m per sec (rounded) $\longleftarrow$ Rate
Time $\longrightarrow$

NOW TRY

EXAMPLE 7 **Solving a Distance-Rate-Time Problem**

Two cars leave Iowa City, Iowa, at the same time and travel east on Interstate 80. One travels at a constant rate of 55 mph. The other travels at a constant rate of 63 mph. In how many hours will the distance between them be 24 mi?

Step 1 **Read** the problem carefully.

Step 2 **Assign a variable.** We are looking for time.

Let $t =$ the number of hours until the distance between them is 24 mi.

The sketch in **FIGURE 18** shows what is happening in the problem.

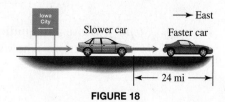

FIGURE 18

**NOW TRY
EXERCISE 7**

From a point on a straight road, two bicyclists ride in the same direction. One travels at a rate of 18 mph. The other travels at a rate of 20 mph. In how many hours will they be 5 mi apart?

To construct a table, we fill in the rates given in the problem, using t for the time traveled by each car. Because $d = rt$, or $rt = d$, we multiply rate by time to find expressions for the distances traveled.

	Rate	Time	Distance
Faster Car	63	t	$63t$
Slower Car	55	t	$55t$

← The quantities $63t$ and $55t$ represent the two distances.

Step 3 **Write an equation.**

$$63t - 55t = 24$$

The *difference* between the larger distance and the smaller distance is 24 mi.

Step 4 **Solve.** $\qquad 8t = 24$ Combine like terms.

$\qquad\qquad\qquad t = 3$ Divide by 8.

Step 5 **State the answer.** It will take the cars 3 hr to be 24 mi apart.

Step 6 **Check.** After 3 hr, the faster car will have traveled $63 \cdot 3 = 189$ mi and the slower car will have traveled $55 \cdot 3 = 165$ mi. The difference is

$$189 - 165 = 24, \quad \text{as required.}$$ **NOW TRY**

PROBLEM-SOLVING HINT In distance-rate-time problems, once we have filled in two pieces of information in each row of a table, we can automatically fill in the third piece of information, using the appropriate form of the distance formula. Then we set up the equation based on a sketch and the information in the table.

EXAMPLE 8 **Solving a Distance-Rate-Time Problem**

Two planes leave Memphis at the same time. One heads south to New Orleans. The other heads north to Chicago. The Chicago plane flies 50 mph faster than the New Orleans plane. In $\frac{1}{2}$ hr, the planes are 275 mi apart. What are their rates?

Step 1 **Read** the problem carefully.

Step 2 **Assign a variable.**

Let $\qquad r =$ the rate of the slower plane.

Then $\quad r + 50 =$ the rate of the faster plane.

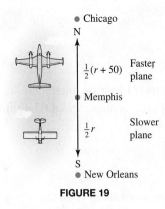

FIGURE 19

	Rate	Time	Distance
Slower Plane	r	$\frac{1}{2}$	$\frac{1}{2}r$
Faster Plane	$r + 50$	$\frac{1}{2}$	$\frac{1}{2}(r + 50)$

Sum is 275 mi.

Step 3 **Write an equation.** As **FIGURE 19** shows, the planes are headed in *opposite* directions. The *sum* of their distances equals 275 mi.

$$\frac{1}{2}r + \frac{1}{2}(r + 50) = 275$$

**NOW TRY
EXERCISE 8**

Two cars leave a parking lot at the same time, one traveling east and the other traveling west. The westbound car travels 6 mph faster than the eastbound car. In $\frac{1}{4}$ hr, they are 35 mi apart. What are their rates?

NOW TRY ANSWER

8. slower car: 67 mph;
   faster car: 73 mph

Step 4 Solve. $\frac{1}{2}r + \frac{1}{2}(r + 50) = 275$ Equation from Step 3

$\frac{1}{2}r + \frac{1}{2}r + 25 = 275$ Distributive property

$r + 25 = 275$ Combine like terms.

Rate of slower plane $\rightarrow r = 250$ Subtract 25.

Step 5 State the answer. The slower plane (headed south) has a rate of 250 mph.

$250 + 50 = 300$ mph $\leftarrow$ Rate of faster plane

Step 6 Check. Verify that $\frac{1}{2}(250) + \frac{1}{2}(300) = 275$ mi, as required.   NOW TRY

2.7 Exercises

FOR EXTRA HELP   ▶ MyMathLab®

▶ *Complete solution available in MyMathLab*

Concept Check   In Exercises 1–7, choose the letter of the correct response.

1. Which expression represents the amount of pure alcohol in x liters of a 75% alcohol solution?

   A. $0.75x$ liters   **B.** $(75 + x)$ liters   **C.** $(75 - x)$ liters   **D.** $75x$ liters

2. Which expression represents the value of x quarters?

   A. $25x$ dollars   **B.** $\frac{25}{x}$ dollars   **C.** $0.25x$ dollars   **D.** $(x + 0.25)$ dollars

3. If a minivan travels at 55 mph for t hours, which expression represents the distance traveled?

   A. $(t + 55)$ miles   **B.** $(t - 55)$ miles   **C.** $55t$ miles   **D.** $\frac{55}{t}$ miles

4. If a car travels at r mph for 6 hr, which expression represents the distance traveled?

   A. $\frac{r}{6}$ miles   **B.** $(r - 6)$ miles   **C.** $(r + 6)$ miles   **D.** $6r$ miles

5. Suppose that a chemist is mixing two acid solutions, one of 20% concentration and the other of 30% concentration. Which concentration could *not* be obtained?

   A. 22%   **B.** 24%   **C.** 28%   **D.** 32%

6. Suppose that water is added to a 24% alcohol mixture. Which concentration could be obtained? (*Hint:* The solution is being diluted.)

   A. 22%   **B.** 26%   **C.** 28%   **D.** 30%

7. Which choice is the best estimate for the average rate of a bus trip of 405 mi that lasted 8.2 hr?

   A. 50 mph   **B.** 30 mph   **C.** 60 mph   **D.** 40 mph

8. *Concept Check*   An automobile averages 45 mph and travels for 30 min. Is the distance traveled $45 \cdot 30 = 1350$ mi? If not, explain why not, and give the correct distance.

*Answer each question. **See Example 1 and the Problem-Solving Hint preceding Example 5.***

▶ 9. How much pure alcohol is in 150 L of a 30% alcohol solution?

10. How much pure acid is in 250 mL of a 14% acid solution?

11. If $25,000 is invested for 1 yr at 3% simple interest, how much interest is earned?

12. If $10,000 is invested for 1 yr at 3.5% simple interest, how much interest is earned?

13. What is the monetary value of 35 half-dollars?

14. What is the monetary value of 283 nickels?

Solve each problem. See Examples 2 and 3.

15. How many liters of 25% acid solution must a chemist add to 80 L of 40% acid solution to obtain a mixture that is 30% acid?

Liters of Solution	Rate (as a decimal)	Liters of Pure Acid
x	0.25	0.25x
80	0.40	0.40(80)
x + 80	0.30	0.30(x + 80)

16. How many gallons of 50% antifreeze must be mixed with 80 gal of 20% antifreeze to obtain a mixture that is 40% antifreeze?

Gallons of Mixture	Rate (as a decimal)	Gallons of Pure Antifreeze
x	0.50	0.50x
80	0.20	0.20(80)
x + 80	0.40	0.40(x + 80)

17. A pharmacist has 20 L of a 10% drug solution. How many liters of 5% drug solution must be added to obtain a mixture that is 8%?

Liters of Solution	Rate (as a decimal)	Liters of Pure Drug
20		20(0.10)
	0.05	
	0.08	

18. A certain metal is 20% tin. How many kilograms of this metal must be mixed with 80 kg of a metal that is 70% tin to obtain a metal that is 50% tin?

Kilograms of Metal	Rate (as a decimal)	Kilograms of Pure Tin
x	0.20	
	0.70	
	0.50	

19. In a chemistry class, 12 L of a 12% alcohol solution must be mixed with a 20% solution to obtain a 14% solution. How many liters of the 20% solution are needed?

20. How many liters of a 10% alcohol solution must be mixed with 40 L of a 50% solution to obtain a 40% solution?

21. Minoxidil is a drug that has proven to be effective in treating male pattern baldness. Water must be added to 20 mL of a 4% minoxidil solution to dilute it to a 2% solution. How many milliliters of water should be used? (*Hint:* Water is 0% minoxidil.)

22. A pharmacist wishes to mix a solution that is 2% minoxidil. She has on hand 50 mL of a 1% solution, and she wishes to add some 4% solution to it to obtain the desired 2% solution. How much 4% solution should she add?

23. How many liters of a 60% acid solution must be mixed with a 75% acid solution to obtain 20 L of a 72% solution?

24. How many gallons of a fruit drink that is 50% real juice must be mixed with a fruit drink that is 20% real juice to obtain 12 gal of a fruit drink that is 40% real juice?

Solve each problem. See Example 4.

25. Arlene is saving money for her college education. She deposited some money in a savings account paying 5% and $1200 less than that amount in a second account paying 4%. The two accounts produced a total of $141 interest in 1 yr. How much did she invest at each rate?

26. Margaret won a prize for her work. She invested part of the money in a certificate of deposit at 2% and $3000 more than that amount in a bond paying 3%. Her annual interest income was $390. How much did Margaret invest at each rate?

27. An artist invests in a tax-free bond paying 6%, and $6000 more than three times as much in mutual funds paying 5%. Her total annual interest income from the investments is $825. How much does she invest at each rate?

28. With income earned by selling the rights to his life story, an actor invests some of the money at 3% and $30,000 more than twice as much at 4%. The total annual interest earned from the investments is $5600. How much is invested at each rate?

29. Jamal had $2500, some of which he deposited in a mutual fund account paying 8%. The rest he deposited in a money market account paying 2%. How much did he deposit in each account if the total annual interest was $152?

Amount Invested (in dollars)	Rate (as a decimal)	Interest for One Year (in dollars)
x	0.08	
	0.02	

30. Carter invested a total of $9000 in two accounts, one paying 1% and the other paying 4%. If he earned total annual interest of $285, how much did he deposit in each account?

Amount Invested (in dollars)	Rate (as a decimal)	Interest for One Year (in dollars)
x	0.01	
	0.04	

*Solve each problem. **See Example 5.***

▶ 31. A coin collector has $1.70 in dimes and nickels. She has two more dimes than nickels. How many nickels does she have?

Number of Coins	Denomination (in dollars)	Total Value (in dollars)
x	0.05	0.05x
	0.10	

32. A bank teller has $725 in $5 bills and $20 bills. The teller has five more twenties than fives. How many $5 bills does the teller have?

Number of Bills	Denomination (in dollars)	Total Value (in dollars)
x	5	
x + 5	20	

33. In January 2013, U.S. first-class mail rates increased to 46 cents for the first ounce, and 20 cents for each additional ounce. If Sabrina spent $15.50 for a total of 45 stamps of these two denominations, how many stamps of each denomination did she buy? (*Source:* U.S. Postal Service.)

34. A movie theater has two ticket prices: $8 for adults and $5 for children. If the box office took in $4116 from the sale of 600 tickets, how many tickets of each kind were sold?

35. Harriet operates a coffee shop. One of her customers wants to buy two kinds of beans: Arabian Mocha and Colombian Decaf. If she wants twice as much Arabian Mocha as Colombian Decaf, how much of each can she buy for a total of $87.50? (Prices are listed on the sign.)

36. See **Exercise 35.** Another one of Harriet's customers wants to buy Italian Espresso beans and Kona Deluxe beans. If he wants four times as much Kona Deluxe as Italian Espresso, how much of each can he buy for a total of $247.50?

Arabian Mocha.........$ 8.50/lb
Chocolate Mint.......$10.50/lb
Colombian Decaf........$ 8.00/lb
French Roast..........$ 7.50/lb
Guatemalan Spice......$ 9.50/lb
Hazelnut Decaf.......$10.00/lb
Italian Espresso........$ 9.00/lb
Kona Deluxe...........$11.50/lb

*Solve each problem. **See Example 6.***

⏵ **37.** A driver averaged 53 mph and took 10 hr to travel from Memphis to Chicago. What is the distance between Memphis and Chicago?

38. A small plane traveled from Warsaw to Rome, averaging 164 mph. The trip took 2 hr. What is the distance from Warsaw to Rome?

39. The winner of the 2013 Indianapolis 500 (mile) race was Tony Kanaan, who drove his Dellara-Chevrolet to victory at a rate of 187.433 mph. What was his time (to the nearest thousandth of an hour)? (*Source: USA Today.*)

40. In 2013, Ryan Newman drove his Chevrolet to victory in the Brickyard 400 (mile) race at a rate of 153.485 mph. What was his time (to the nearest thousandth of an hour)? (*Source: World Almanac and Book of Facts.*)

In Exercises 41–44, find the rate on the basis of the information provided. Round answers to the nearest hundredth. All events were at the 2012 Olympics. (Source: World Almanac and Book of Facts.) ***See Example 6.***

	Event	Participant	Distance	Time
41.	200-m run, women	Allyson Felix, USA	200 m	21.88 sec
42.	400-m run, women	Sanya Richards-Ross, USA	400 m	49.55 sec
43.	110-m hurdles, men	Aries Merritt, USA	110 m	12.92 sec
44.	200-m run, men	Usain Bolt, Jamaica	200 m	19.32 sec

*Solve each problem. **See Examples 7 and 8.***

⏵ **45.** From a point on a straight road, Marco and Celeste ride bicycles in the same direction. Marco rides at 10 mph and Celeste rides at 12 mph. In how many hours will they be 15 mi apart?

46. At a given hour, two steamboats leave a city in the same direction on a straight canal. One travels at 18 mph and the other travels at 24 mph. In how many hours will the boats be 9 mi apart?

47. Atlanta and Cincinnati are 440 mi apart. John leaves Cincinnati, driving toward Atlanta at an average rate of 60 mph. Pat leaves Atlanta at the same time, driving toward Cincinnati in her antique auto, averaging 28 mph. How long will it take them to meet?

	r	t	d
John	60	t	$60t$
Pat	28	t	$28t$

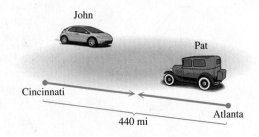

48. St. Louis and Portland are 2060 mi apart. A small plane leaves Portland, traveling toward St. Louis at an average rate of 90 mph. Another plane leaves St. Louis at the same time, traveling toward Portland and averaging 116 mph. How long will it take them to meet?

	r	t	d
Plane Leaving Portland	90	t	$90t$
Plane Leaving St. Louis	116	t	$116t$

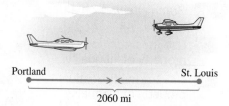

Portland   St. Louis

2060 mi

49. A train leaves Kansas City, Kansas, and travels north at 85 km per hr. Another train leaves at the same time and travels south at 95 km per hour. How long will it take before they are 315 km apart?

50. Two steamers leave a port on a river at the same time, traveling in opposite directions. Each is traveling at 22 mph. How long will it take for them to be 110 mi apart?

▶ **51.** Two planes leave an airport at the same time, one flying east, the other flying west. The eastbound plane travels 150 mph slower. They are 2250 mi apart after 3 hr. Find the rate of each plane.

52. Two trains leave a city at the same time. One travels north, and the other travels south 20 mph faster. In 2 hr, the trains are 280 mi apart. Find their rates.

	r	t	d
Eastbound	$x - 150$	3	
Westbound	x	3	

	r	t	d
Northbound	x	2	
Southbound	$x + 20$	2	

53. Two cars start from towns 400 mi apart and travel toward each other. They meet after 4 hr. Find the rate of each car if one travels 20 mph faster than the other.

54. Two cars leave towns 230 km apart at the same time, traveling directly toward one another. One car travels 15 km per hr slower than the other. They pass one another 2 hr later. What are their rates?

Extending Skills   *Solve each problem.*

55. Kevin is three times as old as Bob. Three years ago the sum of their ages was 22 yr. How old is each now? (*Hint:* Write an expression first for the age of each now and then for the age of each three years ago.)

56. A store has 39 qt of milk, some in pint cartons and some in quart cartons. There are six times as many quart cartons as pint cartons. How many quart cartons are there? (*Hint:* 1 qt = 2 pt)

57. A table is three times as long as it is wide. If it were 3 ft shorter and 3 ft wider, it would be square (with all sides equal). How long and how wide is the table?

58. Elena works for $8 an hour. A total of 25% of her salary is deducted for taxes and insurance. How many hours must she work to take home $450?

59. Paula received a paycheck for $585 for her weekly wages less 10% deductions. How much was she paid before the deductions were made?

60. At the end of a day, the owner of a gift shop had $2394 in the cash register. This amount included sales tax of 5% on all sales. Find the amount of the sales.

2.8   Solving Linear Inequalities

VOCABULARY

☐ inequality
☐ linear inequality in one variable
☐ interval
☐ three-part inequality

An **inequality** relates algebraic expressions using the symbols

$<$ "is less than," $\leq$ "is less than or equal to,"

$>$ "is greater than," $\geq$ "is greater than or equal to."

Linear Inequality in One Variable

A **linear inequality in one variable** (here x) can be written in the form

$$Ax + B < C, \quad Ax + B \leq C, \quad Ax + B > C, \quad \text{or} \quad Ax + B \geq C,$$

where A, B, and C represent real numbers and $A \neq 0$.

Examples: $x + 5 < 2$, $z - \dfrac{3}{4} \geq 5$, and $2k + 5 \leq 10$ Linear inequalities

---

We solve a linear inequality by finding all real number solutions of it. For example, the solution set $\{x \mid x \leq 2\}$ includes *all real numbers* that are less than or equal to 2, not just the *integers* less than or equal to 2.

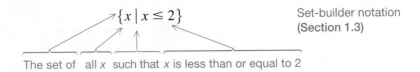

Set-builder notation
(Section 1.3)

The set of all x such that x is less than or equal to 2

OBJECTIVE 1 Graph intervals on a number line.

Graphing is a good way to show the solution set of an inequality. To graph all real numbers belonging to the set

$$\{x \mid x \leq 2\},$$

we place a square bracket at 2 on a number line and draw an arrow extending from the bracket to the left (because all numbers *less than* 2 are also part of the graph). See **FIGURE 20**.

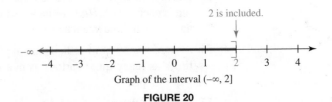

Graph of the interval $(-\infty, 2]$

FIGURE 20

The set of numbers less than or equal to 2 is an example of an **interval** on a number line. We can write this interval using **interval notation** as follows.

$$(-\infty, 2]$$ Interval notation

The **negative infinity symbol** $-\infty$ does not indicate a number, but shows that the interval includes *all* real numbers less than 2. Again, the square bracket indicates that 2 is part of the solution. Intervals that continue indefinitely in the positive direction are written with the **positive infinity symbol** ∞.

NOW TRY
EXERCISE 1

Write each inequality in interval notation, and graph the interval.

(a) $x < -1$ **(b)** $-2 \le x$

EXAMPLE 1 Graphing Intervals on a Number Line

Write each inequality in interval notation, and graph the interval.

(a) $x > -5$

The statement $x > -5$ says that x can represent any number greater than -5 but cannot equal -5. The interval is written $(-5, \infty)$. We graph this interval by placing a parenthesis at -5 and drawing an arrow to the right, as in **FIGURE 21**. The parenthesis at -5 indicates that -5 is *not* part of the graph.

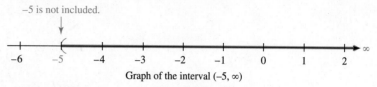

Graph of the interval $(-5, \infty)$

FIGURE 21

(b) $3 > x$

The statement $3 > x$ means the same as $x < 3$. *The inequality symbol continues to point toward the lesser number.* The graph of $x < 3$, written in interval notation as $(-\infty, 3)$, is shown in **FIGURE 22**.

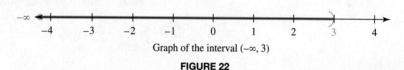

Graph of the interval $(-\infty, 3)$

FIGURE 22

CHECK To confirm that the interval in **FIGURE 22** is graphed in the proper direction, select a value that is part of the graph and substitute it into the given inequality $3 > x$. For example, we select 0 and substitute to obtain $3 > 0$, a true statement. ✓

NOW TRY

Important Concepts Regarding Interval Notation

1. A parenthesis indicates that an endpoint is *not included* in a solution set.

2. A bracket indicates that an endpoint is *included* in a solution set.

3. A parenthesis is *always* used next to an infinity symbol, $-\infty$ or ∞.

4. The set of all real numbers is written in interval notation as $(-\infty, \infty)$.

▼ **Methods of Expressing Solution Sets of Linear Inequalities**

Set-Builder Notation	Interval Notation	Graph
$\{x \mid x < a\}$	$(-\infty, a)$	
$\{x \mid x \le a\}$	$(-\infty, a]$	
$\{x \mid x > a\}$	(a, ∞)	
$\{x \mid x \ge a\}$	$[a, \infty)$	
$\{x \mid x \text{ is a real number}\}$	$(-\infty, \infty)$	

NOW TRY ANSWERS
1. (a) $(-\infty, -1)$

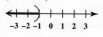

(b) $[-2, \infty)$

> **NOTE** Some texts use a solid circle ● rather than a square bracket to indicate that an endpoint is included in a number line graph. An open circle ○ is used to indicate noninclusion, rather than a parenthesis.

OBJECTIVE 2 Use the addition property of inequality.

Consider the true inequality $2 < 5$. Add 4 to each side.

$$2 < 5$$
$$2 + 4 < 5 + 4 \qquad \text{Add 4.}$$
$$6 < 9 \qquad \text{True}$$

The result is a true statement. This suggests the **addition property of inequality.**

Addition Property of Inequality

If A, B, and C represent real numbers, then the inequalities

$$A < B \quad \text{and} \quad A + C < B + C \quad \text{are equivalent.}^*$$

That is, the same number may be added to each side of an inequality without changing the solution set.

*This also applies to $A \leq B$, $A > B$, and $A \geq B$.

Consider the inequality $2 < 5$ again. This time subtract 4 from each side.

$$2 < 5$$
$$2 - 4 < 5 - 4 \qquad \text{Subtract 4.}$$
$$-2 < 1 \qquad \text{True}$$

Again, a true statement results. *As with the addition property of equality, the same number may be **subtracted** from each side of an inequality.*

NOW TRY
EXERCISE 2

Solve the inequality, and graph the solution set.

$$5 + 5x \geq 4x + 3$$

EXAMPLE 2 Using the Addition Property of Inequality

Solve $7 + 3x \geq 2x - 5$, and graph the solution set.

$$7 + 3x \geq 2x - 5 \qquad \text{As with equations, our goal is to isolate } x.$$
$$7 + 3x - 2x \geq 2x - 5 - 2x \qquad \text{Subtract } 2x.$$
$$7 + x \geq -5 \qquad \text{Combine like terms.}$$
$$7 + x - 7 \geq -5 - 7 \qquad \text{Subtract 7.}$$
$$x \geq -12 \qquad \text{Combine like terms.}$$

The solution set is $[-12, \infty)$. Its graph is shown in **FIGURE 23**.

NOW TRY ANSWER
2. $[-2, \infty)$

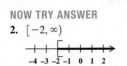

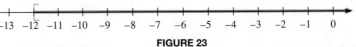

FIGURE 23

NOW TRY

NOTE Because an inequality has many solutions, we cannot check all of them by substitution as we did with the single solution of an equation. To check the solutions in the interval $[-12, \infty)$ in **Example 2,** we first substitute -12 for x in the related *equation*.

CHECK

$$7 + 3x = 2x - 5 \qquad \text{Related equation}$$

$$7 + 3(-12) \stackrel{?}{=} 2(-12) - 5 \qquad \text{Let } x = -12.$$

$$7 - 36 \stackrel{?}{=} -24 - 5 \qquad \text{Multiply.}$$

$$-29 = -29 \ \checkmark \qquad \text{True}$$

A true statement results, so -12 is indeed the "boundary" point. Next we test a number other than -12 from the interval $[-12, \infty)$. We choose 0.

CHECK

$$7 + 3x \geq 2x - 5 \qquad \text{Original inequality}$$

$$7 + 3(0) \stackrel{?}{\geq} 2(0) - 5 \qquad \text{Let } x = 0.$$

0 is easy to substitute.

$$7 + 0 \stackrel{?}{\geq} 0 - 5 \qquad \text{Multiply.}$$

$$7 \geq -5 \ \checkmark \qquad \text{True}$$

Again, a true statement results. The checks confirm that solutions to the inequality are in the interval $[-12, \infty)$. Any number "outside" the interval $[-12, \infty)$—that is, any number in $(-\infty, -12)$—will give a false statement when tested. (Try this.)

OBJECTIVE 3 Use the multiplication property of inequality.

Consider the true inequality $3 < 7$. Multiply each side by the positive number 2.

$$3 < 7$$

$$2(3) < 2(7) \qquad \text{Multiply each side by 2.}$$

$$6 < 14 \qquad \text{True}$$

The result is a true statement. Now multiply each side of $3 < 7$ by the negative number -5.

$$3 < 7$$

$$-5(3) < -5(7) \qquad \text{Multiply each side by } -5.$$

$$-15 < -35 \qquad \text{False}$$

To obtain a true statement when multiplying each side by -5, *we must reverse the direction of the inequality symbol.*

$$3 < 7$$

$$-5(3) > -5(7) \qquad \text{Multiply by } -5. \text{ Reverse the direction of the symbol.}$$

$$-15 > -35 \qquad \text{True}$$

NOTE The above illustrations began with the inequality $3 < 7$, a true statement involving two positive numbers. Similar results occur when one or both of the numbers is negative. Verify this by multiplying each of the following inequalities first by 2 and then by -5.

$$-3 < 7, \ 3 > -7, \ \text{and} \ -7 < -3$$

These observations suggest the **multiplication property of inequality.**

Multiplication Property of Inequality

Let A, B, and C represent real numbers, where $C \neq 0$.

1. If C is *positive,* then the inequalities

$$A < B \quad \text{and} \quad AC < BC \quad \text{are equivalent.}^*$$

2. If C is *negative,* then the inequalities

$$A < B \quad \text{and} \quad AC > BC \quad \text{are equivalent.}^*$$

That is, each side of an inequality may be multiplied by the same positive number without changing the direction of the inequality symbol. *If the multiplier is negative, we must reverse the direction of the inequality symbol.*

*This also applies to $A \leq B$, $A > B$, and $A \geq B$.

As with the multiplication property of equality, the same nonzero number may be divided into each side of an inequality.

Note the following differences for positive and negative numbers.

1. When each side of an inequality is multiplied or divided by a *positive number,* the direction of the inequality symbol *does not change.*

2. *Reverse the direction of the inequality symbol ONLY when multiplying or dividing each side of an inequality by a NEGATIVE NUMBER.*

EXAMPLE 3   Using the Multiplication Property of Inequality

Solve each inequality, and graph the solution set.

(a) $3x < -18$

We divide each side by 3, a positive number, so the direction of the inequality symbol *does not* change. *(It does not matter that the number on the right side of the inequality is negative.)*

$$3x < -18$$

$$\frac{3x}{3} < \frac{-18}{3} \qquad \text{Divide by 3.}$$

> 3 is *positive.* Do NOT reverse the direction of the symbol.

$$x < -6$$

The solution set is $(-\infty, -6)$. The graph is shown in **FIGURE 24**.

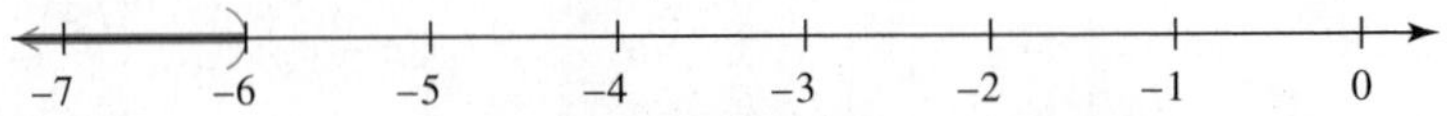

FIGURE 24

(b) $-4x \geq 8$

Here, each side of the inequality must be divided by -4, a negative number, which *does* require changing the direction of the inequality symbol.

$$-4x \geq 8$$

> To avoid errors, show the division as a separate step.

$$\frac{-4x}{-4} \leq \frac{8}{-4} \qquad \begin{array}{l}\text{Divide by } -4. \\ \text{Reverse the symbol.}\end{array}$$

> -4 is *negative.* Change $\geq$ to $\leq$.

$$x \leq -2$$

NOW TRY
EXERCISE 3

Solve the inequality, and graph the solution set.

$$-5k \geq 15$$

The solution set $(-\infty, -2]$ is graphed in **FIGURE 25**.

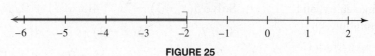

FIGURE 25

NOW TRY

OBJECTIVE 4 Solve linear inequalities using both properties of inequality.

> ### Solving a Linear Inequality in One Variable
>
> **Step 1 Simplify each side separately.** Use the distributive property as needed.
> - Clear any parentheses.
> - Clear any fractions or decimals.
> - Combine like terms.
>
> **Step 2 Isolate the variable terms on one side.** Use the addition property of inequality so that all terms with variables are on one side of the inequality and all constants (numbers) are on the other side.
>
> **Step 3 Isolate the variable.** Use the multiplication property of inequality to obtain an inequality in one of the following forms, where k is a constant (number).
>
> $$\text{variable} < k, \quad \text{variable} \leq k, \quad \text{variable} > k, \quad \text{or} \quad \text{variable} \geq k$$
>
> ***Remember:*** *Reverse the direction of the inequality symbol only when multiplying or dividing each side of an inequality by a negative number.*

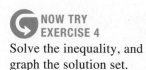

NOW TRY
EXERCISE 4

Solve the inequality, and graph the solution set.

$$6 - 2t + 5t \leq 8t - 4$$

EXAMPLE 4 Solving a Linear Inequality

Solve $3x + 2 - 5 > -x + 7 + 2x$, and graph the solution set.

Step 1 Combine like terms and simplify.

$$3x + 2 - 5 > -x + 7 + 2x$$

$$3x - 3 > x + 7$$

Step 2 Use the addition property of inequality.

$$3x - 3 - x > x + 7 - x \qquad \text{Subtract } x.$$

$$2x - 3 > 7 \qquad \text{Combine like terms.}$$

$$2x - 3 + 3 > 7 + 3 \qquad \text{Add 3.}$$

$$2x > 10 \qquad \text{Combine like terms.}$$

Step 3 Use the multiplication property of inequality.

Because 2 is positive, keep the symbol $>$.

$$\frac{2x}{2} > \frac{10}{2} \qquad \text{Divide by 2.}$$

$$x > 5$$

NOW TRY ANSWERS

3. $(-\infty, -3]$

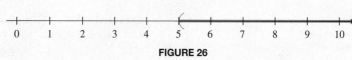

4. $[2, \infty)$

The solution set is $(5, \infty)$. Its graph is shown in **FIGURE 26**.

FIGURE 26

NOW TRY

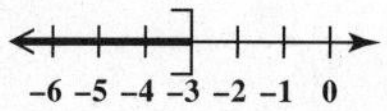

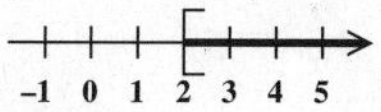

NOW TRY EXERCISE 5

Solve the inequality, and graph the solution set.

$$2x - 3(x - 6) < 4(x + 7)$$

EXAMPLE 5 Solving a Linear Inequality

Solve $5(x - 3) - 7x \geq 4(x - 3) + 9$, and graph the solution set.

Step 1 $\quad\quad 5(x - 3) - 7x \geq 4(x - 3) + 9$

$\quad\quad\quad 5x - 15 - 7x \geq 4x - 12 + 9$ Distributive property

$\quad\quad\quad -2x - 15 \geq 4x - 3$ Combine like terms.

Step 2 $\quad -2x - 15 - 4x \geq 4x - 3 - 4x$ Subtract $4x$.

$\quad\quad\quad -6x - 15 \geq -3$ Combine like terms.

$\quad\quad -6x - 15 + 15 \geq -3 + 15$ Add 15.

$\quad\quad\quad -6x \geq 12$ Combine like terms.

Step 3 $\quad\quad \dfrac{-6x}{-6} \leq \dfrac{12}{-6}$ Divide by -6. Reverse the symbol.

> Because -6 is negative, change $\geq$ to $\leq$.

$\quad\quad\quad x \leq -2$

The solution set is $(-\infty, -2]$. Its graph is shown in **FIGURE 27**.

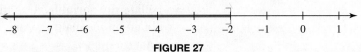

FIGURE 27

NOW TRY

NOW TRY EXERCISE 6

Solve the inequality, and graph the solution set.

$$\frac{1}{8}(x + 4) \geq \frac{1}{6}(2x + 8)$$

EXAMPLE 6 Solving a Linear Inequality with Fractions

Solve $\frac{3}{4}(x - 6) < \frac{2}{3}(5x + 1)$, and graph the solution set.

Step 1 $\quad\quad \dfrac{3}{4}(x - 6) < \dfrac{2}{3}(5x + 1)$

> Clear the parentheses first. Then clear the fractions.

$\quad\quad\quad \dfrac{3}{4}x - \dfrac{9}{2} < \dfrac{10}{3}x + \dfrac{2}{3}$ Distributive property

$\quad 12\left(\dfrac{3}{4}x - \dfrac{9}{2}\right) < 12\left(\dfrac{10}{3}x + \dfrac{2}{3}\right)$ Multiply each side by the LCD, 12.

$\quad\quad\quad 9x - 54 < 40x + 8$ Distributive property

Step 2 $\quad 9x - 54 - 40x < 40x + 8 - 40x$ Subtract $40x$.

$\quad\quad\quad -31x - 54 < 8$ Combine like terms.

$\quad\quad -31x - 54 + 54 < 8 + 54$ Add 54.

$\quad\quad\quad -31x < 62$ Combine like terms.

Step 3 $\quad\quad \dfrac{-31x}{-31} > \dfrac{62}{-31}$ Divide by -31. Reverse the symbol.

$\quad\quad\quad x > -2$

The solution set is $(-2, \infty)$. Its graph is shown in **FIGURE 28**.

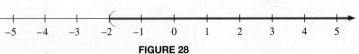

FIGURE 28

NOW TRY

NOW TRY ANSWERS

5. $(-2, \infty)$

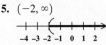

6. $(-\infty, -4]$

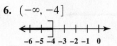

OBJECTIVE 5 Solve applied problems using inequalities.

▼ **Words and Phrases That Indicate Inequality**

Word or Phrase	Example	Inequality
Is more than	A number *is more than* 4	$x > 4$
Is less than	A number *is less than* −12	$x < -12$
Exceeds	A number *exceeds* 3.5	$x > 3.5$
Is at least	A number *is at least* 6	$x \geq 6$
Is at most	A number *is at most* 8	$x \leq 8$

❶ CAUTION   Do not confuse statements such as "5 is more than a number" with the phrase "5 more than a number." The first of these is expressed as $5 > x$, while the second is expressed as $x + 5$, or $5 + x$.

The next example uses the idea of finding the average of a number of scores. ***To find the average of n numbers, add the numbers and divide by n.*** We use the six problem-solving steps from **Section 2.4,** changing Step 3 to "Write an inequality."

NOW TRY EXERCISE 7

Will has grades of 98 and 85 on his first two tests in algebra. If he wants an average of at least 90 after his third test, what score must he make on that test?

EXAMPLE 7   Finding an Average Test Score

John has grades of 86, 88, and 78 on his first three tests in geometry. If he wants an average of at least 80 after his fourth test, what are the possible scores he can make on that test?

Step 1   **Read** the problem again.

Step 2   **Assign a variable.** Let x = John's score on his fourth test.

Step 3   **Write an inequality.**

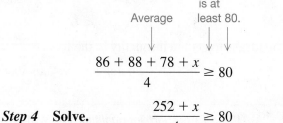

$$\frac{86 + 88 + 78 + x}{4} \geq 80$$

To find his average after four tests, add the test scores and divide by 4.

Step 4   **Solve.**
$$\frac{252 + x}{4} \geq 80$$ Add in the numerator.

$$4\left(\frac{252 + x}{4}\right) \geq 4(80)$$ Multiply by 4.

$$252 + x \geq 320$$

$$252 + x - 252 \geq 320 - 252$$ Subtract 252.

$$x \geq 68$$ Combine like terms.

Step 5   **State the answer.** He must score 68 or more on the fourth test to have an average of *at least* 80.

Step 6   **Check.** $$\frac{86 + 88 + 78 + 68}{4} = \frac{320}{4} = 80$$

To complete the check, also show that any number greater than 68 (but less than or equal to 100) makes the average greater than 80.

NOW TRY ANSWER
7. 87 or more

NOW TRY 🔄

> ⓘ **CAUTION** In applied problems, remember that
>
> is at least   translates as   is greater than or equal to
>
> and   is at most   translates as   is less than or equal to.

OBJECTIVE 6 Solve linear inequalities with three parts.

An inequality that says that one number is *between* two other numbers is a **three-part inequality.** For example,

$$-3 < 5 < 7 \quad \text{says that} \quad 5 \text{ is } between -3 \text{ and } 7.$$

NOW TRY EXERCISE 8

Write the inequality in interval notation, and graph the interval.

$$0 \leq x < 2$$

EXAMPLE 8   Graphing a Three-Part Inequality

Write the inequality in interval notation, and graph the interval.

$$-1 \leq x < 3$$

The statement is read "-1 is less than or equal to x *and* x is less than 3." We want the set of numbers *between* -1 and 3, with -1 included and 3 excluded. In interval notation, we write $[-1, 3)$, using a square bracket at -1 because -1 is part of the graph and a parenthesis at 3 because 3 is not part of the graph. See **FIGURE 29**.

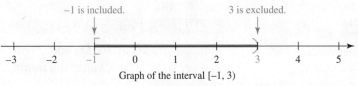

Graph of the interval $[-1, 3)$

FIGURE 29   NOW TRY ↺

The three-part inequality $3 < x + 2 < 8$ says that $x + 2$ is between 3 and 8. We solve this inequality as follows.

$$3 - 2 < x + 2 - 2 < 8 - 2 \qquad \text{Subtract 2 from } each \text{ part.}$$

$$1 < \quad x \quad < 6$$

The idea is to get the inequality in the form

$$\text{a number} < x < \text{another number.}$$

> ⓘ **CAUTION**   *Three-part inequalities are written so that the symbols point in the same direction and both point toward the lesser number.* It would be *wrong* to write $8 < x + 2 < 3$, which would imply that $8 < 3$, a false statement.

EXAMPLE 9   Solving Three-Part Inequalities

Solve each inequality, and graph the solution set.

(a) $4 < \quad 3x - 5 \quad \leq 10$ ◁ Work with all three parts at the same time.

$$4 + 5 < 3x - 5 + 5 \leq 10 + 5 \qquad \text{Add 5 to each part.}$$

$$9 < \quad 3x \quad \leq 15 \qquad \text{Combine like terms.}$$

Remember to divide all *three* parts by 3. ▷ $\dfrac{9}{3} < \quad \dfrac{3x}{3} \quad \leq \dfrac{15}{3} \qquad$ Divide each part by 3.

$$3 < \quad x \quad \leq 5$$

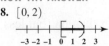

NOW TRY
EXERCISE 9

Solve the inequality, and graph the solution set.

$$-4 \le \frac{3}{2}x - 1 \le 0$$

The solution set is $(3, 5]$. Its graph is shown in **FIGURE 30**.

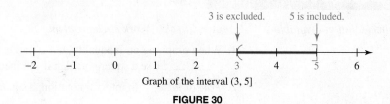

Graph of the interval $(3, 5]$

FIGURE 30

(b)
$$-4 \le \frac{2}{3}m - 1 < 8$$

$$3(-4) \le 3\left(\frac{2}{3}m - 1\right) < 3(8) \qquad \text{Multiply each part by 3 to clear the fraction.}$$

$$-12 \le 2m - 3 < 24 \qquad \text{Distributive property}$$

$$-12 + 3 \le 2m - 3 + 3 < 24 + 3 \qquad \text{Add 3 to each part.}$$

$$-9 \le 2m < 27 \qquad \text{Combine like terms.}$$

$$\frac{-9}{2} \le \frac{2m}{2} < \frac{27}{2} \qquad \text{Divide each part by 2.}$$

$$-\frac{9}{2} \le m < \frac{27}{2}$$

The solution set is $\left[-\frac{9}{2}, \frac{27}{2}\right)$. Its graph is shown in **FIGURE 31**.

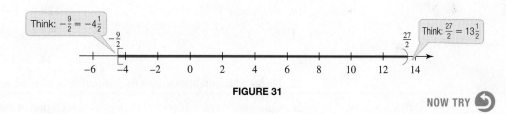

FIGURE 31

NOW TRY

⚠ **CAUTION** Be especially careful of whether to use parentheses or square brackets when writing and graphing solution sets of three-part inequalities. The following table illustrates the four possibilities that may occur.

▼ **Methods of Expressing Solution Sets of Three-Part Inequalities**

Set-Builder Notation	Interval Notation	Graph
$\{x \mid a < x < b\}$	(a, b)	$\overset{}{\underset{a \qquad b}{\longleftrightarrow}}$
$\{x \mid a < x \le b\}$	$(a, b]$	$\overset{}{\underset{a \qquad b}{\longleftrightarrow}}$
$\{x \mid a \le x < b\}$	$[a, b)$	$\overset{}{\underset{a \qquad b}{\longleftrightarrow}}$
$\{x \mid a \le x \le b\}$	$[a, b]$	$\overset{}{\underset{a \qquad b}{\longleftrightarrow}}$

NOW TRY ANSWER

9. $\left[-2, \frac{2}{3}\right]$

$$\underset{-3 \ -2 \ -1 \quad 0 \quad 1 \quad 2 \quad 3}{\overset{\frac{2}{3}}{\longleftrightarrow}}$$

2.8 Exercises

     ▶ MyMathLab®

▶ *Complete solution available in MyMathLab*

Concept Check   *Work each problem.*

1. When graphing an inequality, use a parenthesis if the inequality symbol is _____ or _____ . Use a square bracket if the inequality symbol is _____ or _____ .

2. *True* or *false?*   In interval notation, a square bracket is sometimes used next to an infinity symbol.

3. In interval notation, the set $\{x \mid x > 0\}$ is written _____ .

4. In interval notation, the set of all real numbers is written _____ .

Concept Check   *Write an inequality involving the variable x that describes each set of numbers graphed.*

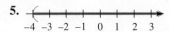

5.

6.

7.

8.

Write each inequality in interval notation, and graph the interval. **See Example 1.**

▶ 9. $k \le 4$ 10. $x \le 3$ 11. $x < -3$ 12. $r < -11$

13. $t > 4$ 14. $m > 5$ 15. $0 \ge x$

16. $1 \ge x$ 17. $-\dfrac{1}{2} \le x$ 18. $-\dfrac{3}{4} \le x$

Solve each inequality. Write the solution set in interval notation, and graph it. **See Example 2.**

19. $z - 8 \ge -7$ 20. $p - 3 \ge -11$

▶ 21. $2k + 3 \ge k + 8$ 22. $3x + 7 \ge 2x + 11$

23. $3n + 5 < 2n - 6$ 24. $5x - 2 < 4x - 5$

25. Under what conditions must the inequality symbol be reversed when solving an inequality?

26. *Concept Check*   If $p < q$ and $r < 0$, which one of the following statements is false?

    A. $pr < qr$ **B.** $pr > qr$ **C.** $p + r < q + r$ **D.** $p - r < q - r$

Solve each inequality. Write the solution set in interval notation, and graph it. **See Example 3.**

27. $3x < 18$ 28. $5x < 35$ 29. $2y \ge -20$

30. $6m \ge -24$ ▶ 31. $-8t > 24$ 32. $-7x > 49$

33. $-x \ge 0$ 34. $-k < 0$ 35. $-\dfrac{3}{4}r < -15$

36. $-\dfrac{7}{8}t < -14$ 37. $-0.02x \le 0.06$ 38. $-0.03v \ge -0.12$

Solve each inequality. Write the solution set in interval notation, and graph it. **See Examples 4–6.**

39. $8x + 9 \le -15$ 40. $6x + 7 \le -17$

41. $-4x - 3 < 1$ 42. $-5x - 4 < 6$

43. $5r + 1 \ge 3r - 9$ 44. $6t + 3 < 3t + 12$

45. $5x - 2 \leq -x + 10$

46. $3x - 9 \geq -2x + 6$

47. $-7x + 4 > -3x - 2$

48. $-8x + 1 < -4x + 11$

▶ **49.** $6x + 3 + x < 2 + 4x + 4$

50. $-4w + 12 + 9w \geq w + 9 + w$

51. $-x + 4 + 7x \leq -2 + 3x + 6$

52. $14x - 6 + 7x > 4 + 10x - 10$

53. $5(t - 1) > 3(t - 2)$

54. $7(m - 2) < 4(m - 4)$

▶ **55.** $5(x + 3) - 6x \leq 3(2x + 1) - 4x$

56. $2(x - 5) + 3x < 4(x - 6) + 1$

57. $\frac{1}{3}(5x - 4) \geq \frac{2}{5}(x + 3)$

58. $\frac{5}{12}(5x - 7) < \frac{5}{6}(x - 5)$

59. $\frac{2}{3}(p + 3) > \frac{5}{6}(p - 4)$

60. $\frac{7}{9}(x - 4) \leq \frac{4}{3}(x + 5)$

61. $\frac{4}{5}x - \frac{1}{2}(x + 3) \leq \frac{3}{10}$

62. $\frac{1}{6}x + \frac{1}{3}(x - 1) > \frac{1}{2}$

63. $4x - (6x + 1) \leq 8x + 2(x - 3)$

64. $2z - (4z + 3) > 6z + 3(z + 4)$

65. $5(2k + 3) - 2(k - 8) > 3(2k + 4) + k - 2$

66. $2(3z - 5) + 4(z + 6) \geq 2(3z + 2) + 3z - 15$

Concept Check   *Translate each statement into an inequality. Use x as the variable.*

67. You must be at least 18 yr old to vote.

68. Less than 1 in. of rain fell.

69. Chicago received more than 5 in. of snow.

70. A full-time student must take at least 12 credits.

71. Tracy could spend at most $20 on a gift.

72. The car's speed exceeded 60 mph.

Solve each problem. ***See Example 7.***

▶ **73.** Christy has scores of 76 and 81 on her first two algebra tests. If she wants an average of at least 80 after her third test, what possible scores can she make on that test?

74. Joseph has scores of 96 and 86 on his first two geometry tests. What possible scores can he make on his third test so that his average is at least 90?

75. The average monthly precipitation in Houston, TX, for October, November, and December is 4.6 in. If 5.7 in. falls in October and 4.3 in. falls in November, how many inches must fall in December so that the average monthly precipitation for these months exceeds 4.6 in.? (*Source:* National Climatic Data Center.)

76. The average monthly precipitation in New Orleans, LA, for June, July, and August is 6.7 in. If 8.1 in. falls in June and 5.7 in. falls in July, how many inches must fall in August so that the average monthly precipitation for these months exceeds 6.7 in.? (*Source:* National Climatic Data Center.)

77. When 2 is added to the difference between six times a number and 5, the result is greater than 13 added to five times the number. Find all such numbers.

78. When 8 is subtracted from the sum of three times a number and 6, the result is less than 4 more than the number. Find all such numbers.

79. The formula for converting Fahrenheit temperature to Celsius is

$$C = \frac{5}{9}(F - 32).$$

If the Celsius temperature on a certain winter day in Minneapolis is never less than $-25°$, how would we describe the corresponding Fahrenheit temperatures? (*Source:* National Climatic Data Center.)

80. The formula for converting Celsius temperature to Fahrenheit is

$$F = \frac{9}{5}C + 32.$$

The Fahrenheit temperature of Phoenix has never exceeded 122°. How would we describe this using Celsius temperature? (*Source:* National Climatic Data Center.)

81. For what values of x would the rectangle have a perimeter of at least 400?

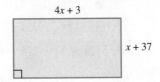

82. For what values of x would the triangle have a perimeter of at least 72?

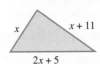

83. An international phone call costs $2.00, plus $0.30 per minute or fractional part of a minute. If x represents the number of minutes of the length of the call, then $2 + 0.30x$ represents the cost of the call. If Alan has $5.60 to spend on a call, what is the maximum total time he can use the phone?

84. At the Speedy Gas'n Go, a car wash costs $3.00 and gasoline is selling for $3.60 per gallon. Carla has $48.00 to spend, and her car is so dirty that she must have it washed. What is the maximum number of gallons of gasoline that she can purchase?

A company that produces DVDs has found that revenue from sales of DVDs is $5 per DVD, less sales costs of $100. Production costs are $125, plus $4 per DVD. Profit (P) is given by revenue (R) less cost (C), so the company must find the production level x that makes

$$P > 0, \quad \text{that is,} \quad R - C > 0. \qquad P = R - C$$

85. Write an expression for revenue R, letting x represent the production level (number of DVDs to be produced).

86. Write an expression for production costs C in terms of x.

87. Write an expression for profit P, and then solve the inequality $P > 0$.

88. Describe the solution in terms of the problem.

Concept Check Write a three-part inequality involving the variable x that describes each set of numbers graphed.

89.

90.

91.

92.

Write each inequality in interval notation, and graph the interval. *See Example 8.*

93. $8 \leq x \leq 10$ **94.** $3 \leq x \leq 5$ **95.** $0 < y \leq 10$

96. $-3 \leq x < 0$ **97.** $4 > x > -3$ **98.** $6 \geq x \geq -4$

Solve each inequality. Write the solution set in interval notation, and graph it. *See Example 9.*

99. $-8 < 4x \leq 4$ **100.** $-3 \leq 3x < 12$

▶ **101.** $-5 \leq 2x - 3 \leq 9$ **102.** $-7 \leq 3x - 4 \leq 8$

103. $10 < 7p + 3 < 24$ **104.** $-8 \leq 3r - 1 \leq -1$

105. $-4 < -2x < 12$ **106.** $9 < -3x < 15$

107. $5 < 1 - 6m < 12$ **108.** $-1 \leq 1 - 5q \leq 16$

109. $6 \leq 3(x - 1) < 18$ **110.** $-4 < 2(x + 1) \leq 6$

111. $-12 \leq \dfrac{1}{2}z + 1 \leq 4$ **112.** $-6 \leq \dfrac{1}{3}x + 3 \leq 5$

113. $1 \leq 3 + \dfrac{2}{3}p \leq 7$ **114.** $2 < 6 + \dfrac{3}{4}x < 12$

115. $-7 \leq \dfrac{5}{4}r - 1 \leq -1$ **116.** $-12 \leq \dfrac{3}{7}x + 2 \leq -4$

RELATING CONCEPTS For Individual or Group Work (Exercises 117–120)

Work Exercises 117–120 in order, to see the connection between the solution of an equation and the solutions of the corresponding inequalities.

117. Solve the following equation, and graph the solution set on a number line.

$$3x + 2 = 14$$

118. Solve the following inequality, and graph the solution set on a number line.

$$3x + 2 > 14$$

119. Solve the following inequality, and graph the solution set on a number line.

$$3x + 2 < 14$$

120. If we were to graph all the solution sets from **Exercises 117–119** on the same number line, describe the graph that we would obtain. (This is the **union** of all the solution sets.)

STUDY SKILLS

Taking Math Tests

Techniques to Improve Your Test Score	Comments
Come prepared with a pencil, eraser, paper, and calculator, if allowed.	Working in pencil lets you erase, keeping your work neat.
Scan the entire test, note the point values of different problems, and plan your time accordingly.	To do 20 problems in 50 minutes, allow $50 \div 20 = 2.5$ minutes per problem. Spend less time on the easier problems.
Do a "knowledge dump" when you get the test. Write important notes, such as formulas, in a corner of the test.	Writing down tips and information that you've learned at the beginning allows you to relax later.
Read directions carefully, and circle any significant words. When you finish a problem, reread the directions. Did you do what was asked?	Pay attention to any announcements written on the board or made by your instructor. Ask if you don't understand something.
Show all your work. Many teachers give partial credit if some steps are correct, even if the final answer is wrong. ***Write neatly.***	If your teacher can't read your writing, you won't get credit for it. If you need more space to work, ask to use extra paper.
Write down anything that might help solve a problem: a formula, a diagram, etc. If necessary, circle the problem and come back to it later. Do *not* erase anything you wrote down.	If you know even a little bit about a problem, write it down. The answer may come to you as you work on it, or you may get partial credit. Don't spend too long on any one problem.
If you can't solve a problem, make a guess. Do not change it unless you find an obvious mistake.	Have a good reason for changing an answer. Your first guess is usually your best bet.
Check that the answer to an application problem is reasonable and makes sense. Reread the problem to make sure you've answered the question.	Use common sense. Can the father really be seven years old? Would a month's rent be $32,140? Remember to label your answer if needed: $, years, inches, etc.
Check for careless errors. Rework each problem without looking at your previous work. Then compare the two answers.	Reworking a problem from the beginning forces you to rethink it. If possible, use a different method to solve the problem.

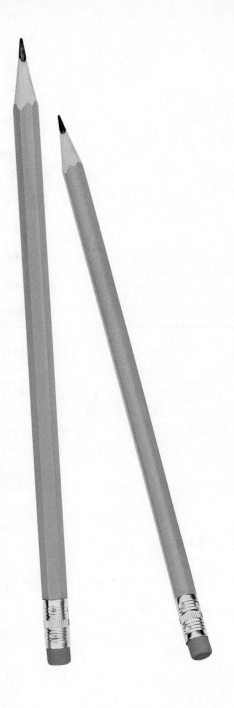

Think through and answer each question.

1. What two or three tips will you try when you take your next math test?

2. How did the tips you selected work for you when you took your math test?

3. What will you do differently when taking your next math test?

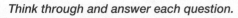

Chapter 2	Summary

Key Terms

2.1
equation
linear equation in one
  variable
solution
solution set
equivalent equations

2.3
conditional equation
identity

contradiction
empty (null) set

2.4
consecutive integers
consecutive even
  (or odd) integers
degree
complementary angles
right angle
supplementary angles
straight angle

2.5
formula
area
perimeter
vertical angles

2.6
ratio
proportion
terms of a proportion
extremes
means

cross products of
  a proportion
percent
percentage
base

2.8
inequality
linear inequality in
  one variable
interval
three-part inequality

New Symbols

$\varnothing$ empty set

$1°$ one degree

$\lnot$ right angle

a **to** b, $a : b$, or $\dfrac{a}{b}$
  ratio of a to b

∞ infinity

$-\infty$ negative infinity

$(-\infty, \infty)$ set of all real
  numbers

(a, b) interval notation
  for $a < x < b$

$[a, b]$ interval notation
  for $a \le x \le b$

Test Your Word Power

See how well you have learned the vocabulary in this chapter.

1. A **solution set** is the set of numbers that
   A. make an expression undefined
   B. make an equation false
   C. make an equation true
   D. make an expression equal to 0.

2. **Complementary angles** are angles
   A. formed by two parallel lines
   B. whose sum is 90°
   C. whose sum is 180°
   D. formed by perpendicular lines.

3. **Supplementary angles** are angles
   A. formed by two parallel lines
   B. whose sum is 90°
   C. whose sum is 180°
   D. formed by perpendicular lines.

4. A **ratio**
   A. compares two quantities using a quotient
   B. says that two quotients are equal
   C. is a product of two quantities
   D. is a difference of two quantities.

5. A **proportion**
   A. compares two quantities using a quotient
   B. says that two quotients are equal
   C. is a product of two quantities
   D. is a difference of two quantities.

6. An **inequality** is
   A. a statement that two algebraic expressions are equal
   B. a point on a number line
   C. an equation with no solutions
   D. a statement that relates algebraic expressions using $<$, $\le$, $>$, or $\ge$.

ANSWERS

1. C; *Example:* {8} is the solution set of $2x + 5 = 21$.   2. B; *Example:* Angles with measures 35° and 55° are complementary angles.

3. C; *Example:* Angles with measures 112° and 68° are supplementary angles.   4. A; *Example:* $\dfrac{7 \text{ in.}}{12 \text{ in.}}$, or $\dfrac{7}{12}$ 5. B; *Example:* $\dfrac{2}{3} = \dfrac{8}{12}$

6. D; *Examples:* $x < 5$, $7 + 2y \ge 11$, $-5 < 2z - 1 \le 3$

Quick Review

CONCEPTS	EXAMPLES

2.1 The Addition Property of Equality

The same number may be added to (or subtracted from) each side of an equation without changing the solution set.

Solve. $x - 6 = 12$

$x - 6 + 6 = 12 + 6$ Add 6.

$x = 18$ Combine like terms.

Solution set: $\{18\}$

2.2 The Multiplication Property of Equality

Each side of an equation may be multiplied (or divided) by the same nonzero number without changing the solution set.

Solve. $\dfrac{3}{4}x = -9$

$\dfrac{4}{3} \cdot \dfrac{3}{4}x = \dfrac{4}{3} \cdot (-9)$ Multiply by $\frac{4}{3}$, the reciprocal of $\frac{3}{4}$.

$x = -12$

Solution set: $\{-12\}$

2.3 More on Solving Linear Equations

Step 1 Simplify each side separately.
- Clear any parentheses.
- Clear any fractions or decimals.
- Combine like terms.

Step 2 Isolate the variable terms on one side.

Step 3 Isolate the variable.

Step 4 Check.

Solve.

$2x + 2(x + 1) = 14 + x$

$2x + 2x + 2 = 14 + x$ Distributive property

$4x + 2 = 14 + x$ Combine like terms.

$4x + 2 - x - 2 = 14 + x - x - 2$ Subtract x. Subtract 2.

$3x = 12$ Combine like terms.

$\dfrac{3x}{3} = \dfrac{12}{3}$ Divide by 3.

$x = 4$

CHECK $2(4) + 2(4 + 1) \overset{?}{=} 14 + 4$ Let $x = 4$.

$18 = 18$ ✓   True

Solution set: $\{4\}$

2.4 Applications of Linear Equations

Step 1 Read.

Step 2 Assign a variable.

Step 3 Write an equation.

Step 4 Solve the equation.

Step 5 State the answer.

Step 6 Check.

One number is five more than another. Their sum is 21. What are the numbers?

Let $x =$ the lesser number.
Then $x + 5 =$ the greater number.

$x + (x + 5) = 21$

$2x + 5 = 21$ Combine like terms.

$2x = 16$ Subtract 5.

$x = 8$ Divide by 2.

The numbers are 8 and 13.

13 is five more than 8, and $8 + 13 = 21$. The answer checks.

CONCEPTS	**EXAMPLES**

2.5 Formulas and Additional Applications from Geometry

To find the value of one of the variables in a formula, given values for the others, substitute the known values into the formula.

Find L if $\mathcal{A} = LW$, given that $\mathcal{A} = 24$ and $W = 3$.

$$\mathcal{A} = LW$$
$$24 = L \cdot 3 \qquad \mathcal{A} = 24, W = 3$$
$$\frac{24}{3} = \frac{L \cdot 3}{3} \qquad \text{Divide by 3.}$$
$$8 = L$$

To solve a formula for one of the variables, isolate that variable by treating the other variables as numbers and using the steps for solving equations.

Solve $P = 2a + 2b$ for b.

$$P = 2a + 2b$$
$$P - 2a = 2a + 2b - 2a \qquad \text{Subtract } 2a.$$
$$P - 2a = 2b \qquad \text{Combine like terms.}$$
$$\frac{P - 2a}{2} = \frac{2b}{2} \qquad \text{Divide by 2.}$$
$$\frac{P - 2a}{2} = b, \quad \text{or} \quad b = \frac{P - 2a}{2}$$

2.6 Ratio, Proportion, and Percent

To write a ratio, express quantities in the same units.

4 ft to 8 in. can be written 48 in. to 8 in., which is the ratio

$$\frac{48}{8}, \quad \text{or} \quad \frac{6}{1}.$$

To solve a proportion, use the method of cross products.

Solve.

$$\frac{x}{12} = \frac{35}{60}$$

$$60x = 12 \cdot 35 \qquad \text{Cross products}$$
$$60x = 420 \qquad \text{Multiply.}$$
$$x = 7 \qquad \text{Divide by 60.}$$

Solution set: $\{7\}$

To solve a percent problem, use the percent equation.

amount = percent (as a decimal) · base

65   is   what percent   of   325?

$$65 = p \cdot 325$$

$$\frac{65}{325} = p$$

$$0.2 = p, \quad \text{or} \quad 20\% = p$$

65 is 20% of 325.

CONCEPTS	EXAMPLES

2.7 Further Applications of Linear Equations

Step 1   Read.

Step 2   Assign a variable. Make a table and/or draw a sketch to help solve the problem.

The three forms of the formula relating distance, rate, and time are

$$d = rt, \quad r = \frac{d}{t}, \quad \text{and} \quad t = \frac{d}{r}.$$

Step 3   Write an equation.

Step 4   Solve the equation.

Steps 5 and 6   State the answer and check the solution.

Two cars leave from the same point, traveling in opposite directions. One travels at 45 mph and the other at 60 mph. How long will it take them to be 210 mi apart?

Let t = time it takes for the two cars to be 210 mi apart.

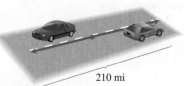

210 mi

	Rate	Time	Distance	
One Car	45	t	$45t$	The sum of the
Other Car	60	t	$60t$	distances is 210 mi.

$$45t + 60t = 210$$
$$105t = 210 \qquad \text{Combine like terms.}$$
$$t = 2 \qquad \text{Divide by 105.}$$

It will take them 2 hr to be 210 mi apart.

2.8 Solving Linear Inequalities

Step 1   Simplify each side separately.

- Clear any parentheses.
- Clear any fractions or decimals.
- Combine like terms.

Step 2   Isolate the variable terms on one side.

Step 3   Isolate the variable.

Be sure to reverse the direction of the inequality symbol when multiplying or dividing by a negative number.

To solve a three-part inequality such as

$$4 < 2x + 6 \le 8,$$

work with all three parts at the same time.

Solve the inequality, and graph the solution set.

$$3(1 - x) + 5 - 2x > 9 - 6$$
$$3 - 3x + 5 - 2x > 9 - 6 \qquad \text{Distributive property}$$
$$8 - 5x > 3 \qquad \text{Combine like terms.}$$
$$8 - 5x - 8 > 3 - 8 \qquad \text{Subtract 8.}$$
$$-5x > -5 \qquad \text{Combine like terms.}$$
$$\frac{-5x}{-5} < \frac{-5}{-5} \qquad \begin{array}{l}\text{Divide by } -5.\\ \text{Change } > \text{ to } <.\end{array}$$
$$x < 1$$

Solution set: $(-\infty, 1)$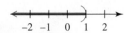

Solve the inequality, and graph the solution set.

$$4 < \quad 2x + 6 \quad \le 8$$
$$4 - 6 < \quad 2x + 6 - 6 \le 8 - 6 \qquad \text{Subtract 6.}$$
$$-2 < \quad 2x \quad \le 2 \qquad \text{Combine like terms.}$$
$$\frac{-2}{2} < \quad \frac{2x}{2} \quad \le \frac{2}{2} \qquad \text{Divide by 2.}$$
$$-1 < \quad x \quad \le 1$$

Solution set: $(-1, 1]$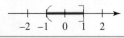

Chapter 2 Review Exercises

Answers (left column)

1. $\{6\}$ 2. $\{-12\}$

3. $\{7\}$ 4. $\left\{\dfrac{2}{3}\right\}$

5. $\{11\}$ 6. $\{17\}$

7. $\{5\}$ 8. $\{-4\}$

9. $\{5\}$ 10. $\{-12\}$

11. $\left\{\dfrac{64}{5}\right\}$ 12. $\{4\}$

13. {all real numbers}

14. $\{-19\}$

15. {all real numbers}

16. $\{20\}$

17. $\varnothing$

18. $\{-1\}$

19. $-\dfrac{7}{2}$

20. Democrats: 71;
Republicans: 47

21. Hawaii: 6425 mi²;
Rhode Island: 1212 mi²

22. Seven Falls: 300 ft;
Twin Falls: 120 ft

23. 80°   24. 11, 13

25. $h = 11$ 26. $\mathcal{A} = 28$

27. $r = 4.75$ 28. $V = 904.32$

29. $h = \dfrac{\mathcal{A}}{b}$ 30. $h = \dfrac{2\mathcal{A}}{b + B}$

For Exercises 31 and 32, there are
other correct forms.

31. $y = -x + 11$

32. $y = \dfrac{3}{2}x - 6$

33. 135°; 45°   34. 100°; 100°

2.1–2.3 *Solve each equation.*

1. $x - 5 = 1$ 2. $x + 8 = -4$ 3. $3t + 1 = 2t + 8$

4. $5z = 4z + \dfrac{2}{3}$ 5. $(4r - 2) - (3r + 1) = 8$ 6. $3(2x - 5) = 2 + 5x$

7. $7x = 35$ 8. $12r = -48$ 9. $2p - 7p + 8p = 15$

10. $\dfrac{x}{12} = -1$ 11. $\dfrac{5}{8}q = 8$ 12. $12m + 11 = 59$

13. $3(2x + 6) - 5(x + 8) = x - 22$ 14. $5x + 9 - (2x - 3) = 2x - 7$

15. $\dfrac{1}{2}r - \dfrac{r}{3} = \dfrac{r}{6}$ 16. $0.1(x + 80) + 0.2x = 14$

17. $3x - (-2x + 6) = 4(x - 4) + x$ 18. $\dfrac{1}{2}(x + 3) - \dfrac{2}{3}(x - 2) = 3$

2.4 *Solve each problem.*

19. If 7 is added to five times a number, the result is equal to three times the number. Find the number.

20. In 2013, Illinois had 118 members in its House of Representatives, consisting of only Democrats and Republicans. There were 24 more Democrats than Republicans. How many representatives from each party were there? (*Source:* www.ilga.gov)

21. The land area of Hawaii is 5213 mi² greater than the area of Rhode Island. Together, the areas total 7637 mi². What is the area of each of the two states?

22. The height of Seven Falls in Colorado is $\dfrac{5}{2}$ the height of Twin Falls in Idaho. The sum of the heights is 420 ft. Find the height of each. (*Source: World Almanac and Book of Facts.*)

23. The supplement of an angle measures 10 times the measure of its complement. What is the measure of the angle?

24. Find two consecutive odd integers such that when the lesser is added to twice the greater, the result is 24 more than the greater integer.

2.5 *A formula is given along with the values of all but one of the variables. Find the value of the variable that is not given. Use 3.14 as an approximation for π.*

25. $\mathcal{A} = \dfrac{1}{2}bh;$ $\mathcal{A} = 44, b = 8$ 26. $\mathcal{A} = \dfrac{1}{2}h(b + B);$ $h = 8, b = 3, B = 4$

27. $C = 2\pi r;$ $C = 29.83$ 28. $V = \dfrac{4}{3}\pi r^3;$ $r = 6$

Solve each formula for the specified variable.

29. $\mathcal{A} = bh$ for h 30. $\mathcal{A} = \dfrac{1}{2}h(b + B)$ for h

Solve each equation for y.

31. $x + y = 11$ 32. $3x - 2y = 12$

Find the measure of each marked angle.

33. $(8x - 1)°$ / $(3x - 6)°$

34.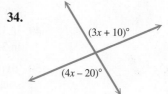
$(3x + 10)°$
$(4x - 20)°$

35. 2 cm   **36.** 42.2°; 92.8°

37. $\dfrac{3}{2}$ **38.** $\dfrac{3}{4}$

39. $\left\{\dfrac{7}{2}\right\}$ **40.** $\left\{-\dfrac{8}{3}\right\}$

41. $3.06   **42.** 375 km

43. 8 gold medals

44. 18 oz; $0.249

45. 175%   **46.** 2500

47. 3.75 L

48. $5000 at 5%; $5000 at 3%

49. 8.2 mph   **50.** $2\dfrac{1}{2}$ hr

Solve each problem.

35. The perimeter of a certain rectangle is 16 times the width. The length is 12 cm more than the width. Find the width of the rectangle.

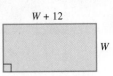

36. A baseball diamond is a square with a side of 90 ft. The pitcher's mound is located 60.5 ft from home plate, as shown in the figure. Find the measures of the angles marked in the figure. (*Hint:* Recall that the sum of the measures of the angles of any triangle is 180°.)

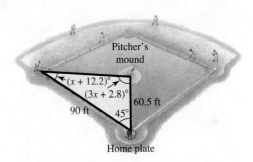

2.6 *Write a ratio for each word phrase. Express fractions in lowest terms.*

37. 60 cm to 40 cm

38. 90 in. to 10 ft

Solve each equation.

39. $\dfrac{p}{21} = \dfrac{5}{30}$

40. $\dfrac{5 + x}{3} = \dfrac{2 - x}{6}$

Solve each problem.

41. The tax on a $24.00 item is $2.04. How much tax would be paid on a $36.00 item?

42. The distance between two cities on a road map is 32 cm. The two cities are actually 150 km apart. The distance on the map between two other cities is 80 cm. How far apart are these cities?

43. In the 2012 Summer Olympics in London, England, Italian athletes earned 28 medals. Two of every 7 medals were gold. How many gold medals did Italy earn? (*Source: World Almanac and Book of Facts.*)

44. Find the best buy. Give the unit price to the nearest thousandth for that size. (*Source: Jewel-Osco.*)

Cereal

Size	Price
9 oz	$ 3.49
14 oz	$ 3.99
18 oz	$ 4.49

45. What percent of 12 is 21?

46. 36% of what number is 900?

2.7 *Solve each problem.*

47. A nurse must mix 15 L of a 10% solution of a drug with some 60% solution to obtain a 20% mixture. How many liters of the 60% solution will be needed?

48. Robert invested $10,000, from which he earns an annual income of $400 per year. He invested part of the $10,000 at 5% annual interest and the remainder in bonds paying 3% interest. How much did he invest at each rate?

49. In 1846, the vessel *Yorkshire* traveled from Liverpool to New York, a distance of 3150 mi, in 384 hr. What was the *Yorkshire's* average rate? Round the answer to the nearest tenth.

50. Two planes leave St. Louis at the same time. One flies north at 350 mph and the other flies south at 420 mph. In how many hours will they be 1925 mi apart?

51. $[-4, \infty)$ ⊢⊢⊢⊢⟦⊢⊢⊢⊢→
            −4   0

52. $(-\infty, 7)$ ←⊢⊢⊢⊢⊢⊢⊢⟧⊢→
          0   7

53. $[-5, 6)$ ⊢⊢⟦⊢⊢⊢⊢⊢⊢⟧→
         −5 0   6

54. B

55. $[-3, \infty)$ ⊢⊢⊢⟦⊢⊢⊢⊢⊢→
         −3   0

56. $(-\infty, 2)$ ←⊢⊢⟧⊢⊢⊢⊢→
          0   2

57. $[3, \infty)$ ⊢⊢⊢⊢⊢⟦⊢⊢→
        0   3

58. $[46, \infty)$ ⊢⊢⊢⊢⊢⊢⊢⟦⊢→
       0 10   40 46

59. $(-\infty, -5)$ ←⊢⊢⊢⟧⊢⊢⊢⊢⊢→
           −5 0

60. $(-\infty, -4)$ ←⊢⟧⊢⊢⊢⊢⊢→
         −4   0

61. $\left[-2, \dfrac{3}{2}\right]$ ⊢⟦⊢⊢⊢⊢⟧⊢→
        −2 0 1 2

62. $\left(\dfrac{4}{3}, 5\right]$ ⊢⊢⟦⊢⊢⊢⊢⟧→
       0 1 2   5

63. 88 or more

64. all numbers less than or equal to $-\dfrac{1}{3}$

2.8 *Write each inequality in interval notation, and graph the interval.*

51. $x \geq -4$ **52.** $x < 7$ **53.** $-5 \leq x < 6$

54. Which inequality requires reversing the inequality symbol when it is solved?

    A. $4x \geq -36$ **B.** $-4x \leq 36$ **C.** $4x < 36$ **D.** $4x > 36$

Solve each inequality. Write the solution set in interval notation, and graph it.

55. $x + 6 \geq 3$ **56.** $5x < 4x + 2$

57. $-6x \leq -18$ **58.** $8(x - 5) - (2 + 7x) \geq 4$

59. $4x - 3x > 10 - 4x + 7x$ **60.** $3(2x + 5) + 4(8 + 3x) < 5(3x + 7)$

61. $-3 \leq 2x + 1 \leq 4$ **62.** $9 < 3x + 5 \leq 20$

Solve each problem.

63. Awilda has grades of 94 and 88 on her first two calculus tests. What possible scores on a third test will give her an average of at least 90?

64. If nine times a number is added to 6, the result is at most 3. Find all such numbers.

Chapter 2   Mixed Review Exercises

Solve.

1. $\{7\}$ 2. $r = \dfrac{I}{pt}$

3. $(-\infty, 2)$ 4. $\{-9\}$

5. $\{70\}$ 6. $\left\{\dfrac{13}{4}\right\}$

7. $\varnothing$

8. $\{$ all real numbers $\}$

9. 4000 calories

10. DiGiorno: $668.7 million; Red Baron: $268.8 million

11. 160 oz; $0.062

12. $24°, 66°$

1. $\dfrac{x}{7} = \dfrac{x - 5}{2}$ **2.** $I = prt$ for r

3. $-2x > -4$ **4.** $2k - 5 = 4k + 13$

5. $0.05x + 0.02x = 4.9$ **6.** $2 - 3(x - 5) = 4 + x$

7. $9x - (7x + 2) = 3x + (2 - x)$ **8.** $\dfrac{1}{3}s + \dfrac{1}{2}s + 7 = \dfrac{5}{6}s + 5 + 2$

9. Athletes in vigorous training programs can eat 50 calories per day for every 2.2 lb of body weight. To the nearest hundred, how many calories can a 175-lb athlete consume per day? (*Source: The Gazette.*)

10. In a recent year, the top-selling frozen pizza brands, DiGiorno and Red Baron, together had sales of $937.5 million. Red Baron's sales were $399.9 million less than DiGiorno's. What were sales in millions for each brand? (*Source:* www.aibonline.org)

11. Find the best buy. Give the unit price to the nearest thousandth for that size. (*Source:* Jewel-Osco.)

Laundry Detergent

Size	Price
50 oz	$ 3.99
100 oz	$ 7.29
160 oz	$ 9.99

12. Find the measure of each marked angle.

$(3x)°$ $(8x + 2)°$

13. 13 hr

14. faster train: 80 mph;
slower train: 50 mph

15. 44 m

16. 50 m or less

13. Janet drove from Louisville to Dallas, a distance of 819 mi, averaging 63 mph. What was her driving time?

14. Two trains are 390 mi apart. They start at the same time and travel toward one another, meeting 3 hr later. If the rate of one train is 30 mph more than the rate of the other train, find the rate of each train.

15. The perimeter of a triangle is 96 m. One side is twice as long as another, and the third side is 30 m long. What is the length of the longest side?

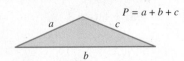

$$P = a + b + c$$

16. The perimeter of a certain square cannot be greater than 200 m. Find the possible values for the length of a side.

| Chapter 2 | Test | FOR EXTRA HELP | Step-by-step test solutions are found on the Chapter Test Prep Videos available in MyMathLab® or on YouTube™. |

▶ *View the complete solutions to all Chapter Test exercises in MyMathLab.*

[2.1–2.3]

1. $\{-6\}$ **2.** $\{21\}$

3. $\varnothing$ **4.** $\{30\}$

5. {all real numbers}

6. $\left\{ \dfrac{13}{4} \right\}$

[2.4]

7. wins: 97; losses: 65

8. Hawaii: 4021 mi²;
Maui: 728 mi²;
Kauai: 551 mi²

9. 50°   **10.** 24, 26

[2.5]

11. (a) $W = \dfrac{P - 2L}{2}$

(b) 18

12. $y = \dfrac{5}{4}x - 2$

(There are other correct forms.)

13. 75°, 75°

Solve each equation.

1. $5x + 9 = 7x + 21$

2. $-\dfrac{4}{7}x = -12$

3. $7 - (x - 4) = -3x + 2(x + 1)$

4. $0.06(x + 20) + 0.08(x - 10) = 4.6$

5. $-8(2x + 4) = -4(4x + 8)$

6. $2 - 3(x - 5) = 3 + (x + 1)$

Solve each problem.

7. In the 2013 baseball season, the St. Louis Cardinals won 33 less than twice as many games as they lost. They played 162 regular-season games. How many wins and losses did the Cardinals have? (*Source:* www.MLB.com)

8. Three islands in the Hawaiian island chain are Hawaii (the Big Island), Maui, and Kauai. Together, their areas total 5300 mi². The island of Hawaii is 3293 mi² larger than the island of Maui, and Maui is 177 mi² larger than Kauai. What is the area of each island?

9. Find the measure of an angle if its supplement measures 10° more than three times its complement.

10. If the lesser of two consecutive even integers is tripled, the result is 20 more than twice the greater integer. Find the two integers.

11. The formula for the perimeter of a rectangle is $P = 2L + 2W$.

(a) Solve for W.

(b) If $P = 116$ and $L = 40$, find the value of W.

12. Solve the equation $5x - 4y = 8$ for y.

13. Find the measure of each marked angle.

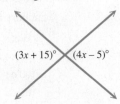

$(3x + 15)°$ $(4x - 5)°$

[2.6]

14. {6}

15. {−29}

16. 40%

17. 16 oz; $0.249

18. 2300 mi

[2.7]

19. $8000 at 3%; $14,000 at 4.5%

20. 4 hr

[2.8]

21. (a) $x < 0$

(b) $-2 < x \le 3$

22. $(-\infty, 11)$

23. $(-2, 6]$

24. $(-\infty, 4]$

25. 83 or more

Solve each equation.

14. $\dfrac{z}{8} = \dfrac{12}{16}$

15. $\dfrac{x+5}{3} = \dfrac{x-3}{4}$

Solve each problem.

16. What percent of 65 is 26?

17. Find the best buy. Give the unit price to the nearest thousandth for that size. (*Source:* Jewel-Osco.)

Cheese Slices

Size	Price
8 oz	$ 2.99
16 oz	$ 3.99
48 oz	$14.69

18. The distance between Milwaukee and Boston is 1050 mi. On a certain map, this distance is represented by 42 in. On the same map, Seattle and Cincinnati are 92 in. apart. What is the actual distance between Seattle and Cincinnati?

19. Carlos invested some money at 3% simple interest and $6000 more than that amount at 4.5% simple interest. After 1 yr, his total interest from the two accounts was $870. How much did he invest at each rate?

20. Two cars leave from the same point, traveling in opposite directions. One travels at a constant rate of 50 mph, while the other travels at a constant rate of 65 mph. How long will it take for them to be 460 mi apart?

21. Write an inequality involving x that describes the numbers graphed.

(a)

(b)

Solve each inequality. Write the solution set in interval notation, and graph it.

22. $-3x > -33$

23. $-10 < 3x - 4 \le 14$

24. $-4x + 2(x - 3) \ge 4x - (3 + 5x) - 7$

25. Susan has grades of 76 and 81 on her first two algebra tests. If she wants an average of at least 80 after her third test, what score must she make on that test?

Chapters R–2   Cumulative Review Exercises

[R.1]

1. $\dfrac{37}{60}$ **2.** $\dfrac{48}{5}$

[R.2]

3. 34.03

[1.2]

4. $\dfrac{1}{2}x - 18$

5. $\dfrac{6}{x+12} = 2$

[1.3]

6. true

[1.4, 1.5]

7. −8   **8.** 28

9. 0

Perform each indicated operation.

1. $\dfrac{5}{6} + \dfrac{1}{4} - \dfrac{7}{15}$

2. $\dfrac{9}{8} \cdot \dfrac{16}{3} \div \dfrac{5}{8}$

3. $4.8 + 12.5 + 16.73$

Translate from words to symbols. Use x as the variable.

4. The difference of half a number and 18

5. The quotient of 6 and 12 more than a number is 2.

6. *True* or *false?* $\dfrac{8(7) - 5(6 + 2)}{3 \cdot 5 + 1} \ge 1$

Perform each indicated operation.

7. $\dfrac{-4(9)(-2)}{-3^2}$

8. $(-7 - 1)(-4) + (-4)$

9. $\dfrac{6(-4) - (-2)(12)}{3^2 + 7^2}$

[1.2–1.5]
10. $-\dfrac{19}{3}$

[1.6]
11. distributive property
12. inverse property
13. identity property

[2.1–2.3]
14. $\{-1\}$
15. $\{-1\}$
16. $\{-12\}$

[2.6]
17. $\{26\}$

[2.5]
18. $y = -\dfrac{3}{4}x + 6$
   (There are other correct forms.)
19. $c = P - a - b - B$

[2.8]
20. $(-\infty, 1]$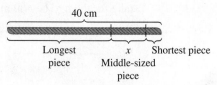
21. $(-1, 2]$

[2.4]
22. 4 cm; 9 cm; 27 cm

[2.5]
23. 12.42 cm

[2.7]
24. slower car: 40 mph;
   faster car: 60 mph

[R.2]
25. (a) 532,000
(b) 504,000
(c) 336,000

10. Find the value of $\dfrac{3x^2 - y^3}{-4z}$ for $x = -2$, $y = -4$, and $z = 3$.

Name the property illustrated by each equation.

11. $7(p + q) = 7p + 7q$ **12.** $7 + (-7) = 0$ **13.** $3.5(1) = 3.5$

Solve each equation, and check the solution.

14. $2r - 6 = 8r$ **15.** $4 - 5(s + 2) = 3(s + 1) - 1$

16. $\dfrac{2}{3}x + \dfrac{3}{4}x = -17$ **17.** $\dfrac{2x + 3}{5} = \dfrac{x - 4}{2}$

18. Solve $3x + 4y = 24$ for y.   **19.** Solve $P = a + b + c + B$ for c.

Solve each inequality. Write the solution set in interval notation, and graph it.

20. $6(r - 1) + 2(3r - 5) \le -4$ **21.** $-18 \le -9z < 9$

Solve each problem.

22. A 40-cm piece of yarn must be cut into three pieces. The longest piece is to be three times as long as the middle-sized piece, and the shortest piece is to be 5 cm shorter than the middle-sized piece. Find the length of each piece.

23. A fully inflated professional basketball has a circumference of 78 cm. What is the radius of a circular cross section through the center of the ball? (Use 3.14 as an approximation for π.) Round the answer to the nearest hundredth.

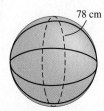

24. Two cars are 400 mi apart. Both start at the same time and travel toward one another. They meet 4 hr later. If the rate of one car is 20 mph faster than the other, what is the rate of each car?

25. The graph shows the breakdown of the colors chosen for new 2012 model-year compact / sports cars sold in the United States. If approximately 2.8 million of these cars were sold, about how many were each color? (*Source:* Ward's Auto Group.)

(a) White   **(b)** Silver   **(c)** Red

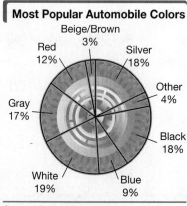

Source: DuPont Automotive Products.

Linear Equations and Inequalities in Two Variables; Functions

We determine location on a map using *coordinates,* a concept that is based on a *rectangular coordinate system,* one of the topics of this chapter.

3.1 Linear Equations and Rectangular Coordinates

OBJECTIVES

1. Interpret graphs.
2. Write a solution as an ordered pair.
3. Decide whether a given ordered pair is a solution of a given equation.
4. Complete ordered pairs for a given equation.
5. Complete a table of values.
6. Plot ordered pairs.

VOCABULARY

☐ line graph
☐ linear equation in two variables
☐ ordered pair
☐ table of values
☐ *x*-axis
☐ *y*-axis
☐ origin
☐ rectangular (Cartesian) coordinate system
☐ quadrant
☐ plane
☐ coordinates
☐ plot
☐ scatter diagram

NOW TRY EXERCISE 1

Refer to the line graph in **FIGURE 1**.

(a) Estimate the average price of a gallon of gasoline in 2010.

(b) About how much did the average price of a gallon of gasoline increase from 2010 to 2012?

NOW TRY ANSWERS
1. (a) about $2.80
   (b) about $0.90

OBJECTIVE 1 Interpret graphs.

A line graph is used to show changes or trends in data over time. To form a **line graph,** we connect a series of points representing data with line segments.

EXAMPLE 1 Interpreting a Line Graph

The line graph in **FIGURE 1** shows average prices of a gallon of regular unleaded gasoline in the United States for the years 2005 through 2012.

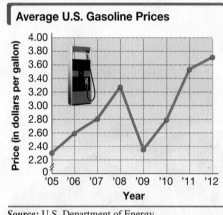

Source: U.S. Department of Energy.

FIGURE 1

(a) Between which years did the average price of a gallon of gasoline decrease?

The line between 2008 and 2009 *falls* from left to right, so the average price of a gallon of gasoline decreased from 2008 to 2009.

(b) What was the general trend in the average price of a gallon of gasoline from 2009 through 2012?

The line graph *rises* from left to right from 2009 to 2012, so the average price of a gallon of gasoline increased over those years.

(c) Estimate the average price of a gallon of gasoline in 2009 and 2012. About how much did the price increase between 2009 and 2012?

Move up from 2009 on the horizontal scale to the point plotted for 2009. This point is about three-fourths of the way between the lines on the vertical scale for $2.20 and $2.40—that is, about $2.35 per gallon.

Locate the point plotted for 2012. Moving across to the vertical scale, this point is about halfway between the lines for $3.60 and $3.80—that is, about $3.70 per gallon.

Between 2009 and 2012, the average price increased about

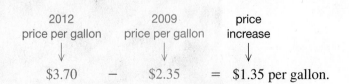

2012 price per gallon − 2009 price per gallon = price increase

$3.70 − $2.35 = $1.35 per gallon. **NOW TRY**

Year	Average Price (in dollars per gallon)
2005	2.30
2006	2.59
2007	2.80
2008	3.27
2009	2.35
2010	2.79
2011	3.53
2012	3.71

Source: U.S. Department of Energy.

The line graph in **FIGURE 1** relates years to average prices for a gallon of gasoline. We can also represent these two related quantities using a table of data, as shown in the margin. In table form, we can see more precise data rather than estimating it. Trends in the data are easier to see from the graph, which gives a "picture" of the data.

We can extend these ideas to the subject of this chapter, *linear equations in two variables*. A linear equation in two variables, one for each of the quantities being related, can be used to represent the data in a table or graph. ***The graph of a linear equation in two variables is a line.***

Linear Equation in Two Variables

A **linear equation in two variables** (here x and y) can be written in the form

$$Ax + By = C,$$

where A, B, and C are real numbers and A and B are not both 0. This form is called *standard form*.

Examples: $3x + 4y = 9$, $x - y = 0$,   and $x + 2y = -8$ Linear equations in two variables

---

NOTE Other linear equations in two variables, such as

$$y = 4x + 5 \quad \text{and} \quad 3x = 7 - 2y,$$

are not written in standard form, but could be algebraically rewritten in this form. We discuss the forms of linear equations in more detail in **Sections 3.4 and 3.5.**

OBJECTIVE 2 Write a solution as an ordered pair.

Recall from **Section 1.2** that a *solution* of an equation is a number that makes the equation true when it replaces the variable. For example, the linear equation in *one* variable

$$x - 2 = 5$$

has solution 7 because replacing x with 7 gives a true statement.

A solution of a linear equation in **two** *variables requires* **two** *numbers, one for each variable.* For example, a true statement results when we replace x with 2 and y with 13 in the equation $y = 4x + 5$ because

$$13 = 4(2) + 5. \quad \text{Let } x = 2 \text{ and } y = 13.$$

The pair of numbers $x = 2$ and $y = 13$ gives a solution of the equation $y = 4x + 5$. The phrase "$x = 2$ and $y = 13$" is abbreviated

$$\underset{\text{Ordered pair}}{\underbrace{(\overset{x\text{-value}}{2}, \overset{y\text{-value}}{13})}}$$

with the x-value, 2, and the y-value, 13, given as a pair of numbers written inside parentheses. ***The x-value is always given first.*** A pair of numbers such as (2, 13) is an **ordered pair**.

> ❗ **CAUTION**   The ordered pairs $(2, 13)$ and $(13, 2)$ are *not* the same. In the first pair, $x = 2$ and $y = 13$. In the second pair, $x = 13$ and $y = 2$. ***The order in which the numbers are written in an ordered pair is important.***

OBJECTIVE 3 Decide whether a given ordered pair is a solution of a given equation.

We substitute the x- and y-values of an ordered pair into a linear equation in two variables to see whether the ordered pair is a solution. An ordered pair that is a solution of an equation is said to *satisfy* the equation.

🔄 **NOW TRY EXERCISE 2**

Decide whether each ordered pair is a solution of the equation.

$$3x - 7y = 19$$

(a) $(3, 4)$ **(b)** $(-3, -4)$

EXAMPLE 2   Deciding Whether Ordered Pairs Are Solutions of an Equation

Decide whether each ordered pair is a solution of the equation $2x + 3y = 12$.

(a) $(3, 2)$

Substitute 3 for x and 2 for y in the given equation.

$$2x + 3y = 12$$
$$2(3) + 3(2) \stackrel{?}{=} 12 \qquad \text{Let } x = 3 \text{ and } y = 2.$$
$$6 + 6 \stackrel{?}{=} 12 \qquad \text{Multiply.}$$
$$12 = 12 \checkmark \qquad \text{True}$$

This result is true, so $(3, 2)$ is a solution of $2x + 3y = 12$.

(b) $(-2, -7)$

$$2x + 3y = 12$$
$$2(-2) + 3(-7) \stackrel{?}{=} 12 \qquad \text{Let } x = -2 \text{ and } y = -7.$$

> Use parentheses to avoid errors.

$$-4 + (-21) \stackrel{?}{=} 12 \qquad \text{Multiply.}$$
$$-25 = 12 \qquad \text{False}$$

This result is false, so $(-2, -7)$ is *not* a solution of $2x + 3y = 12$.   **NOW TRY** 🔄

OBJECTIVE 4 Complete ordered pairs for a given equation.

Substituting a number for one variable in a linear equation makes it possible to find the value of the other variable.

EXAMPLE 3   Completing Ordered Pairs

Complete each ordered pair for the equation $y = 4x + 5$.

(a) $(7, \underline{})$ > The x-value always comes first.

In this ordered pair, $x = 7$. To find the corresponding value of y, replace x with 7 in the given equation.

$$y = 4x + 5$$

> Solve for the value of y.

$$y = 4(7) + 5 \qquad \text{Let } x = 7.$$
$$y = 28 + 5 \qquad \text{Multiply.}$$
$$y = 33 \qquad \text{Add.}$$

NOW TRY ANSWERS

2. (a) no   **(b)** yes

The ordered pair is $(7, 33)$.

**NOW TRY
EXERCISE 3**

Complete each ordered pair for the equation.

$$y = 3x - 12$$

(a) $(4, \_\_)$ **(b)** $(\_\_, 3)$

(b) $(\_\_, -3)$

In this ordered pair, $y = -3$. Find the corresponding value of x by replacing y with -3 in the given equation.

$$y = 4x + 5 \qquad \text{Solve for the value of } x.$$
$$-3 = 4x + 5 \qquad \text{Let } y = -3.$$
$$-8 = 4x \qquad \text{Subtract 5 from each side.}$$
$$-2 = x \qquad \text{Divide each side by 4.}$$

The ordered pair is $(-2, -3)$.

NOW TRY

OBJECTIVE 5 Complete a table of values.

Ordered pairs are often displayed in a **table of values.** Although we usually write tables of values vertically, they may be written horizontally.

EXAMPLE 4 Completing Tables of Values

Complete the table of values for each equation. Write the results as ordered pairs.

(a) $x - 2y = 8$

x	y	Ordered Pairs
2		$(2, \_\_)$
10		$(10, \_\_)$
	0	$(\_\_, 0)$
	-2	$(\_\_, -2)$

From the first row of the table, let $x = 2$ in the equation. From the second row of the table, let $x = 10$.

	If $\quad x = 2$,		If $\quad x = 10$,
then	$x - 2y = 8$	then	$x - 2y = 8$
becomes	$2 - 2y = 8$	becomes	$10 - 2y = 8$
	$-2y = 6$		$-2y = -2$
	$y = -3.$		$y = 1.$

The first two ordered pairs are $(2, -3)$ and $(10, 1)$. From the third and fourth rows of the table, let $y = 0$ and $y = -2$, respectively.

	If $\quad y = 0$,		If $\quad y = -2$,
then	$x - 2y = 8$	then	$x - 2y = 8$
becomes	$x - 2(0) = 8$	becomes	$x - 2(-2) = 8$
	$x - 0 = 8$		$x + 4 = 8$
	$x = 8.$		$x = 4.$

The last two ordered pairs are $(8, 0)$ and $(4, -2)$. The completed table of values and corresponding ordered pairs follow.

Write y-values in the second column.

Write x-values in the first column.

x	y	Ordered Pairs
2	-3	$(2, -3)$
10	1	$(10, 1)$
8	0	$(8, 0)$
4	-2	$(4, -2)$

Each ordered pair is a solution of the given equation $x - 2y = 8$.

NOW TRY
EXERCISE 4

Complete the table of values for the equation. Write the results as ordered pairs.

$$5x - 4y = 20$$

x	y
0	
	0
2	

(b) $x = 5$

x	y
	−2
	6
	3

The given equation is $x = 5$. No matter which value of y is chosen, the value of x is always 5.

x	y	Ordered Pairs
5	−2	→ (5, −2)
5	6	→ (5, 6)
5	3	→ (5, 3)

NOW TRY

NOTE   We can think of $x = 5$ in **Example 4(b)** as an equation in two variables by rewriting $x = 5$ as

$$x + 0y = 5.$$

This form of the equation shows that, for any value of y, the value of x is 5. Similarly, $y = 4$ can be written

$$0x + y = 4.$$

OBJECTIVE 6 Plot ordered pairs.

In **Section 2.3,** we saw that linear equations in *one* variable had either one, zero, or an infinite number of real number solutions. These solutions could be graphed on *one* number line. For example, the linear equation in one variable $x - 2 = 5$ has solution 7, which is graphed on the number line in **FIGURE 2**.

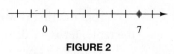

0   7

FIGURE 2

Every linear equation in *two* variables has an infinite number of ordered pairs (x, y) as solutions. To graph these solutions, we need *two* number lines, one for each variable, drawn at right angles as in **FIGURE 3**. The horizontal number line is the **x-axis,** and the vertical line is the **y-axis.** The point at which the x-axis and y-axis intersect is the **origin.** Together, the x-axis and y-axis form a **rectangular coordinate system.**

The rectangular coordinate system is divided into four regions, or **quadrants.** These quadrants are numbered counterclockwise, as shown in **FIGURE 3**.

René Descartes (1596–1650)

The rectangular coordinate system is also called the **Cartesian coordinate system,** in honor of René Descartes, the French mathematician credited with its invention.

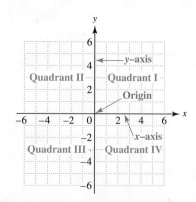

FIGURE 3 Rectangular Coordinate System

NOW TRY ANSWER

4.

x	y
0	−5
4	0
2	$-\frac{5}{2}$

$(0, -5), (4, 0), \left(2, -\frac{5}{2}\right)$

The x-axis and y-axis determine a **plane**—a flat surface illustrated by a sheet of paper. By referring to the two axes, we can associate every point in the plane with an ordered pair. The numbers in the ordered pair are the **coordinates** of the point.

NOTE In a plane, *both* numbers in the ordered pair are needed to locate a point. The ordered pair is a name for the point.

**NOW TRY
EXERCISE 5**

Plot each point in a rectangular coordinate system.

$(-3, 1), (2, -4), (0, -1),$
$\left(\frac{5}{2}, 3\right), (-4, -3), (-4, 0)$

EXAMPLE 5 Plotting Ordered Pairs

Plot each point in a rectangular coordinate system.

(a) $(2, 3)$ **(b)** $(-1, -4)$ **(c)** $(-2, 3)$ **(d)** $(3, -2)$ **(e)** $\left(\frac{3}{2}, 2\right)$

(f) $(4, -3.75)$ **(g)** $(5, 0)$ **(h)** $(0, -3)$ **(i)** $(0, 0)$

The point $(2, 3)$ from part (a) is **plotted** (graphed) in **FIGURE 4**. The other points are plotted in **FIGURE 5**. In each case, we begin at the origin.

Step 1   Move right or left the number of units that corresponds to the x-coordinate in the ordered pair—*right if the x-coordinate is positive or left if the x-coordinate is negative.*

Step 2   Then turn and move up or down the number of units that corresponds to the y-coordinate in the ordered pair—*up if the y-coordinate is positive or down if the y-coordinate is negative.*

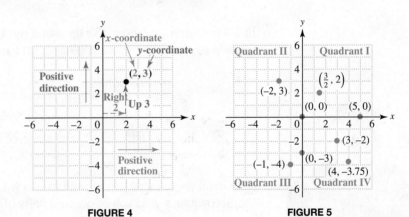

FIGURE 4   **FIGURE 5**

Notice in **FIGURE 5** that the point $(-2, 3)$ is in quadrant II, whereas the point $(3, -2)$ is in quadrant IV. *The order of the coordinates is important. The x-coordinate is always given first in an ordered pair.*

To plot the point $\left(\frac{3}{2}, 2\right)$, think of the improper fraction $\frac{3}{2}$ as the mixed number $1\frac{1}{2}$ and move $\frac{3}{2}$ $\left(\text{or } 1\frac{1}{2}\right)$ units to the right along the x-axis. Then turn and go 2 units up, parallel to the y-axis. The point $(4, -3.75)$ is plotted similarly, by approximating the location of the decimal y-coordinate.

The point $(5, 0)$ lies on the x-axis because the y-coordinate is 0. The point $(0, -3)$ lies on the y-axis because the x-coordinate is 0. The point $(0, 0)$ is at the origin.

Points on the axes themselves are not in any quadrant. **NOW TRY**

NOW TRY ANSWER

5. [graph showing plotted points: $(-3, 1)$, $\left(\frac{5}{2}, 3\right)$, $(-4, 0)$, $(0, -1)$, $(-4, -3)$, $(2, -4)$]

We can use a linear equation in two variables to mathematically describe, or *model,* certain real-life situations.

NOW TRY
EXERCISE 6

Use the linear equation in **Example 6** to approximate the number of twin births, to the nearest thousand, in 2010. Interpret the results.

NOW TRY ANSWER

6. $y \approx 133$; There were approximately 133 thousand (or 133,000) twin births in the U.S. in 2010.

EXAMPLE 6   Using a Linear Equation to Model Twin Births

The annual number of twin births in the United States from 2006 through 2011 can be approximated by the linear equation

Number of twin births ⎤   ⎡ Year

$$y = -1.421x + 2989,$$

which relates x, the year, and y, the number of twin births in thousands. (*Source: National Center for Health Statistics.*)

(a) Complete the table of values for the given linear equation.

x (Year)	y (Number of Twin Births, in thousands)
2006	
2009	
2011	

To find y when $x = 2006$, we substitute into the equation.

$$y = -1.421x + 2989$$

$\approx$ means "is approximately equal to." $y = -1.421(2006) + 2989$ Let $x = 2006$.

$$y \approx 138$$ Use a calculator.

In 2006, there were about 138 thousand (or 138,000) twin births.

We substitute the years 2009 and 2011 in the same way to complete the table.

x (Year)	y (Number of Twin Births, in thousands)	Ordered Pairs (x, y)
2006	138	⟶ (2006, 138)
2009	134	⟶ (2009, 134)
2011	131	⟶ (2011, 131)

Here each year x is paired with a number of twin births y (in thousands).

(b) Graph the ordered pairs found in part (a).

See **FIGURE 6**. A graph of ordered pairs of data is a **scatter diagram.**

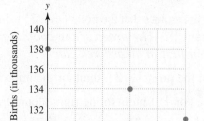

Number of Twin Births

Notice the axis labels and scales. Each grid square represents 1 unit in the horizontal direction and 2 units in the vertical direction. We show a break in the y-axis, to indicate the jump from 0 to 130.

FIGURE 6

A scatter diagram enables us to describe how the two quantities are related to each other. In **FIGURE 6**, the plotted points could be connected to approximate a straight ***line,*** so the variables x (year) and y (number of twin births) have a ***line***ar relationship. The decrease in the number of twin births is also reflected.

NOW TRY

> ⚠ **CAUTION** The equation in **Example 6** is valid only for the years 2006 through 2011. *Do not assume that it would provide reliable data for other years.*

3.1 Exercises

FOR EXTRA HELP   MyMathLab®

▶ *Complete solution available in MyMathLab*

Concept Check *Complete each statement.*

1. The symbol (x, y) (*does/does not*) represent an ordered pair, while the symbols $[x, y]$ and $\{x, y\}$ (*do/do not*) represent ordered pairs.

2. The origin is represented by the ordered pair ____ .

3. The point whose graph has coordinates $(-4, 2)$ is in quadrant ____ .

4. The point whose graph has coordinates $(0, 5)$ lies on the ____ -axis.

5. The ordered pair $(4, \underline{\quad})$ is a solution of the equation $y = 3$.

6. The ordered pair $(\underline{\quad}, -2)$ is a solution of the equation $x = 6$.

Concept Check *Fill in each blank with the word* positive *or the word* negative.

The point with coordinates (x, y) is in

7. quadrant III if x is _____ and y is _____ .

8. quadrant II if x is _____ and y is _____ .

9. quadrant IV if x is _____ and y is _____ .

10. quadrant I if x is _____ and y is _____ .

11. A point (x, y) has the property that $xy < 0$. In which quadrant(s) must the point lie? Explain.

12. A point (x, y) has the property that $xy > 0$. In which quadrant(s) must the point lie? Explain.

The line graph shows the overall unemployment rate in the U.S. civilian labor force for the years 2006 through 2012. Use the graph to work Exercises 13–16. ***See Example 1.***

13. Between which pairs of consecutive years did the unemployment rate decrease?

14. What was the general trend in the unemployment rate between 2007 and 2010?

15. Estimate the overall unemployment rate in 2011 and 2012. About how much did the unemployment rate decline between 2011 and 2012?

16. During which year(s)

  (a) was the unemployment rate greater than 9%, but less than 10%?

  (b) did the unemployment rate stay the same?

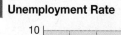

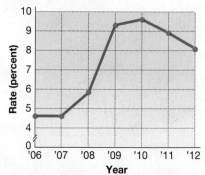

Year

Source: Bureau of Labor Statistics.

Decide whether the given ordered pair is a solution of the given equation. ***See Example 2.***

▶ **17.** $x + y = 8$; $(0, 8)$ **18.** $x + y = 9$; $(0, 9)$ **19.** $2x + y = 5$; $(3, -1)$

20. $2x - y = 6$; $(4, 2)$ ▶ **21.** $5x - 3y = 15$; $(5, 2)$ **22.** $4x - 3y = 6$; $(2, 1)$

23. $x = -4y$; $(-8, 2)$ **24.** $y = 3x$; $(2, 6)$ **25.** $y = 2$; $(4, 2)$

26. $x = -6$; $(-6, 5)$ **27.** $x - 6 = 0$; $(4, 2)$ **28.** $x + 4 = 0$; $(-6, 2)$

Complete each ordered pair for the equation $y = 2x + 7$. ***See Example 3.***

29. $(5, \underline{\hspace{0.3cm}})$ **30.** $(2, \underline{\hspace{0.3cm}})$ **31.** $(\underline{\hspace{0.3cm}}, -3)$ **32.** $(\underline{\hspace{0.3cm}}, 0)$

Complete each ordered pair for the equation $y = -4x - 4$. ***See Example 3.***

33. $(\underline{\hspace{0.3cm}}, 0)$ **34.** $(0, \underline{\hspace{0.3cm}})$ **35.** $(\underline{\hspace{0.3cm}}, 24)$ **36.** $(\underline{\hspace{0.3cm}}, 16)$

Complete the table of values for each equation. Write the results as ordered pairs. ***See Example 4.***

37. $4x + 3y = 24$

x	y
0	
	0
	4

38. $2x + 3y = 12$

x	y
0	
	0
	8

39. $4x - 9y = -36$

x	y
	0
0	
	8

40. $3x - 5y = -15$

x	y
0	
	0
	-6

41. $x = 12$

x	y
	3
	8
	0

42. $x = -9$

x	y
	6
	2
	-3

43. $y = -10$

x	y
4	
0	
-4	

44. $y = -6$

x	y
8	
4	
-2	

45. $y + 2 = 0$

x	y
9	
2	
0	

46. $y + 6 = 0$

x	y
6	
3	
0	

47. $x - 4 = 0$

x	y
	4
	0
	-4

48. $x - 8 = 0$

x	y
	8
	3
	0

49. Do $(3, 4)$ and $(4, 3)$ correspond to the same point in the plane? Explain.

50. Do $(4, -1)$ and $(-1, 4)$ represent the same ordered pair? Explain.

Give ordered pairs for the points labeled A–H in the figure. (Coordinates of the points shown are integers.) Identify the quadrant in which each point is located. ***See Example 5.***

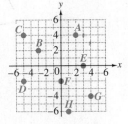

51. A   **52.** B   **53.** C   **54.** D

55. E   **56.** F   **57.** G   **58.** H

Plot and label each point in a rectangular coordinate system. ***See Example 5.***

59. $(6, 2)$ **60.** $(5, 3)$ **61.** $(-4, 2)$ **62.** $(-3, 5)$

63. $\left(-\dfrac{4}{5}, -1\right)$ **64.** $\left(-\dfrac{3}{2}, -4\right)$ **65.** $(3, -1.75)$ **66.** $(5, -4.25)$

67. $(0, 4)$ **68.** $(0, -3)$ **69.** $(4, 0)$ **70.** $(-3, 0)$

Complete the table of values for each equation. Then plot and label the ordered pairs. ***See Examples 4 and 5.***

71. $x - 2y = 6$

x	y
0	
	0
2	
	-1

72. $2x - y = 4$

x	y
0	
	0
1	
	-6

73. $3x - 4y = 12$

x	y
0	
	0
-4	
	-4

74. $2x - 5y = 10$

x	y
0	
	0
−5	
	−3

75. $y + 4 = 0$

x	y
0	
5	
−2	
−3	

76. $x - 5 = 0$

x	y
	1
	0
	6
	−4

77. Look at the graphs of the ordered pairs in **Exercises 71–76**. Describe the pattern indicated by the plotted points.

78. Answer each question.

(a) A line through the plotted points in **Exercise 75** would be horizontal. What do you notice about the *y*-coordinates of the ordered pairs?

(b) A line through the plotted points in **Exercise 76** would be vertical. What do you notice about the *x*-coordinates of the ordered pairs?

*Solve each problem. **See Example 6.***

79. Suppose that it costs a flat fee of $20 plus $5 per day to rent a pressure washer. Therefore, the cost *y* in dollars to rent the pressure washer for *x* days is given by the linear equation

$$y = 5x + 20.$$

Express each of the following as an ordered pair.

(a) When the washer is rented for 5 days, the cost is $45.

(b) We paid $50 when we returned the washer, so we must have rented it for 6 days.

80. Suppose that it costs $5000 to start up a business selling snow cones. Furthermore, it costs $0.50 per cone in labor, ice, syrup, and overhead. Then the cost *y* in dollars to make *x* snow cones is given by the linear equation

$$y = 0.50x + 5000.$$

Express each of the following as an ordered pair.

(a) When 100 snow cones are made, the cost is $5050.

(b) When the cost is $6000, the number of snow cones made is 2000.

81. The table shows the rate (in percent) at which 2-year college students (public) completed a degree within 3 years.

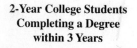

2-Year College Students Completing a Degree within 3 Years

Year	Percent
2008	29.3
2009	28.3
2010	28.0
2011	26.9
2012	25.4
2013	22.5

Source: ACT.

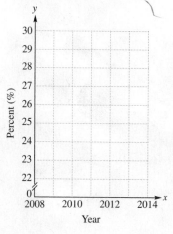

(a) Write the data from the table as ordered pairs (x, y), where *x* represents the year and *y* represents the percent.

(b) What does the ordered pair (2013, 22.5) mean in the context of this problem?

(c) Make a scatter diagram of the data, using the ordered pairs from part (a) and the given grid.

(d) Describe the pattern indicated by the points on the scatter diagram. What happened to the rates at which 2-year college students complete a degree within 3 years?

82. The table shows the number of U.S. students who studied abroad (in thousands).

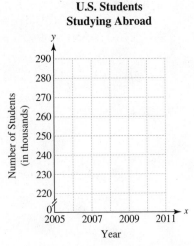

U.S. Students Studying Abroad

Academic Year	Number of Students (in thousands)
2005	224
2006	242
2007	262
2008	260
2009	270
2010	274
2011	283

Source: Institute of International Education.

(a) Write the data from the table as ordered pairs (x, y), where x represents the year and y represents the number of U.S. students (in thousands) studying abroad.

(b) What does the ordered pair $(2011, 283)$ mean in the context of this problem?

(c) Make a scatter diagram of the data, using the ordered pairs from part (a) and the given grid.

(d) Describe the pattern indicated by the points on the scatter diagram. What was the trend in the number of U.S. students studying abroad during these years?

83. The maximum benefit for the heart from exercising occurs if the heart rate is in the target heart rate zone. The lower limit of this target zone can be approximated by the linear equation

$$y = -0.65x + 143,$$

where x represents age and y represents heartbeats per minute. (*Source: The Gazette.*)

Age	Heartbeats (per minute)
20	
40	
60	
80	

(a) Complete the table of values for this linear equation.

(b) Write the data from the table of values as ordered pairs.

(c) Make a scatter diagram of the data. Do the points lie in an approximately linear pattern?

84. (See **Exercise 83.**) The upper limit of the target heart rate zone can be approximated by the linear equation

$$y = -0.85x + 187,$$

where x represents age and y represents heartbeats per minute. (*Source: The Gazette.*)

Age	Heartbeats (per minute)
20	
40	
60	
80	

(a) Complete the table of values for this linear equation.

(b) Write the data from the table of values as ordered pairs.

(c) Make a scatter diagram of the data. Describe the pattern indicated by the data.

85. See **Exercises 83 and 84.** What is the target heart rate zone for age 20?   Age 40?

86. See **Exercises 83 and 84.** What is the target heart rate zone for age 60?   Age 80?

Analyzing Your Test Results

An exam is a learning opportunity—learn from your mistakes. After a test is returned, do the following:

- **Note what you got wrong and why you had points deducted.**

- **Figure out how to solve the problems you missed.** Check your textbook or notes, or ask your instructor. Rework the problems correctly.

- **Keep all quizzes and tests that are returned to you.** Use them to study for future tests and the final exam.

Typical Reasons for Errors on Math Tests

These are test taking errors. They are easy to correct if you read carefully, show all your work, proofread, and double-check units and labels.

1. You read the directions wrong.

2. You read the question wrong or skipped over something.

3. You made a computation error.

4. You made a careless error. (For example, you incorrectly copied a correct answer onto a separate answer sheet.)

5. Your answer was not complete.

6. You labeled your answer wrong. (For example, you labeled an answer "ft" instead of "ft^2.")

7. You didn't show your work.

These are test preparation errors. Be sure to practice all the kinds of problems that you will see on tests.

8. You didn't understand a concept.

9. You were unable to set up the problem (in an application).

10. You were unable to apply a procedure.

Below are sample charts for tracking your test taking progress. Refer to the tests you have taken so far in your course, and use the charts to find out if you tend to make certain kinds of errors. Check the appropriate box when you've made an error in a particular category.

Test Taking Errors

Test	Read directions wrong	Read question wrong	Computation error	Not exact or accurate	Not complete	Labeled wrong	Didn't show work
1							
2							
3							

Test Preparation Errors

Test	Didn't understand concept	Didn't set up problem correctly	Couldn't apply concept to new situation
1			
2			
3			

What will you do to avoid these kinds of errors on your next test?

3.2 Graphing Linear Equations in Two Variables

VOCABULARY

☐ graph, graphing
☐ *x*-intercept
☐ *y*-intercept
☐ horizontal line
☐ vertical line

OBJECTIVE 1   Graph linear equations by plotting ordered pairs.

There are infinitely many ordered pairs that satisfy a linear equation in two variables. We find these ordered-pair solutions by choosing as many values of *x* (or *y*) as we wish and then completing each ordered pair.

For example, consider the equation

$$x + 2y = 7.$$

If we choose $x = 1$, then we can substitute to find the corresponding value of *y*.

$x + 2y = 7$	Given equation.
$1 + 2y = 7$	Let $x = 1$.
$2y = 6$	Subtract 1.
$y = 3$	Divide by 2.

If $x = 1$, then $y = 3$, so the ordered pair $(1, 3)$ is a solution.

$$1 + 2(3) = 7 \qquad \text{(1, 3) is a solution.}$$

This ordered pair and other solutions of $x + 2y = 7$ are graphed in **FIGURE 7**.

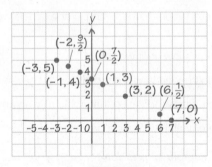

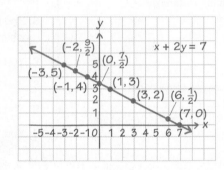

FIGURE 7   **FIGURE 8**

Notice that the points plotted in **FIGURE 7** all appear to lie on a straight line, as shown in **FIGURE 8**. In fact, the following is true.

Every point on the line represents a solution of the equation $x + 2y = 7$, and every solution of the equation corresponds to a point on the line.

The line gives a "picture" of all the solutions of the equation

$$x + 2y = 7.$$

The line extends indefinitely in both directions, as suggested by the arrowhead on each end, and is the **graph** of the equation $x + 2y = 7$. **Graphing** is the process of plotting ordered pairs and drawing a line through the corresponding points.

> **Graph of a Linear Equation**
>
> The graph of any linear equation in two variables is a straight line.

Notice that the word *line* appears in the name "*line*ar equation."

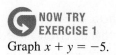

**NOW TRY
EXERCISE 1**

Graph $x + y = -5$.

EXAMPLE 1   Graphing a Linear Equation

Graph $x - y = -3$.

   At least two different ordered pairs are needed to draw the graph. To find them, we arbitrarily choose values for x or y and substitute them into the equation. We choose $x = 0$ to find one ordered pair and $y = 0$ to find another.

$x - y = -3$			$x - y = -3$	
$0 - y = -3$	0 is easy to substitute.		$x - 0 = -3$	0 is easy to substitute.
$-y = -3$	Subtract.		$x = -3$	Subtract.
$y = 3$	Multiply by -1.			

One ordered pair is $(0, 3)$. | One ordered pair is $(-3, 0)$.

   We find a third ordered pair (as a check) by choosing some other value for x or y. We let $x = 2$.

$$x - y = -3$$
$$2 - y = -3 \quad \text{We arbitrarily let } x = 2. \text{ Other numbers could be used for } x, \text{ or for } y, \text{ instead.}$$
$$-y = -5 \quad \text{Subtract 2.}$$
$$y = 5 \quad \text{Multiply by } -1.$$

This gives the ordered pair $(2, 5)$. We plot the three ordered-pair solutions and draw a line through them. See **FIGURE 9**.

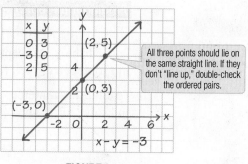

All three points should lie on the same straight line. If they don't "line up," double-check the ordered pairs.

FIGURE 9

 NOW TRY

EXAMPLE 2   Graphing a Linear Equation

Graph $4x - 5y = 20$.

   To find three ordered pairs that are solutions of $4x - 5y = 20$, we choose three arbitrary values for x or y that we think will be easy to substitute.

Let $x = 0$.	Let $y = 0$.	Let $y = 2$.
$4x - 5y = 20$	$4x - 5y = 20$	$4x - 5y = 20$
$4(0) - 5y = 20$	$4x - 5(0) = 20$	$4x - 5(2) = 20$
$0 - 5y = 20$	$4x - 0 = 20$	$4x - 10 = 20$
$-5y = 20$	$4x = 20$	$4x = 30$
$y = -4$	$x = 5$	$\frac{30}{4} = 7\frac{1}{2} \Rightarrow x = 7\frac{1}{2}$
Ordered pair: $(0, -4)$	Ordered pair: $(5, 0)$	Ordered pair: $\left(7\frac{1}{2}, 2\right)$

NOW TRY ANSWER

1.

**NOW TRY
EXERCISE 2**
Graph $2x - 4y = 8$.

We plot the three ordered-pair solutions and draw a line through them. See **FIGURE 10**. Two points determine the line, and the third point is used to check that no errors have been made.

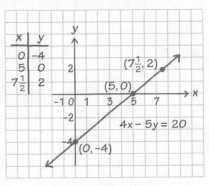

FIGURE 10

NOW TRY

OBJECTIVE 2 Find intercepts.

In **FIGURE 10**, the graph intersects (crosses) the x-axis at $(5, 0)$ and the y-axis at $(0, -4)$. For this reason, $(5, 0)$ is the **x-intercept** and $(0, -4)$ is the **y-intercept** of the graph. The intercepts are often convenient points to use when graphing linear equations.

Finding Intercepts

To find the x-intercept, let $y = 0$ in the given equation and solve for x. Then $(x, 0)$ is the x-intercept.

To find the y-intercept, let $x = 0$ in the given equation and solve for y. Then $(0, y)$ is the y-intercept.

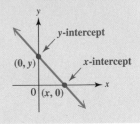

EXAMPLE 3   Graphing a Linear Equation Using Intercepts

Find the intercepts for the graph of $2x + y = 4$. Then draw the graph.

To find the intercepts, we first let $x = 0$ and then let $y = 0$. To find a third point, we arbitrarily let $x = 4$.

Let $x = 0$.	Let $y = 0$.	Let $x = 4$.
$2x + y = 4$	$2x + y = 4$	$2x + y = 4$
$2(0) + y = 4$	$2x + 0 = 4$	$2(4) + y = 4$
$0 + y = 4$	$2x = 4$	$8 + y = 4$
$y = 4$	$x = 2$	$y = -4$
y-intercept: $(0, 4)$	x-intercept: $(2, 0)$	Third point: $(4, -4)$

NOW TRY ANSWER
2.

The graph, with the two intercepts in red, and a table of values is shown in **FIGURE 11** on the next page.

NOW TRY EXERCISE 3

Find the intercepts for the graph of $x + 2y = 2$. Then draw the graph.

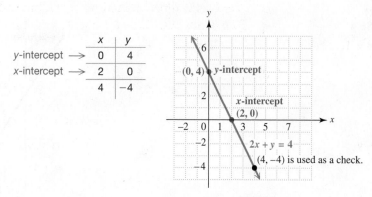

	x	y
y-intercept $\rightarrow$	0	4
x-intercept $\rightarrow$	2	0
	4	−4

$(4, −4)$ is used as a check.

FIGURE 11

NOW TRY

> ⓘ **CAUTION** *When choosing x- or y-values to find ordered pairs to plot, be careful to choose so that the resulting points are not too close together.* For example, using $(−1, −1)$, $(0, 0)$, and $(1, 1)$ to graph $x − y = 0$ may result in an inaccurate line. It is better to choose points whose x-values differ by at least 2.

EXAMPLE 4 **Graphing a Linear Equation Using Intercepts**

Graph $y = -\frac{3}{2}x + 3$.

Although this linear equation is not in standard form $(Ax + By = C)$, it *could* be written in that form. To find the intercepts, we first let $x = 0$ and then let $y = 0$.

$$y = -\frac{3}{2}x + 3$$

$$y = -\frac{3}{2}(0) + 3 \quad \text{Let } x = 0.$$

$$y = 0 + 3 \quad \text{Multiply.}$$

$$y = 3 \quad \text{Add.}$$

y-intercept: $(0, 3)$

$$y = -\frac{3}{2}x + 3$$

$$0 = -\frac{3}{2}x + 3 \quad \text{Let } y = 0.$$

$$\frac{3}{2}x = 3 \quad \text{Add } \frac{3}{2}x.$$

$$x = 2 \quad \text{Multiply by } \frac{2}{3}.$$

x-intercept: $(2, 0)$

To find a third point, we arbitrarily let $x = −2$.

$$y = -\frac{3}{2}x + 3$$

Choosing a multiple of 2 makes multiplying by $-\frac{3}{2}$ easier.

$$y = -\frac{3}{2}(-2) + 3 \quad \text{Let } x = -2.$$

$$y = 3 + 3 \quad \text{Multiply.}$$

$$y = 6 \quad \text{Add.}$$

Third point: $(−2, 6)$

We plot the three ordered-pair solutions and draw a line through them, as shown in **FIGURE 12** on the next page.

NOW TRY ANSWER

3. x-intercept: $(2, 0)$; y-intercept: $(0, 1)$

NOW TRY EXERCISE 4

Graph $y = \frac{1}{3}x + 1$.

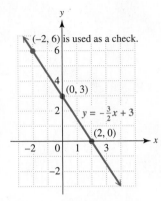

x	y
0	3
2	0
−2	6

$(−2, 6)$ is used as a check.

$y = -\frac{3}{2}x + 3$

FIGURE 12

NOW TRY

OBJECTIVE 3 Graph linear equations of the form $Ax + By = 0$.

NOW TRY EXERCISE 5

Graph $2x + y = 0$.

EXAMPLE 5   Graphing an Equation with x- and y-Intercepts $(0, 0)$

Graph $x - 3y = 0$.

To find the y-intercept, let $x = 0$.	To find the x-intercept, let $y = 0$.
$x - 3y = 0$	$x - 3y = 0$
$0 - 3y = 0$ Let $x = 0$.	$x - 3(0) = 0$ Let $y = 0$.
$-3y = 0$ Subtract.	$x - 0 = 0$ Multiply.
$y = 0$ Divide by -3.	$x = 0$ Subtract.
y-intercept: $(0, 0)$	x-intercept: $(0, 0)$

The x- and y-intercepts are the *same* point, $(0, 0)$. We must select *two other values* for x or y to find two other points on the graph. We choose $x = 6$ and $x = -3$.

> Choosing a multiple of 3 for x makes dividing by -3 in the last step easier.

$x - 3y = 0$	$x - 3y = 0$
$6 - 3y = 0$ Let $x = 6$.	$-3 - 3y = 0$ Let $x = -3$.
$-3y = -6$ Subtract 6.	$-3y = 3$ Add 3.
$y = 2$ Divide by -3.	$y = -1$ Divide by -3.
Ordered pair: $(6, 2)$	Ordered pair: $(-3, -1)$

We use the three ordered-pair solutions to draw the graph in **FIGURE 13**.

4.

5.

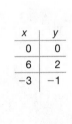

x	y
0	0
6	2
−3	−1

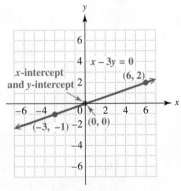

x-intercept and y-intercept

$x - 3y = 0$

$(6, 2)$

$(0, 0)$

$(-3, -1)$

FIGURE 13

NOW TRY

Line through the Origin

The graph of a linear equation of the form

$$Ax + By = 0,$$

where A and B are nonzero real numbers, passes through the origin $(0, 0)$.

OBJECTIVE 4 Graph linear equations of the form $y = b$ or $x = a$.

Consider the following linear equations.

$$y = -4 \quad \text{can be written as} \quad 0x + y = -4.$$

$$x = 3 \quad \text{can be written as} \quad x + 0y = 3.$$

When the coefficient of x or y is 0, the graph is a horizontal or vertical line.

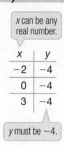

**NOW TRY
EXERCISE 6**

Graph $y = 2$.

EXAMPLE 6 Graphing a Horizontal Line ($y = b$)

Graph $y = -4$.

For any value of x, the value of y is always -4. Three ordered-pair solutions of the equation are shown in the table of values. Drawing a line through these points gives the **horizontal line** in **FIGURE 14**.

The y-intercept is $(0, -4)$.

There is no x-intercept.

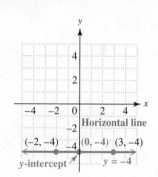

x can be any real number.

x	y
−2	−4
0	−4
3	−4

y must be −4.

FIGURE 14

NOW TRY

**NOW TRY
EXERCISE 7**

Graph $x + 4 = 0$.

EXAMPLE 7 Graphing a Vertical Line ($x = a$)

Graph $x - 3 = 0$.

First we add 3 to each side of the equation $x - 3 = 0$ to obtain $x = 3$. All ordered-pair solutions of this equation have x-coordinate 3. Any number can be used for y. Three ordered pairs that satisfy the equation are given in the table of values. The graph is the **vertical line** in **FIGURE 15**.

The x-intercept is $(3, 0)$.

There is no y-intercept.

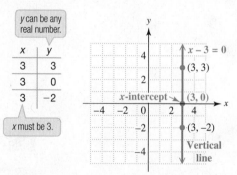

y can be any real number.

x	y
3	3
3	0
3	−2

x must be 3.

FIGURE 15

NOW TRY

Horizontal and Vertical Lines

The graph of **$y = b$,** where b is a real number, is a **horizontal line** with y-intercept $(0, b)$ and no x-intercept (unless the horizontal line is the x-axis itself).

The graph of **$x = a$,** where a is a real number, is a **vertical line** with x-intercept $(a, 0)$ and no y-intercept (unless the vertical line is the y-axis itself).

Keep the following in mind regarding the x- and y-axes.

- *The x-axis is the horizontal line given by the equation $y = 0$.*
- *The y-axis is the vertical line given by the equation $x = 0$.*

⚠ CAUTION The equations of horizontal and vertical lines are often confused with each other. The graph of $y = b$ is parallel to the x-axis and the graph of $x = a$ is parallel to the y-axis (for $a \neq 0$ and $b \neq 0$).

▼ **Forms of Linear Equations**

Equation	To Graph	Example
Ax + By = C (where A, B, and $C \neq 0$)	Find any two points on the line. A good choice is to find the intercepts. Let $x = 0$, and find the corresponding value of y. Then let $y = 0$, and find x. As a check, find a third point by choosing a value for x or y that has not yet been used.	
Ax + By = 0	The graph passes through the point $(0, 0)$. To find additional points that lie on the graph, choose any values for x or y, except 0.	
y = b	Draw a horizontal line, through the point $(0, b)$.	
x = a	Draw a vertical line, through the point $(a, 0)$.	

OBJECTIVE 5 Use a linear equation to model data.

EXAMPLE 8 Using a Linear Equation to Model Internet Use

In the United States, the weekly time spent online y in hours can be modeled by the linear equation

$$y = 0.96x + 9.1,$$

where $x = 0$ represents 2000, $x = 1$ represents 2001, and so on. (*Source: The 2013 Digital Future Report,* USC.)

(a) Use the equation to approximate weekly time spent online in the years 2000, 2006, and 2012.

**NOW TRY
EXERCISE 8**

Use **(a)** the graph and **(b)** the equation in **Example 8** to approximate weekly time spent online in 2009. (Round the answer in part (b) to the nearest tenth.)

Substitute the appropriate value for each year x to find weekly time spent online that year.

$$y = 0.96x + 9.1 \qquad \text{Given linear equation}$$

For 2000: $\quad y = 0.96(0) + 9.1 \qquad$ Replace x with 0.

$\qquad\qquad\ y = 9.1 \text{ hr} \qquad$ Multiply, and then add.

For 2006: $\quad y = 0.96(6) + 9.1 \qquad$ 2006 − 2000 = 6

$\qquad\qquad\ y \approx 14.9 \text{ hr} \qquad$ Replace x with 6.

For 2012: $\quad y = 0.96(12) + 9.1 \qquad$ 2012 − 2000 = 12

$\qquad\qquad\ y \approx 20.6 \text{ hr} \qquad$ Replace x with 12.

(b) Write the information from part (a) as three ordered pairs, and use them to graph the given linear equation.

Because x represents the year and y represents the time, the ordered pairs are

$$(0, 9.1), \quad (6, 14.9), \quad \text{and} \quad (12, 20.6).$$

See **FIGURE 16**. (Arrowheads are not included with the graphed line because the data are for the years 2000 to 2012 only—that is, from $x = 0$ to $x = 12$.)

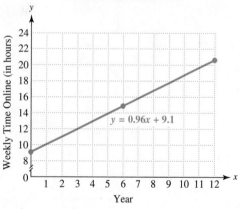

FIGURE 16

(c) Use the graph and then the equation to approximate weekly time spent online in 2010.

For 2010, $x = 10$. On the graph, find 10 on the horizontal axis, move up to the graphed line and then across to the vertical axis. It appears that weekly time spent online in 2010 was about 19 hr. To use the equation, substitute 10 for x.

$$y = 0.96x + 9.1 \qquad \text{Given linear equation}$$

$$y = 0.96(10) + 9.1 \qquad \text{Let } x = 10.$$

$$y = 18.7 \qquad \text{Multiply, and then add.}$$

This result for 2010 is close to our estimate of 19 hr from the graph. **NOW TRY**

NOW TRY ANSWERS
8. (a) about 18 hr
   (b) 17.7 hr

3.2 Exercises

FOR EXTRA HELP ▶ MyMathLab®

▶ *Complete solution available
in MyMathLab*

Concept Check   *Fill in each blank with the correct response.*

1. A linear equation in two variables x and y can be written in the form $Ax +$ _____ = _____ , where A, B, and C are real numbers and A and B are not both _____ .

2. The graph of any linear equation in two variables is a straight _____ . Every point on the line represents a _____ of the equation.

3. *Concept Check*   Match the information about each graph in Column I with the correct linear equation in Column II.

I	II
(a) The graph of the equation has y-intercept $(0, -4)$.	**A.** $3x + y = -4$
(b) The graph of the equation has $(0, 0)$ as x-intercept and y-intercept.	**B.** $x - 4 = 0$
	C. $y = 4x$
(c) The graph of the equation does not have an x-intercept.	**D.** $y = 4$
(d) The graph of the equation has x-intercept $(4, 0)$.	

4. *Concept Check*   Which of these equations have a graph with only one intercept?

    A. $x + 8 = 0$ **B.** $x - y = 3$ **C.** $x + y = 0$ **D.** $y = 4$

Concept Check   *Identify the intercepts of each graph. (Coordinates of the points shown are integers.)*

5. **6.** **7.** **8.**

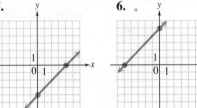

9. *Concept Check*   Match each equation in (a)–(d) with its graph in A–D.

    (a) $x = -2$ **(b)** $y = -2$ **(c)** $x = 2$ **(d)** $y = 2$

A. **B.** **C.** **D.**

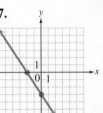
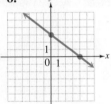

10. *Concept Check*   What is the equation of the x-axis? What is the equation of the y-axis?

Complete the given ordered-pair solutions for each equation. Then graph each equation by plotting the points and drawing a line through them. See Examples 1–4.

▶ **11.** $x + y = 5$

    $(0, \text{\_\_}), (\text{\_\_}, 0), (2, \text{\_\_})$

12. $x - y = 2$

    $(0, \text{\_\_}), (\text{\_\_}, 0), (5, \text{\_\_})$

▶ **13.** $y = \dfrac{2}{3}x + 1$

    $(0, \text{\_\_}), (3, \text{\_\_}), (-3, \text{\_\_})$

14. $y = -\dfrac{3}{4}x + 2$

    $(0, \text{\_\_}), (4, \text{\_\_}), (-4, \text{\_\_})$

15. $3x = -y - 6$

    $(0, \text{\_\_}), (\text{\_\_}, 0), \left(-\dfrac{1}{3}, \text{\_\_}\right)$

16. $x = 2y + 3$

    $(\text{\_\_}, 0), (0, \text{\_\_}), \left(\text{\_\_}, \dfrac{1}{2}\right)$

Find the x- and y-intercepts for the graph of each equation. See Examples 1–7.

17. $x - y = 8$ **18.** $x - y = 7$ **19.** $5x - 2y = 20$ **20.** $-3x + 2y = 12$

21. $x + 6y = 0$ **22.** $3x + y = 0$ **23.** $y = -2x + 4$ **24.** $y = 3x + 6$

25. $y = \frac{1}{3}x - 2$ **26.** $y = \frac{1}{4}x - 1$ **27.** $2x - 3y = 0$ **28.** $4x - 5y = 0$

29. $x - 4 = 0$ **30.** $x - 5 = 0$ **31.** $y = 2.5$ **32.** $y = -1.5$

Graph each linear equation. See Examples 1–7.

33. $x - y = 4$ **34.** $x - y = 5$ **35.** $2x + y = 6$

36. $-3x + y = -6$ **37.** $y = 2x - 5$ **38.** $y = 4x + 3$

39. $x = y + 2$ **40.** $x = -y + 6$ **41.** $2x - 5y = 10$

42. $3x + 2y = 6$ **43.** $3x + 7y = 14$ **44.** $6x - 5y = 18$

45. $y = -\frac{3}{4}x + 3$ **46.** $y = -\frac{2}{3}x - 2$ ▶ **47.** $y - 2x = 0$

48. $y + 3x = 0$ **49.** $y = -6x$ **50.** $y = 4x$

▶ **51.** $y = -1$ **52.** $y = 3$ **53.** $x = 5$

54. $x = -1$ ▶ **55.** $x + 2 = 0$ **56.** $x - 4 = 0$

57. $-3y = 15$ **58.** $-2y = 12$ **59.** $x + 2 = 8$ **60.** $x - 1 = -4$

Concept Check   *Describe what the graph of each linear equation will look like in the coordinate plane. (Hint: Rewrite the equation if necessary so that it is in a more recognizable form.)*

61. $3x = y - 9$ **62.** $2x = y - 4$ **63.** $x - 10 = 1$ **64.** $x + 4 = 3$

65. $3y = -6$ **66.** $5y = -15$ **67.** $2x = 4y$ **68.** $3x = 9y$

Extending Skills   *Plot each set of points, and draw a line through them. Then give an equation of the line.*

69. $(3, 5)$, $(3, 0)$, and $(3, -3)$ **70.** $(1, 3)$, $(1, 0)$, and $(1, -1)$

71. $(-3, -3)$, $(0, -3)$, and $(4, -3)$ **72.** $(-5, 5)$, $(0, 5)$, and $(3, 5)$

Solve each problem. See Example 8.

73. The weight y (in pounds) of a man taller than 60 in. can be approximated by the linear equation

$$y = 5.5x - 220,$$

where x is the height of the man in inches.

(a) Use the equation to approximate the weights of men whose heights are 62 in., 66 in., and 72 in.

(b) Write the information from part (a) as three ordered pairs.

(c) Graph the equation for $x \geq 62$, using the data from part (b).

(d) Use the graph to estimate the height of a man who weighs 155 lb. Then use the equation to find the height of this man to the nearest inch.

74. The height y (in centimeters) of a woman can be approximated by the linear equation

$$y = 3.9x + 73.5,$$

where x is the length of her radius bone (from the wrist to the elbow) in centimeters.

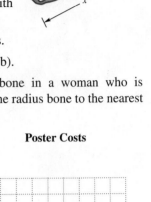

(a) Use the equation to approximate the heights of women with radius bones of lengths 20 cm, 26 cm, and 22 cm.

(b) Write the information from part (a) as three ordered pairs.

(c) Graph the equation for $x \geq 20$, using the data from part (b).

(d) Use the graph to estimate the length of the radius bone in a woman who is 167 cm tall. Then use the equation to find the length of the radius bone to the nearest centimeter.

75. As a fundraiser, a club is selling posters. The printer charges a $25 set-up fee, plus $0.75 for each poster. The cost y in dollars to print x posters is given by the linear equation

$$y = 0.75x + 25.$$

(a) What is the cost y in dollars to print 50 posters? To print 100 posters?

(b) Find the number of posters x if the printer billed the club for costs of $175.

(c) Write the information from parts (a) and (b) as three ordered pairs.

(d) Use the data from part (c) and the given grid to graph the equation for $x \geq 0$.

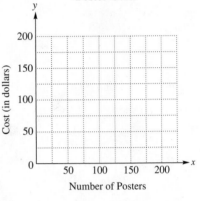

Poster Costs

76. A gas station is selling gasoline for $3.50 per gallon and charges $7 for a car wash. The cost y in dollars for x gallons of gasoline and a car wash is given by the linear equation

$$y = 3.50x + 7.$$

(a) What is the cost y in dollars for 9 gal of gasoline and a car wash? For 4 gal of gasoline and a car wash?

(b) Find the number of gallons of gasoline x if the cost for gasoline and a car wash is $35.

(c) Write the information from parts (a) and (b) as three ordered pairs.

(d) Use the data from part (c) and the given grid to graph the equation for $x \geq 0$.

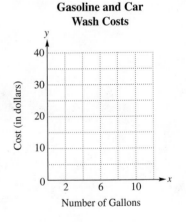

Gasoline and Car Wash Costs

77. The graph shows the value of a sport-utility vehicle (SUV) over the first 5 yr of ownership. Use the graph to do the following.

(a) Determine the initial value of the SUV.

(b) Find the **depreciation** (loss in value) from the original value after the first 3 yr.

(c) What is the annual or yearly depreciation in each of the first 5 yr?

(d) What does the ordered pair $(5, 5000)$ mean in the context of this problem?

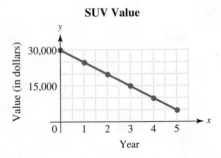

SUV Value

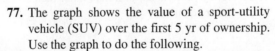

78. Demand for an item is often closely related to its price. As price increases, demand decreases, and as price decreases, demand increases. Suppose demand for a video game is 2000 units when the price is $40, and demand is 2500 units when the price is $30.

(a) Let *x* be the price and *y* be the demand for the game. Graph the two given pairs of prices and demands on the given grid.

(b) Assume that the relationship is linear. Draw a line through the two points from part (a). From the graph, estimate the demand if the price drops to $20.

(c) Use the graph to estimate the price if the demand is 3500 units.

(d) Write the prices and demands from parts (b) and (c) as ordered pairs.

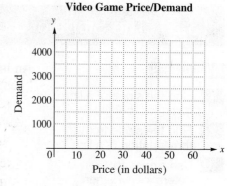

79. U.S. per capita consumption *y* of cheese in pounds from 2000 through 2012 is shown in the graph and modeled by the linear equation

$$y = 0.307x + 30.1,$$

where *x* = 0 represents 2000, *x* = 2 represents 2002, and so on.

(a) Use the equation to approximate cheese consumption (to the nearest tenth) in 2000, 2008, and 2012.

(b) Use the graph to estimate cheese consumption for the same years.

(c) How do the approximations using the equation compare with the estimates from the graph?

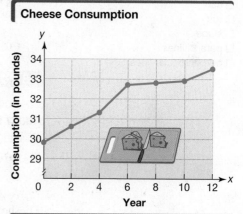

Source: U.S. Department of Agriculture.

(d) The USDA projects that per capita consumption of cheese in 2022 will be 36.8 lb. Use the equation to approximate per capita cheese consumption (to the nearest tenth) in 2022. How does the approximation using the equation compare to the USDA projection?

80. The number of U.S. marathon finishers *y* in thousands from 1990 through 2010 are shown in the graph and modeled by the linear equation

$$y = 13.36x + 220.8,$$

where *x* = 0 represents 1990, *x* = 5 represents 1995, and so on.

(a) Use the equation to approximate the number of U.S. marathon finishers in 1990, 2000, and 2010 to the nearest thousand.

(b) Use the graph to estimate the number of U.S. marathon finishers for the same years.

(c) How do the approximations using the equation compare to the estimates from the graph?

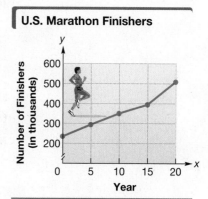

Source: Running U.S.A.

3.3 The Slope of a Line

VOCABULARY

☐ rise
☐ run
☐ slope
☐ parallel lines
☐ perpendicular lines

An important characteristic of the lines we graphed in **Section 3.2** is their slant, or "steepness" as viewed from *left to right*. See **FIGURE 17**.

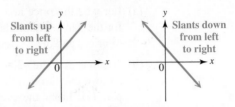

FIGURE 17

One way to measure the steepness of a line is to compare the vertical change in the line with the horizontal change while moving along the line from one fixed point to another. This measure of steepness is the *slope* of the line.

OBJECTIVE 1 Find the slope of a line given two points.

To find the steepness, or slope, of the line in **FIGURE 18**, we begin at point Q and move to point P. The vertical change, or **rise,** is the change in the y-values, which is the difference

$$6 - 1 = 5 \text{ units.}$$

The horizontal change, or **run,** is the change in the x-values, which is the difference

$$5 - 2 = 3 \text{ units.}$$

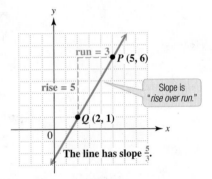

FIGURE 18

Remember from **Section 2.6** that one way to compare two numbers is by using a ratio. **Slope** is a ratio of the vertical change in y to the horizontal change in x. The line in **FIGURE 18** has

$$\text{slope} = \frac{\text{vertical change in } y \text{ (rise)}}{\text{horizontal change in } x \text{ (run)}} = \frac{5}{3}.$$

To confirm this ratio, we can count grid squares. We start at point Q in **FIGURE 18** and count *up* 5 grid squares to find the vertical change (rise). To find the horizontal change (run) and arrive at point P, we count to the *right* 3 grid squares. The slope is $\frac{5}{3}$, as found above.

We can summarize this discussion as follows.

Slope of a Line

Slope is a single number that allows us to determine the direction in which a line is slanting from left to right, as well as how much slant there is to the line.

NOW TRY
EXERCISE 1
Find the slope of the line.

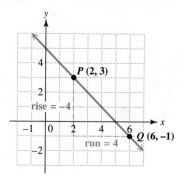

EXAMPLE 1 Finding the Slope of a Line

Find the slope of the line in **FIGURE 19**.

We use the coordinates of the two points shown on the line. The vertical change is the difference in the *y*-values.

$$-1 - 3 = -4$$

The horizontal change is the difference in the *x*-values.

$$6 - 2 = 4$$

Thus, the line has

$$\text{slope} = \frac{\text{change in } y \text{ (rise)}}{\text{change in } x \text{ (run)}} = \frac{-4}{4}, \text{ or } -1.$$

FIGURE 19

Counting grid squares, we begin at point *P* and count *down* 4 grid squares. Then we count to the *right* 4 grid squares to reach point *Q*. Because we counted down, we write the vertical change as a negative number, -4 here. The slope is $\frac{-4}{4}$, or -1.

NOW TRY

NOTE *The slope of a line is the same for any two points on the line.* In **FIGURE 19**, find the points $(3, 2)$ and $(5, 0)$ on the line. If we start at $(3, 2)$ and count *down* 2 units and then to the *right* 2 units, we arrive at $(5, 0)$. The slope is $\frac{-2}{2}$, or -1, the same slope we found in **Example 1**.

The concept of slope is used in many everyday situations. See **FIGURE 20**.

- A highway with a 10%, or $\frac{1}{10}$, grade (or slope) rises 1 m for every 10 m horizontally.

- A roof with pitch (or slope) $\frac{5}{12}$ rises 5 ft for every 12 ft that it runs horizontally.

- A stairwell with slope $\frac{8}{12}$ $\left(\text{or } \frac{2}{3}\right)$ indicates a vertical rise of 8 ft for a horizontal run of 12 ft.

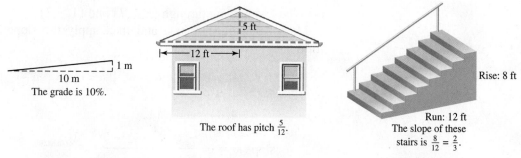

FIGURE 20

We can generalize the preceding discussion and find the slope of a line through two nonspecific points (x_1, y_1) and (x_2, y_2). (This notation is called **subscript notation.** Read x_1 as "**x-sub-one**" and x_2 as "**x-sub-two.**") See **FIGURE 21** on the next page.

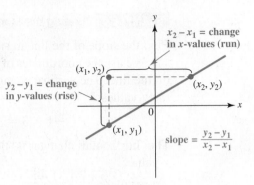

FIGURE 21

Moving along the line from the point (x_1, y_1) to the point (x_2, y_2), we see that y changes by $y_2 - y_1$ units. This is the vertical change (rise). Similarly, x changes by $x_2 - x_1$ units, which is the horizontal change (run). The slope of the line is the ratio of $y_2 - y_1$ to $x_2 - x_1$.

Slope Formula

The **slope m** of the line passing through the points (x_1, y_1) and (x_2, y_2) is defined as follows. (Traditionally, the letter m represents slope.)

$$m = \frac{\text{change in } y}{\text{change in } x} = \frac{y_2 - y_1}{x_2 - x_1} \quad (\text{where } x_1 \neq x_2)$$

The slope gives the change in y for each unit of change in x.

NOTE   Subscript notation is used to identify a point. It does *not* indicate an operation. **Note the difference between x_2, which represents a nonspecific value, and x^2, which means $x \cdot x$.** Read x_2 as "*x*-sub-two," *not* "*x* squared."

EXAMPLE 2   Finding Slopes of Lines

Find the slope of each line.

(a) The line passing through $(-4, 7)$ and $(1, -2)$

Label the given points, and then apply the slope formula.

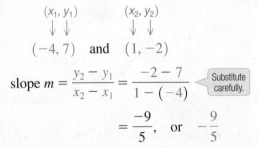

$$\text{slope } m = \frac{y_2 - y_1}{x_2 - x_1} = \frac{-2 - 7}{1 - (-4)} \quad \text{Substitute carefully.}$$

$$= \frac{-9}{5}, \quad \text{or} \quad -\frac{9}{5}$$

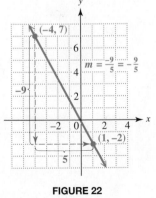

FIGURE 22

Begin at $(-4, 7)$ and count grid squares in **FIGURE 22** to confirm that the slope is $\frac{-9}{5}$, or $-\frac{9}{5}$.

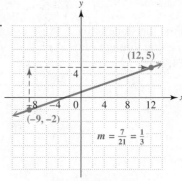

**NOW TRY
EXERCISE 2**

Find the slope of the line passing through $(4, -5)$ and $(-2, -4)$.

(b) The line passing through $(-9, -2)$ and $(12, 5)$

Label the points, and then apply the slope formula.

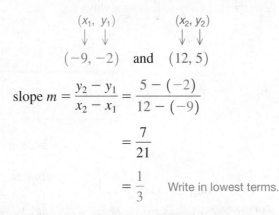

$$(x_1, \ y_1) \qquad (x_2, y_2)$$
$$(-9, -2) \quad \text{and} \quad (12, 5)$$

$$\text{slope } m = \frac{y_2 - y_1}{x_2 - x_1} = \frac{5 - (-2)}{12 - (-9)}$$

$$= \frac{7}{21}$$

$$= \frac{1}{3} \qquad \text{Write in lowest terms.}$$

Confirm this calculation using **FIGURE 23**. (Note the scale on the x- and y-axes.)

The same slope is obtained if we label the points in reverse order. ***It makes no difference which point is identified as*** (x_1, y_1) ***or*** (x_2, y_2)***.***

$$(x_2, \ y_2) \qquad (x_1, y_1)$$
$$(-9, -2) \quad \text{and} \quad (12, 5)$$

$$\text{slope } m = \frac{y_2 - y_1}{x_2 - x_1} = \frac{-2 - 5}{-9 - 12} \qquad \text{Substitute.}$$

$$= \frac{-7}{-21} \qquad \text{Subtract.}$$

$$= \frac{1}{3} \qquad \text{The same slope results.}$$

NOW TRY

The slopes of the lines in **FIGURES 22** and **23** suggest the following.

> **Orientation of Lines with Positive and Negative Slopes**
>
> A line with positive slope rises (slants up) from left to right.
>
> A line with negative slope falls (slants down) from left to right.

EXAMPLE 3 Finding the Slope of a Horizontal Line

Find the slope of the line passing through $(-5, 4)$ and $(2, 4)$.

$$(x_1, \ y_1) \qquad (x_2, y_2)$$
$$(-5, 4) \quad \text{and} \quad (2, 4) \qquad \text{Label the points.}$$

$$m = \frac{y_2 - y_1}{x_2 - x_1} = \frac{4 - 4}{2 - (-5)} \qquad \text{Substitute in the slope formula.}$$

$$= \frac{0}{7} \qquad \text{Subtract.}$$

$$= 0 \qquad \text{Slope 0}$$

NOW TRY ANSWER

2. $-\frac{1}{6}$

NOW TRY
EXERCISE 3

Find the slope of the line passing through $(1, -3)$ and $(4, -3)$.

As shown in **FIGURE 24**, the line passing through the two points $(-5, 4)$ and $(2, 4)$ is horizontal, with equation $y = 4$. ***All horizontal lines have slope 0*** because the difference in their y-values is always 0.

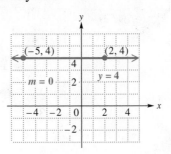

FIGURE 24

NOW TRY

NOW TRY
EXERCISE 4

Find the slope of the line passing through $(-2, 1)$ and $(-2, -4)$.

EXAMPLE 4 Applying the Slope Concept to a Vertical Line

Find the slope of the line passing through $(6, 2)$ and $(6, -4)$.

$$(x_1, y_1) \qquad (x_2, y_2)$$
$$\downarrow\downarrow \qquad\qquad \downarrow\downarrow$$
$$(6, 2) \quad \text{and} \quad (6, -4) \qquad \text{Label the points.}$$

$$m = \frac{y_2 - y_1}{x_2 - x_1} = \frac{-4 - 2}{6 - 6} \qquad \text{Substitute in the slope formula.}$$

$$= \frac{-6}{0} \qquad \text{Undefined slope}$$

Because division by 0 is undefined, this line has undefined slope. (This is why the slope formula has the restriction $x_1 \neq x_2$.)

The graph in **FIGURE 25** shows that this line is vertical, with equation $x = 6$. All points on a vertical line have the same x-value, so ***the slope of any vertical line is undefined.***

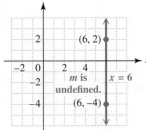

FIGURE 25

NOW TRY

Slopes of Horizontal and Vertical Lines

A **horizontal line,** which has an equation of the form $y = b$ (where b is a constant (number)), has **slope 0.**

A **vertical line,** which has an equation of the form $x = a$ (where a is a constant (number)), has **undefined slope.**

FIGURE 26 summarizes the four cases for slopes of lines.

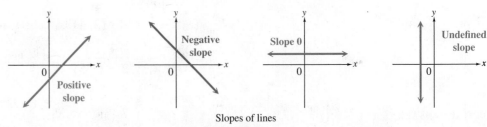

Slopes of lines

FIGURE 26

OBJECTIVE 2 Find the slope from the equation of a line.

Consider this linear equation.

$$y = -3x + 5$$

We can find the slope of this line using any two points on the line. Because the equation is solved for y, we find two points by choosing two different values of x and then finding the corresponding values of y. We arbitrarily choose $x = -2$ and $x = 4$.

$y = -3x + 5$			$y = -3x + 5$	
$y = -3(-2) + 5$	Let $x = -2.$		$y = -3(4) + 5$	Let $x = 4.$
$y = 6 + 5$	Multiply.		$y = -12 + 5$	Multiply.
$y = 11$	Add.		$y = -7$	Add.

The ordered pairs are $(-2, 11)$ and $(4, -7)$. Now we apply the slope formula.

$$m = \frac{-7 - 11}{4 - (-2)} = \frac{-18}{6} = -3$$

The slope, -3, is the same number as the coefficient of x in the given equation $y = -3x + 5$. It can be shown that this always happens, *as long as the equation is solved for y.* This fact is used to find the slope of a line from its equation.

Finding the Slope of a Line from Its Equation

Step 1   Solve the equation for y. (See **Section 2.5.**)

Step 2   The slope is given by the coefficient of x.

EXAMPLE 5   **Finding Slopes from Equations**

Find the slope of each line.

(a) $2x - 5y = 4$

   Step 1   Solve the equation for y.

$$2x - 5y = 4 \quad \text{← Isolate } y \text{ on one side.}$$

$$-5y = 4 - 2x \qquad \text{Subtract } 2x.$$

$$-5y = -2x + 4 \qquad \text{Commutative property}$$

$$\frac{-2x}{-5} = \frac{-2}{-5}x = \frac{2}{5}x \quad\Rightarrow\quad y = \frac{2}{5}x - \frac{4}{5} \qquad \text{Divide } each \text{ term by } -5.$$

$$\uparrow$$
$$\text{Slope}$$

   Step 2   The slope is given by the coefficient of x, so the slope is $\frac{2}{5}$.

(b)
$$8x + 4y = 1$$

$$\boxed{\text{Solve for } y.} \quad 4y = 1 - 8x \qquad \text{Subtract } 8x.$$

$$4y = -8x + 1 \qquad \text{Commutative property}$$

$$y = -2x + \frac{1}{4} \qquad \text{Divide } each \text{ term by } 4.$$

The slope is given by the coefficient of x, which is -2.

**NOW TRY
EXERCISE 5**
Find the slope of the line.

$$3x + 5y = -1$$

(c)

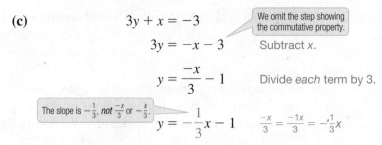

$$3y + x = -3$$

> We omit the step showing the commutative property.

$$3y = -x - 3 \qquad \text{Subtract } x.$$

$$y = \frac{-x}{3} - 1 \qquad \text{Divide \textit{each} term by 3.}$$

> The slope is $-\frac{1}{3}$, *not* $\frac{-x}{3}$ or $\frac{x}{3}$.

$$y = -\frac{1}{3}x - 1 \qquad \frac{-x}{3} = \frac{-1x}{3} = -\frac{1}{3}x$$

The coefficient of x is $-\frac{1}{3}$, so the slope of this line is $-\frac{1}{3}$.

NOW TRY

NOTE   We can solve the linear equation $Ax + By = C$ (where $B \neq 0$) for y to show that, in general, the slope of a line is $m = -\frac{A}{B}$.

OBJECTIVE 3   Use slopes to determine whether two lines are parallel, perpendicular, or neither.

Two lines in a plane that never intersect are **parallel.** We use slopes to tell whether two lines are parallel.

FIGURE 27 on the next page shows the graphs of $x + 2y = 4$ and $x + 2y = -6$. These lines appear to be parallel. We solve each equation for y to find the slope.

$x + 2y = 4$	$x + 2y = -6$
$2y = -x + 4 \qquad \text{Subtract } x.$	$2y = -x - 6 \qquad \text{Subtract } x.$
$y = \dfrac{-x}{2} + 2 \qquad \text{Divide by 2.}$	$y = \dfrac{-x}{2} - 3 \qquad \text{Divide by 2.}$
$y = -\dfrac{1}{2}x + 2 \qquad \frac{-x}{2} = \frac{-1x}{2} = -\frac{1}{2}x$	$y = -\dfrac{1}{2}x - 3 \qquad \frac{-x}{2} = \frac{-1x}{2} = -\frac{1}{2}x$
↑ Slope	↑ Slope

> The slope is $-\frac{1}{2}$, not $-\frac{x}{2}$.

Both lines have slope $-\frac{1}{2}$. *Nonvertical parallel lines always have equal slopes.*

FIGURE 28 shows the graphs of $x + 2y = 4$ and $2x - y = 6$. These lines appear to be **perpendicular** (that is, they intersect at a 90° angle). As shown above, solving $x + 2y = 4$ for y gives $y = -\frac{1}{2}x + 2$, with slope $-\frac{1}{2}$. We solve $2x - y = 6$ for y.

$$2x - y = 6$$

$$-y = -2x + 6 \qquad \text{Subtract } 2x.$$

$$y = 2x - 6 \qquad \text{Multiply by } -1.$$

↑
Slope

The product of the slopes of the two lines is

$$-\frac{1}{2}(2) = -1.$$

NOW TRY ANSWER

5. $-\frac{3}{5}$

▼ Negative Reciprocals

Number	Negative Reciprocal
$\frac{3}{4}$	$-\frac{4}{3}$
$\frac{1}{2}$	$-\frac{2}{1}$, or -2
-6, or $-\frac{6}{1}$	$\frac{1}{6}$
-0.4, or $-\frac{4}{10}$	$\frac{10}{4}$, or 2.5

The product of each number and its negative reciprocal is -1.

It can be proved that the product of the slopes of two perpendicular lines, neither of which is vertical, is always -1. This means that the slopes of perpendicular lines are negative (or opposite) reciprocals—if one slope is the nonzero number a, then the other is $-\frac{1}{a}$.

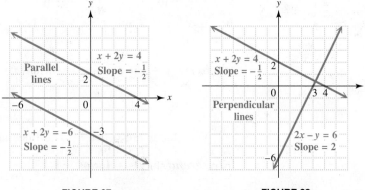

FIGURE 27   FIGURE 28

Slopes of Parallel and Perpendicular Lines

Two lines with the same slope are parallel.

Two lines whose slopes have a product of -1 are perpendicular.

EXAMPLE 6 Deciding Whether Two Lines Are Parallel or Perpendicular

Decide whether each pair of lines is *parallel, perpendicular,* or *neither.*

(a) $x + 3y = 7$

$-3x + \ y = 3$

Find the slope of each line by first solving each equation for y.

$$x + 3y = 7 \qquad\qquad\qquad -3x + y = 3$$
$$3y = -x + 7 \quad \text{Subtract } x. \qquad\qquad y = 3x + 3 \quad \text{Add } 3x.$$
$$y = -\frac{1}{3}x + \frac{7}{3} \quad \text{Divide by 3.}$$

The slope is $-\frac{1}{3}$.   The slope is 3.

The slopes are not equal, so the lines are not parallel. Find the product of the slopes.

$$-\frac{1}{3}(3) = -1 \qquad \text{The slopes are negative reciprocals.}$$

The two lines are perpendicular because the product of their slopes is -1.

(b) $4x - \ y = 4$

$8x - 2y = -12$

Solve each equation for y, and identify the slope.

$$4x - y = 4 \qquad\qquad\qquad 8x - 2y = -12$$
$$-y = -4x + 4 \quad \text{Subtract } 4x. \qquad -2y = -8x - 12 \quad \text{Subtract } 8x.$$
$$y = 4x - 4 \quad \text{Multiply by } -1. \qquad y = 4x + 6 \quad \text{Divide by } -2.$$

Both lines have slope 4, so the lines are parallel.

**NOW TRY
EXERCISE 6**

Decide whether the pair of
lines is *parallel*, *perpendicular*,
or *neither*.

$$2x - 3y = 1$$
$$4x + 6y = 5$$

(c) $4x + 3y = 6$ Solve for *y*. $y = -\dfrac{4}{3}x + 2$

$2x - y = 5$ $\longrightarrow$ $y = 2x - 5$

The slopes are $-\dfrac{4}{3}$ and 2. These two lines are neither parallel nor perpendicular,
because $-\dfrac{4}{3} \neq 2$ and $-\dfrac{4}{3} \cdot 2 \neq -1$.

(d) $6x - y = 1$ Solve for *y*. $y = 6x - 1$

$x - 6y = -12$ $\longrightarrow$ $y = \dfrac{1}{6}x + 2$

The slopes are 6 and $\dfrac{1}{6}$. The lines are not parallel, nor are they perpendicular.

$\left(\textit{Be careful. } \mathbf{6\left(\dfrac{1}{6}\right) = 1, \textit{not} -1.}\right)$

NOW TRY

NOW TRY ANSWER
6. neither

3.3 Exercises

FOR EXTRA HELP   ▶ MyMathLab®

▶ *Complete solution available
in MyMathLab*

Concept Check   *Work each problem involving slope.*

1. Slope is a measure of the _____ of a line. Slope is the (*horizontal / vertical*) change
compared to the (*horizontal / vertical*) change while moving along the line from one point
to another.

2. Slope is the _____ of the vertical change in _____, called the (*rise / run*), to the
horizontal change in _____, called the (*rise / run*).

3. Look at the graph at the right.

   (a) Start at the point $(-1, -4)$ and count vertically up to
   the horizontal line that goes through the other plotted
   point. What is this vertical change? (Remember: "up"
   means positive, "down" means negative.)

   (b) From this new position, count horizontally to the other
   plotted point. What is this horizontal change? (Remember: "right" means positive, "left" means negative.)

   (c) What is the ratio (quotient) of the numbers found in
   parts (a) and (b)? What do we call this number?

4. See **Exercise 3**. If we were to *start* at the point $(3, 2)$ and *end* at the point $(-1, -4)$
would the answer to **Exercise 3(c)** be the same? Explain.

5. Match the graph of each line in (a)–(d) with its slope in A–D. (Coordinates of the points
shown are integers.)

(a)   **(b)**   **(c)**   **(d)**

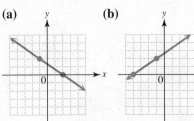   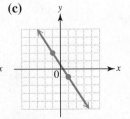   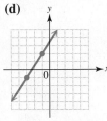

A. $\dfrac{2}{3}$ B. $\dfrac{3}{2}$ C. $-\dfrac{2}{3}$ D. $-\dfrac{3}{2}$

6. Decide whether the line with the given slope rises from left to right, falls from left to right, is horizontal, or is vertical.

(a) $m = -4$ (b) $m = 0$ (c) m is undefined.   (d) $m = \dfrac{3}{7}$

Concept Check On a pair of axes similar to the one shown, sketch the graph of a straight line having the indicated slope.

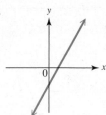

7. Negative   **8.** Positive

9. Undefined   **10.** Zero

Concept Check The figure at the right shows a line that has a positive slope (because it rises from left to right) and a positive y-value for the y-intercept (because it intersects the y-axis above the origin).

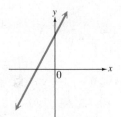

For each line in Exercises 11–16, decide whether

(a) the slope is positive, negative, or zero and

(b) the y-value of the y-intercept is positive, negative, or zero.

11.

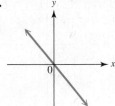

12.

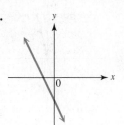

13.

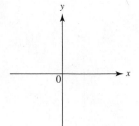

14.

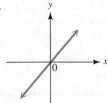

15.

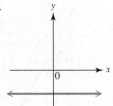

16.

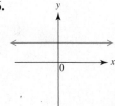

17. *Concept Check* A student was asked to find the slope of the line through the points $(2, 5)$ and $(-1, 3)$. His answer, $-\dfrac{2}{3}$, was incorrect. He showed his work as

$$\frac{3 - 5}{2 - (-1)} = \frac{-2}{3} = -\frac{2}{3}.$$

WHAT WENT WRONG? Give the correct slope.

18. *Concept Check* A student was asked to find the slope of the line through the points $(-2, 4)$ and $(6, -1)$. Her answer, $-\dfrac{8}{5}$, was incorrect. She showed her work as

$$\frac{6 - (-2)}{-1 - 4} = \frac{8}{-5} = -\frac{8}{5}.$$

WHAT WENT WRONG? Give the correct slope.

Concept Check *Find each slope.*

19. What is the slope (or grade) of this hill?

32 m

108 m

20. What is the slope (or pitch) of this roof?

6 ft

20 ft

21. What is the slope of the slide? (*Hint:* The slide *drops* 8 ft vertically as it extends 12 ft horizontally.)

−8 ft

12 ft

22. What is the slope (or grade) of this ski slope? (*Hint:* The ski slope *drops* 25 ft vertically for every 100 horizontal feet.)

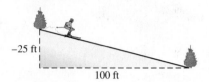

−25 ft

100 ft

Use the coordinates of the indicated points to find the slope of each line. (Coordinates of the points shown are integers.) *See Example 1.*

▶ 23.

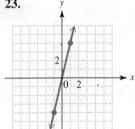

24.

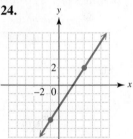

25.

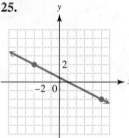

26.

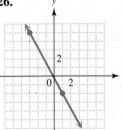

27.

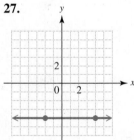

28.

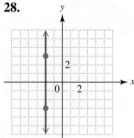

Find the slope of the line passing through each pair of points. *See Examples 2–4.*

▶ 29. $(1, -2)$ and $(-3, -7)$ **30.** $(4, -1)$ and $(-2, -8)$ **31.** $(0, 3)$ and $(-2, 0)$

32. $(8, 0)$ and $(0, -5)$ **▶ 33.** $(4, 3)$ and $(-6, 3)$ **34.** $(6, 5)$ and $(-12, 5)$

35. $(-2, 4)$ and $(-3, 7)$ **36.** $(-4, 5)$ and $(-5, 8)$

▶ 37. $(-12, 3)$ and $(-12, -7)$ **38.** $(-8, 6)$ and $(-8, -1)$

39. $(4.8, 2.5)$ and $(3.6, 2.2)$ **40.** $(3.1, 2.6)$ and $(1.6, 2.1)$

41. $\left(-\frac{7}{5}, \frac{3}{10}\right)$ and $\left(\frac{1}{5}, -\frac{1}{2}\right)$ **42.** $\left(-\frac{4}{3}, \frac{1}{2}\right)$ and $\left(\frac{1}{3}, -\frac{5}{6}\right)$

Find the slope of each line. See Example 5.

43. $y = 5x + 12$

44. $y = 2x + 3$

45. $4y = x + 1$

46. $2y = x + 4$

▶ **47.** $3x - 2y = 3$

48. $6x - 4y = 4$

49. $-3x + 2y = 5$

50. $-2x + 4y = 5$

51. $x + y = -4$

52. $x - y = -2$

53. $y = -5$

54. $y = 4$

55. $x = 6$

56. $x = -2$

Find the slope of each line in two ways by doing the following.

(a) *Give any two points that lie on the line, and use them to determine the slope.*

(b) *Solve the equation for y, and identify the slope from the equation.*

See Objective 2 and Example 5.

57. $2x + y = 10$

58. $-4x + y = -8$

59. $5x - 3y = 15$

60. $3x + 2y = 12$

Each table of values gives several points that lie on a line.

(a) *Use any two of the ordered pairs to find the slope of the line.*

(b) *What is the x-intercept of the line?   The y-intercept?*

(c) *Graph the line.*

61.

x	y
−4	0
−2	2
0	4
1	5

62.

x	y
−4	3
−1	0
0	−1
2	−3

63.

x	y
3	−3
0	−2
−3	−1
−6	0

64.

x	y
−1	−6
0	−4
2	0
5	6

Concept Check   Answer each question.

65. What is the slope of a line whose graph is

 (a) parallel to the graph of $3x + y = 7$?

 (b) perpendicular to the graph of $3x + y = 7$?

66. What is the slope of a line whose graph is

 (a) parallel to the graph of $-5x + y = -3$?

 (b) perpendicular to the graph of $-5x + y = -3$?

67. If two lines are both vertical or both horizontal, which of the following are they?

 A. Parallel   **B.** Perpendicular   **C.** Neither parallel nor perpendicular

68. If a line is vertical, what is true of any line that is perpendicular to it?

For each pair of equations, give the slopes of the lines and then determine whether the two lines are parallel, perpendicular, *or* neither. *See Example 6.*

▶ **69.** $2x + 5y = 4$
   $4x + 10y = 1$

70. $-4x + 3y = 4$
   $-8x + 6y = 0$

71. $8x - 9y = 6$
   $8x + 6y = -5$

72. $5x - 3y = -2$
   $3x - 5y = -8$

73. $3x - 2y = 6$
   $2x + 3y = 3$

74. $3x - 5y = -1$
   $5x + 3y = 2$

75. $5x - y = 1$
   $x - 5y = -10$

76. $3x - 4y = 12$
   $4x + 3y = 12$

The graph shows album sales (which include CD, vinyl, cassette, and digital albums) and music purchases (which include digital tracks, albums, singles, and music videos) in millions of units from 2004 through 2012. Use the graph to work Exercises 77 and 78.

77. Locate the line on the graph that represents music purchases.

    (a) Write two ordered pairs (x, y), where x is the year and y is purchases in millions of units, to represent the data for the years 2004 and 2012.

    (b) Use the ordered pairs from part (a) to find the slope of the line.

    (c) Interpret the meaning of the slope in the context of this problem.

78. Locate the line on the graph that represents album sales. Repeat parts (a)–(c) of **Exercise 77.** For part (a), x is the year and y is sales in millions of units.

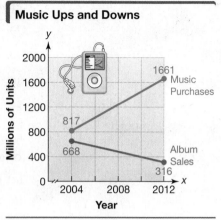

Music Ups and Downs

Source: Nielsen SoundScan.

RELATING CONCEPTS   For Individual or Group Work (Exercises 79–84)

FIGURE A *gives the percent of freshmen at 4-year colleges and universities who planned to major in the Biological Sciences.* **FIGURE B** *shows the percent of the same group of students who planned to major in Business.* **Work Exercises 79–84 in order.**

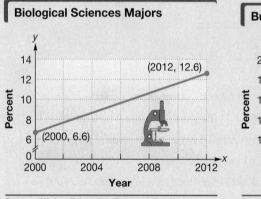

Biological Sciences Majors

Source: Higher Education Research Institute.

FIGURE A

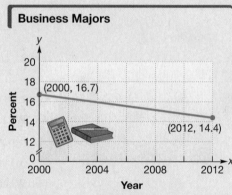

Business Majors

Source: Higher Education Research Institute.

FIGURE B

79. Use the given ordered pairs to find the slope of the line in **FIGURE A**.

80. The slope of the line in **FIGURE A** is (*positive / negative*). This means that during the period represented, the percent of freshmen planning to major in the Biological Sciences (*increased / decreased*).

81. The slope of a line represents the *rate of change*. Based on **FIGURE A**, what was the increase in the percent of freshmen *per year* who planned to major in the Biological Sciences during the period shown?

82. Use the given ordered pairs to find the slope of the line in **FIGURE B** to the nearest tenth.

83. The slope of the line in **FIGURE B** is (*positive / negative*). This means that during the period represented, the percent of freshmen planning to major in Business (*increased / decreased*).

84. Based on **FIGURE B**, what was the decrease in the percent of freshmen *per year* who planned to major in Business?

STUDY SKILLS

Preparing for Your Math Final Exam

Your math final exam is likely to be a comprehensive exam, which means it will cover material from the entire term. **One way to prepare for it now is by working a set of Cumulative Review Exercises** each time your class finishes a chapter. This continual review will help you remember concepts and procedures as you progress through the course.

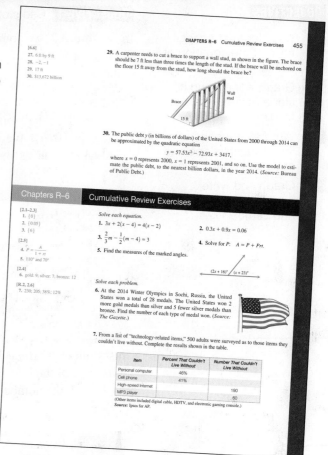

Final Exam Preparation Suggestions

1. **Figure out the grade you need to earn on the final exam to get the course grade you want.** Check your course syllabus for grading policies, or ask your instructor if you are not sure.

   How many points do you need to earn on your math final exam to get the grade you want?

2. **Create a final exam week plan.** Set priorities that allow you to spend extra time studying. This may mean making adjustments, in advance, in your work schedule or enlisting extra help with family responsibilities.

   What adjustments do you need to make for final exam week? List two or three here.

3. **Use the following suggestions to guide your studying.**

   - **Begin reviewing several days before the final exam.** DON'T wait until the last minute.

   - **Know exactly which chapters and sections will be covered on the exam.**

   - **Divide up the chapters.** Decide how much you will review each day.

   - **Keep returned quizzes and tests.** Use them to review.

   - **Practice all types of problems. Use the Cumulative Review Exercises** at the end of each chapter in your textbook beginning in Chapter 2. All answers, with section references, are given in the margins.

   - **Review or rewrite your notes** to create summaries of important information.

   - **Make study cards for all types of problems.** Carry the cards with you, and review them whenever you have a few minutes.

   - **Take plenty of short breaks as you study to reduce physical and mental stress.** Exercising, listening to music, and enjoying a favorite activity are effective stress busters.

   Finally, *DON'T* stay up all night the night before an exam—*get a good night's sleep.*

   Which of these suggestions will you use as you study for your math final exam? List two or three here.

3.4 Slope-Intercept Form of a Linear Equation

OBJECTIVES

1 Use slope-intercept form of the equation of a line.

2 Graph a line using its slope and a point on the line.

3 Write an equation of a line using its slope and any point on the line.

4 Graph and write equations of horizontal and vertical lines.

OBJECTIVE 1 Use slope-intercept form of the equation of a line.

In **Section 3.3,** we found the slope (steepness) of a line by solving the equation of the line for y. In that form, the slope is the coefficient of x. For example, the line with equation

$$y = 2x + 3 \quad \text{has slope} \quad 2.$$

What does the number 3 represent? To find out, suppose a line has slope m and y-intercept $(0, b)$. We can find an equation of this line by choosing another point (x, y) on the line, as shown in **FIGURE 29**. Then we apply the slope formula.

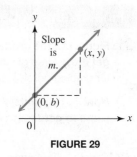

$$m = \frac{y - b}{x - 0} \quad \begin{array}{l} \leftarrow \text{Change in } y\text{-values} \\ \leftarrow \text{Change in } x\text{-values} \end{array}$$

$$m = \frac{y - b}{x} \quad \begin{array}{l} \text{Subtract in the} \\ \text{denominator.} \end{array}$$

$$mx = y - b \quad \text{Multiply by } x.$$

$$mx + b = y \quad \text{Add } b.$$

$$y = mx + b \quad \text{Rewrite.}$$

FIGURE 29

This result is the *slope-intercept form* of the equation of a line. Both the *slope* and the *y-intercept* of the line can be read directly from this form. For the line with equation $y = 2x + 3$, the number 3 gives the y-intercept $(0, 3)$.

> **Slope-Intercept Form**
>
> The **slope-intercept form** of the equation of a line with slope m and y-intercept $(0, b)$ is
>
> $$y = mx + b.$$
> Slope ⤴   ⤴ $(0, b)$ is the y-intercept.
>
> ***The intercept given is the y-intercept.***

NOW TRY EXERCISE 1

Identify the slope and y-intercept of the line with each equation.

(a) $y = -\dfrac{3}{5}x - 9$

(b) $y = -\dfrac{x}{3} + \dfrac{7}{3}$

EXAMPLE 1 Identifying Slopes and y-Intercepts

Identify the slope and y-intercept of the line with each equation.

(a) $y = -4x + 1$
Slope ⤴   ⤴ y-intercept $(0, 1)$

(b) $y = x - 8$ can be written as $y = 1x + (-8)$.
                Slope ⤴   ⤴ y-intercept $(0, -8)$

(c) $y = 6x$ can be written as $y = 6x + 0$.
                Slope ⤴   ⤴ y-intercept $(0, 0)$

NOW TRY ANSWERS

1. (a) slope: $-\frac{3}{5}$; y-intercept: $(0, -9)$

   (b) slope: $-\frac{1}{3}$; y-intercept: $\left(0, \frac{7}{3}\right)$

(d) $y = \dfrac{x}{4} - \dfrac{3}{4}$ can be written as $y = \dfrac{1}{4}x + \left(-\dfrac{3}{4}\right)$.
                Slope ⤴   ⤴ y-intercept $\left(0, -\dfrac{3}{4}\right)$ **NOW TRY**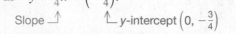

NOTE Slope-intercept form is an especially useful form for a linear equation because of the information we can determine from it. It is also the form used by graphing calculators and the one that describes a *linear function*.

OBJECTIVE 2 Graph a line using its slope and a point on the line.

We can use the slope and the point represented by the y-intercept to graph a line.

Graphing a Line Using the Slope and y-Intercept

Step 1 Write the equation in slope-intercept form $y = mx + b$, if necessary, by solving for y.

Step 2 Identify the y-intercept. Plot the point $(0, b)$.

Step 3 Identify the slope m of the line. Use the geometric interpretation of slope ("*rise over run*") to find another point on the graph by counting from the y-intercept.

Step 4 Join the two points with a line to obtain the graph. (If desired, obtain a third point, such as the x-intercept, as a check.)

EXAMPLE 2 Graphing Lines Using Slopes and y-Intercepts

Graph the equation of each line using the slope and y-intercept.

(a) $y = \dfrac{2}{3}x - 1$

Step 1 The equation is in slope-intercept form.

$$y = \underset{\uparrow}{\dfrac{2}{3}}x \underset{\uparrow}{- 1}$$

Slope   Value of b in y-intercept $(0, b)$

Step 2 The y-intercept is $(0, -1)$. Plot this point. See **FIGURE 30**.

Step 3 The slope is $\dfrac{2}{3}$. By definition,

$$\text{slope } m = \dfrac{\text{change in } y \text{ (rise)}}{\text{change in } x \text{ (run)}} = \dfrac{2}{3}.$$

From the y-intercept, count up 2 units and to the right 3 units to the point $(3, 1)$.

Step 4 Draw the line through the points $(0, -1)$ and $(3, 1)$ to obtain the graph in **FIGURE 30**.

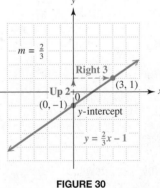

FIGURE 30

(b) $3x + 4y = 8$

Step 1 Solve for y to write the equation in slope-intercept form.

$$3x + 4y = 8$$

Isolate y on one side.

$$4y = -3x + 8 \qquad \text{Subtract } 3x.$$

Slope-intercept form $\rightarrow y = -\dfrac{3}{4}x + 2 \qquad$ Divide by 4.

**NOW TRY
EXERCISE 2**
Graph $3x + 2y = 8$ using the slope and y-intercept.

Step 2 The y-intercept in $y = -\frac{3}{4}x + 2$ is $(0, 2)$. Plot this point. See **FIGURE 31**.

Step 3 The slope is $-\frac{3}{4}$, which can be written as either $\frac{-3}{4}$ or $\frac{3}{-4}$. We use $\frac{-3}{4}$ here.

$$m = \frac{\text{change in } y \text{ (rise)}}{\text{change in } x \text{ (run)}} = \frac{-3}{4}$$

From the y-intercept, count *down* 3 units (because of the negative sign) and to the right 4 units to the point $(4, -1)$.

Step 4 Draw the line through the two points $(0, 2)$ and $(4, -1)$ to obtain the graph in **FIGURE 31**.

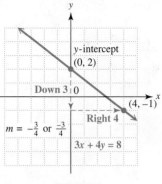

FIGURE 31

NOW TRY

NOTE In Step 3 of **Example 2(b)**, we could use $\frac{3}{-4}$ for the slope. From the y-intercept $(0, 2)$ in **FIGURE 31**, count up 3 units and to the *left* 4 units (because of the negative sign) to the point $(-4, 5)$. Verify that this produces the same line.

**NOW TRY
EXERCISE 3**
Graph the line passing through the point $(-3, -4)$, with slope $\frac{5}{2}$.

EXAMPLE 3   Graphing a Line Using Its Slope and a Point

Graph the line passing through the point $(-2, 3)$, with slope -4.

First, plot the point $(-2, 3)$. See **FIGURE 32**. Then write the slope -4 as

$$m = \frac{\text{change in } y \text{ (rise)}}{\text{change in } x \text{ (run)}} = \frac{-4}{1}.$$

Locate another point on the line by counting *down* 4 units from $(-2, 3)$ and then to the right 1 unit. Finally, draw the line through this new point P and the given point $(-2, 3)$. See **FIGURE 32**.

We could have written the slope as $\frac{4}{-1}$ instead. In this case, we would move up 4 units from $(-2, 3)$ and then to the *left* 1 unit. Verify that this produces the same line.

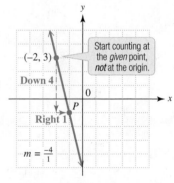

FIGURE 32

NOW TRY

OBJECTIVE 3 Write an equation of a line using its slope and any point on the line.

We can use the slope-intercept form to write the equation of a line if we know the slope and any point on the line.

NOW TRY ANSWERS
2.
3.

EXAMPLE 4   Using Slope-Intercept Form to Write Equations

Write an equation in slope-intercept form of the line passing through the given point and having the given slope.

(a) $(0, -1)$, $m = \frac{2}{3}$

Because the point $(0, -1)$ is the y-intercept, $b = -1$. We can substitute this value for b and the given slope $m = \frac{2}{3}$ directly into slope-intercept form $y = mx + b$ to write an equation.

**NOW TRY
EXERCISE 4**

Write an equation in slope-intercept form of the line passing through the given point and having the given slope.

(a) $(0, 2)$, $m = -4$

(b) $(-2, 1)$, $m = 3$

Slope $\searrow$ $\swarrow$ y-intercept is $(0, b)$.

$$y = mx + b \qquad \text{Slope-intercept form}$$

$$y = \frac{2}{3}x + (-1) \qquad \text{Substitute.}$$

$$y = \frac{2}{3}x - 1 \qquad \text{Definition of subtraction}$$

(b) $(2, 5)$, $m = 4$

This line passes through the point $(2, 5)$, which is *not* the y-intercept because the x-coordinate is 2, *not* 0. *We cannot substitute directly as in part (a).* We can find the y-intercept by substituting $x = 2$ and $y = 5$ from the given point and the given slope $m = 4$ into $y = mx + b$ and solving for b.

$$y = mx + b \qquad \text{Slope-intercept form}$$

$$5 = 4(2) + b \qquad \text{Let } x = 2, y = 5, \text{ and } m = 4.$$

> $(0, b)$ is the y-intercept. Don't stop here.

$$5 = 8 + b \qquad \text{Multiply.}$$

$$-3 = b \qquad \text{Subtract 8.}$$

Now substitute the values of m and b into slope-intercept form.

$$y = mx + b \qquad \text{Slope-intercept form}$$

$$y = 4x - 3 \qquad \text{Let } m = 4 \text{ and } b = -3. \qquad \textbf{NOW TRY} \ \circlearrowleft$$

OBJECTIVE 4 Graph and write equations of horizontal and vertical lines.

EXAMPLE 5 Graphing Horizontal and Vertical Lines Using Slope and a Point

Graph each line passing through the given point and having the given slope.

(a) $(4, -2)$, $m = 0$

Recall from **Section 3.3** that horizontal lines have slope 0. To graph this line, plot the point $(4, -2)$ and draw the horizontal line through it. See **FIGURE 33**.

**NOW TRY
EXERCISE 5**

Graph each line passing through the given point and having the given slope.

(a) $(-3, 3)$, undefined slope

(b) $(3, -3)$, slope 0

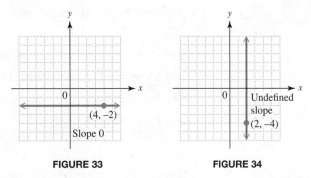

FIGURE 33   **FIGURE 34**

(b) $(2, -4)$, undefined slope

Vertical lines have undefined slope. To graph this line, plot the point $(2, -4)$ and draw the vertical line through it. See **FIGURE 34**.   **NOW TRY** $\circlearrowleft$

NOW TRY ANSWERS

4. **(a)** $y = -4x + 2$
   (b) $y = 3x + 7$

5. **(a)**

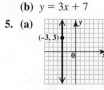

   (b)

NOW TRY
EXERCISE 6

Write an equation of the line passing through the point $(-1, 1)$ and having the given slope.

(a) Undefined slope

(b) $m = 0$

NOW TRY ANSWERS
6. **(a)** $x = -1$ **(b)** $y = 1$

EXAMPLE 6 Writing Equations of Horizontal and Vertical Lines

Write an equation of the line passing through the point $(2, -2)$ and having the given slope.

(a) Slope 0

This line is horizontal because it has slope 0. Recall that a horizontal line through the point (a, b) has equation $y = b$. The y-coordinate of the point $(2, -2)$ is -2, so the equation is $y = -2$. See **FIGURE 35**.

(b) Undefined slope

This line is vertical because it has undefined slope. A vertical line through the point (a, b) has equation $x = a$. The x-coordinate of $(2, -2)$ is 2, so the equation is $x = 2$. See **FIGURE 35**.

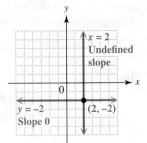

FIGURE 35

NOW TRY

3.4 Exercises

FOR EXTRA HELP

▶ MyMathLab®

▶ *Complete solution available in MyMathLab*

Concept Check   *Fill in each blank with the correct response.*

1. In slope-intercept form $y = mx + b$ of the equation of a line, the slope is _____ and the y-intercept is _____.

2. The line with equation $y = -\frac{x}{2} - 3$ has slope _____ and y-intercept _____.

3. *Concept Check*   Match each equation in parts (a)–(d) with the graph in A–D that would most closely resemble its graph.

(a) $y = x + 3$ **(b)** $y = -x + 3$ **(c)** $y = x - 3$ **(d)** $y = -x - 3$

A.   **B.**   **C.**   **D.**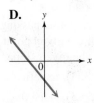

4. *Concept Check*   Match the description in Column I with the correct equation in Column II.

I	II
(a) Slope $= -2$, passes through $(4, 1)$	**A.** $y = 4x$
(b) Slope $= -2$, y-intercept $(0, 1)$	**B.** $y = \frac{1}{4}x$
(c) Passes through $(0, 0)$ and $(4, 1)$	**C.** $y = -4x$
(d) Passes through $(0, 0)$ and $(1, 4)$	**D.** $y = -2x + 1$
	E. $2x + y = 9$

*Identify the slope and y-intercept of the line with each equation. **See Example 1.***

5. $y = \frac{5}{2}x - 4$ **6.** $y = \frac{7}{3}x - 6$ **7.** $y = -x + 9$

8. $y = x + 1$ **9.** $y = \frac{x}{5} - \frac{3}{10}$ **10.** $y = \frac{x}{7} - \frac{5}{14}$

Graph the equation of each line using the slope and y-intercept. See Example 2.

11. $y = 3x + 2$ **12.** $y = 4x - 4$ **13.** $y = -\dfrac{1}{3}x + 4$

14. $y = -\dfrac{1}{2}x + 2$ ▶ **15.** $2x + y = -5$ **16.** $3x + y = -2$

17. $4x - 5y = 20$ **18.** $6x - 5y = 30$

Graph each line passing through the given point and having the given slope. See Examples 3 and 5.

19. $(0, 1), m = 4$ **20.** $(0, -5), m = -2$ ▶ **21.** $(1, -5), m = -\dfrac{2}{5}$

22. $(2, -1), m = -\dfrac{1}{3}$ **23.** $(-1, 4), m = \dfrac{2}{5}$ **24.** $(-2, 2), m = \dfrac{3}{2}$

25. $(0, 0), m = -2$ **26.** $(0, 0), m = -3$ **27.** $(-2, 3), m = 0$

28. $(3, 2), m = 0$ **29.** $(2, 4),$ undefined slope   **30.** $(3, -2),$ undefined slope

31. $(5, -5),$ slope 0 **32.** $(-4, 4),$ slope 0

Concept Check   *Answer each question.*

33. What is the common name given to a vertical line whose *x*-intercept is the origin?

34. What is the common name given to a line with slope 0 whose *y*-intercept is the origin?

*Use the geometric interpretation of slope ("rise over run," from **Section 3.3**) to find the slope of each line. Then, by identifying the y-intercept from the graph, write the slope-intercept form of the equation of the line. (Coordinates of the points shown are integers.)*

35.
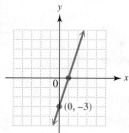

36.
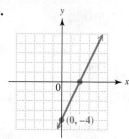

37.
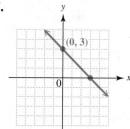

38.
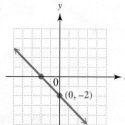

39.
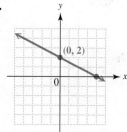

40.
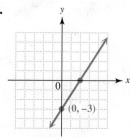

Write an equation in slope-intercept form (if possible) of the line passing through the given point and having the given slope. See Examples 4 and 6.

▶ **41.** $(0, -3), m = 4$ **42.** $(0, 6), m = -5$ **43.** $(0, -7), m = -1$

44. $(0, -9), m = 1$ **45.** $(4, 1), m = 2$ **46.** $(2, 7), m = 3$

▶ **47.** $(-1, 3), m = -4$ **48.** $(-3, 1), m = -2$ **49.** $(9, 3), m = 1$

50. $(8, 4), m = 1$ **51.** $(-4, 1), m = \dfrac{3}{4}$ **52.** $(2, 1), m = \dfrac{5}{2}$

53. $(0, 3), m = 0$ **54.** $(0, -4), m = 0$ **55.** $(2, -6)$, undefined slope

56. $(-1, 7)$, undefined slope   **57.** $(0, -2)$, undefined slope   **58.** $(0, 5)$, undefined slope

59. $(6, -6)$, slope 0   **60.** $(-3, 3)$, slope 0

Each table of values gives several points that lie on a line.

(a) *Use any two of the ordered pairs to find the slope of the line.*

(b) *Identify the y-intercept of the line.*

(c) *Use the slope and y-intercept from parts (a) and (b) to write an equation of the line in slope-intercept form.*

(d) *Graph the equation.*

61.

x	y
0	-1
3	5
5	9

62.

x	y
0	4
2	2
4	0

63.

x	y
-9	1
-6	0
0	-2

64.

x	y
-10	-1
0	3
5	5

Extending Skills   *Solve each problem.*

65. In his job, Andrew earns 5% commission on his sales, plus a base salary of $2000 per month. This is illustrated in the graph and can be modeled by the linear equation

$$y = 0.05x + 2000,$$

where *y* is his monthly salary in dollars and *x* is his sales, also in dollars.

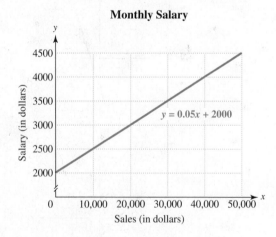

Monthly Salary

(a) What is the slope? With what does the slope correspond in the problem?

(b) What is the *y*-intercept? With what does the *y*-value of the *y*-intercept correspond in the problem?

(c) Use the equation to determine Andrew's monthly salary if his sales are $10,000. Confirm this using the graph.

(d) Use the graph to determine his sales if he wants to earn a monthly salary of $3500. Confirm this using the equation.

66. The cost to rent a moving van is $0.50 per mile, plus a flat fee of $100. This is illustrated in the graph and can be modeled by the linear equation

$$y = 0.50x + 100,$$

where y is the total rental cost in dollars and x is the number of miles driven.

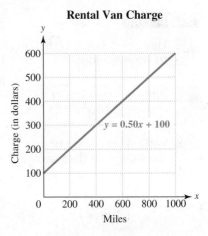

Rental Van Charge

(a) What is the slope? With what does the slope correspond in the problem?

(b) What is the y-intercept? With what does the y-value of the y-intercept correspond in the problem?

(c) Use the equation to determine the total charge if 400 mi are driven. Confirm this using the graph.

(d) Use the graph to determine the number of miles driven if the charge is $500. Confirm this using the equation.

Extending Skills   *The cost y of producing x items is, in some cases, expressed in the form*

$$y = mx + b.$$

*The value of b gives the **fixed cost** (the cost that is the same no matter how many items are produced), and the value of m is the **variable cost** (the cost of producing an additional item). Use this information to work Exercises 67 and 68.*

67. It costs $400 to start up a business selling snow cones. Each snow cone costs $0.25 to produce.

(a) What is the fixed cost?   **(b)** What is the variable cost?

(c) Write the cost equation.

(d) What will be the cost of producing 100 snow cones, based on the cost equation?

(e) How many snow cones will be produced if the total cost is $775?

68. It costs $2000 to purchase a copier, and each copy costs $0.02 to make.

(a) What is the fixed cost?   **(b)** What is the variable cost?

(c) Write the cost equation.

(d) What will be the cost of producing 10,000 copies, based on the cost equation?

(e) How many copies will be produced if the total cost is $2600?

RELATING CONCEPTS   For Individual or Group Work   (Exercises 69–72)

*A line with equation written in slope-intercept form $y = mx + b$ has slope m and y-intercept $(0, b)$. Recall from **Section 3.1** that the standard form of a linear equation in two variables is*

$$Ax + By = C, \quad \text{Standard form} \quad .$$

*where A, B, and C are real numbers and A and B are not both 0. **Work Exercises 69–72 in order.***

69. Write the standard form of a linear equation in slope-intercept form—that is, solved for y—to show that, in general, the slope is given by $-\frac{A}{B}$ (where $B \neq 0$).

70. Use the fact that $m = -\frac{A}{B}$ to find the slope of the line with each equation.

(a) $2x + 3y = 18$ **(b)** $4x - 2y = -1$ **(c)** $3x - 7y = 21$

71. Refer to the slope-intercept form found in **Exercise 69**. What is the y-intercept?

72. Use the result of **Exercise 71** to find the y-intercept of each line in **Exercise 70**.

3.5 Point-Slope Form of a Linear Equation and Modeling

OBJECTIVE 1   Use point-slope form to write an equation of a line.

There is another form that can be used to write an equation of a line. To develop this form, let m represent the slope of a line and let (x_1, y_1) represent a given point on the line. Let (x, y) represent any other point on the line. See **FIGURE 36**.

$$m = \frac{y - y_1}{x - x_1} \qquad \text{Definition of slope}$$

$$m(x - x_1) = y - y_1 \qquad \text{Multiply each side by } x - x_1.$$

$$y - y_1 = m(x - x_1) \qquad \text{Rewrite.}$$

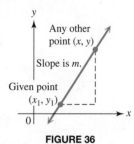

FIGURE 36

This result is the *point-slope form* of the equation of a line.

> **Point-Slope Form**
>
> The **point-slope form** of the equation of a line with slope m passing through the point (x_1, y_1) is
>
> $$y - y_1 = m(x - x_1).$$
>
> with Slope, Given point labeled.

**NOW TRY
EXERCISE 1**

Write an equation of the line passing through $(3, -1)$, with slope $-\frac{2}{5}$. Give the final answer in slope-intercept form.

EXAMPLE 1   Using Point-Slope Form to Write Equations

Write an equation of each line. Give the final answer in slope-intercept form.

(a) The line passing through $(-2, 4)$, with slope -3

The given point is $(-2, 4)$, so $x_1 = -2$ and $y_1 = 4$. Also, $m = -3$. Substitute these values into the point-slope form.

$$y - y_1 = m(x - x_1) \qquad \text{Point-slope form}$$

Only y_1, m, and x_1 are replaced with numbers.

$$y - 4 = -3[x - (-2)] \qquad \text{Let } y_1 = 4, m = -3, x_1 = -2.$$

$$y - 4 = -3(x + 2) \qquad \text{Definition of subtraction}$$

$$y - 4 = -3x - 6 \qquad \text{Distributive property}$$

The answer is in $y = mx + b$ form as specified.

$$y = -3x - 2 \qquad \text{Add 4.}$$

(b) The line passing through $(4, 2)$, with slope $\frac{3}{5}$

$$y - y_1 = m(x - x_1) \qquad \text{Point-slope form}$$

$$y - 2 = \frac{3}{5}(x - 4) \qquad \text{Let } y_1 = 2, m = \frac{3}{5}, x_1 = 4.$$

$$y - 2 = \frac{3}{5}x - \frac{12}{5} \qquad \text{Distributive property}$$

Do not clear fractions here because we want the answer in slope-intercept form—that is, solved for y.

$$y = \frac{3}{5}x - \frac{12}{5} + \frac{10}{5} \qquad \text{Add } 2 = \frac{10}{5} \text{ to each side.}$$

$$y = \frac{3}{5}x - \frac{2}{5} \qquad \text{Combine like terms.} \qquad \textbf{NOW TRY}$$

NOW TRY ANSWER

1. $y = -\frac{2}{5}x + \frac{1}{5}$

OBJECTIVE 2 Write an equation of a line using two points on the line.

Many of the linear equations in **Sections 3.1–3.4** were given in **standard form**

$$Ax + By = C, \quad \text{Standard form}$$

where A, B, and C are real numbers and A and B are not both 0. In most cases, A, B, and C are rational numbers. For consistency in this book, we give answers so that A, B, and C are integers with greatest common factor 1 and $A \geq 0$. (If $A = 0$, then we give $B > 0$.)

NOTE The definition of standard form is not the same in all texts. A linear equation can be written in many different, equally correct, ways. For example,

$$3x + 4y = 12, \quad 6x + 8y = 24, \quad \text{and} \quad -9x - 12y = -36$$

all represent the same set of ordered pairs. When giving answers in standard form, let us agree that $3x + 4y = 12$ is preferable to the other forms because the greatest common factor of 3, 4, and 12 is 1 and $A \geq 0$.

NOW TRY
EXERCISE 2
Write an equation of the line passing through the points $(4, 1)$ and $(6, -2)$. Give the final answer in

(a) slope-intercept form and

(b) standard form.

EXAMPLE 2 Writing an Equation of a Line Using Two Points

Write an equation of the line passing through the points $(-2, 5)$ and $(3, 4)$. Give the final answer in slope-intercept form and then in standard form.

First, find the slope of the line.

$$\overset{(x_1, y_1)}{\underset{\downarrow\downarrow}{}} \qquad \overset{(x_2, y_2)}{\underset{\downarrow\downarrow}{}}$$

$$(3, 4) \quad \text{and} \quad (-2, 5) \qquad \text{Label the points.}$$

$$\text{slope } m = \frac{y_2 - y_1}{x_2 - x_1} = \frac{5 - 4}{-2 - 3} \qquad \text{Apply the slope formula.}$$

$$= \frac{1}{-5}, \quad \text{or} \quad -\frac{1}{5} \qquad \text{Simplify the fraction.}$$

Now use $m = -\frac{1}{5}$ and either $(-2, 5)$ or $(3, 4)$ as (x_1, y_1) in the point-slope form.

$$y - y_1 = m(x - x_1) \qquad \text{We choose } (3, 4).$$

$$y - 4 = -\frac{1}{5}(x - 3) \qquad \text{Let } y_1 = 4, m = -\frac{1}{5}, x_1 = 3.$$

$$y - 4 = -\frac{1}{5}x + \frac{3}{5} \qquad \text{Distributive property}$$

$$y = -\frac{1}{5}x + \frac{3}{5} + \frac{20}{5} \qquad \text{Add } 4 = \frac{20}{5} \text{ to each side.}$$

Slope-intercept form $\longrightarrow$ $$y = -\frac{1}{5}x + \frac{23}{5} \qquad \text{Combine like terms.}$$

$$5y = -x + 23 \qquad \text{Multiply by 5 to clear fractions.}$$

Standard form $\longrightarrow$ $x + 5y = 23$ \qquad Add x.

NOW TRY

NOW TRY ANSWERS
2. (a) $y = -\frac{3}{2}x + 7$
 (b) $3x + 2y = 14$

NOTE   In **Example 2,** the same result would be found using $(-2, 5)$ for (x_1, y_1). We could also substitute the slope and either given point in slope-intercept form $y = mx + b$ and then solve for b, as in **Section 3.4, Example 4(b).**

▼ **Summary of the Forms of Linear Equations**

Equation	Description	Example
$y = mx + b$	**Slope-intercept form** Slope is m. y-intercept is $(0, b)$.	$y = \frac{3}{2}x - 6$
$y - y_1 = m(x - x_1)$	**Point-slope form** Slope is m. Line passes through (x_1, y_1).	$y + 3 = \frac{3}{2}(x - 2)$
$Ax + By = C$ (where A, B, and C are real numbers and A and B are not both 0)	**Standard form** Slope is $-\frac{A}{B}$ $(B \neq 0)$. x-intercept is $\left(\frac{C}{A}, 0\right)$ $(A \neq 0)$. y-intercept is $\left(0, \frac{C}{B}\right)$ $(B \neq 0)$.	$3x - 2y = 12$
$x = a$	**Vertical line** Slope is undefined. x-intercept is $(a, 0)$.	$x = 3$
$y = b$	**Horizontal line** Slope is 0. y-intercept is $(0, b)$.	$y = 3$

OBJECTIVE 3   Write an equation of a line that fits a data set.

If a given set of data fits a linear pattern—that is, if its graph consists of points lying close to a straight line—we can write a linear equation that models the data.

EXAMPLE 3   **Writing an Equation of a Line That Models Data**

The table lists the average annual cost (in dollars) of tuition and fees for in-state students at public 4-year colleges and universities for selected years. Year 1 represents 2001, year 3 represents 2003, and so on. Plot the data and write an equation that approximates it.

Letting y represent the cost in year x, we plot the data as shown in **FIGURE 37.**

Year	Cost (in dollars)
1	3766
3	4645
5	5491
7	6185
9	7050
11	8244

Source: The College Board.

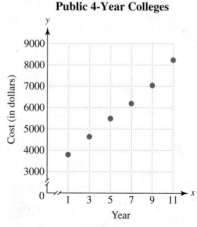

Average Annual Costs at Public 4-Year Colleges

FIGURE 37

NOW TRY EXERCISE 3

Use the points $(3, 4645)$ and $(5, 5491)$ to write an equation in slope-intercept form that approximates the data of **Example 3.** How well does this equation approximate the cost in 2011?

The points appear to lie approximately in a straight line. We choose the ordered pairs $(5, 5491)$ and $(7, 6185)$ from the table and find the slope of the line through these points.

$$m = \frac{y_2 - y_1}{x_2 - x_1} = \frac{6185 - 5491}{7 - 5} = 347 \qquad \begin{array}{l} \text{Let } (7, 6185) = (x_2, y_2) \\ \text{and } (5, 5491) = (x_1, y_1). \end{array}$$

The slope, 347, is positive, indicating that tuition and fees *increased* \$347 each year. Now substitute this slope and the point $(5, 5491)$ in the point-slope form to find an equation of the line.

$$y - y_1 = m(x - x_1) \qquad \text{Point-slope form}$$

$$y - 5491 = 347(x - 5) \qquad \text{Let } (x_1, y_1) = (5, 5491), m = 347.$$

$$y - 5491 = 347x - 1735 \qquad \text{Distributive property}$$

$$y = 347x + 3756 \qquad \text{Add 5491.}$$

Thus, the equation $y = 347x + 3756$ can be used to model the data.

To see how well this equation approximates the ordered pairs in the data table, let $x = 9$ (for 2009) and find y.

$$y = 347x + 3756 \qquad \text{Equation of the line}$$

$$y = 347(9) + 3756 \qquad \text{Substitute 9 for } x.$$

$$y = 3123 + 3756 \qquad \text{Multiply.}$$

$$y = 6879 \qquad \text{Add.}$$

The corresponding value in the table for $x = 9$ is 7050, so the equation approximates the data reasonably well.   NOW TRY

NOW TRY ANSWER

3. $y = 423x + 3376$; The equation gives $y = 8029$ when $x = 11$, which approximates the data reasonably well.

NOTE In **Example 3,** if we had chosen two different data points, we would have obtained a slightly different equation. See **Now Try Exercise 3.**

Also, we could have used slope-intercept form $y = mx + b$ (instead of point-slope form) to write an equation that models the data.

3.5 Exercises

 FOR EXTRA HELP   MyMathLab®

▶ *Complete solution available in MyMathLab*

Concept Check   *Work each problem.*

  1. Match each form or description in Column I with the corresponding equation in Column II.

I	II
(a) Point-slope form	**A.** $x = a$
(b) Horizontal line	**B.** $y = mx + b$
(c) Slope-intercept form	**C.** $y = b$
(d) Standard form	**D.** $y - y_1 = m(x - x_1)$
(e) Vertical line	**E.** $Ax + By = C$

2. Write the equation $y + 1 = -2(x - 5)$ first in slope-intercept form and then in standard form.

3. Which equations are equivalent to $2x - 3y = 6$?

A. $y = \dfrac{2}{3}x - 2$ **B.** $-2x + 3y = -6$

C. $y = -\dfrac{3}{2}x + 3$ **D.** $y - 2 = \dfrac{2}{3}(x - 6)$

4. In the summary box following **Example 2,** we give the equations

$$y = \dfrac{3}{2}x - 6 \quad \text{and} \quad y + 3 = \dfrac{3}{2}(x - 2)$$

as examples of equations in slope-intercept form and point-slope form, respectively. Write each of these equations in standard form. What do you notice?

*Write an equation of the line passing through the given point and having the given slope. Give the final answer in slope-intercept form. See **Example 1.***

5. $(1, 7), m = 5$ **6.** $(2, 9), m = 6$ **7.** $(6, -3), m = 1$

8. $(-4, 4), m = 1$ **9.** $(1, -7), m = -3$ **10.** $(1, -5), m = -7$

11. $(3, -2), m = -1$ **12.** $(-5, 4), m = -1$ ▶ **13.** $(-2, 5), m = \dfrac{2}{3}$

14. $(4, 2), m = -\dfrac{1}{3}$ **15.** $(6, -3), m = -\dfrac{4}{5}$ **16.** $(7, -2), m = -\dfrac{7}{2}$

*Write an equation of the line passing through the given pair of points. Give the final answer in **(a)** slope-intercept form and **(b)** standard form. See **Example 2.***

17. $(4, 10)$ and $(6, 12)$ **18.** $(8, 5)$ and $(9, 6)$ **19.** $(-4, 0)$ and $(0, 2)$

20. $(0, -2)$ and $(-3, 0)$ **21.** $(-2, -1)$ and $(3, -4)$ **22.** $(-1, -7)$ and $(-8, -2)$

23. $\left(-\dfrac{2}{3}, \dfrac{8}{3}\right)$ and $\left(\dfrac{1}{3}, \dfrac{7}{3}\right)$ **24.** $\left(\dfrac{1}{2}, \dfrac{3}{2}\right)$ and $\left(-\dfrac{1}{4}, \dfrac{5}{4}\right)$

*Write an equation of the given line through the given points. Give the final answer in **(a)** slope-intercept form and **(b)** standard form.*

25.

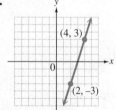

26.

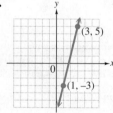

27.

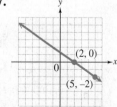

28.

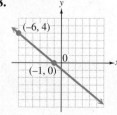

Extending Skills   *Write an equation of the line satisfying the given conditions. Give the final answer in slope-intercept form. (Hint: Recall the relationships among slopes of parallel and perpendicular lines in Section 3.3.)*

29. Perpendicular to $x - 2y = 7$; y-intercept $(0, -3)$

30. Parallel to $5x - y = 10$; y-intercept $(0, -2)$

31. Through $(2, 3)$;   parallel to $4x - y = -2$

32. Through $(4, 2)$;   perpendicular to $x - 3y = 7$

33. Through $(2, -3)$;   parallel to $3x = 4y + 5$

34. Through $(-1, 4)$;   perpendicular to $2x = -3y + 8$

Solve each problem. **See Example 3.**

35. The table lists the average annual cost y (in dollars) of tuition and fees at 2-year colleges for selected years x, where year 1 represents 2008, year 2 represents 2009, and so on.

Year	Cost (in dollars)
1	2530
2	2790
3	2940
4	3070
5	3220

Source: The College Board.

  (a) Write five ordered pairs (x, y) for the data.

  (b) Plot the ordered pairs (x, y). Do the points lie approximately in a straight line?

  (c) Use the ordered pairs $(1, 2530)$ and $(4, 3070)$ to write an equation of a line that approximates the data. Give the final equation in slope-intercept form.

  (d) Use the equation from part (c) to estimate the average annual cost at 2-year colleges in 2013 to the nearest dollar. (*Hint:* What is the value of x for 2013?)

36. The table gives heavy-metal nuclear waste y (in thousands of metric tons) from spent reactor fuel awaiting permanent storage. (*Source: Scientific American.*)

Year x	Waste y
1995	32
2000	42
2010	61
2020*	76

*Estimate by the U.S. Department of Energy.

Let $x = 0$ represent 1995, $x = 5$ represent 2000 (since $2000 - 1995 = 5$), and so on.

  (a) For 1995, the ordered pair is $(0, 32)$. Write ordered pairs (x, y) for the data for the other years given in the table.

  (b) Plot the ordered pairs (x, y). Do the points lie approximately in a straight line?

  (c) Use the ordered pairs $(0, 32)$ and $(25, 76)$ to write an equation of a line that approximates the data. Give the final equation in slope-intercept form.

  (d) Use the equation from part (c) to estimate the amount of nuclear waste in 2015. (*Hint:* What is the value of x for 2015?)

The points on the graph show the number of colleges y that teamed up with banks to issue student ID cards which doubled as debit cards for recent years x. The graph of a linear equation that models the data is also shown.

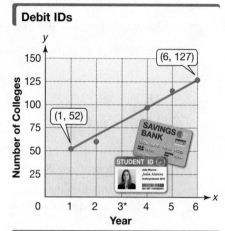

Debit IDs

Source: CR80News.
* Data for year 3 is unavailable.

37. Use the ordered pairs shown on the graph to write an equation of the line that models the data. Give the final equation in slope-intercept form.

38. Use the equation from **Exercise 37** to estimate the number of colleges that teamed up with banks to offer debit IDs in year 3 when data was unavailable.

The points on the graph indicate years of life expected at birth y in the United States for selected years x. The graph of a linear equation that models the data is also shown.

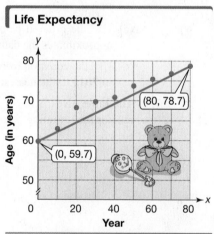

Life Expectancy

Source: National Center for Health Statistics.

Here x = 0 represents 1930, x = 10 represents 1940, and so on.

39. Use the ordered pairs shown on the graph to write an equation of the line that models the data. Give the final equation in slope-intercept form.

40. Use the equation from **Exercise 39** to do the following.

(a) Find years of life expected at birth in 2000. (*Hint:* What is the value of x for 2000?) Round the answer to the nearest tenth.

(b) How does the answer in part (a) compare to the actual value of 76.8 yr?

(c) Project years of life expected at birth in 2020. (*Hint:* What is the value of x for 2020?) Round the answer to the nearest tenth. Does the answer seem reasonable?

1. (a) B (b) D
   (c) A (d) C

2. A, B

3.

4.

5.

6.

7.

8.

9.

10.

11.

12.

13.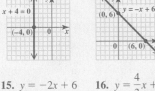

14.

15. $y = -2x + 6$ **16.** $y = \dfrac{4}{3}x + 8$

17. $y = \dfrac{1}{2}x - 2$

18. $y = -\dfrac{2}{3}x + 5$

19. $y = -3x - 6$ **20.** $y = \dfrac{3}{2}x + 12$

21. $y = -4x - 3$ **22.** $x = 0$

23. $y = \dfrac{2}{3}x$ **24.** $y = -x - 4$

25. $y = x - 5$ **26.** $y = 0$

27. $y = \dfrac{5}{3}x + 5$

28. $y = -5x - 8$

---

1. *Concept Check* Match the description in Column I with the correct equation in Column II.

I	II
(a) Slope $= -0.5, b = -2$	**A.** $y = -\dfrac{1}{2}x$
(b) x-intercept $(4, 0)$, y-intercept $(0, 2)$	**B.** $y = -\dfrac{1}{2}x - 2$
(c) Passes through $(4, -2)$ and $(0, 0)$	**C.** $x - 2y = 2$
(d) $m = \dfrac{1}{2}$, passes through $(-2, -2)$	**D.** $x + 2y = 4$
	E. $x = 2y$

2. *Concept Check* Which equations are equivalent to $2x + 5y = 20$?

A. $y = -\dfrac{2}{5}x + 4$ **B.** $y - 2 = -\dfrac{2}{5}(x - 5)$

C. $y = \dfrac{5}{2}x - 4$ **D.** $2x = 5y - 20$

Graph each line, using the given information or equation.

3. $x - 2y = -4$ **4.** $2x + 3y = 12$

5. $m = 1$, y-intercept $(0, -2)$ **6.** $y - 4 = -9$

7. $m = -\dfrac{2}{3}$, passes through $(3, -4)$ **8.** $8x = 6y + 24$

9. $x - 4y = 0$ **10.** $m = -\dfrac{3}{4}$, passes through $(4, -4)$

11. $5x + 2y = 10$ **12.** $x + 5y = 0$

13. $x + 4 = 0$ **14.** $y = -x + 6$

Write an equation in slope-intercept form of each line represented by the table of ordered pairs or the graph.

15.

x	y
3	0
1	4
-1	8

16.

x	y
-6	0
0	8
3	12

17.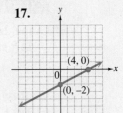

18.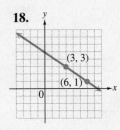

Write an equation of each line. Give the final answer in slope-intercept form if possible.

19. $m = -3$, $b = -6$ **20.** $m = \dfrac{3}{2}$, through $(-4, 6)$

21. Through $(1, -7)$ and $(-2, 5)$ **22.** Through $(0, 0)$, undefined slope

23. Through $(0, 0)$ and $(3, 2)$ **24.** $m = -1$, $b = -4$

25. Through $(5, 0)$ and $(0, -5)$ **26.** Through $(0, 0)$, $m = 0$

27. $m = \dfrac{5}{3}$, through $(-3, 0)$ **28.** Through $(1, -13)$ and $(-2, 2)$

3.6 Graphing Linear Inequalities in Two Variables

OBJECTIVES

1 Graph linear inequalities in two variables.

2 Graph an inequality with a boundary line through the origin.

VOCABULARY
☐ linear inequality in two variables
☐ boundary line

In **Section 3.2,** we graphed linear equations such as

$$2x + 3y = 6.$$

We extend this work to *linear inequalities in two variables,* such as

$$2x + 3y \leq 6.$$

Linear Inequality in Two Variables

A **linear inequality in two variables** (here x and y) can be written in the form

$$Ax + By < C, \quad Ax + By \leq C, \quad Ax + By > C, \quad \text{or} \quad Ax + By \geq C,$$

where A, B, and C are real numbers and A and B are not both 0.

OBJECTIVE 1 Graph linear inequalities in two variables.

Consider the graph in **FIGURE 38**. The graph of the line $x + y = 5$ divides the points in the rectangular coordinate system into three sets.

1. Those points that lie *on* the line itself and satisfy the equation $x + y = 5$ [such as $(0, 5)$, $(2, 3)$, and $(5, 0)$]

2. Those points that lie in the region *above* the line and satisfy the inequality $x + y > 5$ [such as $(5, 3)$ and $(2, 4)$]

3. Those points that lie in the region *below* the line and satisfy the inequality $x + y < 5$ [such as $(0, 0)$ and $(-3, -1)$].

The graph of the line $x + y = 5$ is the **boundary line** for the inequalities

$$x + y > 5 \quad \text{and} \quad x + y < 5.$$

Graphs of linear inequalities in two variables are regions in the real number plane that may or may not include boundary lines.

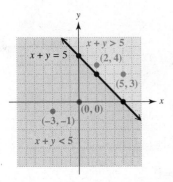

FIGURE 38

EXAMPLE 1 Graphing a Linear Inequality

Graph $2x + 3y \leq 6$.

The inequality $2x + 3y \leq 6$ means that

$$2x + 3y < 6 \quad \text{or} \quad 2x + 3y = 6.$$

We begin by graphing the equation $2x + 3y = 6$, a line with intercepts $(0, 2)$ and $(3, 0)$, as shown in **FIGURE 39** on the next page. This boundary line divides the plane into two regions, one of which satisfies the inequality. To find the correct region, we choose a test point *not* on the boundary line and substitute it into the inequality to see whether the resulting statement is true or false. The point $(0, 0)$ is a convenient choice.

$$2x + 3y < 6 \quad \text{We are testing the region.}$$

$$2(0) + 3(0) \overset{?}{<} 6 \quad \text{Let } x = 0 \text{ and } y = 0.$$

$$\text{Use } (0, 0) \text{ as a test point.} \quad 0 + 0 \overset{?}{<} 6 \quad \text{Multiply.}$$

$$0 < 6 \quad \text{True}$$

NOW TRY
EXERCISE 1
Graph $x + 3y \leq 6$.

Because a true statement results, we shade the region that includes the test point $(0, 0)$. See **FIGURE 39**. The shaded region, along with the boundary line because $\leq$ includes equality, is the desired graph.

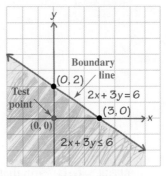

FIGURE 39

 NOW TRY

NOTE Alternatively in **Example 1**, we can find the required region by solving the given inequality for y.

$$2x + 3y \leq 6 \qquad \text{Inequality from \textbf{Example 1}}$$

$$3y \leq -2x + 6 \qquad \text{Subtract } 2x.$$

$$y \leq -\frac{2}{3}x + 2 \qquad \text{Divide by 3.}$$

Ordered pairs in which y is equal to $-\frac{2}{3}x + 2$ are on the boundary line, so pairs in which *y is less than* $-\frac{2}{3}x + 2$ will be *below* that line. (As we move *down* vertically, the y-values *decrease*.) This gives the same region that we shaded in **FIGURE 39**. (Ordered pairs in which y is *greater than* $-\frac{2}{3}x + 2$ will be *above* the boundary line.)

EXAMPLE 2   Graphing a Linear Inequality

Graph $x - y > 5$.

This inequality does *not* involve equality. Therefore, the points on the line

$$x - y = 5$$

do *not* belong to the graph. However, the line still serves as a boundary for two regions, one of which satisfies the inequality.

To graph the inequality, first graph the equation $x - y = 5$ using the intercepts $(5, 0)$ and $(0, -5)$. Draw a *dashed line* to show that the points on the line are *not* solutions of the inequality $x - y > 5$. See **FIGURE 40** on the next page. Then choose a test point to see which region satisfies the inequality.

$$x - y > 5$$

(0, 0) is a convenient test point. → $$0 - 0 \overset{?}{>} 5 \qquad \text{Let } x = 0 \text{ and } y = 0.$$

$$0 > 5 \qquad \text{False}$$

NOW TRY ANSWER

1.

Because $0 > 5$ is false, the graph of the inequality is the region that *does not* contain $(0, 0)$. Shade the *other* region, as shown in **FIGURE 40**.

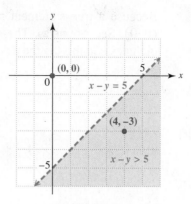

NOW TRY
EXERCISE 2
Graph $2x - 4y > 8$.

FIGURE 40

CHECK   To confirm that the proper region is shaded, we test a point in the shaded region. We arbitrarily use $(4, -3)$.

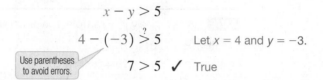

$$x - y > 5$$

$$4 - (-3) \overset{?}{>} 5 \qquad \text{Let } x = 4 \text{ and } y = -3.$$

Use parentheses to avoid errors.

$$7 > 5 \ \checkmark \quad \text{True}$$

This true statement verifies that the correct region is shaded in **FIGURE 40**.

NOW TRY

Graphing a Linear Inequality in Two Variables

Step 1 **Draw the graph of the straight line that is the boundary.** Make the line solid if the inequality involves $\leq$ or $\geq$. Make the line dashed if the inequality involves $<$ or $>$.

Step 2 **Choose a test point.** Choose any point not on the line, and substitute the coordinates of that point in the inequality.

Step 3 **Shade the appropriate region.** Shade the region that includes the test point if it satisfies the original inequality. Otherwise, shade the region on the other side of the boundary line.

EXAMPLE 3   **Graphing a Linear Inequality (Vertical Boundary Line)**

Graph $x < 3$.

First, we graph $x = 3$, a vertical line passing through the point $(3, 0)$. We use a dashed line because $<$ does not include equality and choose $(0, 0)$ as a test point.

$$x < 3$$

$$0 \overset{?}{<} 3 \qquad \text{Let } x = 0.$$

$$0 < 3 \qquad \text{True}$$

NOW TRY ANSWER

2.

Because $0 < 3$ is true, we shade the region containing $(0, 0)$, as in **FIGURE 41** on the next page. Intuitively this makes sense—all values of x along the x-axis in the shaded region are indeed less than 3.

NOW TRY
EXERCISE 3
Graph $x > 2$.

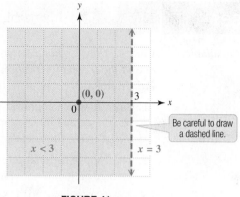

FIGURE 41

 NOW TRY

NOW TRY
EXERCISE 4
Graph $y \leq -2x$.

OBJECTIVE 2 Graph an inequality with a boundary line through the origin.

If the graph of an inequality has a boundary line that goes through the origin, $(0, 0)$ *cannot be used as a test point.*

EXAMPLE 4 Graphing a Linear Inequality (Boundary Line through the Origin)

Graph $x \leq 2y$.

We graph $x = 2y$ using a solid line by determining several ordered pairs that satisfy the equation.

$x = 2y$	$x = 2y$	$x = 2y$
$0 = 2y$ Let $x = 0$.	$6 = 2y$ Let $x = 6$.	$x = 2(2)$ Let $y = 2$.
$0 = y$	$3 = y$	$x = 4$
Ordered pair: $(0, 0)$	Ordered pair: $(6, 3)$	Ordered pair: $(4, 2)$

The line through these three ordered pairs is shown in **FIGURE 42**. Because the point $(0, 0)$ is *on* the line $x = 2y$, it cannot be used as a test point. Instead, we choose a test point *off* the line, such as $(1, 3)$.

$$x < 2y$$

$$1 \overset{?}{<} 2(3) \quad \text{Let } x = 1 \text{ and } y = 3.$$

$$1 < 6 \quad \text{True}$$

Because a true statement results, we shade the region containing the test point $(1, 3)$. See **FIGURE 42**.

NOW TRY ANSWERS

3.

4.

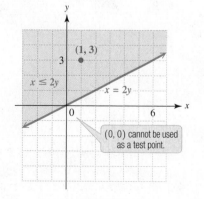

FIGURE 42

NOW TRY

3.6 Exercises

FOR EXTRA HELP MyMathLab®

● *Complete solution available in MyMathLab*

Concept Check *Each statement includes one or more phrases that can be symbolized with one of the inequality symbols* $<, \leq, >,$ *or* $\geq$. *Give the inequality symbol for the boldface italic words.*

1. Since it was recognized in 1981, HIV/AIDS has killed **more than** 25 million people worldwide and infected **more than** 60 million, about two-thirds of whom live in Africa. (*Source*: The President's Emergency Plan for AIDS Relief.)

2. The average national automobile insurance premium in 2012 was $153 **less than** the 2013 average premium. (*Source:* www.finance.yahoo.com)

3. As of December 2013, Southwest Airlines passengers were allowed one carry-on bag, with dimensions totaling **at most** 50 in. (*Source: USA Today.*)

4. A tornado must have winds of **at least** 65 mph to be rated using the Enhanced Fujita Scale. (*Source:* National Weather Service.)

5. By 1937, a population of as many as a million Attwater's prairie chickens had been cut to ***less than*** 9000. (*Source: National Geographic.*)

6. Easter Island's nearly 1000 statues, some ***almost*** 30 feet tall and weighing ***as much as*** 80 tons, are still an enigma. (*Source: Smithsonian.*)

Concept Check *Answer* true *or* false *to each of the following. If false, explain why.*

7. The point $(4, 0)$ lies on the graph of $3x - 4y < 12$.

8. The point $(4, 0)$ lies on the graph of $3x - 4y \leq 12$.

9. The point $(0, 0)$ can be used as a test point to determine which region to shade when graphing the linear inequality $x + 4y > 0$.

10. When graphing the linear inequality $3x + 2y \geq 12$, use a dashed line for the boundary line.

11. The points $(4, 1)$ and $(0, 0)$ lie on the graph of $3x - 2y \geq 0$.

12. The graph of $y > x$ does not contain points in quadrant IV.

In Exercises 13–20, the straight-line boundary has been drawn. Complete the graph by shading the correct region. **See Examples 1–4.**

13. $x + 2y \geq 7$

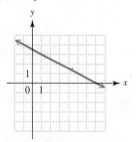

14. $2x + y \geq 5$

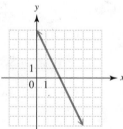

15. $-3x + 4y > 12$

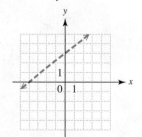

16. $4x - 5y < 20$

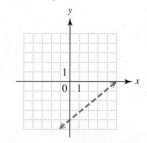

17. $x \le -y$

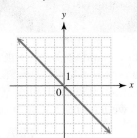

18. $x \le 3y$

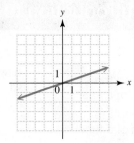

19. $y < -1$

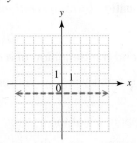

20. $x > 4$

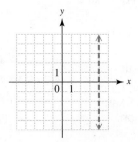

21. Explain how to determine whether to use a dashed line or a solid line when graphing a linear inequality in two variables.

22. Explain why the point $(0, 0)$ is not an appropriate choice for a test point when graphing an inequality whose boundary goes through the origin.

Graph each linear inequality. **See Examples 1–4.**

23. $x + y \le 5$ **24.** $x + y \ge 3$

▶ **25.** $2x + 3y > -6$ **26.** $3x + 4y < 12$

27. $y \ge 2x + 1$ **28.** $y < -3x + 1$

▶ **29.** $x < -2$ **30.** $x > 1$

31. $y \le 5$ **32.** $y \le -3$

▶ **33.** $y \ge 4x$ **34.** $y \le 2x$

35. $x < -2y$ **36.** $x > -5y$

Extending Skills   *For the given information,* **(a)** *graph the inequality (here, $x \ge 0$ and $y \ge 0$, so graph only the part of the inequality in quadrant I) and* **(b)** *give two ordered pairs that satisfy the inequality.*

37. A company will ship x units of merchandise to outlet I and y units of merchandise to outlet II. The company must ship a total of at least 500 units to these two outlets. The preceding information can be expressed using the linear inequality

$$x + y \ge 500.$$

38. A toy manufacturer makes stuffed bears and geese. It takes 20 min to sew a bear and 30 min to sew a goose. There is a total of 480 min of sewing time available to make x bears and y geese. These restrictions lead to the linear inequality

$$20x + 30y \le 480.$$

3.7 Introduction to Functions

VOCABULARY

☐ components
☐ relation
☐ domain
☐ range
☐ function

NOW TRY EXERCISE 1

Identify the domain and range of the relation.

$\{(-2, 3), (0, 7), (2, 8), (2, 10)\}$

If gasoline costs \$4.00 per gal and we buy 1 gal, then we must pay \$4.00(1) = \$4.00. If we buy 2 gal, then the cost is \$4.00(2) = \$8.00. If we buy 3 gal, then the cost is \$4.00(3) = \$12.00, and so on. Generalizing, if x represents the number of gallons, then the cost is \$4.00x. If we let y represent the cost, then the equation

$$y = 4.00x$$

relates the number of gallons, x, to the cost in dollars, y. The set of ordered pairs (x, y) that satisfy this equation forms a *relation*.

OBJECTIVE 1   Understand the definition of a relation.

In an ordered pair (x, y), x and y are the **components.** Any set of ordered pairs is a **relation.**

• The set of all first components of the ordered pairs is the **domain** of the relation.

• The set of all second components of the ordered pairs is the **range** of the relation.

EXAMPLE 1   Identifying Domains and Ranges of Relations

Identify the domain and range of each relation.

(a) $\{(0, 1), (2, 5), (3, 8), (4, 2)\}$

Domain: $\{0, 2, 3, 4\}$ ← Set of first components

Range: $\{1, 5, 8, 2\}$ ← Set of second components

The correspondence between the elements of the domain and the elements of the range is shown in **FIGURE 43.**

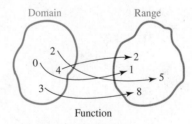

FIGURE 43

(b) $\{(3, 5), (3, 6), (3, 7), (3, 8)\}$

This relation has domain $\{3\}$ and range $\{5, 6, 7, 8\}$.

NOW TRY

OBJECTIVE 2   Understand the definition of a function.

A *function* is a special type of relation.

Function

A **function** is a set of ordered pairs in which each distinct first component corresponds to exactly one second component. See **FIGURE 43.**

NOW TRY ANSWER
1. domain: $\{-2, 0, 2\}$; range: $\{3, 7, 8, 10\}$

Example 1(b)

Domain   Range

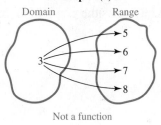

Not a function

FIGURE 44

Based on this definition, the relation in **Example 1(a)**

$$\{(0, 1), (2, 5), (3, 8), (4, 2)\} \text{ is a function.}$$

The relation in **Example 1(b)**

$$\{(3, 5), (3, 6), (3, 7), (3, 8)\} \text{ is } \textbf{\textit{not}} \text{ a function.}$$

See **FIGURE 44**. The *same* first component, 3, corresponds to more than one second component. If the ordered pairs in **Example 1(b)** were interchanged, giving the relation

$$\{(5, 3), (6, 3), (7, 3), (8, 3)\},$$

then this relation *would* be a function. ***In that case, each domain element (first component) is distinct and corresponds to exactly one range element (second component).***

NOW TRY
EXERCISE 2

Determine whether each relation is a function.

(a) $\{(-1, 2), (0, 1), (1, 0), (4, 3), (4, 5)\}$

(b) $\{(-1, -3), (0, 2), (3, 1), (8, 1)\}$

EXAMPLE 2 Determining Whether Relations Are Functions

Determine whether each relation is a function.

(a) $\{(-2, 4), (-1, 1), (0, 0), (1, 1), (2, 4)\}$

Each first component appears once and only once. The relation is a function.

(b) $\{(9, 3), (9, -3), (4, 2)\}$

The first component 9 appears in two ordered pairs and corresponds to two different second components. Therefore, this relation is not a function.   **NOW TRY**

Most functions have an infinite number of ordered pairs and are usually defined with equations that tell how to get the second components (**outputs**), given the first components (**inputs**). Here are some everyday examples of functions.

1. The cost y in dollars charged by an express mail company is a function of the weight x in pounds determined by the equation $y = 1.5(x - 1) + 9$.

2. In Cedar Rapids, Iowa, the sales tax is 7% of the price of an item. The tax y on a particular item is a function of the price x, because $y = 0.07x$.

3. The distance d traveled by a car moving at a constant rate of 45 mph is a function of the time t. Thus, $d = 45t$.

The function concept can be illustrated by an input-output "machine," as seen in **FIGURE 45**. The express mail company equation $y = 1.5(x - 1) + 9$ provides an output (cost y in dollars) for a given input (weight x in pounds).

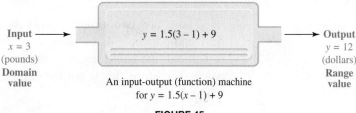

Input ⟶
$x = 3$
(pounds)
Domain value

$y = 1.5(3 - 1) + 9$

⟶ Output
$y = 12$
(dollars)
Range value

An input-output (function) machine
for $y = 1.5(x - 1) + 9$

FIGURE 45

OBJECTIVE 3 Decide whether an equation defines a function.

Given the graph of an equation, the definition of a function can be used to decide whether or not the graph represents a function. By definition, each x-value of a function must lead to exactly one y-value.

NOW TRY ANSWERS
2. (a) not a function
(b) function

In **FIGURE 46(a)**, the indicated *x*-value leads to two *y*-values, so this graph is *not* the graph of a function. A vertical line can be drawn that intersects the graph in more than one point.

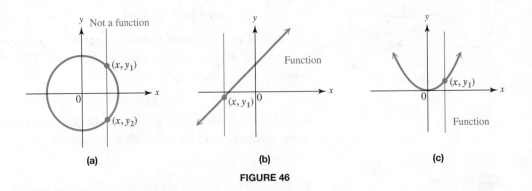

FIGURE 46

By contrast, in **FIGURE 46(b)** and **FIGURE 46(c)** any vertical line will intersect each graph in no more than one point, so these graphs are graphs of functions. This idea leads to the **vertical line test** for a function.

Vertical Line Test

If a vertical line intersects a graph in more than one point, then the graph is not the graph of a function.

As **FIGURE 46(b)** suggests, any *nonvertical* line is the graph of a function.

Thus, any linear equation of the form y = mx + b defines a function.

(Recall that a vertical line has undefined slope.)

EXAMPLE 3   **Determining Whether Relations Define Functions**

Determine whether each relation is a function.

(a)

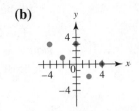

There are two ordered pairs with first component −4, as shown in red. A vertical line could intersect the graph twice. Therefore, this is not the graph of a function.

(b)

Every first component is matched with one and only one second component, and as a result, no vertical line intersects the graph in more than one point. This is the graph of a function.

(c) $y = 2x - 9$

This linear equation is in the form $y = mx + b$. Because the graph of this equation is a line that is not vertical, the equation defines a function.

(d) $x = 4$

The graph of $x = 4$ is a vertical line, so the equation does *not* define a function.

**NOW TRY
EXERCISE 3**

Determine whether each relation is a function.

(a) $y = x - 5$ **(b)** $x = -3$

(c)

(e)

Use the vertical line test. Any vertical line will intersect the graph just once, so this is the graph of a function.

(f)

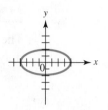

The vertical line test shows that this graph is not the graph of a function. A vertical line could intersect the graph twice.

NOW TRY

NOTE   An equation in which y is squared, such as $x = y^2$, does not usually define a function—most x-values will lead to two y-values. This is true for *any even* power of y, such as

$$y^2, \quad y^4, \quad y^6, \quad \text{and so on.}$$

Similarly, an equation involving $|y|$, such as $x = |y|$, does not usually define a function—most x-values will lead to more than one y-value.

OBJECTIVE 4 Find domains and ranges.

The set of all numbers that can be used as replacements for x in a function is the domain of the function, and the set of all possible values of y is the range of the function.

**NOW TRY
EXERCISE 4**

Find the domain and range of each function.

(a) $y = -x$ **(b)** $y = x^2 - 2$

EXAMPLE 4   Finding Domains and Ranges of Functions

Find the domain and range of each function.

(a) $y = 2x - 4$

Any number may be input for x, so the domain is the set of all real numbers. Also, any number may be output for y, so the range is also the set of all real numbers. As indicated in **FIGURE 47**, the graph of the equation is a straight line that extends infinitely in both directions, confirming that both the domain and range are $(-\infty, \infty)$.

(b) $y = x^2$

Any number can be squared, so the domain is the set of all real numbers. However, y equals the *square* of a real number, so the values of y cannot be negative. Thus, the range is the set of all nonnegative numbers, or $[0, \infty)$ in interval notation. See **FIGURE 48**.

FIGURE 47

x	y
0	0
1	1
−1	1
2	4
−2	4
3	9
−3	9

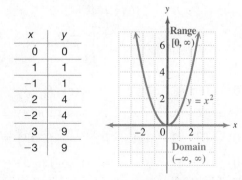

The ordered pairs in the table are used to obtain the graph of the function. While x can take any real number value, notice that y is always greater than or equal to 0.

FIGURE 48

NOW TRY

NOW TRY ANSWERS

3. **(a)** function
   (b) not a function
   (c) not a function
4. **(a)** domain: $(-\infty, \infty)$;
   range: $(-\infty, \infty)$
   (b) domain: $(-\infty, \infty)$;
   range: $[-2, \infty)$

OBJECTIVE 5   Use function notation.

The letters f, g, and h are commonly used to name functions. The function $y = 3x + 5$ may be named f and written

$$f(x) = 3x + 5,$$

where $f(x)$, which represents the value of f at x, is read **"f of x."** The notation $f(x)$, called **function notation,** is another way of expressing the range element y for a function f. For the function $f(x) = 3x + 5$, if $x = 7$, then we find $f(7)$ as follows.

$$f(x) = 3x + 5$$
$$f(7) = 3(7) + 5 \quad \text{Let } x = 7.$$
$$f(7) = 21 + 5 \quad \text{Multiply.}$$
$$f(7) = 26 \quad \text{Add.}$$

We read $f(7) = 26$ as "f of 7 equals 26." The notation $f(7)$ represents the value of y when x is 7. The statement $f(7) = 26$ says that the value of y is 26 when the value of x is 7. It also indicates that the point $(7, 26)$ lies on the graph of f.

⚠ CAUTION   The notation $f(x)$ does *not* mean f times x. *It represents the y-value that corresponds to x in function f.*

Function Notation

In the notation $f(x)$, remember the following.

	f	is the name of the function.
	x	is the domain value.
and	$f(x)$	is the range value y for the domain value x.

NOW TRY EXERCISE 5

Find $f(-2)$ and $f(0)$ for the function f.

$$f(x) = x^3 - 7$$

EXAMPLE 5   Using Function Notation

For the function $f(x) = x^2 - 3$, find each function value.

(a) $f(4)$

$$f(x) = x^2 - 3 \quad \text{Given function}$$
$$f(4) = 4^2 - 3 \quad \text{Let } x = 4.$$

4^2 means $4 \cdot 4$.

$$= 16 - 3 \quad \text{Apply the exponent.}$$
$$f(4) = 13 \quad \text{Subtract.}$$

(b) $f(0)$

$$f(x) = x^2 - 3$$
$$f(0) = 0^2 - 3 \quad \text{Let } x = 0.$$
$$f(0) = 0 - 3$$
$$f(0) = -3$$

(c) $f(-3)$

$$f(x) = x^2 - 3$$
$$f(-3) = (-3)^2 - 3 \quad \text{Let } x = -3.$$
$$f(-3) = 9 - 3$$

$(-3)^2$ means $(-3)(-3)$.

$$f(-3) = 6$$

NOW TRY

NOW TRY ANSWER
5. -15; -7

NOW TRY
EXERCISE 6

Refer to **Example 6.**

(a) Find $f(2006)$ and interpret this result.

(b) For what x-value does $f(x)$ equal 17.3 million?

OBJECTIVE 6 Apply the function concept in an application.

EXAMPLE 6 Applying the Function Concept to Population

Asian-American populations (in millions) are shown in the table.

Year	Population (in millions)
2000	11.2
2004	13.1
2006	14.9
2010	17.3

Source: U.S. Census Bureau.

(a) Use the table to write a set of ordered pairs that defines a function f.

If we choose the years as the domain elements and the populations in millions as the range elements, the information in the table can be written as a set of four ordered pairs. In set notation, the function f is defined as follows.

$$f = \{(2000, 11.2), (2004, 13.1), (2006, 14.9), (2010, 17.3)\}$$ *y*-values are in millions.

(b) What is the domain of f?   What is the range?

The domain is the set of years, or x-values.

$$\{2000, 2004, 2006, 2010\}$$

The range is the set of populations in millions, or y-values.

$$\{11.2, 13.1, 14.9, 17.3\}$$

(c) Find $f(2000)$ and $f(2004)$.

We refer to the table or the ordered pairs found in part (a).

$$f(2000) = 11.2 \text{ million} \quad \text{and} \quad f(2004) = 13.1 \text{ million}$$

NOW TRY ANSWERS
6. (a) $f(2006) = 14.9$;
The population of Asian-Americans was 14.9 million in 2006.
(b) 2010

(d) For what x-value does $f(x)$ equal 14.9 million?   11.2 million?

We use the table or the ordered pairs found in part (a).

$$f(2006) = 14.9 \text{ million} \quad \text{and} \quad f(2000) = 11.2 \text{ million}$$

NOW TRY

3.7 Exercises

FOR EXTRA HELP   ● MyMathLab®

● *Complete solution available in MyMathLab*

Concept Check   *Complete each statement.*

1. Any set of ordered pairs is a(n) _____. The set of first components of the ordered pairs of a relation is the _____. The set of second components of the ordered pairs of a relation is the _____.

2. A(n) _____ is a special type of relation in which each first component corresponds to exactly (*one / two / unlimited*) second _____.

Concept Check   *Complete the table for the linear function* $f(x) = x + 2$.

	x	x + 2	f(x)	(x, y)
	0	2	2	(0, 2)
3.	1			
4.	2			
5.	3			
6.	4			

7. Describe the graph of function f in the table for **Exercises 3–6** if the domain is $\{0, 1, 2, 3, 4\}$.

8. Describe the graph of function f in the table for **Exercises 3–6** if the domain is the set of real numbers $(-\infty, \infty)$.

*Determine whether each relation is a function. Give the domain and range. **See Examples 1–4.***

9. $\{(3, 7), (1, 4), (0, -2), (-1, -1)\}$

10. $\{(-2, 6), (0, 5), (2, 4), (3, 3)\}$

11. $\{(1, -1), (2, -2), (3, -1)\}$

12. $\{(4, 2), (0, 0), (-4, 2)\}$

▶ **13.** $\{(-4, 3), (-2, 1), (0, 5), (-2, -8)\}$

14. $\{(6, 0), (3, -2), (3, -6), (0, -4)\}$

15.
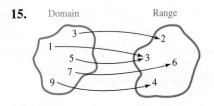

16.
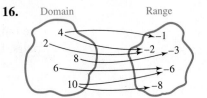

▶ **17.**
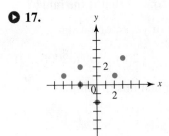

18.
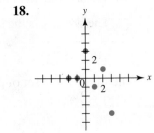

19.
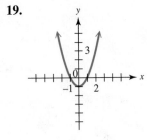

20.
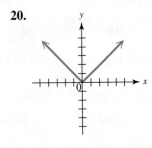

21.
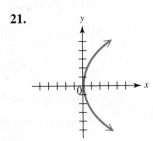

22.
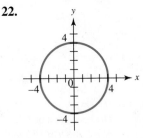

Decide whether each equation defines y as a function of x. (Remember that, to be a function, every value of x must give one and only one value of y.) **See Example 3.**

23. $y = 5x + 3$

24. $y = -7x + 12$

25. $x = -7$

26. $x = -4$

27. $y = 1$

28. $y = -2$

29. $y = x^2$

30. $y = -3x^2$

31. $x = y^2$

32. $x = |y|$

Find the domain and the range of each function. **See Example 4.**

▶ **33.** $y = 3x - 2$

34. $y = -x + 3$

35. $y = x$

36. $y = -5x$

37. $y = x^2 + 2$

38. $y = x^2 - 3$

39. $y = -x^2$

40. $y = -2x^2$

*For each function f, find **(a)** $f(2)$, **(b)** $f(0)$, **(c)** $f(-3)$, **(d)** $f\left(\frac{1}{2}\right)$, and **(e)** $f\left(-\frac{1}{3}\right)$.* **See Example 5.**

41. $f(x) = 4x + 3$

42. $f(x) = 3x + 5$

43. $f(x) = -x - 2$

44. $f(x) = -2x - 1$ ▶ **45.** $f(x) = x^2 - x + 2$ **46.** $f(x) = x^3 + x$

47. $f(x) = |x|$ **48.** $f(x) = -|x|$ **49.** $f(x) = |x + 7|$ **50.** $f(x) = |x - 5|$

The table shows the number of U.S. foreign-born residents (in millions) for selected years. Use the information to work Exercises 51–56. See Example 6.

Year	Population (in millions)
1970	9.6
1980	14.1
1990	19.8
2000	28.4
2010	40.0

Source: U.S. Census Bureau.

51. Write the information in the table as a set of ordered pairs. Does this set represent a function?

52. Suppose that g is the name given to this relation. Give the domain and range of g.

53. Find $g(1980)$ and $g(2000)$.

54. For what value of x does $g(x) = 19.8$ (million)?

55. For what value of x does $g(x) = 40.0$ (million)?

56. Suppose $g(2008) = 38.0$ (million). What does this indicate in the context of the application?

*A calculator can be thought of as a function machine. (See **FIGURE 45**.) We input a number value (from the domain), and then, by pressing the appropriate key, we obtain an output value (from the range). Use a calculator, follow the directions, and then answer each question.*

57. Suppose we enter the value 4 and then take the square root (that is, activate the square root function, using the key labeled $\sqrt{x}$).

   (a) What is the domain value here?   **(b)** What range value is obtained?

58. Enter the value -8 and then activate the squaring function, using the key labeled x^2.

   (a) What is the domain value here?   **(b)** What range value is obtained?

RELATING CONCEPTS For Individual or Group Work (Exercises 59–62)

*A **linear function** is a function defined by the equation of a line, such as*

$$f(x) = 3x - 4.$$

*It can be graphed by replacing $f(x)$ with y and then using the methods described earlier in this chapter. Let us assume that some function is written in the form $f(x) = mx + b$, for particular values of m and b. **Work Exercises 59–62 in order.***

59. Name the coordinates of the point on the line that corresponds to the statement $f(2) = 4$.

60. Name the coordinates of the point on the line that corresponds to the statement $f(-1) = -4$.

61. Use the results of **Exercises 59 and 60** to find the slope of the line.

62. Use the slope-intercept form of the equation of a line to write the function in the form $f(x) = mx + b$.

Chapter 3	Summary

Key Terms

3.1

line graph
linear equation in two
  variables
ordered pair
table of values
x-axis
y-axis
origin
rectangular (Cartesian)
  coordinate system
quadrant

plane
coordinates
plot
scatter diagram

3.2

graph, graphing
x-intercept
y-intercept
horizontal line
vertical line

3.3

rise
run
slope
parallel lines
perpendicular lines

3.6

linear inequality in two
  variables
boundary line

3.7

components
relation
domain
range
function

New Symbols

(x, y) ordered pair

m slope

(x_1, y_1) subscript notation
  (read "*x*-sub-one,
  y-sub-one")

$f(x)$ function notation;
  function of *x*
  (read "*f* of *x*")

Test Your Word Power

See how well you have learned the vocabulary in this chapter.

1. An **ordered pair** is a pair of
   numbers written
   A. in numerical order between
      brackets
   B. between parentheses or brackets
   C. between parentheses in which
      order is important
   D. between parentheses in which
      order does not matter.

2. An **intercept** is
   A. the point where the *x*-axis and
      y-axis intersect
   B. a pair of numbers written in
      parentheses in which order
      matters
   C. one of the four regions
      determined by a rectangular
      coordinate system
   D. the point where a graph intersects
      the *x*-axis or the *y*-axis.

3. The **slope** of a line is
   A. the measure of the run over the
      rise of the line
   B. the distance between two points
      on the line
   C. the ratio of the change in *y* to the
      change in *x* along the line
   D. the horizontal change compared
      to the vertical change of two
      points on the line.

4. Two lines in a plane are **parallel** if
   A. they represent the same line
   B. they never intersect
   C. they intersect at a 90° angle
   D. one has a positive slope and one
      has a negative slope.

5. Two lines in a plane are
   perpendicular if
   A. they represent the same line
   B. they never intersect
   C. they intersect at a 90° angle
   D. one has a positive slope and one
      has a negative slope.

6. A **function** is
   A. any set of ordered pairs
   B. a set of ordered pairs in which
      each distinct first component
      corresponds to exactly one
      second component
   C. two sets of ordered pairs that are
      related
   D. a graph of ordered pairs.

ANSWERS

1. C; *Examples:* $(0, 3)$, $(-3, 8)$, $(4, 0)$ **2.** D; *Example:* The graph of the equation $4x - 3y = 12$ has *x*-intercept $(3, 0)$ and *y*-intercept $(0, -4)$.
3. C; *Example:* The line through $(3, 6)$ and $(5, 4)$ has slope $\frac{4-6}{5-3} = \frac{-2}{2} = -1$.   **4.** B; *Example:* See **FIGURE 27** in **Section 3.3**.
5. C; *Example:* See **FIGURE 28** in **Section 3.3**.   **6.** B; *Example:* The set of ordered pairs $\{(0, 2), (2, 4), (3, 6)\}$ is a function because each *x*-value corresponds to exactly one *y*-value.

Quick Review

CONCEPTS	EXAMPLES

3.1 Linear Equations and Rectangular Coordinates

An ordered pair is a solution of an equation if it satisfies the equation.

Decide whether $(2, -5)$ and $(0, -6)$ are solutions of $4x - 3y = 18$.

$$4(2) - 3(-5) \overset{?}{=} 18 \qquad\qquad 4(0) - 3(-6) \overset{?}{=} 18$$
$$8 + 15 \overset{?}{=} 18 \qquad\qquad 0 + 18 \overset{?}{=} 18$$
$$23 = 18 \quad \text{False} \qquad\qquad 18 = 18 \ \checkmark \ \text{True}$$

$(2, -5)$ is not a solution. $(0, -6)$ is a solution.

If a value of either variable in an equation is given, then the value of the other variable can be found by substitution.

Complete the ordered pair $(0, \underline{\quad})$ for the given equation.

$$3x = y + 4$$
$$3(0) = y + 4 \qquad \text{Let } x = 0.$$
$$0 = y + 4 \qquad \text{Multiply.}$$
$$-4 = y \qquad \text{Subtract 4.}$$

The ordered pair is $(0, -4)$.

To plot an ordered pair, begin at the origin.

Step 1   Move right or left the number of units corresponding to the *x*-coordinate—right if it is positive or left if it is negative.

Step 2   Then turn and move up or down the number of units corresponding to the *y*-coordinate—up if it is positive or down if it is negative.

Plot the ordered pair $(-3, 4)$.

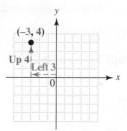

3.2 Graphing Linear Equations in Two Variables

To graph a linear equation, follow these steps.

Step 1   Find at least two ordered pairs that satisfy the equation. (It is good practice to find a third ordered pair as a check.)

Step 2   Plot the corresponding points.

Step 3   Draw a straight line through the points.

Graph $x - 2y = 4$.

	x	y
y-intercept →	0	-2
x-intercept →	4	0
	-2	-3

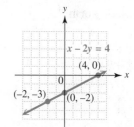

The graph of $Ax + By = 0$ passes through the origin. Find and plot two other points that satisfy the equation. Then draw a straight line through the points.

The graph of $y = b$ is a horizontal line through $(0, b)$.

The graph of $x = a$ is a vertical line through $(a, 0)$.

Graph $2x + 3y = 0$.

x	y
-3	2
0	0
3	-2

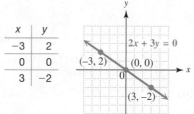

Graph $y = -3$ and $x = -3$.

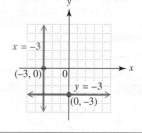

CONCEPTS	**EXAMPLES**
3.3 **The Slope of a Line**	
The slope m of the line passing through the points (x_1, y_1) and (x_2, y_2) is defined as follows.	The line passing through $(-2, 3)$ and $(4, -5)$ has slope
$$m = \frac{\text{change in } y}{\text{change in } x} = \frac{y_2 - y_1}{x_2 - x_1} \quad (\text{where } x_1 \neq x_2)$$	$$m = \frac{-5 - 3}{4 - (-2)} = \frac{-8}{6} = -\frac{4}{3}.$$
Horizontal lines have slope 0.	The line $y = -2$ has slope 0.
Vertical lines have undefined slope.	The line $x = 4$ has undefined slope.
To find the slope of a line from its equation, solve for y. The slope is the coefficient of x.	Find the slope of the line with the following equation. $$3x - 4y = 12$$ $$-4y = -3x + 12 \qquad \text{Subtract } 3x.$$ $$y = \frac{3}{4}x - 3 \qquad \text{Divide by } -4.$$ $$\text{Slope} \nearrow$$
Parallel lines have the same slope.	The lines $y = 3x - 1$ and $y = 3x + 4$ are parallel because both have slope 3.
The slopes of perpendicular lines, neither of which is vertical, are negative reciprocals (that is, their product is -1).	The lines $y = -3x - 1$ and $y = \frac{1}{3}x + 4$ are perpendicular because their slopes are -3 and $\frac{1}{3}$, and $-3\left(\frac{1}{3}\right) = -1$.
3.4 **Slope-Intercept Form of a Linear Equation**	
Slope-Intercept Form $y = mx + b$ m is the slope. $(0, b)$ is the y-intercept.	Write an equation of the line with slope 2 and y-intercept $(0, -5)$. $$y = 2x - 5$$
3.5 **Point-Slope Form of a Linear Equation and Modeling**	
Point-Slope Form $y - y_1 = m(x - x_1)$ m is the slope. (x_1, y_1) is a point on the line.	Write an equation of the line passing through $(-4, 5)$ with slope $-\frac{1}{2}$. $$y - 5 = -\frac{1}{2}[x - (-4)] \qquad \begin{array}{l}\text{Substitute for } m \text{ and } (x_1, y_1) \\ \text{in the point-slope form.}\end{array}$$ $$y - 5 = -\frac{1}{2}(x + 4) \qquad \text{Definition of subtraction}$$ $$y - 5 = -\frac{1}{2}x - 2 \qquad \text{Distributive property}$$ $$y = -\frac{1}{2}x + 3 \qquad \text{Add 5.}$$
Standard Form $Ax + By = C$ A, B, and C are real numbers and A and B are not both 0. (In answers, we give A, B, and C as integers with greatest common factor 1 and $A \geq 0$.)	Write the equation $y = -\frac{1}{2}x + 3$ in standard form. $$y = -\frac{1}{2}x + 3$$ $$-2y = x - 6 \qquad \text{Multiply each term by } -2.$$ $$x + 2y = 6 \qquad A = 1, B = 2, C = 6$$

CONCEPTS	EXAMPLES

3.6 Graphing Linear Inequalities in Two Variables

Step 1   Draw the graph of the straight line that is the boundary. Make it solid if the inequality involves $\leq$ or $\geq$. Make it dashed if the inequality involves $<$ or $>$.

Step 2   Choose a test point not on the line, and substitute the coordinates of that point in the inequality.

Step 3   Shade the region that includes the test point if it satisfies the original inequality. Otherwise, shade the region on the other side of the boundary line.

Graph $2x + y \leq 5$.
Graph the boundary line $2x + y = 5$.
Make it solid because the symbol $\leq$ includes equality.

Use $(0, 0)$ as a test point.
$$2(0) + 0 \overset{?}{<} 5$$
$$0 < 5 \qquad \text{True}$$

Shade the region that includes $(0, 0)$.

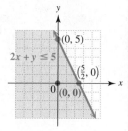

3.7 Introduction to Functions

Vertical Line Test
If a vertical line intersects a graph in more than one point, then the graph is not the graph of a function.

By the vertical line test, the graph shown is not the graph of a function.

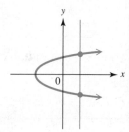

The set of all first components in the ordered pairs of a relation is the **domain** of the relation.

The set of all second components in the ordered pairs of a relation is the **range** of the relation.

The function
$$\{(10, 5), (20, 15), (30, 25)\}$$
has domain $\{10, 20, 30\}$ and range $\{5, 15, 25\}$.

To find $f(x)$ for a specific value of x, replace x by that value in the expression for function f.

If $f(x) = 2x + 7$, find $f(3)$.
$$f(x) = 2x + 7$$
$$f(3) = 2(3) + 7 \qquad \text{Let } x = 3.$$
$$f(3) = 13 \qquad \text{Multiply. Add.}$$

Chapter 3   Review Exercises

3.1   *The line graph shows the number, in millions, of real Christmas trees purchased for the years 2007 through 2012.*

1. Between which years did the number of real trees purchased increase?

2. Between which years did the number of real trees purchased remain the same?

3. Estimate the number of real trees purchased in 2011 and 2012.

4. By about how much did the number of real trees purchased between 2011 and 2012 decrease?

1. from 2010 to 2011
2. from 2008 to 2009
3. 2011: about 31 million;
    2012: about 24 million
4. about 7 million

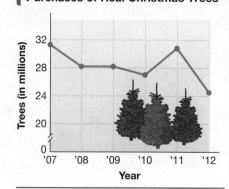

Purchases of Real Christmas Trees

Source: National Christmas Tree Association.

5. $-1; 2; 1$ **6.** $2; \dfrac{3}{2}; \dfrac{14}{3}$

7. $0; \dfrac{8}{3}; -9$ **8.** $7; 7; 7$

9. yes   **10.** no

11. yes   **12.** no

13. I   **14.** II

15. none   **16.** none

Graph for Exercises 13–16

17.   **18.**

$\left(-\dfrac{5}{2}, 0\right); (0, 5)$ $\left(\dfrac{8}{3}, 0\right); (0, 4)$

19. $(-4, 0);$ **20.** $(-6, 0);$
$(0, -2)$ no y-intercept

21. $-\dfrac{1}{2}$ **22.** undefined

23. 3   **24.** 0

25. undefined   **26.** $\dfrac{3}{2}$

27. $-\dfrac{1}{3}$ **28.** $\dfrac{3}{2}$

29. (a) 2   **(b)** $\dfrac{1}{3}$

30. parallel
31. perpendicular
32. neither

33. $y = -x + \dfrac{2}{3}$

34. $y = -\dfrac{1}{2}x + 4$

35. $y = x - 7$

36. $y = \dfrac{2}{3}x + \dfrac{14}{3}$

37. $y = -\dfrac{3}{4}x - \dfrac{1}{4}$

38. $y = -\dfrac{1}{4}x + \dfrac{3}{2}$

39. $y = 1$ **40.** $x = -4$

41. $y = \dfrac{3}{2}x - 2$ **42.** $y = -\dfrac{1}{3}x + 1$

Complete the given ordered pairs for each equation.

5. $y = 3x + 2;$ $(-1, \underline{\quad}), (0, \underline{\quad}), (\underline{\quad}, 5)$

6. $4x + 3y = 6;$ $(0, \underline{\quad}), (\underline{\quad}, 0), (-2, \underline{\quad})$

7. $x = 3y;$ $(0, \underline{\quad}), (8, \underline{\quad}), (\underline{\quad}, -3)$

8. $x - 7 = 0;$ $(\underline{\quad}, -3), (\underline{\quad}, 0), (\underline{\quad}, 5)$

Decide whether the given ordered pair is a solution of the given equation.

9. $x + y = 7;$ $(2, 5)$ **10.** $2x + y = 5;$ $(-1, 3)$

11. $3x - y = 4;$ $\left(\dfrac{1}{3}, -3\right)$ **12.** $x = -1;$ $(0, -1)$

Identify the quadrant in which each point is located. Then plot and label each point in a rectangular coordinate system.

13. $(2, 3)$ **14.** $(-4, 2)$ **15.** $(3, 0)$ **16.** $(0, -6)$

3.2   *Find the x- and y-intercepts for the graph of each equation. Then draw the graph.*

17. $y = 2x + 5$ **18.** $3x + 2y = 8$ **19.** $x + 2y = -4$ **20.** $x = -6$

3.3   *Find the slope of each line. (In Exercises 26 and 27, coordinates of the points shown are integers.)*

21. Through $(2, 3)$ and $(-4, 6)$ **22.** Through $(2, 5)$ and $(2, 8)$

23. $y = 3x - 4$ **24.** $y = 5$ **25.** $x = -7$

26. 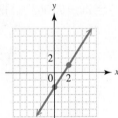   **27.** 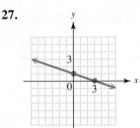   **28.** The line passing through these points

x	y
0	1
2	4
6	10

29. Find each slope.

(a) A line whose graph is parallel to the graph of $y = 2x + 3$

(b) A line whose graph is perpendicular to the graph of $y = -3x + 3$

Decide whether each pair of lines is parallel, perpendicular, *or* neither.

30. $3x + 2y = 6$
   $6x + 4y = 8$

31. $x - 3y = 1$
   $3x + y = 4$

32. $x - 2y = 8$
   $x + 2y = 8$

3.4, 3.5   *Write an equation of each line. Give the final answer in slope-intercept form (if possible).*

33. $m = -1, b = \dfrac{2}{3}$ **34.** Through $(2, 3)$ and $(-4, 6)$

35. Through $(4, -3), m = 1$ **36.** Through $(-1, 4), m = \dfrac{2}{3}$

37. Through $(1, -1), m = -\dfrac{3}{4}$ **38.** $m = -\dfrac{1}{4}, b = \dfrac{3}{2}$

39. Slope 0, through $(-4, 1)$ **40.** Through $(-4, 1)$, undefined slope

41. The line in **Exercise 26**   **42.** The line in **Exercise 27**

43.

44.

45.

46.

3.6 *Graph each linear inequality.*

43. $3x + 5y > 9$ **44.** $2x - 3y > -6$ **45.** $x - 2y \geq 0$ **46.** $y \geq -1$

3.7 *Decide whether each relation is a function. Give the domain and range.*

47. $\{(-2, 4), (0, 8), (2, 5), (2, 3)\}$ **48.** $\{(8, 3), (7, 4), (6, 5), (5, 6), (4, 7)\}$

49.

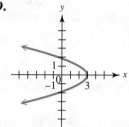

50.

47. not a function;
domain: $\{-2, 0, 2\}$;
range: $\{4, 8, 5, 3\}$

48. function;
domain: $\{8, 7, 6, 5, 4\}$;
range: $\{3, 4, 5, 6, 7\}$

49. not a function;
domain: $(-\infty, 3]$;
range: $(-\infty, \infty)$

50. function;
domain: $(-\infty, \infty)$;
range: $(-\infty, 3]$

51. $2x + 3y = 12$ **52.** $y = x^2$

For each function f, find (a) $f(2)$ and (b) $f(-1)$.

53. $f(x) = 3x + 2$ **54.** $f(x) = 2x^2 - 1$ **55.** $f(x) = |x + 3|$

51. function;
domain: $(-\infty, \infty)$;
range: $(-\infty, \infty)$

52. function;
domain: $(-\infty, \infty)$;
range: $[0, \infty)$

53. (a) 8 (b) -1
54. (a) 7 (b) 1
55. (a) 5 (b) 2

Chapter 3 Mixed Review Exercises

1. A **2.** C, D
3. A, B, D **4.** D
5. C **6.** B

7. $\left(-\dfrac{5}{2}, 0\right)$; $(0, -5)$; -2

8. $(0, 0)$; $(0, 0)$; $-\dfrac{1}{3}$

9. no x-intercept; $(0, 5)$; 0

10. $(-5, 0)$; no y-intercept;
undefined slope

In Exercises 1–6, match each statement to the appropriate graph or graphs in A–D. Graphs may be used more than once.

A. **B.** **C.** **D.**

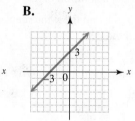

1. The line shown in the graph has undefined slope.

2. The graph of the equation has y-intercept $(0, -3)$.

3. The graph of the equation has x-intercept $(-3, 0)$.

4. The line shown in the graph has negative slope.

5. The graph is that of the equation $y = -3$.

6. The line shown in the graph has slope 1.

Find the x- and y-intercepts and the slope of each line. Then graph the line.

7. $y = -2x - 5$ **8.** $x + 3y = 0$

9. $y - 5 = 0$ **10.** $x = -5$

Write an equation of each line. Give the final answer in (a) slope-intercept form and (b) standard form (if possible).

11. $m = -\dfrac{1}{4}, b = -\dfrac{5}{4}$ **12.** Through $(8, 6)$, $m = -3$

13. Through $(3, -5)$ and $(-4, -1)$ **14.** Slope 0, through $(5, -5)$

11. (a) $y = -\frac{1}{4}x - \frac{5}{4}$

(b) $x + 4y = -5$

12. (a) $y = -3x + 30$

(b) $3x + y = 30$

13. (a) $y = -\frac{4}{7}x - \frac{23}{7}$

(b) $4x + 7y = -23$

14. (a) $y = -5$ (b) $y = -5$

15. **16.**

17. Because the graph falls from left to right, the slope is negative.

18. $(0, 39.6), (3, 36.0)$

19. $y = -1.2x + 39.6$

20. 37.2% $(x = 2)$;

It is a little high, as we might expect. The actual data point lies slightly below the graph of the line.

Graph each linear inequality.

15. $x - 2y \leq 6$

16. $y < -4x$

The points on the graph indicate the percent y of 4-year college students in public schools who earned a degree within 5 years of entry for selected years x. The graph of a linear equation that models the data is also shown. Here x = 0 represents 2010, x = 1 represents 2011, and so on.

17. Because the points of the graph lie approximately in a linear pattern, a straight line can be used to model the data. Will this line have positive or negative slope? Explain.

18. Write two ordered pairs (x, y) for the data for 2010 and 2013.

19. Use the two ordered pairs from **Exercise 18** to write an equation of a line that models the data. Give the final equation in slope-intercept form.

20. Use the equation from **Exercise 19** to approximate the percent for 2012. (What is the value of x for 2012?) How does the answer compare to the actual value of 36.6%?

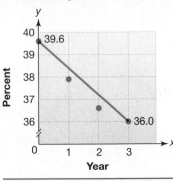

Percent of Students Graduating Within 5 Years

Source: ACT.

Chapter 3 Test

FOR EXTRA HELP *Step-by-step test solutions are found on the Chapter Test Prep Videos available in* MyMathLab® *or on* YouTube™.

▶ *View the complete solutions to all Chapter Test exercises in MyMathLab.*

[3.1]

1. $-6; -10; -5$ **2.** no

[3.2]

3. To find the x-intercept, let $y = 0$, and to find the y-intercept, let $x = 0$.

4. $(2, 0); (0, 6)$

5. $(0, 0); (0, 0)$

6. $(-3, 0)$; no y-intercept

1. Complete the ordered pairs $(0, \underline{\ \ }), (\underline{\ \ }, 0), (\underline{\ \ }, -3)$ for the equation $3x + 5y = -30$.

2. Is $(4, -1)$ a solution of $4x - 7y = 9$?

3. How do we find the x-intercept of the graph of a linear equation in two variables? How do we find the y-intercept?

Graph each linear equation. Give the x- and y-intercepts.

4. $3x + y = 6$

5. $y - 2x = 0$

6. $x + 3 = 0$

7. $y = 1$

8. $x - y = 4$

Find the slope of each line. (In Exercise 13, coordinates of the points shown are integers.)

9. Through $(-4, 6)$ and $(-1, -2)$

10. $2x + y = 10$

11. $x + 12 = 0$

12. A line whose graph is parallel to the graph of $y - 4 = 6$

13.

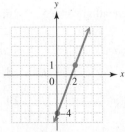

Write an equation of each line. Give the final answer in slope-intercept form.

14. Through $(-1, 4)$, $m = 2$

15. The line in **Exercise 13**

16. Through $(2, -6)$ and $(1, 3)$

7. no *x*-intercept;
(0, 1)

8. (4, 0);
(0, −4)

[3.3]

9. $-\dfrac{8}{3}$ **10.** −2

11. undefined **12.** 0

13. $\dfrac{5}{2}$

[3.4, 3.5]

14. $y = 2x + 6$ **15.** $y = \dfrac{5}{2}x - 4$

16. $y = -9x + 12$

[3.1, 3.5]

17. The slope is negative because sales are decreasing.

18. (0, 209), (13, 145);
 (a) −4.9
 (b) $y = -4.9x + 209$

19. 184.5 thousand;
The equation gives an approximation that is a little high.

20. In 2013, worldwide snowmobile sales were 145 thousand.

[3.6]

21.

22.

[3.7]

23. (a) not a function
 (b) function;
 domain: {0, 1, 2};
 range: {2}

24. not a function

25. 1

The graph shows worldwide snowmobile sales y for selected years x, where x = 0 represents 2000, x = 1 represents 2001, and so on. Use the graph to work Exercises 17–20.

17. Is the slope of the line in the graph positive or negative? Explain.

18. Write two ordered pairs (x, y) for the data points shown in the graph.

 (a) Use the ordered pairs to find the slope of the line to the nearest tenth.

 (b) Write an equation of a line that models the data. Give the final equation in slope-intercept form.

19. Use the equation from **Exercise 18(b)** to approximate worldwide snowmobile sales for 2005. How does the answer compare to the actual sales of 173.7 thousand?

20. What does the ordered pair (13, 145) mean in the context of this problem?

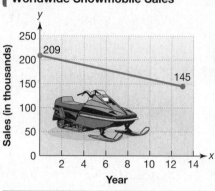

Worldwide Snowmobile Sales

Source: www.snowmobile.org

Graph each linear inequality.

21. $x + y \le 3$ **22.** $3x - y > 0$

Work each problem.

23. Decide whether each relation represents a function. If it does, give the domain and the range.

 (a) $\{(2, 3), (2, 4), (2, 5)\}$ **(b)** $\{(0, 2), (1, 2), (2, 2)\}$

24. Use the vertical line test to determine whether the graph is that of a function.

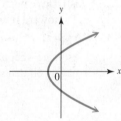

25. If $f(x) = 3x + 7$, find $f(-2)$.

Chapters R–3 Cumulative Review Exercises

[R.1]

1. $\dfrac{301}{40}$, or $7\dfrac{21}{40}$ **2.** 6

[1.4] **[1.5]**

3. 7 **4.** $\dfrac{73}{18}$, or $4\dfrac{1}{18}$

[1.1–1.5]

5. true **6.** −43

[1.6] **[1.7]**

7. distributive property **8.** $-p + 2$

Perform each indicated operation.

1. $10\dfrac{5}{8} - 3\dfrac{1}{10}$ **2.** $\dfrac{3}{4} \div \dfrac{1}{8}$ **3.** $5 - (-4) + (-2)$

4. $\dfrac{(-3)^2 - (-4)(2^4)}{5(2) - (-2)^3}$ **5.** *True* or *false?* $\dfrac{4(3 - 9)}{2 - 6} \ge 6$

6. Find the value of $xz^3 - 5y^2$ for $x = -2$, $y = -3$, and $z = -1$.

7. What property does $3(-2 + x) = -6 + 3x$ illustrate?

8. Simplify $-4p - 6 + 3p + 8$ by combining like terms.

[2.5]

9. $h = \dfrac{3V}{\pi r^2}$

[2.3]

10. $\{-1\}$

11. $\{2\}$

[2.6]

12. $\{-13\}$

[2.8]

13. $(-2.6, \infty)$

14. $(0, \infty)$

15. $(-\infty, -4]$

[2.4]

16. 13 mi

[R.2]

17. (a) $7000   **(b)** $10,000

18. about $30,000

[3.2]   **[3.3]**

19. $(-4, 0); (0, 3)$ **20.** $\dfrac{3}{4}$

[3.2]

21.

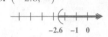

[3.3]

22. perpendicular

[3.4, 3.5]

23. $y = 3x - 11$ **24.** $y = 4$

Solve.

9. $V = \dfrac{1}{3}\pi r^2 h$ for h

10. $6 - 3(1 + x) = 2(x + 5) - 2$

11. $-(m - 3) = 5 - 2m$

12. $\dfrac{x - 2}{3} = \dfrac{2x + 1}{5}$

Solve each inequality, and graph the solution set.

13. $-2.5x < 6.5$ **14.** $4(x + 3) - 5x < 12$ **15.** $\dfrac{2}{3}x - \dfrac{1}{6}x \le -2$

Solve each problem.

16. Mount Mayon in the Philippines is the most perfectly shaped conical volcano in the world. Its base is a circle with circumference 80 mi, and it has a height of about 8100 ft. (One mile is 5280 ft.) Find the radius of the circular base to the nearest mile. (*Source:* www.britannica.hk)

Circumference = 80 mi

17. Over the next 45 yr, baby boomers are expected to inherit $10.4 trillion from their parents, an average of $50,000 each. The circle graph shows how they plan to spend their inheritances.

(a) How much of the $50,000 is expected to go toward a home purchase?

(b) How much of the $50,000 is expected to go toward retirement?

18. Use the answer from **Exercise 17(b)** to estimate the amount expected to go toward paying off debts or funding children's education.

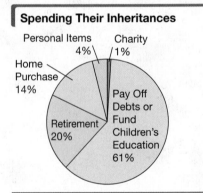

Spending Their Inheritances

Personal Items 4%
Charity 1%
Home Purchase 14%
Retirement 20%
Pay Off Debts or Fund Children's Education 61%

Source: First Interstate Bank Trust and Private Banking Group.

Consider the linear equation $-3x + 4y = 12$. *Find the following.*

19. The *x*- and *y*-intercepts   **20.** The slope   **21.** The graph

22. Are the lines with equations $x + 5y = -6$ and $y = 5x - 8$ *parallel, perpendicular,* or *neither*?

Write an equation of each line. Give the final answer in slope-intercept form if possible.

23. Through $(2, -5)$, slope 3   **24.** Through $(0, 4)$ and $(2, 4)$

4

Systems of Linear Equations and Inequalities

The point of intersection of two lines can be found using a *system of linear equations,* the subject of this chapter.

4.1   Solving Systems of Linear Equations by Graphing

VOCABULARY

☐ system of linear equations (linear system)
☐ solution of a system
☐ solution set of a system
☐ consistent system
☐ inconsistent system
☐ independent equations
☐ dependent equations

NOW TRY EXERCISE 1

Decide whether the ordered pair $(5, 2)$ is a solution of each system.

(a) $2x + 5y = 20$
   $x - y = 7$

(b) $3x - y = 13$
   $2x + y = 12$

A **system of linear equations,** or **linear system,** consists of two or more linear equations with the same variables.

$$2x + 3y = 4 \qquad x + 3y = 1 \qquad x - y = 1$$
$$3x - y = -5 \qquad -y = 4 - 2x \qquad y = 3$$

Linear systems

In the system on the right, think of $y = 3$ as an equation in two variables by writing it as $0x + y = 3$.

OBJECTIVE 1   Decide whether a given ordered pair is a solution of a system.

A **solution of a system** of linear equations is an ordered pair that makes both equations true at the same time. A solution of an equation is said to *satisfy* the equation.

EXAMPLE 1   Determining Whether an Ordered Pair Is a Solution

Decide whether the ordered pair $(4, -3)$ is a solution of each system.

(a) $x + 4y = -8$
   $3x + 2y = 6$

To decide whether $(4, -3)$ is a solution of the system, substitute 4 for x and -3 for y in each equation.

$x + 4y = -8$		$3x + 2y = 6$	
$4 + 4(-3) \overset{?}{=} -8$	Substitute.	$3(4) + 2(-3) \overset{?}{=} 6$	Substitute.
$4 + (-12) \overset{?}{=} -8$	Multiply.	$12 + (-6) \overset{?}{=} 6$	Multiply.
$-8 = -8$ ✓ True		$6 = 6$ ✓ True	

Because $(4, -3)$ satisfies both equations, it is a solution of the system.

(b) $2x + 5y = -7$
   $3x + 4y = 2$

Again, substitute 4 for x and -3 for y in each equation.

$2x + 5y = -7$		$3x + 4y = 2$	
$2(4) + 5(-3) \overset{?}{=} -7$	Substitute.	$3(4) + 4(-3) \overset{?}{=} 2$	Substitute.
$8 + (-15) \overset{?}{=} -7$	Multiply.	$12 + (-12) \overset{?}{=} 2$	Multiply.
$-7 = -7$ ✓ True		$0 = 2$	False

The ordered pair $(4, -3)$ is not a solution of this system because it does not satisfy the second equation.

NOW TRY

OBJECTIVE 2   Solve linear systems by graphing.

The set of all ordered pairs that are solutions of a system is its **solution set.** One way to find the solution set of a system of two linear equations is to graph both equations on the same axes. Any intersection point would be on both lines and would therefore be a solution of *both* equations. *Thus, the coordinates of any point at which the lines intersect give a solution of the system.*

NOW TRY ANSWERS
1. (a) no   **(b)** yes

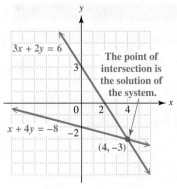

FIGURE 1

The graph in **FIGURE 1** shows that the solution of the system in **Example 1(a)** is the intersection point $(4, -3)$. Because two *different* straight lines can intersect at no more than one point, there can never be more than one solution for such a system.

EXAMPLE 2 Solving a System by Graphing

Solve the system by graphing.

$$2x + 3y = 4$$
$$3x - y = -5$$

We graph these two lines by plotting several points for each line. Recall from **Section 3.2** that the intercepts are often convenient choices. We show finding the intercepts for $2x + 3y = 4$.

To find the y-intercept, let $x = 0$.	To find the x-intercept, let $y = 0$.
$2x + 3y = 4$	$2x + 3y = 4$
$2(0) + 3y = 4$ Let $x = 0$.	$2x + 3(0) = 4$ Let $y = 0$.
$3y = 4$	$2x = 4$
$y = \dfrac{4}{3}$ y-intercept $\left(0, \frac{4}{3}\right)$	$x = 2$ x-intercept $(2, 0)$

The tables show the intercepts and a check point for each graph.

$2x + 3y = 4$

x	y	
0	$\frac{4}{3}$	← y-intercept
2	0	← x-intercept
-2	$\frac{8}{3}$	

Find a third ordered pair as a check.

$3x - y = -5$

x	y	
0	5	← y-intercept
$-\frac{5}{3}$	0	← x-intercept
-2	-1	

The lines in **FIGURE 2** suggest that the graphs intersect at the point $(-1, 2)$. We check by substituting -1 for x and 2 for y in *both* equations.

CHECK

$2x + 3y = 4$	First equation
$2(-1) + 3(2) \overset{?}{=} 4$	Substitute.
$4 = 4$ ✓	True
$3x - y = -5$	Second equation
$3(-1) - 2 \overset{?}{=} -5$	Substitute.
$-5 = -5$ ✓	True

FIGURE 2

Because $(-1, 2)$ satisfies both equations, the solution set of the system is $\{(-1, 2)\}$.

NOW TRY

NOTE We can also write each equation in a system in slope-intercept form and use the slope and y-intercept to graph each line. See **Example 2.**

$2x + 3y = 4$ becomes $y = -\frac{2}{3}x + \frac{4}{3}$.   y-intercept $\left(0, \frac{4}{3}\right)$; slope $-\frac{2}{3}$

$3x - y = -5$ becomes $y = 3x + 5$.   y-intercept $(0, 5)$; slope 3, or $\frac{3}{1}$

Confirm that graphing these equations gives the same results shown in **FIGURE 2.**

**NOW TRY
EXERCISE 2**

Solve the system by graphing.

$$x - 2y = 4$$
$$2x + y = 3$$

NOW TRY ANSWER
2. $\{(2, -1)\}$

Solving a Linear System by Graphing

Step 1 **Graph each equation** of the system on the same coordinate axes.

Step 2 **Find the coordinates of the point of intersection** of the graphs if possible, and write it as an ordered pair.

Step 3 **Check** that the ordered pair is the solution by substituting it in *both* of the *original* equations. If it satisfies *both* equations, write the solution set.

⚠ **CAUTION** We recommend using graph paper and a straightedge when solving systems of equations graphically. It may not be possible to determine from the graph the exact coordinates of the point that represents the solution, particularly if those coordinates are not integers. The graphing method does, however, show geometrically how solutions are found and is useful when approximate answers will suffice.

OBJECTIVE 3 Solve special systems by graphing.

EXAMPLE 3 Solving Special Systems by Graphing

Solve each system by graphing.

(a) $2x + y = 2$

$\quad\ 2x + y = 8$

See the graphs in **FIGURE 3**. The two lines are parallel and have no points in common. For such a system, there is no solution. The solution set is $\varnothing$.

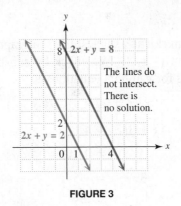

The lines do not intersect. There is no solution.

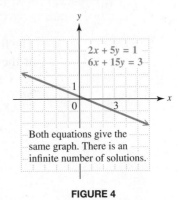

Both equations give the same graph. There is an infinite number of solutions.

FIGURE 3 **FIGURE 4**

(b) $2x + \ 5y = 1$

$\quad\ 6x + 15y = 3$

The graphs of these two equations are the same line. See **FIGURE 4**. We can obtain the second equation by multiplying each side of the first equation by 3. In this case, every point on the line is a solution of the system, and the solution set contains an infinite number of ordered pairs, each of which satisfies both equations of the system. We write the solution set as

$$\{(x, y) \mid 2x + 5y = 1\},$$

> This is the first equation in the system. See the Note on the next page.

read "the set of ordered pairs (x, y) such that $2x + 5y = 1$." Recall from **Section 1.3** that this notation is called **set-builder notation**.

NOW TRY ↻

**NOW TRY
EXERCISE 3**

Solve each system by graphing.

(a) $\quad 5x - 3y = 2$

$\quad\ 10x - 6y = 4$

(b) $\quad 4x + \ y = 7$

$\quad\ 12x + 3y = 10$

NOW TRY ANSWERS

3. (a) $\{(x, y) \mid 5x - 3y = 2\}$

$\quad$ **(b)** $\varnothing$

NOTE When a system has an infinite number of solutions, as in **Example 3(b),** either equation of the system could be used to write the solution set. *We prefer to use the equation in standard form with integer coefficients having greatest common factor 1.* If neither of the given equations is in this form, we use an *equivalent* equation that is in standard form with integer coefficients having greatest common factor 1.

The system in **Example 2** has exactly one solution. A system with at least one solution is a **consistent system.** A system with no solution, such as the one in **Example 3(a),** is an **inconsistent system.**

The equations in **Example 2** are **independent equations** with different graphs. The equations of the system in **Example 3(b)** have the same graph and are equivalent. Because they are different forms of the same equation, these equations are **dependent equations.**

Examples 2 and 3 illustrate the three cases that may occur when solving a system of equations with two variables.

Three Cases for Solutions of Linear Systems with Two Variables

Case 1 The graphs intersect at exactly one point, which gives the (single) ordered-pair solution of the system. The **system is consistent** and the **equations are independent.** See **FIGURE 5(a).**

Case 2 The graphs are parallel lines, so there is no solution and the solution set is ∅. The **system is inconsistent** and the **equations are independent.** See **FIGURE 5(b).**

Case 3 The graphs are the same line. There is an infinite number of solutions, and the solution set is written in set-builder notation as

$$\{(x, y) \mid \underline{\qquad}\},$$

where one of the equations follows the | symbol. The **system is consistent** and the **equations are dependent.** See **FIGURE 5(c).**

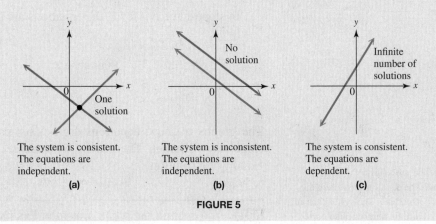

The system is consistent. The equations are independent.	The system is inconsistent. The equations are independent.	The system is consistent. The equations are dependent.
(a)	(b)	(c)

FIGURE 5

OBJECTIVE 4 Identify special systems without graphing.

We can recognize special systems without graphing by comparing their slopes and *y*-intercepts. We do this by writing each equation in slope-intercept form, solving for *y*.

**NOW TRY
EXERCISE 4**

Describe each system without graphing. State the number of solutions.

(a) $5x - 8y = 4$

$\quad x - \dfrac{8}{5}y = \dfrac{4}{5}$

(b) $2x + y = 7$

$\quad 3y = -6x - 12$

(c) $y - 3x = 7$

$\quad 3y - x = 0$

EXAMPLE 4   **Identifying the Three Cases Using Slopes**

Describe each system without graphing. State the number of solutions.

(a) $3x + 2y = 6$

$\quad -2y = 3x - 5$

Write each equation in slope-intercept form $y = mx + b$.

$3x + 2y = 6$ ⟨ Solve for y. ⟩

$\quad 2y = -3x + 6$ Subtract $3x$.

$\quad y = -\dfrac{3}{2}x + 3$ Divide *each* term by 2.

$-2y = 3x - 5$ ⟨ Solve for y. ⟩

$\quad y = -\dfrac{3}{2}x + \dfrac{5}{2}$ Divide *each* term by -2.

Both equations have slope $-\dfrac{3}{2}$, but they have different y-intercepts, $(0, 3)$ and $\left(0, \dfrac{5}{2}\right)$. Such lines are parallel (**Section 3.3**), so these equations have graphs that are parallel lines, which do not intersect. Thus, the system has no solution.

(b) $2x - y = 4$

$\quad x = \dfrac{y}{2} + 2$

Again, write the equations in slope-intercept form.

$2x - y = 4$

$\quad -y = -2x + 4$ Subtract $2x$.

$\quad y = 2x - 4$ Multiply by -1.

$\dfrac{y}{2} + 2 = x$ Interchange sides.

$\dfrac{y}{2} = x - 2$ Subtract 2.

$y = 2x - 4$ Multiply by 2.

The equations are exactly the same—their graphs are the same line. Any ordered-pair solution of one equation is also a solution of the other equation. Thus, the system has an infinite number of solutions.

(c) $x - 3y = 5$

$\quad 2x + \ y = 8$

In slope-intercept form, the equations are as follows.

$x - 3y = 5$

$\quad -3y = -x + 5$ Subtract x.

$\quad y = \dfrac{1}{3}x - \dfrac{5}{3}$ Divide by -3.

$2x + y = 8$

$\quad y = -2x + 8$ Subtract $2x$.

The graphs of these equations are neither parallel nor the same line because the slopes are different. The graphs will intersect in one point—thus, the system has exactly one solution.

NOW TRY

NOW TRY ANSWERS

4. (a) The equations represent the same line. The system has an infinite number of solutions.
(b) The equations represent parallel lines. The system has no solution.
(c) The equations represent lines that are neither parallel nor the same line. The system has exactly one solution.

NOTE   The solution set of the system in **Example 4(a)** is $\varnothing$ because the graphs of the equations of the system are parallel lines. The solution set of the system in **Example 4(b)**, written using set-builder notation and the first equation, is

$$\{(x, y) \mid 2x - y = 4\}.$$

If we try to solve the system in **Example 4(c)** by graphing, we will have difficulty identifying the point of intersection of the graphs. We introduce an algebraic method for solving systems like this in **Section 4.2**.

4.1 Exercises

 MyMathLab®

▶ *Complete solution available in MyMathLab*

Concept Check *Complete each statement. The following terms may be used once, more than once, or not at all.*

consistent	system of linear equations	inconsistent	solution
ordered pair	independent	linear equation	dependent

1. A(n) _____ consists of two or more linear equations with the (*same/different*) variables.

2. A solution of a system of linear equations is a(n) _____ that makes all equations of the system (*true/false*) at the same time.

3. The equations of two parallel lines form a(n) _____ system that has (*one/no/infinitely many*) solution(s). The equations are _____ because their graphs are different.

4. If the graphs of a linear system intersect in one point, the point of intersection is the _____ of the system. The system is _____ and the equations are independent.

5. If two equations of a linear system have the same graph, the equations are _____. The system is _____ and has (*one/no/infinitely many*) solution(s).

6. If a linear system is inconsistent, the graphs of the two equations are (*intersecting/parallel/the same*) line(s). The system has no _____.

Concept Check *Work each problem.*

7. A student determined that the ordered pair $(1, -2)$ is a solution of the following system. His reasoning was that the ordered pair satisfies the equation $x + y = -1$ because $1 + (-2) = -1$. **WHAT WENT WRONG?**

$$x + y = -1$$
$$2x + y = 4$$

8. The following system has infinitely many solutions. Write its solution set using set-builder notation as described in **Example 3(b).**

$$6x - 4y = 8$$
$$3x - 2y = 4$$

9. Which ordered pair could not be a solution of the system graphed? Why is it the only valid choice?

A. $(-4, -4)$ **B.** $(-2, 2)$

C. $(-4, 4)$ **D.** $(-3, 3)$

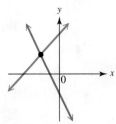

10. Which ordered pair could be a solution of the system graphed? Why is it the only valid choice?

A. $(2, 0)$ **B.** $(0, 2)$

C. $(-2, 0)$ **D.** $(0, -2)$

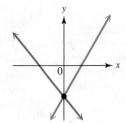

11. Each ordered pair in parts (a)–(d) is a solution of one of the systems graphed in choices A–D. Use the location of the point of intersection to determine the correct system for each solution. Match each system from A–D with its solution from (a)–(d).

(a) $(3, 4)$ **A.**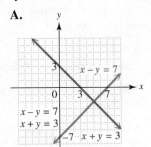

(b) $(-2, 3)$ **B.**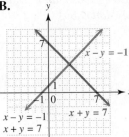

(c) $(-3, 2)$ **C.**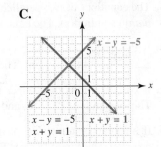

(d) $(5, -2)$ **D.**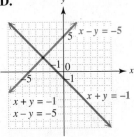

12. Solve the system by graphing. Can the solution be checked? Why or why not?

$$2x + 3y = 6$$
$$x - 3y = 5$$

*Decide whether the given ordered pair is a solution of the given system. **See Example 1.***

13. $(2, -3)$
$x + y = -1$
$2x + 5y = 19$

14. $(4, 3)$
$x + 2y = 10$
$3x + 5y = 3$

15. $(-1, -3)$
$3x + 5y = -18$
$4x + 2y = -10$

16. $(-9, -2)$
$2x - 5y = -8$
$3x + 6y = -39$

17. $(7, -2)$
$4x = 26 - y$
$3x = 29 + 4y$

18. $(9, 1)$
$2x = 23 - 5y$
$3x = 24 + 3y$

19. $(6, -8)$
$-2y = x + 10$
$3y = 2x + 30$

20. $(-5, 2)$
$5y = 3x + 20$
$3y = -2x - 4$

21. $(0, 0)$
$4x + 2y = 0$
$x + y = 0$

22. $(-1, -1)$
$-4x + 4y = 0$
$x - y = 0$

*Solve each system by graphing. If the system is inconsistent or the equations are dependent, say so. **See Examples 2 and 3.***

23. $x - y = 2$
$x + y = 6$

24. $x - y = 3$
$x + y = -1$

25. $x + y = 4$
$y - x = 4$

26. $x + y = -5$
$y - x = -5$

27. $x - 2y = 6$
$x + 2y = 2$

28. $2x - y = 4$
$4x + y = 2$

29. $3x - 2y = -3$
$-3x - y = -6$

30. $2x - y = 4$
$2x + 3y = 12$

31. $3x + y = 5$
$6x + 2y = 10$

32. $2x - y = 4$
$4x - 2y = 8$

33. $2x - 3y = -6$
$y = -3x + 2$

34. $-3x + y = -3$
$y = x - 3$

35. $2x - y = 6$
$4x - 2y = 8$

36. $x + 2y = 4$
$2x + 4y = 12$

37. $2y - 6x = 12$
$3x - y = -6$

38. $-8y - 2x = -8$
$x + 4y = 4$

39. $3x - 4y = 24$
$y = -\dfrac{3}{2}x + 3$

40. $4x + y = 5$
$y = \dfrac{3}{2}x - 6$

41. $2x = y - 4$
$4x + 4 = 2y$

42. $3x = y + 5$
$6x - 5 = 2y$

Without graphing, answer the following questions for each linear system. **See Example 4.**

(**a**) *Is the system inconsistent, are the equations dependent, or neither?*
(**b**) *Is the graph a pair of intersecting lines, a pair of parallel lines, or one line?*
(**c**) *Does the system have one solution, no solution, or an infinite number of solutions?*

43. $y - x = -5$
$x + y = 1$

44. $y + 2x = 6$
$x - 3y = -4$

45. $x + 2y = 0$
$4y = -2x$

46. $2x - y = 4$
$y + 4 = 2x$

47. $x - 3y = 5$
$2x + y = 8$

48. $2x + 3y = 12$
$2x - y = 4$

49. $5x + 4y = 7$
$10x + 8y = 4$

50. $3x + 2y = 5$
$6x + 4y = 3$

Work each problem using the graph provided.

51. The numbers of daily morning and evening newspapers in the United States in selected years over the period 1980–2012 are shown in the graph.

(**a**) For which years were there more evening dailies than morning dailies?

(**b**) Estimate the year in which the number of evening and morning dailies was closest to the same. About how many newspapers of each type were there in that year?

(**c**) Express the point of intersection of the two graphs as an ordered pair written in the form (year, number of newspapers).

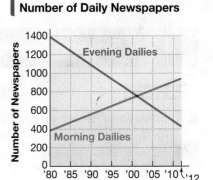

Number of Daily Newspapers

Source: Editor & Publisher International Year Book.

52. The graph shows how sales of music CDs and digital downloads of single songs (in millions) in the United States have changed over the years 2004 through 2012.

(**a**) In what year did Americans purchase about the same number of CDs as single digital downloads? How many units was this?

(**b**) Express the point of intersection of the two graphs as an ordered pair written in the form (year, units sold in millions).

(**c**) Describe the trend in sales of music CDs over the years 2004 to 2012. If a straight line were used to approximate its graph, would the slope of the line be positive, negative, or zero? Explain.

(**d**) If a straight line were used to approximate the graph of sales of digital downloads over the years 2004 to 2012, would the slope of the line be positive, negative, or zero? Explain.

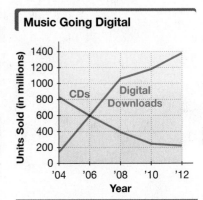

Music Going Digital

Source: Recording Industry Association of America.

Economics deals with **supply** and **demand**. Typically, as the price of an item increases, the demand for the item decreases and the supply increases. If supply and demand can be described by straight-line equations, the point at which the lines intersect determines the **equilibrium supply** and **equilibrium demand**.

The price per unit, p, and the demand, x, for a particular aluminum siding are related by the linear equation

$$p = 60 - \frac{3}{4}x,$$

while the price and the supply are related by

$$p = \frac{3}{4}x.$$

Supply and Demand

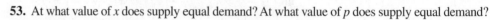

Use the graph to work Exercises 53–56.

53. At what value of *x* does supply equal demand? At what value of *p* does supply equal demand?

54. Express the equilibrium supply and equilibrium demand as an ordered pair of the form (quantity, price).

55. When *x* > 40, does demand exceed supply or does supply exceed demand?

56. When *x* < 40, does demand exceed supply or does supply exceed demand?

4.2   Solving Systems of Linear Equations by Substitution

OBJECTIVES

1. Solve linear systems by substitution.
2. Solve special systems by substitution.
3. Solve linear systems with fractions and decimals.

OBJECTIVE 1   Solve linear systems by substitution.

Graphing to solve a system of equations has a serious drawback. For example, consider the system graphed in **FIGURE 6**. It is difficult to determine an accurate solution of the system from the graph.

As a result, there are algebraic methods for solving systems of equations. The **substitution method,** which gets its name from the fact that an expression in one variable is *substituted* for the other variable, is one such method.

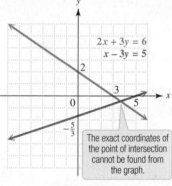

FIGURE 6

EXAMPLE 1   **Using the Substitution Method**

Solve the system by the substitution method.

$$3x + 5y = 26 \qquad (1)$$
$$y = 2x \qquad (2)$$

We number the equations for reference in our discussion.

Equation (2) is already solved for *y*. This equation says that *y* is equal to 2*x*, so we substitute 2*x* for *y* in equation (1).

$3x + 5y = 26$	(1)
$3x + 5(2x) = 26$	Let $y = 2x$.
$3x + 10x = 26$	Multiply.
$13x = 26$	Combine like terms.
$x = 2$	Divide by 13.

Don't stop here.

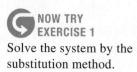

**NOW TRY
EXERCISE 1**

Solve the system by the substitution method.

$$2x - 4y = 28$$
$$y = -3x$$

Now we can find the value of y by substituting 2 for x in either equation. We choose equation (2) because the substitution is easier.

$$y = 2x \qquad (2)$$
$$y = 2(2) \qquad \text{Let } x = 2.$$
$$y = 4 \qquad \text{Multiply.}$$

We check that the ordered pair $(2, 4)$ is the solution by substituting 2 for x and 4 for y in *both* equations.

CHECK

$3x + 5y = 26$	(1)		$y = 2x$	(2)
$3(2) + 5(4) \stackrel{?}{=} 26$	Substitute.		$4 \stackrel{?}{=} 2(2)$	Substitute.
$6 + 20 \stackrel{?}{=} 26$	Multiply.		$4 = 4 \ \checkmark$	True
$26 = 26 \ \checkmark$	True			

Because $(2, 4)$ satisfies both equations, the solution set of the system is $\{(2, 4)\}$.

NOW TRY

! **CAUTION** *A system is not completely solved until values for both x and y are found.* Write the solution set as a set containing an ordered pair.

**NOW TRY
EXERCISE 2**

Solve the system by the substitution method.

$$4x + 9y = 1$$
$$x = y - 3$$

EXAMPLE 2 Using the Substitution Method

Solve the system by the substitution method.

$$2x + 5y = 7 \qquad (1)$$
$$x = -1 - y \qquad (2)$$

Equation (2) gives x in terms of y. Substitute $-1 - y$ for x in equation (1).

$2x + 5y = 7$	(1)	*Be sure to substitute in the other equation.*
$2(-1 - y) + 5y = 7$	Let $x = -1 - y$.	
$-2 - 2y + 5y = 7$	Distributive property	
$-2 + 3y = 7$	Combine like terms.	
$3y = 9$	Add 2.	
$y = 3$	Divide by 3.	

Distribute 2 to both -1 and $-y$.

To find x, substitute 3 for y in equation (2).

$$x = -1 - y \qquad (2)$$
$$x = -1 - 3 \qquad \text{Let } y = 3.$$
$$x = -4 \qquad \text{Subtract.}$$

Check that $(-4, 3)$ is the solution.

CHECK

$2x + 5y = 7$	(1)		$x = -1 - y$	(2)
$2(-4) + 5(3) \stackrel{?}{=} 7$	Substitute.		$-4 \stackrel{?}{=} -1 - 3$	Substitute.
$-8 + 15 \stackrel{?}{=} 7$	Multiply.		$-4 = -4 \ \checkmark$	True
$7 = 7 \ \checkmark$	True			

Write the x-coordinate first.

NOW TRY ANSWERS
1. $\{(2, -6)\}$ **2.** $\{(-2, 1)\}$

Both results are true. The solution set of the system is $\{(-4, 3)\}$. **NOW TRY**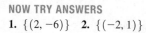

🛑 **CAUTION**   Even though we found y first in **Example 2,** *the x-coordinate is always written first in the ordered-pair solution of a system.* The ordered pair $(-4, 3)$ is *not* the same as $(3, -4)$.

Solving a Linear System by Substitution

Step 1   **Solve one equation for either variable.** If one of the equations has a variable term with coefficient 1 or -1, choose it because the substitution method is usually easier.

Step 2   **Substitute** for that variable in the other equation. The result should be an equation with just one variable.

Step 3   **Solve** the equation from Step 2.

Step 4   **Find the other value.** Substitute the result from Step 3 into the equation from Step 1 and solve for the other variable.

Step 5   **Check** the values in *both* of the *original* equations. Then write the solution set as a set containing an ordered pair.

EXAMPLE 3   **Using the Substitution Method**

Solve the system by the substitution method.

$$2x = 4 - y \qquad (1)$$
$$5x + 3y = 10 \qquad (2)$$

Step 1   We must solve one of the equations for either x or y. Because the coefficient of y in equation (1) is -1, we avoid fractions by solving this equation for y.

$$2x = 4 - y \qquad (1)$$

$$y + 2x = 4 \qquad \text{Add } y.$$

$$y = -2x + 4 \qquad \text{Subtract } 2x.$$

Step 2   Now substitute $-2x + 4$ for y in equation (2).

$$5x + 3y = 10 \qquad (2)$$

$$5x + 3(-2x + 4) = 10 \qquad \text{Let } y = -2x + 4.$$

Step 3   Solve the equation from Step 2.

$$5x - 6x + 12 = 10 \qquad \text{Distributive property}$$

Distribute 3 to both $-2x$ and 4.

$$-x + 12 = 10 \qquad \text{Combine like terms.}$$

$$-x = -2 \qquad \text{Subtract 12.}$$

$$x = 2 \qquad \text{Multiply by } -1.$$

Step 4   We solved equation (1) for y in Step 1. Substitute 2 for x in this equation to find y.

$$y = -2x + 4 \qquad \text{Equation (1) solved for } y$$

$$y = -2(2) + 4 \qquad \text{Let } x = 2.$$

$$y = 0 \qquad \text{Multiply, and then add.}$$

NOW TRY
EXERCISE 3

Solve the system by the substitution method.

$$2y = x - 2$$
$$4x - 5y = -4$$

Step 5 Check that $(2, 0)$ is the solution.

CHECK

$2x = 4 - y$ (1)	$5x + 3y = 10$ (2)
$2(2) \overset{?}{=} 4 - 0$ Substitute.	$5(2) + 3(0) \overset{?}{=} 10$ Substitute.
$4 = 4$ ✓ True	$10 = 10$ ✓ True

Because both results are true, the solution set of the system is $\{(2, 0)\}$.

NOW TRY

OBJECTIVE 2 Solve special systems by substitution.

EXAMPLE 4 Solving an Inconsistent System Using Substitution

Solve the system by the substitution method.

$$x = 5 - 2y \qquad (1)$$
$$2x + 4y = 6 \qquad (2)$$

NOW TRY
EXERCISE 4

Solve the system by the substitution method.

$$8x - 2y = 1$$
$$y = 4x - 8$$

Because equation (1) is solved for x, we substitute $5 - 2y$ for x in equation (2).

$2x + 4y = 6$ (2)	
$2(5 - 2y) + 4y = 6$	Let $x = 5 - 2y$.
$10 - 4y + 4y = 6$	Distributive property
$10 = 6$	False

The false result means that the equations in the system have graphs that are parallel lines. The system is inconsistent and has no solution, so the solution set is $\varnothing$.

CHECK We can confirm the solution set by writing each equation in slope-intercept form—that is, solved for y. (See **Section 4.1, Example 4.**)

$x = 5 - 2y$ (1)	$2x + 4y = 6$ (2)
$2y = -x + 5$	$4y = -2x + 6$
$y = -\dfrac{1}{2}x + \dfrac{5}{2}$	$y = -\dfrac{1}{2}x + \dfrac{3}{2}$

The two lines have the same slope, but different y-intercepts. Therefore, they are parallel and do not intersect, confirming that the solution set is $\varnothing$. See **FIGURE 7.** ✓

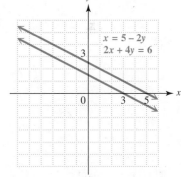

FIGURE 7

NOW TRY

EXAMPLE 5 Solving a System with Dependent Equations Using Substitution

Solve the system by the substitution method.

$$3x - y = 4 \qquad (1)$$
$$-9x + 3y = -12 \qquad (2)$$

Begin by solving equation (1) for y to obtain

$$y = 3x - 4. \qquad \text{Equation (1) solved for } y$$

We substitute $3x - 4$ for y in equation (2) and solve the resulting equation.

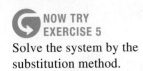

**NOW TRY
EXERCISE 5**

Solve the system by the substitution method.

$$5x - y = 6$$
$$-10x + 2y = -12$$

$$-9x + 3y = -12 \qquad (2)$$
$$-9x + 3(3x - 4) = -12 \qquad \text{Let } y = 3x - 4.$$
$$-9x + 9x - 12 = -12 \qquad \text{Distributive property}$$
$$0 = 0 \qquad \text{Add 12. Combine like terms.}$$

This true result means that every solution of one equation is also a solution of the other, so the system has an infinite number of solutions. The solution set, written in set-builder notation using equation (1), is

$$\{(x, y) \mid 3x - y = 4\}.$$

CHECK If we multiply equation (1) by -3, we obtain equation (2). Therefore,

$$3x - y = 4 \quad \text{and} \quad -9x + 3y = -12$$

are equivalent equations. They represent the same line. All of the ordered pairs corresponding to points that lie on the common graph are solutions. See **FIGURE 8**. ✓

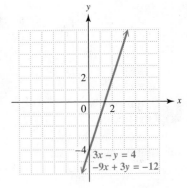

FIGURE 8

NOW TRY

⚠ CAUTION Avoid these common mistakes.

1. Do not give "false" as the solution of an inconsistent system. The correct response is ∅. (See **Example 4**.)

2. Do not give "true" as the solution of a system of dependent equations. In this book, we write the solution set in set-builder notation using the equation in the system (or an equivalent equation) that is in standard form with integer coefficients having greatest common factor 1. (See **Example 5**.)

OBJECTIVE 3 Solve linear systems with fractions and decimals.

EXAMPLE 6 Using the Substitution Method (Fractional Coefficients)

Solve the system by the substitution method.

$$3x + \frac{1}{4}y = 2 \qquad (1)$$

$$\frac{1}{2}x + \frac{3}{4}y = -\frac{5}{2} \qquad (2)$$

Clear equation (1) of fractions by multiplying each side by 4.

$$3x + \frac{1}{4}y = 2 \qquad (1)$$

$$4\left(3x + \frac{1}{4}y\right) = 4(2) \qquad \text{Multiply by 4.}$$

$$4(3x) + 4\left(\frac{1}{4}y\right) = 4(2) \qquad \text{Distributive property}$$

$$12x + y = 8 \qquad (3)$$

**NOW TRY
EXERCISE 6**
Solve the system by the
substitution method.

$$x + \frac{1}{2}y = \frac{1}{2}$$

$$\frac{1}{6}x - \frac{1}{3}y = \frac{4}{3}$$

Now clear equation (2) of fractions by multiplying each side by 4.

$$\frac{1}{2}x + \frac{3}{4}y = -\frac{5}{2} \qquad \text{(2)}$$

$$4\left(\frac{1}{2}x + \frac{3}{4}y\right) = 4\left(-\frac{5}{2}\right) \qquad \text{Multiply by 4, the common denominator.}$$

$$4\left(\frac{1}{2}x\right) + 4\left(\frac{3}{4}y\right) = 4\left(-\frac{5}{2}\right) \qquad \text{Distributive property}$$

$$2x + 3y = -10 \qquad \text{(4)}$$

The given system of equations has been simplified to an equivalent system.

$$12x + \ y = 8 \qquad \text{(3)}$$

$$2x + 3y = -10 \qquad \text{(4)}$$

To solve this system by substitution, solve equation (3) for y.

$$12x + y = 8 \qquad \text{(3)}$$

$$y = -12x + 8 \qquad \text{Subtract 12x.}$$

Now substitute this result for y in equation (4).

$$2x + 3y = -10 \qquad \text{(4)}$$

$$2x + 3(-12x + 8) = -10 \qquad \text{Let } y = -12x + 8.$$

$$2x - 36x + 24 = -10 \qquad \text{Distributive property}$$

$$-34x = -34 \qquad \text{Combine like terms. Subtract 24.}$$

$$x = 1 \qquad \text{Divide by } -34.$$

Substitute 1 for x in $y = -12x + 8$ (equation (3) solved for y).

$$y = -12(1) + 8 \qquad \text{Let } x = 1.$$

$$y = -4 \qquad \text{Multiply, and then add.}$$

Check $(1, -4)$ in both of the original equations. The solution set is $\{(1, -4)\}$.

NOW TRY

EXAMPLE 7 Using the Substitution Method (Decimal Coefficients)

Solve the system by the substitution method.

$$0.5x + 2.4y = 4.2 \qquad \text{(1)}$$

$$-0.1x + 1.5y = 5.1 \qquad \text{(2)}$$

Clear each equation of decimals by multiplying by 10.

$$10(0.5x + 2.4y) = 10(4.2) \qquad \text{Multiply equation (1) by 10.}$$

$$10(0.5x) + 10(2.4y) = 10(4.2) \qquad \text{Distributive property}$$

$$5x + 24y = 42 \qquad \text{(3)}$$

$$10(-0.1x + 1.5y) = 10(5.1) \qquad \text{Multiply equation (2) by 10.}$$

$$10(-0.1x) + 10(1.5y) = 10(5.1) \qquad \text{Distributive property}$$

NOW TRY ANSWER
6. $\{(2, -3)\}$

 $10(-0.1x) = -1x = -x$ $\longrightarrow$ $-x + 15y = 51 \qquad \text{(4)}$

**NOW TRY
EXERCISE 7**

Solve the system by the substitution method.

$$0.2x + 0.3y = 0.5$$
$$0.3x - 0.1y = 1.3$$

Now solve the equivalent system of equations by substitution.

$$5x + 24y = 42 \qquad (3)$$
$$-x + 15y = 51 \qquad (4)$$

Equation (4) can be solved for x.

$$x = 15y - 51 \qquad \text{Equation (4) solved for } x$$

Substitute this result for x in equation (3).

$$5x + 24y = 42 \qquad (3)$$
$$5(15y - 51) + 24y = 42 \qquad \text{Let } x = 15y - 51.$$
$$75y - 255 + 24y = 42 \qquad \text{Distributive property}$$
$$99y = 297 \qquad \text{Combine like terms. Add 255.}$$
$$y = 3 \qquad \text{Divide by 99.}$$

Equation (4) solved for x is $x = 15y - 51$. Substitute 3 for y.

$$x = 15(3) - 51 \qquad \text{Let } y = 3.$$
$$x = -6 \qquad \text{Multiply, and then subtract.}$$

Check $(-6, 3)$ in both of the original equations. The solution set is $\{(-6, 3)\}$.

NOW TRY ANSWER

7. $\{(4, -1)\}$

NOW TRY

4.2 Exercises

FOR EXTRA HELP   ▶ MyMathLab®

▶ *Complete solution available
in MyMathLab*

Concept Check   *Work each problem.*

1. A student solves the following system and finds that $x = 3$, which is correct. The student gives the solution set as $\{3\}$. *WHAT WENT WRONG?*

$$5x - y = 15$$
$$7x + y = 21$$

2. A student solves the following system and obtains the equation $0 = 0$. The student gives the solution set as $\{(0, 0)\}$. *WHAT WENT WRONG?*

$$x + y = 4$$
$$2x + 2y = 8$$

3. When we use the substitution method, how can we tell that a system has no solution?

4. When we use the substitution method, how can we tell that a system has an infinite number of solutions?

Solve each system by the substitution method. Check each solution. ***See Examples 1–5.***

5. $x + y = 12$
$\quad y = 3x$

6. $x + 3y = -28$
$\quad y = -5x$

7. $3x + 2y = 27$
$\quad x = y + 4$

8. $4x + 3y = -5$
$\quad x = y - 3$

9. $3x + 4 = -y$
$\quad 2x + y = 0$

10. $2x - 5 = -y$
$\quad x + 3y = 0$

11. $7x + 4y = 13$
$\quad x + y = 1$

12. $3x - 2y = 19$
$\quad x + y = 8$

13. $3x + 5y = 25$
$\quad x - 2y = -10$

14. $5x + 2y = -15$
$2x - y = -6$

15. $3x - y = 5$
$y = 3x - 5$

16. $4x - y = -3$
$y = 4x + 3$

17. $2x + y = 0$
$4x - 2y = 2$

18. $x + y = 0$
$4x + 2y = 3$

19. $2x + 8y = 3$
$x = 8 - 4y$

20. $2x + 10y = 3$
$x = 1 - 5y$

21. $2y = 4x + 24$
$2x - y = -12$

22. $2y = 14 - 6x$
$3x + y = 7$

23. $y = 6 - x$
$y = 2x + 3$

24. $y = 4x - 4$
$y = -3x - 11$

25. $x = y - 4$
$x - y = 1$

26. $x = 2 - y$
$x + y = -5$

27. $x + y = 0$
$3x - 3y = 0$

28. $5x + y = 0$
$x - y = 0$

Solve each system by the substitution method. Check each solution. **See Examples 6 and 7.**

29. $\frac{1}{2}x + \frac{1}{3}y = 3$
$y = 3x$

30. $\frac{1}{4}x - \frac{1}{5}y = 9$
$y = 5x$

31. $\frac{1}{2}x + \frac{1}{3}y = -\frac{1}{3}$
$\frac{1}{2}x + 2y = -7$

32. $\frac{1}{6}x + \frac{1}{6}y = 1$
$-\frac{1}{2}x - \frac{1}{3}y = -5$

33. $\frac{x}{5} + 2y = \frac{8}{5}$
$\frac{3x}{5} + \frac{y}{2} = -\frac{7}{10}$

34. $\frac{x}{2} + \frac{y}{3} = \frac{7}{6}$
$\frac{x}{4} - \frac{3y}{2} = \frac{9}{4}$

35. $\frac{1}{6}x + \frac{1}{3}y = 8$
$\frac{1}{4}x + \frac{1}{2}y = 12$

36. $\frac{1}{2}x - \frac{1}{8}y = -\frac{1}{4}$
$\frac{1}{3}x - \frac{1}{12}y = -\frac{1}{6}$

37. $0.2x - 1.3y = -3.2$
$-0.1x + 2.7y = 9.8$

38. $0.1x + 0.9y = -2$
$0.5x - 0.2y = 4.1$

39. $0.3x - 0.1y = 2.1$
$0.6x + 0.3y = -0.3$

40. $0.8x - 0.1y = 1.3$
$2.2x + 1.5y = 8.9$

RELATING CONCEPTS For Individual or Group Work (Exercises 41–44)

A system of linear equations can be used to model the cost and the revenue of a business. ***Work Exercises 41–44 in order.***

41. Suppose that it costs $5000 to start a business manufacturing and selling bicycles. Each bicycle will cost $400 to manufacture. Explain why the linear equation

$$y_1 = 400x + 5000 \quad (y_1 \text{ in dollars})$$

gives the *total* cost to manufacture x bicycles.

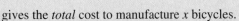

42. We decide to sell each bicycle for $600. Write an equation using y_2 (in dollars) to express the revenue when we sell x bicycles.

43. Form a system from the two equations in **Exercises 41 and 42.** Solve the system.

44. The value of x from **Exercise 43** is the number of bicycles it takes to *break even*. Fill in the blanks: When _____ bicycles are sold, the break-even point is reached. At that point, we have spent _____ dollars and taken in _____ dollars.

4.3 Solving Systems of Linear Equations by Elimination

NOW TRY
EXERCISE 1

Solve the system by the elimination method.

$$x - y = 4$$
$$3x + y = 8$$

OBJECTIVE 1 Solve linear systems by elimination.

Adding the same quantity to each side of an equation results in equal sums.

$$\text{If}\quad A = B,\quad \text{then}\quad A + C = B + C.$$

We can take this addition a step further. Adding *equal* quantities, rather than the *same* quantity, to each side of an equation also results in equal sums.

$$\text{If}\quad A = B\quad \text{and}\quad C = D,\quad \text{then}\quad A + C = B + D.$$

The **elimination method** uses the addition property of equality to solve systems of equations.

EXAMPLE 1 Using the Elimination Method

Solve the system by the elimination method.

$$x + y = 5 \quad (1)$$
$$x - y = 3 \quad (2)$$

Each equation in this system is a statement of equality, so the sum of the left sides equals the sum of the right sides. Adding vertically in this way gives the following.

$$
\begin{array}{ll}
x + y = 5 & (1) \\
\underline{x - y = 3} & (2) \\
2x\quad\ = 8 & \text{Add left sides and add right sides.} \\
x = 4 & \text{Divide by 2.}
\end{array}
$$

Notice that y has been eliminated. The result, $x = 4$, gives the x-value of an ordered pair. To find the corresponding y-value, substitute 4 for x in either of the two equations of the system. We choose equation (1).

$$
\begin{array}{ll}
x + y = 5 & (1) \\
4 + y = 5 & \text{Let } x = 4. \\
y = 1 & \text{Subtract 4.}
\end{array}
$$

Check the ordered pair $(4, 1)$ in both equations of the given system.

$$
\begin{array}{llll}
\textbf{CHECK} \quad x + y = 5 & (1) & \qquad x - y = 3 & (2) \\
4 + 1 \overset{?}{=} 5 & \text{Substitute.} & \qquad 4 - 1 \overset{?}{=} 3 & \text{Substitute.} \\
5 = 5 \ \checkmark & \text{True} & \qquad 3 = 3 \ \checkmark & \text{True}
\end{array}
$$

Both results are true, so the solution set of the system is $\{(4, 1)\}$.   **NOW TRY**

With the elimination method, the idea is to *eliminate* one of the two variables in a system.

To do this, one pair of variable terms in the two equations must have coefficients that are opposites (additive inverses).

NOW TRY ANSWER
1. $\{(3, -1)\}$

Solving a Linear System by Elimination

Step 1 **Write both equations in standard form** $Ax + By = C$.

Step 2 **Transform the equations as needed so that the coefficients of one pair of variable terms are opposites.** Multiply one or both equations by appropriate numbers so that the sum of the coefficients of either the x- or y-terms is 0.

Step 3 **Add** the new equations to eliminate a variable. The sum should be an equation with just one variable.

Step 4 **Solve** the equation from Step 3 for the remaining variable.

Step 5 **Find the other value.** Substitute the result from Step 4 into either of the original equations, and solve for the other variable.

Step 6 **Check** the values in *both* of the *original* equations. Then write the solution set as a set containing an ordered pair.

It does not matter which variable is eliminated first. Usually, we choose the one that is more convenient to work with.

NOW TRY
EXERCISE 2

Solve the system.

$$2x - 6 = -3y$$
$$5x - 3y = -27$$

| EXAMPLE 2 | Using the Elimination Method |

Solve the system.

$$y + 11 = 2x \qquad (1)$$
$$5x = y + 26 \qquad (2)$$

Step 1 Write both equations in standard form $Ax + By = C$.

$$-2x + y = -11 \qquad \text{Subtract } 2x \text{ and } 11 \text{ in equation (1).}$$
$$5x - y = 26 \qquad \text{Subtract } y \text{ in equation (2).}$$

Step 2 Because the coefficients of y are 1 and -1, adding will eliminate y. It is not necessary to multiply either equation by a number.

Step 3 Add the two equations.

$$-2x + y = -11$$
$$\underline{5x - y = 26}$$
$$3x = 15 \qquad \text{Add in columns.}$$

Step 4 Solve. $\qquad x = 5 \qquad$ Divide by 3.

Step 5 Find the value of y by substituting 5 for x in either of the original equations.

$$y + 11 = 2x \qquad (1)$$
$$y + 11 = 2(5) \qquad \text{Let } x = 5.$$
$$y + 11 = 10 \qquad \text{Multiply.}$$
$$y = -1 \qquad \text{Subtract 11.}$$

Step 6 Check the ordered pair $(5, -1)$ in both of the original equations.

CHECK $\qquad y + 11 = 2x \qquad (1) \qquad\qquad\quad 5x = y + 26 \qquad (2)$

$$(-1) + 11 \stackrel{?}{=} 2(5) \quad \text{Substitute.} \qquad 5(5) = -1 + 26 \quad \text{Substitute.}$$
$$10 = 10 \ \checkmark \quad \text{True} \qquad\qquad\quad 25 = 25 \ \checkmark \qquad \text{True}$$

Because $(5, -1)$ is a solution of *both* equations, the solution set is $\{(5, -1)\}$.

NOW TRY

OBJECTIVE 2   Multiply when using the elimination method.

Sometimes we need to multiply each side of one or both equations in a system by a number before adding will eliminate a variable.

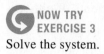

NOW TRY
EXERCISE 3

Solve the system.

$$3x - 5y = 25$$
$$2x + 8y = -6$$

EXAMPLE 3   Using the Elimination Method

Solve the system.

$$2x + 3y = -15 \qquad (1)$$
$$5x + 2y = 1 \qquad (2)$$

Adding the two equations gives $7x + 5y = -14$, which does not eliminate either variable. However, we can multiply each equation by a suitable number so that the coefficients of one of the two variables are opposites. For example, to eliminate x, we multiply each side of $2x + 3y = -15$ (equation (1)) by 5 and each side of $5x + 2y = 1$ (equation (2)) by -2.

$$10x + 15y = -75 \qquad \text{Multiply both sides of equation (1) by 5.}$$
$$\underline{-10x - 4y = -2} \qquad \text{Multiply both sides of equation (2) by } -2.$$

> The coefficients of x are opposites.

$$11y = -77 \qquad \text{Add.}$$
$$y = -7 \qquad \text{Divide by 11.}$$

Find the value of x by substituting -7 for y in either equation (1) or (2).

$$5x + 2y = 1 \qquad (2)$$
$$5x + 2(-7) = 1 \qquad \text{Let } y = -7.$$
$$5x - 14 = 1 \qquad \text{Multiply.}$$
$$5x = 15 \qquad \text{Add 14.}$$
$$x = 3 \qquad \text{Divide by 5.}$$

Check that the solution set of the system is $\{(3, -7)\}$. ◁ Write the x-value first.   **NOW TRY**

NOTE   In **Example 3,** we eliminated the variable x. Alternatively, we could multiply each equation of the system by a suitable number so that the variable y is eliminated.

$$2x + 3y = -15 \quad (1) \xrightarrow{\text{Multiply by 2.}} 4x + 6y = -30$$

$$5x + 2y = 1 \quad (2) \xrightarrow{\text{Multiply by } -3.} -15x - 6y = -3$$

Complete this approach and confirm that the same solution results.

⚠ **CAUTION**   When using the elimination method, remember to *multiply both sides* of an equation by the same nonzero number.

OBJECTIVE 3   Use an alternative method to find the second value in a solution.

Sometimes it is easier to find the value of the second variable in a solution using the elimination method twice.

NOW TRY ANSWER
3. $\{(5, -2)\}$

**NOW TRY
EXERCISE 4**

Solve the system.

$$4x + 9y = 3$$
$$5y = 6 - 3x$$

EXAMPLE 4 Finding the Second Value Using an Alternative Method

Solve the system.

$$4x = 9 - 3y \qquad \text{(1)}$$
$$5x - 2y = 8 \qquad \text{(2)}$$

Write equation (1) in standard form by adding $3y$ to each side.

$$4x + 3y = 9 \qquad \text{(3)}$$
$$5x - 2y = 8 \qquad \text{(2)}$$

One way to proceed is to eliminate y by multiplying each side of equation (3) by 2 and each side of equation (2) by 3 and then adding.

$$8x + 6y = 18 \qquad \text{Multiply equation (3) by 2.}$$
$$\underline{15x - 6y = 24} \qquad \text{Multiply equation (2) by 3.}$$
$$23x \qquad = 42 \qquad \text{Add.}$$

The coefficients of y are opposites.

$$x = \frac{42}{23} \qquad \text{Divide by 23.}$$

Substituting $\frac{42}{23}$ for x in one of the given equations would give y, but the arithmetic would be complicated. Instead, solve for y by starting over again with the original equations written in standard form (equations (3) and (2)) and eliminating x.

$$20x + 15y = 45 \qquad \text{Multiply equation (3) by 5.}$$
$$\underline{-20x + 8y = -32} \qquad \text{Multiply equation (2) by } -4.$$
$$23y = 13 \qquad \text{Add.}$$

The coefficients of x are opposites.

$$y = \frac{13}{23} \qquad \text{Divide by 23.}$$

Check that the solution set is $\left\{\left(\frac{42}{23}, \frac{13}{23}\right)\right\}$.

NOW TRY

NOTE When the value of the first variable is a fraction, the method used in **Example 4** helps avoid arithmetic errors. This method could be used to solve any system.

OBJECTIVE 4 Solve special systems by elimination.

EXAMPLE 5 Solving Special Systems Using the Elimination Method

Solve each system by the elimination method.

(a)

$$2x + 4y = 5 \qquad \text{(1)}$$
$$4x + 8y = -9 \qquad \text{(2)}$$

Multiply each side of equation (1) by -2. Then add the two equations.

$$-4x - 8y = -10 \qquad \text{Multiply equation (1) by } -2.$$
$$\underline{4x + 8y = -9} \qquad \text{(2)}$$
$$0 = -19 \qquad \text{False}$$

The false statement $0 = -19$ indicates that the given system has solution set $\varnothing$.

NOW TRY ANSWER

4. $\left\{\left(\frac{39}{7}, -\frac{15}{7}\right)\right\}$

**NOW TRY
EXERCISE 5**

Solve each system by the elimination method.

(a) $x - y = 2$
  $5x - 5y = 10$

(b) $4x + 3y = 0$
  $-4x - 3y = -1$

NOW TRY ANSWERS

5. (a) $\{(x, y) \mid x - y = 2\}$
  (b) $\varnothing$

(b)
$$3x - \ y = 4 \qquad (1)$$
$$-9x + 3y = -12 \qquad (2)$$

Multiply each side of equation (1) by 3. Then add the two equations.

$$9x - 3y = \quad 12 \qquad \text{Multiply equation (1) by 3.}$$
$$\underline{-9x + 3y = -12} \qquad (2)$$
$$0 = \quad 0 \qquad \text{True}$$

A true statement occurs when the equations are equivalent. This indicates that every solution of one equation is also a solution of the other. The solution set is

$$\{(x, y) \mid 3x - y = 4\}.$$

(See **Section 4.2, Example 5,** where this same system was solved by substitution.)

NOW TRY

4.3 Exercises

FOR EXTRA HELP

▶ MyMathLab®

▶ *Complete solution available in MyMathLab*

Concept Check   *Answer* true *or* false *for each statement. If false, tell why.*

  1. If the elimination method leads to $0 = -1$, the solution set of the system is $\{(0, -1)\}$.

  2. A system that includes the equation $5x - 4y = 0$ cannot have $(4, -5)$ as a solution.

Solve each system by the elimination method. Check each solution. **See Examples 1 and 2.**

▶ **3.** $x - y = -2$
  $x + y = 10$

4. $x + y = 10$
  $x - y = -6$

5. $2x + y = -5$
  $x - y = 2$

6. $2x + y = -15$
  $-x - y = 10$

▶ **7.** $2y = -3x$
  $-3x - y = 3$

8. $5x = y + 5$
  $-5x + 2y = 0$

9. $6x - y = -1$
  $5y = 17 + 6x$

10. $y = 9 - 6x$
  $-6x + 3y = 15$

Solve each system by the elimination method. Check each solution. **See Examples 3–5.**

11. $2x - \ y = 12$
  $3x + 2y = -3$

12. $x + \ y = 3$
  $-3x + 2y = -19$

13. $x + 4y = 16$
  $3x + 5y = 20$

14. $2x + \ y = 8$
  $5x - 2y = -16$

15. $2x - 8y = 0$
  $4x + 5y = 0$

16. $3x - 15y = 0$
  $6x + 10y = 0$

▶ **17.** $3x + 3y = 33$
  $5x - 2y = 27$

18. $4x - 3y = -19$
  $3x + 2y = 24$

19. $5x + 4y = 12$
  $3x + 5y = 15$

20. $2x + 3y = 21$
  $5x - 2y = -14$

21. $5x - 4y = 15$
  $-3x + 6y = -9$

22. $4x + 5y = -16$
  $5x - 6y = -20$

▶ **23.** $-x + 3y = 4$
  $-2x + 6y = 8$

24. $6x - 2y = 24$
  $-3x + \ y = -12$

25. $5x - 2y = 3$
  $10x - 4y = 5$

26. $3x - \ 5y = 1$
  $6x - 10y = 4$

27. $6x - 2y = -22$
  $-3x + 4y = 17$

28. $5x - 4y = -1$
  $x + 8y = -9$

29. $3x = 3 + 2y$

$-\dfrac{4}{3}x + y = \dfrac{1}{3}$

30. $3x = 27 + 2y$

$x - \dfrac{7}{2}y = -25$

31. $\dfrac{1}{5}x + \quad y = \dfrac{6}{5}$

$\dfrac{1}{10}x + \dfrac{1}{3}y = \dfrac{5}{6}$

32. $\dfrac{1}{3}x + \dfrac{1}{2}y = \dfrac{13}{6}$

$\dfrac{1}{2}x - \dfrac{1}{4}y = -\dfrac{3}{4}$

33. $2.4x + 1.7y = 7.6$

$1.2x - 0.5y = 9.2$

34. $0.5x + 3.4y = 13$

$1.5x - 2.6y = -25$

35. $x + 3y = 6$

$-2x + 12 = 6y$

36. $7x + 2y = 0$

$4y = -14x$

37. $4x - 3y = 1$

$8x = 3 + 6y$

38. $5x + 8y = 10$

$24y = -15x - 10$

▶ 39. $4x = 3y - 2$

$5x + 3 = 2y$

40. $2x + 3y = 0$

$4x + 12 = 9y$

41. $24x + 12y = -7$

$16x - 18y = 17$

42. $9x + 4y = -3$

$6x + 6y = -7$

RELATING CONCEPTS   For Individual or Group Work   (Exercises 43–46)

The graph shows average U.S. movie theater ticket prices from 2004 through 2012. In 2004, the average price was $6.21, as represented on the graph by the point $P(2004, 6.21)$. *In 2012, the average price was $7.96, as represented on the graph by the point* $Q(2012, 7.96)$. **Work Exercises 43–46 in order.**

Average Movie Ticket Price

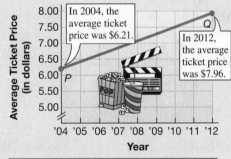

In 2004, the average ticket price was $6.21.

In 2012, the average ticket price was $7.96.

Source: Motion Picture Association of America.

43. Line segment PQ has an equation that can be written in the form $y = ax + b$. Using the coordinates of point P with $x = 2004$ and $y = 6.21$, write an equation in the variables a and b.

44. Using the coordinates of point Q with $x = 2012$ and $y = 7.96$, write a second equation in the variables a and b.

45. Write the system of equations formed from the two equations in **Exercises 43 and 44.** Solve the system, giving the values of a and b. (*Hint:* Eliminate b using the elimination method.)

46. Answer each of the following.

(a) What is the equation of the line on which segment PQ lies?

(b) Let $x = 2011$ in the equation from part (a), and solve for y (to two decimal places). How does the result compare with the actual figure of $7.93?

SUMMARY EXERCISES Applying Techniques for Solving Systems of Linear Equations

Guidelines for Choosing a Method to Solve a System of Linear Equations

1. If one of the equations of the system is already solved for one of the variables, as indicated by the arrows in the following systems, the substitution method is the better choice.

$$3x + 4y = 9 \qquad \text{and} \qquad \rightarrow x = 3y - 7$$
$$\rightarrow y = 2x - 6 \qquad \qquad -5x + 3y = 9$$

2. If both equations are in standard form $Ax + By = C$ and none of the variables has coefficient -1 or 1, as in the following system, the elimination method is the better choice.

$$4x - 11y = 3$$
$$-2x + 3y = 4$$

3. If one or both of the equations are in standard form and the coefficient of one of the variables is -1 or 1, as indicated by the arrows in the following systems, either method is appropriate.

$$\rightarrow 3x + y = -2 \qquad \text{and} \qquad 3x - 2y = 8$$
$$-5x + 2y = 4 \qquad \qquad \rightarrow -x + 3y = -4$$

1. (a) Use substitution because the second equation is solved for y.
  (b) Use elimination because the coefficients of the y-terms are opposites.
  (c) Use elimination because the equations are in standard form with no coefficients of 1 or -1. Solving by substitution would involve fractions.

2. System B is easier to solve by substitution because the second equation is already solved for y.

3. (a) $\{(1, 4)\}$ **(b)** $\{(1, 4)\}$
  (c) Answers will vary.
4. (a) $\{(-5, 2)\}$
  (b) $\{(-5, 2)\}$
  (c) Answers will vary.

5. $\{(3, 12)\}$ **6.** $\{(-3, 2)\}$
7. $\left\{\left(\dfrac{1}{3}, \dfrac{1}{2}\right)\right\}$ **8.** $\varnothing$
9. $\{(3, -2)\}$ **10.** $\{(-1, -11)\}$
11. $\{(x, y) \mid 2x - 3y = 5\}$
12. $\{(9, 4)\}$
13. $\left\{\left(\dfrac{45}{31}, \dfrac{4}{31}\right)\right\}$

Concept Check *Use the preceding guidelines to solve each problem.*

1. To minimize the amount of work required, tell whether you would use the substitution or elimination method to solve each system, and why. *Do not actually solve.*

  (a) $3x + 5y = 69$ **(b)** $3x + y = -7$ **(c)** $3x - 2y = 0$
  $y = 4x$ $x - y = -5$ $9x + 8y = 7$

2. Which system would be easier to solve with the substitution method? Why?

  System A: $5x - 3y = 7$ *System B:* $7x + 2y = 4$
  $2x + 8y = 3$ $y = -3x + 1$

In Exercises 3 and 4, (a) solve the system by the elimination method, (b) solve the system by the substitution method, and (c) tell which method you prefer for that particular system and why.

3. $4x - 3y = -8$ **4.** $2x + 5y = 0$
  $x + 3y = 13$ $x = -3y + 1$

*Solve each system by any method. (For Exercises 5–7, see the answers to **Exercise 1**.)*

5. $3x + 5y = 69$ **6.** $3x + y = -7$ **7.** $3x - 2y = 0$
  $y = 4x$ $x - y = -5$ $9x + 8y = 7$

8. $x + y = 7$ **9.** $6x + 7y = 4$ **10.** $6x - y = 5$
  $x = -3 - y$ $5x + 8y = -1$ $y = 11x$

11. $4x - 6y = 10$ **12.** $3x - 5y = 7$ **13.** $5x = 7 + 2y$
  $-10x + 15y = -25$ $2x + 3y = 30$ $5y = 5 - 3x$

14. $\{(4, -5)\}$
15. $\varnothing$ **16.** $\{(-4, 6)\}$
17. $\{(0, 0)\}$ **18.** $\left\{\left(\dfrac{22}{13}, -\dfrac{23}{13}\right)\right\}$

19. $\{(2, -3)\}$ **20.** $\{(24, -12)\}$
21. $\{(3, 2)\}$ **22.** $\{(10, -12)\}$
23. $\{(-4, 2)\}$ **24.** $\{(5, 3)\}$

14. $4x + 3y = 1$
$\quad\;\; 3x + 2y = 2$

15. $\quad 2x - 3y = 7$
$\quad -4x + 6y = 14$

16. $\quad 2x + 3y = 10$
$\quad -3x + \;\; y = 18$

17. $7x - 4y = 0$
$\quad\; 3x = 2y$

18. $\quad x - 3y = 7$
$\quad 4x + \;\; y = 5$

Solve each system by any method. First clear all fractions or decimals.

19. $\dfrac{1}{5}x + \dfrac{2}{3}y = -\dfrac{8}{5}$
$\quad\; 3x - \;\;\; y = 9$

20. $\dfrac{1}{6}x + \dfrac{1}{6}y = 2$
$\quad -\dfrac{1}{2}x - \dfrac{1}{3}y = -8$

21. $\dfrac{x}{3} - \dfrac{3y}{4} = -\dfrac{1}{2}$
$\quad \dfrac{x}{6} + \dfrac{y}{8} = \dfrac{3}{4}$

22. $\dfrac{x}{2} - \dfrac{y}{3} = 9$
$\quad \dfrac{x}{5} - \dfrac{y}{4} = 5$

23. $0.1x + \;\;\; y = 1.6$
$\quad 0.6x + 0.5y = -1.4$

24. $0.2x - 0.3y = 0.1$
$\quad 0.3x - 0.2y = 0.9$

4.4 Applications of Linear Systems

OBJECTIVES

1 Solve problems about unknown numbers.
2 Solve problems about quantities and their costs.
3 Solve problems about mixtures.
4 Solve problems about distance, rate (or speed), and time.

Recall from **Section 2.4** the six-step method for solving applied problems. We modify those steps slightly to allow for two variables and two equations.

Solving an Applied Problem Using a System of Equations

Step 1 **Read** the problem carefully. *What information is given? What is to be found?*

Step 2 **Assign variables** to represent the unknown values. Make a sketch, diagram, or table, as needed. Write down what each variable represents.

Step 3 **Write two equations** using both variables.

Step 4 **Solve** the system of two equations.

Step 5 **State the answer.** Label it appropriately. *Does it seem reasonable?*

Step 6 **Check** the answer in the words of the *original* problem.

OBJECTIVE 1 Solve problems about unknown numbers.

EXAMPLE 1 Solving a Problem about Two Unknown Numbers

In a recent year, consumer sales of sports equipment were $307 million more for snow skiing than for snowboarding. Together, total equipment sales for these two sports were $931 million. What were equipment sales for each sport? (*Source:* National Sporting Goods Association.)

Step 1 **Read** the problem carefully. We must find equipment sales (in millions of dollars) for snow skiing and for snowboarding. We know how much more equipment sales were for snow skiing than for snowboarding. Also, we know the total sales.

NOW TRY EXERCISE 1
Marina pays a total of $1150 per month for rent and electricity. It costs $650 more for rent per month than for electricity. What are the costs for each?

Step 2 **Assign variables.**

Let x = equipment sales for skiing (in millions of dollars),

and y = equipment sales for snowboarding (in millions of dollars).

Step 3 **Write two equations.** If we use two variables, then we must write two equations.

$x = 307 + y$ Equipment sales for skiing were $307 million more than equipment sales for snowboarding.   (1)

$x + y = 931$ Total sales were $931 million.   (2)

Step 4 **Solve** the system from Step 3. We use the substitution method because the first equation is already solved for x.

$$x + y = 931 \qquad (2)$$
$$(307 + y) + y = 931 \qquad \text{Let } x = 307 + y.$$
$$307 + 2y = 931 \qquad \text{Combine like terms.}$$
$$2y = 624 \qquad \text{Subtract 307.}$$
$$\boxed{\text{Don't stop here.}} \; y = 312 \qquad \text{Divide by 2.}$$

To find the value of x, we substitute 312 for y in either equation (1) or (2).

$$x = 307 + y \qquad (1)$$
$$x = 307 + 312 \qquad \text{Let } y = 312.$$
$$x = 619 \qquad \text{Add.}$$

Step 5 **State the answer.** Equipment sales for skiing were $619 million, and equipment sales for snowboarding were $312 million.

Step 6 **Check** the answer in the original problem. Because

$$619 = 307 + 312 \quad \text{and} \quad 619 + 312 = 931$$

are both true, the answer satisfies the information given in the problem.

NOW TRY ↺

❗ **CAUTION** If an applied problem asks for *two* values, as in **Example 1,** be sure to give both of them in the answer.

OBJECTIVE 2 Solve problems about quantities and their costs.

EXAMPLE 2 Solving a Problem about Quantities and Costs

For a production of the musical *Wicked* at the Ford Center in Chicago, main floor tickets cost $148, while the best balcony tickets cost $65. The members of a club spent a total of $2614 for 30 tickets. How many tickets of each kind did they buy? (*Source:* www.ticketmaster.com)

Step 1 **Read** the problem, several times as needed.

Step 2 **Assign variables.**

Let x = the number of main floor tickets,

and y = the number of balcony tickets.

NOW TRY ANSWER
1. rent: $900; electricity: $250

General admission rates at a local water park are $19 for adults and $16 for children. If a group of 27 people paid $462 for admission, how many adults and how many children were there?

Summarize the information given in the problem in a table.

	Number of Tickets	Price per Ticket (in dollars)	Total Value (in dollars)
Main Floor	x	148	$148x$
Balcony	y	65	$65y$
Total	30		2614

The entries in the first two rows of the Total Value column were found by multiplying the Number of Tickets by the Price per Ticket.

Step 3 **Write two equations.**

$$x + y = 30 \qquad \text{Total number of tickets was 30.} \quad (1)$$

$$148x + 65y = 2614 \qquad \text{Total value of tickets was \$2614.} \quad (2)$$

Step 4 **Solve** the system from Step 3 using the elimination method.

$$-65x - 65y = -1950 \qquad \text{Multiply equation (1) by } -65.$$

$$\underline{148x + 65y = \quad 2614} \qquad (2)$$

$$83x \qquad = \quad 664 \qquad \text{Add.}$$

Main floor tickets $\longrightarrow$ $x = 8$ \qquad Divide by 83.

Substitute 8 for x in equation (1).

$$x + y = 30 \qquad (1)$$

$$8 + y = 30 \qquad \text{Let } x = 8.$$

Balcony tickets $\longrightarrow$ $y = 22$ \qquad Subtract 8.

Step 5 **State the answer.** They bought 8 main floor tickets and 22 balcony tickets.

Step 6 **Check.** The sum of 8 and 22 is 30, so the total number of tickets is correct. The total of all the ticket prices is

$$\$148(8) + \$65(22) = \$2614, \quad \text{as required.} \qquad \text{NOW TRY} \ \text{↺}$$

OBJECTIVE 3 Solve problems about mixtures.

In **Section 2.7,** we solved mixture problems using one variable. Many mixture problems can also be solved using a system of two equations in two variables.

EXAMPLE 3 Solving a Mixture Problem Involving Percent

A pharmacist has a 30% alcohol solution and an 80% alcohol solution, which he can mix. How many liters of each will be required to make 100 L of a 50% alcohol solution?

Step 1 **Read** the problem. Note the percentage of each solution and of the mixture.

Step 2 **Assign variables.**

Let x = the number of liters of 30% alcohol needed,

and y = the number of liters of 80% alcohol needed.

Liters of Solution	Percent (as a decimal)	Liters of Pure Alcohol
x	0.30	$0.30x$
y	0.80	$0.80y$
100	0.50	$0.50(100)$

Summarize the information in a table. Percents are written as decimals.

**NOW TRY
EXERCISE 3**

A biologist needs 80 L of a 30% saline solution. He has a 10% saline solution and a 35% saline solution with which to work. How many liters of each will be required to make the 80 L of a 30% solution?

FIGURE 9 gives an idea of what is happening in this problem.

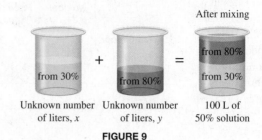

Unknown number of liters, *x*   Unknown number of liters, *y*   100 L of 50% solution

FIGURE 9

Step 3   **Write two equations.** The total number of liters in the final mixture will be 100, which gives one equation.

$$x + y = 100 \qquad \text{Refer to the first column of the table.}$$

To find the amount of pure alcohol in each mixture, multiply the number of liters by the concentration. The amount of pure alcohol in the 30% solution added to the amount of pure alcohol in the 80% solution will equal the amount of pure alcohol in the final 50% solution. This gives a second equation.

$$0.30x + 0.80y = 0.50(100) \qquad \begin{array}{l}\text{Refer to the last column} \\ \text{of the table.}\end{array}$$

$$0.30x + 0.80y = 50 \qquad \text{Multiply.}$$

$$3x + 8y = 500 \qquad \begin{array}{l}\text{Multiply each side by} \\ \text{10 to clear the decimals.}\end{array}$$

This is a simpler, yet equivalent, equation.

These two equations form a system.

Be sure to write *two* equations.

$$x + \ y = 100 \qquad (1)$$

$$3x + 8y = 500 \qquad (2)$$

Step 4   **Solve** the system using the substitution method. Solving equation (1) for *x* gives $x = 100 - y$. Substitute $100 - y$ for *x* in equation (2).

$$3x + 8y = 500 \qquad (2)$$

$$3(100 - y) + 8y = 500 \qquad \text{Let } x = 100 - y.$$

$$300 - 3y + 8y = 500 \qquad \text{Distributive property}$$

$$300 + 5y = 500 \qquad \text{Combine like terms.}$$

$$5y = 200 \qquad \text{Subtract 300.}$$

Liters of 80% solution → $y = 40 \qquad \text{Divide by 5.}$

Equation (1) solved for *x* is $x = 100 - y$. Substitute 40 for *y* to find *x*.

$$x = 100 - 40 \qquad \text{Let } y = 40.$$

Liters of 30% solution → $x = 60 \qquad \text{Subtract.}$

Step 5   **State the answer.** The pharmacist should use 60 L of the 30% solution and 40 L of the 80% solution.

Step 6   **Check** the answer in the original problem. Because

Use the original equations written from the table.

$$60 + 40 = 100 \quad \text{and} \quad 0.30(60) + 0.80(40) = 0.50(100),$$

this mixture will give the 100 L of 50% solution, as required.   **NOW TRY**

NOTE   In **Example 3,** we could have used the elimination method.

OBJECTIVE 4 Solve problems about distance, rate (or speed), and time.

Problems that use the distance formula $d = rt$ were solved in **Section 2.7.**

NOW TRY EXERCISE 4

From a truck stop, two trucks travel in opposite directions on a straight highway. In 3 hr they are 405 mi apart. Find the rate of each truck if one travels 5 mph faster than the other.

EXAMPLE 4 Solving a Problem about Distance, Rate, and Time

Two executives in cities 400 mi apart drive to a business meeting at a location on the line between their cities. They meet after 4 hr. Find the rate (speed) of each car if one travels 20 mph faster than the other.

Step 1 **Read** the problem carefully.

Step 2 **Assign variables.**

Let x = the rate of the faster car,

and y = the rate of the slower car.

Make a table using the formula $d = rt$, and draw a sketch. See **FIGURE 10.**

	r	t	d
Faster Car	x	4	$x \cdot 4$, or $4x$
Slower Car	y	4	$y \cdot 4$, or $4y$

Because each car travels for 4 hr, the time t for each car is 4. We find d using $d = rt$ (or $rt = d$).

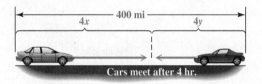

FIGURE 10

Step 3 **Write two equations.** The total distance traveled by both cars is 400 mi, which gives equation (1). The faster car goes 20 mph faster than the slower car, which gives equation (2).

$$4x + 4y = 400 \qquad (1)$$

$$x = 20 + y \qquad (2)$$

Step 4 **Solve** the system using the substitution method. Replace x with $20 + y$ in equation (1) and solve for y.

$$4x + 4y = 400 \qquad (1)$$

$$4(20 + y) + 4y = 400 \qquad \text{Let } x = 20 + y.$$

$$80 + 4y + 4y = 400 \qquad \text{Distributive property}$$

$$80 + 8y = 400 \qquad \text{Combine like terms.}$$

$$8y = 320 \qquad \text{Subtract 80.}$$

$$\text{Slower car} \rightarrow y = 40 \qquad \text{Divide by 8.}$$

To find x, substitute 40 for y in equation (2), $x = 20 + y$.

$$x = 20 + 40 \qquad \text{Let } y = 40.$$

$$\text{Faster car} \rightarrow x = 60 \qquad \text{Add.}$$

Step 5 **State the answer.** The rates of the two cars are 40 mph and 60 mph.

NOW TRY ANSWER

4. faster truck: 70 mph; slower truck: 65 mph

Step 6 **Check.** Each car travels for 4 hr, so the total distance traveled is

$$4(60) + 4(40) = 400 \text{ mi}, \quad \text{as required.} \qquad \text{NOW TRY} \text{ } $$

In 1 hr, Gigi can row 10 mi
with the current or 2 mi
against the current. Find
Gigi's rate in still water and
the rate of the current.

EXAMPLE 5 Solving a Problem about Distance, Rate, and Time

A plane flies 560 mi in 1.75 hr traveling with the wind. The return trip against the same wind takes the plane 2 hr. Find the rate (speed) of the plane in still air and the wind speed.

Step 1 **Read** the problem, several times as needed.

Step 2 **Assign variables.**

Let x = the rate of the plane in still air,

and y = the wind speed.

When the plane is traveling *with* the wind, the wind "pushes" the plane. In this case, the rate (speed) of the plane is the *sum* of the rate of the plane and the wind speed, $(x + y)$ mph. See **FIGURE 11**.

When the plane is traveling *against* the wind, the wind "slows" the plane down. In this case, the rate (speed) of the plane is the *difference* between the rate of the plane and the wind speed, $(x - y)$ mph. Again, see **FIGURE 11**.

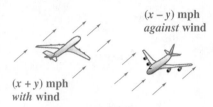

$(x - y)$ mph
against wind

$(x + y)$ mph
with wind

FIGURE 11

	r	t	d
With Wind	$x + y$	1.75	560
Against Wind	$x - y$	2	560

Summarize the information in a table. The distance is the same both ways.

Step 3 **Write two equations.** Refer to the table, and use the formula $d = rt$ (or $rt = d$) to do this.

$(x + y)1.75 = 560$ $\xrightarrow{\text{Divide by 1.75.}}$ $x + y = 320$ (1)

$(x - y)2 = 560$ $\xrightarrow{\text{Divide by 2.}}$ $x - y = 280$ (2)

Step 4 **Solve** the system using the elimination method.

$$x + y = 320 \quad (1)$$
$$\underline{x - y = 280} \quad (2)$$
$$2x \quad\;\; = 600 \quad \text{Add.}$$
$$x = 300 \quad \text{Divide by 2.}$$

Because $x + y = 320$ and $x = 300$, it follows that

$$y = 20.$$

Step 5 **State the answer.** The rate of the plane is 300 mph, and the wind speed is 20 mph.

Step 6 **Check.** The answer seems reasonable. Because

$$300 + 20 = 320 \quad \text{and} \quad 300 - 20 = 280$$

are both true, the answer is correct.

NOW TRY

4.4 Exercises

 FOR EXTRA HELP

 MyMathLab®

Complete solution available in MyMathLab

Concept Check *Choose the correct response.*

1. Which expression represents the monetary value of x 5-dollar bills?

A. $\dfrac{x}{5}$ dollars   **B.** $\dfrac{5}{x}$ dollars   **C.** $(5 + x)$ dollars   **D.** $5x$ dollars

2. Which expression represents the cost of t pounds of candy that sells for $4.95 per lb?

A. $\$4.95t$ **B.** $\dfrac{\$4.95}{t}$ **C.** $\dfrac{t}{\$4.95}$ **D.** $\$4.95 + t$

3. Which expression represents the amount of interest earned on d dollars at an interest rate of 3%?

A. $3d$ dollars   **B.** $0.03d$ dollars   **C.** $0.3d$ dollars   **D.** $300d$ dollars

4. Which expression represents the amount of pure alcohol in x liters of a 25% alcohol solution?

A. $25x$ liters   **B.** $(25 + x)$ liters   **C.** $0.25x$ liters   **D.** $(0.25 + x)$ liters

5. According to *Natural History* magazine, the speed of a cheetah is 70 mph. If a cheetah runs for x hours, how many miles does the cheetah cover?

A. $(70 + x)$ miles   **B.** $(70 - x)$ miles   **C.** $\dfrac{70}{x}$ miles   **D.** $70x$ miles

6. How far does a car travel in 2.5 hr if it travels at an average rate of x miles per hour?

A. $(x + 2.5)$ miles   **B.** $\dfrac{2.5}{x}$ miles   **C.** $\dfrac{x}{2.5}$ miles   **D.** $2.5x$ miles

▶ **7.** What is the rate of a plane that travels at 650 mph *with* a wind of r mph?

A. $\dfrac{r}{650}$ mph   **B.** $(650 - r)$ mph   **C.** $(650 + r)$ mph   **D.** $(r - 650)$ mph

8. What is the rate of a plane that travels at 650 mph *against* a wind of r mph?

A. $(650 + r)$ mph   **B.** $\dfrac{650}{r}$ mph   **C.** $(650 - r)$ mph   **D.** $(r - 650)$ mph

9. Suppose that Ira wants to mix x liters of a 40% acid solution with y liters of a 35% solution to obtain 100 L of a 38% solution. One equation in a system for solving this problem is $x + y = 100$. Which one of the following is the other equation?

A. $0.35x + 0.40y = 0.38(100)$ **B.** $0.40x + 0.35y = 0.38(100)$

C. $35x + 40y = 38$ **D.** $40x + 35y = 0.38(100)$

10. Suppose that two trucks leave a rest stop traveling in opposite directions on a straight highway. One truck travels 15 mph faster than the other, and in 8 hr they are 840 mi apart. If x and y represent the rates of the trucks, one equation in a system for solving this problem is $x = 15 + y$. Which one of the following is the other equation?

A. $8x + 8y = 840$ **B.** $8x + y = 840$

C. $x + y = 840$ **D.** $8x - 8y = 840$

In Exercises 11 and 12, refer to the six-step problem-solving method, fill in the blanks for Steps 2 and 3, and complete the solution by applying Steps 4–6.

11. The sum of two numbers is 98. The difference between them is 48. Find the two numbers.

Step 1 **Read** the problem carefully.

Step 2 **Assign variables.**

   Let x = the first number and let $y =$ _____ .

Step 3 **Write two equations.**

   First equation: $x + y = 98$;   Second equation: _____

12. The sum of two numbers is 201. The difference between them is 11. Find the two numbers.

   Step 1   **Read** the problem carefully.

   Step 2   **Assign variables.**

   Let $x = $ _____ and let $y = $ the second number.

   Step 3   **Write two equations.**

   First equation: $x + y = 201$;   Second equation: _____

*Solve each problem using a system of equations. **See Example 1.***

13. Two of the longest-running shows on Broadway are *The Phantom of the Opera* and *Cats.* As of October 2013, there had been a total of 18,188 performances of the two shows, with 3218 more performances of *The Phantom of the Opera* than *Cats.* How many performances were there of each show? (*Source:* The Broadway League.)

14. During Broadway runs of *A Chorus Line* and *Beauty and the Beast,* there were 676 fewer performances of *Beauty and the Beast* than of *A Chorus Line.* There were a total of 11,598 performances of the two shows. How many performances were there of each show? (*Source:* The Broadway League.)

15. The two domestic top-grossing movies of 2012 were *Marvel's The Avengers* and *The Dark Knight Rises.* The *Dark Knight* movie grossed $175 million less than the *Avengers* movie, and together the two films took in $1071 million. How much did each of these movies earn? (*Source:* Rentrack Corporation.)

16. Two domestic top-grossing movies of 2012 were *The Hunger Games* and *The Twilight Saga: Breaking Dawn Part 2. The Hunger Games* grossed $120 million more than *Breaking Dawn Part 2,* and together the two films took in $696 million. How much did each of these movies earn? (*Source:* Rentrack Corporation.)

17. The Terminal Tower in Cleveland, Ohio, is 239 ft shorter than the Key Tower, also in Cleveland. The total of the heights of the two buildings is 1655 ft. Find the heights of the buildings. (*Source: World Almanac and Book of Facts.*)

18. The total of the heights of the Chase Tower and the One America Tower, both in Indianapolis, Indiana, is 1234 ft. The Chase Tower is 168 ft taller than the One America Tower. Find the heights of the two buildings. (*Source: World Almanac and Book of Facts.*)

In Exercises 19 and 20, refer to the six-step problem-solving method, fill in the blanks for Steps 1–3, and complete the solution by applying Steps 4–6.

19. An official playing field (including end zones) for the Indoor Football League has length 38 yd longer than its width. The perimeter of the rectangular field is 188 yd. Find the length and width of the field. (*Source:* Indoor Football League.)

Steps 1 and 2   **Read** the problem carefully, and **assign** _____.

Let x = the length (in yards), and y = the _____ (in yards).

Step 3   **Write two equations.**

Equation (1): See the first sentence in the problem. Express the length in terms of the width.

$$x = \text{\underline{\hspace{1.5cm}}} \qquad (1)$$

Equation (2): See the second sentence in the problem. Perimeter of a rectangle equals twice the _____ plus _____ the width.

$$2x + \text{\underline{\hspace{1.5cm}}} = \text{\underline{\hspace{1.5cm}}} \qquad (2)$$

20. Pickleball is a combination of badminton, tennis, and ping pong. The perimeter of the rectangular-shaped court is 128 ft. The width is 24 ft shorter than the length. Find the length and width of the court. (*Source:* www.sportsknowhow.com)

Steps 1 and 2   **Read** the problem carefully, and **assign** _____.

Let x = the _____ (in feet), and y = the width (in feet).

Step 3   **Write two equations.**

Equation (1): Perimeter of a rectangle equals _____ the length plus twice the _____.

$$\text{\underline{\hspace{1.5cm}}} + 2y = \text{\underline{\hspace{1.5cm}}} \qquad (1)$$

Equation (2): See the third sentence in the problem. Express the width in terms of the length.

$$y = \text{\underline{\hspace{1.5cm}}} \qquad (2)$$

*Suppose that x units of a product cost C dollars to manufacture and earn revenue of R dollars. The value of x at which the expressions for C and R are equal is the **break-even quantity**—the number of units that produce 0 profit.*

In Exercises 21 and 22, (a) find the break-even quantity, and (b) decide whether the product should be produced on the basis of whether it will earn a profit. (Profit = Revenue − Cost.)

21. $C = 85x + 900$; $R = 105x$

No more than 38 units can be sold.

22. $C = 105x + 6000$; $R = 255x$

No more than 400 units can be sold.

*Solve each problem using a system of equations. **See Example 2.***

23. Jonathan counted the money in his piggy bank. He had only quarters and dimes. When he added up his money, he had 39 coins worth a total of $7.50. How many coins of each kind did he have?

Number of Coins	Value per Coin (in dollars)	Total Value (in dollars)
x	0.25	
y	0.10	
39		7.50

24. At the post office, Marilyn spent $24.88 on 56 stamps, made up of a combination of 49-cent and 33-cent stamps. How many stamps of each denomination did she buy? (*Source:* www.usps.com)

Number of Stamps	Denomination (in dollars)	Total Value (in dollars)
x	0.49	
y	0.33	
		24.88

25. Joyce bought each of her seven nephews a DVD of the movie *Iron Man 3* or a Blu-ray disc of *The Hunger Games: Catching Fire*. Each DVD cost $14.95. Each Blu-ray disc cost $16.88. She spent a total of $114.30. How many of each did she buy?

26. Jason bought each of his five nieces a DVD of *Monsters University* or a Blu-ray disc of *Gravity*. Each DVD cost $14.99. Each Blu-ray disc cost $19.99. He spent a total of $94.95. How many of each did he buy?

27. Karen has twice as much money invested at 5% annual simple interest as she does at 4%. If her yearly income from these two investments is $350, how much does she have invested at each rate?

28. Glenmore invested some money in two accounts, one paying 3% annual simple interest and the other paying 2% interest. He earned a total of $880 interest. If he invested three times as much in the 3% account as he did in the 2% account, how much did he invest at each rate?

29. Two top-grossing North American concert tours in 2012 were Madonna and Bruce Springsteen and the E Street Band. Based on average ticket prices, it cost a total of $1300 to buy six tickets for Madonna and five tickets for Springsteen. Three tickets for Madonna and four for Springsteen cost a total of $788. How much did an average ticket cost for each tour? (*Source:* www.businessinsider.com)

30. Two top-grossing North American concert tours in 2012 were Cirque du Soleil and Coldplay. Based on average ticket prices, it cost a total of $1099 to buy eight tickets for Cirque du Soleil and three tickets for Coldplay. Four tickets for Cirque du Soleil and five tickets for Coldplay cost $833. How much did an average ticket cost for each tour? (*Source:* www.businessinsider.com)

Solve each problem using a system of equations. **See Example 3.**

▶ 31. A 40% dye solution is to be mixed with a 70% solution to make 120 L of a 50% solution. How many liters of the 40% and 70% solutions will be needed?

Liters of Solution	Percent (as a decimal)	Liters of Pure Dye
x	0.40	
y	0.70	
120	0.50	

32. A 90% antifreeze solution is to be mixed with a 75% solution to make 120 L of a 78% solution. How many liters of the 90% and 75% solutions will be used?

Liters of Solution	Percent (as a decimal)	Liters of Pure Antifreeze
x	0.90	
y	0.75	
120	0.78	

33. Deoraj wishes to mix coffee worth $6 per lb with coffee worth $3 per lb to obtain 90 lb of a mixture worth $4 per lb. How many pounds of the $6 and the $3 coffees will be needed?

Pounds of Coffee	Cost per Pound (in dollars)	Total Value (in dollars)
x	6	
y		
90		

34. Andrea wishes to blend candy selling for $1.20 per lb with candy selling for $1.80 per lb to obtain 45 lb of a mixture that will be sold for $1.40 per lb. How many pounds of the $1.20 and the $1.80 candies should be used?

Pounds of Candy	Cost per Pound (in dollars)	Total Value (in dollars)
x		
y	1.80	
45		

35. How many pounds of nuts selling for $6 per lb and raisins selling for $3 per lb should Theresa combine to obtain 60 lb of a trail mix selling for $5 per lb?

36. Jasmine is preparing a cheese tray using some cheeses that sell for $8 per lb and others that sell for $12 per lb. How many pounds of each cheese should she use in order for the cheeses on the tray to weigh a total of 56 lb and sell for $10.50 per lb?

*Solve each problem using a system of equations. **See Examples 4 and 5.***

▶ **37.** Two trains start from towns 495 mi apart and travel toward each other on parallel tracks. They pass each other 4.5 hr later. If one train travels 10 mph faster than the other, find the rate of each train.

	r	t	d
Train 1	x		
Train 2			

38. Two trains that are 495 mi apart travel toward each other on parallel tracks. They pass each other 5 hr later. If one train travels half as fast as the other, what are their rates?

	r	t	d
Train 1	x		
Train 2			

39. Kansas City and Denver are 600 mi apart. Two cars start from these cities, traveling toward each other. They pass each other after 6 hr. Find the rate of each car if one travels 30 mph slower than the other.

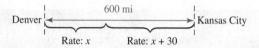

Denver ⟷ 600 mi ⟷ Kansas City
Rate: x   Rate: x + 30

40. Toledo and Cincinnati are 200 mi apart. A car leaves Toledo traveling toward Cincinnati, and another car leaves Cincinnati at the same time, traveling toward Toledo. The car leaving Toledo averages 15 mph faster than the other car, and they meet after 1 hr, 36 min. What are the rates of the cars?

41. RAGBRAI®, the *Des Moines Register's* **A**nnual **G**reat **B**icycle **R**ide **A**cross **I**owa, is the longest and oldest touring bicycle ride in the world. Suppose a cyclist began the 405 mi ride on July 21, 2013, in western Iowa at the same time that a car traveling toward it left eastern Iowa. If the bicycle and the car met after 6 hr and the car traveled 40 mph faster than the bicycle, find the average rate of each. (*Source:* www.ragbrai.com)

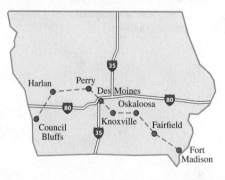

42. Suppose two planes leave Atlanta's Hartsfield-Jackson Airport at the same time, one traveling east and the other traveling west. If the planes are 2100 mi apart after 2 hr and one plane travels 50 mph faster than the other, find the rate of each plane.

43. In 1 hr, a plane can travel 440 mi into the wind and 500 mi with the wind. Find the wind speed and the rate of the plane in still air.

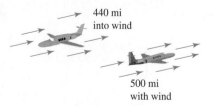

440 mi
into wind

500 mi
with wind

44. In 1 hr, a small plane travels 200 mi with the wind and 120 mi against it. Find the wind speed and the rate of the plane in still air.

45. A boat takes 3 hr to go 24 mi upstream. It can go 36 mi downstream in the same time. Find the rate of the current and the rate of the boat in still water if $x =$ the rate of the boat in still water and $y =$ the rate of the current.

	r	t	d
Downstream	$x + y$		
Upstream		3	

46. It takes a boat $1\frac{1}{2}$ hr to go 12 mi downstream, and 6 hr to return. Find the rate of the boat in still water and the rate of the current. Let $x =$ the rate of the boat in still water and $y =$ the rate of the current.

	r	t	d
Downstream		$\frac{3}{2}$	
Upstream	$x - y$		

Extending Skills   Solve each problem.

47. At the beginning of a bicycle ride for charity, Yady and Dane are 30 mi apart. If they leave at the same time and ride in the same direction, Yady overtakes Dane in 6 hr. If they ride toward each other, they pass each other in 1 hr. What are their rates?

48. Humera left Farmersville in a plane at noon to travel to Exeter. Walter left Exeter in his automobile at 2 P.M. to travel to Farmersville. It is 400 mi from Exeter to Farmersville. If the sum of their rates was 120 mph, and if they crossed paths at 4 P.M., find the rate of each.

4.5　Solving Systems of Linear Inequalities

We graphed the solutions of a linear inequality in **Section 3.6.** For example, recall that to graph the solutions of

$$x + 3y > 12,$$

we first graph $x + 3y = 12$ by finding and plotting a few ordered pairs that satisfy the equation. (The x- and y-intercepts are good choices.) Because the symbol $>$ does not include equality, the points on the line do *not* satisfy the inequality and we graph it using a dashed line. To decide which region includes the points that are solutions, we choose a test point not on the line.

	$x + 3y > 12$ Original inequality
(0, 0) is a convenient test point.	$0 + 3(0) \overset{?}{>} 12$ Let $x = 0$ and $y = 0$.
	$0 > 12$ False

This false result indicates that the solutions are those points in the region that does *not* include $(0, 0)$, as shown in **FIGURE 12.**

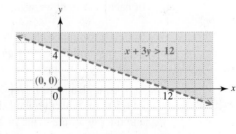

FIGURE 12

We use the same techniques to solve *systems* of linear inequalities.

OBJECTIVE 1 Solve systems of linear inequalities by graphing.

A **system of linear inequalities** consists of two or more linear inequalities. The **solution set of a system of linear inequalities** includes all ordered pairs that make all inequalities of the system true at the same time.

Solving a System of Linear Inequalities

Step 1 **Graph each inequality.** Use the method of **Section 3.6**.

Step 2 **Choose the intersection.** Indicate the solution set by shading the intersection of the graphs—that is, the region where the graphs overlap.

EXAMPLE 1 Solving a System of Linear Inequalities

Graph the solution set of the system.

$$3x + 2y \leq 6$$
$$2x - 5y \geq 10$$

Step 1 To graph $3x + 2y \leq 6$, graph the solid boundary line $3x + 2y = 6$ using the intercepts $(0, 3)$ and $(2, 0)$. Determine the region to shade.

$$3x + 2y < 6 \quad \text{We are testing the region.}$$

$$3(0) + 2(0) \overset{?}{<} 6 \quad \text{Use } (0, 0) \text{ as a test point.}$$

$$0 < 6 \quad \text{True}$$

Shade the region that includes $(0, 0)$. See **FIGURE 13(a)**.

Now graph $2x - 5y \geq 10$ with solid boundary line $2x - 5y = 10$ using the intercepts $(0, -2)$ and $(5, 0)$. Determine the region to shade.

$$2x - 5y > 10$$

$$2(0) - 5(0) \overset{?}{>} 10 \quad \text{Use } (0, 0) \text{ as a test point.}$$

$$0 > 10 \quad \text{False}$$

Shade the region that does *not* include $(0, 0)$. See **FIGURE 13(b)**.

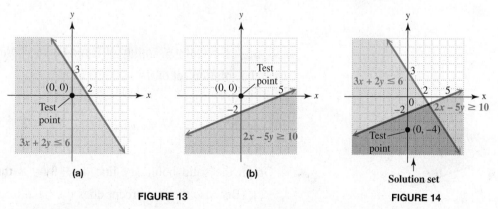

(a)   (b)

FIGURE 13

FIGURE 14

Step 2 The solution set of this system includes all points in the intersection—that is, the overlap—of the graphs of the two inequalities. As shown in **FIGURE 14**, this intersection is the gray shaded region and portions of the two boundary lines that surround it.

NOW TRY EXERCISE 1

Graph the solution set of the system.

$$4x - 2y \le 8$$
$$x + 3y \ge 3$$

CHECK   To confirm the solution set in **FIGURE 14**, select a test point in the gray shaded region, such as $(0, -4)$, and substitute it into *both* inequalities to make sure that true statements result. (Using an ordered pair that has one coordinate 0 makes the substitution easier.)

$3x + 2y < 6$	$2x - 5y > 10$
$3(0) + 2(-4) \overset{?}{<} 6$ Test $(0, -4)$.	$2(0) - 5(-4) \overset{?}{>} 10$ Test $(0, -4)$.
$-8 < 6$ True	$20 > 10$ True

We have shaded the correct region in **FIGURE 14**. Test points selected in the other three regions will satisfy only one of the inequalities or neither of them. (Verify this.) ✓

NOTE   We usually do all the work on one set of axes. In the remaining examples, only one graph is shown. Be sure that the region of the final solution set is clearly indicated.

NOW TRY EXERCISE 2

Graph the solution set of the system.

$$2x + 5y > 10$$
$$x - 2y < 0$$

EXAMPLE 2   Solving a System of Linear Inequalities

Graph the solution set of the system.

$$x - y > 5$$
$$2x + y < 2$$

FIGURE 15 shows the graphs of both $x - y > 5$ and $2x + y < 2$. Dashed lines indicate that the graphs of the inequalities do not include their boundary lines. Use $(0, 0)$ as a test point to determine the region to shade for each inequality.

The solution set of the system is the region with gray shading. Neither boundary line is included. (Use $(0, -6)$ in the gray shaded region as a test point to confirm the solution set.)

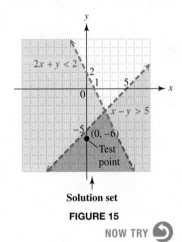

Solution set

FIGURE 15

EXAMPLE 3   Solving a System of Three Linear Inequalities

Graph the solution set of the system.

$$4x - 3y \le 8$$
$$x \ge 2$$
$$y \le 4$$

Graph the solid boundary line $4x - 3y = 8$ through the intercepts $(2, 0)$ and $\left(0, -\frac{8}{3}\right)$. (Because the y-intercept does not have integer coordinates, we also use the point $(-1, -4)$ to help draw an accurate line.) Recall that $x = 2$ is a vertical line through the point $(2, 0)$, and $y = 4$ is a horizontal line through the point $(0, 4)$. See **FIGURE 16** on the next page.

Use $(0, 0)$ as a test point to determine the region to shade for each inequality.

NOW TRY ANSWERS

1.

2.

**NOW TRY
EXERCISE 3**

Graph the solution set of the
system.

$$x - y < 2$$
$$x \geq -2$$
$$y \leq 4$$

NOW TRY ANSWER

3.

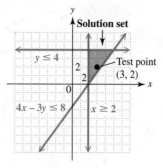

FIGURE 16

The graph of the solution set is the shaded region in **FIGURE 16**, including all
boundary lines. (Use $(3, 2)$ in the gray shaded region as a test point to confirm the
solution set.)

NOW TRY ↻

4.5 Exercises

FOR
EXTRA
HELP

▶ MyMathLab®

▶ *Complete solution available
in MyMathLab*

Concept Check   Match each system of inequalities with the correct graph from choices A–D.

1. $x \geq 5$
   $y \leq -3$

A.
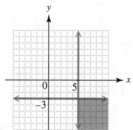

B.
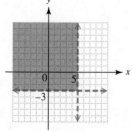

2. $x \leq 5$
   $y \geq -3$

3. $x > 5$
   $y < -3$

C.
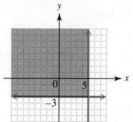

D.
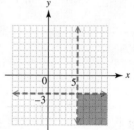

4. $x < 5$
   $y > -3$

*Concept Check   Decide whether each ordered pair is a solution of the given system of
inequalities. Then shade the solution set of each system. Boundary lines are already graphed.*

5. $x - 3y \leq 6$
   $x \geq -4$
   (a) $(-5, -4)$ **(b)** $(0, -4)$ **(c)** $(0, 0)$

6. $x - 2y \geq 4$
   $x \leq -2$
   (a) $(-3, 0)$ **(b)** $(0, 0)$ **(c)** $(-4, -5)$

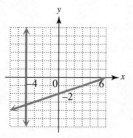

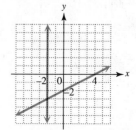

7. $x + y > 4$

$5x - 3y < 15$

(a) $(0, 0)$ **(b)** $(3, 3)$ **(c)** $(5, 0)$

8. $3x - 2y > 12$

$4x + 3y < 12$

(a) $(0, 0)$ **(b)** $(3, -3)$ **(c)** $(6, 0)$

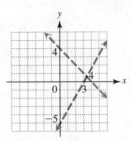   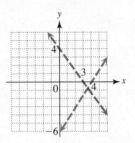

Graph the solution set of each system of linear inequalities. **See Examples 1 and 2.**

9. $x + y \leq 6$

$x - y \geq 1$

10. $x + y \leq 2$

$x - y \geq 3$

11. $4x + 5y \geq 20$

$x - 2y \leq 5$

12. $x + 4y \leq 8$

$2x - y \geq 4$

13. $2x + 3y < 6$

$x - y < 5$

14. $x + 2y < 4$

$x - y < -1$

15. $y \leq 2x - 5$

$x < 3y + 2$

16. $x \geq 2y + 6$

$y > -2x + 4$

17. $4x + 3y < 6$

$x - 2y > 4$

18. $3x + y > 4$

$x + 2y < 2$

19. $x \leq 2y + 3$

$x + y < 0$

20. $x \leq 4y + 3$

$x + y > 0$

21. $x - 3y \leq 6$

$x \geq -5$

22. $x - 2y \geq 2$

$x \leq -3$

23. $-3x + y \geq 1$

$6x - 2y \geq -10$

24. $x + y < 4$

$-2x - 2y < 4$

25. $2x + 3y < 6$

$4x + 6y > 18$

26. $2x - y < -3$

$6x - 3y > 9$

Graph the solution set of each system of linear inequalities. **See Example 3.**

27. $4x + 5y < 8$

$y > -2$

$x > -4$

28. $x - 2y \geq -2$

$y \geq -2$

$x \leq 3$

29. $x + y \geq -3$

$x - y \leq 3$

$y \leq 3$

30. $x + y < 4$

$x - y > -4$

$y > -1$

31. $3x - 2y \geq 6$

$x + y \leq 4$

$x \geq 0$

$y \geq -4$

32. $2x - 3y < 6$

$x + y > 3$

$x < 4$

$y < 4$

Chapter 4   Summary

Key Terms

4.1

system of linear equations
(linear system)

solution of a system
solution set of a system
consistent system

inconsistent system
independent equations
dependent equations

4.5

system of linear inequalities
solution set of a system of
linear inequalities

Test Your Word Power

See how well you have learned the vocabulary in this chapter.

1. A **system of linear equations** consists of
  A. at least two linear equations with different variables
  B. two or more linear equations that have an infinite number of solutions
  C. two or more linear equations with the same variables
  D. two or more linear inequalities.

2. A **consistent system** is a system of equations
  A. with at least one solution
  B. with no solution
  C. with graphs that do not intersect
  D. with solution set $\varnothing$.

3. An **inconsistent system** is a system of equations
  A. with one solution
  B. with no solution

  C. with an infinite number of solutions
  D. that have the same graph.

4. **Dependent equations**
  A. have different graphs
  B. have no solution
  C. have one solution
  D. are different forms of the same equation.

ANSWERS

1. C; *Example:* $2x + y = 7$, $3x - y = 3$ **2.** A; *Example:* The system in **Answer 1** is consistent. The graphs of the equations intersect at exactly one point—in this case, the solution $(2, 3)$.   **3.** B; *Example:* The equations of two parallel lines make up an inconsistent system. Their graphs never intersect, so there is no solution to the system.   **4.** D; *Example:* The equations $4x - y = 8$ and $8x - 2y = 16$ are dependent because their graphs are the same line.

Quick Review

CONCEPTS

EXAMPLES

4.1   Solving Systems of Linear Equations by Graphing

An ordered pair is a solution of a system if it makes all equations of the system true at the same time.

Is $(4, -1)$ a solution of the following system?

$$x + y = 3$$
$$2x - y = 9$$

Yes, because $4 + (-1) = 3$ and $2(4) - (-1) = 9$ are both true, $(4, -1)$ is a solution.

To solve a linear system by graphing, follow these steps.

Step 1   Graph each equation of the system on the same axes.

Step 2   Find the coordinates of the point of intersection.

Step 3   Check. Write the solution set.

Solve the system by graphing.

$$x + y = 5$$
$$2x - y = 4$$

The ordered pair $(3, 2)$ satisfies *both* equations, so $\{(3, 2)\}$ is the solution set.

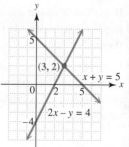

CONCEPTS	EXAMPLES

4.2 Solving Systems of Linear Equations by Substitution

Step 1 Solve one equation for either variable.

Solve the system by substitution.

$$x + 2y = -5 \quad (1)$$
$$y = -2x - 1 \quad (2)$$

Equation (2) is already solved for y.

Step 2 Substitute for that variable in the other equation to obtain an equation in one variable.

Step 3 Solve the equation from Step 2.

Substitute $-2x - 1$ for y in equation (1).

$$x + 2(-2x - 1) = -5 \quad \text{Let } y = -2x - 1 \text{ in (1).}$$
$$x - 4x - 2 = -5 \quad \text{Distributive property}$$
$$-3x - 2 = -5 \quad \text{Combine like terms.}$$
$$-3x = -3 \quad \text{Add 2.}$$
$$x = 1 \quad \text{Divide by } -3.$$

Step 4 Find the other value by substituting the result from Step 3 into the equation from Step 1 and solving for the remaining variable.

To find y, let $x = 1$ in equation (2).

$$y = -2x - 1 \quad (2)$$
$$y = -2(1) - 1 \quad \text{Let } x = 1.$$
$$y = -3 \quad \text{Multiply, and then subtract.}$$

Step 5 Check. Write the solution set.

A check confirms that $\{(1, -3)\}$ is the solution set.

4.3 Solving Systems of Linear Equations by Elimination

Step 1 Write both equations in standard form $Ax + By = C$.

Step 2 Multiply to transform the equations so that the coefficients of one pair of variable terms are opposites.

Step 3 Add the equations to get an equation with only one variable.

Step 4 Solve the equation from Step 3.

Step 5 Find the other value by substituting the result from Step 4 into either of the original equations and solving for the remaining variable.

Step 6 Check. Write the solution set.

If the result of the addition step (Step 3) is a false statement, such as $0 = 4$, the graphs are parallel lines and *there is no solution. The solution set is* $\varnothing$.

If the result is a true statement, such as $0 = 0$, the graphs are the same line, and an *infinite number of ordered pairs are solutions. The solution set is written in set-builder notation as*

$$\{(x, y) | \underline{\qquad}\},$$

where a form of the equation is written in the blank.

Solve the system by elimination.

$$x + 3y = 7 \quad (1)$$
$$3x - y = 1 \quad (2)$$

Multiply equation (1) by -3 to eliminate the x-terms.

$$\begin{array}{rl} -3x - 9y = -21 & \text{Multiply equation (1) by } -3. \\ \underline{3x - y = 1} & (2) \\ -10y = -20 & \text{Add.} \end{array}$$

$$y = 2 \quad \text{Divide by } -10.$$

Substitute to find the value of x.

$$x + 3y = 7 \quad (1)$$
$$x + 3(2) = 7 \quad \text{Let } y = 2.$$
$$x + 6 = 7 \quad \text{Multiply.}$$
$$x = 1 \quad \text{Subtract 6.}$$

A check confirms that $\{(1, 2)\}$ is the solution set.

$$\begin{array}{r} x - 2y = 6 \\ \underline{-x + 2y = -2} \\ 0 = 4 \end{array}$$

Solution set: $\varnothing$

$$\begin{array}{r} x - 2y = 6 \\ \underline{-x + 2y = -6} \\ 0 = 0 \end{array}$$

Solution set:
$$\{(x, y) | x - 2y = 6\}$$

CONCEPTS	EXAMPLES
4.4 **Applications of Linear Systems**	The sum of two numbers is 30. Their difference is 6. Find the numbers.
Use the modified six-step method.	
Step 1 Read.	
Step 2 Assign variables.	Let x = one number, and let y = the other number.
Step 3 Write two equations using both variables.	$$x + y = 30 \quad (1)$$ $$\underline{x - y = 6} \quad (2)$$
Step 4 Solve the system.	$$2x = 36 \quad \text{Add.}$$ $$x = 18 \quad \text{Divide by 2.}$$ Let $x = 18$ in equation (1), $x + y = 30$. $$18 + y = 30 \quad \text{Let } x = 18.$$ $$y = 12 \quad \text{Subtract 18.}$$
Step 5 State the answer.	The two numbers are 18 and 12.
Step 6 Check.	The sum of 18 and 12 is 30, and the difference of 18 and 12 is 6, so the answer checks.
4.5 **Solving Systems of Linear Inequalities**	Graph the solution set of the system.
To solve a system of linear inequalities, follow these steps.	$$2x + 4y \geq 5$$ $$x \geq 1$$
Step 1 Graph each inequality on the same axes. (This was explained in **Section 3.6.**)	First graph the solid boundary lines $2x + 4y = 5$ and $x = 1$. Then use a test point, such as $(0, 0)$, to determine the region to shade for each inequality. The intersection, the gray shaded region, is the solution set of the system. 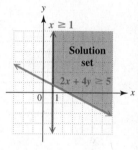
Step 2 Choose the intersection. The solution set of the system is formed by the overlap of the regions of the two graphs.	

Chapter 4   Review Exercises

4.1 *Decide whether the given ordered pair is a solution of the given system.*

1. $(3, 4)$

$4x - 2y = 4$

$5x + y = 19$

2. $(-5, 2)$

$x - 4y = -13$

$2x + 3y = 4$

Solve each system by graphing.

3. $x + y = 4$

$2x - y = 5$

4. $x - 2y = 4$

$2x + y = -2$

5. $2x + 4 = 2y$

$y - x = -3$

6. $x - 2 = 2y$

$2x - 4y = 4$

4.2 *Answer each question.*

7. To solve the following system by substitution, which variable in which equation would be easiest to solve for in the first step?

$$5x - 3y = 7$$

$$-x + 2y = 4$$

Answers (left margin):

1. yes   **2.** no

3. $\{(3, 1)\}$ **4.** $\{(0, -2)\}$

5. $\varnothing$

6. $\{(x, y) \mid x - 2y = 2\}$

7. It would be easiest to solve for x in the second equation because its coefficient is -1. No fractions would be involved.

8. The true statement $0 = 0$ is an indication that the system has an infinite number of solutions. Write the solution set using set-builder notation and the equation of the system that is in standard form with integer coefficients having greatest common factor 1.

9. $\{(2, 1)\}$ **10.** $\{(3, 5)\}$

11. $\{(6, 4)\}$

12. $\{(x, y) \mid x + 3y = 6\}$

13. C

14. (a) 2   (b) 9

15. $\{(7, 1)\}$ **16.** $\{(-4, 3)\}$

17. $\{(x, y) \mid 3x - 4y = 9\}$

18. $\varnothing$

19. $\{(-4, 1)\}$

20. $\{(x, y) \mid 2x - 3y = 0\}$

21. $\{(9, 2)\}$ **22.** $\{(8, 9)\}$

23. $\{(2, 1)\}$ **24.** $\{(-3, 2)\}$

25. Pizza Hut: 6118 locations; Domino's: 4926 locations

26. *Reader's Digest:* 5.2 million; *People:* 3.5 million

27. length: 27 m; width: 18 m

28. twenties: 13; tens: 7

8. After solving a system of linear equations by the substitution method, a student obtained the equation "$0 = 0$." He gave the solution set of the system as $\{(0, 0)\}$. *WHAT WENT WRONG?*

Solve each system by the substitution method.

9. $3x + y = 7$
  $x = 2y$

10. $2x - 5y = -19$
  $y = x + 2$

11. $4x + 5y = 44$
  $x + 2 = 2y$

12. $5x + 15y = 30$
  $x + 3y = 6$

4.3 *Answer each question.*

13. Which system does not require that we multiply one or both equations by a constant to solve the system by the elimination method?

  A. $-4x + 3y = 7$
    $3x - 4y = 4$

  B. $5x + 8y = 13$
    $12x + 24y = 36$

  C. $2x + 3y = 5$
    $x - 3y = 12$

  D. $x + 2y = 9$
    $3x - y = 6$

14. If we were to multiply equation (1) by -3 in the system below, by what number would we have to multiply equation (2) in order to do the following?

$$2x + 12y = 7 \quad (1)$$
$$3x + 4y = 1 \quad (2)$$

  (a) Eliminate the *x*-terms when solving by the elimination method.

  (b) Eliminate the *y*-terms when solving by the elimination method.

Solve each system by the elimination method.

15. $2x - y = 13$
  $x + y = 8$

16. $-4x + 3y = 25$
  $6x - 5y = -39$

17. $3x - 4y = 9$
  $6x - 8y = 18$

18. $2x + y = 3$
  $-4x - 2y = 6$

4.1–4.3 *Solve each system by any method.*

19. $2x + 3y = -5$
  $3x + 4y = -8$

20. $6x - 9y = 0$
  $2x - 3y = 0$

21. $x - 2y = 5$
  $y = x - 7$

22. $\dfrac{x}{2} + \dfrac{y}{3} = 7$

  $\dfrac{x}{4} + \dfrac{2y}{3} = 8$

23. $\dfrac{3}{4}x - \dfrac{1}{3}y = \dfrac{7}{6}$

  $\dfrac{1}{2}x + \dfrac{2}{3}y = \dfrac{5}{3}$

24. $0.4x - 0.5y = -2.2$
  $0.3x + 0.2y = -0.5$

4.4 *Solve each problem using a system of equations.*

25. The two leading pizza chains in the United States are Pizza Hut and Domino's Pizza. In September 2013, Pizza Hut had 1192 more locations than Domino's, and together the two chains had 11,044 locations. How many locations did each chain have? (*Source: PMQ Pizza Magazine.*)

26. Two popular magazines in the United States are *Reader's Digest* and *People*. Together, the average paid circulation for these two magazines during the first six months of 2013 was 8.7 million. The circulation for *People* was 1.7 million less than that of *Reader's Digest*. What were the circulation figures for each magazine? (*Source:* Audit Bureau of Circulations.)

27. The perimeter of a rectangle is 90 m. Its length is $1\frac{1}{2}$ times its width. Find the length and width of the rectangle.

28. Laura has 20 bills, all of which are $10 or $20 bills. The total value of the money is $330. How many of each denomination does she have?

29. 25 lb of $1.30 candy; 75 lb of $0.90 candy
30. 40% solution: 60 L; 70% solution: 30 L
31. $7000 at 3%; $11,000 at 4%
32. plane: 250 mph; wind: 20 mph

33.

34.

35.

36.

29. Sharon has candy that sells for $1.30 per lb, to be mixed with candy selling for $0.90 per lb to obtain 100 lb of a mix that will sell for $1 per lb. How much of each type should she use?

30. A 40% antifreeze solution is to be mixed with a 70% solution to obtain 90 L of a 50% solution. How many liters of the 40% and 70% solutions will be needed?

31. Nancy invested $18,000, part of it at 3% annual simple interest and the rest at 4%. Her interest income for the first year was $650. How much did she invest at each rate?

Liters of Solution	Percent (as a decimal)	Liters of Pure Antifreeze
x	0.40	
y	0.70	
90	0.50	

Amount Invested (in dollars)	Rate (as a decimal)	Interest (in dollars)
x	0.03	
y	0.04	
18,000		

32. A plane flying with the wind travels 540 mi in 2 hr. Flying against the same wind, the plane travels 690 mi in 3 hr. Find the rate of the plane in still air and the wind speed.

4.5 *Graph the solution set of each system of linear inequalities.*

33. $x + y \geq 2$
$x - y \leq 4$

34. $y \geq 2x$
$2x + 3y \leq 6$

35. $x + y < 3$
$2x > y$

36. $3x - y \leq 3$
$x \geq -1$
$y \leq 2$

Chapter 4   Mixed Review Exercises

Solve each system by any method.

1. $\{(4, 8)\}$
2. $\{(x, y) \mid x - y = 6\}$
3. $\{(2, 0)\}$
4. $\{(4, 1)\}$

5. In System B, the second equation is already solved for y.
6. (a) years 0 to 6
(b) year 6; about $650

1. $\dfrac{2x}{3} + \dfrac{y}{4} = \dfrac{14}{3}$

$\dfrac{x}{2} + \dfrac{y}{12} = \dfrac{8}{3}$

2. $x = y + 6$
$2y - 2x = -12$

3. $3x + 4y = 6$
$4x - 5y = 8$

4. $0.4x - 0.9y = 0.7$
$0.3x + 0.2y = 1.4$

Solve each problem.

5. Why would it be easier to solve System B by the substitution method than System A?

System A: $-5x + 6y = -7$
$2x + 5y = -5$

System B: $2x + 9y = 13$
$y = 3x - 2$

6. Patricia compared the monthly payments she would incur for two types of mortgages: fixed rate and variable rate. Her observations led to the graph shown.

(a) For which years would the monthly payment be more for the fixed rate mortgage than for the variable rate mortgage?

(b) In what year would the payments be the same, and what would those payments be?

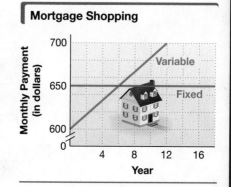

7. 8 in., 8 in., 13 in.
8. Lincoln Memorial: 6.2 million;
World War II Memorial:
4.4 million
9. slower car: 38 mph;
faster car: 68 mph
10. B

11.   **12.**

7. The perimeter of an isosceles triangle measures 29 in. One side of the triangle is 5 in. longer than each of the two equal sides. Find the lengths of the sides of the triangle.

8. In 2012, a total of 10.6 million people visited the Lincoln Memorial and the World War II Memorial, two popular tourist attractions. The World War II Memorial had 1.8 million fewer visitors than the Lincoln Memorial. How many visitors did each of these attractions have? (*Source:* National Park Service.)

9. Two cars leave from the same place and travel in opposite directions. One car travels 30 mph faster than the other. After $2\frac{1}{2}$ hr, they are 265 mi apart. What are the rates of the cars?

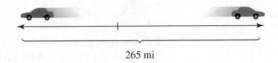

265 mi

10. Which system of linear inequalities is graphed in the figure?

A. $x \le 3$	**B.** $x \le 3$	**C.** $x \ge 3$	**D.** $x \ge 3$
$y \le 1$	$y \ge 1$	$y \le 1$	$y \ge 1$

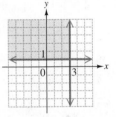

Graph the solution set of each system of linear inequalities.

11. $x + y < 5$
$x - y \ge 2$

12. $y \le 2x$
$x + 2y > 4$

Chapter 4   Test

FOR EXTRA HELP *Step-by-step test solutions are found on the Chapter Test Prep Videos available in* MyMathLab® *or on* YouTube.

▶ *View the complete solutions to all Chapter Test exercises in MyMathLab.*

[4.1]
1. (a) no (b) no (c) yes
2. $\{(4, 1)\}$
3. It has no solution.

[4.2]
4. $\{(1, -6)\}$
5. $\{(-35, 35)\}$

[4.3]
6. $\{(5, 6)\}$
7. $\{(-1, 3)\}$
8. $\{(-1, 3)\}$
9. ∅
10. $\{(0, 0)\}$

1. Decide whether each ordered pair is a solution of the system.
$$2x + y = -3$$
$$x - y = -9$$

(a) $(1, -5)$ **(b)** $(1, 10)$ **(c)** $(-4, 5)$

2. Solve the system $\begin{aligned} x + 2y &= 6 \\ -2x + y &= -7 \end{aligned}$ by graphing.

3. Suppose that the graph of a system of two linear equations consists of lines that have the same slope but different y-intercepts. How many solutions does the system have?

Solve each system by the substitution method.

4. $2x + y = -4$
$x = y + 7$

5. $4x + 3y = -35$
$x + y = 0$

Solve each system by the elimination method.

6. $2x - y = 4$
$3x + y = 21$

7. $4x + 2y = 2$
$5x + 4y = 7$

8. $3x + 4y = 9$
$2x + 5y = 13$

9. $4x + 5y = 2$
$-8x - 10y = 6$

10. $6x - 5y = 0$
$-2x + 3y = 0$

[4.1–4.3]
11. $\{(x, y) \mid 3x + 4y = 5\}$
12. $\{(-15, 6)\}$
13. $\{(-4, 6)\}$

[4.4]
14. Memphis and Atlanta: 394 mi;
Minneapolis and Houston:
1176 mi
15. Magic Kingdom: 17.5 million;
Disneyland: 16.0 million
16. 25% solution: $33\frac{1}{3}$ L;

40% solution: $16\frac{2}{3}$ L
17. slower car: 45 mph;
faster car: 60 mph

[4.5]
18.

19.

20. B

Solve each system by any method.

11. $4y = -3x + 5$
    $6x = -8y + 10$

12. $\dfrac{6}{5}x - \dfrac{1}{3}y = -20$

    $-\dfrac{2}{3}x + \dfrac{1}{6}y = 11$

13. $0.2x + 0.3y = 1.0$
    $-0.3x + 0.1y = 1.8$

Solve each problem using a system of equations.

14. The distance between Memphis and Atlanta is 782 mi less than the distance between Minneapolis and Houston. Together, the two distances total 1570 mi. How far is it between Memphis and Atlanta? How far is it between Minneapolis and Houston?

15. In 2012, the two most popular amusement parks in the United States were Disneyland and the Magic Kingdom at Walt Disney World. Disneyland had 1.5 million fewer visitors than the Magic Kingdom, and together they had 33.5 million visitors. How many visitors did each park have? (*Source:* www.aecom.com)

16. Sohail has a 25% solution of alcohol to mix with a 40% solution to make 50 L of a final mixture that is 30% alcohol. How much of each of the original solutions should be used?

17. Two cars leave from Perham, Minnesota, at the same time and travel in the same direction. One car travels one and one-third times as fast as the other. After 3 hr, they are 45 mi apart. What are the rates of the cars?

Liters of Solution	Percent (as a decimal)	Liters of Pure Alcohol

	r	t	d
Faster Car			
Slower Car			

Graph the solution set of each system of linear inequalities.

18. $2x + 7y \le 14$
    $x - y \ge 1$

19. $2x - y > 6$
    $4y + 12 \ge -3x$

20. Without actually graphing, determine which one of the following systems of inequalities has no solution.

A. $x \ge 4$
    $y \le 3$

B. $x + y > 4$
    $x + y < 3$

C. $x > 2$
    $y < 1$

D. $x + y > 4$
    $x - y < 3$

Chapters R–4   Cumulative Review Exercises

[1.5]
1. $-1, 1, -2, 2, -4, 4, -5, 5, -8,$
    $8, -10, 10, -20, 20, -40, 40$
2. 46

[1.2]
3. 1

[1.6]
4. distributive property

[2.3]
5. $\left\{-\dfrac{13}{11}\right\}$ **6.** $\left\{\dfrac{9}{11}\right\}$

[2.5]

7. $T = \dfrac{PV}{k}$

1. List all integer factors of 40.

2. Evaluate $-2 + 6[3 - (4 - 9)]$.

3. Find the value of the expression

$$\frac{3x^2 + 2y^2}{10y + 3}$$

for $x = 1$ and $y = 5$.

4. Name the property that justifies the statement.

$$r(s - k) = rs - rk$$

Solve each linear equation.

5. $2 - 3(6x + 2) = 4(x + 1) + 18$

6. $\dfrac{3}{2}\left(\dfrac{1}{3}x + 4\right) = 6\left(\dfrac{1}{4} + x\right)$

7. Solve the formula $P = \dfrac{kT}{V}$ for T.

[2.8]

8. $(-18, \infty)$

9. $\left(-\dfrac{11}{2}, \infty\right)$

[R.2, 2.6]

10. 2010; 1813; 62.8%; 57.2%

[2.4]

11. Ravens: 34; 49ers: 31

[2.5]

12. 46°, 46°, 88°

[3.2]

13.　　　　**14.**

[3.3]

15. $-\dfrac{4}{3}$　　**16.** $-\dfrac{1}{4}$

[3.4, 3.5]

17. $y = \dfrac{1}{2}x + 3$

18. $y = 2x + 1$

19. (a) $x = 9$　**(b)** $y = -1$

[4.1–4.3]

20. $\{(-1, 6)\}$

21. $\{(3, -4)\}$

22. $\varnothing$

[4.4]

23. adult tickets: 405;
child tickets: 49

24. 20% solution: 4 L;
50% solution: 8 L

[4.5]

25.

Solve each linear inequality.

8. $-\dfrac{5}{6}x < 15$　　　　**9.** $-8 < 2x + 3$

Solve each problem.

10. A survey of 2500 people measured public recognition of some classic advertising slogans. Complete the results shown in the table.

Slogan (product or company)	Percent Recognition (nearest tenth of a percent)	Actual Number That Recognized Slogan (nearest whole number)
Please Don't Squeeze the . . . (Charmin®)	80.4%	
The Breakfast of Champions (Wheaties)	72.5%	
The King of Beers (Budweiser®)		1570
Like a Good Neighbor (State Farm)		1430

(Other slogans included "You're in Good Hands" (Allstate), "Snap, Crackle, Pop" (Rice Krispies®), and "The Un-Cola" (7-Up).)
Source: Department of Integrated Marketing Communications, Northwestern University.

11. In Super Bowl XLVII, the Baltimore Ravens defeated the San Francisco 49ers by 3 points. The two teams scored 65 total points. What was the final score of the game? (*Source: World Almanac and Book of Facts.*)

12. Two angles of a triangle have the same measure. The measure of the third angle is 4° less than twice the measure of each of the equal angles. Find the measures of the three angles.

Measures are in degrees.

Graph each linear equation.

13. $x - y = 4$　　　　　　**14.** $3x + y = 6$

Find the slope of each line.

15. Through $(-5, 6)$ and $(1, -2)$　　**16.** Perpendicular to the line $y = 4x - 3$

Write an equation of each line. Give the final answer in slope-intercept form.

17. Through $(-4, 1)$, slope $\dfrac{1}{2}$　　**18.** Through the points $(1, 3)$ and $(-2, -3)$

19. Write an equation of the line satisfying the given conditions.

　　(a) Vertical, through $(9, -2)$　　　　**(b)** Horizontal, through $(4, -1)$

Solve each system by any method.

20. $2x - y = -8$　　　**21.** $4x + 5y = -8$　　　**22.** $3x + 4y = 2$
　　　$x + 2y = 11$　　　　　　$3x + 4y = -7$　　　　　　$6x + 8y = 1$

Solve each problem using a system of equations.

23. Admission prices at a high school football game were $6 for adults and $2 for children. The total value of the tickets sold was $2528, and 454 tickets were sold. How many adult and how many child tickets were sold?

Number of Tickets	Cost of Each (in dollars)	Total Value (in dollars)
x	6	$6x$
y		
454	✕✕✕	

24. A chemist needs 12 L of a 40% alcohol solution. She must mix a 20% solution and a 50% solution. How many liters of each will be required to obtain what she needs?

25. Graph the solution set of the system of linear inequalities.

$$x + 2y \le 12$$
$$2x - y \le 8$$

5

Exponents and Polynomials

Numbers such as the one shown here are often written using *scientific notation,* a method of expressing them using a power of 10 as a factor.

17,512,083,047,998

   = 1.7512083047998 × 10^{13}

5.1   The Product Rule and Power Rules for Exponents

VOCABULARY

☐ base
☐ exponent (power)
☐ exponential expression

**NOW TRY
EXERCISE 1**

Write $4 \cdot 4 \cdot 4$ in exponential form and evaluate.

**NOW TRY
EXERCISE 2**

Identify the base and the exponent of each expression. Then evaluate.

(a) $(-3)^4$ **(b)** -3^4

OBJECTIVE 1   Use exponents.

Recall from **Section 1.1** that in the expression 5^2, the number 5 is the **base** and 2 is the **exponent,** or **power.** The expression 5^2 is an **exponential expression.** Although we do not usually write the exponent when it is 1, the following holds for any quantity *a* in general.

$$a^1 = a$$

EXAMPLE 1   Using Exponents

Write $3 \cdot 3 \cdot 3 \cdot 3$ in exponential form and evaluate.

Because 3 occurs as a factor four times, the base is 3 and the exponent is 4. The exponential expression is 3^4, read "3 to the fourth power" or simply "3 to the fourth."

$$\underbrace{3 \cdot 3 \cdot 3 \cdot 3}_{4 \text{ factors of } 3} \quad \text{means} \quad 3^4, \quad \text{which equals} \quad 81.$$

NOW TRY

EXAMPLE 2   Evaluating Exponential Expressions

Identify the base and the exponent of each expression. Then evaluate.

Expression	Base	Exponent	Value
(a) 5^4	5	4	$5 \cdot 5 \cdot 5 \cdot 5$, which equals 625
(b) -5^4	5	4	$-1 \cdot (5 \cdot 5 \cdot 5 \cdot 5)$, which equals -625
(c) $(-5)^4$	-5	4	$(-5)(-5)(-5)(-5)$, which equals 625

NOW TRY

> ⚠️ **CAUTION**   Compare **Examples 2(b) and 2(c).** In the expression -5^4, the absence of parentheses shows that the exponent 4 applies only to the base 5, not -5. In $(-5)^4$, the parentheses show that the exponent 4 applies to the base -5. In summary, $-a^n$ and $(-a)^n$ are not necessarily the same.

Expression	Base	Exponent	Example
$-a^n$	a	n	$-3^2 = -(3 \cdot 3) = -9$
$(-a)^n$	$-a$	n	$(-3)^2 = (-3)(-3) = 9$

OBJECTIVE 2   Use the product rule for exponents.

To develop the product rule, we use the definition of exponents.

$$2^4 \cdot 2^3$$

$$= \overbrace{(2 \cdot 2 \cdot 2 \cdot 2)}^{4 \text{ factors}} \overbrace{(2 \cdot 2 \cdot 2)}^{3 \text{ factors}}$$

$$= \underbrace{2 \cdot 2 \cdot 2 \cdot 2 \cdot 2 \cdot 2 \cdot 2}_{4 + 3 = 7 \text{ factors}}$$

$$= 2^7$$

NOW TRY ANSWERS
1. 4^3; 64
2. **(a)** -3; 4; 81   **(b)** 3; 4; -81

Also,
$$6^2 \cdot 6^3$$
$$= (6 \cdot 6)(6 \cdot 6 \cdot 6)$$
$$= 6 \cdot 6 \cdot 6 \cdot 6 \cdot 6$$
$$= 6^5.$$

Generalizing from these examples, we have the following.

$2^4 \cdot 2^3$ is equal to 2^{4+3},   which equals 2^7.

$6^2 \cdot 6^3$ is equal to 6^{2+3},   which equals 6^5.

In each case, adding the exponents gives the exponent of the product, suggesting the **product rule for exponents.**

Product Rule for Exponents

For any positive integers m and n, $a^m \cdot a^n = a^{m+n}$.
(Keep the same base and add the exponents.)

Example: $6^2 \cdot 6^5 = 6^{2+5} = 6^7$

⚠ CAUTION Do not multiply the bases when using the product rule. ***Keep the same base and add the exponents.*** For example,

$$6^2 \cdot 6^5 = 6^7, \quad \textbf{not} \quad 36^7.$$

🔄 NOW TRY
EXERCISE 3
Use the product rule for exponents to simplify each expression, if possible.
(a) $(-5)^2(-5)^4$
(b) $y^2 \cdot y \cdot y^5$
(c) $(2x^3)(4x^6)$
(d) $2^4 \cdot 5^3$
(e) $3^2 + 3^3$

EXAMPLE 3   **Using the Product Rule**

Use the product rule for exponents to simplify each expression, if possible.

(a) $6^3 \cdot 6^5$

$= 6^{3+5}$ Product rule

$= 6^8$ Add the exponents.

(b) $(-4)^7(-4)^2$

$= (-4)^{7+2}$ Product rule

$= (-4)^9$ Add the exponents.

(c) $x^2 \cdot x$

$= x^2 \cdot x^1$ $a = a^1$, for all a.

$= x^{2+1}$ Product rule

$= x^3$ Add the exponents.

(d) $m^4 m^3 m^5$

$= m^{4+3+5}$ Product rule

$= m^{12}$ Add the exponents.

(e) [Think: 2^3 means $2 \cdot 2 \cdot 2$.] $2^3 \cdot 3^2$ [Think: 3^2 means $3 \cdot 3$.]   The product rule does not apply. ***The bases are different.***

$= 8 \cdot 9$ Evaluate 2^3 and 3^2.

$= 72$ Multiply.

(f) $2^3 + 2^4$ The product rule does not apply. ***This is a sum, not a product.***

$= 8 + 16$ Evaluate 2^3 and 2^4.

$= 24$ Add.

(g) $(2x^3)(3x^7)$ [$2x^3$ means $2 \cdot x^3$ and $3x^7$ means $3 \cdot x^7$.]

$= (2 \cdot 3) \cdot (x^3 \cdot x^7)$ Commutative and associative properties

$= 6x^{3+7}$ Multiply and then use the product rule.

$= 6x^{10}$ Add the exponents.   **NOW TRY 🔄**

NOW TRY ANSWERS
3. (a) $(-5)^6$ **(b)** y^8 **(c)** $8x^9$
   (d) The product rule does not apply; 2000
   (e) The product rule does not apply; 36

> **⊘ CAUTION**   Be sure that you understand the difference between *adding* and *multiplying* exponential expressions.

$$8x^3 + 5x^3 \quad \text{means} \quad (8+5)x^3, \quad \text{which equals} \quad 13x^3.$$
$$(8x^3)(5x^3) \quad \text{means} \quad (8 \cdot 5)x^{3+3}, \quad \text{which equals} \quad 40x^6.$$

OBJECTIVE 3   Use the rule $(a^m)^n = a^{mn}$.

We can simplify an expression such as $(5^2)^4$ with the product rule for exponents, as follows.

$$(5^2)^4$$

$$= 5^2 \cdot 5^2 \cdot 5^2 \cdot 5^2 \qquad \text{Definition of exponent}$$

$$= 5^{2+2+2+2} \qquad \text{Product rule}$$

$$= 5^8 \qquad \text{Add.}$$

Observe that $2 \cdot 4 = 8$. This example suggests **power rule (a) for exponents.**

Power Rule (a) for Exponents

For any positive integers m and n, $(a^m)^n = a^{mn}$.
(Raise a power to a power by multiplying exponents.)

Example: $(3^2)^4 = 3^{2 \cdot 4} = 3^8$

↻ NOW TRY EXERCISE 4

Use power rule (a) for exponents to simplify.

(a) $(4^7)^5$ **(b)** $(y^4)^7$

EXAMPLE 4   Using Power Rule (a)

Use power rule (a) for exponents to simplify.

(a) $(2^5)^3$

$$= 2^{5 \cdot 3}$$

$$= 2^{15}$$

(b) $(5^7)^2$

$$= 5^{7 \cdot 2}$$

$$= 5^{14}$$

(c) $(x^2)^5$

$$= x^{2 \cdot 5} \qquad \text{Power rule (a)}$$

$$= x^{10} \qquad \text{Multiply.}$$

NOW TRY ↻

OBJECTIVE 4   Use the rule $(ab)^m = a^m b^m$.

Consider the following.

$$(4x)^3$$

$$= (4x)(4x)(4x) \qquad \text{Definition of exponent}$$

$$= (4 \cdot 4 \cdot 4)(x \cdot x \cdot x) \qquad \text{Commutative and associative properties}$$

$$= 4^3 x^3 \qquad \text{Definition of exponent}$$

This example suggests **power rule (b) for exponents.**

Power Rule (b) for Exponents

For any positive integer m, $(ab)^m = a^m b^m$.
(Raise a product to a power by raising each factor to the power.)

Example: $(2p)^5 = 2^5 p^5$

NOW TRY ANSWERS
4. (a) 4^{35} **(b)** y^{28}

NOW TRY EXERCISE 5

Use power (b) for exponents to simplify.

(a) $(-5ab)^3$ **(b)** $(4t^3p^5)^2$

EXAMPLE 5 Using Power Rule (b)

Use power rule (b) for exponents to simplify.

(a) $(3xy)^2$

$= 3^2x^2y^2$ Power rule (b)

$= 9x^2y^2$ $3^2 = 3 \cdot 3 = 9$

(b) $5(pq)^2$

$= 5(p^2q^2)$ Power rule (b)

$= 5p^2q^2$ Multiply.

(c) $(2m^2p^3)^4$

$= 2^4(m^2)^4(p^3)^4$ Power rule (b)

$= 2^4m^8p^{12}$ Power rule (a)

$= 16m^8p^{12}$ $2^4 = 2 \cdot 2 \cdot 2 \cdot 2 = 16$

(d) $(-5^6)^3$

$= (-1 \cdot 5^6)^3$ $-a = -1 \cdot a$

$= (-1)^3 \cdot (5^6)^3$ Power rule (b)

Raise -1 to the designated power.

$= -1 \cdot 5^{18}$ Power rule (a)

$= -5^{18}$ Multiply.

NOW TRY

> **! CAUTION** *Power rule (b) does not apply to a sum.*
>
> $$(4x)^2 = 4^2x^2, \quad \text{but} \quad (4 + x)^2 \neq 4^2 + x^2.$$

OBJECTIVE 5 Use the rule $\left(\frac{a}{b}\right)^m = \frac{a^m}{b^m}$.

Because the quotient $\frac{a}{b}$ can be written as $a\left(\frac{1}{b}\right)$, we use this fact, power rule (b), and some properties of real numbers to obtain **power rule (c) for exponents.**

> **Power Rule (c) for Exponents**
>
> For any positive integer m, $\left(\dfrac{a}{b}\right)^m = \dfrac{a^m}{b^m}$ **(where $b \neq 0$).**
>
> (Raise a quotient to a power by raising both numerator and denominator to the power.)
>
> *Example:* $\left(\dfrac{5}{3}\right)^2 = \dfrac{5^2}{3^2}$

NOW TRY EXERCISE 6

Use power (c) for exponents to simplify.

(a) $\left(\dfrac{p}{q}\right)^5$ **(b)** $\left(\dfrac{1}{4}\right)^3$

    $(q \neq 0)$

NOW TRY ANSWERS

5. (a) $-125a^3b^3$ (b) $16t^6p^{10}$

6. (a) $\dfrac{p^5}{q^5}$ (b) $\dfrac{1}{64}$

EXAMPLE 6 Using Power Rule (c)

Use power rule (c) for exponents to simplify.

(a) $\left(\dfrac{2}{3}\right)^5$

$= \dfrac{2^5}{3^5}$

$= \dfrac{32}{243}$

(b) $\left(\dfrac{m}{n}\right)^4$

$= \dfrac{m^4}{n^4}$

$(n \neq 0)$

(c) $\left(\dfrac{1}{5}\right)^4$

$= \dfrac{1^4}{5^4}$ Power rule (c)

$= \dfrac{1}{625}$ Simplify.

NOW TRY

NOTE   In **Example 6(c)**, we used the fact that $1^4 = 1$.

In general, $1^n = 1$, for any integer n.

Rules for Exponents

For positive integers m and n, the following hold.

		Examples
Product rule	$a^m \cdot a^n = a^{m+n}$	$6^2 \cdot 6^5 = 6^{2+5} = 6^7$
Power rules (a)	$(a^m)^n = a^{mn}$	$(3^2)^4 = 3^{2 \cdot 4} = 3^8$
(b)	$(ab)^m = a^m b^m$	$(2p)^5 = 2^5 p^5$
(c)	$\left(\dfrac{a}{b}\right)^m = \dfrac{a^m}{b^m}$ (where $b \neq 0$)	$\left(\dfrac{5}{3}\right)^2 = \dfrac{5^2}{3^2}$

OBJECTIVE 6   Use combinations of the rules for exponents.

EXAMPLE 7   **Using Combinations of the Rules**

Simplify.

(a) $\left(\dfrac{2}{3}\right)^2 \cdot 2^3$

$= \dfrac{2^2}{3^2} \cdot \dfrac{2^3}{1}$ Power rule (c)

$= \dfrac{2^2 \cdot 2^3}{3^2 \cdot 1}$ Multiply fractions.

$= \dfrac{2^{2+3}}{3^2}$ Product rule

$= \dfrac{2^5}{3^2}$ Add.

$= \dfrac{32}{9}$ Apply the exponents.

(b) $(5x)^3 (5x)^4$

$= (5x)^7$ Product rule

$= 5^7 x^7$ Power rule (b)

An equally correct way to simplify this expression follows.

$(5x)^3 (5x)^4$

$= 5^3 x^3 5^4 x^4$ Power rule (b)

$= 5^3 \cdot 5^4 x^3 x^4$ Commutative property

$= 5^7 x^7$ Product rule

(c) $(2x^2 y^3)^4 (3xy^2)^3$

$= 2^4 (x^2)^4 (y^3)^4 \cdot 3^3 x^3 (y^2)^3$ Power rule (b)

$= 2^4 x^8 y^{12} \cdot 3^3 x^3 y^6$ Power rule (a)

$= 2^4 \cdot 3^3 x^8 x^3 y^{12} y^6$ Commutative and associative properties

$= 16 \cdot 27 x^{11} y^{18}$ Apply the exponents; product rule

$= 432 x^{11} y^{18}$ Multiply.

Notice that $(2x^2 y^3)^4$ means $2^4 x^{2 \cdot 4} y^{3 \cdot 4}$,   ***not*** $(2 \cdot 4) x^{2 \cdot 4} y^{3 \cdot 4}$.

**NOW TRY
EXERCISE 7**

Simplify.

(a) $\left(\dfrac{3}{5}\right)^3 \cdot 3^2$ **(b)** $(8k)^5(8k)^4$

(c) $(x^4y)^5(-2x^2y^5)^3$

(d) $(-x^3y)^2(-x^5y^4)^3$ Think of the negative sign as a factor of -1.

$= (-1 \cdot x^3y)^2(-1 \cdot x^5y^4)^3$ $-a = -1 \cdot a$

$= (-1)^2(x^3)^2y^2 \cdot (-1)^3(x^5)^3(y^4)^3$ Power rule (b)

$= (-1)^2(x^6)(y^2)(-1)^3(x^{15})(y^{12})$ Power rule (a)

$= (-1)^5(x^{21})(y^{14})$ Product rule

$= -x^{21}y^{14}$ Simplify. **NOW TRY**

⊘ CAUTION Be aware of the distinction between $(2y)^3$ and $2y^3$.

$$(2y)^3 = 2y \cdot 2y \cdot 2y = 8y^3, \qquad \text{The base is } 2y.$$

while $\qquad\qquad 2y^3 = 2 \cdot y \cdot y \cdot y. \qquad$ The base is y.

OBJECTIVE 7 Use the rules for exponents in a geometry application.

EXAMPLE 8 Using Area Formulas

**NOW TRY
EXERCISE 8**

Find an expression that represents the area of the figure. Assume $x > 0$.

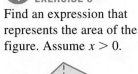
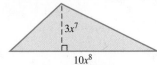

Find an expression that represents the area in **(a) FIGURE 1** and **(b) FIGURE 2**.

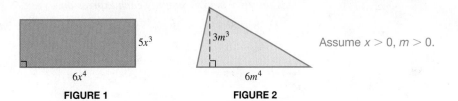

Assume $x > 0$, $m > 0$.

FIGURE 1 **FIGURE 2**

(a) For **FIGURE 1**, use the formula for the area of a rectangle.

$\mathcal{A} = LW$ Area formula

$\mathcal{A} = (6x^4)(5x^3)$ Substitute.

$\mathcal{A} = 6 \cdot 5 \cdot x^{4+3}$ Commutative property; product rule

$\mathcal{A} = 30x^7$ Multiply. Add the exponents.

(b) **FIGURE 2** is a triangle with base $6m^4$ and height $3m^3$.

$\mathcal{A} = \dfrac{1}{2}bh$ Area formula

$\mathcal{A} = \dfrac{1}{2}(6m^4)(3m^3)$ Substitute.

$\mathcal{A} = \dfrac{1}{2}(6 \cdot 3 \cdot m^{4+3})$ Properties of real numbers; product rule

$\mathcal{A} = 9m^7$ Multiply. Add the exponents. **NOW TRY**

NOW TRY ANSWERS

7. (a) $\dfrac{243}{125}$ **(b)** 8^9k^9

   (c) $-8x^{26}y^{20}$

8. $15x^{15}$

5.1 Exercises

FOR EXTRA HELP

 MyMathLab®

▶ *Complete solution available in MyMathLab*

Concept Check   *Decide whether each statement is* true *or* false. *If false, tell why.*

1. $3^3 = 9$

2. $(x^2)^3 = x^5$

3. $(-3)^4 = 3^4$

4. $\left(\dfrac{1}{5}\right)^2 = \dfrac{1}{5^2}$

Write each expression in exponential form. See Example 1.

5. $w \cdot w \cdot w \cdot w \cdot w \cdot w$

6. $t \cdot t \cdot t \cdot t \cdot t \cdot t \cdot t \cdot t$

▶ **7.** $\left(\dfrac{1}{2}\right)\left(\dfrac{1}{2}\right)\left(\dfrac{1}{2}\right)\left(\dfrac{1}{2}\right)\left(\dfrac{1}{2}\right)\left(\dfrac{1}{2}\right)$

8. $\left(\dfrac{1}{4}\right)\left(\dfrac{1}{4}\right)\left(\dfrac{1}{4}\right)\left(\dfrac{1}{4}\right)\left(\dfrac{1}{4}\right)$

9. $(-4)(-4)(-4)(-4)$

10. $(-3)(-3)(-3)(-3)(-3)(-3)$

11. $(-7y)(-7y)(-7y)(-7y)$

12. $(-8p)(-8p)(-8p)(-8p)(-8p)$

13. Explain how the expressions $(-3)^4$ and -3^4 are different.

14. Explain how the expressions $(5x)^3$ and $5x^3$ are different.

Identify the base and the exponent of each expression. In Exercises 15–18, also evaluate. See Example 2.

▶ **15.** 3^5

16. 2^7

▶ **17.** $(-3)^5$

18. $(-2)^7$

19. $(-6x)^4$

20. $(-8x)^4$

21. $-6x^4$

22. $-8x^4$

Concept Check   *Simplify each expression.*

23. $8^2 \cdot 8^5$

$= 8^{\underline{}+\underline{}}$

$= \underline{}$

24. $5m^2 \cdot 2m^6$

$= (5 \cdot \underline{}) \cdot (m^{\underline{}} \cdot m^{\underline{}})$

$= \underline{}m^{\underline{}+\underline{}}$

$= \underline{}$

Use the product rule for exponents to simplify each expression, if possible. Write each answer in exponential form. See Example 3.

▶ **25.** $5^2 \cdot 5^6$

26. $3^6 \cdot 3^7$

27. $4^2 \cdot 4^7 \cdot 4^3$

28. $5^3 \cdot 5^8 \cdot 5^2$

29. $(-7)^3(-7)^6$

30. $(-9)^8(-9)^5$

▶ **31.** $t^3 \cdot t^8 \cdot t^{13}$

32. $n^5 \cdot n^6 \cdot n^9$

33. $(-8r^4)(7r^3)$

34. $(10a^7)(-4a^3)$

▶ **35.** $(-6p^5)(-7p^5)$

36. $(-5w^8)(-9w^8)$

▶ **37.** $(5x^2)(-2x^3)(3x^4)$

38. $(12y^3)(4y)(-3y^5)$

▶ **39.** $3^8 + 3^9$

40. $4^{12} + 4^5$

41. $5^8 \cdot 3^9$

42. $6^3 \cdot 8^9$

Use the power rules for exponents to simplify each expression. Assume that variables in denominators are not zero. Write each answer in exponential form. See Examples 4–6.

▶ **43.** $(4^3)^2$

44. $(8^3)^6$

▶ **45.** $(t^4)^5$

46. $(y^6)^5$

47. $(7r)^3$

48. $(11x)^4$

▶ **49.** $(5xy)^5$

50. $(9pq)^6$

51. $(-5^2)^6$

52. $(-9^4)^8$

53. $(-8^3)^5$

54. $(-7^5)^7$

55. $8(qr)^3$

56. $4(vw)^5$

57. $\left(\dfrac{9}{5}\right)^8$

58. $\left(\dfrac{12}{7}\right)^3$

▶ 59. $\left(\dfrac{1}{2}\right)^3$ **60.** $\left(\dfrac{1}{3}\right)^5$ **61.** $\left(\dfrac{a}{b}\right)^3$ **62.** $\left(\dfrac{r}{t}\right)^4$

63. $\left(\dfrac{x}{2}\right)^3$ **64.** $\left(\dfrac{y}{3}\right)^4$ **65.** $\left(-\dfrac{2x}{y}\right)^5$ **66.** $\left(-\dfrac{4p}{q}\right)^3$

Simplify each expression. ***See Example 7.***

67. $\left(\dfrac{5}{2}\right)^3 \cdot \left(\dfrac{5}{2}\right)^2$ **68.** $\left(\dfrac{3}{4}\right)^5 \cdot \left(\dfrac{3}{4}\right)^6$ **▶ 69.** $\left(\dfrac{9}{8}\right)^3 \cdot 9^2$

70. $\left(\dfrac{8}{5}\right)^4 \cdot 8^3$ **71.** $(2x)^9(2x)^3$ **72.** $(6y)^5(6y)^8$

73. $(-6p)^4(-6p)$ **74.** $(-13q)^3(-13q)$ **75.** $(6x^2y^3)^5$

76. $(5r^5t^6)^7$ **77.** $(x^2)^3(x^3)^5$ **78.** $(y^4)^5(y^3)^5$

79. $(2w^2x^3y)^2(x^4y)^5$ **80.** $(3x^4y^2z)^3(yz^4)^5$ **▶ 81.** $(-r^4s)^2(-r^2s^3)^5$

82. $(-ts^6)^4(-t^3s^5)^3$ **83.** $\left(\dfrac{5a^2b^5}{c^6}\right)^3 \ \ (c \neq 0)$ **84.** $\left(\dfrac{6x^3y^9}{z^5}\right)^4 \ \ (z \neq 0)$

85. *Concept Check*   A student wrote the following as a simplification of $(10^2)^3$.

$$1000^6$$

WHAT WENT WRONG?

86. *Concept Check*   A student wrote the following as a simplification of $(3x^2y^3)^4$.

$$3 \cdot 4x^8y^{12}, \quad \text{or} \quad 12x^8y^{12}$$

WHAT WENT WRONG?

Find an expression that represents the area of each figure. ***See Example 8.*** *(If necessary, refer to the formulas found inside the back cover of this book. The ⌐ in the figures indicates 90° angles.)*

▶ 87.

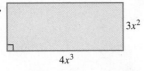

$3x^2$

$4x^3$

88.

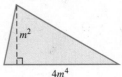

m^2

$4m^4$

89.

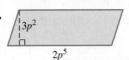

$3p^2$

$2p^5$

90.

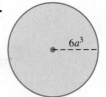

$6a^3$

Find an expression that represents the volume of each figure. (If necessary, refer to the formulas found inside the back cover of this book.)

91.

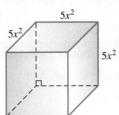

$5x^2$

$5x^2$

$5x^2$

92.

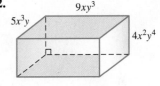

$9xy^3$

$5x^3y$

$4x^2y^4$

Compound interest is interest paid on both the principal and the interest earned earlier. The formula for compound interest, which involves an exponential expression, is

$$A = P(1 + r)^n,$$

where A is the amount accumulated from a principal of P dollars left untouched for n years with an annual interest rate r (expressed as a decimal).

In Exercises 93–96, use the preceding formula and a calculator to find A to the nearest cent.

93. $P = \$250$, $r = 0.04$, $n = 5$ **94.** $P = \$400$, $r = 0.04$, $n = 3$

95. $P = \$1500$, $r = 0.015$, $n = 6$ **96.** $P = \$2000$, $r = 0.015$, $n = 4$

5.2   Integer Exponents and the Quotient Rule

OBJECTIVES

1   Use 0 as an exponent.

2   Use negative numbers as exponents.

3   Use the quotient rule for exponents.

4   Use combinations of the rules for exponents.

Consider the following list.

$$2^4 = 16$$
$$2^3 = 8$$
$$2^2 = 4$$

As exponents decrease by 1, the results are divided by 2 each time.

Each time we decrease the exponent by 1, the value is divided by 2 (the base). Using this pattern, we can continue the list to lesser and lesser integer exponents.

$$2^1 = 2$$
$$2^0 = 1$$
$$2^{-1} = \frac{1}{2}$$
$$2^{-2} = \frac{1}{4}$$
$$2^{-3} = \frac{1}{8}$$

We continue the pattern here.

From the preceding list, it appears that we should define 2^0 as 1 and bases raised to negative exponents as reciprocals of those bases.

OBJECTIVE 1   Use 0 as an exponent.

The definitions of 0 and negative exponents must be consistent with the rules for exponents from **Section 5.1.** For example, if we define 6^0 to be 1, then

$$6^0 \cdot 6^2 = 1 \cdot 6^2 = 6^2 \quad \text{and} \quad 6^0 \cdot 6^2 = 6^{0+2} = 6^2,$$

and we see that the product rule is satisfied. Check that the power rules are also valid for a 0 exponent. Thus, we define a 0 exponent as follows.

Zero Exponent

For any nonzero real number a, $a^0 = 1.$

Example: $17^0 = 1$

**NOW TRY
EXERCISE 1**

Evaluate.

(a) 6^0

(b) -12^0

(c) $(-12x)^0$ $(x \neq 0)$

(d) $14^0 - 12^0$

EXAMPLE 1 Using Zero Exponents

Evaluate.

(a) $60^0 = 1$

(b) $(-60)^0 = 1$

(c) $-60^0 = -(1) = -1$

(d) $y^0 = 1$ $(y \neq 0)$

(e) $6y^0 = 6(1) = 6$ $(y \neq 0)$

(f) $(6y)^0 = 1$ $(y \neq 0)$

(g) $8^0 + 11^0 = 1 + 1 = 2$

(h) $-8^0 - 11^0 = -1 - 1 = -2$

NOW TRY

⚠️ **CAUTION** Look again at **Examples 1(b) and 1(c).** In $(-60)^0$, the base is -60, and because any nonzero base raised to the 0 exponent is 1, $(-60)^0 = 1$. In -60^0, which can be written $-(60)^0$, the base is 60, so $-60^0 = -1$.

OBJECTIVE 2 Use negative numbers as exponents.

From the lists at the beginning of this section, $2^{-2} = \frac{1}{4}$ and $2^{-3} = \frac{1}{8}$. We can make a conjecture that 2^{-n} should equal $\frac{1}{2^n}$. Is the product rule valid in such cases? For example,

$$6^{-2} \cdot 6^2 = 6^{-2+2} = 6^0 = 1.$$

The expression 6^{-2} behaves as if it were the reciprocal of 6^2, because their product is 1. The reciprocal of 6^2 is also $\frac{1}{6^2}$, leading us to define 6^{-2} as $\frac{1}{6^2}$, and generalize accordingly.

Negative Exponents

For any nonzero real number a and any integer n, $\quad a^{-n} = \dfrac{1}{a^n}.$

Example: $3^{-2} = \dfrac{1}{3^2}$

By definition, a^{-n} and a^n are reciprocals.

$$a^n \cdot a^{-n} = a^n \cdot \frac{1}{a^n} = 1$$

Because $1^n = 1$, the definition of a^{-n} can also be written as follows.

$$a^{-n} = \frac{1}{a^n} = \frac{1^n}{a^n} = \left(\frac{1}{a}\right)^n$$

For example, $\quad 6^{-3} = \left(\dfrac{1}{6}\right)^3 \quad$ and $\quad \left(\dfrac{1}{3}\right)^{-2} = 3^2.$

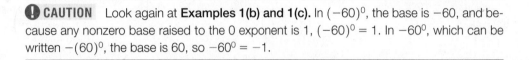

NOW TRY ANSWERS
1. (a) 1 (b) -1 (c) 1 (d) 0

**NOW TRY
EXERCISE 2**

Write with positive exponents and simplify.

(a) 2^{-3} **(b)** $\left(\dfrac{1}{7}\right)^{-2}$

(c) $\left(\dfrac{3}{2}\right)^{-4}$ **(d)** $3^{-2} + 4^{-2}$

(e) p^{-4} $(p \neq 0)$

EXAMPLE 2   **Using Negative Exponents**

Write with positive exponents and simplify. Assume that all variables represent non-zero real numbers.

(a) $3^{-2} = \dfrac{1}{3^2} = \dfrac{1}{9}$ $a^{-n} = \frac{1}{a^n}$ **(b)** $5^{-3} = \dfrac{1}{5^3} = \dfrac{1}{125}$ $a^{-n} = \frac{1}{a^n}$

(c) $\left(\dfrac{1}{2}\right)^{-3} = 2^3 = 8$ $\frac{1}{2}$ and 2 are reciprocals.
(Reciprocals have a product of 1.)

　　Notice that we can change the base to its reciprocal if we also change the sign of the exponent.

(d) $\left(\dfrac{2}{5}\right)^{-4}$

$= \left(\dfrac{5}{2}\right)^4$　$\frac{2}{5}$ and $\frac{5}{2}$ are reciprocals.

$= \dfrac{5^4}{2^4}$　Power rule (c)

$= \dfrac{625}{16}$　Apply the exponents.

(e) $\left(\dfrac{4}{3}\right)^{-5}$

$= \left(\dfrac{3}{4}\right)^5$　$\frac{4}{3}$ and $\frac{3}{4}$ are reciprocals.

$= \dfrac{3^5}{4^5}$　Power rule (c)

$= \dfrac{243}{1024}$　Apply the exponents.

(f) $4^{-1} - 2^{-1}$

$= \dfrac{1}{4} - \dfrac{1}{2}$　Apply the exponents.

$= \dfrac{1}{4} - \dfrac{2}{4}$　Find a common denominator.

$= -\dfrac{1}{4}$　Subtract.

(g) $3p^{-2}$

$= \dfrac{3}{1} \cdot \dfrac{1}{p^2}$　$a^{-n} = \frac{1}{a^n}$

$= \dfrac{3}{p^2}$　Multiply.

(h) $\dfrac{1}{x^{-4}}$

$= \dfrac{1^{-4}}{x^{-4}}$　$1^n = 1$, for any integer n.

$= \left(\dfrac{1}{x}\right)^{-4}$　Power rule (c)

$= x^4$　$\frac{1}{x}$ and x are reciprocals.

(i) $x^3 y^{-4}$

$= \dfrac{x^3}{1} \cdot \dfrac{1}{y^4}$　$a^{-n} = \frac{1}{a^n}$

$= \dfrac{x^3}{y^4}$　Multiply.

In general,　$\dfrac{1}{a^{-n}} = a^n.$　**NOW TRY**

⚠ **CAUTION**　*A negative exponent does not indicate a negative number. Negative exponents lead to reciprocals.*

Expression	Example	
a^{-n}	$3^{-2} = \dfrac{1}{3^2} = \dfrac{1}{9}$	Not negative
$-a^{-n}$	$\uparrow\,-3^{-2} = \,\uparrow-\dfrac{1}{3^2} = \,\uparrow-\dfrac{1}{9}$	Negative

Consider the following.

$$\frac{2^{-3}}{3^{-4}}$$

$$= \frac{\dfrac{1}{2^3}}{\dfrac{1}{3^4}} \qquad \text{Definition of negative exponent}$$

$$= \frac{1}{2^3} \div \frac{1}{3^4} \qquad \frac{a}{b} \text{ means } a \div b.$$

$$= \frac{1}{2^3} \cdot \frac{3^4}{1} \qquad \begin{array}{l} \text{To divide, multiply by the} \\ \text{reciprocal of the divisor.} \end{array}$$

$$= \frac{3^4}{2^3} \qquad \text{Multiply.}$$

Therefore, $\qquad \dfrac{2^{-3}}{3^{-4}} = \dfrac{3^4}{2^3}.$

Changing from Negative to Positive Exponents

For any nonzero numbers a and b and any integers m and n, the following hold.

$$\frac{a^{-m}}{b^{-n}} = \frac{b^n}{a^m} \quad \text{and} \quad \left(\frac{a}{b}\right)^{-m} = \left(\frac{b}{a}\right)^m$$

Examples: $\quad \dfrac{3^{-5}}{2^{-4}} = \dfrac{2^4}{3^5} \quad \text{and} \quad \left(\dfrac{4}{5}\right)^{-3} = \left(\dfrac{5}{4}\right)^3$

NOW TRY EXERCISE 3

Write with positive exponents and simplify. Assume that all variables represent nonzero real numbers.

(a) $\dfrac{5^{-3}}{6^{-2}}$ (b) $m^2 n^{-4}$

(c) $\dfrac{x^2 y^{-3}}{5z^{-4}}$

EXAMPLE 3 Changing from Negative to Positive Exponents

Write with positive exponents and simplify. Assume that all variables represent non-zero real numbers.

(a) $\dfrac{4^{-2}}{5^{-3}} = \dfrac{5^3}{4^2} = \dfrac{125}{16}$

(b) $\dfrac{m^{-5}}{p^{-1}} = \dfrac{p^1}{m^5} = \dfrac{p}{m^5}$

(c) $\dfrac{a^{-2}b}{3d^{-3}} = \dfrac{bd^3}{3a^2} \qquad \begin{array}{l}\text{Notice that } b \text{ in the numerator and 3 in} \\ \text{the denominator are not affected.}\end{array}$

(d) $\left(\dfrac{x}{2y}\right)^{-4}$

$\quad = \left(\dfrac{2y}{x}\right)^4 \qquad \text{Negative-to-positive rule}$

$\quad = \dfrac{2^4 y^4}{x^4} \qquad \text{Power rules (b) and (c)}$

$\quad = \dfrac{16y^4}{x^4} \qquad \text{Apply the exponent.}$

NOW TRY

NOW TRY ANSWERS

3. (a) $\dfrac{36}{125}$ (b) $\dfrac{m^2}{n^4}$ (c) $\dfrac{x^2 z^4}{5y^3}$

⚠ CAUTION   Be careful. We cannot use the rule $\dfrac{a^{-m}}{b^{-n}} = \dfrac{b^n}{a^m}$ to change negative exponents to positive exponents if the exponents occur in a *sum* or *difference* of terms. For example,

$$\frac{5^{-2} + 3^{-1}}{7 - 2^{-3}} \quad \text{would be written with positive exponents as} \quad \frac{\dfrac{1}{5^2} + \dfrac{1}{3}}{7 - \dfrac{1}{2^3}}.$$

OBJECTIVE 3   Use the quotient rule for exponents.

Consider the following.

$$\frac{6^5}{6^3} = \frac{6 \cdot 6 \cdot 6 \cdot 6 \cdot 6}{6 \cdot 6 \cdot 6} = 6^2$$

The difference of the exponents, $5 - 3 = 2$, is the exponent in the quotient.

Also,

$$\frac{6^2}{6^4} = \frac{6 \cdot 6}{6 \cdot 6 \cdot 6 \cdot 6} = \frac{1}{6^2} = 6^{-2}.$$

Here, $2 - 4 = -2$. These examples suggest the **quotient rule for exponents.**

Quotient Rule for Exponents

For any nonzero real number a and any integers m and n,

$$\frac{a^m}{a^n} = a^{m-n}.$$

(Keep the same base and subtract the exponents.)

Example: $\dfrac{5^8}{5^4} = 5^{8-4} = 5^4$

⚠ CAUTION   A common **error** is to write $\dfrac{5^8}{5^4} = 1^{8-4} = 1^4$. ***This is incorrect.*** By the quotient rule, the quotient must have the *same base*, 5, just as in the product rule.

$$\frac{5^8}{5^4} = 5^{8-4} = 5^4$$

We can confirm this by writing out the factors.

$$\frac{5^8}{5^4} = \frac{5 \cdot 5 \cdot 5 \cdot 5 \cdot 5 \cdot 5 \cdot 5 \cdot 5}{5 \cdot 5 \cdot 5 \cdot 5} = 5^4$$

EXAMPLE 4   **Using the Quotient Rule**

Simplify. Assume that all variables represent nonzero real numbers.

(a) $\dfrac{5^8}{5^6} = 5^{8-6} = 5^2 = 25$

Keep the same base.

(b) $\dfrac{4^2}{4^5} = 4^{2-5} = 4^{-3} = \dfrac{1}{4^3} = \dfrac{1}{64}$

**NOW TRY
EXERCISE 4**

Simplify. Assume that all variables represent nonzero real numbers.

(a) $\dfrac{6^3}{6^4}$ **(b)** $\dfrac{t^4}{t^{-5}}$

(c) $\dfrac{(p+q)^{-3}}{(p+q)^{-7}}$ $(p \neq -q)$

(d) $\dfrac{5^2xy^{-3}}{3^{-1}x^{-2}y^2}$

(c) $\dfrac{5^{-3}}{5^{-7}} = 5^{-3-(-7)} = 5^4 = 625$

Be careful with signs.

(d) $\dfrac{q^5}{q^{-3}} = q^{5-(-3)} = q^8$

(e) $\dfrac{3^2x^5}{3^4x^3}$

$= \dfrac{3^2}{3^4} \cdot \dfrac{x^5}{x^3}$

$= 3^{2-4} \cdot x^{5-3}$ Quotient rule

$= 3^{-2}x^2$ Subtract.

$= \dfrac{x^2}{3^2}$ Definition of negative exponent

$= \dfrac{x^2}{9}$ Apply the exponent.

(f) $\dfrac{(m+n)^{-2}}{(m+n)^{-4}}$

$= (m+n)^{-2-(-4)}$

$= (m+n)^{-2+4}$

$= (m+n)^2$ $(m \neq -n)$

The restriction $m \neq -n$ is necessary to prevent a denominator of 0 in the original expression. Division by 0 is undefined.

(g) $3x^{-5}$

Avoid the error of applying −5 to 3.

$= 3 \cdot \dfrac{1}{x^5}$ −5 applies *only* to x.

$= \dfrac{3}{x^5}$ Multiply.

(h) $\dfrac{7x^{-3}y^2}{2^{-1}x^2y^{-5}}$

$= \dfrac{7 \cdot 2^1y^2y^5}{x^2x^3}$ Negative-to-positive rule

$= \dfrac{14y^7}{x^5}$ Multiply; product rule

NOW TRY

The definitions and rules for exponents are summarized here.

Definitions and Rules for Exponents

For any integers m and n, the following hold.   **Examples**

Product rule	$a^m \cdot a^n = a^{m+n}$	$7^4 \cdot 7^5 = 7^{4+5} = 7^9$
Zero exponent	$a^0 = 1$ (where $a \neq 0$)	$(-3)^0 = 1$
Negative exponent	$a^{-n} = \dfrac{1}{a^n}$ (where $a \neq 0$)	$5^{-3} = \dfrac{1}{5^3}$
Quotient rule	$\dfrac{a^m}{a^n} = a^{m-n}$ (where $a \neq 0$)	$\dfrac{2^2}{2^5} = 2^{2-5} = 2^{-3} = \dfrac{1}{2^3}$
Power rules (a)	$(a^m)^n = a^{mn}$	$(4^2)^3 = 4^{2 \cdot 3} = 4^6$
(b)	$(ab)^m = a^mb^m$	$(3k)^4 = 3^4k^4$
(c)	$\left(\dfrac{a}{b}\right)^m = \dfrac{a^m}{b^m}$ (where $b \neq 0$)	$\left(\dfrac{2}{3}\right)^2 = \dfrac{2^2}{3^2}$
Negative-to-positive rules	$\dfrac{a^{-m}}{b^{-n}} = \dfrac{b^n}{a^m}$ (where $a \neq 0$, $b \neq 0$)	$\dfrac{2^{-4}}{5^{-3}} = \dfrac{5^3}{2^4}$
	$\left(\dfrac{a}{b}\right)^{-m} = \left(\dfrac{b}{a}\right)^m$	$\left(\dfrac{4}{7}\right)^{-2} = \left(\dfrac{7}{4}\right)^2$

NOW TRY ANSWERS

4. **(a)** $\dfrac{1}{6}$ **(b)** t^9 **(c)** $(p+q)^4$

   (d) $\dfrac{75x^3}{y^5}$

OBJECTIVE 4 Use combinations of the rules for exponents.

EXAMPLE 5 Using Combinations of Rules

Simplify. Assume that all variables represent nonzero real numbers.

(a) $\dfrac{(4^2)^3}{4^5}$

$= \dfrac{4^6}{4^5}$ Power rule (a)

$= 4^{6-5}$ Quotient rule

$= 4^1$ Subtract.

$= 4$ $a^1 = a$, for all a.

(b) $(2x)^3(2x)^2$

$= (2x)^5$ Product rule

$= 2^5 x^5$ Power rule (b)

$= 32x^5$ $2^5 = 32$

(c) $\left(\dfrac{2x^3}{5}\right)^{-4}$

$= \left(\dfrac{5}{2x^3}\right)^4$ Negative-to-positive rule

$= \dfrac{5^4}{2^4 x^{12}}$ Power rules (a)–(c)

$= \dfrac{625}{16x^{12}}$ Apply the exponents.

(d) $\left(\dfrac{3x^{-2}}{4^{-1}y^3}\right)^{-3}$

$= \dfrac{3^{-3}x^6}{4^3 y^{-9}}$ Power rules (a)–(c)

$= \dfrac{x^6 y^9}{4^3 \cdot 3^3}$ Negative-to-positive rule

$= \dfrac{x^6 y^9}{1728}$ $4^3 \cdot 3^3 = 64 \cdot 27$ $= 1728$

(e) $\dfrac{(4m)^{-3}}{(3m)^{-4}}$

$= \dfrac{4^{-3}m^{-3}}{3^{-4}m^{-4}}$ Power rule (b)

$= \dfrac{3^4 m^4}{4^3 m^3}$ Negative-to-positive rule

$= \dfrac{3^4 m^{4-3}}{4^3}$ Quotient rule

$= \dfrac{3^4 m^1}{4^3}$ Subtract.

$= \dfrac{81m}{64}$ Apply the exponents.

(f) $\dfrac{(7y)^{-3}(7y)^4}{(7y)^{12}(7y)^{-10}}$

$= \dfrac{(7y)^{-3+4}}{(7y)^{12+(-10)}}$ Product rule

$= \dfrac{(7y)^1}{(7y)^2}$ Add.

$= (7y)^{1-2}$ Quotient rule

$= (7y)^{-1}$ Subtract.

$= \dfrac{1}{7y}$ $a^{-1} = \frac{1}{a}$, for $a \neq 0$.

NOW TRY ↻

NOW TRY EXERCISE 5

Simplify. Assume that all variables represent nonzero real numbers.

(a) $\dfrac{3^{15}}{(3^3)^4}$ **(b)** $(4t)^5(4t)^{-3}$

(c) $\left(\dfrac{7y^4}{10}\right)^{-3}$ **(d)** $\dfrac{(a^2 b^{-2}c)^{-3}}{(2ab^3 c^{-4})^5}$

(e) $\dfrac{(5k)^{-6}(5k)^8}{(5k)^7(5k)^{-4}}$

NOW TRY ANSWERS

5. **(a)** 27   **(b)** $16t^2$ **(c)** $\dfrac{1000}{343y^{12}}$

   (d) $\dfrac{c^{17}}{32a^{11}b^9}$ **(e)** $\dfrac{1}{5k}$

NOTE Because the steps can be done in several different orders, there are many equally correct ways to simplify expressions like those in **Examples 5(c)–5(f).**

5.2 Exercises

FOR EXTRA HELP   ▶ MyMathLab®

▶ *Complete solution available in MyMathLab*

Concept Check *Decide whether each expression is* positive, negative, *or* 0.

1. $(-2)^{-3}$ **2.** $(-3)^{-2}$ **3.** -2^4 **4.** -3^6

5. $\left(\dfrac{1}{4}\right)^{-2}$ **6.** $\left(\dfrac{1}{5}\right)^{-2}$ **7.** $1 - 5^0$ **8.** $1 - 7^0$

Decide whether each expression is equal to 0, 1, *or* −1. *See Example 1.*

▶ **9.** 9^0

10. 3^0

11. $(-2)^0$

12. $(-12)^0$

13. -8^0

14. -6^0

15. $-(-6)^0$

16. $-(-13)^0$

17. $(-4)^0 - 4^0$

18. $(-11)^0 - 11^0$

19. $\dfrac{0^{10}}{12^0}$

20. $\dfrac{0^5}{2^0}$

21. $8^0 - 12^0$

22. $6^0 - 13^0$

23. $\dfrac{0^2}{2^0 + 0^2}$

24. $\dfrac{2^0}{0^2 + 2^0}$

Concept Check   In Exercises 25 and 26, match each expression in Column I with the equivalent expression in Column II. Choices in Column II may be used once, more than once, or not at all. (In Exercise 25, $x \neq 0$.)

	I	**II**
25.	(a) x^0	**A.** 0
	(b) $-x^0$	**B.** 1
	(c) $7x^0$	**C.** −1
	(d) $(7x)^0$	**D.** 7
	(e) $-7x^0$	**E.** −7
	(f) $(-7x)^0$	**F.** $\dfrac{1}{7}$

	I	**II**
26.	(a) -2^{-4}	**A.** 8
	(b) $(-2)^{-4}$	**B.** 16
	(c) 2^{-4}	**C.** $-\dfrac{1}{16}$
	(d) $\dfrac{1}{2^{-4}}$	**D.** −8
	(e) $\dfrac{1}{-2^{-4}}$	**E.** −16
	(f) $\dfrac{1}{(-2)^{-4}}$	**F.** $\dfrac{1}{16}$

Evaluate each expression. See Examples 1 and 2.

27. $6^0 + 8^0$

28. $4^0 + 2^0$

▶ **29.** 4^{-3}

30. 5^{-4}

31. $\left(\dfrac{1}{2}\right)^{-4}$

32. $\left(\dfrac{1}{3}\right)^{-3}$

33. $\left(\dfrac{6}{7}\right)^{-2}$

34. $\left(\dfrac{2}{3}\right)^{-3}$

35. $(-3)^{-4}$

36. $(-4)^{-3}$

37. $5^{-1} + 3^{-1}$

38. $6^{-1} + 2^{-1}$

39. $3^{-2} - 2^{-1}$

40. $6^{-2} - 3^{-1}$

41. $\left(\dfrac{1}{2}\right)^{-1} + \left(\dfrac{2}{3}\right)^{-1}$

42. $\left(\dfrac{1}{3}\right)^{-1} + \left(\dfrac{4}{3}\right)^{-1}$

Concept Check   Simplify each expression.

43. $\dfrac{5^{11}}{5^8}$

$= 5^{—-—}$

$= 5^{—}$

$= \underline{\quad}$

44. $\dfrac{6^{-5}}{6^{-2}}$

$= 6^{—-(—)}$

$= 6^{—}$

$= \dfrac{1}{6^{—}}$

$= \underline{\quad}$

Simplify each expression. Assume that all variables represent nonzero real numbers and no denominators are zero. See Examples 2–4.

▶ **45.** $\dfrac{5^8}{5^5}$

46. $\dfrac{11^6}{11^3}$

▶ **47.** $\dfrac{3^{-2}}{5^{-3}}$

48. $\dfrac{4^{-3}}{3^{-2}}$

49. $\dfrac{5}{5^{-1}}$

50. $\dfrac{6}{6^{-2}}$

51. $\dfrac{x^{12}}{x^{-3}}$

52. $\dfrac{y^4}{y^{-6}}$

53. $\dfrac{1}{6^{-3}}$

54. $\dfrac{1}{5^{-2}}$

55. $\dfrac{2}{r^{-4}}$

56. $\dfrac{3}{s^{-8}}$

57. $\dfrac{4^{-3}}{5^{-2}}$

58. $\dfrac{6^{-2}}{5^{-4}}$

59. $p^5 q^{-8}$

60. $x^{-8} y^4$

61. $\dfrac{r^5}{r^{-4}}$

62. $\dfrac{a^6}{a^{-4}}$

63. $\dfrac{x^{-3} y}{4z^{-2}}$

64. $\dfrac{p^{-5} q^{-4}}{9r^{-3}}$

65. $\dfrac{(a+b)^{-3}}{(a+b)^{-4}}$

66. $\dfrac{(x+y)^{-8}}{(x+y)^{-9}}$

67. $\dfrac{(x+2y)^{-3}}{(x+2y)^{-5}}$

68. $\dfrac{(p-3q)^{-2}}{(p-3q)^{-4}}$

Simplify each expression. Assume that all variables represent nonzero real numbers. **See Example 5.**

69. $\dfrac{(7^4)^3}{7^9}$

70. $\dfrac{(5^3)^2}{5^2}$

71. $x^{-3} \cdot x^5 \cdot x^{-4}$

72. $y^{-8} \cdot y^5 \cdot y^{-2}$

73. $\dfrac{(3x)^{-2}}{(4x)^{-3}}$

74. $\dfrac{(2y)^{-3}}{(5y)^{-4}}$

$\blacktriangleright$ **75.** $\left(\dfrac{x^{-1}y}{z^2}\right)^{-2}$

76. $\left(\dfrac{p^{-4}q}{r^{-3}}\right)^{-3}$

77. $(6x)^4 (6x)^{-3}$

78. $(10y)^9 (10y)^{-8}$

79. $\dfrac{(m^7 n)^{-2}}{m^{-4} n^3}$

80. $\dfrac{(m^{-8} n^{-4})^2}{m^2 n^5}$

81. $\dfrac{(x^{-1} y^2 z)^{-2}}{(x^{-3} y^3 z)^{-1}}$

82. $\dfrac{(a^2 b^3 c^4)^{-4}}{(a^{-2} b^{-3} c^{-4})^{-5}}$

83. $\left(\dfrac{xy^{-2}}{x^2 y}\right)^{-3}$

84. $\left(\dfrac{wz^{-5}}{w^{-3} z}\right)^{-2}$

85. $\dfrac{(2r)^{-4} (2r)^5}{(2r)^9 (2r)^{-7}}$

86. $\dfrac{(8x)^{-8} (8x)^9}{(8x)^{13} (8x)^{-11}}$

87. $\dfrac{(-4y)^8 (-4y)^{-8}}{(-4y)^{-26} (-4y)^{27}}$

88. $\dfrac{(-9p)^{16} (-9p)^{-16}}{(-9p)^{-41} (-9p)^{42}}$

89. *Concept Check* A student simplified $\dfrac{16^3}{2^2}$ as shown.

$$\dfrac{16^3}{2^2} = \left(\dfrac{16}{2}\right)^{3-2} = 8^1 = 8$$

WHAT WENT WRONG? Give the correct answer.

90. *Concept Check* A student simplified -5^4 as shown.

$$-5^4 = (-5^4) = 625$$

WHAT WENT WRONG? Give the correct answer.

Extending Skills Simplify each expression. Assume that all variables represent nonzero real numbers.

91. $\dfrac{(4a^2 b^3)^{-2} (2ab^{-1})^3}{(a^3 b)^{-4}}$

92. $\dfrac{(m^6 n)^{-2} (m^2 n^{-2})^3}{m^{-1} n^{-2}}$

93. $\dfrac{(2y^{-1} z^2)^2 (3y^{-2} z^{-3})^3}{(y^3 z^2)^{-1}}$

94. $\dfrac{(3p^{-2} q^3)^2 (5p^{-1} q^{-4})^{-1}}{(p^2 q^{-2})^{-3}}$

95. $\dfrac{(9^{-1} z^{-2} x)^{-1} (4z^2 x^4)^{-2}}{(5z^{-2} x^{-3})^2}$

96. $\dfrac{(4^{-1} a^{-1} b^{-2})^{-2} (5a^{-3} b^4)^{-2}}{(3a^{-3} b^{-5})^2}$

SUMMARY EXERCISES   Applying the Rules for Exponents

Simplify each expression. Use only positive exponents in the answers. Assume that all variables represent nonzero real numbers.

1. $10^5x^7y^{14}$ **2.** $-128a^{10}b^{15}c^4$

3. $\dfrac{729w^3x^9}{y^{12}}$ **4.** $\dfrac{x^4y^6}{16}$

5. c^{22} **6.** $\dfrac{1}{k^4t^{12}}$

7. $\dfrac{11}{30}$ **8.** $y^{12}z^3$

9. $\dfrac{x^6}{y^5}$ **10.** 0

11. $\dfrac{1}{z^2}$ **12.** $\dfrac{9}{r^2s^2t^{10}}$

13. $\dfrac{300x^3}{y^3}$ **14.** $\dfrac{3}{5x^6}$

15. x^8 **16.** $\dfrac{y^{11}}{x^{11}}$

17. $\dfrac{a^6}{b^4}$ **18.** $6ab$

19. $\dfrac{61}{900}$ **20.** 1

21. $\dfrac{343a^6b^9}{8}$ **22.** 1

23. -1 **24.** 0

25. $\dfrac{27y^{18}}{4x^8}$ **26.** $\dfrac{1}{a^8b^{12}c^{16}}$

27. $\dfrac{x^{15}}{216z^9}$ **28.** $\dfrac{q}{8p^6r^3}$

29. x^6y^6 **30.** 0

31. $\dfrac{343}{x^{15}}$ **32.** $\dfrac{9}{x^6}$

33. $5p^{10}q^9$ **34.** $\dfrac{7}{24}$

35. $\dfrac{r^{14}t}{2s^2}$ **36.** 1

37. $8p^{10}q$ **38.** $\dfrac{1}{mn^3p^3}$

39. -1

40. (a) D   **(b)** D
(c) E   **(d)** B
(e) J   **(f)** F
(g) I   **(h)** B
(i) E   **(j)** F

1. $(10x^2y^4)^2(10xy^2)^3$ **2.** $(-2ab^3c)^4(-2a^2b)^3$ **3.** $\left(\dfrac{9wx^3}{y^4}\right)^3$

4. $(4x^{-2}y^{-3})^{-2}$ **5.** $\dfrac{c^{11}(c^2)^4}{(c^3)^3(c^2)^{-6}}$ **6.** $\left(\dfrac{k^4t^2}{k^2t^{-4}}\right)^{-2}$

7. $5^{-1}+6^{-1}$ **8.** $\dfrac{(3y^{-1}z^3)^{-1}(3y^2)}{(y^3z^2)^{-3}}$ **9.** $\dfrac{(2xy^{-1})^3}{2^3x^{-3}y^2}$

10. $-4^0+(-4)^0$ **11.** $(z^4)^{-3}(z^{-2})^{-5}$ **12.** $\left(\dfrac{r^2st^5}{3r}\right)^{-2}$

13. $\dfrac{(3^{-1}x^{-3}y)^{-1}(2x^2y^{-3})^2}{(5x^{-2}y^2)^{-2}}$ **14.** $\left(\dfrac{5x^2}{3x^{-4}}\right)^{-1}$ **15.** $\left(\dfrac{-9x^{-2}}{9x^2}\right)^{-2}$

16. $\dfrac{(x^{-4}y^2)^3(x^2y)^{-1}}{(xy^2)^{-3}}$ **17.** $\dfrac{(a^{-2}b^3)^{-4}}{(a^{-3}b^2)^{-2}(ab)^{-4}}$ **18.** $(2a^{-30}b^{-29})(3a^{31}b^{30})$

19. $5^{-2}+6^{-2}$ **20.** $\left(\dfrac{(x^{43}y^{23})^2}{x^{-26}y^{-42}}\right)^0$ **21.** $\left(\dfrac{7a^2b^3}{2}\right)^3$

22. $-(-19^0)$ **23.** $-(-13)^0$ **24.** $\dfrac{0^{13}}{13^0}$

25. $\dfrac{(2xy^{-3})^{-2}}{(3x^{-2}y^4)^{-3}}$ **26.** $\left(\dfrac{a^2b^3c^4}{a^{-2}b^{-3}c^{-4}}\right)^{-2}$ **27.** $(6x^{-5}z^3)^{-3}$

28. $(2p^{-2}qr^{-3})(2p)^{-4}$ **29.** $\dfrac{(xy)^{-3}(xy)^5}{(xy)^{-4}}$ **30.** $52^0-(-8)^0$

31. $\dfrac{(7^{-1}x^{-3})^{-2}(x^4)^{-6}}{7^{-1}x^{-3}}$ **32.** $\left(\dfrac{3^{-4}x^{-3}}{3^{-3}x^{-6}}\right)^{-2}$ **33.** $(5p^{-2}q)^{-3}(5pq^3)^4$

34. $8^{-1}+6^{-1}$ **35.** $\left(\dfrac{4r^{-6}s^{-2}t}{2r^8s^{-4}t^2}\right)^{-1}$ **36.** $(13x^{-6}y)(13x^{-6}y)^{-1}$

37. $\dfrac{(8pq^{-2})^4}{(8p^{-2}q^{-3})^3}$ **38.** $\left(\dfrac{mn^{-2}p}{m^2np^4}\right)^{-2}\left(\dfrac{mn^{-2}p}{m^2np^4}\right)^3$ **39.** $-(-8^0)^0$

40. *Concept Check* Match each expression (a)–(j) in Column I with the equivalent expression A–J in Column II. Choices in Column II may be used once, more than once, or not at all.

I

(a) 2^0+2^0 **(b)** $2^1\cdot2^0$

(c) 2^0-2^{-1} **(d)** 2^1-2^0

(e) $2^0\cdot2^{-2}$ **(f)** $2^1\cdot2^1$

(g) $2^{-2}-2^{-1}$ **(h)** $2^0\cdot2^0$

(i) $2^{-2}\div2^{-1}$ **(j)** $2^0\div2^{-2}$

II

A. 0 **B.** 1

C. -1 **D.** 2

E. $\dfrac{1}{2}$ **F.** 4

G. -2 **H.** -4

I. $-\dfrac{1}{4}$ **J.** $\dfrac{1}{4}$

5.3 Scientific Notation

OBJECTIVES

1. Express numbers in scientific notation.
2. Convert numbers in scientific notation to standard notation.
3. Use scientific notation in calculations.

VOCABULARY

☐ scientific notation
☐ standard notation

OBJECTIVE 1 Express numbers in scientific notation.

Numbers occurring in science are often extremely large (such as the distance from Earth to the sun, 93,000,000 mi) or extremely small (the wavelength of blue light, approximately 0.000000475 m). Because of the difficulty of working with many zeros, scientists often express such numbers with exponents using *scientific notation*.

Scientific Notation

A number is written in **scientific notation** when it is expressed in the form

$$a \times 10^n,$$

where $1 \le |a| < 10$ and n is an integer.

In scientific notation, there is *always* one nonzero digit before the decimal point.

$3.19 \times 10^1 = 3.19 \times 10 = 31.9$	Decimal point moves 1 place to the right.
$3.19 \times 10^2 = 3.19 \times 100 = 319.$	Decimal point moves 2 places to the right.
$3.19 \times 10^3 = 3.19 \times 1000 = 3190.$	Decimal point moves 3 places to the right.
$3.19 \times 10^{-1} = 3.19 \times 0.1 = 0.319$	Decimal point moves 1 place to the left.
$3.19 \times 10^{-2} = 3.19 \times 0.01 = 0.0319$	Decimal point moves 2 places to the left.
$3.19 \times 10^{-3} = 3.19 \times 0.001 = 0.00319$	Decimal point moves 3 places to the left.

NOTE In scientific notation, the multiplication cross $\times$ is commonly used.

A number in scientific notation is always written with the decimal point after the first nonzero digit and then multiplied by the appropriate power of 10. For example, 56,200 is written 5.62×10^4 because

$$56,200 = 5.62 \times 10,000 = 5.62 \times 10^4.$$

Other examples include the following.

42,000,000	is written	4.2×10^7,
0.000586	is written	5.86×10^{-4},
and 2,000,000,000	is written	2×10^9.

To write a number in scientific notation, follow these steps. For a negative number, follow these steps using the *absolute value* of the number. Then make the result negative.

Converting a Positive Number to Scientific Notation

Step 1 **Position the decimal point.** Place a caret, ^, to the right of the first nonzero digit, where the decimal point will be placed.

Step 2 **Determine the numeral for the exponent.** Count the number of digits from the decimal point to the caret. This number gives the absolute value of the exponent on 10.

Step 3 **Determine the sign for the exponent.** Decide whether multiplying by 10^n should make the result of Step 1 greater or less.

- The exponent should be positive to make the result greater.

- The exponent should be negative to make the result less.

NOW TRY
EXERCISE 1

Write each number in scientific notation.

(a) 12,600,000

(b) 0.00027

(c) −0.0000341

EXAMPLE 1 Using Scientific Notation

Write each number in scientific notation.

(a) 93,000,000

Step 1 Place a caret to the right of the 9 (the first nonzero digit) to mark the new location of the decimal point.

$$9_\wedge 3{,}000{,}000$$

Step 2 Count from the decimal point, which is understood to be after the last 0, to the caret.

$$9.3{,}000{,}000. \leftarrow \text{Decimal point}$$
Count 7 places.

Step 3 Because 9.3 is to be made greater, the exponent on 10 is positive.

$$93{,}000{,}000 = 9.3 \times 10^7$$

(b) $63{,}200{,}000{,}000 = 6.3200000000 = 6.32 \times 10^{10}$
10 places

(c) 0.00462

Move the decimal point to the right of the first nonzero digit, and count the number of places the decimal point was moved.

$$0.00462 \quad \text{3 places}$$

Because 0.00462 is *less* than 4.62, the exponent must be *negative*.

$$0.00462 = 4.62 \times 10^{-3}$$

(d) $-0.0000762 = -7.62 \times 10^{-5}$ Remember the negative sign. **NOW TRY**
5 places

NOTE When writing a positive number in scientific notation, think as follows.

- If the original number is "large," like 93,000,000, use a *positive* exponent on 10 because positive is greater than negative.

- If the original number is "small," like 0.00462, use a *negative* exponent on 10 because negative is less than positive.

NOW TRY ANSWERS
1. **(a)** 1.26×10^7
(b) 2.7×10^{-4}
(c) -3.41×10^{-5}

OBJECTIVE 2 Convert numbers in scientific notation to standard notation.

Multiplying a number by a positive power of 10 will make the number greater. Multiplying by a negative power of 10 will make the number less.

We refer to a number such as 475 as the **standard notation** of 4.75×10^2.

> **NOW TRY EXERCISE 2**
>
> Write each number in standard notation.
>
> **(a)** 5.71×10^4
>
> **(b)** 2.72×10^{-5}
>
> **(c)** -8.81×10^{-4}

EXAMPLE 2 Writing Numbers in Standard Notation

Write each number in standard notation.

(a) 6.2×10^3

Because the exponent is positive, we make 6.2 greater by moving the decimal point 3 places to the right. We attach two zeros.

$$6.2 \times 10^3 = 6.200 = 6200$$

(b) $4.283 \times 10^6 = 4.283000 = 4,283,000$ Move 6 places to the right. Attach zeros as necessary.

(c) $-7.04 \times 10^{-3} = -0.00704$ Move 3 places to the left.

The exponent tells the number of places and the direction in which the decimal point is moved.

NOW TRY

OBJECTIVE 3 Use scientific notation in calculations.

> **NOW TRY EXERCISE 3**
>
> Perform each calculation. Write answers in both scientific and standard notation.
>
> **(a)** $(6 \times 10^7)(7 \times 10^{-4})$
>
> **(b)** $\dfrac{18 \times 10^{-3}}{6 \times 10^4}$

EXAMPLE 3 Multiplying and Dividing with Scientific Notation

Perform each calculation.

(a)

$$(7 \times 10^3)(5 \times 10^4)$$

$$= (7 \times 5)(10^3 \times 10^4) \qquad \text{Commutative and associative properties}$$

$$= 35 \times 10^7 \qquad \text{Multiply. Use the product rule.}$$

> Don't stop. This number is *not* in scientific notation because 35 is not between 1 and 10.

$$= (3.5 \times 10^1) \times 10^7 \qquad \text{Write 35 in scientific notation.}$$

$$= 3.5 \times (10^1 \times 10^7) \qquad \text{Associative property}$$

$$= 3.5 \times 10^8 \qquad \text{Product rule}$$

$$= 350{,}000{,}000 \qquad \text{Write in standard notation.}$$

(b) $\dfrac{4 \times 10^{-5}}{2 \times 10^3}$

$$= \frac{4}{2} \times \frac{10^{-5}}{10^3}$$

$$= 2 \times 10^{-8} \qquad \text{Divide. Use the quotient rule.}$$

$$= 0.00000002 \qquad \text{Write in standard notation.}$$

NOW TRY

NOTE Multiplying or dividing numbers written in scientific notation may produce an answer in the form $a \times 10^0$. Because $10^0 = 1$, $a \times 10^0 = a$. For example,

$$(8 \times 10^{-4})(5 \times 10^4) = 40 \times 10^0 = 40. \qquad 10^0 = 1$$

Also, if $a = 1$, then $a \times 10^n = 10^n$. For example, we could write 1,000,000 as 10^6 instead of 1×10^6.

NOW TRY ANSWERS

2. (a) 57,100   **(b)** 0.0000272
   (c) −0.000881

3. (a) 4.2×10^4, or 42,000
   (b) 3×10^{-7}, or 0.0000003

**NOW TRY
EXERCISE 4**

See **Example 4.** About
how much would 8,000,000
nanometers measure in
inches?

EXAMPLE 4   Using Scientific Notation to Solve an Application

A *nanometer* is a very small unit of measure that is equivalent to about
0.00000003937 in. About how much would 700,000 nanometers measure in inches?
(*Source:* www.conversion-metric.org)

Write each number in scientific notation, and then multiply.

$$700{,}000(0.00000003937)$$

$$= (7 \times 10^5)(3.937 \times 10^{-8}) \qquad \text{Write in scientific notation.}$$

$$= (7 \times 3.937)(10^5 \times 10^{-8}) \qquad \text{Properties of real numbers}$$

$$= 27.559 \times 10^{-3} \qquad \text{Multiply. Use the product rule.}$$

Don't stop here. $\;= (2.7559 \times 10^1) \times 10^{-3} \qquad$ Write 27.559 in scientific notation.

$$= 2.7559 \times 10^{-2} \qquad \text{Product rule}$$

$$= 0.027559 \qquad \text{Write in standard notation.}$$

Thus, 700,000 nanometers would measure

$$2.7559 \times 10^{-2} \text{ in., or } 0.027559 \text{ in.} \qquad \text{NOW TRY} \; \; \circlearrowright$$

**NOW TRY
EXERCISE 5**

The land area of California is
approximately 1.6×10^5 mi²,
and the 2012 population of
California was approximately
3.8×10^7 people. Use this
information to estimate the
number of square miles per
California resident in 2012.
(*Source:* U.S. Census Bureau.)

EXAMPLE 5   Using Scientific Notation to Solve an Application

In 2013, the gross federal debt was about $\$1.7218 \times 10^{13}$ (which is more than
$17 trillion). The population of the United States was approximately 317 million that
year. About how much would each person have had to contribute in order to pay off
the federal debt? (*Source*: www.usgovernmentdebt.us; www.census.gov)

Write the population in scientific notation. Then divide to obtain the per person
contribution.

$$\frac{1.7218 \times 10^{13}}{317{,}000{,}000}$$

$$= \frac{1.7218 \times 10^{13}}{3.17 \times 10^8} \qquad \text{Write 317 million in scientific notation.}$$

$$= \frac{1.7218}{3.17} \times 10^5 \qquad \text{Quotient rule}$$

$$= 0.54315 \times 10^5 \qquad \text{Divide. Round to 5 decimal places.}$$

$$= 54{,}315 \qquad \text{Write in standard notation.}$$

Each person would have to pay about \$54,315.   NOW TRY $\circlearrowright$

NOW TRY ANSWERS
4. 3.1496×10^{-1} in., or 0.31496 in.
5. 4.2×10^{-3} mi², or 0.0042 mi²

5.3 Exercises

  MyMathLab®

▶ *Complete solution available
in MyMathLab*

Concept Check *Match each number written in scientific notation in Column I with the correct
choice from Column II. Not all choices in Column II will be used.*

I	II	I	II
1. (a) 4.6×10^{-4}	**A.** 46,000	**2. (a)** 1×10^9	**A.** 1 billion
(b) 4.6×10^4	**B.** 460,000	**(b)** 1×10^6	**B.** 100 million
(c) 4.6×10^5	**C.** 0.00046	**(c)** 1×10^8	**C.** 1 million
(d) 4.6×10^{-5}	**D.** 0.000046	**(d)** 1×10^{10}	**D.** 10 billion
	E. 4600		**E.** 100 billion

Concept Check   *Determine whether or not each number is written in scientific notation as defined in* **Objective 1.** *If it is not, write it as such.*

3. 4.56×10^4 **4.** 7.34×10^6 **5.** 5,600,000   **6.** 34,000

7. 0.8×10^2 **8.** 0.9×10^3 **9.** 0.004   **10.** 0.0007

11. *Concept Check*   Write each number in scientific notation.

(a) 63,000

The first nonzero digit is _____ . The decimal point should be moved _____ places.

$$63{,}000 = \underline{} \times 10^{\underline{}}$$

(b) 0.0571

The first nonzero digit is _____ . The decimal point should be moved _____ places.

$$0.0571 = \underline{} \times 10^{\underline{}}$$

12. *Concept Check*   Write each number in standard notation.

(a) 4.2×10^3

Move the decimal point _____ places to the _____ .

$$4.2 \times 10^3 = \underline{}$$

(b) 6.42×10^{-3}

Move the decimal point _____ places to the _____ .

$$6.42 \times 10^{-3} = \underline{}$$

Write each number in scientific notation. **See Example 1.**

▶ 13. 5,876,000,000   **14.** 9,994,000,000   **15.** 82,350   **16.** 78,330

17. 0.000007   **18.** 0.0000004   **19.** 0.00203   **20.** 0.0000578

21. −13,000,000   **22.** −25,000,000,000   **23.** −0.006   **24.** −0.01234

Write each number in standard notation. **See Example 2.**

▶ 25. 7.5×10^5 **26.** 8.8×10^6 **27.** 5.677×10^{12} **28.** 8.766×10^9

29. 1×10^{12} **30.** 1×10^7 **31.** 6.21×10^0 **32.** 8.56×10^0

33. 7.8×10^{-4} **34.** 8.9×10^{-5} **35.** 5.134×10^{-9} **36.** 7.123×10^{-10}

37. -4×10^{-3} **38.** -6×10^{-4} **39.** -8.1×10^5 **40.** -9.6×10^6

Perform the indicated operations. Write each answer **(a)** *in scientific notation and* **(b)** *in standard notation.* **See Example 3.**

41. $(2 \times 10^8)(3 \times 10^3)$ **42.** $(4 \times 10^7)(3 \times 10^3)$

▶ 43. $(5 \times 10^4)(3 \times 10^2)$ **44.** $(8 \times 10^5)(2 \times 10^3)$

45. $(3 \times 10^{-4})(-2 \times 10^8)$ **46.** $(4 \times 10^{-3})(-2 \times 10^7)$

47. $(6 \times 10^3)(4 \times 10^{-2})$ **48.** $(7 \times 10^5)(3 \times 10^{-4})$

49. $(9 \times 10^4)(7 \times 10^{-7})$ **50.** $(6 \times 10^4)(8 \times 10^{-8})$

51. $\dfrac{9 \times 10^{-5}}{3 \times 10^{-1}}$ **52.** $\dfrac{12 \times 10^{-4}}{4 \times 10^{-3}}$ **53.** $\dfrac{8 \times 10^3}{-2 \times 10^2}$

54. $\dfrac{15 \times 10^4}{-3 \times 10^3}$ **55.** $\dfrac{2.6 \times 10^{-3}}{2 \times 10^2}$ **56.** $\dfrac{9.5 \times 10^{-1}}{5 \times 10^3}$

57. $\dfrac{4 \times 10^5}{8 \times 10^2}$ **58.** $\dfrac{3 \times 10^9}{6 \times 10^5}$ **59.** $\dfrac{-4.5 \times 10^4}{1.5 \times 10^{-2}}$

60. $\dfrac{-7.2 \times 10^3}{6.0 \times 10^{-1}}$ **61.** $\dfrac{-8 \times 10^{-4}}{-4 \times 10^3}$ **62.** $\dfrac{-5 \times 10^{-6}}{-2 \times 10^2}$

Calculators can express numbers in scientific notation. The displays often use notation such as

$$5.4\text{E}3 \quad \text{to represent} \quad 5.4 \times 10^3.$$

Similarly, 5.4E–3 represents 5.4×10^{-3}. They can also perform operations with numbers entered in scientific notation.

Predict the display the calculator would give for the expression shown in each screen.

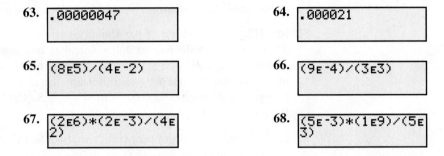

63. `.00000047`

64. `.000021`

65. `(8E5)/(4E-2)`

66. `(9E-4)/(3E3)`

67. `(2E6)*(2E-3)/(4E2)`

68. `(5E-3)*(1E9)/(5E3)`

Extending Skills *Use scientific notation to calculate the result in each expression. Write answers in scientific notation.*

69. $\dfrac{650{,}000{,}000(0.0000032)}{0.00002}$

70. $\dfrac{3{,}400{,}000{,}000(0.000075)}{0.00025}$

71. $\dfrac{0.00000072(0.00023)}{0.000000018}$

72. $\dfrac{0.000000081(0.000036)}{0.00000048}$

73. $\dfrac{0.0000016(240{,}000{,}000)}{0.00002(0.0032)}$

74. $\dfrac{0.000015(42{,}000{,}000)}{0.000009(0.000005)}$

Each statement contains a number in boldface italic type. If the number is in scientific notation, write it in standard notation. If the number is not in scientific notation, write it as such. **See Examples 1 and 2.**

75. A *muon* is an atomic particle closely related to an electron. According to the Web page *Muon basics,* the half-life of a muon is about 2 millionths (**2×10^{-6}**) of a second. (*Source:* www2.fisica.unlp.edu.ar)

76. There are 13 red balls and 39 black balls in a box. Mix them up and draw 13 out one at a time without returning any ball . . . the probability that the 13 drawings each will produce a red ball is . . . **1.6×10^{-12}**. (*Source:* Weaver, W., *Lady Luck.*)

77. An electron and a positron attract each other in two ways: the electromagnetic attraction of their opposite electric charges, and the gravitational attraction of their two masses. The electromagnetic attraction is

4,200,000,000,000,000,000,000,000,000,000,000,000,000

times as strong as the gravitational. (*Source:* Asimov, I., *Isaac Asimov's Book of Facts.*)

78. The name "googol" applies to the number

10,000,000,000,000,000,000,000,000,000,000,000,000,000,000,000,000, 000,000,000,000,000,000,000,000,000,000,000,000,000.

It was created by Edward Kasner and his nephew in 1938. The Web search engine Google honors this number. Sergey Brin, president and cofounder of Google, Inc., was a mathematics major. He chose the name Google to describe the vast reach of this search engine. (*Source: The Gazette.*)

*Use scientific notation to calculate the answer to each problem. **See Examples 3–5.***

79. The Double Helix Nebula, a conglomeration of dust and gas stretching across the center of the Milky Way galaxy, is 25,000 light-years from Earth. If one light-year is about 6,000,000,000,000 mi, about how many miles is the Double Helix Nebula from Earth? (*Source:* www.spitzer.caltech.edu)

80. Pollux, one of the brightest stars in the night sky, is 33.7 light-years from Earth. If one light-year is about 6,000,000,000,000 mi (that is, 6 trillion mi), about how many miles is Pollux from Earth? (*Source: World Almanac and Book of Facts.*)

81. In 2012, the population of the United States was about 313.9 million. To the nearest dollar, calculate how much each person in the United States would have had to contribute in order to make one person a trillionaire (that is, to give that person $1,000,000,000,000). (*Source:* U.S. Census Bureau.)

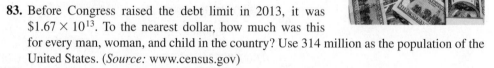

82. In 2011, the U.S. government collected about $4105 per person in individual income taxes. If the population at that time was 310,000,000, how much did the government collect in taxes for 2011? (*Source:* www.usgovernmentrevenue.com)

83. Before Congress raised the debt limit in 2013, it was 1.67×10^{13}. To the nearest dollar, how much was this for every man, woman, and child in the country? Use 314 million as the population of the United States. (*Source:* www.census.gov)

84. In 2010, the state of Minnesota had about 8.1×10^4 farms with an average of 3.44×10^2 acres per farm. What was the total number of acres devoted to farmland in Minnesota that year? (*Source:* U.S. Department of Agriculture.)

85. Light travels at a speed of 1.86×10^5 mi per sec. When Venus is 6.68×10^7 mi from the sun, how long (in seconds) does it take light to travel from the sun to Venus? (*Source: World Almanac and Book of Facts.*)

86. The distance to Earth from Pluto is 4.58×10^9 km. *Pioneer 10* transmitted radio signals from Pluto to Earth at the speed of light, 3.00×10^5 km per sec. About how long (in seconds) did it take for the signals to reach Earth?

87. During the 2012–2013 season, Broadway shows grossed a total of 1.14×10^9. Total attendance for the season was 1.16×10^7. What was the average ticket price (to the nearest cent) for a Broadway show? (*Source:* The Broadway League.)

88. In 2012, 1.08×10^{10} was spent to attend motion pictures in the United States and Canada. The total number of tickets sold was 1.36 billion. What was the average ticket price (to the nearest cent) for a movie? (*Source:* Motion Picture Association of America.)

89. In 2012, the world's fastest computer could perform 10,000,000,000,000,000 calculations per second. How many could it perform per minute? Per hour?
(*Source:* www.japantimes.co.jp)

90. In 2013, it was reported that the world's fastest computer could handle 33.86 quadrillion calculations per second. (*Hint:* 1 quadrillion $= 1 \times 10^{15}$) How many could it perform per minute? Per hour? (*Source*: www.top500.org)

RELATING CONCEPTS For Individual or Group Work (Exercises 91–94)

In 1935, Charles F. Richter devised a scale to compare the intensities of earthquakes. The *intensity* of an earthquake is measured relative to the intensity of a standard *zero-level* earthquake of intensity I_0. The relationship is equivalent to

$$I = I_0 \times 10^R, \quad \text{where } R \text{ is the } \textbf{Richter scale} \text{ measure.}$$

For example, if an earthquake has magnitude 5.0 on the Richter scale, then its intensity is calculated as

$$I = I_0 \times 10^{5.0} = I_0 \times 100,000,$$

which is 100,000 times as intense as a zero-level earthquake.

Intensity	$I_0 \times 10^0$	$I_0 \times 10^1$	$I_0 \times 10^2$	$I_0 \times 10^3$	$I_0 \times 10^4$	$I_0 \times 10^5$	$I_0 \times 10^6$	$I_0 \times 10^7$	$I_0 \times 10^8$
Richter Scale	0	1	2	3	4	5	6	7	8

To compare two earthquakes, such as one that measures 8.0 to one that measures 5.0, calculate the *ratio* of their intensities.

$$\frac{\text{intensity } 8.0}{\text{intensity } 5.0} = \frac{I_0 \times 10^{8.0}}{I_0 \times 10^{5.0}} = \frac{10^8}{10^5} = 10^{8-5} = 10^3 = 1000$$

An earthquake that measures 8.0 is 1000 time as intense as one that measures 5.0.

The table lists information for selected earthquakes with their years, locations, and magnitudes. **Work Exercises 91–94 in order.**

Year	Earthquake Location	Richter Scale Measurement
1960	Chile	9.5
1952	Kamchatka	9.0
2007	Southern Sumatra, Indonesia	8.5
2013	Obihoro, Japan	6.9
2002	Hindu Kush, Afghanistan	5.9

Source: earthquake.usgs.gov

91. Compare the intensity of the 1960 Chile earthquake with the 2007 Southern Sumatra earthquake.

92. Compare the intensity of the 2013 Obihoro earthquake with the 2002 Hindu Kush earthquake.

93. Compare the intensity of the 1952 Kamchatka earthquake with the 2007 Southern Sumatra earthquake. $\left(\textit{Hint: } 10^{0.5} = \sqrt{10}\right)$

94. Suppose an earthquake measures 7.5 on the Richter scale. How would the intensity of the 1960 Chile earthquake compare to it?

5.4 Adding, Subtracting, and Graphing Polynomials

VOCABULARY

☐ term
☐ leading term
☐ numerical coefficient (coefficient)
☐ like terms
☐ unlike terms
☐ polynomial
☐ descending powers
☐ degree of a term
☐ degree of a polynomial
☐ monomial
☐ binomial
☐ trinomial
☐ parabola
☐ vertex
☐ axis of symmetry (axis)

 NOW TRY
EXERCISE 1

Name the coefficient of each term in the expression. Give the number of terms.

$$t - 10t^2$$

OBJECTIVE 1 Identify terms and coefficients.

In an expression such as

$$4x^3 + 6x^2 + 5x + 8,$$

the quantities $4x^3$, $6x^2$, $5x$, and 8 are **terms.** (See **Section 1.7.**) In the **leading** (or first) **term** $4x^3$, the number 4 is the **numerical coefficient,** or simply the **coefficient,** of x^3. In the same way, 6 is the coefficient of x^2 in the term $6x^2$, and 5 is the coefficient of x in the term $5x$. The constant term 8 can be thought of as

$$8 \cdot 1 = 8x^0 \quad \text{because} \quad x^0 = 1,$$

so 8 is the coefficient in the term 8.

EXAMPLE 1 Identifying Coefficients

Name the coefficient of each term in each expression. Give the number of terms.

(a) $x - 6x^4 + 3$ can be written as $1x + (-6x^4) + 3x^0$.

The coefficients are 1, −6, and 3.

There are three terms: x, $-6x^4$, and 3.

(b) $5 - v^3$ can be written as $5v^0 + (-1v^3)$.

There are two terms.

The coefficients are 5 and −1.

NOW TRY

OBJECTIVE 2 Combine like terms.

Recall from **Section 1.7** that **like terms** have exactly the same variables, with the same exponents on the variables. *Only the coefficients may differ.*

$19m^5$ and $14m^5$			$7x$ and $7y$		
$6y^9$, $-37y^9$, and y^9	Examples of like terms		z^4 and z	Examples of unlike terms	
$3pq$ and $-2pq$			$2pq$ and $2p$		
$2xy^2$ and $-xy^2$			$-4xy^2$ and $5x^2y$		

Using the distributive property, we combine like terms by adding or subtracting their coefficients.

EXAMPLE 2 Combining Like Terms

Simplify by combining like terms.

(a) $-4x^3 + 6x^3$

$ac + bc$
$= (a + b)c$

$= (-4 + 6)x^3$

$= 2x^3$

(b) $9x^6 - 14x^6 + x^6$ $x^6 = 1x^6$

$= (9 - 14 + 1)x^6$

$= -4x^6$

NOW TRY EXERCISE 2

Simplify by combining like terms.

(a) $x - \dfrac{2}{5}x$

(b) $3x^2 - x^2 + 2x$

(c) $y + \dfrac{2}{3}y$

$= 1y + \dfrac{2}{3}y \qquad y = 1y$

$= \left(\dfrac{3}{3} + \dfrac{2}{3}\right)y \qquad 1 = \frac{3}{3};$ Distributive property

$= \dfrac{5}{3}y \qquad$ Add the fractions.

(e) $12m^2 + 5m + 4m^2$

$= (12 + 4)m^2 + 5m$

$= 16m^2 + 5m$ ◁ Stop here. These are unlike terms.

(d) $8rs - 13rs + 9rs$

$= (8 - 13 + 9)rs$

$= 4rs$

(f) $5u + 11v$

These are unlike terms. They cannot be combined.

NOW TRY ↺

⚠ **CAUTION** In **Example 2(e)**, we cannot combine $16m^2$ and $5m$ because the exponents on the variables are different. *Unlike terms have different variables or different exponents on the same variables.*

OBJECTIVE 3 Know the vocabulary for polynomials.

Polynomial in x

A **polynomial in x** is a term or the sum of a finite number of terms of the form

ax^n, for any real number a and any whole number n.

For example,

$$16x^8 - 7x^6 + 5x^4 - 3x^2 + 4$$ Polynomial in x
(The 4 can be written as $4x^0$.)

is a polynomial in x. This polynomial is written in **descending powers** of the variable because the exponents on x decrease from left to right. By contrast,

$$2x^3 - x^2 + \dfrac{4}{x}, \quad \text{or} \quad 2x^3 - x^2 + 4x^{-1}, \qquad \text{Not a polynomial}$$

is not a polynomial in x. A variable appears in a denominator or as a factor to a negative power in a numerator.

NOTE We can define *polynomial* using any variable and not just x, as in **Example 2(e)**. Polynomials may have terms with more than one variable, as in **Example 2(d)**.

The **degree of a term** is the sum of the exponents on the variables. The **degree of a polynomial** is the greatest degree of any nonzero term of the polynomial.

▼ **Degrees of Terms and Polynomials**

Term	Degree	Polynomial	Degree
$3x^4$	4	$3x^4 - 5x^2 + 6$	4
$5x$, or $5x^1$	1	$5x + 7$	1
-7, or $-7x^0$	0	$x^5 + 3x^6 - 7$	6
$2x^2y$, or $2x^2y^1$	$2 + 1 = 3$	$2x^2y + xy - 5y^2$	3

NOW TRY ANSWERS

2. **(a)** $\dfrac{3}{5}x$ **(b)** $2x^2 + 2x$

A polynomial with only one term is a **monomial.** *(Mono-* means "one," as in *mono*rail.) A polynomial with exactly two terms is a **binomial.** *(Bi-* means "two," as in *bi*cycle.) A polynomial with exactly three terms is a **trinomial.** *(Tri-* means "three," as in *tri*angle.)

$$9m, \quad -6y^5, \quad a^2, \quad \text{and} \quad 6 \qquad \text{Monomials}$$

$$-9x^4 + 9x^3, \quad 8m^2 + 6m, \quad \text{and} \quad 3m^5 - 9m^2 \qquad \text{Binomials}$$

$$9m^3 - 4m^2 + 6, \quad \frac{19}{3}y^2 + \frac{8}{3}y + 5, \quad \text{and} \quad -3m^5 - 9m^2 + 2 \qquad \text{Trinomials}$$

NOW TRY EXERCISE 3

Simplify, give the degree, and tell whether the simplified polynomial is a *monomial,* a *binomial,* a *trinomial,* or *none of these.*

(a) $3x^2 + 2x - 4$

(b) $x^3 + 4x^3$

(c) $x^8 - x^7 + 2x^8$

EXAMPLE 3 Classifying Polynomials

For each polynomial, first simplify, if possible. Then give the degree and tell whether the simplified polynomial is a *monomial,* a *binomial,* a *trinomial,* or *none of these.*

(a) $2x^3 + 5$ The polynomial cannot be simplified. It is a binomial of degree 3.

(b) $6x - 8x + 13x$

$= 11x$ Combine like terms to simplify.

The degree is 1 (because $x = x^1$). The simplified polynomial is a monomial.

(c) $4xy - 5xy + 2xy$

$= xy$ Combine like terms to simplify.

The degree is 2 (because $xy = x^1y^1$, and $1 + 1 = 2$). The simplified polynomial is a monomial.   **NOW TRY**

OBJECTIVE 4 Evaluate polynomials.

A polynomial usually represents different numbers for different values of the variable.

NOW TRY EXERCISE 4

Find the value for $t = -3$.

$$4t^3 - t^2 - t$$

EXAMPLE 4 Evaluating a Polynomial

Find the value of $3x^4 + 5x^3 - 4x - 4$ for **(a)** $x = -2$ and **(b)** $x = 3$.

(a) $3x^4 + 5x^3 - 4x - 4$

$= 3(-2)^4 + 5(-2)^3 - 4(-2) - 4$ Substitute -2 for x.

[Use parentheses to avoid errors.] $= 3(16) + 5(-8) - 4(-2) - 4$ Apply the exponents.

$= 48 - 40 + 8 - 4$ Multiply.

$= 12$ Add and subtract.

(b) $3x^4 + 5x^3 - 4x - 4$

[Replace x with 3.] $= 3(3)^4 + 5(3)^3 - 4(3) - 4$ Let $x = 3$.

$= 3(81) + 5(27) - 4(3) - 4$ Apply the exponents.

$= 243 + 135 - 12 - 4$ Multiply.

$= 362$ Add and subtract.   **NOW TRY**

NOW TRY ANSWERS

3. (a) The polynomial cannot be simplified; degree 2; trinomial

  (b) $5x^3$; degree 3; monomial

  (c) $3x^8 - x^7$; degree 8; binomial

4. -114

! CAUTION Use parentheses around the numbers that are substituted for the variable, as in **Example 4.** *Be particularly careful when substituting a negative number for a variable that is raised to a power, or a sign error may result.*

OBJECTIVE 5 Add and subtract polynomials.

Adding Polynomials

To add two polynomials, add like terms.

NOW TRY EXERCISE 5

Find each sum.

(a) Add $7y^3 - 4y^2 + 2$ and $-6y^3 + 5y^2 - 3$.

(b) Add $-5x^4 - 2x + 3$ and $x^3 - 5x$.

EXAMPLE 5   Adding Polynomials Vertically

Find each sum.

(a) Add $6x^3 - 4x^2 + 3$ and $-2x^3 + 7x^2 - 5$.

$$\begin{array}{c} 6x^3 - 4x^2 + 3 \\ \underline{-2x^3 + 7x^2 - 5} \end{array}$$ Write like terms in columns.

Now add, column by column.

Add the coefficients only. Do *not* add the exponents.

$$\begin{array}{c} 6x^3 \\ \underline{-2x^3} \\ 4x^3 \end{array} \qquad \begin{array}{c} -4x^2 \\ \underline{7x^2} \\ 3x^2 \end{array} \qquad \begin{array}{c} 3 \\ \underline{-5} \\ -2 \end{array}$$

Add the three sums together to obtain the answer.

$$4x^3 + 3x^2 + (-2) = 4x^3 + 3x^2 - 2 \;\leftarrow \text{Final sum}$$

(b) Add $2x^2 - 4x + 3$ and $x^3 + 5x$.

Write like terms in columns and add column by column.

$$\begin{array}{c} 2x^2 - 4x + 3 \\ \underline{x^3 \qquad\quad + 5x} \\ x^3 + 2x^2 + \;\; x + 3 \end{array}$$ Leave spaces for missing terms.

NOW TRY

The polynomials in **Example 5** also can be added horizontally.

NOW TRY EXERCISE 6

Add $10x^4 - 3x^2 - x$ and $x^4 - 3x^2 + 5x$ horizontally.

EXAMPLE 6   Adding Polynomials Horizontally

Find each sum.

(a) Add $6x^3 - 4x^2 + 3$ and $-2x^3 + 7x^2 - 5$.

$$(6x^3 - 4x^2 + 3) + (-2x^3 + 7x^2 - 5) = 4x^3 + 3x^2 - 2$$ Same answer as found in **Example 5(a)**

(b) Add $2x^2 - 4x + 3$ and $x^3 + 5x$.

$$(2x^2 - 4x + 3) + (x^3 + 5x)$$
$$= x^3 + 2x^2 - 4x + 5x + 3 \qquad \text{Commutative property}$$
$$= x^3 + 2x^2 + x + 3 \qquad \text{See Example 5(b).} \qquad \text{NOW TRY}$$

In **Section 1.4,** the difference $x - y$ was defined as $x + (-y)$. (We find the difference $x - y$ by adding x and the opposite of y.)

$$7 - 2 \quad \text{is equivalent to} \quad 7 + (-2), \quad \text{which equals} \quad 5.$$

$$-8 - (-2) \quad \text{is equivalent to} \quad -8 + 2, \quad \text{which equals} \quad -6.$$

NOW TRY ANSWERS
5. (a) $y^3 + y^2 - 1$
   (b) $-5x^4 + x^3 - 7x + 3$
6. $11x^4 - 6x^2 + 4x$

A similar method is used to subtract polynomials.

To subtract two polynomials, change all the signs of the subtrahend (second polynomial) and add the result to the minuend (first polynomial).

**NOW TRY
EXERCISE 7**

Perform each subtraction.

(a) $(3x - 8) - (5x - 9)$

(b) $(4t^4 - t^2 + 7)$
$\quad - (5t^4 - 3t^2 + 1)$

EXAMPLE 7   Subtracting Polynomials Horizontally

Perform each subtraction.

(a) $(5x - 2) - (3x - 8)$

$\quad = (5x - 2) + [-(3x - 8)]$ Definition of subtraction

$\quad = (5x - 2) + [-1(3x - 8)]$ $-a = -1a$

$\quad = (5x - 2) + (-3x + 8)$ Distributive property

$\quad = 2x + 6$ Combine like terms.

CHECK   To check a subtraction problem, use the fact that

$$\text{if}\quad a - b = c, \quad \text{then}\quad a = b + c.$$

Here, add $3x - 8$ and $2x + 6$.

$$(3x - 8) + (2x + 6)$$
$$= 5x - 2 \;\checkmark$$

(b) Subtract $6x^3 - 4x^2 + 2$ from $11x^3 + 2x^2 - 8$.

> Be careful to write the problem in the correct order.

$$(11x^3 + 2x^2 - 8) - (6x^3 - 4x^2 + 2)$$

$$= (11x^3 + 2x^2 - 8) + (-6x^3 + 4x^2 - 2)$$

$$= 5x^3 + 6x^2 - 10 \quad \text{Combine like terms.}$$

CHECK   Add $6x^3 - 4x^2 + 2$ and $5x^3 + 6x^2 - 10$.

$$(6x^3 - 4x^2 + 2) + (5x^3 + 6x^2 - 10)$$

$$= 11x^3 + 2x^2 - 8 \;\checkmark$$ NOW TRY

Subtraction can also be done in columns. We use vertical subtraction in **Section 5.7** when we divide polynomials.

**NOW TRY
EXERCISE 8**

Subtract by columns.

$\quad (12x^2 - 9x + 4)$
$\quad - (-10x^2 - 3x + 7)$

EXAMPLE 8   Subtracting Polynomials Vertically

Subtract by columns: $(14y^3 - 6y^2 + 2y - 5) - (2y^3 - 7y^2 - 4y + 6)$.

$$14y^3 - 6y^2 + 2y - 5$$
$$\underline{2y^3 - 7y^2 - 4y + 6}$$ Arrange like terms in columns.

Change all signs in the second row (the subtrahend), and then add.

$$14y^3 - 6y^2 + 2y - 5$$
$$\underline{-2y^3 + 7y^2 + 4y - 6}$$ Change all signs.
$$12y^3 + y^2 + 6y - 11$$ Add.   NOW TRY

NOW TRY ANSWERS
7. (a) $-2x + 1$
   (b) $-t^4 + 2t^2 + 6$
8. $22x^2 - 6x - 3$

**NOW TRY
EXERCISE 9**

Perform the indicated
operations.

$$(6p^4 - 8p^3 + 2p - 1)$$
$$- (-7p^4 + 6p^2 - 12)$$
$$+ (p^4 - 3p + 8)$$

**NOW TRY
EXERCISE 10**

Subtract.

$$(4x^2 - 2xy + y^2)$$
$$- (6x^2 - 7xy + 2y^2)$$

EXAMPLE 9 Adding and Subtracting More Than Two Polynomials

Perform the indicated operations.

$$(4 - x + 3x^2) - (2 - 3x + 5x^2) + (8 + 2x - 4x^2)$$

Rewrite, using the definition of subtraction.

$$(4 - x + 3x^2) - (2 - 3x + 5x^2) + (8 + 2x - 4x^2)$$

$$= (4 - x + 3x^2) + (-2 + 3x - 5x^2) + (8 + 2x - 4x^2)$$

$$= (2 + 2x - 2x^2) + (8 + 2x - 4x^2) \quad \text{Combine like terms.}$$

$$= 10 + 4x - 6x^2 \quad \text{Combine like terms.}$$

NOW TRY

EXAMPLE 10 Adding and Subtracting Multivariable Polynomials

Add or subtract as indicated.

(a) $(4a + 2ab - b) + (3a - ab + b)$

$$= 4a + 2ab - b + 3a - ab + b$$

$$= 7a + ab \quad \text{Combine like terms.}$$

(b) $(2x^2y + 3xy + y^2) - (3x^2y - xy - 2y^2)$

$$= 2x^2y + 3xy + y^2 - 3x^2y + xy + 2y^2 \quad \text{Be careful with signs. The coefficient of } xy \text{ is 1.}$$

$$= -x^2y + 4xy + 3y^2$$

NOW TRY

OBJECTIVE 6 Graph equations defined by polynomials of degree 2.

In **Chapter 3,** we graphed linear equations (which are actually polynomial equations of degree 1). By plotting points, we can graph polynomial equations of degree 2.

EXAMPLE 11 Graphing Equations Defined by Polynomials of Degree 2

Graph each equation.

(a) $y = x^2$

It is easier to select values for x and find corresponding y-values. Selecting $x = 2$ and substituting in $y = x^2$ gives

$$y = 2^2 = 4.$$

The point $(2, 4)$ is on the graph of $y = x^2$. (Recall that in an ordered pair such as $(2, 4)$, *the x-value comes first and the y-value second.*) We show some ordered pairs that satisfy $y = x^2$ in the table with **FIGURE 3** on the next page. If we plot the ordered pairs from the table on a coordinate system and draw a smooth curve through them, we obtain the graph shown in **FIGURE 3**.

The graph of $y = x^2$ is the graph of a function, because each input x is related to just one output y. The curve in **FIGURE 3** is a **parabola.** The point $(0, 0)$, the *lowest* point on this graph, is the **vertex** of the parabola. The vertical line through the vertex (the y-axis here) is the **axis of symmetry,** or simply the **axis,** of the parabola. This axis is a line of symmetry for the graph. If the graph is folded on this line, the two halves will coincide.

NOW TRY ANSWERS
9. $14p^4 - 8p^3 - 6p^2 - p + 19$
10. $-2x^2 + 5xy - y^2$

**NOW TRY
EXERCISE 11**
Graph $y = -x^2 - 1$.

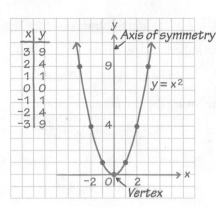

FIGURE 3

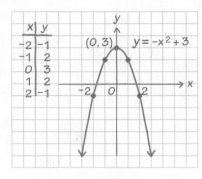

FIGURE 4

(b) $y = -x^2 + 3$

Plot points to obtain the graph. For example, let $x = -2$ and $x = 0$.

$$y = -(-2)^2 + 3 \qquad y = -0^2 + 3$$
$$y = -4 + 3 \qquad y = 0 + 3$$
$$y = -1 \qquad y = 3$$

The points $(-2, -1)$, $(0, 3)$, and several others are shown in the table that accompanies the graph in **FIGURE 4**. The vertex $(0, 3)$ is the *highest* point on this graph. The graph opens downward because x^2 has a negative coefficient.   **NOW TRY**

NOW TRY ANSWER
11.

NOTE *All polynomials of degree 2 have parabolas as their graphs.* When graphing, find points until the vertex and points on either side of it are located. (In this section, all parabolas have their vertices on the *x*-axis or the *y*-axis.)

5.4 Exercises

FOR EXTRA HELP   **MyMathLab®**

 Complete solution available in MyMathLab

Concept Check   Complete each statement.

1. In the term $4x^6$, the coefficient of x^6 is _____ and the exponent is _____.

2. The expression $4x^3 - 5x^2$ has exactly (*one / two / three*) term(s).

3. The degree of the term $-3x^9$ is _____.

4. The polynomial $4x^2 + y^2$ (*is / is not*) an example of a trinomial.

5. When $x^2 + 10$ is evaluated for $x = 3$, the result is _____.

6. $5x^{\underline{\quad}} + 3x^3 - 7x$ is a trinomial of degree 6.

7. Combining like terms in $-3xy - 2xy + 5xy$ gives _____.

8. _____ is an example of a monomial with coefficient 8, in the variable x, having degree 5.

For each expression, determine the number of terms and name the coefficients of the terms. See Example 1.

9. $6x^4$ **10.** $-9y^5$ **11.** t^4 **12.** s^7

13. $-19r^2 - r$ **14.** $2y^3 - y$ **15.** $x + 8x^2 + 5x^3$ **16.** $v - 2v^3 - v^7$

*In each polynomial, simplify by combining like terms whenever possible. In Exercises 17–26, write the result in descending powers of the variable. **See Example 2 and Objective 3.***

▶ 17. $-3m^5 + 5m^5$ **18.** $-4y^3 + 3y^3$ **19.** $2r^5 + (-3r^5)$

20. $9y^2 + (-19y^2)$ **21.** $0.2m^5 - 0.5m^2$ **22.** $-0.9y + 0.9y^2$

23. $-3x^5 + 3x^5 - 5x^5$ **24.** $6x^3 - 9x^3 + 10x^3$ **25.** $-4p^7 + 8p^7 + 5p^9$

26. $-3a^8 + 4a^8 - 3a^2$ **27.** $-1.5x^2 + 5.3x^2 - 3.8x^2$ **28.** $8.6y^4 - 10.3y^4 + 1.7y^4$

29. $-4xy^2 + 3xy^2 - 2xy^2 + xy^2$ **30.** $3pr^5 - 8pr^5 + pr^5 + 2pr^5$

31. $-\dfrac{1}{3}tu^7 + \dfrac{2}{5}tu^7 + \dfrac{1}{15}tu^7 - \dfrac{8}{5}tu^7$ **32.** $-\dfrac{3}{4}p^2q - \dfrac{1}{3}p^2q + \dfrac{7}{12}p^2q - \dfrac{1}{6}p^2q$

*For each polynomial, first simplify, if possible, and write the result in descending powers of the variable. Then give the degree and tell whether the simplified polynomial is a monomial, a binomial, a trinomial, or none of these. **See Example 3.***

▶ 33. $6x^4 - 9x$ **34.** $7t^3 - 3t$

35. $5m^4 - 3m^2 + 6m^4 - 7m^3$ **36.** $6p^5 + 4p^3 - 8p^5 + 10p^2$

37. $\dfrac{5}{3}x^4 - \dfrac{2}{3}x^4$ **38.** $\dfrac{4}{5}r^6 + \dfrac{1}{5}r^6$

39. $0.8x^4 - 0.3x^4 - 0.5x^4 + 7$ **40.** $1.2t^3 - 0.9t^3 - 0.3t^3 + 9$

41. $-11ab + 2ab - 4ab$ **42.** $5xy + 13xy - 12xy$

*Find the value of each polynomial for **(a)** $x = 2$ and **(b)** $x = -1$. **See Example 4.***

▶ 43. $2x^2 - 3x - 5$ **44.** $x^2 + 5x - 10$ **45.** $-3x^2 + 14x - 2$

46. $-2x^2 + 5x - 1$ **47.** $2x^5 - 4x^4 + 5x^3 - x^2$ **48.** $x^4 - 6x^3 + x^2 - x$

*Add. **See Examples 5 and 6.***

49. $\begin{array}{l} 2x^2 - 4x \\ \underline{3x^2 + 2x} \end{array}$ **50.** $\begin{array}{l} -5y^3 + 3y \\ \underline{8y^3 - 4y} \end{array}$ **▶ 51.** $\begin{array}{l} 3m^2 + 5m + 6 \\ \underline{2m^2 - 2m - 4} \end{array}$

52. $\begin{array}{l} 4a^3 - 4a^2 - 4 \\ \underline{6a^3 + 5a^2 - 8} \end{array}$ **53.** $\begin{array}{l} \frac{2}{3}x^2 + \frac{1}{5}x + \frac{1}{6} \\ \underline{\frac{1}{2}x^2 - \frac{1}{3}x + \frac{2}{3}} \end{array}$ **54.** $\begin{array}{l} \frac{4}{7}y^2 - \frac{1}{5}y + \frac{7}{9} \\ \underline{\frac{1}{3}y^2 - \frac{1}{3}y + \frac{2}{5}} \end{array}$

55. $9m^3 - 5m^2 + 4m - 8$ and $-3m^3 + 6m^2 - 6$

56. $12r^5 + 11r^4 - 7r^3 - 2r^2$ and $-8r^5 + 3r^3 + 2r^2$

*Subtract. **See Example 8.***

57. $\begin{array}{l} 5y^3 - 3y^2 \\ \underline{2y^3 + 8y^2} \end{array}$ **58.** $\begin{array}{l} -6t^3 + 4t^2 \\ \underline{8t^3 - 6t^2} \end{array}$

59. $\begin{array}{l} 12x^4 - x^2 + x \\ \underline{8x^4 + 3x^2 - 3x} \end{array}$ **60.** $\begin{array}{l} 13y^5 - y^3 - 8y^2 \\ \underline{7y^5 + 5y^3 + y^2} \end{array}$

▶ 61. $\begin{array}{l} 12m^3 - 8m^2 + 6m + 7 \\ \underline{-3m^3 + 5m^2 - 2m - 4} \end{array}$ **62.** $\begin{array}{l} 5a^4 - 3a^3 + 2a^2 - a + 6 \\ \underline{-6a^4 + a^3 - a^2 + a - 1} \end{array}$

*Perform each indicated operation. **See Examples 6, 7, and 9.***

▶ 63. $(8m^2 - 7m) - (3m^2 + 7m - 6)$ **64.** $(x^2 + x) - (3x^2 + 2x - 1)$

▶ **65.** $(16x^3 - x^2 + 3x) + (-12x^3 + 3x^2 + 2x)$

66. $(-2b^6 + 3b^4 - b^2) + (b^6 + 2b^4 + 2b^2)$

67. Subtract $18y^4 - 5y^2 + y$ from $7y^4 + 3y^2 + 2y$.

68. Subtract $19t^5 - 6t^3 + t$ from $8t^5 + 3t^3 + 5t$.

69. $(9a^4 - 3a^2 + 2) + (4a^4 - 4a^2 + 2) + (-12a^4 + 6a^2 - 3)$

70. $(4m^2 - 3m + 2) + (5m^2 + 13m - 4) + (-16m^2 - 4m + 3)$

71. $[(8m^2 + 4m - 7) - (2m^2 - 5m + 2)] - (m^2 + m + 1)$

72. $[(9b^3 - 4b^2 + 3b + 2) - (-2b^3 - 3b^2 + b)] - (8b^3 + 6b + 4)$

73. $[(3x^2 - 2x + 7) - (4x^2 + 2x - 3)] - [(9x^2 + 4x - 6) + (-4x^2 + 4x + 4)]$

74. $[(6t^2 - 3t + 1) - (12t^2 + 2t - 6)] - [(4t^2 - 3t - 8) + (-6t^2 + 10t - 12)]$

75. *Concept Check* Without actually performing the operations, determine mentally the coefficient of the x^2-term in the simplified form of
$$(-4x^2 + 2x - 3) - (-2x^2 + x - 1) + (-8x^2 + 3x - 4).$$

76. *Concept Check* Without actually performing the operations, determine mentally the coefficient of the x-term in the simplified form of
$$(-8x^2 - 3x + 2) - (4x^2 - 3x + 8) - (-2x^2 - x + 7).$$

Add or subtract as indicated. **See Example 10.**

▶ **77.** $(6b + 3c) + (-2b - 8c)$ **78.** $(-5t + 13s) + (8t - 3s)$

79. $(4x + 2xy - 3) - (-2x + 3xy + 4)$ **80.** $(8ab + 2a - 3b) - (6ab - 2a + 3b)$

81. $(5x^2y - 2xy + 9xy^2) - (8x^2y + 13xy + 12xy^2)$

82. $(16t^3s^2 + 8t^2s^3 + 9ts^4) - (-24t^3s^2 + 3t^2s^3 - 18ts^4)$

Find a polynomial that represents the perimeter of each rectangle, square, or triangle.

83.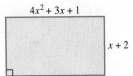
$4x^2 + 3x + 1$
$x + 2$

84.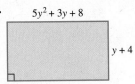
$5y^2 + 3y + 8$
$y + 4$

85.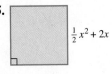
$\frac{1}{2}x^2 + 2x$

86.
$\frac{3}{4}x^2 + x$

87.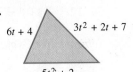
$6t + 4$ $3t^2 + 2t + 7$
$5t^2 + 2$

88.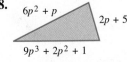
$6p^2 + p$ $2p + 5$
$9p^3 + 2p^2 + 1$

Find **(a)** *a polynomial that represents the perimeter of each triangle and* **(b)** *the degree measures of the angles of the triangle. (Hint: The sum of the measures of the angles of any triangle is* $180°$.)

89.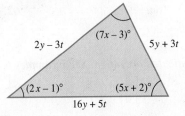
$2y - 3t$ $(7x - 3)°$ $5y + 3t$
$(2x - 1)°$ $(5x + 2)°$
$16y + 5t$

90.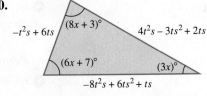
$-t^2s + 6ts$ $(8x + 3)°$ $4t^2s - 3ts^2 + 2ts$
$(6x + 7)°$ $(3x)°$
$-8t^2s + 6ts^2 + ts$

Extending Skills *Perform each indicated operation.*

91. Find the difference of the sum of $5x^2 + 2x - 3$ and $x^2 - 8x + 2$ and the sum of $7x^2 - 3x + 6$ and $-x^2 + 4x - 6$.

92. Subtract the sum of $9t^3 - 3t + 8$ and $t^2 - 8t + 4$ from the sum of $12t + 8$ and $t^2 - 10t + 3$.

Graph each equation by completing the table of values. ***See Example 11.***

▶ **93.** $y = x^2 - 4$

x	y
-2	
-1	
0	
1	
2	

94. $y = x^2 - 6$

x	y
-2	
-1	
0	
1	
2	

95. $y = 2x^2 - 1$

x	y
-2	
-1	
0	
1	
2	

96. $y = 2x^2 + 2$

x	y
-2	
-1	
0	
1	
2	

97. $y = -x^2 + 4$

x	y
-2	
-1	
0	
1	
2	

98. $y = -x^2 + 2$

x	y
-2	
-1	
0	
1	
2	

99. $y = (x + 3)^2$

x	-5	-4	-3	-2	-1
y					

100. $y = (x - 4)^2$

x	2	3	4	5	6
y					

RELATING CONCEPTS For Individual or Group Work (Exercises 101–104)

The age of a dog in human years y is given by the polynomial equation

$$y = -0.1133x^2 + 6.966x + 4.915,$$

where x represents age in dog years (based on data from www.dogyears.com*). For example if a dog is 4 in dog years, then we let x = 4 to find that y ≈ 31. (Verify this.) The dog is about 31 yr old in human years. This illustrates the concept of a function—for each input value x, we obtain one and only one output value y.*

Exercises 101–104 further illustrate the function concept with polynomials. ***Work them in order.***

101. It used to be thought that each dog year was about 7 human years, so that $y = 7x$ gave the number of human years for x dog years. Evaluate y for $x = 9$, and interpret the result.

102. Use the polynomial equation in the directions above to find the number of human years equivalent to each given number of dog years. Round to the nearest whole number.

(a) 5 **(b)** 11 **(c)** 14

103. If an object is projected upward under certain conditions, its height y in feet is given by

$$y = -16x^2 + 60x + 80,$$

where x is in seconds. Evaluate y for $x = 2.5$. Use the result to fill in the blanks: If _____ seconds have elapsed, the height of the object is _____ feet.

104. If it costs \$15 to rent a chain saw, plus \$2 per day, the equation

$$y = 2x + 15$$

gives the cost y in dollars to rent the chain saw for x days. Evaluate y for $x = 6$. Use the result to fill in the blanks: If the saw is rented for _____ days, the cost is _____ .

5.5 Multiplying Polynomials

VOCABULARY

☐ FOIL method
☐ outer product
☐ inner product

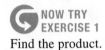 **NOW TRY EXERCISE 1**

Find the product.

$$-3x^5(2x^3 - 5x^2 + 10)$$

OBJECTIVE 1 Multiply a monomial and a polynomial.

In **Section 5.1,** we found the product of two monomials as follows.

$$(-8m^6)(-9n^6)$$
$$= (-8)(-9)(m^6)(n^6) \quad \text{Commutative and associative properties}$$
$$= 72m^6n^6 \quad \text{Multiply.}$$

> ⚠ **CAUTION** *Do not confuse addition of terms with multiplication of terms.*

$7q^5 + 2q^5$		$(7q^5)(2q^5)$	Commutative property;
$= (7 + 2)q^5$	Distributive property	$= 7 \cdot 2q^{5+5}$	product rule
$= 9q^5$	Add.	$= 14q^{10}$	Multiply. Add.

To find the product of a monomial and a polynomial with more than one term, we use the distributive property and multiplication of monomials.

> **EXAMPLE 1** Multiplying Monomials and Polynomials

Find each product.

(a) $4x^2(3x + 5)$ $a(b + c) = ab + ac$

$$= 4x^2(3x) + 4x^2(5) \quad \text{Distributive property}$$
$$= 12x^3 + 20x^2 \quad \text{Multiply monomials.}$$

(b) $-8m^3(4m^3 + 3m^2 + 2m - 1)$

$$= -8m^3(4m^3) + (-8m^3)(3m^2)$$
$$+ (-8m^3)(2m) + (-8m^3)(-1) \quad \text{Distributive property}$$
$$= -32m^6 - 24m^5 - 16m^4 + 8m^3 \quad \text{Multiply monomials.} \quad \text{NOW TRY} \; ↺$$

OBJECTIVE 2 Multiply two polynomials.

To find the product of the polynomials $x^2 + 3x + 5$ and $x - 4$, we can think of $x - 4$ as a single quantity and use the distributive property as follows.

$$(x^2 + 3x + 5)(x - 4)$$
$$= x^2(x - 4) + 3x(x - 4) + 5(x - 4) \quad \text{Distributive property}$$
$$= x^2(x) + x^2(-4) + 3x(x) + 3x(-4) + 5(x) + 5(-4)$$
$$\quad \text{Distributive property again}$$
$$= x^3 - 4x^2 + 3x^2 - 12x + 5x - 20 \quad \text{Multiply monomials.}$$
$$= x^3 - x^2 - 7x - 20 \quad \text{Combine like terms.}$$

Multiplying Polynomials

To multiply two polynomials, multiply each term of the second polynomial by each term of the first polynomial and add the products.

NOW TRY
EXERCISE 2

Multiply.

$(x^2 - 4)(2x^2 - 5x + 3)$

| EXAMPLE 2 | **Multiplying Two Polynomials** |

Multiply $(m^2 + 5)(4m^3 - 2m^2 + 4m)$.

$(m^2 + 5)(4m^3 - 2m^2 + 4m)$ Multiply each term of the second polynomial by each term of the first.

$= m^2(4m^3) + m^2(-2m^2) + m^2(4m) + 5(4m^3) + 5(-2m^2) + 5(4m)$

                                               Distributive properly

$= 4m^5 - 2m^4 + 4m^3 + 20m^3 - 10m^2 + 20m$ Multiply monomials.

$= 4m^5 - 2m^4 + 24m^3 - 10m^2 + 20m$ Combine like terms.

NOW TRY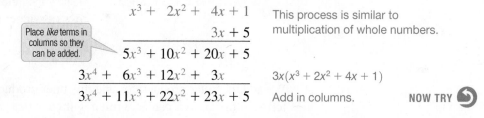

NOW TRY
EXERCISE 3

Multiply.

$$5t^2 - 7t + 4$$
$$\underline{\quad\quad\ 2t - 6}$$

| EXAMPLE 3 | **Multiplying Polynomials Vertically** |

Multiply $(x^3 + 2x^2 + 4x + 1)(3x + 5)$ vertically.

$$x^3 + 2x^2 + 4x + 1$$
$$\underline{\quad\quad\quad\quad\quad\ 3x + 5}$$

Write the polynomials vertically.

Begin by multiplying each of the terms in the top row by 5.

$$x^3 + \ 2x^2 + \ 4x + 1$$
$$\underline{\quad\quad\quad\quad\quad\quad\ 3x + 5}$$
$$5x^3 + 10x^2 + 20x + 5 \quad\quad 5(x^3 + 2x^2 + 4x + 1)$$

Now multiply each term in the top row by $3x$. Then add like terms.

Place *like* terms in columns so they can be added.

$$x^3 + \ 2x^2 + \ 4x + 1$$
$$\underline{\quad\quad\quad\quad\quad\quad\ 3x + 5}$$
$$5x^3 + 10x^2 + 20x + 5$$
$$\underline{3x^4 + \ 6x^3 + 12x^2 + \ 3x}$$
$$3x^4 + 11x^3 + 22x^2 + 23x + 5$$

This process is similar to multiplication of whole numbers.

$3x(x^3 + 2x^2 + 4x + 1)$

Add in columns.   **NOW TRY**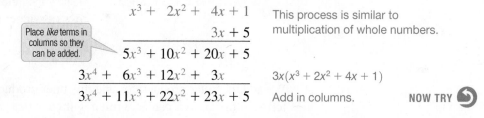

NOW TRY
EXERCISE 4

Find the product of

$9x^3 - 12x^2 + 3$ and $\frac{1}{3}x^2 - \frac{2}{3}$.

| EXAMPLE 4 | **Multiplying Polynomials Vertically (Fractional Coefficients)** |

Find the product of $4m^3 - 2m^2 + 4m$ and $\frac{1}{2}m^2 + \frac{5}{2}$.

$$4m^3 - 2m^2 + 4m$$
$$\underline{\quad\quad\quad \frac{1}{2}m^2 + \frac{5}{2}}$$
$$10m^3 - 5m^2 + 10m \quad\quad \text{Terms of top row are multiplied by } \tfrac{5}{2}.$$
$$\underline{2m^5 - m^4 + \ 2m^3} \quad\quad\quad\ \text{Terms of top row are multiplied by } \tfrac{1}{2}m^2.$$
$$2m^5 - m^4 + 12m^3 - 5m^2 + 10m \quad\quad \text{Add in columns.} \quad\quad \textbf{NOW TRY} $$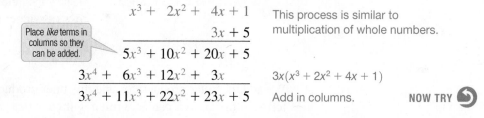

We can use a rectangle to model polynomial multiplication. For example, to find

$$(2x + 1)(3x + 2),$$

we label a rectangle with each term, as shown below on the left. Then put the product of each pair of monomials in the appropriate box, as shown on the right.

NOW TRY ANSWERS
2. $2x^4 - 5x^3 - 5x^2 + 20x - 12$
3. $10t^3 - 44t^2 + 50t - 24$
4. $3x^5 - 4x^4 - 6x^3 + 9x^2 - 2$

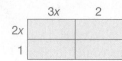

The product of the binomials is the sum of the four monomial products.

	$3x$	2
$2x$	$6x^2$	$4x$
1	$3x$	2

$$(2x + 1)(3x + 2)$$
$$= 6x^2 + 4x + 3x + 2$$
$$= 6x^2 + 7x + 2 \qquad \text{Combine like terms.}$$

This approach can be extended to polynomials with any number of terms.

OBJECTIVE 3 Multiply binomials by the FOIL method.

When multiplying binomials, the **FOIL method** reduces the rectangle method to a systematic approach without the rectangle. Consider this example.

$$(x + 3)(x + 5)$$
$$= (x + 3)x + (x + 3)5 \qquad \text{Distributive property}$$
$$= x(x) + 3(x) + x(5) + 3(5) \qquad \text{Distributive property again}$$
$$= x^2 + 3x + 5x + 15 \qquad \text{Multiply.}$$
$$= x^2 + 8x + 15 \qquad \text{Combine like terms.}$$

The letters of the word FOIL refer to the positions of the terms.

$(x + 3)(x + 5)$ Multiply the **First terms:** $x(x)$.   **F**

$(x + 3)(x + 5)$ Multiply the **Outer terms:** $x(5)$.   **O**
This is the **outer product.**

$(x + 3)(x + 5)$ Multiply the **Inner terms:** $3(x)$.   **I**
This is the **inner product.**

$(x + 3)(x + 5)$ Multiply the **Last terms:** $3(5)$.   **L**

We add the outer product, $5x$, and the inner product, $3x$, to obtain $8x$ so that the three terms of the answer can be written without extra steps.

$$(x + 3)(x + 5)$$
$$= x^2 + 8x + 15$$

Multiplying Binomials by the FOIL Method

Step 1 Multiply the two **First** terms of the binomials to obtain the first term of the product.

Step 2 Find the **Outer** product and the **Inner** product and combine them (when possible) to obtain the middle term of the product.

Step 3 Multiply the two **Last** terms of the binomials to obtain the last term of the product.

$$\mathbf{F} = x^2 \qquad \mathbf{L} = 15$$

$(x + 3)(x + 5)$ $\qquad$ $(x + 3)(x + 5)$
$$= x^2 + 8x + 15$$

I $\quad 3x$
O $\quad \underline{5x}$
$\quad\quad 8x$ $\qquad$ Combine like terms.

Add the terms found in Steps 1–3.

NOW TRY
EXERCISE 5
Use the FOIL method to find the product.

$$(t - 6)(t + 5)$$

EXAMPLE 5 Using the FOIL Method

Use the FOIL method to find the product $(x + 8)(x - 6)$.

Step 1   **F**   Multiply the First terms: $x(x) = x^2$.

Step 2   **O**   Find the Outer product: $x(-6) = -6x$.

　　　　　I   Find the Inner product: $8(x) = 8x$.

　　　　　　　Combine the outer and inner products mentally: $-6x + 8x = 2x$.

Step 3   **L**   Multiply the Last terms: $8(-6) = -48$.

　　　　　　　The product $(x + 8)(x - 6)$ is $x^2 + 2x - 48$.　　Add the terms found in Steps 1–3.

Shortcut:

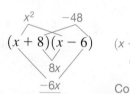

$$(x + 8)(x - 6)$$
$$= x^2 + 2x - 48$$

Combine like terms.   **NOW TRY**

EXAMPLE 6 Using the FOIL Method

Multiply $(9x - 2)(3y + 1)$.

First	$(9x - 2)(3y + 1)$	$27xy$
Outer	$(9x - 2)(3y + 1)$	$9x$
Inner	$(9x - 2)(3y + 1)$	$-6y$
Last	$(9x - 2)(3y + 1)$	-2

These unlike terms *cannot* be combined.

　　　　　　　　　　　　　F　　O　　I　　L
The product $(9x - 2)(3y + 1)$　is　$27xy + 9x - 6y - 2$.　　**NOW TRY**

NOW TRY
EXERCISE 6
Multiply.

$$(7y - 3)(2x + 5)$$

NOW TRY
EXERCISE 7
Find each product.

(a) $(3p - 5q)(4p - q)$

(b) $5x^2(3x + 1)(x - 5)$

EXAMPLE 7 Using the FOIL Method

Find each product.

(a) $(2k + 5y)(k + 3y)$

$$\begin{aligned} & \; \overset{\text{F}}{2k(k)} + \overset{\text{O}}{2k(3y)} + \overset{\text{I}}{5y(k)} + \overset{\text{L}}{5y(3y)} \\ &= 2k^2 + 6ky + 5ky + 15y^2 \qquad \text{Multiply.} \\ &= 2k^2 + 11ky + 15y^2 \qquad \text{Combine like terms.} \end{aligned}$$

(b) $(7p + 2q)(3p - q)$

$$= 21p^2 - pq - 2q^2 \quad \begin{array}{l}\text{FOIL}\\ \text{method}\end{array}$$

(c) $2x^2(x - 3)(3x + 4)$

$$\begin{aligned} &= 2x^2(3x^2 - 5x - 12) \quad \text{FOIL method} \\ &= 6x^4 - 10x^3 - 24x^2 \quad \begin{array}{l}\text{Distributive}\\ \text{property}\end{array} \end{aligned}$$

NOW TRY

NOW TRY ANSWERS
5. $t^2 - t - 30$
6. $14yx + 35y - 6x - 15$
7. (a) $12p^2 - 23pq + 5q^2$
　　(b) $15x^4 - 70x^3 - 25x^2$

NOTE   Alternatively, the factors in **Example 7(c)** can be multiplied as follows.

$$2x^2(x - 3)(3x + 4) \qquad \text{Multiply } 2x^2 \text{ and } x - 3 \text{ first.}$$

$$= (2x^3 - 6x^2)(3x + 4) \qquad \text{Multiply that product and } 3x + 4.$$

$$= 6x^4 - 10x^3 - 24x^2 \qquad \text{The same answer results.}$$

5.5 Exercises

   MyMathLab®

▶ *Complete solution available in MyMathLab*

Concept Check   In Exercises 1 and 2, match each product in Column I with the correct polynomial in Column II.

	I	II		I	II
1.	(a) $5x^3(6x^7)$	**A.** $125x^{21}$	**2.**	(a) $(x-5)(x+4)$	**A.** $x^2 + 9x + 20$
	(b) $-5x^7(6x^3)$	**B.** $30x^{10}$		(b) $(x+5)(x+4)$	**B.** $x^2 - 9x + 20$
	(c) $(5x^7)^3$	**C.** $-216x^9$		(c) $(x-5)(x-4)$	**C.** $x^2 - x - 20$
	(d) $(-6x^3)^3$	**D.** $-30x^{10}$		(d) $(x+5)(x-4)$	**D.** $x^2 + x - 20$

Concept Check   Fill in each blank with the correct response.

3. In multiplying a monomial by a polynomial, such as in

$$4x(3x^2 + 7x^3) = 4x(3x^2) + 4x(7x^3),$$

the first property that is used is the _____ property.

4. The FOIL method can only be used to multiply two polynomials when both polynomials are _____ .

5. The product $2x^2(-3x^5)$ has exactly _____ term(s) after the multiplication is performed.

6. The product $(a + b)(c + d)$ has exactly _____ term(s) after the multiplication is performed.

Find each product. **See Objective 1.**

7. $5y^4(3y^7)$

8. $10p^2(5p^3)$

9. $-15a^4(-2a^5)$

10. $-3m^6(-5m^4)$

11. $5p(3q^2)$

12. $4a^3(3b^2)$

13. $-6m^3(3n^2)$

14. $9r^3(-2s^2)$

15. $y^5 \cdot 9y \cdot y^4$

16. $x^2 \cdot 3x^3 \cdot 2x$

17. $(4x^3)(2x^2)(-x^5)$

18. $(7t^5)(3t^4)(-t^8)$

Find each product. **See Example 1.**

▶ **19.** $2m(3m + 2)$

20. $4x(5x + 3)$

21. $3p(-2p^3 + 4p^2)$

22. $4x(3 + 2x + 5x^3)$

23. $-8z(2z + 3z^2 + 3z^3)$

24. $-7y(3 + 5y^2 - 2y^3)$

25. $2y^3(3 + 2y + 5y^4)$

26. $2m^4(6 + 5m + 3m^2)$

27. $-4r^3(-7r^2 + 8r - 9)$

28. $-9a^5(-3a^6 - 2a^4 + 8a^2)$

29. $3a^2(2a^2 - 4ab + 5b^2)$

30. $4z^3(8z^2 + 5zy - 3y^2)$

Concept Check   Multiply.

31. $7m^3n^2(3m^2 + 2mn - n^3)$

$= 7m^3n^2(\underline{\hspace{0.5cm}}) + 7m^3n^2(\underline{\hspace{0.5cm}})$

$+ 7m^3n^2(\underline{\hspace{0.5cm}})$

$= \underline{\hspace{2cm}}$

32. $2p^2q(3p^2q^2 - 5p + 2q^2)$

$= \underline{\hspace{0.5cm}}(3p^2q^2) + \underline{\hspace{0.5cm}}(-5p)$

$+ \underline{\hspace{0.5cm}}(2q^2)$

$= \underline{\hspace{2cm}}$

Find each product. **See Examples 2–4.**

▶ **33.** $(6x + 1)(2x^2 + 4x + 1)$

34. $(9a + 2)(9a^2 + a + 1)$

35. $(9y - 2)(8y^2 - 6y + 1)$

36. $(2r - 1)(3r^2 + 4r - 4)$

▶ **37.** $(4m + 3)(5m^3 - 4m^2 + m - 5)$ **38.** $(2y + 8)(3y^4 - 2y^2 + 1)$

39. $(2x - 1)(3x^5 - 2x^3 + x^2 - 2x + 3)$ **40.** $(2a + 3)(a^4 - a^3 + a^2 - a + 1)$

41. $(5x^2 + 2x + 1)(x^2 - 3x + 5)$ **42.** $(2m^2 + m - 3)(m^2 - 4m + 5)$

▶ **43.** $(6x^4 - 4x^2 + 8x)\left(\dfrac{1}{2}x + 3\right)$ **44.** $(8y^6 + 4y^4 - 12y^2)\left(\dfrac{3}{4}y^2 + 2\right)$

Find each product using the rectangle method shown in the text. Determine the individual terms that should appear on the blanks or in the rectangles, and then give the final product.

45. $(x + 3)(x + 4)$ **46.** $(x + 5)(x + 2)$

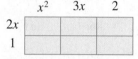

Product: _____   Product: _____

47. $(2x + 1)(x^2 + 3x + 2)$ **48.** $(x + 4)(3x^2 + 2x + 1)$

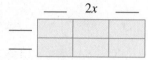
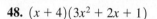

Product: _____   Product: _____

49. *Concept Check*   For the product

$$(2p - 5)(3p + 7),$$

find and simplify the following.

(a) Product of first terms   **(b)** Outer product   **(c)** Inner product

    ___ (___)   ___ (___)   ___ (___)

    = ___   = ___   = ___

(d) Product of last terms   **(e)** Complete product in simplified form

    ___ (___)   _____

    = ___

50. *Concept Check*   Repeat the process of **Exercise 49** for $(2p - 5)(2p + 5)$, and compare the results in parts (b) and (c). What do you notice? What is the complete product in simplified form?

Find each product. See Examples 5–7.

▶ **51.** $(m + 7)(m + 5)$ **52.** $(n + 9)(n + 3)$ **53.** $(n - 1)(n + 4)$

54. $(t - 3)(t + 8)$ **55.** $(2x + 3)(6x - 4)$ **56.** $(3y + 5)(8y - 6)$

57. $(9 + t)(9 - t)$ **58.** $(10 + r)(10 - r)$ **59.** $(3x - 2)(3x - 2)$

60. $(4m + 3)(4m + 3)$ **61.** $(5a + 1)(2a + 7)$ **62.** $(b + 8)(6b - 2)$

63. $(6 - 5m)(2 + 3m)$ **64.** $(8 - 3a)(2 + a)$ **65.** $(5 - 3x)(4 + x)$

66. $(6 - 5x)(2 + x)$ **67.** $(3t - 4s)(t + 3s)$ **68.** $(2m - 3n)(m + 5n)$

▶ **69.** $(4x + 3)(2y - 1)$ **70.** $(5x + 7)(3y - 8)$ ▶ **71.** $(3x + 2y)(5x - 3y)$

72. $(5a + 3b)(5a - 4b)$ **73.** $3y^3(2y + 3)(y - 5)$ **74.** $2x^2(2x - 5)(x + 3)$

75. $-8r^3(5r^2 + 2)(5r^2 - 2)$ **76.** $-5t^4(2t^4 + 1)(2t^4 - 1)$

Find polynomials that represent **(a)** *the area and* **(b)** *the perimeter of each square or rectangle. (If necessary, refer to the formulas found inside the back cover of this book.)*

77.

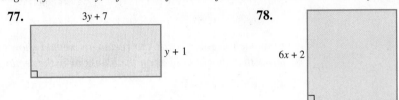

78.

Extending Skills   *Find each product. In Exercises 87–90, 95, and 96, apply the meaning of exponents.*

79. $(x + 7)^2$ **80.** $(m + 6)^2$ **81.** $(a - 4)(a + 4)$

82. $(b - 10)(b + 10)$ **83.** $(2p - 5)^2$ **84.** $(3m - 1)^2$

85. $(5k + 3q)^2$ **86.** $(8m + 3n)^2$ **87.** $(m - 5)^3$

88. $(p - 3)^3$ **89.** $(2a + 1)^3$ **90.** $(3m + 1)^3$

91. $-3a(3a + 1)(a - 4)$ **92.** $-4r(3r + 2)(2r - 5)$

93. $7(4m - 3)(2m + 1)$ **94.** $5(3k - 7)(5k + 2)$

95. $(3r - 2s)^4$ **96.** $(2z - 5y)^4$

97. $3p^3(2p^2 + 5p)(p^3 + 2p + 1)$ **98.** $5k^2(k^3 - 3)(k^2 - k + 4)$

99. $-2x^5(3x^2 + 2x - 5)(4x + 2)$ **100.** $-4x^3(3x^4 + 2x^2 - x)(-2x + 1)$

101. $\left(3p^2 + \dfrac{5}{4}q\right)\left(2p^2 - \dfrac{5}{3}q\right)$ **102.** $\left(2x^2 + \dfrac{2}{3}y\right)\left(3x^2 - \dfrac{3}{4}y\right)$

The figures in Exercises 103–106 are composed of triangles, squares, rectangles, and circles. Find a polynomial that represents the area of each shaded region. In Exercises 105 and 106, leave π in the answers. (If necessary, refer to the formulas found inside the back cover of this book.)

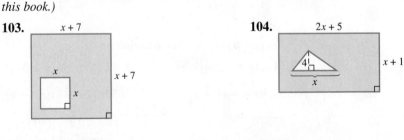

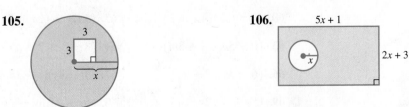

RELATING CONCEPTS For Individual or Group Work (Exercises 107–112)

Work Exercises 107–112 in order. (All units are in feet.)

107. Find a polynomial that represents the area, in square feet, of the rectangle.

108. Suppose we know that the area of the rectangle is 600 ft^2. Use this information and the polynomial from **Exercise 107** to write an equation in x, and solve it.

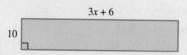

109. Refer to **Exercise 108.** What are the dimensions of the rectangle?

110. Use the result of **Exercise 109** to find the perimeter of the rectangle.

111. Suppose the rectangle represents a strip of lawn and it costs $0.75 per square foot to lay sod on the lawn. How much will it cost to sod the entire lawn?

112. Suppose it costs $20.50 per linear foot for fencing. How much will it cost to fence the entire lawn?

5.6 Special Products

OBJECTIVES

1 Square binomials.
2 Find the product of the sum and difference of two terms.
3 Find greater powers of binomials.

VOCABULARY

☐ conjugates

**NOW TRY
EXERCISE 1**
Find $(x + 5)^2$.

OBJECTIVE 1 Square binomials.

EXAMPLE 1 **Squaring a Binomial**

Find $(m + 3)^2$.

$$(m + 3)(m + 3)$$ *($(m + 3)^2$ means $(m + 3)(m + 3)$.)*

$$= m^2 + 3m + 3m + 9$$ FOIL method

This is the answer. $$= m^2 + 6m + 9$$ Combine like terms.

This result has the squares of the first and the last terms of the binomial.

$$m^2 = m^2 \quad \text{and} \quad 3^2 = 9$$

The middle term, 6m, is twice the product of the two terms of the binomial, because the outer and inner products are $m(3)$ and $3(m)$. Then we find their sum.

$$m(3) + 3(m)$$
$$= 2(m)(3)$$
$$= 6m$$

NOW TRY

Example 1 suggests the following rules.

Square of a Binomial

The square of a binomial is a trinomial consisting of

the square of
the first term $+$ twice the product
of the two terms $+$ the square of
the last term.

For x and y, the following hold.

$$(x + y)^2 = x^2 + 2xy + y^2$$
$$(x - y)^2 = x^2 - 2xy + y^2$$

NOW TRY ANSWER
1. $x^2 + 10x + 25$

Square each binomial.

(a) $(3x - 1)^2$

(b) $(4p - 5q)^2$

(c) $\left(6t - \frac{1}{3}\right)^2$

(d) $-(3y + 2)^2$

(e) $m(2m + 3)^2$

EXAMPLE 2 **Squaring Binomials**

Square each binomial.

$$(x - y)^2 = x^2 - 2 \cdot x \cdot y + y^2$$
$$\downarrow \quad \downarrow \qquad \downarrow \quad \downarrow \quad \downarrow \quad \downarrow \qquad \downarrow$$

(a) $(5z - 1)^2 = (5z)^2 - 2(5z)(1) + (1)^2$

$\qquad\qquad = 25z^2 - 10z + 1$ $\qquad$ $(5z)^2 = 5^2z^2 = 25z^2$

(b) $(3b + 5r)^2$

> Be careful to square $3b$ and $5r$ correctly.

$\qquad = (3b)^2 + 2(3b)(5r) + (5r)^2$

$\qquad = 9b^2 + 30br + 25r^2$

(c) $(2a - 9x)^2$

$\qquad = (2a)^2 - 2(2a)(9x) + (9x)^2$

$\qquad = 4a^2 - 36ax + 81x^2$

(d) $\left(4m + \frac{1}{2}\right)^2$

$\qquad = (4m)^2 + 2(4m)\left(\frac{1}{2}\right) + \left(\frac{1}{2}\right)^2$

$\qquad = 16m^2 + 4m + \frac{1}{4}$

(e) $-(2x - 3)^2$

$\qquad = -\left[(2x)^2 - 2(2x)(3) + 3^2\right]$

$\qquad = -(4x^2 - 12x + 9)$

$\qquad = -4x^2 + 12x - 9$

(f) $x(4x - 3)^2$ $\qquad$ Remember the middle term.

$\qquad = x(16x^2 - 24x + 9)$ $\qquad$ Square the binomial.

$\qquad = 16x^3 - 24x^2 + 9x$ $\qquad$ Distributive property $\qquad$ NOW TRY

In the square of a sum, all of the terms are positive, as in Examples 2(b) and (d). *In the square of a difference, the middle term is negative,* as in Examples 2(a), (c), and (f).

⚠ CAUTION A common error when squaring a binomial is to forget the middle term of the product. In general, remember the following.

$$(x + y)^2 = x^2 + 2xy + y^2, \quad not \quad x^2 + y^2,$$

and $\qquad\qquad (x - y)^2 = x^2 - 2xy + y^2, \quad not \quad x^2 - y^2.$

OBJECTIVE 2 Find the product of the sum and difference of two terms.

In binomial products of the form $(x + y)(x - y)$, one binomial is a sum of two terms. The other is a difference of the *same* two terms. Consider the following.

$$(x + 2)(x - 2)$$

$\qquad = x^2 - 2x + 2x - 4$ $\qquad$ FOIL method

$\qquad = x^2 - 4$ $\qquad$ Combine like terms.

Thus, the product of $x + y$ and $x - y$ is a difference of two squares.

2. (a) $9x^2 - 6x + 1$
(b) $16p^2 - 40pq + 25q^2$
(c) $36t^2 - 4t + \frac{1}{9}$
(d) $-9y^2 - 12y - 4$
(e) $4m^3 + 12m^2 + 9m$

Product of a Sum and Difference of Two Terms

$$(x + y)(x - y) = x^2 - y^2$$

NOTE The expressions $x + y$ and $x - y$, a sum and difference of the *same* two terms, are **conjugates.** In the preceding example, $x + 2$ and $x - 2$ are conjugates.

**NOW TRY
EXERCISE 3**

Find the product.

$$(t + 10)(t - 10)$$

EXAMPLE 3 Finding the Product of a Sum and Difference of Two Terms

Find each product.

(a) $(x + 4)(x - 4)$ ← This is a sum and difference of two terms.

$= x^2 - 4^2$ $(x + y)(x - y) = x^2 - y^2$

$= x^2 - 16$ Square 4.

(b) $\left(\dfrac{2}{3} - w\right)\left(\dfrac{2}{3} + w\right)$

$= \left(\dfrac{2}{3} + w\right)\left(\dfrac{2}{3} - w\right)$ Commutative property

$= \left(\dfrac{2}{3}\right)^2 - w^2$ $(x + y)(x - y) = x^2 - y^2$

$= \dfrac{4}{9} - w^2$ Square $\frac{2}{3}$.

(c) $x(x + 2)(x - 2)$

$= x(x^2 - 4)$ Find the product of the sum and difference of two terms.

$= x^3 - 4x$ Distributive property   **NOW TRY** ↻

**NOW TRY
EXERCISE 4**

Find each product.

(a) $(4x - 6)(4x + 6)$

(b) $\left(5r - \dfrac{4}{5}\right)\left(5r + \dfrac{4}{5}\right)$

(c) $y(3y + 1)(3y - 1)$

(d) $-5(p + q^2)(p - q^2)$

EXAMPLE 4 Finding the Product of a Sum and Difference of Two Terms

Find each product.

(a) $(x + y)\ (x - y)$
    ↓   ↓   ↓   ↓

$(5m + 3)(5m - 3)$ ← This indicates the product of a sum and difference of two terms.

$= (5m)^2 - 3^2$ $(x + y)(x - y) = x^2 - y^2$

Be careful to square $5m$ correctly.

$= 25m^2 - 9$ Apply the exponents.

(b) $(4x + y)(4x - y)$

$= (4x)^2 - y^2$

$= 16x^2 - y^2$

(c) $\left(z - \dfrac{1}{4}\right)\left(z + \dfrac{1}{4}\right)$

$= z^2 - \dfrac{1}{16}$

(d) $p(2p + 1)(2p - 1)$

$= p(4p^2 - 1)$

$= 4p^3 - p$ Distributive property

(e) $-3(x + y^2)(x - y^2)$

$= -3(x^2 - y^4)$

$= -3x^2 + 3y^4$

NOW TRY ↻

NOW TRY ANSWERS
3. $t^2 - 100$
4. **(a)** $16x^2 - 36$
  (b) $25r^2 - \dfrac{16}{25}$
  (c) $9y^3 - y$
  (d) $-5p^2 + 5q^4$

OBJECTIVE 3 Find greater powers of binomials.

The methods used in the previous section and this section can be combined to find greater powers of binomials.

**NOW TRY
EXERCISE 5**

Find the product.

$$(2m - 1)^3$$

EXAMPLE 5   Finding Greater Powers of Binomials

Find each product.

(a) $(x + 5)^3$

$$= (x + 5)(x + 5)^2 \qquad\qquad a^3 = a \cdot a^2$$

$$= (x + 5)(x^2 + 10x + 25) \qquad \text{Square the binomial.}$$

$$= x^3 + 10x^2 + 25x + 5x^2 + 50x + 125 \qquad \text{Multiply polynomials.}$$

$$= x^3 + 15x^2 + 75x + 125 \qquad \text{Combine like terms.}$$

(b) $(2y - 3)^4$

$$= (2y - 3)^2(2y - 3)^2 \qquad\qquad a^4 = a^2 \cdot a^2$$

$$= (4y^2 - 12y + 9)(4y^2 - 12y + 9) \qquad \text{Square each binomial.}$$

$$= 16y^4 - 48y^3 + 36y^2 - 48y^3 + 144y^2 \qquad \text{Multiply polynomials.}$$

$$\quad - 108y + 36y^2 - 108y + 81$$

$$= 16y^4 - 96y^3 + 216y^2 - 216y + 81 \qquad \text{Combine like terms.}$$

(c) $-2r(r + 2)^3$

$$= -2r(r + 2)(r + 2)^2 \qquad\qquad a^3 = a \cdot a^2$$

$$= -2r(r + 2)(r^2 + 4r + 4) \qquad \text{Square the binomial.}$$

$$= -2r(r^3 + 4r^2 + 4r + 2r^2 + 8r + 8) \qquad \text{Multiply polynomials.}$$

$$= -2r(r^3 + 6r^2 + 12r + 8) \qquad \text{Combine like terms.}$$

$$= -2r^4 - 12r^3 - 24r^2 - 16r \qquad \text{Multiply.} \qquad \textbf{NOW TRY} \ \circlearrowright$$

NOW TRY ANSWER

5. $8m^3 - 12m^2 + 6m - 1$

5.6 Exercises

FOR
EXTRA
HELP

▶ MyMathLab®

● *Complete solution available
in MyMathLab*

1. *Concept Check*   Consider the square of the binomial $4x + 3$:

$$(4x + 3)^2.$$

(a) What is the first term of the binomial? Square it.

(b) Find twice the product of the two terms of the binomial: $2(\underline{\quad})(\underline{\quad}) = \underline{\quad}$.

(c) What is the last term of the binomial? Square it.

(d) Use the results of parts (a)–(c) to find $(4x + 3)^2$.

2. *Concept Check*   Consider the product of $(7x + 3y)$ and $(7x - 3y)$:

$$(7x + 3y)(7x - 3y).$$

(a) What is the first term of each binomial factor? Square it.

(b) What is the product of the outer terms? The inner terms? Add them.

(c) What are the last terms of the binomial factors? Multiply them.

(d) Use the results of parts (a)–(c) to find $(7x + 3y)(7x - 3y)$.

Find each product. See Examples 1 and 2.

3. $(m + 2)^2$ **4.** $(x + 8)^2$ **5.** $(r - 3)^2$ **6.** $(z - 5)^2$

 7. $(x + 2y)^2$ **8.** $(p - 3m)^2$ **9.** $(5p + 2q)^2$ **10.** $(8a + 3b)^2$

11. $(4x - 3)^2$ **12.** $(9x - 4)^2$ **13.** $(4a + 5b)^2$ **14.** $(9y + 4z)^2$

15. $\left(6m - \dfrac{4}{5}n\right)^2$ **16.** $\left(5x + \dfrac{2}{5}y\right)^2$ **17.** $\left(\dfrac{1}{2}x + \dfrac{1}{3}\right)^2$ **18.** $\left(\dfrac{1}{4}x + \dfrac{1}{5}\right)^2$

19. $2(x + 6)^2$ **20.** $4(x + 3)^2$ **21.** $t(3t - 1)^2$ **22.** $x(2x + 5)^2$

23. $3t(4t + 1)^2$ **24.** $2x(7x - 2)^2$ **25.** $-(4r - 2)^2$ **26.** $-(3y - 8)^2$

Find each product. ***See Examples 3 and 4.***

27. $(k + 5)(k - 5)$ **28.** $(a + 8)(a - 8)$ **29.** $(4 - 3t)(4 + 3t)$

30. $(7 - 2x)(7 + 2x)$ **31.** $(5x + 2)(5x - 2)$ **32.** $(2m + 5)(2m - 5)$

33. $(5y + 3x)(5y - 3x)$ **34.** $(3x + 4y)(3x - 4y)$ **35.** $(10x + 3y)(10x - 3y)$

36. $(13r + 2z)(13r - 2z)$ **37.** $(2x^2 - 5)(2x^2 + 5)$ **38.** $(9y^2 - 2)(9y^2 + 2)$

39. $\left(\dfrac{3}{4} - x\right)\left(\dfrac{3}{4} + x\right)$ **40.** $\left(\dfrac{2}{3} + r\right)\left(\dfrac{2}{3} - r\right)$ **41.** $\left(9y + \dfrac{2}{3}\right)\left(9y - \dfrac{2}{3}\right)$

42. $\left(7x + \dfrac{3}{7}\right)\left(7x - \dfrac{3}{7}\right)$ **43.** $q(5q - 1)(5q + 1)$ **44.** $p(3p + 7)(3p - 7)$

45. $-5(a - b^3)(a + b^3)$ **46.** $-6(r - s^4)(r + s^4)$

47. $\dfrac{1}{2}(2k - 1)(2k + 1)$ **48.** $\dfrac{1}{3}(3m - 5)(3m + 5)$

49. $-\dfrac{1}{100}(10x + 10)(10x - 10)$ **50.** $-\dfrac{1}{200}(20y + 20)(20y - 20)$

Find each product. ***See Example 5.***

51. $(x + 1)^3$ **52.** $(y + 2)^3$ **53.** $(t - 3)^3$ **54.** $(m - 5)^3$

55. $(r + 5)^3$ **56.** $(p + 3)^3$ **57.** $(2a + 1)^3$ **58.** $(3m + 1)^3$

59. $(4x - 1)^4$ **60.** $(2x - 1)^4$ **61.** $(3r - 2t)^4$ **62.** $(2z + 5y)^4$

63. $2x(x + 1)^3$ **64.** $3y(y + 2)^3$ **65.** $-4t(t + 3)^3$

66. $-5r(r + 1)^3$ **67.** $(x + y)^2(x - y)^2$ **68.** $(s + 2)^2(s - 2)^2$

69. *Concept Check* Does $(a + b)^n = a^n + b^n$ hold true in general?

70. *Concept Check* Give values for a, b, and n for which $(a + b)^n = a^n + b^n$ is true.

The special product $(x + y)(x - y) = x^2 - y^2$ *can be used to perform some multiplication problems. Here are two examples.*

$$
\begin{aligned}
51 \times 49 &= (50 + 1)(50 - 1) \\
&= 50^2 - 1^2 \\
&= 2500 - 1 \\
&= 2499
\end{aligned}
\qquad
\begin{aligned}
102 \times 98 &= (100 + 2)(100 - 2) \\
&= 100^2 - 2^2 \\
&= 10{,}000 - 4 \\
&= 9996
\end{aligned}
$$

Once these patterns are recognized, multiplications of this type can be done mentally. Use this method to calculate each product mentally.

71. 101×99 **72.** 103×97 **73.** 201×199

74. 301×299 **75.** $20\dfrac{1}{2} \times 19\dfrac{1}{2}$ **76.** $30\dfrac{1}{3} \times 29\dfrac{2}{3}$

Find a polynomial that represents the area of each figure. (If necessary, refer to the formulas found inside the back cover of this book.)

77.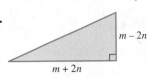
$m + 2n$
$m - 2n$

78.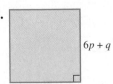
$6p + q$
$6p + q$

79.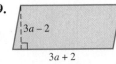
$3a - 2$
$3a + 2$

80.
$4b - 1$
$4b + 1$

81.
$x + 2$

82.
$3x + 1$
4
$5x + 3$

In Exercises 83 and 84, refer to the figure shown here.

83. Find a polynomial that represents the volume of the cube (in cubic units).

84. If the value of x is 6, what is the volume of the cube (in cubic units)?

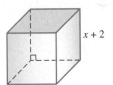
$x + 2$

RELATING CONCEPTS For Individual or Group Work (Exercises 85–94)

*Special products can be illustrated by using areas of rectangles. Use the figure, and **work Exercises 85–90 in order** to justify*

$$(a + b)^2 = a^2 + 2ab + b^2.$$

85. Express the area of the large square as the square of a binomial.

86. Give the monomial that represents the area of the red square.

87. Give the monomial that represents the sum of the areas of the blue rectangles.

88. Give the monomial that represents the area of the yellow square.

89. What is the sum of the monomials obtained in **Exercises 86–88**?

90. Explain why the binomial square found in **Exercise 85** must equal the polynomial found in **Exercise 89**.

*To understand how the special product $(a + b)^2 = a^2 + 2ab + b^2$ can be applied to a purely numerical problem, **work Exercises 91–94 in order**.*

91. Evaluate 35^2, using either traditional paper-and-pencil methods or a calculator.

92. The number 35 can be written as $30 + 5$. Therefore, $35^2 = (30 + 5)^2$. Use the special product for squaring a binomial with $a = 30$ and $b = 5$ to write an expression for $(30 + 5)^2$. Do not simplify at this time.

93. Use the order of operations to simplify the expression found in **Exercise 92**.

94. How do the answers in **Exercises 91 and 94** compare?

5.7 Dividing Polynomials

OBJECTIVES

1. Divide a polynomial by a monomial.
2. Divide a polynomial by a polynomial.
3. Use division in a geometry application.

OBJECTIVE 1 Divide a polynomial by a monomial.

We add two fractions with a common denominator as follows.

$$\frac{a}{c} + \frac{b}{c} = \frac{a + b}{c}$$

In reverse, this statement gives a rule for dividing a polynomial by a monomial.

Dividing a Polynomial by a Monomial

To divide a polynomial by a monomial, divide each term of the polynomial by the monomial.

$$\frac{a+b}{c} = \frac{a}{c} + \frac{b}{c} \quad (\text{where } c \neq 0)$$

Examples: $\quad \dfrac{2+5}{3} = \dfrac{2}{3} + \dfrac{5}{3} \quad$ and $\quad \dfrac{x+3z}{2y} = \dfrac{x}{2y} + \dfrac{3z}{2y}$

The parts of a division problem are named as follows.

Dividend $\longrightarrow$ $\dfrac{12x^2 + 6x}{6x}$ $= 2x + 1 \longleftarrow$ Quotient
Divisor $\longrightarrow$

**NOW TRY
EXERCISE 1**

Divide $16a^6 - 12a^4$ by $4a^2$.

EXAMPLE 1 Dividing a Polynomial by a Monomial

Divide $5m^5 - 10m^3$ by $5m^2$.

$$\frac{5m^5 - 10m^3}{5m^2} \quad \text{A fraction bar means division.}$$

$$= \frac{5m^5}{5m^2} - \frac{10m^3}{5m^2} \quad \text{Use the preceding rule, with + replaced by }-.$$

$$= m^3 - 2m \quad \text{Quotient rule}$$

CHECK   Multiply $\quad 5m^2 \cdot (m^3 - 2m) = 5m^5 - 10m^3 \quad$ ✓

Divisor   Quotient   Original polynomial (Dividend)

Because division by 0 is undefined, the quotient $\frac{5m^5 - 10m^3}{5m^2}$ is undefined if $5m^2 = 0$, or $m = 0$. From now on, we assume that no denominators are 0.   **NOW TRY**

**NOW TRY
EXERCISE 2**

Divide.

$$\frac{36x^5 + 24x^4 - 12x^3}{6x^4}$$

EXAMPLE 2 Dividing a Polynomial by a Monomial

Divide.

$$\frac{16a^5 - 12a^4 + 8a^2}{4a^3}$$

This becomes $\frac{2}{a}$, **not** 2a.

$$= \frac{16a^5}{4a^3} - \frac{12a^4}{4a^3} + \frac{8a^2}{4a^3} \quad \text{Divide each term by } 4a^3.$$

$$= 4a^2 - 3a + \frac{2}{a} \quad \text{Quotient rule}$$

The quotient $4a^2 - 3a + \frac{2}{a}$ is *not* a polynomial because $\frac{2}{a}$ has a variable in the denominator. While the sum, difference, and product of two polynomials are always polynomials, the quotient of two polynomials may not be a polynomial.

CHECK $4a^3 \left(4a^2 - 3a + \dfrac{2}{a} \right)$ Divisor × Quotient should equal Dividend.

$$= 4a^3(4a^2) + 4a^3(-3a) + 4a^3 \left(\frac{2}{a} \right) \quad \text{Distributive property}$$

$$= 16a^5 - 12a^4 + 8a^2 \quad ✓ \quad \text{Dividend} \qquad \text{NOW TRY}$$

NOW TRY ANSWERS

1. $4a^4 - 3a^2$

2. $6x + 4 - \dfrac{2}{x}$

⚠ **CAUTION**   The most frequent error in a problem like that in **Example 2** is with the last term of the quotient.

$$\frac{8a^2}{4a^3} = \frac{8}{4}a^{2-3} = 2a^{-1} = 2\left(\frac{1}{a}\right) = \frac{2}{a}$$

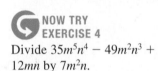
NOW TRY EXERCISE 3
Divide $7y^4 - 40y^5 + 100y^2$ by $-5y^2$.

EXAMPLE 3   Dividing a Polynomial by a Monomial (Negative Coefficient)

Divide $-7x^3 + 12x^4 - 4x$ by $-4x$.

Write the dividend polynomial in descending powers as $12x^4 - 7x^3 - 4x$.

$$\underbrace{\frac{12x^4 - 7x^3 - 4x}{-4x}}_{\text{Write in descending powers before dividing.}}$$

$$= \frac{12x^4}{-4x} - \frac{7x^3}{-4x} - \frac{4x}{-4x} \qquad \text{Divide each term by } -4x.$$

$$= -3x^3 - \frac{7x^2}{-4} - (-1) \qquad \text{Quotient rule}$$

$$= -3x^3 + \frac{7x^2}{4} + 1 \qquad \boxed{\text{Be careful with signs, and be sure to include 1 in the answer.}}$$

Check by multiplying this quotient by $-4x$ to obtain the dividend.   **NOW TRY** ↻

NOW TRY EXERCISE 4
Divide $35m^5n^4 - 49m^2n^3 + 12mn$ by $7m^2n$.

EXAMPLE 4   Dividing a Polynomial by a Monomial

Divide $180x^4y^{10} - 150x^3y^8 + 120x^2y^6 - 90xy^4 + 100y$ by $30xy^2$.

$$\frac{180x^4y^{10} - 150x^3y^8 + 120x^2y^6 - 90xy^4 + 100y}{30xy^2}$$

$$= \frac{180x^4y^{10}}{30xy^2} - \frac{150x^3y^8}{30xy^2} + \frac{120x^2y^6}{30xy^2} - \frac{90xy^4}{30xy^2} + \frac{100y}{30xy^2}$$

$$= 6x^3y^8 - 5x^2y^6 + 4xy^4 - 3y^2 + \frac{10}{3xy} \qquad \text{**NOW TRY** ↻}$$

OBJECTIVE 2   Divide a polynomial by a polynomial.

We use a method of "long division" to divide a polynomial by a polynomial (other than a monomial). *Both polynomials must first be written in descending powers.*

Dividing Whole Numbers	Dividing Polynomials
Step 1 Divide 6696 by 27. $27\overline{)6696}$	Divide $8x^3 - 4x^2 - 14x + 15$ by $2x + 3$. $2x + 3\overline{)8x^3 - 4x^2 - 14x + 15}$
Step 2 66 divided by 27 = 2. $2 \cdot 27 = 54$ 2 $27\overline{)6696}$ 54	$8x^3$ divided by $2x = 4x^2$. $4x^2(2x + 3) = 8x^3 + 12x^2$ $4x^2$ $2x + 3\overline{)8x^3 - 4x^2 - 14x + 15}$ $8x^3 + 12x^2$

Step 3

Subtract. Then bring down the next digit.

$$
\begin{array}{r}
2 \\
27\overline{)6696} \\
54\downarrow \\
\overline{129}
\end{array}
$$

Subtract. Then bring down the next term.

$$
\begin{array}{r}
4x^2 \\
2x+3\overline{)8x^3-4x^2-14x+15} \\
8x^3+12x^2\downarrow \\
\overline{-16x^2-14x}
\end{array}
$$

(To subtract two polynomials, change the signs of the second and then add.)

Step 4

129 divided by 27 = 4.
4 · 27 = 108

$$
\begin{array}{r}
24 \\
27\overline{)6696} \\
54 \\
\overline{129} \\
108
\end{array}
$$

$-16x^2$ divided by $2x = -8x$.
$-8x(2x+3) = -16x^2-24x$

$$
\begin{array}{r}
4x^2-8x \\
2x+3\overline{)8x^3-4x^2-14x+15} \\
8x^3+12x^2 \\
\overline{-16x^2-14x} \\
-16x^2-24x
\end{array}
$$

Step 5

Subtract. Then bring down the next digit.

$$
\begin{array}{r}
24 \\
27\overline{)6696} \\
54\big| \\
\overline{129}\big| \\
108\big| \\
\overline{216}
\end{array}
$$

Subtract. Then bring down the next term.

$$
\begin{array}{r}
4x^2-8x \\
2x+3\overline{)8x^3-4x^2-14x+15} \\
8x^3+12x^2 \\
\overline{-16x^2-14x} \\
\underline{-16x^2-24x} \\
10x+15
\end{array}
$$

Step 6

216 divided by 27 = 8.
8 · 27 = 216

$$
\begin{array}{r}
248 \\
27\overline{)6696} \\
54 \\
\overline{129} \\
108 \\
\overline{216} \\
216
\end{array}
$$

Remainder $\longrightarrow 0$

10x divided by 2x = 5.
5(2x + 3) = 10x + 15

$$
\begin{array}{r}
4x^2-8x+5 \\
2x+3\overline{)8x^3-4x^2-14x+15} \\
8x^3+12x^2 \\
\overline{-16x^2-14x} \\
\underline{-16x^2-24x} \\
10x+15 \\
10x+15
\end{array}
$$

Remainder $\longrightarrow 0$

6696 divided by 27 is 248. The remainder is 0.

$8x^3 - 4x^2 - 14x + 15$ divided by $2x + 3$ is $4x^2 - 8x + 5$. The remainder is 0.

Step 7

CHECK Multiply.

$$27 \cdot 248 = 6696 \ ✓$$

CHECK Multiply.

$$(2x+3)(4x^2-8x+5)$$
$$= 8x^3 - 4x^2 - 14x + 15 \ ✓$$

**NOW TRY
EXERCISE 5**

Divide.

$$\frac{4x^2 + x - 18}{x - 2}$$

EXAMPLE 5   Dividing a Polynomial by a Polynomial

Divide. $\dfrac{3x^2 - 5x - 28}{x - 4}$

Step 1 $3x^2$ divided by x is $3x$.
$3x(x - 4) = 3x^2 - 12x$

Step 2   Subtract $3x^2 - 12x$ from
$3x^2 - 5x$. Bring down -28.

Step 3 $7x$ divided by x is 7.
$7(x - 4) = 7x - 28$

Step 4   Subtract $7x - 28$ from $7x - 28$. The remainder is 0.

$$\begin{array}{r}
3x + 7 \quad \leftarrow \text{Quotient} \\
x - 4 \overline{)\, 3x^2 - 5x - 28\,} \leftarrow \text{Dividend} \\
\underline{3x^2 - 12x} \\
7x - 28 \\
\underline{7x - 28} \\
0
\end{array}$$

Divisor $\downarrow$

CHECK   Multiply the divisor, $x - 4$, by the quotient, $3x + 7$. The product must be the original dividend, $3x^2 - 5x - 28$.

$$(x - 4)(3x + 7) = 3x^2 + 7x - 12x - 28$$
$$= 3x^2 - 5x - 28 \;\checkmark$$

Divisor   Quotient

Dividend

NOW TRY

EXAMPLE 6   Dividing a Polynomial by a Polynomial

Divide. $\dfrac{5x + 4x^3 - 8 - 4x^2}{2x - 1}$

The first polynomial must be written in descending powers as $4x^3 - 4x^2 + 5x - 8$. Then divide by $2x - 1$.

$$\begin{array}{r}
2x^2 - x + 2 \\
2x - 1 \overline{)\, 4x^3 - 4x^2 + 5x - 8\,} \\
\underline{4x^3 - 2x^2} \\
-2x^2 + 5x \\
\underline{-2x^2 + x} \\
4x - 8 \\
\underline{4x - 2} \\
-6 \quad \leftarrow \text{Remainder}
\end{array}$$

Write in descending powers.

In each subtraction, add the opposite.

Step 1 $4x^3$ divided by $2x$ is $2x^2$. $2x^2(2x - 1) = 4x^3 - 2x^2$

Step 2   Subtract. Bring down the next term.

Step 3 $-2x^2$ divided by $2x$ is $-x$. $-x(2x - 1) = -2x^2 + x$

Step 4   Subtract. Bring down the next term.

Step 5 $4x$ divided by $2x$ is 2. $2(2x - 1) = 4x - 2$

Step 6   Subtract. The remainder is -6. Write the remainder as the numerator of a fraction that has $2x - 1$ as its denominator. Because there is a nonzero remainder, the answer is not a polynomial.

Remember to add $\frac{\text{remainder}}{\text{divisor}}$. Don't forget the $+$ sign.

Dividend $\longrightarrow$
Divisor $\longrightarrow$
$$\dfrac{4x^3 - 4x^2 + 5x - 8}{2x - 1} = \underbrace{2x^2 - x + 2}_{\substack{\text{Quotient} \\ \text{polynomial}}} + \underbrace{\dfrac{-6}{2x - 1}}_{\substack{\text{Fractional part} \\ \text{of quotient}}} \quad \begin{array}{l} \leftarrow \text{Remainder} \\ \leftarrow \text{Divisor} \end{array}$$

NOW TRY ANSWER
5. $4x + 9$

**NOW TRY
EXERCISE 6**

Divide.

$$\frac{6k^3 - 20k - k^2 + 1}{2k - 3}$$

Step 7 CHECK

$$(2x - 1)\left(2x^2 - x + 2 + \frac{-6}{2x - 1}\right) \quad \text{Multiply Divisor} \times \text{(Quotient including the Remainder).}$$

$$= (2x - 1)(2x^2) + (2x - 1)(-x) + (2x - 1)(2) + (2x - 1)\left(\frac{-6}{2x - 1}\right)$$

$$= 4x^3 - 2x^2 - 2x^2 + x + 4x - 2 - 6$$

$$= 4x^3 - 4x^2 + 5x - 8 \ \checkmark \qquad \text{NOW TRY} \ \circlearrowleft$$

**NOW TRY
EXERCISE 7**

Divide $m^3 - 1000$ by $m - 10$.

EXAMPLE 7 Dividing into a Polynomial with Missing Terms

Divide $x^3 - 1$ by $x - 1$.

Here, the dividend, $x^3 - 1$, is missing the x^2-term and the x-term. We use 0 as the coefficient for each missing term. Thus, $x^3 - 1 = x^3 + 0x^2 + 0x - 1$.

$$
\begin{array}{r}
x^2 + x + 1 \\
x - 1 \overline{\smash{\big)}\ x^3 + 0x^2 + 0x - 1} \\
\underline{x^3 - x^2} \\
x^2 + 0x \\
\underline{x^2 - x} \\
x - 1 \\
\underline{x - 1} \\
0
\end{array}
$$

Insert placeholders for the missing terms.

The remainder is 0. The quotient is $x^2 + x + 1$.

CHECK $(x - 1)(x^2 + x + 1)$

$$= x^3 + x^2 + x - x^2 - x - 1$$

$$= x^3 - 1 \ \checkmark \quad \text{Divisor} \times \text{Quotient} = \text{Dividend} \qquad \text{NOW TRY} \ \circlearrowleft$$

**NOW TRY
EXERCISE 8**

Divide

$$y^4 - 5y^3 + 6y^2 + y - 4$$

by $y^2 + 2$.

EXAMPLE 8 Dividing by a Polynomial with Missing Terms

Divide $x^4 + 2x^3 + 2x^2 - x - 1$ by $x^2 + 1$.

Because the divisor, $x^2 + 1$, has a missing x-term, write it as $x^2 + 0x + 1$.

$$
\begin{array}{r}
x^2 + 2x + 1 \\
x^2 + 0x + 1 \overline{\smash{\big)}\ x^4 + 2x^3 + 2x^2 - x - 1} \\
\underline{x^4 + 0x^3 + x^2} \\
2x^3 + x^2 - x \\
\underline{2x^3 + 0x^2 + 2x} \\
x^2 - 3x - 1 \\
\underline{x^2 + 0x + 1} \\
-3x - 2 \ \leftarrow \text{Remainder}
\end{array}
$$

Insert a placeholder for the missing term.

When the result of subtracting ($-3x - 2$ here) is a constant or a polynomial of degree less than the divisor ($x^2 + 0x + 1$), that constant or polynomial is the remainder. We write the answer as follows.

$$x^2 + 2x + 1 + \frac{-3x - 2}{x^2 + 1}$$

Remember to write "$+ \frac{\text{remainder}}{\text{divisor}}$".

Multiply to check that this is correct. NOW TRY

NOW TRY ANSWERS

6. $3k^2 + 4k - 4 + \dfrac{-11}{2k - 3}$

7. $m^2 + 10m + 100$

8. $y^2 - 5y + 4 + \dfrac{11y - 12}{y^2 + 2}$

**NOW TRY
EXERCISE 9**
Divide $10x^3 + 21x^2 + 5x - 8$
by $2x + 4$.

| EXAMPLE 9 | Dividing a Polynomial When the Quotient Has Fractional Coefficients |

Divide $4x^3 + 2x^2 + 3x + 2$ by $4x - 4$.

$$\frac{6x^2}{4x} = \frac{3}{2}x$$
$$x^2 + \frac{3}{2}x + \frac{9}{4} \leftarrow \frac{9x}{4x} = \frac{9}{4}$$

$$
\begin{array}{r}
4x - 4 \overline{)4x^3 + 2x^2 + 3x + 2} \\
\underline{4x^3 - 4x^2} \\
6x^2 + 3x \\
\underline{6x^2 - 6x} \\
9x + 2 \\
\underline{9x - 9} \\
11
\end{array}
$$

The answer is $x^2 + \frac{3}{2}x + \frac{9}{4} + \frac{11}{4x - 4}$.

NOW TRY

OBJECTIVE 3 Use division in a geometry application.

| EXAMPLE 10 | Using an Area Formula |

**NOW TRY
EXERCISE 10**
The area of a rectangle is given
by $(x^3 + 7x^2 + 17x + 20)$ sq.
units. The width is given by
$(x + 4)$ units. What is its
length?

The area of the rectangle in **FIGURE 5** is given
by $(x^3 + 4x^2 + 8x + 8)$ sq. units. The width is
given by $(x + 2)$ units. What is its length?

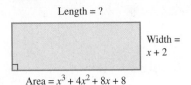

Length = ?

Width = $x + 2$

Area = $x^3 + 4x^2 + 8x + 8$

FIGURE 5

For a rectangle, $\mathcal{A} = LW$. Solving for L gives $L = \frac{\mathcal{A}}{W}$. Divide the area,
$x^3 + 4x^2 + 8x + 8$, by the width, $x + 2$, to find the length.

$$
\begin{array}{r}
x^2 + 2x + 4 \\
x + 2 \overline{)x^3 + 4x^2 + 8x + 8} \\
\underline{x^3 + 2x^2} \\
2x^2 + 8x \\
\underline{2x^2 + 4x} \\
4x + 8 \\
\underline{4x + 8} \\
0
\end{array}
$$

NOW TRY ANSWERS
9. $5x^2 + \frac{1}{2}x + \frac{3}{2} + \frac{-14}{2x + 4}$
10. $(x^2 + 3x + 5)$ units

The quotient $(x^2 + 2x + 4)$ units represents the length.

NOW TRY

5.7 Exercises

FOR EXTRA HELP

 MyMathLab®

▶ *Complete solution available
in MyMathLab*

Concept Check   *Complete each statement.*

1. In the statement $\frac{10x^2 + 8}{2} = 5x^2 + 4$, _____ is the dividend, _____ is the divisor, and
_____ is the quotient.

2. The expression $\frac{3x + 13}{x}$ is undefined for $x =$ _____.

3. To check the division shown in **Exercise 1,** multiply _____ by _____ and show that the
product is _____.

4. The expression $5x^2 - 4x + 6 + \frac{2}{x}$ (*is / is not*) a polynomial.

Concept Check   *Divide.*

5. $\dfrac{6p^4 + 18p^7}{3p^2}$

$= \dfrac{\overline{}}{3p^2} + \dfrac{\overline{}}{3p^2}$

$= \underline{}$

6. $\dfrac{20x^4 - 25x^3 + 5x}{5x^2}$

$= \dfrac{\overline{}}{\underline{}} - \dfrac{\overline{}}{\underline{}} + \dfrac{\overline{}}{\underline{}}$

$= \underline{}$

Perform each division. **See Examples 1–3.**

7. $\dfrac{60x^4 - 20x^2 + 10x}{2x}$

8. $\dfrac{120x^6 - 60x^3 + 80x^2}{2x}$

9. $\dfrac{20m^5 - 10m^4 + 5m^2}{5m^2}$

10. $\dfrac{12t^5 - 6t^3 + 6t^2}{6t^2}$

11. $\dfrac{8t^5 - 4t^3 + 4t^2}{2t}$

12. $\dfrac{8r^4 - 4r^3 + 6r^2}{2r}$

▶ **13.** $\dfrac{4a^5 - 4a^2 + 8}{4a}$

14. $\dfrac{5t^8 + 5t^7 + 15}{5t}$

15. $\dfrac{18p^5 + 12p^3 - 6p^2}{-6p^3}$

16. $\dfrac{32x^8 + 24x^5 - 8x}{-8x^2}$

17. $\dfrac{-7r^7 + 6r^5 - r^4}{-r^5}$

18. $\dfrac{-13t^9 + 8t^6 - t^5}{-t^6}$

Divide each polynomial by $3x^2$. **See Examples 1–3.**

▶ **19.** $12x^5 - 9x^4 + 6x^3$

20. $24x^6 - 12x^5 + 30x^4$

21. $3x^2 + 15x^3 - 27x^4$

22. $3x^2 - 18x^4 + 30x^5$

23. $36x + 24x^2 + 6x^3$

24. $9x - 12x^2 + 9x^3$

25. $4x^4 + 3x^3 + 2x$

26. $5x^4 - 6x^3 + 8x$

27. $-81x^5 + 30x^4 + 12x^2$

28. *Concept Check*   If $-60x^5 - 30x^4 + 20x^3$ is divided by $3x^2$, what is the sum of the coefficients of the third- and second-degree terms in the quotient?

Perform each division. **See Examples 1–4.**

▶ **29.** $\dfrac{-27r^4 + 36r^3 - 6r^2 - 26r + 2}{-3r}$

30. $\dfrac{-8k^4 + 12k^3 + 2k^2 - 7k + 3}{-2k}$

31. $\dfrac{2m^5 - 6m^4 + 8m^2}{-2m^3}$

32. $\dfrac{6r^5 - 8r^4 + 10r^2}{-2r^3}$

33. $(20a^4 - 15a^5 + 25a^3) \div (5a^4)$

34. $(36y^2 - 12y^3 + 20y) \div (4y^2)$

35. $(120x^{11} - 60x^{10} + 140x^9 - 100x^8) \div (10x^{12})$

36. $(45y^7 + 9y^6 - 6y^5 + 12y^4) \div (3y^8)$

▶ **37.** $(120x^5y^4 - 80x^2y^3 + 40x^2y^4 - 20x^5y^3) \div (20xy^2)$

38. $(200a^5b^6 - 160a^4b^7 - 120a^3b^9 + 40a^2b^2) \div (40a^2b)$

Perform each division using the "long division" process. **See Examples 5 and 6.**

39. $\dfrac{x^2 - x - 6}{x - 3}$

40. $\dfrac{m^2 - 2m - 24}{m - 6}$

41. $\dfrac{2y^2 + 9y - 35}{y + 7}$

42. $\dfrac{2y^2 + 9y + 7}{y + 1}$

43. $\dfrac{p^2 + 2p + 20}{p + 6}$

44. $\dfrac{x^2 + 11x + 16}{x + 8}$

45. $\dfrac{12m^2 - 20m + 3}{2m - 3}$

46. $\dfrac{12y^2 + 20y + 7}{2y + 1}$

47. $\dfrac{4a^2 - 22a + 32}{2a + 3}$

48. $\dfrac{9w^2 + 6w + 10}{3w - 2}$

▶ **49.** $\dfrac{8x^3 - 10x^2 - x + 3}{2x + 1}$

50. $\dfrac{12t^3 - 11t^2 + 9t + 18}{4t + 3}$

51. $\dfrac{8k^4 - 12k^3 - 2k^2 + 7k - 6}{2k - 3}$

52. $\dfrac{27r^4 - 36r^3 - 6r^2 + 26r - 24}{3r - 4}$

53. $\dfrac{5y^4 + 5y^3 + 2y^2 - y - 8}{y + 1}$

54. $\dfrac{2r^3 - 5r^2 - 6r + 15}{r - 3}$

▶ **55.** $\dfrac{3k^3 - 4k^2 - 6k + 10}{k - 2}$

56. $\dfrac{5z^3 - z^2 + 10z + 2}{z + 2}$

57. $\dfrac{6p^4 - 16p^3 + 15p^2 - 5p + 10}{3p + 1}$

58. $\dfrac{6r^4 - 11r^3 - r^2 + 16r - 8}{2r - 3}$

Perform each division. **See Examples 6–9.**

▶ **59.** $(x^3 + 2x^2 - 3) \div (x - 1)$

60. $(x^3 - 2x^2 - 9) \div (x - 3)$

61. $(2x^3 + x + 2) \div (x + 3)$

62. $(3x^3 + x + 5) \div (x + 1)$

63. $\dfrac{5 - 2r^2 + r^4}{r^2 - 4}$

64. $\dfrac{4t^2 + t^4 + 7}{t^2 - 4}$

65. $\dfrac{-4x + 3x^3 + 2}{x - 1}$

66. $\dfrac{-5x + 6x^3 + 5}{x - 1}$

67. $\dfrac{y^3 + 27}{y + 3}$

68. $\dfrac{y^3 - 64}{y - 4}$

69. $\dfrac{a^4 - 25}{a^2 - 5}$

70. $\dfrac{a^4 - 36}{a^2 + 6}$

▶ **71.** $\dfrac{x^4 - 4x^3 + 5x^2 - 3x + 2}{x^2 + 3}$

72. $\dfrac{3t^4 + 5t^3 - 8t^2 - 13t + 2}{t^2 - 5}$

73. $\dfrac{2x^5 + 9x^4 + 8x^3 + 10x^2 + 14x + 5}{2x^2 + 3x + 1}$

74. $\dfrac{4t^5 - 11t^4 - 6t^3 + 5t^2 - t + 3}{4t^2 + t - 3}$

75. $(3a^2 - 11a + 17) \div (2a + 6)$

76. $(4x^2 + 11x - 8) \div (3x + 6)$

77. $\dfrac{3x^3 + 5x^2 - 9x + 5}{3x - 3}$

78. $\dfrac{5x^3 + 4x^2 + 10x + 20}{5x + 5}$

*In Exercises 79–84, if necessary, refer to the formulas found inside the back cover of this book. **See Example 10.***

79. The area of the rectangle is given by the polynomial

$$5x^3 + 7x^2 - 13x - 6.$$

What polynomial expresses the length (in appropriate units)?

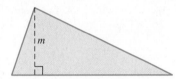

80. The area of the rectangle is given by the polynomial

$$15x^3 + 12x^2 - 9x + 3.$$

What polynomial expresses the length (in appropriate units)?

81. The area of the triangle is given by the polynomial

$$24m^3 + 48m^2 + 12m.$$

What polynomial expresses the length of the base (in appropriate units)?

82. The area of the parallelogram is given by the polynomial

$$2x^3 + 2x^2 - 3x - 1.$$

What polynomial expresses the length of the base (in appropriate units)?

83. If the distance traveled is $(5x^3 - 6x^2 + 3x + 14)$ miles and the rate is $(x + 1)$ mph, write an expression, in hours, for the time traveled.

84. If it costs $(4x^5 + 3x^4 + 2x^3 + 9x^2 - 29x + 2)$ dollars to fertilize a garden, and fertilizer costs $(x + 2)$ dollars per square yard, write an expression, in square yards, for the area of the garden.

Chapter 5   Summary

Key Terms

5.1
base
exponent (power)
exponential expression

5.3
scientific notation
standard notation

5.4
term
leading term
numerical coefficient
 (coefficient)
like terms
unlike terms
polynomial

descending powers
degree of a term
degree of a polynomial
monomial
binomial
trinomial
parabola
vertex
axis of symmetry (axis)

5.5
FOIL method
outer product
inner product

5.6
conjugates

New Symbols

x^{-n} x to the negative n
      power

Test Your Word Power

See how well you have learned the vocabulary in this chapter.

1. A **polynomial** is an algebraic expression made up of
   A. a term or a finite product of terms with positive coefficients and exponents
   B. a term or a finite sum of terms with real coefficients and whole number exponents
   C. the product of two or more terms with positive exponents
   D. the sum of two or more terms with whole number coefficients and exponents.

2. The **degree of a term** is
   A. the number of variables in the term
   B. the product of the exponents on the variables
   C. the least exponent on the variables
   D. the sum of the exponents on the variables.

3. The **FOIL** method is used when
   A. adding two binomials
   B. adding two trinomials
   C. multiplying two binomials
   D. multiplying two trinomials.

4. A **binomial** is a polynomial with
   A. only one term
   B. exactly two terms
   C. exactly three terms
   D. more than three terms.

5. A **monomial** is a polynomial with
   A. only one term
   B. exactly two terms
   C. exactly three terms
   D. more than three terms.

6. A **trinomial** is a polynomial with
   A. only one term
   B. exactly two terms
   C. exactly three terms
   D. more than three terms.

ANSWERS

1. B; *Example:* $5x^3 + 2x^2 - 7$ 2. D; *Examples:* The term 6 has degree 0, $3x$ has degree 1, $-2x^8$ has degree 8, and $5x^2y^4$ has degree 6.

3. C; *Example:* $(m + 4)(m - 3) = m(m) - 3m + 4m + 4(-3) = m^2 + m - 12$ 4. B; *Example:* $3t^3 + 5t$ 5. A; *Examples:* -5 and $4xy^5$
6. C; *Example:* $2a^2 - 3ab + b^2$

Quick Review

CONCEPTS

| **EXAMPLES** |

5.1 The Product Rule and Power Rules for Exponents

For any integers m and n, the following hold.

Product Rule $a^m \cdot a^n = a^{m+n}$

Power Rules   (a) $(a^m)^n = a^{mn}$

(b) $(ab)^m = a^m b^m$

(c) $\left(\dfrac{a}{b}\right)^m = \dfrac{a^m}{b^m}$ **(where $b \neq 0$)**

Simplify by using the rules for exponents.

$$2^4 \cdot 2^5 = 2^{4+5} = 2^9$$

$$(3^4)^2 = 3^{4 \cdot 2} = 3^8$$

$$(6a)^5 = 6^5 a^5$$

$$\left(\frac{2}{3}\right)^4 = \frac{2^4}{3^4}$$

5.2 Integer Exponents and the Quotient Rule

If $a \neq 0$, then for integers m and n, the following hold.

Zero Exponent $a^0 = 1$

Negative Exponent $a^{-n} = \dfrac{1}{a^n}$

Quotient Rule $\dfrac{a^m}{a^n} = a^{m-n}$

Negative-to-Positive Rules $\dfrac{a^{-m}}{b^{-n}} = \dfrac{b^n}{a^m}$ **(where $b \neq 0$)**

$\left(\dfrac{a}{b}\right)^{-m} = \left(\dfrac{b}{a}\right)^m$ **(where $b \neq 0$)**

Simplify by using the rules for exponents.

$$15^0 = 1$$

$$5^{-2} = \frac{1}{5^2} = \frac{1}{25}$$

$$\frac{4^8}{4^3} = 4^{8-3} = 4^5$$

$$\frac{4^{-2}}{3^{-5}} = \frac{3^5}{4^2}$$

$$\left(\frac{6}{5}\right)^{-3} = \left(\frac{5}{6}\right)^3$$

CONCEPTS	**EXAMPLES**

5.3 Scientific Notation

To write a positive number in scientific notation

$$a \times 10^n, \quad \text{where} \quad 1 \le |a| < 10,$$

move the decimal point to follow the first nonzero digit.

1. If moving the decimal point makes the number less, then n is positive.
2. If it makes the number greater, n is negative.
3. If the decimal point is not moved, then n is 0.

For a negative number, follow these steps using the absolute value of the number. Then make the result negative.

Write in scientific notation.

$$247 = 2.47 \times 10^2$$
$$0.0051 = 5.1 \times 10^{-3}$$
$$-4.8 = -4.8 \times 10^0$$

Write in standard notation.

$$3.25 \times 10^5 = 325{,}000$$
$$8.44 \times 10^{-6} = 0.00000844$$

5.4 Adding, Subtracting, and Graphing Polynomials

Adding Polynomials
Add like terms.

Add.
$$\begin{array}{r} 2x^2 + 5x - 3 \\ 5x^2 - 2x + 7 \\ \hline 7x^2 + 3x + 4 \end{array}$$

Subtracting Polynomials
Change the signs of the terms in the subtrahend (second polynomial) and add the result to the minuend (first polynomial).

Subtract. $\quad (2x^2 + 5x - 3) - (5x^2 - 2x + 7)$

$$= (2x^2 + 5x - 3) + (-5x^2 + 2x - 7)$$
$$= -3x^2 + 7x - 10$$

Graphing Polynomials
To graph a simple polynomial equation such as

$$y = x^2 - 2,$$

plot points near the vertex. (In this chapter, all parabolas have a vertex on the x-axis or the y-axis.)

Graph $y = x^2 - 2$.

x	y
−2	2
−1	−1
0	−2
1	−1
2	2

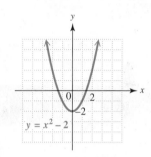

5.5 Multiplying Polynomials

General Method for Multiplying Polynomials
Multiply each term of the first polynomial by each term of the second polynomial. Then add like terms.

Multiply.
$$\begin{array}{r} 3x^3 - 4x^2 + 2x - 7 \\ 4x + 3 \\ \hline 9x^3 - 12x^2 + 6x - 21 \\ 12x^4 - 16x^3 + 8x^2 - 28x \\ \hline 12x^4 - 7x^3 - 4x^2 - 22x - 21 \end{array}$$

FOIL Method for Multiplying Binomials

Step 1 Multiply the two **F**irst terms to obtain the first term of the product.

Step 2 Find the **O**uter product and the **I**nner product and combine them (when possible) to obtain the middle term of the product.

Step 3 Multiply the two **L**ast terms to obtain the last term of the product.

Add the terms found in Steps 1–3.

Multiply. $\quad (2x + 3)(5x - 4)$

$$2x(5x) = 10x^2 \quad \textbf{F}$$

$$2x(-4) + 3(5x) = 7x \quad \textbf{O, I}$$

$$3(-4) = -12 \quad \textbf{L}$$

The product is $10x^2 + 7x - 12$.

CONCEPTS	EXAMPLES

5.6 Special Products

Square of a Binomial

$$(x + y)^2 = x^2 + 2xy + y^2$$
$$(x - y)^2 = x^2 - 2xy + y^2$$

Multiply.

$$(3x + 1)^2$$
$$= (3x)^2 + 2(3x)(1) + 1^2$$
$$= 9x^2 + 6x + 1$$

$$(2m - 5n)^2$$
$$= (2m)^2 - 2(2m)(5n) + (5n)^2$$
$$= 4m^2 - 20mn + 25n^2$$

Product of a Sum and Difference of Two Terms

$$(x + y)(x - y) = x^2 - y^2$$

$$(4a + 3)(4a - 3)$$
$$= (4a)^2 - 3^2$$
$$= 16a^2 - 9$$

5.7 Dividing Polynomials

Dividing a Polynomial by a Monomial
Divide each term of the polynomial by the monomial.

$$\frac{a + b}{c} = \frac{a}{c} + \frac{b}{c} \quad \text{(where } c \neq 0\text{)}$$

Dividing a Polynomial by a Polynomial
Use "long division."

Divide.

$$\frac{4x^3 - 2x^2 + 6x - 9}{2x} = 2x^2 - x + 3 - \frac{9}{2x} \quad \begin{array}{l}\text{Divide each}\\\text{term in the}\\\text{numerator by } 2x.\end{array}$$

$$\begin{array}{r} 2x - 5 \\ 3x + 4 \overline{)6x^2 - 7x - 21} \\ \underline{6x^2 + 8x} \\ -15x - 21 \\ \underline{-15x - 20} \\ -1 \leftarrow \text{Remainder} \end{array}$$

The answer is $2x - 5 + \frac{-1}{3x + 4}$.

Chapter 5   Review Exercises

5.1 *Use the product rule, power rules, or both to simplify each expression. Write each answer in exponential form.*

1. $4^3 \cdot 4^8$

2. $(-5)^6(-5)^5$

3. $(-8x^4)(9x^3)$

4. $(2x^2)(5x^3)(x^9)$

5. $(19x)^5$

6. $(-4y)^7$

7. $5(pt)^4$

8. $\left(\dfrac{7}{5}\right)^6$

9. $(3x^2y^3)^3$

10. $(t^4)^8(t^2)^5$

11. $(6x^2z^4)^2(x^3yz^2)^4$

12. $\left(\dfrac{2m^3n}{p^2}\right)^3$

5.2 *Evaluate each expression.*

13. -10^0

14. $-(-23)^0$

15. $6^0 + (-6)^0$

16. $-3^0 - 2^0$

Simplify each expression. Assume that all variables represent nonzero real numbers.

17. -7^{-2}

18. $\left(\dfrac{5}{8}\right)^{-2}$

19. $(2^{-2})^{-3}$

Answers column:

1. 4^{11}
2. $(-5)^{11}$
3. $-72x^7$
4. $10x^{14}$
5. 19^5x^5
6. $(-4)^7y^7$
7. $5p^4t^4$
8. $\dfrac{7^6}{5^6}$
9. $3^3x^6y^9$
10. t^{42}
11. $6^2x^{16}y^4z^{16}$
12. $\dfrac{2^3m^9n^3}{p^6}$
13. -1
14. -1
15. 2
16. -2
17. $-\dfrac{1}{49}$
18. $\dfrac{64}{25}$
19. 64

20. $\dfrac{1}{81}$

21. $\dfrac{3}{4}$ **22.** $\dfrac{1}{36}$

23. x^2 **24.** y^7

25. $\dfrac{r^8}{81}$ **26.** $\dfrac{3^5}{p^3}$

27. $\dfrac{1}{a^3b^5}$ **28.** $72r^5$

29. 4.8×10^7 **30.** 2.8988×10^{10}
31. 8.24×10^{-8} **32.** -4.82×10^6
33. $24{,}000$ **34.** $78{,}300{,}000$
35. 0.000000897 **36.** -0.00076
37. 800 **38.** 5
39. $4{,}000{,}000$ **40.** 0.025
41. $81{,}887{,}000{,}000{,}000{,}000$
42. $37{,}217{,}400$
43. 1.0086×10^{15}
44. 6.78×10^{13}

45. $20m^2$; degree 2; monomial
46. $p^3 - p^2 - 4p$; degree 3; trinomial
47. $-8y^5 - 7y^4$; degree 5; binomial
48. $-r^3 - 2r + 7$
49. $13x^3y^2 - 5xy^5 + 21x^2$
50. $a^3 + 4a^2$
51. $y^2 - 10y + 9$
52. $-13k^4 - 15k^2 + 18k$

20. $9^3 \cdot 9^{-5}$ **21.** $2^{-1} + 4^{-1}$ **22.** $\dfrac{6^{-5}}{6^{-3}}$

23. $\dfrac{x^{-7}}{x^{-9}}$ **24.** $\dfrac{y^4 \cdot y^{-2}}{y^{-5}}$ **25.** $\left(3r^{-2}\right)^{-4}$

26. $(3p)^4(3p^{-7})$ **27.** $\dfrac{ab^{-3}}{a^4b^2}$ **28.** $\dfrac{(6r^{-1})^2(2r^{-4})}{r^{-5}(r^2)^{-3}}$

5.3   *Write each number in scientific notation.*

29. $48{,}000{,}000$ **30.** $28{,}988{,}000{,}000$

31. 0.0000000824 **32.** $-4{,}820{,}000$

Write each number in standard notation.

33. 2.4×10^4 **34.** 7.83×10^7 **35.** 8.97×10^{-7} **36.** -7.6×10^{-4}

Perform the indicated operations. Write each answer in standard notation.

37. $(2 \times 10^{-3})(4 \times 10^5)$ **38.** $(2.5 \times 10^{-51})(2.0 \times 10^{51})$

39. $\dfrac{8 \times 10^4}{2 \times 10^{-2}}$ **40.** $\dfrac{60 \times 10^{-1}}{24 \times 10}$

Write each boldface italic number in standard form.

41. In a recent year, China was the world's largest energy producer. China accounted for ***8.1887 × 10¹⁶*** Btu. (*Source: World Almanac and Book of Facts.*)

42. The 2011 population of Tokyo, Japan, was ***3.72174 × 10⁷***. (*Source: World Almanac and Book of Facts.*)

Write each boldface italic number in scientific notation.

43. As of July 2013, Japan's outstanding public debt was ***1,008,600,000,000,000*** yen. (*Source:* www.bloomberg.com)

44. In 2011, the budget of the U.S. Department of Defense was ***67,800,000,000,000*** dollars. (*Source:* www.defense.gov)

5.4   *In Exercises 45–47, simplify by combining like terms whenever possible. Write the result in descending powers of the variable. Then give the degree and tell whether the simplified polynomial is a* monomial, *a* binomial, *a* trinomial, *or* none of these.

45. $9m^2 + 11m^2$ **46.** $-4p + p^3 - p^2$ **47.** $-7y^5 - 8y^4 - y^5 + y^4$

Add or subtract as indicated.

48. $(12r^4 - 7r^3 + 2r^2) - (5r^4 - 3r^3 + 2r^2 - 1) - (7r^4 - 3r^3 + 2r - 6)$

49. $(5x^3y^2 - 3xy^5 + 12x^2) - (-9x^2 - 8x^3y^2 + 2xy^5)$

50. Add.

$$-2a^3 + 5a^2$$
$$\underline{3a^3 - a^2}$$

51. Subtract.

$$6y^2 - 8y + 2$$
$$\underline{5y^2 + 2y - 7}$$

52. Subtract.

$$-12k^4 - 8k^2 + 7k$$
$$\underline{k^4 + 7k^2 - 11k}$$

53. 1, 4, 5, 4, 1

54. 10, 1, −2, 1, 10

55. $a^3 - 2a^2 - 7a + 2$
56. $6r^3 + 8r^2 - 17r + 6$
57. $5p^5 - 2p^4 - 3p^3 + 25p^2 + 15p$
58. $m^2 - 7m - 18$
59. $6k^2 - 9k - 6$
60. $2a^2 + 5ab - 3b^2$
61. $12k^2 - 32kq - 35q^2$
62. $s^3 - 3s^2 + 3s - 1$

63. $a^2 + 8a + 16$
64. $4r^2 + 20rt + 25t^2$
65. $36m^2 - 25$
66. $25a^2 - 36b^2$
67. $r^3 + 6r^2 + 12r + 8$
68. $25t^3 - 30t^2 + 9t$
69. (a) Answers will vary. For example, let $x = 1$ and $y = 2$.
$$(1 + 2)^2 \neq 1^2 + 2^2,$$
because $9 \neq 5$.
(b) Answers will vary. For example, let $x = 1$ and $y = 2$.
$$(1 + 2)^3 \neq 1^3 + 2^3,$$
because $27 \neq 9$.
70. Find the third power of a binomial, such as $(a + b)^3$, as follows.
$(a + b)^3$
$\quad = (a + b)(a + b)^2$
$\quad = (a + b)(a^2 + 2ab + b^2)$
$\quad = a^3 + 2a^2b + ab^2 + a^2b +$
$\qquad 2ab^2 + b^3$
$\quad = a^3 + 3a^2b + 3ab^2 + b^3$
71. $x^6 + 6x^4 + 12x^2 + 8$
72. $\dfrac{4}{3}\pi x^3 + 4\pi x^2 + 4\pi x + \dfrac{4}{3}\pi$

73. $\dfrac{-5y^2}{3}$

74. $-2m^2n + mn + \dfrac{6n^3}{5}$

75. $y^3 - 2y + 3$

76. $-6r^5s - 3r^4 + \dfrac{2}{r^2s^5}$

77. $2mn + 3m^4n^2 - 4n$
78. The friend wrote the second term of the quotient as $-12x$ rather than $-2x$.
$$\dfrac{6x^2 - 12x}{6} = \dfrac{6x^2}{6} - \dfrac{12x}{6}$$
$$= x^2 - 2x$$

Graph each equation by completing the table of values.

53. $y = -x^2 + 5$

x	−2	−1	0	1	2
y					

54. $y = 3x^2 - 2$

x	−2	−1	0	1	2
y					

5.5 *Find each product.*

55. $(a + 2)(a^2 - 4a + 1)$

56. $(3r - 2)(2r^2 + 4r - 3)$

57. $(5p^2 + 3p)(p^3 - p^2 + 5)$

58. $(m - 9)(m + 2)$

59. $(3k - 6)(2k + 1)$

60. $(a + 3b)(2a - b)$

61. $(6k + 5q)(2k - 7q)$

62. $(s - 1)^3$

5.6 *Find each product.*

63. $(a + 4)^2$ **64.** $(2r + 5t)^2$ **65.** $(6m - 5)(6m + 5)$

66. $(5a + 6b)(5a - 6b)$ **67.** $(r + 2)^3$ **68.** $t(5t - 3)^2$

69. Choose values for x and y to show that, in general, the following hold true.
   (a) $(x + y)^2 \neq x^2 + y^2$ **(b)** $(x + y)^3 \neq x^3 + y^3$

70. Write an explanation on how to raise a binomial to the third power. Give an example.

In Exercises 71 and 72, refer to the formulas found inside the back cover of this book, if necessary.

71. Find a polynomial that represents, in cubic centimeters, the volume of a cube with one side having length $(x^2 + 2)$ centimeters.

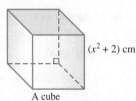

A cube

72. Find a polynomial that represents, in cubic inches, the volume of a sphere with radius $(x + 1)$ inches.

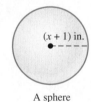

A sphere

5.7 *Perform each division.*

73. $\dfrac{-15y^4}{9y^2}$

74. $(-10m^4n^2 + 5m^3n^2 + 6m^2n^4) \div (5m^2n)$

75. $\dfrac{6y^4 - 12y^2 + 18y}{6y}$

76. $\dfrac{24r^8s^6 + 12r^7s^5 - 8r}{-4r^3s^5}$

77. What polynomial, when multiplied by $6m^2n$, gives the product
$$12m^3n^2 + 18m^6n^3 - 24m^2n^2?$$

78. One of your friends in class simplified
$$\dfrac{6x^2 - 12x}{6} \quad \text{as} \quad x^2 - 12x.$$

WHAT WENT WRONG? Give the correct answer.

79. $2r + 7$

80. $2a^2 + 3a - 1 + \dfrac{6}{5a - 3}$

81. $x^2 + 3x - 4$

82. $m^2 + 4m - 2$

83. $4x - 5$

84. $5y - 10$

85. $y^2 + 2y + 4$

86. $100x^4 - 10x^2 + 1$

87. $2y^2 - 5y + 4 + \dfrac{-5}{3y^2 + 1}$

88. $x^3 - 2x^2 + 4 + \dfrac{-3}{4x^2 - 3}$

Perform each division.

79. $\dfrac{2r^2 + 3r - 14}{r - 2}$

80. $\dfrac{10a^3 + 9a^2 - 14a + 9}{5a - 3}$

81. $\dfrac{x^4 - 5x^2 + 3x^3 - 3x + 4}{x^2 - 1}$

82. $\dfrac{m^4 + 4m^3 - 12m - 5m^2 + 6}{m^2 - 3}$

83. $\dfrac{16x^2 - 25}{4x + 5}$

84. $\dfrac{25y^2 - 100}{5y + 10}$

85. $\dfrac{y^3 - 8}{y - 2}$

86. $\dfrac{1000x^6 + 1}{10x^2 + 1}$

87. $\dfrac{6y^4 - 15y^3 + 14y^2 - 5y - 1}{3y^2 + 1}$

88. $\dfrac{4x^5 - 8x^4 - 3x^3 + 22x^2 - 15}{4x^2 - 3}$

Chapter 5   Mixed Review Exercises

Perform each indicated operation, or simplify each expression. Assume that all variables represent nonzero real numbers.

1. 2

2. $\dfrac{216r^6p^3}{5^3}$

3. $144a^2 - 1$

4. $\dfrac{1}{16}$

5. $\dfrac{1}{256}$

6. $p - 3 + \dfrac{5}{2p}$

7. $\dfrac{2}{3m^3}$

8. $6k^3 - 21k - 6$

9. r^{13}

10. $4r^2 + 20rs + 25s^2$

11. $y^2 + 5y + 1$

12. $10r^2 + 21r - 10$

13. $-y^2 - 4y + 4$

14. $\dfrac{5}{2} - \dfrac{4}{5xy} + \dfrac{3x}{2y^2}$

15. $10p^2 - 3p - 5$

16. $3x^2 + 9x + 25 + \dfrac{80}{x - 3}$

17. $49 - 28k + 4k^2$

18. $\dfrac{1}{x^4 y^{12}}$

19. (a) $6x - 2$

   (b) $2x^2 + x - 6$

20. (a) $20x^4 + 8x^2$

   (b) $25x^8 + 20x^6 + 4x^4$

1. $5^0 + 7^0$

2. $\left(\dfrac{6r^2p}{5}\right)^3$

3. $(12a + 1)(12a - 1)$

4. 2^{-4}

5. $(4^{-2})^2$

6. $\dfrac{2p^3 - 6p^2 + 5p}{2p^2}$

7. $\dfrac{(2m^{-5})(3m^2)^{-1}}{m^{-2}(m^{-1})^2}$

8. $(3k - 6)(2k^2 + 4k + 1)$

9. $\dfrac{r^9 \cdot r^{-5}}{r^{-2} \cdot r^{-7}}$

10. $(2r + 5s)^2$

11. $\dfrac{2y^3 + 17y^2 + 37y + 7}{2y + 7}$

12. $(2r + 5)(5r - 2)$

13. $(-5y^2 + 3y - 11) + (4y^2 - 7y + 15)$

14. $(25x^2y^3 - 8xy^2 + 15x^3y) \div (10x^2y^3)$

15. $(6p^2 - p - 8) - (-4p^2 + 2p - 3)$

16. $\dfrac{3x^3 - 2x + 5}{x - 3}$

17. $(-7 + 2k)^2$

18. $\left(\dfrac{x}{y^{-3}}\right)^{-4}$

Find polynomials that represent (a) the perimeter and (b) the area of each square or rectangle.

19.

20.

| Chapter 5 | Test | FOR EXTRA HELP | *Step-by-step test solutions are found on the Chapter Test Prep Videos available in* MyMathLab®, *or on* You Tube™. |

▶ *View the complete solutions to all Chapter Test exercises in MyMathLab.*

[5.1, 5.2]

1. $\dfrac{1}{625}$

2. 2

3. $\dfrac{7}{12}$

4. $9x^3y^5$

5. 8^5

6. x^2y^6

7. (a) positive   (b) positive
   (c) negative   (d) positive
   (e) zero   (f) negative

[5.3]

8. (a) 4.5×10^{10}
   (b) 0.0000036
   (c) 0.00019

9. (a) 1×10^3; 5.89×10^{12}
   (b) 5.89×10^{15} mi

[5.4]

10. $-7x^2 + 8x$; 2; binomial

11. $4n^4 + 13n^3 - 10n^2$; 4; trinomial

12. $4, -2, -4, -2, 4$

$y = 2x^2 - 4$

13. $-2y^2 - 9y + 17$

14. $-21a^3b^2 + 7ab^5 - 5a^2b^2$

15. $16r^2 - 19$

16. $-12t^2 + 5t + 8$

[5.5]

17. $-27x^5 + 18x^4 - 6x^3 + 3x^2$

18. $t^2 - 5t - 24$

19. $8x^2 + 2xy - 3y^2$

[5.6]

20. $25x^2 - 20xy + 4y^2$

21. $100v^2 - 9w^2$

[5.5]

22. $2r^3 + r^2 - 16r + 15$

Evaluate each expression.

1. 5^{-4}

2. $(-3)^0 + 4^0$

3. $4^{-1} + 3^{-1}$

4. Simplify $\dfrac{(3x^2y)^2(xy^3)^2}{(xy)^3}$. Assume that x and y represent nonzero numbers.

Simplify each expression. Assume that all variables represent nonzero real numbers.

5. $\dfrac{8^{-1} \cdot 8^4}{8^{-2}}$

6. $\dfrac{(x^{-3})^{-2}(x^{-1}y)^2}{(xy^{-2})^2}$

7. Determine whether each expression represents a number that is *positive*, *negative*, or *zero*.

   (a) 3^{-4} (b) $(-3)^4$ (c) -3^4 (d) 3^0 (e) $(-3)^0 - 3^0$ (f) $(-3)^{-3}$

8. (a) Write $45{,}000{,}000{,}000$ using scientific notation.

   (b) Write 3.6×10^{-6} using standard notation.

   (c) Write the quotient $\dfrac{9.5 \times 10^{-1}}{5 \times 10^3}$ using standard notation.

9. A satellite galaxy of the Milky Way, known as the Large Magellanic Cloud, is *1000* light-years across. A *light-year* is equal to *5,890,000,000,000* mi. (*Source:* "Images of Brightest Nebula Unveiled," *USA Today.*)

   (a) Write the two boldface italic numbers in scientific notation.

   (b) How many miles across is the Large Magellanic Cloud?

For each polynomial, simplify by combining like terms whenever possible. Write the result in descending powers of the variable. Then give the degree and tell whether the simplified polynomial is a monomial, *a* binomial, *a* trinomial, *or* none of these.

10. $5x^2 + 8x - 12x^2$

11. $13n^3 - n^2 + n^4 + 3n^4 - 9n^2$

12. Graph the equation $y = 2x^2 - 4$ by completing the table of values.

x	-2	-1	0	1	2
y					

Perform each indicated operation.

13. $(2y^2 - 8y + 8) + (-3y^2 + 2y + 3) - (y^2 + 3y - 6)$

14. $(-9a^3b^2 + 13ab^5 + 5a^2b^2) - (6ab^5 + 12a^3b^2 + 10a^2b^2)$

15. Add.
$$-6r^5 + 4r^2 - 3$$
$$\underline{6r^5 + 12r^2 - 16}$$

16. Subtract.
$$9t^3 - 4t^2 + 2t + 2$$
$$\underline{9t^3 + 8t^2 - 3t - 6}$$

17. $3x^2(-9x^3 + 6x^2 - 2x + 1)$

18. $(t - 8)(t + 3)$

19. $(4x + 3y)(2x - y)$

20. $(5x - 2y)^2$

21. $(10v + 3w)(10v - 3w)$

22. $(2r - 3)(r^2 + 2r - 5)$

[5.6]

23. $12x + 36$

24. $9x^2 + 54x + 81$

[5.7]

25. $4y^2 - 3y + 2 + \dfrac{5}{y}$

26. $-3xy^2 + 2x^3y^2 + 4y^2$

27. $x - 2$

28. $3x^2 + 6x + 11 + \dfrac{26}{x - 2}$

Refer to the square below. Find polynomials that represent the following.

23. The perimeter

24. The area

$3x + 9$

Perform each division.

25. $\dfrac{8y^3 - 6y^2 + 4y + 10}{2y}$

26. $(-9x^2y^3 + 6x^4y^3 + 12xy^3) \div (3xy)$

27. $\dfrac{5x^2 - x - 18}{5x + 9}$

28. $(3x^3 - x + 4) \div (x - 2)$

Chapters R–5 Cumulative Review Exercises

[R.1]

1. $\dfrac{7}{4}$ **2.** 5

3. $31\dfrac{1}{4}$ yd^3

[R.2]

4. $1836

[1.4, 1.5]

5. 1, 3, 5, 9, 15, 45

6. -8

7. $\dfrac{1}{2}$ **8.** -4

[1.6]

9. associative property

10. distributive property

[1.7]

11. $-10x^2 + 21x - 29$

[2.1–2.3]

12. $\left\{ \dfrac{13}{4} \right\}$

13. $\varnothing$

[2.5]

14. $r = \dfrac{d}{t}$

[2.6]

15. $\{-5\}$

[2.1–2.3]

16. $\{0\}$

17. $\{20\}$ **18.** $\{-12\}$

19. $\{$ all real numbers $\}$

Write each fraction in lowest terms.

1. $\dfrac{28}{16}$

2. $\dfrac{55}{11}$

3. A contractor installs sheds. Each requires $1\dfrac{1}{4}$ yd^3 of concrete. How much concrete would be needed for 25 sheds?

4. A retailer has $34,000 invested in her business. She finds that last year she earned 5.4% on this investment. How much did she earn?

5. List all positive integer factors of 45.

6. Find the value of $\dfrac{4x - 2y}{x + y}$ for $x = -2$ and $y = 4$.

Perform each indicated operation.

7. $\dfrac{(-13 + 15) - (3 + 2)}{6 - 12}$

8. $-7 - 3[2 + (5 - 8)]$

Name the property illustrated.

9. $(9 + 2) + 3 = 9 + (2 + 3)$

10. $6(4 + 2) = 6(4) + 6(2)$

11. Simplify the expression $-3(2x^2 - 8x + 9) - (4x^2 + 3x + 2)$.

Solve each equation.

12. $2 - 3(t - 5) = 4 + t$

13. $2(5x + 1) = 10x + 4$

14. $d = rt$ for r

15. $\dfrac{x}{5} = \dfrac{x - 2}{7}$

16. $3x - (4 + 2x) = -4$

17. $0.05x + 0.15(50 - x) = 5.50$

18. $\dfrac{1}{3}p - \dfrac{1}{6}p = -2$

19. $4 - (3x + 12) = -7 - (3x + 1)$

[2.4]

20. exertion: 9443 calories; regulating body temperature: 1757 calories

[2.8]

21. 11 ft and 22 ft

22. $\left(-\infty, -\dfrac{14}{5}\right)$

23. $[-4, 2)$

[3.2]

24.

[3.3–3.5]

25. (a) 1   (b) $y = x + 6$

[3.7]

26. -1

[4.2]

27. $\{(-3, -1)\}$

[4.3]

28. $\{(4, -5)\}$

[5.1, 5.2]

29. $\dfrac{5}{4}$ 30. 1

31. $\dfrac{2b}{a^{10}}$

[5.3]

32. 10,800,000 km

[5.4]

33.

34. $11x^3 - 14x^2 - x + 14$

[5.5]

35. $63x^2 + 57x + 12$

[5.7]

36. $y^2 - 2y + 6$

Solve each problem.

20. A husky running the Iditarod in Alaska burns $5\dfrac{3}{8}$ calories in exertion for every 1 calorie burned in thermoregulation in extreme cold. According to one scientific study, a husky in top condition burns an amazing total of 11,200 calories per day. How many calories are burned for exertion, and how many are burned for regulation of body temperature? Round answers to the nearest whole number.

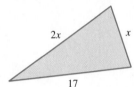

21. One side of a triangle is twice as long as a second side. The third side of the triangle is 17 ft long. The perimeter of the triangle cannot be more than 50 ft. Find the longest possible values for the other two sides of the triangle.

Solve each inequality.

22. $-2(x + 4) > 3x + 6$ 23. $-3 \le 2x + 5 < 9$

24. Graph $y = -3x + 6$.

25. Consider the two points $(-1, 5)$ and $(2, 8)$.

   (a) Find the slope of the line passing through them.

   (b) Find the equation of the line passing through them.

26. If $f(x) = x + 7$, find $f(-8)$.

Solve each system of equations using the method indicated.

27. $y = 2x + 5$ 28. $3x + 2y = 2$

   $x + y = -4$ (Substitution) $2x + 3y = -7$ (Elimination)

Evaluate each expression.

29. $4^{-1} + 3^0$ 30. $\dfrac{8^{-5} \cdot 8^7}{8^2}$

31. Write $\dfrac{(a^{-3}b^2)^2}{(2a^{-4}b^{-3})^{-1}}$ with positive exponents only.

32. It takes about 3.6×10^1 sec at a speed of 3.0×10^5 km per sec for light from the sun to reach Venus. How far is Venus from the sun? (*Source: World Almanac and Book of Facts.*)

33. Graph $y = (x + 4)^2$, using the x-values -6, -5, -4, -3, and -2 to obtain a set of points.

Perform each indicated operation.

34. $(7x^3 - 12x^2 - 3x + 8) + (6x^2 + 4) - (-4x^3 + 8x^2 - 2x - 2)$

35. $(7x + 4)(9x + 3)$ 36. $\dfrac{y^3 - 3y^2 + 8y - 6}{y - 1}$

6

Factoring and Applications

Formulas associated with the mathematicians Pythagoras (c. 380–300 B.C.) and Galileo (1564–1642) are used in applications that involve *factoring* polynomials, the subject of this chapter.

6.1   The Greatest Common Factor; Factoring by Grouping

OBJECTIVES

1 Find the greatest common factor of a list of terms.

2 Factor out the greatest common factor.

3 Factor by grouping.

VOCABULARY

☐ factor
☐ factored form
☐ common factor
☐ greatest common factor (GCF)

To **factor** a number means to write it as a product of two or more numbers. The product is a **factored form** of the number. Consider an example.

$$12 = \overset{\text{Factors}}{\underset{\text{Factored form}}{6 \cdot 2}}$$

Factoring is a process that "undoes" multiplying. We multiply $6 \cdot 2$ to obtain 12, but we factor 12 by writing it as $6 \cdot 2$. Other factored forms of 12 are

$$-6(-2), \quad 3 \cdot 4, \quad -3(-4), \quad 12 \cdot 1, \quad -12(-1), \quad \text{and} \quad 2 \cdot 2 \cdot 3.$$

OBJECTIVE 1 Find the greatest common factor of a list of terms.

An integer that is a factor of two or more integers is a **common factor** of those integers. For example, 6 is a common factor of 18 and 24 because 6 is a factor of both 18 and 24. Other common factors of 18 and 24 are 1, 2, and 3.

The **greatest common factor (GCF)** of a list of integers is the largest common factor of those integers. Thus, 6 is the greatest common factor of 18 and 24 because it is the largest of their common factors.

Finding the Greatest Common Factor (GCF)

Step 1 **Factor.** Write each number in prime factored form.

Step 2 **List common factors.** List each prime number or each variable that is a factor of every term in the list. (If a prime does not appear in one of the prime factored forms, it *cannot* appear in the greatest common factor.)

Step 3 **Choose least exponents.** Use as exponents on the common prime factors the *least* exponents from the prime factored forms.

Step 4 **Multiply** the primes from Step 3. If there are no primes left after Step 3, the greatest common factor is 1.

NOTE *Factors* of a number are also *divisors* of the number. The *greatest common factor* is the same as the *greatest common divisor.* Divisibility tests are useful for deciding what numbers divide into a given number.

▼ **Divisibility Tests**

A Whole Number Divisible by	Must Have the Following Property:
2	Ends in 0, 2, 4, 6, or 8
3	Sum of digits divisible by 3
4	Last two digits form a number divisible by 4
5	Ends in 0 or 5
6	Divisible by both 2 and 3
8	Last three digits form a number divisible by 8
9	Sum of digits divisible by 9
10	Ends in 0

**NOW TRY
EXERCISE 1**

Find the greatest common factor for each list of numbers.

(a) 24, 36

(b) 54, 90, 108

(c) 15, 19, 25

EXAMPLE 1 Finding the Greatest Common Factor (Numbers)

Find the greatest common factor for each list of numbers.

(a) 30, 45

$$30 = 2 \cdot 3 \cdot 5$$
$$45 = 3 \cdot 3 \cdot 5$$

Write the prime factored form of each number.

Use each prime the **least** *number of times it appears in* **all** *the factored forms.* There is no 2 in the prime factored form of 45, so there will be no 2 in the greatest common factor. The least number of times 3 appears in all the factored forms is 1. The least number of times 5 appears is also 1.

$$\text{GCF} = 3^1 \cdot 5^1 = 15 \qquad 3^1 = 3 \text{ and } 5^1 = 5.$$

(b) 72, 120, 432

$$72 = 2 \cdot 2 \cdot 2 \cdot 3 \cdot 3$$
$$120 = 2 \cdot 2 \cdot 2 \cdot 3 \cdot 5$$
$$432 = 2 \cdot 2 \cdot 2 \cdot 2 \cdot 3 \cdot 3 \cdot 3$$

Write the prime factored form of each number.

The least number of times 2 appears in all the factored forms is 3, and the least number of times 3 appears is 1. There is no 5 in the prime factored form of either 72 or 432.

$$\text{GCF} = 2^3 \cdot 3^1 = 24 \qquad 2^3 = 8 \text{ and } 3^1 = 3.$$

(c) 10, 11, 14

$$10 = 2 \cdot 5$$
$$11 = 11$$
$$14 = 2 \cdot 7$$

Write the prime factored form of each number.

There are no primes common to all three numbers, so the GCF is 1.   **NOW TRY**

The greatest common factor can also be found for a list of variable terms. For example, the terms x^4, x^5, x^6, and x^7 have x^4 as the greatest common factor because the least exponent on the variable x in the factored forms is 4.

$$x^4 = 1 \cdot x^4, \quad x^5 = x \cdot x^4, \quad x^6 = x^2 \cdot x^4, \quad x^7 = x^3 \cdot x^4$$

$$\text{GCF} = x^4$$

NOTE *The exponent on a variable in the GCF is the* **least** *exponent that appears on that variable in* **all** *the terms.*

EXAMPLE 2 Finding the Greatest Common Factor (Variable Terms)

Find the greatest common factor for each list of terms.

(a) $21m^7$, $18m^6$, $45m^8$, $24m^5$

$$21m^7 = 3 \cdot 7 \cdot m^7$$
$$18m^6 = 2 \cdot 3 \cdot 3 \cdot m^6$$
$$45m^8 = 3 \cdot 3 \cdot 5 \cdot m^8$$
$$24m^5 = 2 \cdot 2 \cdot 2 \cdot 3 \cdot m^5$$

Here, 3 is the greatest common factor of the coefficients 21, 18, 45, and 24. The least exponent on m is 5.

$$\text{GCF} = 3m^5$$

NOW TRY ANSWERS
1. (a) 12   **(b)** 18   **(c)** 1

NOW TRY
EXERCISE 2
Find the greatest common factor for each list of terms.

(a) $25k^3, 15k^2, 35k^5$

(b) m^3n^5, m^4n^4, m^5n^2

(b) $x^4y^2, \quad x^7y^5, \quad x^3y^7, \quad y^{15}$

$$x^4y^2 = x^4 \cdot y^2$$
$$x^7y^5 = x^7 \cdot y^5$$
$$x^3y^7 = x^3 \cdot y^7$$
$$y^{15} = y^{15}$$

There is no x in the last term, y^{15}, so x will not appear in the greatest common factor. There is a y in each term, however, and 2 is the least exponent on y.

$$\text{GCF} = y^2$$

NOW TRY

OBJECTIVE 2   Factor out the greatest common factor.

Factoring a polynomial is the process of writing a polynomial sum in factored form as a product. For example, the polynomial

$$3m + 12$$

has two terms, $3m$ and 12. The greatest common factor of these two terms is 3. We can write $3m + 12$ so that each term is a product with 3 as one factor.

$$3m + 12$$
$$= 3 \cdot m + 3 \cdot 4 \qquad \text{GCF} = 3$$
$$= 3(m + 4) \qquad \text{Distributive property,}$$
$$\qquad\qquad\qquad a \cdot b + a \cdot c = a(b + c)$$

The factored form of $3m + 12$ is $3(m + 4)$. This process is called **factoring out the greatest common factor.**

> ⚠ **CAUTION**   The polynomial $3m + 12$ is *not* in factored form when written as
>
> $$3 \cdot m + 3 \cdot 4. \qquad \text{Not in factored form}$$
>
> **The terms are factored, but the polynomial is not.** The factored form of $3m + 12$ is the *product*
>
> The factors here are 3 and $m + 4$. $3(m + 4)$.   In factored form

---

EXAMPLE 3   Factoring Out the Greatest Common Factor

Write in factored form by factoring out the greatest common factor.

(a) $5y^2 + 10y$

$$= 5y(y) + 5y(2) \qquad \text{GCF} = 5y$$
$$= 5y(y + 2) \qquad \text{Distributive property}$$

CHECK $5y(y + 2)$ Multiply the factored form.

$$= 5y(y) + 5y(2) \qquad \text{Distributive property}$$
$$= 5y^2 + 10y \checkmark \qquad \text{Original polynomial}$$

(b) $20m^5 + 10m^4 - 15m^3$

$$= 5m^3(4m^2) + 5m^3(2m) - 5m^3(3) \qquad \text{GCF} = 5m^3$$
$$= 5m^3(4m^2 + 2m - 3) \qquad \text{Factor out } 5m^3.$$

**NOW TRY
EXERCISE 3**

Write in factored form by factoring out the greatest common factor.

(a) $7t^4 - 14t^3$

(b) $8x^6 - 20x^5 + 28x^4$

(c) $30m^4n^3 - 42m^2n^2$

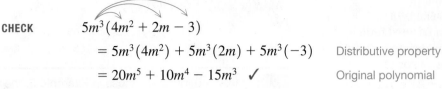

CHECK

$$5m^3(4m^2 + 2m - 3)$$

$$= 5m^3(4m^2) + 5m^3(2m) + 5m^3(-3) \qquad \text{Distributive property}$$

$$= 20m^5 + 10m^4 - 15m^3 \checkmark \qquad \text{Original polynomial}$$

(c) $x^5 + x^3$

$$= x^3(x^2) + x^3(1) \qquad \text{GCF} = x^3$$

$$= x^3(x^2 + 1) \quad \longleftarrow \boxed{\text{Don't forget the 1.}}$$

Check mentally by distributing x^3 over each term inside the parentheses.

(d) $20m^7p^2 - 36m^3p^4$

$$= 4m^3p^2(5m^4) - 4m^3p^2(9p^2) \qquad \text{GCF} = 4m^3p^2$$

$$= 4m^3p^2(5m^4 - 9p^2) \qquad \text{Factor out } 4m^3p^2.$$

NOW TRY

⚠ **CAUTION**   Be sure to include the 1 in a problem like **Example 3(c).** *Check that the factored form can be multiplied out to give the original polynomial.*

**NOW TRY
EXERCISE 4**

Write

$$-14b^2 - 21b^3 + 7b$$

in factored form by factoring out a negative common factor.

EXAMPLE 4   Factoring Out a Negative Common Factor

Write $-8x^4 + 16x^3 - 4x^2$ in factored form.

We can factor out either $4x^2$ or $-4x^2$ here. So that the coefficient of the leading (first) term in the trinomial factor will be positive, we factor out $-4x^2$.

$$-8x^4 + 16x^3 - 4x^2 \qquad \boxed{\text{Be careful with signs.}}$$

$$= -4x^2(2x^2) - 4x^2(-4x) - 4x^2(1) \qquad -4x^2 \text{ is a common factor.}$$

$$= -4x^2(2x^2 - 4x + 1) \qquad \text{Factor out } -4x^2.$$

$$\text{Positive coefficient}$$

CHECK

$$-4x^2(2x^2 - 4x + 1)$$

$$= -4x^2(2x^2) - 4x^2(-4x) - 4x^2(1) \qquad \text{Distributive property}$$

$$= -8x^4 + 16x^3 - 4x^2 \checkmark \qquad \text{Original polynomial}$$

NOW TRY

NOTE   Whenever we factor a polynomial in which the coefficient of the leading term is negative, we will factor out the negative common factor, even if it is just -1. However, it would also be correct to factor out $4x^2$ in **Example 4** to obtain

$$4x^2(-2x^2 + 4x - 1).$$

$$\text{Negative coefficient}$$

EXAMPLE 5   Factoring Out the Greatest Common Factor

Write in factored form by factoring out the greatest common factor.

$$\text{Same}$$

(a) $a(a + 3) + 4(a + 3) \qquad$ The binomial $a + 3$ is the greatest common factor.

$$= (a + 3)(a + 4) \qquad \text{Factor out } a + 3.$$

NOW TRY ANSWERS

3. (a) $7t^3(t - 2)$

(b) $4x^4(2x^2 - 5x + 7)$

(c) $6m^2n^2(5m^2n - 7)$

4. $-7b(2b + 3b^2 - 1)$

**NOW TRY
EXERCISE 5**

Write in factored form by factoring out the greatest common factor.

(a) $x(x + 2) + 5(x + 2)$

(b) $a(t + 10) - b(t + 10)$

(b) $x^2(x + 1) - 5(x + 1)$

$= (x + 1)(x^2 - 5)$ Factor out $x + 1$.

NOW TRY

NOTE In factored forms like those in **Example 5,** the order of the factors does not matter because of the commutative property of multiplication, $ab = ba$.

$$(a + 3)(a + 4) \quad \text{can also be written} \quad (a + 4)(a + 3).$$

OBJECTIVE 3 Factor by grouping.

When a polynomial has four terms, common factors can sometimes be used to factor by grouping.

EXAMPLE 6 Factoring by Grouping

Factor by grouping.

(a) $2x + 6 + ax + 3a$

Group the first two terms and the last two terms because the first two terms have a common factor of 2 and the last two terms have a common factor of a.

$$2x + 6 + ax + 3a$$

$$= (2x + 6) + (ax + 3a) \qquad \text{Group the terms.}$$

$$= 2(x + 3) + a(x + 3) \qquad \text{Factor each group.}$$

The expression is still not in factored form because it is the *sum* of two terms. Now, however, $x + 3$ is a common factor and can be factored out.

$$= 2(x + 3) + a(x + 3) \qquad x + 3 \text{ is a common factor.}$$

$(2 + a)(x + 3)$ is also correct.

$$= (x + 3)(2 + a) \qquad \text{Factor out } x + 3.$$

The final result $(x + 3)(2 + a)$ is in factored form because it is a ***product.***

CHECK $(x + 3)(2 + a)$

$$= x(2) + x(a) + 3(2) + 3(a) \qquad \begin{array}{l}\text{Multiply using the FOIL method.}\\\text{(Section 5.5)}\end{array}$$

$$= 2x + ax + 6 + 3a \qquad \text{Simplify.}$$

$$= 2x + 6 + ax + 3a \ \checkmark \qquad \begin{array}{l}\text{Rearrange terms to obtain the original}\\\text{polynomial.}\end{array}$$

(b) $6ax + 24x + a + 4$

$$= (6ax + 24x) + (a + 4) \qquad \text{Group the terms.}$$

$$= 6x(a + 4) + 1(a + 4) \qquad \text{Factor each group.}$$

Remember the 1.

$$= (a + 4)(6x + 1) \qquad \text{Factor out } a + 4.$$

CHECK $(a + 4)(6x + 1)$

$$= 6ax + a + 24x + 4 \qquad \text{FOIL method}$$

$$= 6ax + 24x + a + 4 \ \checkmark \qquad \begin{array}{l}\text{Rearrange terms to obtain the}\\\text{original polynomial.}\end{array}$$

NOW TRY ANSWERS
5. (a) $(x + 2)(x + 5)$
  (b) $(t + 10)(a - b)$

NOW TRY
EXERCISE 6
Factor by grouping.
(a) $ab + 3a + 5b + 15$
(b) $12xy + 3x + 4y + 1$
(c) $x^3 + 5x^2 - 8x - 40$

(c) $2x^2 - 10x + 3xy - 15y$

$= (2x^2 - 10x) + (3xy - 15y)$ Group the terms.

$= 2x(x - 5) + 3y(x - 5)$ Factor each group.

$= (x - 5)(2x + 3y)$ Factor out $x - 5$.

CHECK $(x - 5)(2x + 3y)$

$= 2x^2 + 3xy - 10x - 15y$ FOIL method

$= 2x^2 - 10x + 3xy - 15y$ ✓   Original polynomial

(d) $t^3 + 2t^2 - 3t - 6$ [Write a + sign between the groups.]

$= (t^3 + 2t^2) + (-3t - 6)$ Group the terms.

$= t^2(t + 2) - 3(t + 2)$ Factor out -3 so there is a common factor, $t + 2$. Check: $-3(t + 2) = -3t - 6$

[Be careful with signs.]

$= (t + 2)(t^2 - 3)$ Factor out $t + 2$.

Check by multiplying using the FOIL method.   **NOW TRY**

⚠ **CAUTION**   *Be careful with signs when grouping* in a problem like **Example 6(d).** It is wise to check the factoring in the second step, as shown in the example side comment, before continuing.

> **Factoring a Polynomial with Four Terms by Grouping**
>
> **Step 1**   **Group the terms.** Collect the terms into two groups so that each group has a common factor.
>
> **Step 2**   **Factor within the groups.** Factor out the greatest common factor from each group.
>
> **Step 3**   **If possible, factor the entire polynomial.** Factor out a common binomial factor from the results of Step 2.
>
> **Step 4**   **If necessary, rearrange terms.** If Step 2 does not result in a common binomial factor, try a different grouping.
>
> *Always check the factored form by multiplying.*

EXAMPLE 7   Rearranging Terms before Factoring by Grouping

Factor by grouping.

(a) $10x^2 - 12y + 15x - 8xy$

Factoring out a common factor of 2 from the first two terms and a common factor of x from the last two terms gives the following.

$(10x^2 - 12y) + (15x - 8xy)$ Group the terms.

$= 2(5x^2 - 6y) + x(15 - 8y)$ Factor each group.

This does not lead to a common factor, so we try rearranging the terms. There is usually more than one way to do this.

NOW TRY ANSWERS
6. (a) $(b + 3)(a + 5)$
   (b) $(4y + 1)(3x + 1)$
   (c) $(x + 5)(x^2 - 8)$

NOW TRY
EXERCISE 7
Factor by grouping.
(a) $12p^2 - 28q - 16pq + 21p$
(b) $5xy - 6 - 15x + 2y$

We try the following.

$$10x^2 - 12y + 15x - 8xy \qquad \text{Original polynomial}$$
$$= 10x^2 - 8xy - 12y + 15x \qquad \text{Commutative property}$$
$$= (10x^2 - 8xy) + (-12y + 15x) \qquad \text{Group the terms.}$$
$$= 2x(5x - 4y) + 3(-4y + 5x) \qquad \text{Factor each group.}$$
$$= 2x(5x - 4y) + 3(5x - 4y) \qquad \text{Rewrite } -4y + 5x.$$
$$= (5x - 4y)(2x + 3) \qquad \text{Factor out } 5x - 4y.$$

CHECK $(5x - 4y)(2x + 3)$

$$= 10x^2 + 15x - 8xy - 12y \qquad \text{FOIL method}$$
$$= 10x^2 - 12y + 15x - 8xy \; \checkmark \qquad \text{Original polynomial}$$

(b) $2xy + 12 - 3y - 8x$

We must rearrange these terms to obtain two groups that each have a common factor. Trial and error suggests the following grouping.

$$2xy + 12 - 3y - 8x \quad \boxed{\text{Write a + sign between the groups.}}$$
$$= (2xy - 3y) + (-8x + 12) \qquad \text{Rearrange and group the terms.}$$
$$= y(2x - 3) - 4(2x - 3) \qquad \begin{array}{l}\text{Factor each group.}\\ \textit{Check: } -4(2x - 3) = -8x + 12\end{array}$$

$\boxed{\text{Be careful with signs.}}$

$$= (2x - 3)(y - 4) \qquad \text{Factor out } 2x - 3.$$

Because the quantities in parentheses in the second step must be the same, we factored out -4 rather than 4.

CHECK $(2x - 3)(y - 4)$

$$= 2xy - 8x - 3y + 12 \qquad \text{FOIL method}$$
$$= 2xy + 12 - 3y - 8x \; \checkmark \qquad \text{Original polynomial} \qquad \text{NOW TRY} \; \text{⟳}$$

NOW TRY ANSWERS
7. (a) $(3p - 4q)(4p + 7)$
   (b) $(5x + 2)(y - 3)$

⚠ CAUTION Use negative signs carefully when grouping, as in **Example 7(b),** or a sign error will occur. **_Always check by multiplying._**

6.1 Exercises

 FOR EXTRA HELP ▶ MyMathLab®

▶ *Complete solution available in MyMathLab*

Concept Check   *Complete each statement.*

1. To factor a number or quantity means to write it as a(n) _____. Factoring is the opposite, or inverse, process of _____.

2. An integer or variable expression that is a factor of two or more terms is a(n) _____. For example, 12 (*is/is not*) a common factor of both 36 and 72 because it _____ evenly into both integers.

*Find the greatest common factor for each list of numbers. **See Example 1.***

3. 12, 16   **4.** 18, 24   ▶ **5.** 40, 20, 4   **6.** 50, 30, 5

7. 18, 24, 36, 48   **8.** 15, 30, 45, 75   **9.** 6, 8, 9   **10.** 20, 22, 23

Find the greatest common factor for each list of terms. ***See Examples 1 and 2.***

11. $16y, 24$

12. $18w, 27$

13. $30x^3, 40x^6, 50x^7$

14. $60z^4, 70z^8, 90z^9$

15. x^4y^3, xy^2

16. a^4b^5, a^3b

17. $42ab^3, 36a, 90b, 48ab$

18. $45c^3d, 75c, 90d, 105cd$

19. $12m^3n^2, 18m^5n^4, 36m^8n^3$

20. $25p^5r^7, 30p^7r^8, 50p^5r^3$

Concept Check   An expression is factored when it is written as a product, not a sum. Decide whether each expression is factored *or* not factored.

21. $2k^2(5k)$

22. $2k^2(5k + 1)$

23. $2k^2 + (5k + 1)$

24. $(2k^2 + 5k) + 1$

25. *Concept Check*   A student factored as follows.

$$18x^3y^2 + 9xy$$

$$= 9xy(2x^2y)$$

WHAT WENT WRONG?   Factor correctly.

26. How can we check an answer when we factor a polynomial?

Complete each factoring by writing each polynomial as the product of two factors. ***See Example 3.***

27. $9m^4$

$= 3m^2(\underline{\hspace{1cm}})$

28. $12p^5$

$= 6p^3(\underline{\hspace{1cm}})$

29. $-8z^9$

$= -4z^5(\underline{\hspace{1cm}})$

30. $-15k^{11}$

$= -5k^8(\underline{\hspace{1cm}})$

31. $6m^4n^5$

$= 3m^3n(\underline{\hspace{1cm}})$

32. $27a^3b^2$

$= 9a^2b(\underline{\hspace{1cm}})$

33. $12y + 24$

$= 12(\underline{\hspace{1cm}})$

34. $18p + 36$

$= 18(\underline{\hspace{1cm}})$

35. $10a^2 - 20a$

$= 10a(\underline{\hspace{1cm}})$

36. $15x^2 - 30x$

$= 15x(\underline{\hspace{1cm}})$

37. $8x^2y + 12x^3y^2$

$= 4x^2y(\underline{\hspace{1cm}})$

38. $18s^3t^2 + 10st$

$= 2st(\underline{\hspace{1cm}})$

Write in factored form by factoring out the greatest common factor. ***See Examples 3–5.***

39. $x^2 - 4x$

40. $m^2 - 7m$

41. $6t^2 + 15t$

42. $8x^2 + 6x$

43. $27m^3 - 9m$

44. $12p^3 - 4p$

45. $m^3 - m^2$

46. $p^3 - p^2$

▶ **47.** $16z^4 + 24z^2$

48. $25k^4 + 15k^2$

49. $-12x^3 - 6x^2$

50. $-21b^3 - 7b^2$

51. $65y^{10} + 35y^6$

52. $100a^5 + 16a^3$

53. $11w^3 - 100$

54. $13z^5 - 80$

55. $8mn^3 + 24m^2n^3$

56. $19p^2y + 38p^2y^3$

57. $13y^8 + 26y^4 - 39y^2$

58. $5x^5 + 25x^4 - 20x^3$

59. $-4x^3 + 10x^2 - 6x$

60. $-9z^3 + 6z^2 - 12z$

61. $36p^6q + 45p^5q^4 + 81p^3q^2$

62. $125a^3z^5 + 60a^4z^4 + 85a^5z^2$

63. $a^5 + 2a^3b^2 - 3a^5b^2 + 4a^4b^3$

64. $x^6 + 5x^4y^3 - 6xy^4 + 10xy$

▶ **65.** $c(x + 2) - d(x + 2)$

66. $r(x + 5) - t(x + 5)$

67. $m(m + 2n) + n(m + 2n)$

68. $q(q + 4p) + p(q + 4p)$

69. $q^2(p - 4) + 1(p - 4)$

70. $y^2(x - 9) + 1(x - 9)$

Students often have difficulty when factoring by grouping because they are not able to tell when a polynomial is completely factored. For example,

$$5y(2x - 3) + 8t(2x - 3)   \text{Not in factored form}$$

is not in factored form, because it is the *sum* of two terms: $5y(2x - 3)$ and $8t(2x - 3)$. However, because $2x - 3$ is a common factor of these two terms, the expression can now be factored.

$$(2x - 3)(5y + 8t)   \text{In factored form}$$

The factored form is a *product* of the two factors $2x - 3$ and $5y + 8t$.

Concept Check   *Determine whether each expression is* in factored form *or is* not in factored form. *If it is not in factored form, factor it if possible.*

71. $8(7t + 4) + x(7t + 4)$

72. $3r(5x - 1) + 7(5x - 1)$

73. $(8 + x)(7t + 4)$

74. $(3r + 7)(5x - 1)$

75. $18x^2(y + 4) + 7(y - 4)$

76. $12k^3(s - 3) + 7(s + 3)$

77. *Concept Check*   A student factored as follows.

$$x^3 + 4x^2 - 2x - 8$$
$$= (x^3 + 4x^2) + (-2x - 8)$$
$$= x^2(x + 4) + 2(-x - 4)$$

The student could not find a common factor of the two terms. **WHAT WENT WRONG?** Complete the factoring.

78. *Concept Check*   A student factored as follows.

$$10xy + 18 + 12x + 15y$$
$$= (10xy + 18) + (12x + 15y)$$
$$= 2(5xy + 9) + 3(4x + 5y)$$

The student could not find a common factor of the two terms. **WHAT WENT WRONG?** Complete the factoring.

Factor by grouping. **See Examples 6 and 7.**

79. $p^2 + 4p + pq + 4q$

80. $m^2 + 2m + mn + 2n$

▶ **81.** $a^2 - 2a + ab - 2b$

82. $y^2 - 6y + yw - 6w$

83. $7z^2 + 14z - az - 2a$

84. $5m^2 + 15mp - 2mr - 6pr$

85. $18r^2 + 12ry - 3xr - 2xy$

86. $8s^2 + 6sy - 4st - 3yt$

87. $3a^3 + 3ab^2 + 2a^2b + 2b^3$

88. $4x^3 + 4xy^2 + 3x^2y + 3y^3$

89. $12 - 4a - 3b + ab$

90. $6 - 3x - 2y + xy$

91. $16m^3 - 4m^2p^2 - 4mp + p^3$

92. $10t^3 - 2t^2s^2 - 5ts + s^3$

93. $y^2 + 3x + 3y + xy$

94. $m^2 + 14p + 7m + 2mp$

▶ **95.** $5m - 6p - 2mp + 15$

96. $7y - 9x - 3xy + 21$

97. $18r^2 - 2ty + 12ry - 3rt$

98. $12a^2 - 4bc + 16ac - 3ab$

99. $a^5 - 3 + 2a^5b - 6b$

100. $b^3 - 2 + 5ab^3 - 10a$

Extending Skills   *Factor each polynomial.* (*Hint: As the first step, factor out the greatest common factor.*)

101. $16a^2 + 40ab^2 + 16ab + 40b^3$

102. $18x^2 + 12xy^2 + 18xy + 12y^3$

103. $2p^2q^2 - 2p^2q + 2p^3 - 2pq^3$

104. $4m^2n^2 - 4mn^2 - 4m^3n + 4n^3$

6.2 Factoring Trinomials

OBJECTIVES

1 Factor trinomials with coefficient 1 for the second-degree term.

2 Factor such trinomials after factoring out the greatest common factor.

VOCABULARY

☐ prime polynomial

Using the FOIL method, we can find the product of the binomials $k - 3$ and $k + 1$.

$$(k - 3)(k + 1) = k^2 - 2k - 3 \qquad \text{Multiplying}$$

Suppose instead that we are given the polynomial $k^2 - 2k - 3$ and want to write it as the product $(k - 3)(k + 1)$.

$$k^2 - 2k - 3 = (k - 3)(k + 1) \qquad \text{Factoring}$$

Recall that *factoring* is a process that reverses, or "undoes," multiplying.

OBJECTIVE 1 Factor trinomials with coefficient 1 for the second-degree term.

When factoring polynomials with integer coefficients, we use only integers in the factors. For example, we can factor $x^2 + 5x + 6$ by finding integers m and n such that

$$x^2 + 5x + 6 \quad \text{is written as} \quad (x + m)(x + n).$$

To find these integers m and n, we multiply the two binomials on the right.

$$(x + m)(x + n)$$
$$= x^2 + nx + mx + mn \qquad \text{FOIL method}$$
$$= x^2 + (n + m)x + mn \qquad \text{Distributive property}$$

Comparing this result with $x^2 + 5x + 6$ shows that we must find integers m and n having a sum of 5 and a product of 6.

Product of *m* and *n* is 6.

$$x^2 + 5x + 6 = x^2 + (n + m)x + mn$$

Sum of *m* and *n* is 5.

Because many pairs of integers have a sum of 5, it is best to begin by listing those pairs of integers whose product is 6. Both 5 and 6 are positive, so we consider only pairs in which both integers are positive.

Factors of 6	Sums of Factors
6, 1	6 + 1 = 7
3, 2	3 + 2 = 5

Sum is 5.

Both pairs have a product of 6, but only the pair 3 and 2 has a sum of 5. So 3 and 2 are the required integers.

$$x^2 + 5x + 6 \quad \text{factors as} \quad (x + 3)(x + 2).$$

Check by using the FOIL method to multiply the binomials. ***Make sure that the sum of the outer and inner products produces the correct middle term.***

CHECK $\qquad (x + 3)(x + 2) = x^2 + 5x + 6 \quad \checkmark \quad$ Correct

$$3x$$
$$\underline{2x}$$
$$5x \qquad \text{Add.}$$

This method can be used only to factor trinomials that have 1 as the coefficient of the second-degree (squared variable) term.

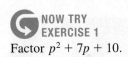

Factor $p^2 + 7p + 10$.

EXAMPLE 1 Factoring a Trinomial (All Positive Terms)

Factor $m^2 + 9m + 14$.

Look for two integers whose product is 14 and whose sum is 9. List pairs of integers whose product is 14, and examine the sums. Only positive integers are needed because all signs in $m^2 + 9m + 14$ are positive.

Factors of 14	Sums of Factors
14, 1	$14 + 1 = 15$
7, 2	$7 + 2 = 9$

Sum is 9.

The required integers are 7 and 2 because $7 \cdot 2 = 14$ and $7 + 2 = 9$.

$$m^2 + 9m + 14 \quad \text{factors as} \quad (m + 7)(m + 2).$$

$(m + 2)(m + 7)$ is also correct.

CHECK $(m + 7)(m + 2)$

$= m^2 + 2m + 7m + 14$ FOIL method

$= m^2 + 9m + 14$ ✓ Original polynomial NOW TRY

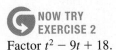

Factor $t^2 - 9t + 18$.

EXAMPLE 2 Factoring a Trinomial (Negative Middle Term)

Factor $x^2 - 9x + 20$.

We must find two integers whose product is 20 and whose sum is -9. Because the numbers we are looking for have a *positive product* and a *negative sum*, we consider only pairs of negative integers.

Factors of 20	Sums of Factors
$-20, -1$	$-20 + (-1) = -21$
$-10, -2$	$-10 + (-2) = -12$
$-5, -4$	$-5 + (-4) = -9$

Sum is -9.

The required integers are -5 and -4.

$$x^2 - 9x + 20 \quad \text{factors as} \quad (x - 5)(x - 4).$$

The order of the factors does not matter.

CHECK $(x - 5)(x - 4)$

$= x^2 - 4x - 5x + 20$ FOIL method

$= x^2 - 9x + 20$ ✓ Original polynomial NOW TRY

Factor $x^2 + x - 42$.

EXAMPLE 3 Factoring a Trinomial (Negative Last (Constant) Term)

Factor $x^2 + x - 6$.

We must find two integers whose product is -6 and whose sum is 1 (because the coefficient of x, or $1x$, is 1). To obtain a *negative product*, the pairs of integers must have different signs.

Factors of -6	Sums of Factors
$6, -1$	$6 + (-1) = 5$
$-6, 1$	$-6 + 1 = -5$
$3, -2$	$3 + (-2) = 1$

Once we find the required pair, we can stop listing factors.

Sum is 1.

NOW TRY ANSWERS
1. $(p + 5)(p + 2)$
2. $(t - 6)(t - 3)$
3. $(x + 7)(x - 6)$

The required integers are 3 and -2.

To check, multiply the factored form.

$$x^2 + x - 6 \quad \text{factors as} \quad (x + 3)(x - 2).$$

NOW TRY

NOW TRY
EXERCISE 4

Factor $x^2 - 4x - 21$.

EXAMPLE 4 Factoring a Trinomial (Two Negative Terms)

Factor $p^2 - 2p - 15$.

Find two integers whose product is -15 and whose sum is -2. Because the constant term, -15, is negative, list pairs of integers with different signs.

Factors of −15	Sums of Factors
15, −1	$15 + (-1) = 14$
−15, 1	$-15 + 1 = -14$
5, −3	$5 + (-3) = 2$
−5, 3	$-5 + 3 = -2$

Sum is -2.

The required integers are -5 and 3.

$$p^2 - 2p - 15 \quad \text{factors as} \quad (p - 5)(p + 3).$$

To check, multiply the factored form.

NOW TRY

> **NOTE** In **Examples 1–4,** we listed factors in descending order (disregarding their signs) when we were looking for the required pair of integers. This helps avoid skipping the correct combination.

Trinomials that cannot be factored using only integers are **prime polynomials.**

NOW TRY
EXERCISE 5

Factor each trinomial if possible.

(a) $m^2 + 5m + 8$

(b) $t^2 + 11t - 24$

EXAMPLE 5 Deciding Whether Polynomials Are Prime

Factor each trinomial if possible.

(a) $x^2 - 5x + 12$

As in **Example 2,** both factors must be negative to give a positive product and a negative sum. List pairs of negative integers whose product is 12, and examine the sums.

Factors of 12	Sums of Factors
−12, −1	$-12 + (-1) = -13$
−6, −2	$-6 + (-2) = -8$
−4, −3	$-4 + (-3) = -7$

No sum is -5.

None of the pairs of integers has a sum of -5. Therefore, the trinomial $x^2 - 5x + 12$ *cannot be factored using only integers.* It is a prime polynomial.

(b) $k^2 - 8k + 11$

There is no pair of integers whose product is 11 and whose sum is -8, so $k^2 - 8k + 11$ is a prime polynomial.

NOW TRY

Guidelines for Factoring $x^2 + bx + c$

Find two integers whose product is c and whose sum is b.

1. Both integers must be positive if b and c are positive. (See **Example 1.**)

2. Both integers must be negative if c is positive and b is negative. (See **Example 2.**)

3. One integer must be positive and one must be negative if c is negative. (See **Examples 3 and 4.**)

NOW TRY ANSWERS

4. $(x - 7)(x + 3)$

5. (a) prime **(b)** prime

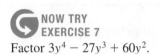

**NOW TRY
EXERCISE 6**

Factor $a^2 + 2ab - 15b^2$.

EXAMPLE 6 Factoring a Multivariable Trinomial

Factor $z^2 - 2bz - 3b^2$.

Here, the coefficient of z in the middle term is $-2b$, so we need to find two expressions whose product is $-3b^2$ and whose sum is $-2b$.

Factors of $-3b^2$	Sums of Factors
$3b, -b$	$3b + (-b) = 2b$
$-3b, b$	$-3b + b = -2b$

Sum is $-2b$.

$z^2 - 2bz - 3b^2$ factors as $(z - 3b)(z + b)$.

CHECK

$$(z - 3b)(z + b)$$
$$= z^2 + zb - 3bz - 3b^2 \qquad \text{FOIL method}$$
$$= z^2 + 1bz - 3bz - 3b^2 \qquad \text{Identity and commutative properties}$$
$$= z^2 - 2bz - 3b^2 \ \checkmark \qquad \text{Combine like terms.} \qquad \text{NOW TRY} $$

OBJECTIVE 2 Factor such trinomials after factoring out the greatest common factor.

**NOW TRY
EXERCISE 7**

Factor $3y^4 - 27y^3 + 60y^2$.

EXAMPLE 7 Factoring a Trinomial with a Common Factor

Factor $4x^5 - 28x^4 + 40x^3$.

$$4x^5 - 28x^4 + 40x^3 \quad \boxed{\text{The terms have a common factor.}}$$

$$= 4x^3(x^2 - 7x + 10) \qquad \text{Factor out the greatest common factor, } 4x^3.$$

Factor $x^2 - 7x + 10$. The integers -5 and -2 have a product of 10 and a sum of -7.

$$\boxed{\text{Include } 4x^3.} \quad = 4x^3(x - 5)(x - 2) \qquad \text{Completely factored form}$$

CHECK $\quad 4x^3(x - 5)(x - 2)$

$$= 4x^3(x^2 - 2x - 5x + 10) \qquad \text{FOIL method}$$
$$= 4x^3(x^2 - 7x + 10) \qquad \text{Combine like terms.}$$
$$= 4x^5 - 28x^4 + 40x^3 \ \checkmark \qquad \text{Distributive property} \qquad \text{NOW TRY} $$

⚠ **CAUTION** *When factoring, always look for a common factor first.* Remember to include the common factor as part of the answer. Check by multiplying out the completely factored form.

NOW TRY ANSWERS
6. $(a + 5b)(a - 3b)$
7. $3y^2(y - 5)(y - 4)$

6.2 Exercises

FOR EXTRA HELP

 MyMathLab®

▶ *Complete solution available
in MyMathLab*

Concept Check   Answer each question.

1. When factoring a trinomial in x as $(x + a)(x + b)$, what must be true of a and b if the coefficient of the constant term of the trinomial is negative?

2. In **Exercise 1,** what must be true of a and b if the coefficient of the constant term is positive?

3. Which is the correct factored form of $x^2 - 12x + 32$?

   A. $(x - 8)(x + 4)$ **B.** $(x + 8)(x - 4)$

   C. $(x - 8)(x - 4)$ **D.** $(x + 8)(x + 4)$

4. What is the suggested first step in factoring

$$2x^3 + 8x^2 - 10x? \quad \text{(See Example 7.)}$$

5. What polynomial can be factored as $(a + 9)(a + 4)$?

6. What polynomial can be factored as $(y - 7)(y + 3)$?

List all pairs of integers with the given product. Then find the pair whose sum is given. **See the tables in Examples 1–4.**

  7. Product: 48;   Sum: -19 **8.** Product: 18;   Sum: 9

  9. Product: -24;   Sum: -5 **10.** Product: -36;   Sum: -16

Concept Check   *Complete each factoring.*

11. To factor $y^2 + 12y + 20$, find two integers whose product is _____ and whose sum is _____. Complete the table.

Factors of 20	Sums of Factors
20, 1	20 + 1 = 21
10, ___	10 + ___ = ___
5, ___	5 + ___ = ___

Which pair of factors has the required sum? _____
Now factor the trinomial.

12. To factor $t^2 - 12t + 32$, find two integers whose product is _____ and whose sum is _____. Complete the table.

Factors of 32	Sums of Factors
$-32, -1$	$-32 + (-1) = -33$
$-16,$ ___	$-16 + (\text{___}) =$ ___
$-8,$ ___	$-8 + (\text{___}) =$ ___

Which pair of factors has the required sum? _____
Now factor the trinomial.

Complete each factoring. **See Examples 1–4.**

13. $p^2 + 11p + 30$
   $= (p + 5)(\text{_____})$

14. $x^2 + 10x + 21$
   $= (x + 7)(\text{_____})$

15. $x^2 + 15x + 44$
   $= (x + 4)(\text{_____})$

16. $r^2 + 15r + 56$
   $= (r + 7)(\text{_____})$

17. $x^2 - 9x + 8$
   $= (x - 1)(\text{_____})$

18. $t^2 - 14t + 24$
   $= (t - 2)(\text{_____})$

19. $y^2 - 2y - 15$
   $= (y + 3)(\text{_____})$

20. $t^2 - t - 42$
   $= (t + 6)(\text{_____})$

21. $x^2 + 9x - 22$
   $= (x - 2)(\text{_____})$

22. $x^2 + 6x - 27$
   $= (x - 3)(\text{_____})$

23. $y^2 - 7y - 18$
   $= (y + 2)(\text{_____})$

24. $y^2 - 2y - 24$
   $= (y + 4)(\text{_____})$

Factor completely. If a polynomial cannot be factored, write prime. **See Examples 1–5.**
(Hint: In Exercises 43 and 44, first write the trinomial in descending powers and then factor.)

25. $y^2 + 9y + 8$ **26.** $a^2 + 9a + 20$

▶ **27.** $b^2 + 8b + 15$ **28.** $x^2 + 6x + 8$

29. $m^2 + m - 20$

30. $p^2 + 4p - 5$

31. $y^2 - 8y + 15$

32. $y^2 - 6y + 8$

33. $x^2 + 4x + 5$

34. $t^2 + 11t + 12$

35. $z^2 - 15z + 56$

36. $x^2 - 13x + 36$

37. $r^2 - r - 30$

38. $q^2 - q - 42$

39. $a^2 - 8a - 48$

40. $d^2 - 4d - 45$

41. $x^2 + 3x - 39$

42. $m^2 + 10m - 30$

43. $-32 + 14x + x^2$

44. $-39 + 10x + x^2$

*Factor completely. **See Example 6.***

45. $r^2 + 3ra + 2a^2$

46. $x^2 + 5xa + 4a^2$

47. $x^2 + 4xy + 3y^2$

48. $p^2 + 9pq + 8q^2$

49. $t^2 - tz - 6z^2$

50. $a^2 - ab - 12b^2$

51. $v^2 - 11vw + 30w^2$

52. $v^2 - 11vx + 24x^2$

53. $m^2 + 4mn - 12n^2$

54. $x^2 + 6xy - 16y^2$

55. $a^2 - 9ab + 18b^2$

56. $h^2 - 11hk + 28k^2$

*Factor completely. **See Example 7.***

57. $4x^2 + 12x - 40$

58. $5y^2 + 5y - 30$

59. $2t^3 + 8t^2 + 6t$

60. $3t^3 + 27t^2 + 24t$

61. $2x^6 + 8x^5 - 42x^4$

62. $4y^5 + 12y^4 - 40y^3$

63. $6z^4 - 24z^3 + 18z^2$

64. $5x^4 - 35x^3 + 30x^2$

65. $5m^5 - 25m^4 + 40m^2$

66. $12k^5 - 6k^3 + 10k^2$

67. $x^3 - 7x^2y + 12xy^2$

68. $p^3 - 8p^2q + 15pq^2$

69. $a^5 + 3a^4b - 4a^3b^2$

70. $k^7 - 2k^6m - 15k^5m^2$

71. $z^{10} - 4z^9y - 21z^8y^2$

72. $x^9 + 5x^8w - 24x^7w^2$

73. $m^3n - 10m^2n^2 + 24mn^3$

74. $y^3z + 3y^2z^2 - 54yz^3$

75. $y^3z + y^2z^2 - 6yz^3$

76. $m^3n - 2m^2n^2 - 3mn^3$

Extending Skills   Factor each polynomial.

77. $(a + b)x^2 + (a + b)x - 12(a + b)$

78. $(x + y)n^2 + (x + y)n - 20(x + y)$

79. $(2p + q)r^2 - 12(2p + q)r + 27(2p + q)$

80. $(3m - n)k^2 - 13(3m - n)k + 40(3m - n)$

6.3   More on Factoring Trinomials

OBJECTIVES

1 Factor trinomials by grouping when the coefficient of the second-degree term is not 1.

2 Factor trinomials using the FOIL method.

NOW TRY EXERCISE 1

Factor $2m^2 + 7m + 3$.

OBJECTIVE 1   Factor trinomials by grouping when the coefficient of the second-degree term is not 1.

We factor a trinomial in which the coefficient of the second-degree term is *not* 1, such as

$$2x^2 + 7x + 6,$$

by extending our work from the previous sections.

EXAMPLE 1   Factoring by Grouping (Coefficient of the Second-Degree Term Not 1)

Factor $2x^2 + 7x + 6$.

To factor this trinomial, we look for two positive integers whose product is $2 \cdot 6 = 12$ and whose sum is 7.

Sum is 7.

$$2x^2 + 7x + 6$$

Product is $2 \cdot 6 = 12$.

The required integers are 3 and 4. We use these integers to write the middle term $7x$ as $3x + 4x$.

$$2x^2 + 7x + 6$$
$$= 2x^2 + \underbrace{3x + 4x}_{7x} + 6$$
$$= (2x^2 + 3x) + (4x + 6) \qquad \text{Group the terms.}$$
$$= x(2x + 3) + 2(2x + 3) \qquad \text{Factor each group.}$$

Must be the same factor

$$= (2x + 3)(x + 2) \qquad \text{Factor out } 2x + 3.$$

CHECK   Multiply $(2x + 3)(x + 2)$ to obtain $2x^2 + 7x + 6$.   ✓   NOW TRY

NOTE   In **Example 1,** we could have written $7x$ as $4x + 3x$, rather than as $3x + 4x$. Factoring by grouping would give the same answer. Try this.

EXAMPLE 2   Factoring Trinomials by Grouping

Factor each trinomial.

(a) $6r^2 + r - 1$

We must find two integers with a product of $6(-1) = -6$ and a sum of 1.

Sum is 1.

$$6r^2 + 1r - 1 \qquad \text{The coefficient of } r, \text{ or } 1r, \text{ is } 1.$$

Product is $6(-1) = -6$.

NOW TRY ANSWER
1. $(2m + 1)(m + 3)$

**NOW TRY
EXERCISE 2**

Factor.

(a) $2z^2 + 5z + 3$

(b) $15m^2 + m - 2$

(c) $8x^2 - 2xy - 3y^2$

The integers -2 and 3 have a product of -6 and a sum of 1. We write the middle term r as $-2r + 3r$.

$$6r^2 + r - 1$$
$$= 6r^2 - 2r + 3r - 1 \qquad r = -2r + 3r$$
$$= (6r^2 - 2r) + (3r - 1) \qquad \text{Group the terms.}$$
$$= 2r(3r - 1) + 1(3r - 1) \qquad \text{The binomials must be the same.}$$

> Remember the 1.

$$= (3r - 1)(2r + 1) \qquad \text{Factor out } 3r - 1.$$

CHECK Multiply $(3r - 1)(2r + 1)$ to obtain $6r^2 + r - 1.$ ✓

(b) $12z^2 - 5z - 2$

Look for two integers whose product is $12(-2) = -24$ and whose sum is -5. The required integers are 3 and -8.

$$12z^2 - 5z - 2$$
$$= 12z^2 + 3z - 8z - 2 \qquad -5z = 3z - 8z$$
$$= (12z^2 + 3z) + (-8z - 2) \qquad \text{Group the terms.}$$
$$= 3z(4z + 1) - 2(4z + 1) \qquad \text{Factor each group.}$$

> Be careful with signs.

$$= (4z + 1)(3z - 2) \qquad \text{Factor out } 4z + 1.$$

CHECK Multiply $(4z + 1)(3z - 2)$ to obtain $12z^2 - 5z - 2.$ ✓

(c) $10m^2 + mn - 3n^2$

Two integers whose product is $10(-3) = -30$ and whose sum is 1 are -5 and 6.

$$10m^2 + mn - 3n^2$$
$$= 10m^2 - 5mn + 6mn - 3n^2 \qquad mn = -5mn + 6mn$$
$$= (10m^2 - 5mn) + (6mn - 3n^2) \qquad \text{Group the terms.}$$
$$= 5m(2m - n) + 3n(2m - n) \qquad \text{Factor each group.}$$
$$= (2m - n)(5m + 3n) \qquad \text{Factor out } 2m - n.$$

CHECK Multiply $(2m - n)(5m + 3n)$ to obtain $10m^2 + mn - 3n^2.$ ✓

NOW TRY

EXAMPLE 3 Factoring a Trinomial with a Common Factor by Grouping

Factor $28x^5 - 58x^4 - 30x^3$.

$$28x^5 - 58x^4 - 30x^3$$
$$= 2x^3(14x^2 - 29x - 15) \qquad \text{Factor out the greatest common factor, } 2x^3.$$

To factor $14x^2 - 29x - 15$, find two integers whose product is $14(-15) = -210$ and whose sum is -29. Factoring 210 into prime factors helps find these integers.

$$210 = 2 \cdot 3 \cdot 5 \cdot 7$$

NOW TRY ANSWERS
2. (a) $(2z + 3)(z + 1)$
(b) $(3m - 1)(5m + 2)$
(c) $(4x - 3y)(2x + y)$

Combine the prime factors into pairs in different ways, using one positive factor and one negative factor to obtain -210. The factors 6 and -35 have the correct sum, -29.

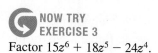

NOW TRY
EXERCISE 3
Factor $15z^6 + 18z^5 - 24z^4$.

$$28x^5 - 58x^4 - 30x^3$$

$$= 2x^3(14x^2 - 29x - 15) \qquad \text{Factor out the GCF.}$$

> Remember the common factor.

$$= 2x^3(14x^2 + 6x - 35x - 15) \qquad -29x = 6x - 35x$$

$$= 2x^3[(14x^2 + 6x) + (-35x - 15)] \qquad \text{Group the terms.}$$

$$= 2x^3[2x(7x + 3) - 5(7x + 3)] \qquad \text{Factor each group.}$$

$$= 2x^3[(7x + 3)(2x - 5)] \qquad \text{Factor out } 7x + 3.$$

$$= 2x^3(7x + 3)(2x - 5)$$

CHECK One way to check is to first multiply $2x^3(7x + 3)$ to obtain $(14x^4 + 6x^3)$. Then multiply

$$(14x^4 + 6x^3)(2x - 5) \quad \text{to obtain} \quad 28x^5 - 58x^4 - 30x^3. \quad \checkmark \quad \text{NOW TRY} \;\circlearrowleft$$

OBJECTIVE 2 Factor trinomials using the FOIL method.

There is an alternative method of factoring trinomials that uses trial and error.

EXAMPLE 4 Factoring Using FOIL (Coefficient of the Second-Degree Term Not 1)

Factor $2x^2 + 7x + 6$. (We factored this trinomial by grouping in **Example 1.**)
 We want to write $2x^2 + 7x + 6$ as the product of two binomials.

$$2x^2 + 7x + 6$$
$$= (\underline{\qquad})(\underline{\qquad})$$

> We use the FOIL method in reverse.

The product of the two first terms of the binomials must be $2x^2$. The possible factors of $2x^2$ are $2x$ and x, or $-2x$ and $-x$. Because all terms of the trinomial are positive, we consider only positive factors. Thus, we have the following.

$$2x^2 + 7x + 6$$
$$= (2x\underline{\qquad})(x\underline{\qquad})$$

The product of the two last terms of the binomials must be 6. It can be factored as $1 \cdot 6$, $6 \cdot 1$, $2 \cdot 3$, or $3 \cdot 2$. Beginning with 1 and 6, we try each pair of factors in $(2x\underline{\qquad})(x\underline{\qquad})$ to find the pair that gives the correct middle term, $7x$.

$$(2x + 1)(x + 6) \qquad \text{Incorrect}$$
$$x$$
$$12x$$
$$13x \qquad \text{Add. (Wrong middle term)}$$

Now try the pair 6 and 1 in $(2x\underline{\qquad})(x\underline{\qquad})$.

$$(2x + 6)(x + 1) \qquad \text{Incorrect}$$
$$6x$$
$$2x$$
$$8x \qquad \text{Add. (Wrong middle term)}$$

Because $2x + 6 = 2(x + 3)$, the terms of the binomial $2x + 6$ have a common factor of 2, while the terms of $2x^2 + 7x + 6$ have no common factor other than 1. The product $(2x + 6)(x + 1)$ cannot be correct.

NOW TRY ANSWER
3. $3z^4(5z - 4)(z + 2)$

If the terms of the original polynomial have greatest common factor 1, then each of its factors will also have terms with GCF 1.

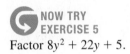
**NOW TRY
EXERCISE 4**
Factor $2p^2 + 9p + 9$.

We try the pair 2 and 3 in $(2x\_\_\_\_\_)(x\_\_\_\_\_)$. Because of the common factor 2 in the terms of $2x + 2$, the product $(2x + 2)(x + 3)$ will not work. Finally, we try the pair 3 and 2 in $(2x\_\_\_\_\_)(x\_\_\_\_\_)$.

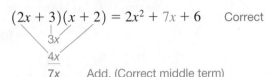

$$(2x + 3)(x + 2) = 2x^2 + 7x + 6 \quad \text{Correct}$$

$3x$

$4x$

$7x$ Add. (Correct middle term)

Thus, $2x^2 + 7x + 6$ factors as $(2x + 3)(x + 2)$.

CHECK   Multiply $(2x + 3)(x + 2)$ to obtain $2x^2 + 7x + 6$. ✓   NOW TRY

**NOW TRY
EXERCISE 5**
Factor $8y^2 + 22y + 5$.

EXAMPLE 5   Factoring a Trinomial Using FOIL (All Positive Terms)

Factor $8p^2 + 14p + 5$.

The number 8 has several possible pairs of factors, but 5 has only 1 and 5 or -1 and -5, so we begin by considering the factors of 5. We ignore the negative factors because all coefficients in the trinomial are positive. If $8p^2 + 14p + 5$ can be factored, the factors will have this form.

$$(\_\_\_\_ + 5)(\_\_\_\_ + 1)$$

The possible pairs of factors of $8p^2$ are $8p$ and p, or $4p$ and $2p$. We try various combinations, checking to see if the middle term is $14p$.

$(8p + 5)(p + 1)$ Incorrect	$(p + 5)(8p + 1)$ Incorrect	$(4p + 5)(2p + 1)$ Correct
$5p$	$40p$	$10p$
$8p$	p	$4p$
$13p$ Add.	$41p$ Add.	$14p$ Add.

The combination on the right produces $14p$, the correct middle term.

$$8p^2 + 14p + 5 \quad \text{factors as} \quad (4p + 5)(2p + 1).$$

CHECK   Multiply $(4p + 5)(2p + 1)$ to obtain $8p^2 + 14p + 5$. ✓   NOW TRY

**NOW TRY
EXERCISE 6**
Factor $10x^2 - 9x + 2$.

EXAMPLE 6   Factoring a Trinomial Using FOIL (Negative Middle Term)

Factor $6x^2 - 11x + 3$.

Because 3 has only 1 and 3 or -1 and -3 as factors, it is better here to begin by factoring 3. The last (constant) term of the trinomial $6x^2 - 11x + 3$ is positive and the middle term has a negative coefficient, so we consider only negative factors. We need two negative factors, because the *product* of two negative factors is positive and their *sum* is negative, as required.

We try -3 and -1 as factors of 3.

$$(\_\_\_\_ - 3)(\_\_\_\_ - 1)$$

The factors of $6x^2$ may be either $6x$ and x, or $2x$ and $3x$.

$(6x - 3)(x - 1)$ Incorrect	$(2x - 3)(3x - 1)$ Correct
$-3x$	$-9x$
$-6x$	$-2x$
$-9x$ Add.	$-11x$ Add.

The factors $2x$ and $3x$ produce $-11x$, the correct middle term.

> Check by multiplying.

$$6x^2 - 11x + 3 \quad \text{factors as} \quad (2x - 3)(3x - 1).$$

NOW TRY

NOW TRY ANSWERS
4. $(2p + 3)(p + 3)$
5. $(4y + 1)(2y + 5)$
6. $(5x - 2)(2x - 1)$

NOTE In **Example 6,** our initial attempt to factor $6x^2 - 11x + 3$ as $(6x - 3)(x - 1)$ *cannot* be correct because the terms of $6x - 3$ have a common factor of 3, while those of the original polynomial do not.

NOW TRY EXERCISE 7
Factor $10a^2 + 31a - 14$.

EXAMPLE 7 Factoring a Trinomial Using FOIL (Negative Constant Term)

Factor $8x^2 + 6x - 9$.

The integer 8 has several possible pairs of factors, as does -9. Because the constant term is negative, one positive factor and one negative factor of -9 are needed. The coefficient of the middle term is relatively small, so we avoid large factors such as 8 or 9. We try $4x$ and $2x$ as factors of $8x^2$, and 3 and -3 as factors of -9.

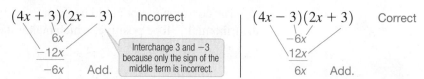

The combination on the right produces $6x$, the correct middle term.

$$8x^2 + 6x - 9 \quad \text{factors as} \quad (4x - 3)(2x + 3).$$

Check by multiplying.

NOW TRY

EXAMPLE 8 Factoring a Multivariable Trinomial

NOW TRY EXERCISE 8
Factor $8z^2 + 2wz - 15w^2$.

Factor $12a^2 - ab - 20b^2$.

There are several pairs of factors of $12a^2$, including

$$12a \text{ and } a, \quad 6a \text{ and } 2a, \quad \text{and} \quad 3a \text{ and } 4a.$$

There are also many pairs of factors of $-20b^2$, including

$$20b \text{ and } -b, \quad -20b \text{ and } b, \quad 10b \text{ and } -2b, \quad -10b \text{ and } 2b,$$

$$4b \text{ and } -5b, \quad \text{and} \quad -4b \text{ and } 5b.$$

Once again, because the coefficient of the desired middle term is relatively small, avoid the larger factors. Try the factors $6a$ and $2a$, and $4b$ and $-5b$.

$$(6a + 4b)(2a - 5b)$$

This cannot be correct because the terms of $6a + 4b$ have 2 as a common factor, while the terms of the given trinomial do not. Try $3a$ and $4a$ with $4b$ and $-5b$.

$$(3a + 4b)(4a - 5b)$$
$$= 12a^2 + ab - 20b^2 \quad \text{Incorrect}$$

Here the middle term is ab rather than $-ab$, so we interchange the signs of the last two terms in the factors.

Check by multiplying.

$$12a^2 - ab - 20b^2 \quad \text{factors as} \quad (3a - 4b)(4a + 5b).$$

NOW TRY

EXAMPLE 9 Factoring Trinomials with Common Factors

Factor each trinomial.

(a) $15y^3 + 55y^2 + 30y$

$$= 5y(3y^2 + 11y + 6) \quad \text{Factor out the greatest common factor, 5y.}$$

To factor $3y^2 + 11y + 6$, try $3y$ and y as factors of $3y^2$, and 2 and 3 as factors of 6.

$$(3y + 2)(y + 3)$$
$$= 3y^2 + 11y + 6 \quad \text{Correct}$$

NOW TRY ANSWERS
7. $(5a - 2)(2a + 7)$
8. $(4z - 5w)(2z + 3w)$

**NOW TRY
EXERCISE 9**
Factor $-10x^3 - 45x^2 + 90x$.

This leads to the completely factored form.

$$15y^3 + 55y^2 + 30y$$

> Remember the common factor.

$$= 5y(3y + 2)(y + 3)$$

CHECK $5y(3y + 2)(y + 3)$

$= 5y(3y^2 + 9y + 2y + 6)$ FOIL method

$= 5y(3y^2 + 11y + 6)$ Combine like terms.

$= 15y^3 + 55y^2 + 30y$ ✓   Distributive property

(b) $-24a^3 - 42a^2 + 45a$

The common factor could be $3a$ or $-3a$. If we factor out $-3a$, the leading term of the trinomial will be positive, which makes it easier to factor the remaining trinomial.

$$-24a^3 - 42a^2 + 45a$$

$$= -3a(8a^2 + 14a - 15)$$ Factor out $-3a$.

$$= -3a(4a - 3)(2a + 5)$$ Factor the trinomial.

CHECK $-3a(4a - 3)(2a + 5)$

> We can multiply $-3a(4a - 3)$ first.

$= (-12a^2 + 9a)(2a + 5)$ Distributive property

$= -24a^3 - 42a^2 + 45a$ ✓   FOIL method

 NOW TRY ↺

NOW TRY ANSWER
9. $-5x(2x - 3)(x + 6)$

6.3 Exercises

> FOR EXTRA HELP

 MyMathLab®

▶ *Complete solution available in MyMathLab*

Concept Check   *The middle term of each trinomial has been rewritten. Now factor by grouping.*

1. $10t^2 + 9t + 2$
$= 10t^2 + 5t + 4t + 2$

2. $6x^2 + 13x + 6$
$= 6x^2 + 9x + 4x + 6$

3. $15z^2 - 19z + 6$
$= 15z^2 - 10z - 9z + 6$

4. $12p^2 - 17p + 6$
$= 12p^2 - 9p - 8p + 6$

5. $8s^2 + 2st - 3t^2$
$= 8s^2 - 4st + 6st - 3t^2$

6. $3x^2 - xy - 14y^2$
$= 3x^2 - 7xy + 6xy - 14y^2$

Concept Check   *Complete the steps to factor each trinomial by grouping.*

7. $2m^2 + 11m + 12$

(a) Find two integers whose product is
_____ · _____ = _____ and whose
sum is _____.

(b) The required integers are _____ and
_____.

(c) Now write the middle term, $11m$, as
_____ + _____.

(d) Rewrite the given trinomial using
four terms as _____.

(e) Factor the polynomial in part (d) by
grouping.

(f) Check by multiplying.

8. $6y^2 - 19y + 10$

(a) Find two integers whose product is
_____ · _____ = _____ and whose
sum is _____.

(b) The required integers are _____ and
_____.

(c) Now write the middle term, $-19y$, as
_____ + _____.

(d) Rewrite the given trinomial using
four terms as _____.

(e) Factor the polynomial in part (d) by
grouping.

(f) Check by multiplying.

9. *Concept Check* Which pair of integers would be used to rewrite the middle term when factoring $12y^2 + 5y - 2$ by grouping?

A. $-8, 3$ **B.** $8, -3$ **C.** $-6, 4$ **D.** $6, -4$

10. *Concept Check* Which pair of integers would be used to rewrite the middle term when factoring $20b^2 - 13b + 2$ by grouping?

A. $10, 3$ **B.** $-10, -3$ **C.** $8, 5$ **D.** $-8, -5$

Concept Check *Which is the correct factored form of the given polynomial?*

11. $2x^2 - x - 1$

    A. $(2x - 1)(x + 1)$

    B. $(2x + 1)(x - 1)$

12. $3a^2 - 5a - 2$

    A. $(3a + 1)(a - 2)$

    B. $(3a - 1)(a + 2)$

13. $4y^2 + 17y - 15$

    A. $(y + 5)(4y - 3)$

    B. $(2y - 5)(2y + 3)$

14. $12c^2 - 7c - 12$

    A. $(6c - 2)(2c + 6)$

    B. $(4c + 3)(3c - 4)$

15. *Concept Check* A student factoring the trinomial

$$12x^2 + 7x - 12$$

wrote $(4x + 4)$ as one binomial factor. ***WHAT WENT WRONG?*** Factor correctly.

16. *Concept Check* Another student factored $3k^3 - 12k^2 - 15k$ by first factoring out the common factor $3k$ to obtain $3k(k^2 - 4k - 5)$. Then she wrote the following.

$$k^2 - 4k - 5$$
$$= k^2 - 5k + k - 5$$
$$= k(k - 5) + 1(k - 5)$$
$$= (k - 5)(k + 1) \qquad \text{Her answer}$$

WHAT WENT WRONG? What is the correct factored form?

Complete each factoring. ***See Examples 1–9.***

17. $6a^2 + 7ab - 20b^2$

    $= (3a - 4b)(\underline{\hspace{1cm}})$

18. $9m^2 + 6mn - 8n^2$

    $= (3m - 2n)(\underline{\hspace{1cm}})$

19. $2x^2 + 6x - 8$

    $= 2(\underline{\hspace{2cm}})$

    $= 2(\underline{\hspace{1cm}})(\underline{\hspace{1cm}})$

20. $3x^2 + 9x - 30$

    $= 3(\underline{\hspace{2cm}})$

    $= 3(\underline{\hspace{1cm}})(\underline{\hspace{1cm}})$

21. $4z^3 - 10z^2 - 6z$

    $= 2z(\underline{\hspace{2cm}})$

    $= 2z(\underline{\hspace{1cm}})(\underline{\hspace{1cm}})$

22. $15r^3 - 39r^2 - 18r$

    $= 3r(\underline{\hspace{2cm}})$

    $= 3r(\underline{\hspace{1cm}})(\underline{\hspace{1cm}})$

Factor each trinomial completely. ***See Examples 1–9.*** *(Hint: In Exercises 57 and 58, first write the trinomial in descending powers and then factor.)*

▶ **23.** $3a^2 + 10a + 7$ **24.** $7r^2 + 8r + 1$ ▶ **25.** $2y^2 + 7y + 6$

26. $5z^2 + 12z + 4$ **27.** $15m^2 + m - 2$ **28.** $6x^2 + x - 1$

29. $12s^2 + 11s - 5$ **30.** $20x^2 + 11x - 3$ ▶ **31.** $10m^2 - 23m + 12$

32. $6x^2 - 17x + 12$ **33.** $8w^2 - 14w + 3$ **34.** $9p^2 - 18p + 8$

▶ **35.** $20y^2 - 39y - 11$ **36.** $10x^2 - 11x - 6$ **37.** $3x^2 - 15x + 16$

38. $2t^2 - 14t + 15$

39. $20x^2 + 22x + 6$

40. $36y^2 + 81y + 45$

41. $24x^2 - 42x + 9$

42. $48b^2 - 74b - 10$

43. $-40m^2q - mq + 6q$

44. $-15a^2b - 22ab - 8b$

▶ **45.** $15n^4 - 39n^3 + 18n^2$

46. $24a^4 + 10a^3 - 4a^2$

▶ **47.** $15x^2y^2 - 7xy^2 - 4y^2$

48. $14a^2b^3 + 15ab^3 - 9b^3$

49. $5a^2 - 7ab - 6b^2$

50. $6x^2 - 5xy - y^2$

▶ **51.** $12s^2 + 11st - 5t^2$

52. $25a^2 + 25ab + 6b^2$

53. $6m^6n + 7m^5n^2 + 2m^4n^3$

54. $12k^3q^4 - 4k^2q^5 - kq^6$

55. $x^2 - 6x - 5$

56. $x^2 - 8x - 7$

57. $16 + 16x + 3x^2$

58. $18 + 65x + 7x^2$

59. $-10x^3 + 5x^2 + 140x$

60. $-18k^3 - 48k^2 + 66k$

61. $12x^2 - 7x - 4$

62. $12x^2 - 9x - 10$

63. $24y^2 - 41xy - 14x^2$

64. $24x^2 + 19xy - 5y^2$

65. $36x^4 - 64x^2y + 15y^2$

66. $36x^4 + 59x^2y + 24y^2$

67. $48a^2 - 94ab - 4b^2$

68. $48t^2 - 147ts + 9s^2$

69. $10x^4y^5 + 39x^3y^5 - 4x^2y^5$

70. $14x^7y^4 - 31x^6y^4 + 6x^5y^4$

71. $36a^3b^2 - 104a^2b^2 - 12ab^2$

72. $36p^4q + 129p^3q - 60p^2q$

73. $24x^2 - 46x + 15$

74. $24x^2 - 94x + 35$

75. $24x^4 + 55x^2 - 24$

76. $24x^4 + 17x^2 - 20$

77. $24x^2 + 38xy + 15y^2$

78. $24x^2 + 62xy + 33y^2$

If a trinomial has a negative coefficient for the second-degree term, such as $-2x^2 + 11x - 12$, it is usually easier to factor by first factoring out the common factor -1.

$$-2x^2 + 11x - 12$$
$$= -1(2x^2 - 11x + 12) \qquad \text{Factor out } -1.$$
$$= -1(2x - 3)(x - 4) \qquad \text{Factor the trinomial.}$$

Use this method to factor each trinomial. ***See Example 9(b).***

79. $-x^2 - 4x + 21$

80. $-x^2 + x + 72$

81. $-3x^2 - x + 4$

82. $-5x^2 + 2x + 16$

83. $-2a^2 - 5ab - 2b^2$

84. $-3p^2 + 13pq - 4q^2$

Extending Skills *Factor each polynomial. (Hint: As the first step, factor out the greatest common factor.)*

85. $25q^2(m + 1)^3 - 5q(m + 1)^3 - 2(m + 1)^3$

86. $18x^2(y - 3)^2 - 21x(y - 3)^2 - 4(y - 3)^2$

87. $9x^2(r + 3)^3 + 12xy(r + 3)^3 + 4y^2(r + 3)^3$

88. $4t^2(k + 9)^7 + 20ts(k + 9)^7 + 25s^2(k + 9)^7$

Extending Skills *Find all integers k so that the trinomial can be factored by the methods of this section.*

89. $5x^2 + kx - 1$

90. $2x^2 + kx - 3$

91. $2m^2 + km + 5$

92. $3y^2 + ky + 4$

6.4 Special Factoring Techniques

VOCABULARY

☐ perfect square
☐ perfect square trinomial
☐ perfect cube

By reversing the rules for multiplication of binomials from **Section 5.6,** we obtain rules for factoring polynomials in certain forms.

OBJECTIVE 1 Factor a difference of squares.

The rule for finding the product of a sum and difference of the same two terms is

$$(x + y)(x - y) = x^2 - y^2.$$

Reversing this rule leads to the following special factoring rule.

Factoring a Difference of Squares

$$x^2 - y^2 = (x + y)(x - y)$$

For example,

$$m^2 - 4$$
$$= m^2 - 2^2$$
$$= (m + 2)(m - 2).$$

Two conditions must be true for a binomial to be a difference of squares.

1. Both terms of the binomial must be **perfect squares,** such as

$$x^2, \quad 9y^2 = (3y)^2, \quad m^4 = (m^2)^2, \quad 1 = 1^2, \quad 25 = 5^2, \quad 144 = 12^2.$$

2. The terms of the binomial must have different signs (one positive and one negative).

NOW TRY EXERCISE 1

Factor each binomial if possible.

(a) $x^2 - 100$

(b) $x^2 + 100$

(c) $x^2 - 32$

EXAMPLE 1 Factoring Binomials

Factor each binomial if possible.

$$x^2 - y^2 = (x + y)(x - y)$$
$$\downarrow \quad \downarrow \quad\quad \downarrow \quad \downarrow \quad \downarrow \quad \downarrow$$
(a) $p^2 - 16 = p^2 - 4^2 = (p + 4)(p - 4)$

(b) $x^2 - 8$
    Because 8 is not the square of an integer, this binomial does not satisfy Condition 1 above. It cannot be factored, so it is a prime polynomial.

(c) $p^2 + 16$
    The binomial $p^2 + 16$ does not satisfy Condition 2 above. It is a *sum* of squares— it is *not* equal to $(p + 4)(p - 4)$. (See part (a).) We can use the FOIL method and try the following.

$$(p - 4)(p - 4) \qquad\qquad\qquad (p + 4)(p + 4)$$
$$= p^2 - 8p + 16, \quad \text{not} \quad p^2 + 16. \qquad = p^2 + 8p + 16, \quad \text{not} \quad p^2 + 16.$$

Thus, $p^2 + 16$ is a prime polynomial.

NOW TRY ANSWERS
1. (a) $(x + 10)(x - 10)$
    (b) prime   (c) prime

NOW TRY

> **Sum of Squares**
>
> **If x and y have no common factors,** then the following holds.
>
> **A *sum of squares* $x^2 + y^2$ cannot be factored using real numbers.**
>
> That is, $x^2 + y^2$ is prime. (See **Example 1(c).**)

NOW TRY
EXERCISE 2

Factor each binomial.

(a) $9t^2 - 100$

(b) $36a^2 - 49b^2$

EXAMPLE 2 Factoring Differences of Squares

Factor each binomial.

$$x^2 \ - \ y^2 \ = \ (x \ + \ y) \ (x \ - \ y)$$

(a) $25m^2 - 4 = (5m)^2 - 2^2 = (5m + 2)(5m - 2)$

(b) $49z^2 - 64t^2$

$\qquad = (7z)^2 - (8t)^2$ Write each term as a square.

$\qquad = (7z + 8t)(7z - 8t)$ Factor the difference of squares.

CHECK $(7z + 8t)(7z - 8t)$

$\qquad = 49z^2 - 56zt + 56tz - 64t^2$ FOIL method

$\qquad = 49z^2 - 64t^2$ ✓ Commutative property; Combine like terms.

NOW TRY

> **NOTE** *Always check a factored form by multiplying.*

NOW TRY
EXERCISE 3

Factor each binomial
completely.

(a) $16k^2 - 64$

(b) $m^4 - 144$

(c) $v^4 - 625$

EXAMPLE 3 Factoring More Complex Differences of Squares

Factor each binomial completely.

(a) $81y^2 - 36$ ⟵ Always check for a common factor first.

$\qquad = 9(9y^2 - 4)$ Factor out the GCF, 9.

$\qquad = 9[(3y)^2 - 2^2]$ Write each term as a square.

$\qquad = 9(3y + 2)(3y - 2)$ Factor the difference of squares.

(b) $\qquad\qquad p^4 - 36$

$\qquad\qquad = (p^2)^2 - 6^2$ Write each term as a square.

Neither binomial can be factored further. ⟶ $= (p^2 + 6)(p^2 - 6)$ Factor the difference of squares.

(c) $\qquad\quad m^4 - 16$

$\qquad = (m^2)^2 - 4^2$ Write each term as a square.

$\qquad = (m^2 + 4)(m^2 - 4)$ Factor the difference of squares.

Don't stop here. ⟶ $= (m^2 + 4)(m + 2)(m - 2)$ Factor the difference of squares again.

NOW TRY

NOW TRY ANSWERS

2. (a) $(3t + 10)(3t - 10)$

 (b) $(6a + 7b)(6a - 7b)$

3. (a) $16(k + 2)(k - 2)$

 (b) $(m^2 + 12)(m^2 - 12)$

 (c) $(v^2 + 25)(v + 5)(v - 5)$

> ❗ **CAUTION** *Factor again when any of the factors is a difference of squares,* as in
> **Example 3(c).** Check by multiplying.

OBJECTIVE 2   Factor a perfect square trinomial.

Recall the rules for squaring binomials from **Section 5.6.**

$(x + y)^2$ Squared binomial	$(x - y)^2$ Squared binomial
$= (x + y)(x + y)$	$= (x - y)(x - y)$
$= x^2 + 2xy + y^2$ Perfect square trinomial	$= x^2 - 2xy + y^2$ Perfect square trinomial

A **perfect square trinomial** is a trinomial that is the square of a binomial. For example, $x^2 + 8x + 16$ is a perfect square trinomial because it is the square of the binomial $x + 4$.

$$(x + 4)^2 \qquad \text{Squared binomial}$$
$$= (x + 4)(x + 4)$$
$$= x^2 + 8x + 16 \qquad \text{Perfect square trinomial}$$

Two conditions must be true for a trinomial to be a perfect square trinomial.

1. Two of its terms must be perfect squares. In the perfect square trinomial $x^2 + 8x + 16$, the terms x^2 and $16 = 4^2$ are perfect squares.

2. *The remaining (middle) term of a perfect square trinomial is always twice the product of the two terms in the squared binomial.* For example,

$$x^2 + 8x + 16$$
$$= x^2 + 2(x)(4) + 4^2 \qquad 8x = 2(x)(4)$$
$$= (x + 4)^2. \qquad \text{Factor.}$$

The following are *not* perfect square trinomials.

$16x^2 + 4x + 15$ 　　Violates Condition 1 (Only $16x^2 = (4x)^2$ is a perfect square; 15 is not.)

$x^2 + 6x + 36$ 　　Violates Condition 2 (x^2 and $36 = 6^2$ are perfect squares, but $2(x)(6) = 12x$, *not* $6x$.)

Reversing the rules for squaring binomials leads to the following special factoring rules.

Factoring Perfect Square Trinomials

$$x^2 + 2xy + y^2 = (x + y)^2$$
$$x^2 - 2xy + y^2 = (x - y)^2$$

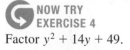
NOW TRY
EXERCISE 4
Factor $y^2 + 14y + 49$.

EXAMPLE 4   Factoring a Perfect Square Trinomial

Factor $x^2 + 10x + 25$.

The x^2-term is a perfect square, and so is 25, which equals 5^2.

Try to factor $x^2 + 10x + 25$ as the squared binomial $(x + 5)^2$.

To check, take twice the product of the two terms in the squared binomial.

$$2 \cdot x \cdot 5 = 10x \leftarrow \text{Middle term of } x^2 + 10x + 25$$

Twice　First term　　　　　Last term
　　　of binomial　　　　of binomial

NOW TRY ANSWER
4. $(y + 7)^2$

Because $10x$ is the middle term of the trinomial, the trinomial is a perfect square.

$$x^2 + 10x + 25 \quad \text{factors as} \quad (x + 5)^2. \qquad \text{NOW TRY}$$

NOW TRY
EXERCISE 5

Factor each trinomial.

(a) $t^2 - 18t + 81$

(b) $4p^2 - 28p + 49$

(c) $9x^2 + 6x + 4$

(d) $80x^3 + 120x^2 + 45x$

EXAMPLE 5 **Factoring Perfect Square Trinomials**

Factor each trinomial.

(a) $x^2 - 22x + 121$

The first and last terms are perfect squares ($121 = 11^2$ or $(-11)^2$). Check to see whether the middle term of $x^2 - 22x + 121$ is twice the product of the first and last terms of the binomial $x - 11$.

$$2 \cdot x \cdot (-11) = -22x \leftarrow \text{Middle term of } x^2 - 22x + 121$$

Twice ⎯⎯ First term Last term

Thus, $x^2 - 22x + 121$ is a perfect square trinomial.

$x^2 - 22x + 121$ factors as $(x - 11)^2.$

Same sign

> Check by squaring the binomial.

Notice that the sign of the second term in the squared binomial is the same as the sign of the middle term in the trinomial.

(b) $9m^2 - 24m + 16 = (3m)^2 + 2(3m)(-4) + (-4)^2$

⎯Perfect squares⎯   Twice ⎯ First term   Last term

$$= (3m - 4)^2$$

(c) $25y^2 + 20y + 16$

The first and last terms are perfect squares.

$$25y^2 = (5y)^2 \quad \text{and} \quad 16 = 4^2$$

Twice the product of the first and last terms of the binomial $5y + 4$ is

$$2 \cdot 5y \cdot 4 = 40y,$$

which is *not* the middle term of

$$25y^2 + 20y + 16.$$

This trinomial is not a perfect square. In fact, the trinomial cannot be factored even with the methods of the previous sections. It is a prime polynomial.

(d) $12z^3 + 60z^2 + 75z$

$$= 3z(4z^2 + 20z + 25) \qquad \text{Factor out the common factor, } 3z.$$

$$= 3z[(2z)^2 + 2(2z)(5) + 5^2] \qquad 4z^2 + 20z + 25 \text{ is a perfect square trinomial.}$$

$$= 3z(2z + 5)^2 \qquad \text{Factor.} \qquad \textbf{NOW TRY}$$

NOTE Keep the following in mind when factoring perfect square trinomials.

1. The sign of the second term in the squared binomial is always the same as the sign of the middle term in the trinomial.

2. The first and last terms of a perfect square trinomial must be *positive* because they are squares. For example, $x^2 - 2x - 1$ cannot be a perfect square trinomial because the last term is negative.

3. Perfect square trinomials can also be factored by grouping or the FOIL method. Using the method of this section is often easier.

NOW TRY ANSWERS

5. **(a)** $(t - 9)^2$
   (b) $(2p - 7)^2$
   (c) prime
   (d) $5x(4x + 3)^2$

OBJECTIVE 3   Factor a difference of cubes.

In a difference of cubes $x^3 - y^3$, both terms of the binomial must be **perfect cubes,** such as

$$x^3, \quad 8p^3 = (2p)^3, \quad s^6 = (s^2)^3, \quad 1 = 1^3, \quad 27 = 3^3, \quad 216 = 6^3.$$

We can factor a **difference of cubes** using the following rule.

Factoring a Difference of Cubes

$$x^3 - y^3 = (x - y)(x^2 + xy + y^2)$$

This rule for factoring a difference of cubes should be memorized. To see that the rule is correct, multiply $(x - y)(x^2 + xy + y^2)$.

$$
\begin{array}{ll}
x^2 + xy \;+ y^2 & \text{Multiply vertically.} \\
\underline{ x \;- y} & \text{(Section 5.5)} \\
-x^2 y - xy^2 - y^3 & -y(x^2 + xy + y^2) \\
\underline{x^3 + x^2 y + xy^2 } & x(x^2 + xy + y^2) \\
x^3 - y^3 & \text{Add.}
\end{array}
$$

Notice the pattern of the terms in the factored form of $x^3 - y^3$.

- $x^3 - y^3$ factors as (a binomial factor) $\cdot$ (a trinomial factor).

- The binomial factor has the difference of the cube roots of the given terms. (*Note:* A cube root of 1 is 1 because $1^3 = 1$, a cube root of 8 is 2 because $2^3 = 8$, and so on.)

- The terms in the trinomial factor are all positive.

- The terms in the binomial factor help to determine the trinomial factor.

$$
\begin{array}{ccccc}
& \text{First term} & & \overset{\text{positive}}{\text{product of}} & \text{second term} \\
& \text{squared} & + & \text{the terms} & + & \text{squared} \\
x^3 - y^3 = (x - y)(& x^2 & + & xy & + & y^2 \;)
\end{array}
$$

⚠ **CAUTION**   The polynomial $x^3 - y^3$ is *not* equivalent to $(x - y)^3$.

$$
\begin{array}{l|l}
x^3 - y^3 & (x - y)^3 \\
= (x - y)(x^2 + xy + y^2) & = (x - y)(x - y)(x - y) \\
& = (x - y)(x^2 - 2xy + y^2)
\end{array}
$$

EXAMPLE 6   Factoring Differences of Cubes

Factor each binomial.

$$x^3 \;-\; y^3 \;=\; (x \;-\; y) \;(x^2 \;+\; xy \;+\; y^2)$$

(a) $m^3 - 125 = m^3 - 5^3 = (m - 5)(m^2 + 5m + 5^2)$ Let $x = m$ and $y = 5$.

$$= (m - 5)(m^2 + 5m + 25) \qquad 5^2 = 25$$

**NOW TRY
EXERCISE 6**

Factor each binomial.

(a) $a^3 - 27$

(b) $8t^3 - 125$

(c) $3k^3 - 192$

(d) $125x^3 - 343y^6$

(b) $8p^3 - 27$

$= (2p)^3 - 3^3$ $8p^3 = (2p)^3$ and $27 = 3^3$.

$= (2p - 3)[(2p)^2 + (2p)3 + 3^2]$ Let $x = 2p$ and $y = 3$.

$= (2p - 3)(4p^2 + 6p + 9)$ Apply the exponents. Multiply.

> $(2p)^2 = 2^2p^2 = 4p^2$,
> **not** $2p^2$.

(c) $4m^3 - 32$

$= 4(m^3 - 8)$ Factor out the common factor, 4.

$= 4(m^3 - 2^3)$ $8 = 2^3$

$= 4(m - 2)(m^2 + 2m + 4)$ Factor the difference of cubes.

(d) $125t^3 - 216s^6$

$= (5t)^3 - (6s^2)^3$ Write each term as a cube.

$= (5t - 6s^2)[(5t)^2 + 5t(6s^2) + (6s^2)^2]$ Factor the difference of cubes.

$= (5t - 6s^2)(25t^2 + 30ts^2 + 36s^4)$ Apply the exponents. Multiply.

> Square carefully.
> $(6s^2)^2 = 6^2(s^2)^2 = 36s^4$

NOW TRY

⚠ CAUTION   A common error when factoring $x^3 - y^3 = (x - y)(x^2 + xy + y^2)$ is to try to factor $x^2 + xy + y^2$. This is usually not possible.

OBJECTIVE 4   **Factor a sum of cubes.**

A *sum of squares,* such as $m^2 + 25$, *cannot* be factored using real numbers, but a **sum of cubes** can.

Factoring a Sum of Cubes

$$x^3 + y^3 = (x + y)(x^2 - xy + y^2)$$

Compare the rule for the *sum* of cubes with that for the *difference* of cubes.

┌─Positive─┐

$x^3 - y^3 = (x - y)(x^2 + xy + y^2)$ Difference of cubes

└─Same─┘   Opposite
  sign   sign

The only difference between the rules is the positive and negative signs.

NOW TRY ANSWERS

6. (a) $(a - 3)(a^2 + 3a + 9)$

   (b) $(2t - 5)(4t^2 + 10t + 25)$

   (c) $3(k - 4)(k^2 + 4k + 16)$

   (d) $(5x - 7y^2) \cdot$
           $(25x^2 + 35xy^2 + 49y^4)$

┌─Positive─┐

$x^3 + y^3 = (x + y)(x^2 - xy + y^2)$ Sum of cubes

└─Same─┘   Opposite
  sign   sign

NOW TRY
EXERCISE 7
Factor each binomial.

(a) $x^3 + 125$

(b) $27a^3 + 8b^3$

EXAMPLE 7 Factoring Sums of Cubes

Factor each binomial.

(a) $k^3 + 27$

$$= k^3 + 3^3 \qquad\qquad 27 = 3^3$$

$$= (k + 3)(k^2 - 3k + 3^2) \qquad \text{Factor the sum of cubes.}$$

$$= (k + 3)(k^2 - 3k + 9) \qquad \text{Apply the exponent.}$$

(b) $8m^3 + 125n^3$

$$= (2m)^3 + (5n)^3 \qquad\qquad 8m^3 = (2m)^3 \text{ and } 125n^3 = (5n)^3.$$

$$= (2m + 5n)[(2m)^2 - 2m(5n) + (5n)^2] \qquad \text{Factor the sum of cubes.}$$

$$= (2m + 5n)(4m^2 - 10mn + 25n^2) \qquad \boxed{\begin{array}{l}\text{Be careful:} \\ (2m)^2 = 2^2m^2 \\ \text{and } (5n)^2 = 5^2n^2.\end{array}}$$

(c) $1000a^6 + 27b^3$

$$= (10a^2)^3 + (3b)^3$$

$$= (10a^2 + 3b)[(10a^2)^2 - (10a^2)(3b) + (3b)^2] \qquad \text{Factor the sum of cubes.}$$

$$= (10a^2 + 3b)(100a^4 - 30a^2b + 9b^2) \qquad (10a^2)^2 = 10^2(a^2)^2 = 100a^4$$

NOW TRY

The methods of factoring discussed in this section are summarized here.

Special Factoring Rules

Difference of squares	$x^2 - y^2 = (x + y)(x - y)$
Perfect square trinomials	$x^2 + 2xy + y^2 = (x + y)^2$
	$x^2 - 2xy + y^2 = (x - y)^2$
Difference of cubes	$x^3 - y^3 = (x - y)(x^2 + xy + y^2)$
Sum of cubes	$x^3 + y^3 = (x + y)(x^2 - xy + y^2)$

A sum of squares can be factored only if the terms have a common factor.

NOW TRY ANSWERS
7. **(a)** $(x + 5)(x^2 - 5x + 25)$
(b) $(3a + 2b)(9a^2 - 6ab + 4b^2)$

6.4 Exercises

 FOR EXTRA HELP MyMathLab®

▶ *Complete solution available*
in MyMathLab

Concept Check Work each problem.

1. To help factor differences of squares, complete the following list of perfect squares.

$1^2 = \rule{1cm}{0.4pt}$ $2^2 = \rule{1cm}{0.4pt}$ $3^2 = \rule{1cm}{0.4pt}$ $4^2 = \rule{1cm}{0.4pt}$ $5^2 = \rule{1cm}{0.4pt}$

$6^2 = \rule{1cm}{0.4pt}$ $7^2 = \rule{1cm}{0.4pt}$ $8^2 = \rule{1cm}{0.4pt}$ $9^2 = \rule{1cm}{0.4pt}$ $10^2 = \rule{1cm}{0.4pt}$

$11^2 = \rule{1cm}{0.4pt}$ $12^2 = \rule{1cm}{0.4pt}$ $13^2 = \rule{1cm}{0.4pt}$ $14^2 = \rule{1cm}{0.4pt}$ $15^2 = \rule{1cm}{0.4pt}$

$16^2 = \rule{1cm}{0.4pt}$ $17^2 = \rule{1cm}{0.4pt}$ $18^2 = \rule{1cm}{0.4pt}$ $19^2 = \rule{1cm}{0.4pt}$ $20^2 = \rule{1cm}{0.4pt}$

2. The following powers of x are all perfect squares:

$$x^2, \quad x^4, \quad x^6, \quad x^8, \quad x^{10}.$$

On the basis of this observation, we may make a conjecture (an educated guess) that if the power of a variable is divisible by \rule{1cm}{0.4pt} (with 0 remainder), then we have a perfect square.

3. Which of the following are differences of squares?

    A. $x^2 - 4$ **B.** $y^2 + 9$ **C.** $2a^2 - 25$ **D.** $9m^2 - 1$

4. Which of the following binomial sums can be factored?

    A. $x^2 + 36$ **B.** $x^3 + x$ **C.** $3x^2 + 12$ **D.** $25x^2 + 49$

5. On a quiz, a student indicated *prime* when asked to factor $4x^2 + 16$ because she said that a sum of a squares cannot be factored. **WHAT WENT WRONG?**

6. When a student was directed to factor $k^4 - 81$ completely, his teacher did not give him full credit.

$$(k^2 + 9)(k^2 - 9) \qquad \text{His answer}$$

The student argued that since his answer does indeed give $k^4 - 81$ when multiplied out, he should be given full credit. **WHAT WENT WRONG?** Give the correct factored form.

Factor each binomial completely. If the binomial is prime, say so. Use the answers from Exercises 1 and 2 as necessary. **See Examples 1–3.**

▶ **7.** $y^2 - 25$ **8.** $t^2 - 36$ **9.** $x^2 - 144$ **10.** $x^2 - 400$

11. $m^2 - 12$ **12.** $k^2 - 18$ **13.** $m^2 + 64$ **14.** $k^2 + 49$

15. $4m^2 + 16$ **16.** $9x^2 + 81$ ▶ **17.** $9r^2 - 4$

18. $4x^2 - 9$ ▶ **19.** $36x^2 - 16$ **20.** $32a^2 - 8$

21. $196p^2 - 225$ **22.** $361q^2 - 400$ **23.** $16r^2 - 25a^2$

24. $49m^2 - 100p^2$ **25.** $81x^2 - 49y^2$ **26.** $36y^2 - 121z^2$

27. $54x^2 - 6y^2$ **28.** $48m^2 - 75n^2$ **29.** $100x^2 + 49$

30. $81w^2 + 16$ **31.** $4 - x^2$ **32.** $25 - x^2$

33. $36 - 25t^2$ **34.** $16 - 49p^2$ **35.** $x^3 + 4x$

36. $z^3 + 25z$ **37.** $x^4 - x^2$ **38.** $y^4 - 9y^2$

39. $p^4 - 49$ **40.** $r^4 - 25$ **41.** $x^4 - 1$

42. $y^4 - 10,000$ **43.** $p^4 - 256$ **44.** $k^4 - 81$

Concept Check   *Work each problem.*

45. Which of the following are perfect square trinomials?

    A. $y^2 - 13y + 36$ **B.** $x^2 + 6x + 9$ **C.** $4z^2 - 4z + 1$ **D.** $16m^2 + 10m + 1$

46. In the polynomial $9y^2 + 14y + 25$, the first and last terms are perfect squares. Can the polynomial be factored? If it can, factor it. If it cannot, explain why it is not a perfect square trinomial.

Concept Check   *Find the value of the indicated variable.*

47. Find b so that $x^2 + bx + 25$ factors as $(x + 5)^2$.

48. Find c so that $4m^2 - 12m + c$ factors as $(2m - 3)^2$.

49. Find a so that $ay^2 - 12y + 4$ factors as $(3y - 2)^2$.

50. Find b so that $100a^2 + ba + 9$ factors as $(10a + 3)^2$.

Factor each trinomial completely. **See Examples 4 and 5.**

▶ **51.** $w^2 + 2w + 1$ **52.** $p^2 + 4p + 4$ ▶ **53.** $x^2 - 8x + 16$

54. $x^2 - 10x + 25$ **55.** $x^2 - 10x + 100$ **56.** $x^2 - 18x + 36$

57. $2x^2 + 24x + 72$

58. $3y^2 + 48y + 192$

59. $4x^2 + 12x + 9$

60. $25x^2 + 10x + 1$

61. $16x^2 - 40x + 25$

62. $36y^2 - 60y + 25$

63. $49x^2 - 28xy + 4y^2$

64. $4z^2 - 12zw + 9w^2$

65. $64x^2 + 48xy + 9y^2$

66. $9t^2 + 24tr + 16r^2$

67. $50h^2 - 40hy + 8y^2$

68. $18x^2 - 48xy + 32y^2$

69. $4k^3 - 4k^2 + 9k$

70. $9r^3 - 6r^2 + 16r$

71. $25z^4 + 5z^3 + z^2$

72. $4x^4 + 2x^3 + x^2$

Concept Check   *Work each problem.*

73. To help factor sums or differences of cubes, complete the following list of perfect cubes.

$1^3 = \underline{\hspace{0.5cm}}$ $2^3 = \underline{\hspace{0.5cm}}$ $3^3 = \underline{\hspace{0.5cm}}$ $4^3 = \underline{\hspace{0.5cm}}$ $5^3 = \underline{\hspace{0.5cm}}$

$6^3 = \underline{\hspace{0.5cm}}$ $7^3 = \underline{\hspace{0.5cm}}$ $8^3 = \underline{\hspace{0.5cm}}$ $9^3 = \underline{\hspace{0.5cm}}$ $10^3 = \underline{\hspace{0.5cm}}$

74. The following powers of x are all perfect cubes:

$$x^3, \quad x^6, \quad x^9, \quad x^{12}, \quad x^{15}.$$

On the basis of this observation, we may make a conjecture that if the power of a variable is divisible by $\underline{\hspace{0.5cm}}$ (with 0 remainder), then we have a perfect cube.

75. Which of the following are differences of cubes?

A. $9x^3 - 125$ **B.** $x^3 - 16$ **C.** $x^3 - 1$ **D.** $8x^3 - 27y^3$

76. Which of the following are sums of cubes?

A. $x^3 + 1$ **B.** $x^3 + 36$ **C.** $12x^3 + 27$ **D.** $64x^3 + 216y^3$

77. Identify each monomial as a *perfect square*, a *perfect cube*, *both of these*, or *neither of these*.

(a) $4x^3$ **(b)** $8y^6$ **(c)** $49x^{12}$ **(d)** $81r^{10}$ **(e)** $64x^6y^{12}$ **(f)** $125t^6$

78. What must be true for x^n to be both a perfect square and a perfect cube?

Factor each binomial completely. Use the answers from **Exercises 73 and 74** *as necessary.* **See Examples 6 and 7.**

▶ **79.** $a^3 - 1$

80. $m^3 - 8$

▶ **81.** $m^3 + 8$

82. $b^3 + 1$

83. $y^3 - 216$

84. $x^3 - 343$

85. $k^3 + 1000$

86. $p^3 + 512$

87. $27x^3 - 64$

88. $64y^3 - 27$

89. $6p^3 + 6$

90. $81x^3 + 3$

91. $5x^3 + 40$

92. $128y^3 + 54$

93. $y^3 - 8x^3$

94. $w^3 - 216z^3$

95. $2x^3 - 16y^3$

96. $27w^3 - 216z^3$

97. $8p^3 + 729q^3$

98. $64x^3 + 125y^3$

99. $27a^3 + 64b^3$

100. $125m^3 + 8p^3$

101. $125t^3 + 8s^3$

102. $27r^3 + 1000s^3$

103. $8x^3 - 125y^6$

104. $27t^3 - 64s^6$

105. $27m^6 + 8n^3$

106. $1000r^6 + 27s^3$

107. $x^9 + y^9$

108. $x^9 - y^9$

Extending Skills   *Although we usually factor polynomials using integers, we can apply the same concepts to factoring using fractions and decimals.*

$$z^2 - \frac{9}{16}$$

$$= z^2 - \left(\frac{3}{4}\right)^2 \qquad \frac{9}{16} = \left(\frac{3}{4}\right)^2$$

$$= \left(z + \frac{3}{4}\right)\left(z - \frac{3}{4}\right) \qquad \text{Factor the difference of squares.}$$

Apply the special factoring rules of this section to factor each polynomial.

109. $p^2 - \dfrac{1}{9}$

110. $q^2 - \dfrac{1}{4}$

111. $36m^2 - \dfrac{16}{25}$

112. $100b^2 - \dfrac{4}{49}$

113. $x^2 - 0.64$

114. $y^2 - 0.36$

115. $t^2 + t + \dfrac{1}{4}$

116. $m^2 + \dfrac{2}{3}m + \dfrac{1}{9}$

117. $x^2 - 1.8x + 0.81$

118. $y^2 - 1.4y + 0.49$

119. $x^3 + \dfrac{1}{8}$

120. $x^3 + \dfrac{1}{64}$

Extending Skills   *Factor each polynomial completely.*

121. $(m + n)^2 - (m - n)^2$

122. $(a - b)^3 - (a + b)^3$

123. $m^2 - p^2 + 2m + 2p$

124. $3r - 3k + 3r^2 - 3k^2$

SUMMARY EXERCISES   Recognizing and Applying Factoring Strategies

When factoring a polynomial, ask these questions to decide on a suitable factoring technique.

Factoring a Polynomial

Question 1   **Is there a common factor other than 1?** If so, factor it out.

Question 2   **How many terms are in the polynomial?**

Two terms: Is it a difference of squares or a sum or difference of cubes? If so, factor as in **Section 6.4.**

Three terms: Is it a perfect square trinomial? In this case, factor as in **Section 6.4.**

If the trinomial is not a perfect square trinomial, what is the coefficient of the second-degree term?

• If it is 1, use the factoring method of **Section 6.2.**

• If it is not 1, use the general factoring methods of **Section 6.3.**

Four terms: Try to factor by grouping, as in **Section 6.1.**

Question 3   **Can any factors be factored further?** If so, factor them.

! **CAUTION** Be careful when checking the answer to a factoring problem.

1. Check that the product of all the factors does indeed yield the original polynomial.

2. Check that the original polynomial has been factored **completely**.

NOW TRY
EXERCISE

Factor completely.

$24m^2 - 42my + 9y^2$

EXAMPLE Applying Factoring Strategies

Factor $12x^2 + 26xy + 12y^2$ completely.

Question 1 **Is there a common factor other than 1?**
Yes, 2 is a common factor, so factor it out.

$$12x^2 + 26xy + 12y^2$$
$$= 2(6x^2 + 13xy + 6y^2)$$

Question 2 **How many terms are in the polynomial?**
The polynomial $6x^2 + 13xy + 6y^2$ has three terms. It is not a perfect square trinomial. To factor by grouping, we find two integers with a product of $6 \cdot 6$, or 36, and a sum of 13. These integers are 4 and 9.

$$12x^2 + 26xy + 12y^2$$

$= 2(6x^2 + 13xy + 6y^2)$	Factor out the GCF, 2.
$= 2(6x^2 + 4xy + 9xy + 6y^2)$	$4 \cdot 9 = 36; 4 + 9 = 13$
$= 2[(6x^2 + 4xy) + (9xy + 6y^2)]$	Group the terms.
$= 2[2x(3x + 2y) + 3y(3x + 2y)]$	Factor each group.
$= 2(3x + 2y)(2x + 3y)$	Factor out the common factor, $3x + 2y$.

We could also have factored the trinomial $6x^2 + 13xy + 6y^2$ by trial and error, using the FOIL method in reverse, as in **Section 6.3.**

Question 3 **Can any factors be factored further?**
No. The original polynomial has been factored completely.

NOW TRY ↩

NOW TRY ANSWER
$3(4m - y)(2m - 3y)$

Match each polynomial in Column I with the best choice for factoring it in Column II. The choices in Column II may be used once, more than once, or not at all.

<table>
<tr><td align="center">**I**</td><td align="center">**II**</td></tr>
</table>

1. $12x^2 + 20x + 8$

A. Factor out the GCF. No further factoring is possible.

2. $x^2 - 17x + 72$

B. Factor a difference of squares.

3. $16m^2n + 24mn - 40mn^2$

C. Factor a difference of cubes.

4. $64a^2 - 121b^2$

D. Factor a sum of cubes.

5. $36p^2 - 60pq + 25q^2$

E. Factor a perfect square trinomial.

6. $z^2 - 4z + 6$

F. Factor by grouping.

7. $8r^3 - 125$

G. Factor out the GCF. Then factor a trinomial by grouping or trial and error.

8. $x^6 + 4x^4 - 3x^2 - 12$

H. Factor into two binomials by finding two integers whose product is the constant in the trinomial and whose sum is the coefficient of the middle term.

9. $4w^2 + 49$

10. $z^2 - 24z + 144$

I. The polynomial is prime.

1. G   **2.** H
3. A   **4.** B
5. E   **6.** I
7. C   **8.** F
9. I   **10.** E

11. $(a - 6)(a + 2)$
12. $(a + 8)(a + 9)$
13. $6(y - 2)(y + 1)$
14. $7y^4(y + 6)(y - 4)$
15. $6(a + 2b + 3c)$
16. $(m - 4n)(m + n)$
17. $(p - 11)(p - 6)$
18. $(z + 7)(z - 6)$
19. $(5z - 6)(2z + 1)$
20. $2(m - 8)(m + 3)$
21. $17xy(x^2y + 3)$
22. $5(3y + 1)$
23. $8a^3(a - 3)(a + 2)$
24. $(4k + 1)(2k - 3)$
25. $(z - 5a)(z + 2a)$
26. $50(z^2 - 2)$
27. $(x - 5)(x - 4)$
28. prime
29. $(3n - 2)(2n - 5)$
30. $(3y - 1)(3y + 5)$
31. $4(4x + 5)$
32. $(m + 5)(m - 3)$
33. $(3y - 4)(2y + 1)$
34. $(m + 9)(m - 9)$
35. $(6z + 1)(z + 5)$
36. $(12x - 1)(x + 4)$
37. $(2k - 3)^2$
38. $(8p - 1)(p + 3)$
39. $6(3m + 2z)(3m - 2z)$
40. $(4m - 3)(2m + 1)$
41. $(3k - 2)(k + 2)$
42. $(2a - 3)(4a^2 + 6a + 9)$
43. $7k(2k + 5)(k - 2)$
44. $(5 + r)(1 - s)$
45. $(y^2 + 4)(y + 2)(y - 2)$
46. prime
47. $8m(1 - 2m)$
48. $(k + 4)(k - 4)$
49. $(z - 2)(z^2 + 2z + 4)$
50. $(y - 8)(y + 7)$
51. prime
52. $9p^8(3p + 7)(p - 4)$
53. $8m^3(4m^6 + 2m^2 + 3)$
54. $(2m + 5)(4m^2 - 10m + 25)$
55. $(4r + 3m)^2$ **56.** $(z - 6)^2$
57. $(5h + 7g)(3h - 2g)$
58. $5z(z - 7)(z - 2)$
59. $(k - 5)(k - 6)$
60. $4(4p - 5m)(4p + 5m)$
61. $3k(k - 5)(k + 1)$
62. $(y - 6k)(y + 2k)$
63. $(10p + 3)(100p^2 - 30p + 9)$
64. $(4r - 7)(16r^2 + 28r + 49)$
65. $(2 + m)(3 + p)$
66. $(2m - 3n)(m + 5n)$
67. $(4z - 1)^2$
68. $(a^2 + 25)(a + 5)(a - 5)$
69. $3(6m - 1)^2$
70. $(10a + 9y)(10a - 9y)$
71. prime
72. $(2y + 5)(2y - 5)$
73. $8z(4z - 1)(z + 2)$
74. $5(2m - 3)(m + 4)$

Factor each polynomial completely.

11. $a^2 - 4a - 12$

12. $a^2 + 17a + 72$

13. $6y^2 - 6y - 12$

14. $7y^6 + 14y^5 - 168y^4$

15. $6a + 12b + 18c$

16. $m^2 - 3mn - 4n^2$

17. $p^2 - 17p + 66$

18. $z^2 - 6z + 7z - 42$

19. $10z^2 - 7z - 6$

20. $2m^2 - 10m - 48$

21. $17x^3y^2 + 51xy$

22. $15y + 5$

23. $8a^5 - 8a^4 - 48a^3$

24. $8k^2 - 10k - 3$

25. $z^2 - 3za - 10a^2$

26. $50z^2 - 100$

27. $x^2 - 4x - 5x + 20$

28. $x^2 + 2x + 16$

29. $6n^2 - 19n + 10$

30. $9y^2 + 12y - 5$

31. $16x + 20$

32. $m^2 + 2m - 15$

33. $6y^2 - 5y - 4$

34. $m^2 - 81$

35. $6z^2 + 31z + 5$

36. $12x^2 + 47x - 4$

37. $4k^2 - 12k + 9$

38. $8p^2 + 23p - 3$

39. $54m^2 - 24z^2$

40. $8m^2 - 2m - 3$

41. $3k^2 + 4k - 4$

42. $8a^3 - 27$

43. $14k^3 + 7k^2 - 70k$

44. $5 + r - 5s - rs$

45. $y^4 - 16$

46. $9z^2 + 64$

47. $8m - 16m^2$

48. $k^2 - 16$

49. $z^3 - 8$

50. $y^2 - y - 56$

51. $k^2 + 9$

52. $27p^{10} - 45p^9 - 252p^8$

53. $32m^9 + 16m^5 + 24m^3$

54. $8m^3 + 125$

55. $16r^2 + 24rm + 9m^2$

56. $z^2 - 12z + 36$

57. $15h^2 + 11hg - 14g^2$

58. $5z^3 - 45z^2 + 70z$

59. $k^2 - 11k + 30$

60. $64p^2 - 100m^2$

61. $3k^3 - 12k^2 - 15k$

62. $y^2 - 4yk - 12k^2$

63. $1000p^3 + 27$

64. $64r^3 - 343$

65. $6 + 3m + 2p + mp$

66. $2m^2 + 7mn - 15n^2$

67. $16z^2 - 8z + 1$

68. $a^4 - 625$

69. $108m^2 - 36m + 3$

70. $100a^2 - 81y^2$

71. $x^2 - xy + y^2$

72. $4y^2 - 25$

73. $32z^3 + 56z^2 - 16z$

74. $10m^2 + 25m - 60$

75. $(8m - 5n)^2$
76. $(2 - q)(2 - 3p)$
77. $2(3a - 1)(a + 2)$
78. $6y^4(3y + 4)(2y - 5)$
79. prime
80. $4(2k - 3)^2$
81. $(4 + m)(5 + 3n)$
82. $12y^2(6yz^2 + 1 - 2y^2z^2)$
83. $(4k - 3h)(2k + h)$
84. $(2a + 5)(a - 6)$
85. $2(x + 4)(x^2 - 4x + 16)$
86. $15a^3b^2(3b^3 - 4a + 5a^3b^2)$
87. $(5y - 6z)(2y + z)$
88. $(m - 2)^2$
89. $(8a - b)(a + 3b)$
90. $5m^2(5m - 3n)(5m - 13n)$

75. $64m^2 - 80mn + 25n^2$

76. $4 - 2q - 6p + 3pq$

77. $6a^2 + 10a - 4$

78. $36y^6 - 42y^5 - 120y^4$

79. $36x^2 + 32x + 9$

80. $16k^2 - 48k + 36$

81. $20 + 5m + 12n + 3mn$

82. $72y^3z^2 + 12y^2 - 24y^4z^2$

83. $8k^2 - 2kh - 3h^2$

84. $2a^2 - 7a - 30$

85. $2x^3 + 128$

86. $45a^3b^5 - 60a^4b^2 + 75a^6b^4$

87. $10y^2 - 7yz - 6z^2$

88. $m^2 - 4m + 4$

89. $8a^2 + 23ab - 3b^2$

90. $125m^4 - 400m^3n + 195m^2n^2$

6.5 Solving Quadratic Equations Using the Zero-Factor Property

OBJECTIVES

1. Solve quadratic equations using the zero-factor property.

2. Solve other equations using the zero-factor property.

VOCABULARY

☐ quadratic equation
☐ double solution

Galileo Galilei (1564–1642)

Galileo Galilei developed theories to explain physical phenomena. According to legend, Galileo dropped objects of different weights from the Leaning Tower of Pisa to disprove the belief that heavier objects fall faster than lighter objects. He developed a formula that describes the motion of freely falling objects,

$$d = 16t^2,$$

where d is the distance in feet that an object falls (disregarding air resistance) in t seconds, regardless of weight. The equation $d = 16t^2$ is a *quadratic equation*.

Quadratic Equation

A **quadratic equation** (in x here) can be written in the form

$$ax^2 + bx + c = 0,$$

where a, b, and c are real numbers and $a \neq 0$. The given form is called **standard form.**

Examples: $x^2 + 5x + 6 = 0$, $2x^2 - 5x = 3$, $x^2 = 4$ Quadratic equations

A quadratic equation has a second-degree term and no terms of greater degree. Of the above examples, only $x^2 + 5x + 6 = 0$ is in standard form.

We have factored many quadratic *expressions* of the form $ax^2 + bx + c$. In this section, we use factored quadratic expressions to solve quadratic *equations*.

OBJECTIVE 1 Solve quadratic equations using the zero-factor property.

We use the following property to solve some quadratic equations.

Zero-Factor Property

If a and b are real numbers and if $ab = 0$, then $a = 0$ or $b = 0$.

That is, if the product of two numbers is 0, then at least one of the numbers must be 0. One number *must* be 0, but both *may* be 0.

**NOW TRY
EXERCISE 1**

Solve each equation.

(a) $(x - 4)(3x + 1) = 0$

(b) $y(4y - 5) = 0$

EXAMPLE 1   Using the Zero-Factor Property

Solve each equation.

(a) $(x + 3)(2x - 1) = 0$

   The product $(x + 3)(2x - 1)$ is equal to 0. By the zero-factor property, the product of these two factors will equal 0 only if at least one of the factors equals 0. Therefore, either $x + 3 = 0$ or $2x - 1 = 0$.

$$x + 3 = 0 \quad \text{or} \quad 2x - 1 = 0 \qquad \text{Zero-factor property}$$
$$x = -3 \qquad\qquad 2x = 1 \qquad \text{Solve each equation.}$$
$$x = \frac{1}{2}. \qquad \text{Divide each side by 2.}$$

Check these values by substituting -3 for x in the original equation. ***Then start over*** and substitute $\frac{1}{2}$ for x.

CHECK   Let $x = -3$.

$$(x + 3)(2x - 1) = 0$$
$$(-3 + 3)[2(-3) - 1] \overset{?}{=} 0$$
$$0(-7) \overset{?}{=} 0$$
$$0 = 0 \ \checkmark \ \text{True}$$

Let $x = \frac{1}{2}$.

$$(x + 3)(2x - 1) = 0$$
$$\left(\frac{1}{2} + 3\right)\left(2 \cdot \frac{1}{2} - 1\right) \overset{?}{=} 0$$
$$\frac{7}{2}(0) \overset{?}{=} 0$$
$$0 = 0 \ \checkmark \ \text{True}$$

Because true statements result, the solution set is $\left\{-3, \frac{1}{2}\right\}$. ⟵ Include *both* solutions in the solution set.

(b)
$$y(3y - 4) = 0$$
$$y = 0 \quad \text{or} \quad 3y - 4 = 0 \qquad \text{Zero-factor property}$$

Don't forget that 0 is a solution.

$$3y = 4 \qquad \text{Add 4.}$$
$$y = \frac{4}{3} \qquad \text{Divide by 3.}$$

CHECK   Let $y = 0$.

$$y(3y - 4) = 0$$
$$0(3 \cdot 0 - 4) \overset{?}{=} 0$$
$$0(-4) \overset{?}{=} 0$$
$$0 = 0 \ \checkmark \ \text{True}$$

Let $y = \frac{4}{3}$.

$$y(3y - 4) = 0$$
$$\frac{4}{3}\left(3 \cdot \frac{4}{3} - 4\right) \overset{?}{=} 0$$
$$\frac{4}{3}(0) \overset{?}{=} 0$$
$$0 = 0 \ \checkmark \ \text{True}$$

True statements result. The solution set is $\left\{0, \frac{4}{3}\right\}$.

NOW TRY

NOTE   The word *or* as used in **Example 1** means "one or the other or both."

NOW TRY ANSWERS

1. (a) $\left\{-\frac{1}{3}, 4\right\}$ **(b)** $\left\{0, \frac{5}{4}\right\}$

If the polynomial in an equation is not already factored, first make sure that the equation is in standard form. Then factor and solve.

**NOW TRY
EXERCISE 2**
Solve $t^2 = -3t + 18$.

EXAMPLE 2 Solving Quadratic Equations

Solve each equation.

(a) $x^2 - 5x = -6$

First, write the equation in standard form $ax^2 + bx + c = 0$.

> *Don't* factor x out at this step.

$$x^2 - 5x = -6$$
$$x^2 - 5x + 6 = 0 \qquad \text{Add 6 to each side.}$$

Now factor $x^2 - 5x + 6$. Find two numbers whose product is 6 and whose sum is -5. These two numbers are -2 and -3, so we factor as follows.

$$(x - 2)(x - 3) = 0 \qquad \text{Factor.}$$
$$x - 2 = 0 \quad \text{or} \quad x - 3 = 0 \qquad \text{Zero-factor property}$$
$$x = 2 \quad \text{or} \qquad x = 3 \qquad \text{Solve each equation.}$$

CHECK   Let $x = 2$.

$$x^2 - 5x = -6$$
$$2^2 - 5(2) \overset{?}{=} -6$$
$$4 - 10 \overset{?}{=} -6$$
$$-6 = -6 \quad \checkmark \quad \text{True}$$

Let $x = 3$.

$$x^2 - 5x = -6$$
$$3^2 - 5(3) \overset{?}{=} -6$$
$$9 - 15 \overset{?}{=} -6$$
$$-6 = -6 \quad \checkmark \quad \text{True}$$

Both values check, so the solution set is $\{2, 3\}$.

(b) $y^2 = y + 20$ ← Write this equation in standard form.

Standard form ⟶ $y^2 - y - 20 = 0$ Subtract y and 20.

$$(y - 5)(y + 4) = 0 \qquad \text{Factor.}$$
$$y - 5 = 0 \quad \text{or} \quad y + 4 = 0 \qquad \text{Zero-factor property}$$
$$y = 5 \quad \text{or} \qquad y = -4 \qquad \text{Solve each equation.}$$

Check each result to verify that the solution set is $\{-4, 5\}$.   NOW TRY

Solving a Quadratic Equation Using the Zero-Factor Property

Step 1   **Write the equation in standard form**—that is, with all terms on one side of the equality symbol in descending powers of the variable and 0 on the other side.

Step 2   **Factor** completely.

Step 3   **Apply the zero-factor property.** Set each factor with a variable equal to 0.

Step 4   **Solve** the resulting equations.

Step 5   **Check** each result in the original equation. Write the solution set.

NOW TRY ANSWER
2. $\{-6, 3\}$

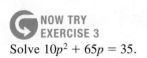

NOW TRY
EXERCISE 3
Solve $10p^2 + 65p = 35$.

EXAMPLE 3 Solving a Quadratic Equation (Common Factor)

Solve $4x^2 + 40 = 26x$.

$$4x^2 + 40 = 26x \quad \text{⟵ Write this equation in the form } ax^2 + bx + c = 0.$$

Step 1 $\qquad 4x^2 - 26x + 40 = 0 \qquad$ Standard form

This 2 is *not* a solution of the equation. $\quad 2(2x^2 - 13x + 20) = 0 \qquad$ Factor out 2.

$$2x^2 - 13x + 20 = 0 \qquad \text{Divide each side by 2.}$$

Step 2 $\qquad (2x - 5)(x - 4) = 0 \qquad$ Factor.

Step 3 $\quad 2x - 5 = 0 \quad$ or $\quad x - 4 = 0 \qquad$ Zero-factor property

Step 4 $\qquad 2x = 5 \qquad\qquad x = 4 \qquad$ Solve each equation.

$$x = \frac{5}{2}$$

Step 5 *Check* each result to verify that the solution set is $\left\{\frac{5}{2}, 4\right\}$. **NOW TRY**

⚠ **CAUTION** A common error is to include the common factor 2 as a solution in **Example 3**. *Only factors containing variables lead to solutions,* such as the factor y in the equation $y(3y - 4) = 0$ in **Example 1(b)**.

EXAMPLE 4 Solving Quadratic Equations

Solve each equation.

(a) $\qquad\qquad\qquad 16m^2 - 25 = 0 \quad \text{⟵ This equation is in standard form } ax^2 + bx + c = 0. \text{ There is no first-degree term because } b = 0.$

$$(4m + 5)(4m - 5) = 0 \qquad \text{Factor the difference of squares. (Section 6.4)}$$

$$4m + 5 = 0 \quad \text{or} \quad 4m - 5 = 0 \qquad \text{Zero-factor property}$$

$$4m = -5 \quad \text{or} \qquad 4m = 5 \qquad \text{Solve each equation.}$$

$$m = -\frac{5}{4} \quad \text{or} \qquad m = \frac{5}{4}$$

Check $-\frac{5}{4}$ and $\frac{5}{4}$ in the original equation. The solution set is $\left\{-\frac{5}{4}, \frac{5}{4}\right\}$.

(b) $\qquad\qquad\qquad\qquad y^2 = 2y \quad \text{⟵ This equation is in the form } ax^2 + bx + c = 0. \text{ Here, } c = 0.$

$$y^2 - 2y = 0$$

$$y(y - 2) = 0 \qquad \text{Factor.}$$

Don't forget to set the variable factor y equal to 0. $\quad y = 0 \quad$ or $\quad y - 2 = 0 \qquad$ Zero-factor property

$$y = 2 \qquad \text{Solve.}$$

A check confirms that the solution set is $\{0, 2\}$.

(c) $\qquad\qquad\qquad k(2k + 1) = 3 \quad \text{⟵ To be in standard form, 0 must be on the right side.}$

$$2k^2 + k = 3 \qquad \text{Distributive property}$$

Standard form $\longrightarrow 2k^2 + k - 3 = 0 \qquad$ Subtract 3.

$$(2k + 3)(k - 1) = 0 \qquad \text{Factor.}$$

NOW TRY ANSWER
3. $\left\{-7, \frac{1}{2}\right\}$

**NOW TRY
EXERCISE 4**

Solve each equation.

(a) $9x^2 - 64 = 0$

(b) $m^2 = 5m$

(c) $p(6p - 1) = 2$

$$2k + 3 = 0 \quad \text{or} \quad k - 1 = 0 \qquad \text{Zero-factor property}$$

$$2k = -3 \qquad\qquad k = 1 \qquad \text{Solve each equation.}$$

$$k = -\frac{3}{2}$$

A check confirms that the solution set is $\left\{-\frac{3}{2}, 1\right\}$.

 NOW TRY

⚠ **CAUTION** In **Example 4(b),** it is tempting to begin by dividing both sides of

$$y^2 = 2y$$

by y to obtain $y = 2$. Note, however, that we do not find the other solution, 0, if we divide by a variable. (We *may* divide each side of an equation by a *nonzero* real number, however. In **Example 3** we divided each side by 2.)

In **Example 4(c),** we cannot directly apply the zero-factor property to solve

$$k(2k + 1) = 3$$

in its given form because of the 3 on the right side of the equation. ***We can apply the zero-factor property only to a product that equals 0.***

EXAMPLE 5 Solving Quadratic Equations (Double Solutions)

Solve each equation.

(a)
$$z^2 - 22z + 121 = 0 \qquad \boxed{\text{This is a perfect square trinomial.}}$$

$$(z - 11)^2 = 0 \qquad \text{Factor.}$$

$$(z - 11)(z - 11) = 0 \qquad a^2 = a \cdot a$$

$$z - 11 = 0 \quad \text{or} \quad z - 11 = 0 \qquad \text{Zero-factor property}$$

Because the two factors are identical, they both lead to the same solution, called a **double solution.**

$$z = 11 \qquad \text{Add 11.}$$

CHECK
$$z^2 - 22z + 121 = 0 \qquad \text{Original equation}$$

$$11^2 - 22(11) + 121 \stackrel{?}{=} 0 \qquad \text{Let } z = 11.$$

$$121 - 242 + 121 \stackrel{?}{=} 0 \qquad \text{Apply the exponent. Multiply.}$$

$$0 = 0 \checkmark \qquad \text{True}$$

The solution set is $\{11\}$.

(b)
$$9t^2 - 30t = -25$$

$$9t^2 - 30t + 25 = 0 \qquad \text{Standard form}$$

$$(3t - 5)^2 = 0 \qquad \text{Factor the perfect square trinomial.}$$

$$3t - 5 = 0 \quad \text{or} \quad 3t - 5 = 0 \qquad \text{Zero-factor property}$$

$$3t = 5 \qquad \text{Solve the equation.}$$

$$t = \frac{5}{3} \qquad \tfrac{5}{3} \text{ is a double solution.}$$

NOW TRY ANSWERS

4. **(a)** $\left\{-\frac{8}{3}, \frac{8}{3}\right\}$ **(b)** $\{0, 5\}$

(c) $\left\{-\frac{1}{2}, \frac{2}{3}\right\}$

**NOW TRY
EXERCISE 5**

Solve.

$$4x^2 - 4x + 1 = 0$$

CHECK

$$9t^2 - 30t = -25 \qquad \text{Original equation}$$

$$9\left(\frac{5}{3}\right)^2 - 30\left(\frac{5}{3}\right) \stackrel{?}{=} -25 \qquad \text{Let } t = \frac{5}{3}.$$

$$9\left(\frac{25}{9}\right) - 30\left(\frac{5}{3}\right) \stackrel{?}{=} -25 \qquad \text{Apply the exponent.}$$

$$25 - 50 \stackrel{?}{=} -25 \qquad \text{Multiply.}$$

$$-25 = -25 \quad \checkmark \quad \text{True}$$

The solution set is $\left\{\frac{5}{3}\right\}$.

NOW TRY

> ⊘ **CAUTION**   Each of the equations in **Example 5** has only *one* distinct solution. ***There is no need to write the same number more than once in a solution set.***

OBJECTIVE 2   Solve other equations using the zero-factor property.

We can also use the zero-factor property to solve equations that involve more than two factors with variables. (These equations will have at least one term greater than second degree. They are *not* quadratic equations.)

**NOW TRY
EXERCISE 6**

Solve each equation.

(a) $3x^3 - 27x = 0$

(b) $(3a - 1) \cdot$
$(2a^2 - 5a - 12) = 0$

EXAMPLE 6   Solving Equations with More Than Two Variable Factors

Solve each equation.

(a)

$$6z^3 - 6z = 0$$

$$6z(z^2 - 1) = 0 \qquad \text{Factor out 6z.}$$

$$6z(z + 1)(z - 1) = 0 \qquad \text{Factor } z^2 - 1.$$

By an extension of the zero-factor property, this product can equal 0 only if at least one of the factors equals 0. Write and solve three equations, one for each factor with a variable.

$$6z = 0 \quad \text{or} \quad z + 1 = 0 \quad \text{or} \quad z - 1 = 0$$

$$z = 0 \quad \text{or} \qquad z = -1 \quad \text{or} \qquad z = 1$$

Check by substituting, in turn, 0, −1, and 1 into the original equation. The solution set is $\{-1, 0, 1\}$.

(b)

$$(3x - 1)(x^2 - 9x + 20) = 0 \qquad \begin{array}{l}\text{The product of the} \\ \text{factors is 0, as required.} \\ \text{Do } \textit{not} \text{ multiply.}\end{array}$$

$$(3x - 1)(x - 5)(x - 4) = 0 \qquad \text{Factor the trinomial.}$$

$$3x - 1 = 0 \quad \text{or} \quad x - 5 = 0 \quad \text{or} \quad x - 4 = 0 \qquad \text{Zero-factor property}$$

$$x = \frac{1}{3} \quad \text{or} \qquad x = 5 \quad \text{or} \qquad x = 4 \qquad \text{Solve each equation.}$$

Check to verify that the solution set is $\left\{\frac{1}{3}, 4, 5\right\}$.

NOW TRY

5. $\left\{\frac{1}{2}\right\}$

6. (a) $\{-3, 0, 3\}$

   (b) $\left\{-\frac{3}{2}, \frac{1}{3}, 4\right\}$

> ⊘ **CAUTION**   In **Example 6(b),** it would be unproductive to begin by multiplying the two factors together. The zero-factor property requires the *product* of two or more factors to equal 0. *Always consider first whether an equation is given in an appropriate form to apply the zero-factor property.*

NOW TRY
EXERCISE 7

Solve.

$$x(4x - 9) = (x - 2)^2 + 24$$

EXAMPLE 7   Solving an Equation Requiring Multiplication before Factoring

Solve $(3x + 1)x = (x + 1)^2 + 5$.

The zero-factor property requires the *product* of two or more factors to equal 0.

$$(3x + 1)x = (x + 1)^2 + 5 \quad \text{This equation is } not \text{ in the correct form.}$$

$$3x^2 + x = x^2 + 2x + 1 + 5 \qquad \text{Multiply on the left.}$$
$$\text{Square } x + 1 \text{ on the right.}$$

$$3x^2 + x = x^2 + 2x + 6 \qquad \text{Combine like terms.}$$

$$2x^2 - x - 6 = 0 \qquad \text{Standard form}$$

The product of the factors is now 0. $\quad (2x + 3)(x - 2) = 0 \qquad$ Factor.

$$2x + 3 = 0 \quad \text{or} \quad x - 2 = 0 \qquad \text{Zero-factor property}$$

$$x = -\frac{3}{2} \quad \text{or} \qquad x = 2 \qquad \text{Solve each equation.}$$

NOW TRY ANSWER
7. $\left\{ -\frac{7}{3}, 4 \right\}$

Check to verify that the solution set is $\left\{ -\frac{3}{2}, 2 \right\}$.   **NOW TRY**

NOTE   Not all quadratic equations can be solved using the zero-factor property. A more general method for solving such equations is given in **Chapter 9**.

6.5 Exercises

FOR EXTRA HELP   ▶   MyMathLab®

● *Complete solution available in MyMathLab*

Concept Check   *Fill in each blank with the correct response.*

1. A quadratic equation in x can be written in the form _____ = 0.

2. The form $ax^2 + bx + c = 0$ is called _____ form.

3. If the product of two numbers is 0, then at least one of the numbers is _____. This is the _____ property.

4. If a quadratic equation is in standard form, to solve the equation we should begin by attempting to _____ the polynomial.

5. The equation $x^3 + x^2 + x = 0$ is not a quadratic equation because _____.

6. If a quadratic equation $ax^2 + bx + c = 0$ has $c = 0$, then _____ *must* be a solution because _____ is a factor of the polynomial.

Concept Check   *Work each problem.*

7. Identify each equation as *linear* or *quadratic*.

    (a) $2x - 5 = 6$ **(b)** $x^2 - 5 = -4$

    (c) $x^2 + 2x - 3 = 2x^2 - 2$ **(d)** $5^2x + 2 = 0$

8. The number 9 is a *double solution* of the following equation. Why is this so?

$$(x - 9)^2 = 0$$

9. Look at this "solution."
   WHAT WENT WRONG?

$$x(7x - 1) = 0$$

$$7x - 1 = 0 \quad \text{Zero-factor property}$$

$$x = \frac{1}{7}$$

The solution set is $\left\{\frac{1}{7}\right\}$.

10. Look at this "solution."
   WHAT WENT WRONG?

$$3x(5x - 4) = 0$$

$$x = 3 \quad \text{or} \quad x = 0 \quad \text{or} \quad 5x - 4 = 0$$

$$x = \frac{4}{5}$$

The solution set is $\left\{3, 0, \frac{4}{5}\right\}$.

*Solve each equation, and check the solutions. **See Example 1.***

11. $(x + 5)(x - 2) = 0$

12. $(x - 1)(x + 8) = 0$

▶ **13.** $(2m - 7)(m - 3) = 0$

14. $(6x + 5)(x + 4) = 0$

15. $(2x + 1)(6x - 1) = 0$

16. $(3x + 2)(10x - 1) = 0$

17. $t(6t + 5) = 0$

18. $w(4w + 1) = 0$

19. $2x(3x - 4) = 0$

20. $6y(4y + 9) = 0$

21. $(x - 6)(x - 6) = 0$

22. $(y + 1)(y + 1) = 0$

*Solve each equation, and check the solutions. **See Examples 2–7.***

23. $y^2 + 3y + 2 = 0$

24. $p^2 + 8p + 7 = 0$

25. $y^2 - 3y + 2 = 0$

26. $r^2 - 4r + 3 = 0$

▶ **27.** $x^2 = 24 - 5x$

28. $t^2 = 2t + 15$

29. $x^2 = 3 + 2x$

30. $x^2 = 4 + 3x$

31. $z^2 + 3z = -2$

32. $p^2 - 2p = 3$

33. $m^2 + 8m + 16 = 0$

34. $x^2 - 6x + 9 = 0$

35. $3x^2 + 5x - 2 = 0$

36. $6r^2 - r - 2 = 0$

▶ **37.** $12p^2 = 8 - 10p$

38. $18x^2 = 12 + 15x$

39. $9s^2 + 12s = -4$

40. $36x^2 + 60x = -25$

41. $y^2 - 9 = 0$

42. $m^2 - 100 = 0$

43. $16x^2 - 49 = 0$

44. $4w^2 - 9 = 0$

45. $n^2 = 121$

46. $x^2 = 400$

47. $x^2 + 6x = 0$

48. $x^2 + 4x = 0$

49. $x^2 = 7x$

50. $t^2 = 9t$

51. $6r^2 = 3r$

52. $10y^2 = -5y$

▶ **53.** $x(x - 7) = -10$

54. $r(r - 5) = -6$

55. $3z(2z + 7) = 12$

56. $4x(2x + 3) = 36$

57. $2y(y + 13) = 136$

58. $t(3t - 20) = -12$

59. $(x - 8)(x + 6) = 6x$

60. $(x - 2)(x + 9) = 4x$

61. $(x + 4)(x + 7) = 10$

62. $(x + 2)(x + 5) = 4$

63. $9y^3 - 49y = 0$

64. $16r^3 - 9r = 0$

65. $r^3 - 2r^2 - 8r = 0$

66. $x^3 - x^2 - 6x = 0$

67. $x^3 + x^2 - 20x = 0$

68. $y^3 - 6y^2 + 8y = 0$

69. $4x^3 - 18x^2 + 8x = 0$

70. $9x^3 - 24x^2 + 12x = 0$

71. $r^4 = 2r^3 + 15r^2$

72. $x^4 = 3x^2 + 2x^3$

▶ **73.** $(2r + 5)(3r^2 - 16r + 5) = 0$

74. $(3m + 4)(6m^2 + m - 2) = 0$

75. $(2x + 7)(x^2 + 2x - 3) = 0$

76. $(x + 1)(6x^2 + x - 12) = 0$

77. $3x(x + 1) = (2x + 3)(x + 1)$

78. $2x(x + 3) = (3x + 1)(x + 3)$

79. $x^2 + (x + 1)^2 = (x + 2)^2$

80. $(x - 7)^2 + x^2 = (x + 1)^2$

Extending Skills   Solve each equation, and check the solutions.

81. $(2x)^2 = (2x + 4)^2 - (x + 5)^2$

82. $5 - (x - 1)^2 = (x - 2)^2$

83. $(x + 3)^2 - (2x - 1)^2 = 0$

84. $(4y - 3)^3 - 9(4y - 3) = 0$

85. $6p^2(p + 1) = 4(p + 1) - 5p(p + 1)$

86. $6x^2(2x + 3) = 4(2x + 3) + 5x(2x + 3)$

Galileo's formula describing the motion of freely falling objects is

$$d = 16t^2.$$

*The distance d in feet an object falls depends on the time t elapsed, in seconds. (This is an example of an important mathematical concept, the **function**.)*

87. (a) Use Galileo's formula and complete the following table. (*Hint:* Substitute each given value into the formula and solve for the unknown value.)

t in seconds	0	1	2	3		
d in feet	0	16			256	576

(b) When $t = 0$, we find that $d = 0$. Explain this in the context of the problem.

88. Refer to **Exercise 87.** When 256 was substituted for d and the formula was solved for t, there should have been two solutions, 4 and -4. Why doesn't -4 make sense as an answer?

6.6 Applications of Quadratic Equations

OBJECTIVES

1 Solve problems involving geometric figures.

2 Solve problems involving consecutive integers.

3 Solve problems by applying the Pythagorean theorem.

4 Solve problems using given quadratic models.

VOCABULARY

☐ consecutive integers
☐ consecutive even (odd) integers
☐ legs
☐ hypotenuse

We use the zero-factor property to solve quadratic equations that arise in application problems. We follow the same six problem-solving steps given in **Section 2.4.**

Solving an Applied Problem

Step 1 **Read** the problem carefully. *What information is given? What is to be found?*

Step 2 **Assign a variable** to represent the unknown value. Make a sketch, diagram, or table, as needed. If necessary, express any other unknown values in terms of the variable.

Step 3 **Write an equation** using the variable expression(s).

Step 4 **Solve** the equation.

Step 5 **State the answer.** Label it appropriately. *Does it seem reasonable?*

Step 6 **Check** the answer in the words of the *original* problem.

PROBLEM-SOLVING HINT   Refer to the formulas inside the back cover of the text as needed when solving application problems.

NOW TRY EXERCISE 1

A right triangle has one leg that is 4 ft shorter than the other leg. The area of the triangle is 6 ft². Determine the lengths of the legs.

OBJECTIVE 1 Solve problems involving geometric figures.

EXAMPLE 1 Solving an Area Problem

Abe wants to plant a triangular flower bed in a corner of his garden. One leg of the right-triangular flower bed will be 2 m shorter than the other leg. He wants the bed to have an area of 24 m². Find the lengths of the legs.

Step 1 **Read** the problem. We need to find the lengths of the legs of a right triangle with area 24 m².

Step 2 **Assign a variable.**

Let x = the length of one leg.

Then $x - 2$ = the length of the other leg.

See **FIGURE 1**.

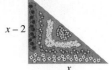

FIGURE 1

Step 3 **Write an equation.** In a right triangle, the legs are the base and height, so we substitute 24 for the area, x for the base, and $x - 2$ for the height in the formula for the area of a triangle.

$$\mathcal{A} = \frac{1}{2}bh \qquad \text{Formula for the area of a triangle}$$

$$24 = \frac{1}{2}x(x - 2) \qquad \text{Let } \mathcal{A} = 24, b = x, h = x - 2.$$

Step 4 **Solve.**

$$24 = \frac{1}{2}x^2 - x \qquad \text{Distributive property}$$

$$48 = x^2 - 2x \qquad \text{Multiply each term by 2.}$$

$$x^2 - 2x - 48 = 0 \qquad \text{Standard form}$$

$$(x + 6)(x - 8) = 0 \qquad \text{Factor.}$$

$$x + 6 = 0 \quad \text{or} \quad x - 8 = 0 \qquad \text{Zero-factor property}$$

$$x = -6 \quad \text{or} \qquad x = 8 \qquad \text{Solve each equation.}$$

Step 5 **State the answer.** The solutions are −6 and 8. Because a triangle cannot have a side of negative length, we discard the solution −6. Then the lengths of the legs will be 8 m and $8 - 2 = 6$ m.

Step 6 **Check.** The length of one leg is 2 m less than the length of the other leg, and the area is

$$\frac{1}{2}(8)(6) = 24 \text{ m}^2, \quad \text{as required.}$$

NOW TRY

⚠ CAUTION *In solving applied problems, always check solutions against physical facts and discard any answers that are not appropriate.*

OBJECTIVE 2 Solve problems involving consecutive integers.

Recall from **Section 2.4** that **consecutive integers** are integers that are next to each other on a number line, such as

3 and 4, or −11 and −10.

See **FIGURE 2**.

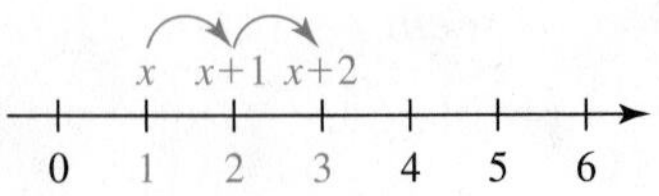

$x \quad x+1 \quad x+2$

0 1 2 3 4 5 6

Consecutive integers

FIGURE 2

NOW TRY ANSWER
1. 2 ft, 6 ft

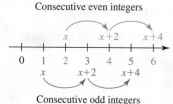

Consecutive even integers

Consecutive odd integers

FIGURE 3

Consecutive even integers are *even* integers that are next to each other on a number line, such as

$$4 \text{ and } 6, \quad \text{or} \quad -10 \text{ and } -8.$$

Consecutive odd integers are defined similarly—for example, 3 and 5 are consecutive *odd* integers, as are -13 and -11. See **FIGURE 3**.

PROBLEM-SOLVING HINT   If x = the lesser (least) integer in a consecutive integer problem, then the following apply.

- For two consecutive integers, use   **x, $x + 1$.**
- For three consecutive integers, use   **x, $x + 1$, $x + 2$.**
- For two consecutive even or odd integers, use   **x, $x + 2$.**
- For three consecutive even or odd integers, use   **x, $x + 2$, $x + 4$.**

In this book, we list consecutive integers in increasing order.

**NOW TRY
EXERCISE 2**

The product of the first and second of three consecutive integers is 2 more than 8 times the third integer. Find the integers.

EXAMPLE 2   Solving a Consecutive Integer Problem

The product of the second and third of three consecutive integers is 2 more than 7 times the first integer. Find the integers.

Step 1   **Read** the problem. Note that the integers are consecutive.

Step 2   **Assign a variable.**

Let x = the first integer.

Then $x + 1$ = the second integer,

and $x + 2$ = the third integer.

Step 3   **Write an equation.**

The product of the second and third   is   2 more than 7 times the first.

$$(x + 1)(x + 2) \qquad = \qquad 7x + 2$$

Step 4   **Solve.**
$$x^2 + 3x + 2 = 7x + 2 \qquad \text{Multiply.}$$
$$x^2 - 4x = 0 \qquad \text{Standard form}$$
$$x(x - 4) = 0 \qquad \text{Factor.}$$
$$x = 0 \quad \text{or} \quad x - 4 = 0 \qquad \text{Zero-factor property}$$
$$x = 4 \qquad \text{Add 4.}$$

Step 5   **State the answer.** The values 0 and 4 each lead to a distinct answer.

If $x = 0$, then $x + 1 = 1$ and $x + 2 = 2$. The integers are $0, 1, 2$.

If $x = 4$, then $x + 1 = 5$ and $x + 2 = 6$. The integers are $4, 5, 6$.

Step 6   **Check.** The product of the second and third integers must equal 2 more than 7 times the first. Because

$$1 \cdot 2 = 7 \cdot 0 + 2 \quad \text{and} \quad 5 \cdot 6 = 7 \cdot 4 + 2 \quad \text{are both true,}$$

both sets of consecutive integers satisfy the statement of the problem.

NOW TRY

OBJECTIVE 3   Solve problems by applying the Pythagorean theorem.

Pythagorean Theorem

If a and b are the lengths of the shorter sides of a right triangle (a triangle with a 90° angle) and c is the length of the longest side, then

$$a^2 + b^2 = c^2.$$

The two shorter sides are the **legs** of the triangle, and the longest side, the **hypotenuse,** is opposite the right angle.

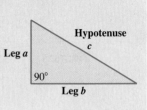

PROBLEM-SOLVING HINT   In solving a problem involving the Pythagorean theorem, be sure that the expressions for the sides are properly placed.

$$(\text{one leg})^2 + (\text{other leg})^2 = \text{hypotenuse}^2$$

EXAMPLE 3   Applying the Pythagorean Theorem

Patricia and Ali leave their office, with Patricia traveling north and Ali traveling east. When Ali is 1 mi farther than Patricia from the office, the distance between them is 2 mi more than Patricia's distance from the office. Find their distances from the office and the distance between them.

Step 1   **Read** the problem again. We must find three distances.

Step 2   **Assign a variable.**

Let x = Patricia's distance from the office.

Then $x + 1$ = Ali's distance from the office,

and $x + 2$ = the distance between them.

Place these expressions on a right triangle, as in **FIGURE 4**.

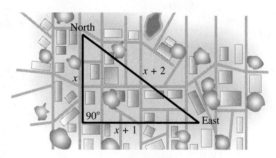

FIGURE 4

Step 3   **Write an equation.** Substitute into the Pythagorean theorem.

$$a^2 + b^2 = c^2$$
$$x^2 + (x + 1)^2 = (x + 2)^2 \quad \boxed{\text{Be careful to substitute properly.}}$$

Step 4   **Solve.** $x^2 + x^2 + 2x + 1 = x^2 + 4x + 4$ Square each binomial.

$\boxed{\text{Remember the middle terms when squaring the binomials.}}$ $x^2 - 2x - 3 = 0$ Standard form

$(x - 3)(x + 1) = 0$ Factor.

$x - 3 = 0$ or $x + 1 = 0$ Zero-factor property

$x = 3$ or $x = -1$ Solve each equation.

NOW TRY EXERCISE 3

The longer leg of a right triangle is 7 ft longer than the shorter leg and the hypotenuse is 8 ft longer than the shorter leg. Find the lengths of the sides of the triangle.

NOW TRY EXERCISE 4

Refer to **Example 4.** How long will it take for the ball to reach a height of 50 ft?

Step 5 **State the answer.** Because -1 cannot represent a distance, 3 is the only possible answer. Patricia's distance is 3 mi, Ali's distance is $3 + 1 = 4$ mi, and the distance between them is $3 + 2 = 5$ mi.

Step 6 **Check.** Because $3^2 + 4^2 = 5^2$ is true, the answers are correct. **NOW TRY**

OBJECTIVE 4 Solve problems using given quadratic models.

In **Examples 1–3,** we wrote quadratic equations to model, or mathematically describe, various situations and then solved the equations. In the remaining examples, we are given quadratic models and must use them to determine data.

EXAMPLE 4 Finding the Height of a Ball

A tennis player's serve travels 180 ft per sec (123 mph). If she hits a ball directly upward, the height h of the ball in feet at time t in seconds is modeled by the quadratic equation

$$h = -16t^2 + 180t + 6.$$

How long will it take for the ball to reach a height of 206 ft?

A height of 206 ft means that $h = 206$, so we substitute 206 for h in the equation and solve for t.

$$h = -16t^2 + 180t + 6$$

$$206 = -16t^2 + 180t + 6 \qquad \text{Let } h = 206.$$

$$-16t^2 + 180t + 6 = 206 \qquad \text{Interchange sides.}$$

$$-16t^2 + 180t - 200 = 0 \qquad \text{Standard form}$$

$$4t^2 - 45t + 50 = 0 \qquad \text{Divide by } -4.$$

$$(4t - 5)(t - 10) = 0 \qquad \text{Factor.}$$

$$4t - 5 = 0 \quad \text{or} \quad t - 10 = 0 \qquad \text{Zero-factor property}$$

$$4t = 5 \quad \text{or} \qquad t = 10 \qquad \text{Solve each equation.}$$

$$t = \frac{5}{4}$$

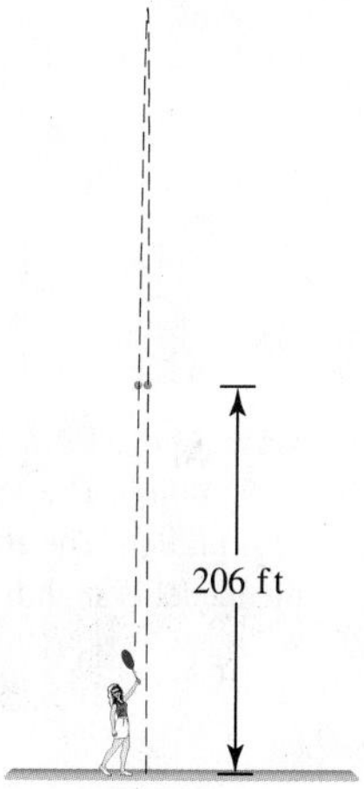

Because we found two acceptable answers, the ball will be 206 ft above the ground twice—once on its way up and once on its way down—at $\frac{5}{4}$ sec and at 10 sec after it is hit. See **FIGURE 5.**

206 ft

FIGURE 5

NOW TRY

EXAMPLE 5 Modeling the Foreign-Born Population of the United States

The foreign-born population of the United States over the years 1930–2010 can be modeled by the quadratic equation

$$y = 0.009665x^2 - 0.4942x + 15.12,$$

where $x = 0$ represents 1930, $x = 10$ represents 1940, and so on, and y is the number of people in millions. (*Source:* U.S. Census Bureau.)

(a) Use the model to find the foreign-born population in 1980 to the nearest tenth of a million.

NOW TRY ANSWERS

3. 5 ft, 12 ft, 13 ft

4. $\frac{1}{4}$ sec and 11 sec

NOW TRY
EXERCISE 5
Use the model in **Example 5** to find the foreign-born population of the United States in the year 2000. Give the answer to the nearest tenth of a million. How does it compare to the actual value from the table?

Because $x = 0$ represents 1930, $x = 50$ represents 1980. Substitute 50 for x in the given equation.

$y = 0.009665x^2 - 0.4942x + 15.12$	Given quadratic model
$y = 0.009665(50)^2 - 0.4942(50) + 15.12$	Let $x = 50$.
$y = 14.6$	Round to the nearest tenth.

In 1980, the foreign-born population of the United States was about 14.6 million.

(b) Repeat part (a) for 2010.

$y = 0.009665(80)^2 - 0.4942(80) + 15.12$	For 2010, let $x = 80$.
$y = 37.4$	Round to the nearest tenth.

In 2010, the U.S. foreign-born population was about 37.4 million.

(c) The model used above was developed from the data in the table. How do the results in parts (a) and (b) compare to the actual data from the table?

Year	Foreign-Born Population (in millions)
1930	14.2
1940	11.6
1950	10.3
1960	9.7
1970	9.6
1980	14.1
1990	19.8
2000	28.4
2010	37.6

NOW TRY ANSWER
5. 27.9 million; The actual value is 28.4 million. The answer using the model is slightly low.

From the table, the actual value for 1980 is 14.1 million. Our answer in part (a), 14.6 million, is slightly high. For 2010, the actual value is 37.6 million, so our answer of 37.4 million in part (b) is slightly low, but a good estimate.

NOW TRY

6.6 Exercises

FOR EXTRA HELP   ▶ MyMathLab®

▶ *Complete solution available in MyMathLab*

1. *Concept Check*   To review the six problem-solving steps first introduced in **Section 2.4,** complete each statement.

Step 1: _____ the problem carefully.

Step 2: Assign a _____ to represent the unknown value.

Step 3: Write a(n) _____ using the variable expression(s).

Step 4: _____ the equation.

Step 5: State the _____ .

Step 6: _____ the answer in the words of the _____ problem.

2. *Concept Check*   A student solves an applied problem and gets 6 or -3 for the length of the side of a square. Which of these answers is reasonable? Why?

*In Exercises 3–6, a geometric figure is given. Write the indicated formula, and using x as the variable, complete Steps 3–6 for each problem. (Refer to the steps in **Exercise 1** as needed.)*

3.

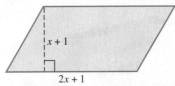

$x + 1$

$2x + 1$

The area of this parallelogram is 45 sq. units. Find its base and height.

Formula for the area of a parallelogram:

_____

Step 3: 45 = _____

Step 4: $x =$ ____ or $x =$ ____

Step 5: base: ____ units;
height: ____ units

Step 6: _____ = 45

4.

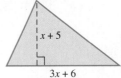

$x + 5$

$3x + 6$

The area of this triangle is 60 sq. units. Find its base and height.

Formula for the area of a triangle:

_____

Step 3: 60 = _____

Step 4: $x =$ ____ or $x =$ ____

Step 5: base: ____ units;
height: ____ units

Step 6: _____ = 60

5.

$x - 8$

$x + 8$

The area of this rug is 80 sq. units. Find its length and width.

Formula for the area of a rectangle:

_____

Step 3: ____ = $(x + 8)$ _____

Step 4: $x =$ ____ or $x =$ ____

Step 5: length: ____ units;
width: ____ units

Step 6: _____ = 80

6.

4

$x + 2$ x

The volume of this box is 192 cu. units. Find its length and width.

Formula for the volume of a rectangular solid: _____

Step 3: ____ = ____ $(x + 2)$

Step 4: $x =$ ____ or $x =$ ____

Step 5: length: ____ units;
width: ____ units

Step 6: _____ · 4 = ____

*Solve each problem. Check answers to be sure that they are reasonable. Refer to the formulas inside the back cover of this book as needed. **See Example 1.***

▶ **7.** The length of a standard jewel case is 2 cm more than its width. The area of the rectangular top of the case is 168 cm^2. Find the length and width of the jewel case.

8. A standard DVD case is 6 cm longer than it is wide. The area of the rectangular top of the case is 247 cm^2. Find the length and width of the case.

9. The area of a triangle is 30 in.^{2}. The base of the triangle measures 2 in. more than twice the height of the triangle. Find the measures of the base and the height.

10. A certain triangle has its base equal in measure to its height. The area of the triangle is 72 m^2. Find the equal base and height measure.

11. A 10-gal aquarium is 3 in. higher than it is wide. Its length is 21 in., and its volume is 2730 in.³. What are the height and width of the aquarium?

12. A toolbox is 2 ft high, and its width is 3 ft less than its length. If its volume is 80 ft³, find the length and width of the box.

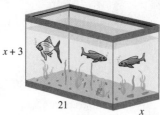

13. The dimensions of a rectangular monitor screen are such that its length is 3 in. more than its width. If the length were doubled and if the width were decreased by 1 in., the area would be increased by 150 in.². What are the length and width of the screen?

14. A computer keyboard is 11 in. longer than it is wide. If the length were doubled and if 2 in. were added to the width, the area would be increased by 198 in.². What are the length and width of the keyboard?

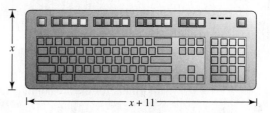

15. A square mirror has sides measuring 2 ft less than the sides of a square painting. If the difference between their areas is 32 ft², find the lengths of the sides of the mirror and the painting.

16. The sides of one square have length 3 m more than the sides of a second square. If the area of the larger square is subtracted from 4 times the area of the smaller square, the result is 36 m². What are the lengths of the sides of each square?

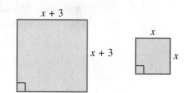

Solve each problem. ***See Example 2.***

17. The product of the numbers on two consecutive volumes of research data is 420. Find the volume numbers.

18. The product of the page numbers on two facing pages of a book is 600. Find the page numbers.

▶ **19.** The product of the second and third of three consecutive integers is 2 more than 10 times the first integer. Find the integers.

20. The product of the first and third of three consecutive integers is 3 more than 3 times the second integer. Find the integers.

21. Find two consecutive odd integers such that their product is 15 more than three times their sum.

22. Find two consecutive odd integers such that five times their sum is 23 less than their product.

23. Find three consecutive odd integers such that 3 times the sum of all three is 18 more than the product of the first and second integers.

24. Find three consecutive odd integers such that the sum of all three is 42 less than the product of the second and third integers.

25. Find three consecutive even integers such that the sum of the squares of the first and second integers is equal to the square of the third integer.

26. Find three consecutive even integers such that the square of the sum of the first and second integers is equal to twice the third integer.

Solve each problem. ***See Example 3.***

▶ **27.** The hypotenuse of a right triangle is 1 cm longer than the longer leg. The shorter leg is 7 cm shorter than the longer leg. Find the length of the longer leg of the triangle.

28. The longer leg of a right triangle is 1 m longer than the shorter leg. The hypotenuse is 1 m shorter than twice the shorter leg. Find the length of the shorter leg of the triangle.

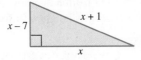

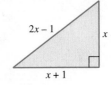

29. The length of a rectangle is 5 in. longer than its width. The diagonal is 5 in. shorter than twice the width. Find the length, width, and diagonal measures of the rectangle.

30. The length of a rectangle is 4 in. longer than its width. The diagonal is 8 in. longer than the width. Find the length, width, and diagonal measures of the rectangle.

31. Tram works due north of home. Her husband Alan works due east. They leave for work at the same time. By the time Tram is 5 mi from home, the distance between them is 1 mi more than Alan's distance from home. How far from home is Alan?

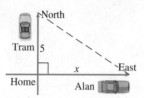

32. Two cars left an intersection at the same time. One traveled north. The other traveled 14 mi farther, but to the east. How far apart were they at that time if the distance between them was 4 mi more than the distance traveled east?

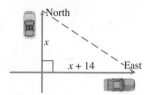

33. A ladder is leaning against a building. The distance from the bottom of the ladder to the building is 4 ft less than the length of the ladder. How high up the side of the building is the top of the ladder if that distance is 2 ft less than the length of the ladder?

34. A lot has the shape of a right triangle with one leg 2 m longer than the other. The hypotenuse is 2 m less than twice the length of the shorter leg. Find the length of the shorter leg.

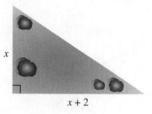

If an object is projected upward with an initial velocity of 128 ft per sec, its height h in feet after t seconds is given by the quadratic equation

$$h = -16t^2 + 128t.$$

*Find the height of the object after each time listed. **See Example 4.***

35. 1 sec   **36.** 2 sec   **37.** 4 sec

38. How long does it take the object just described to return to the ground? (*Hint:* When the object hits the ground, $h = 0$.)

*Solve each problem. **See Example 4.***

▶ **39.** If an object is projected from a height of 48 ft with an initial velocity of 32 ft per sec, its height h in feet after t seconds is given by

$$h = -16t^2 + 32t + 48.$$

(a) After how many seconds is the height 64 ft? (*Hint:* Let $h = 64$ and solve.)

(b) After how many seconds is the height 60 ft?

(c) After how many seconds does the object hit the ground? (*Hint:* When the object hits the ground, $h = 0$.)

(d) The quadratic equation from part (c) has two solutions, yet only one of them is appropriate for answering the question. Why is this so?

40. If an object is projected upward from ground level with an initial velocity of 64 ft per sec, its height h in feet t seconds later is given by

$$h = -16t^2 + 64t.$$

(a) After how many seconds is the height 48 ft?

(b) The object reaches its maximum height 2 sec after it is projected. What is this maximum height?

(c) After how many seconds does the object hit the ground? (*Hint:* When the object hits the ground, $h = 0$.)

(d) The quadratic equation from part (c) has two solutions, yet only one of them is appropriate for answering the question. Why is this so?

(e) Find the number of seconds after which the height is 60 ft.

(f) What is the physical interpretation of why part (e) has two answers?

Solve each problem. **See Example 5.**

41. The table shows the number of cellular phone subscribers (in millions) in the United States.

Year	Subscribers (in millions)
1990	5
1992	11
1994	24
1996	44
1998	69
2000	109
2002	141
2004	182
2006	233
2008	270
2010	296
2012	326

Source: CTIA.

We used the data to develop the quadratic equation

$$y = 0.339x^2 + 8.50x - 8.26,$$

which models the number of cellular phone subscribers y (in millions) in the year x, where $x = 0$ represents 1990, $x = 2$ represents 1992, and so on.

(a) What value of x corresponds to 2000?

(b) Use the model to find the number of subscribers in 2000, to the nearest million. How does the result compare with the actual data in the table?

(c) What value of x corresponds to 2012?

(d) Use the model to find the number of cellular phone subscribers in 2012, to the nearest million. How does the result compare with the actual data in the table?

(e) Assuming that the trend in the data continues, what value of x would correspond to 2014?

(f) Use the model to find the number of cellular phone subscribers in 2014, to the nearest million.

42. World population (in billions) is shown in the table. Using the data, we developed the quadratic equation

$$y = 0.0002x^2 + 0.0593x + 2.501,$$

which models population y (in billions) in the year x, where $x = 0$ represents 1950, $x = 10$ represents 1960, and so on.

Year	Population (in billions)
1950	2.6
1960	3.0
1970	3.7
1980	4.5
1990	5.3
2000	6.1
2010	6.8
2013	7.1

Source: www.worldpopulationstatistics.com

(a) What value of x corresponds to the year 2000? To the year 2013?

(b) Use the model to find world population in 2000 and 2013, to the nearest tenth. How do the results compare with the actual data in the table?

(c) World population is projected to reach 8.0 billion in 2025. What value of x corresponds to the year 2025?

(d) Use the model to project world population in 2025, to the nearest tenth. How does the result compare to the projection given in part (c)?

Chapter 6   Summary

Key Terms

6.1
factor
factored form
common factor
greatest common factor (GCF)

6.2
prime polynomial

6.4
perfect square
perfect square trinomial
perfect cube

6.5
quadratic equation
double solution

6.6
consecutive integers
consecutive even (odd) integers

legs
hypotenuse

Test Your Word Power

See how well you have learned the vocabulary in this chapter.

1. **Factoring** is
   A. a method of multiplying polynomials
   B. the process of writing a polynomial as a product
   C. the answer in a multiplication problem
   D. a way to add the terms of a polynomial.

2. A polynomial is in **factored form** when
   A. it is prime
   B. it is written as a sum

   C. the second-degree term has a coefficient of 1
   D. it is written as a product.

3. A **perfect square trinomial** is a trinomial
   A. that can be factored as the square of a binomial
   B. that cannot be factored
   C. that is multiplied by a binomial
   D. where all terms are perfect squares.

4. A **quadratic equation** is an equation that can be written in the form
   A. $y = mx + b$
   B. $ax^2 + bx + c = 0$ $(a \neq 0)$
   C. $Ax + By = C$
   D. $x = k$.

5. A **hypotenuse** is
   A. either of the two shorter sides of a triangle
   B. the shortest side of a triangle
   C. the side opposite the right angle in a triangle
   D. the longest side in any triangle.

ANSWERS

1. B; *Example:* $x^2 - 5x - 14$ factors as $(x - 7)(x + 2)$.   **2.** D; *Example:* The factored form of $x^2 - 5x - 14$ is $(x - 7)(x + 2)$.   **3.** A; *Example:* $a^2 + 2a + 1$ is a perfect square trinomial. Its factored form is $(a + 1)^2$.   **4.** B; *Examples:* $y^2 - 3y + 2 = 0$, $x^2 - 9 = 0$, $2m^2 = 6m + 8$ **5.** C; *Example:* In **FIGURE 4** of **Section 6.6,** the hypotenuse is the side labeled $x + 2$.

Quick Review

CONCEPTS | **EXAMPLES**

6.1 The Greatest Common Factor; Factoring by Grouping

Finding the Greatest Common Factor (GCF)

Step 1 Write each number in prime factored form.

Step 2 List each prime number or each variable that is a factor of every term in the list.

Step 3 Use as exponents on the common prime factors the *least* exponents from the prime factored forms.

Step 4 Multiply the primes from Step 3.

Find the greatest common factor of $4x^2y$, $6x^2y^3$, and $2xy^2$.

$$4x^2y = 2 \cdot 2 \cdot x^2 \cdot y$$
$$6x^2y^3 = 2 \cdot 3 \cdot x^2 \cdot y^3$$
$$2xy^2 = 2 \cdot x \cdot y^2$$

The greatest common factor is $2xy$.

Factoring by Grouping

Step 1 Group the terms.

Step 2 Factor out the greatest common factor from each group.

Step 3 Factor out a common binomial factor from the results of Step 2.

Step 4 If necessary, rearrange terms and try a different grouping.

Factor by grouping.

$$3x^2 + 5x - 24xy - 40y$$
$$= (3x^2 + 5x) + (-24xy - 40y) \qquad \text{Group the terms.}$$
$$= x(3x + 5) - 8y(3x + 5) \qquad \text{Factor each group.}$$
$$= (3x + 5)(x - 8y) \qquad \text{Factor out } 3x + 5.$$

6.2 Factoring Trinomials

To factor $x^2 + bx + c$, find two integers m and n such that $mn = c$ and $m + n = b$.

$$
\begin{array}{c}
mn = c \\
\downarrow \\
x^2 + bx + c \\
\uparrow \\
m + n = b
\end{array}
$$

Then $x^2 + bx + c$ factors as $(x + m)(x + n)$.
Check by multiplying.

Factor $x^2 + 6x + 8$.

$$
\begin{array}{c}
mn = 8 \\
\downarrow \\
x^2 + 6x + 8 \qquad \text{Here, } m = 2 \text{ and } n = 4. \\
\uparrow \\
m + n = 6
\end{array}
$$

$x^2 + 6x + 8$ factors as $(x + 2)(x + 4)$.

CHECK $(x + 2)(x + 4)$

$$= x^2 + 4x + 2x + 8 \qquad \text{FOIL method}$$
$$= x^2 + 6x + 8 \ \checkmark \qquad \text{Combine like terms.}$$

6.3 More on Factoring Trinomials

To factor $ax^2 + bx + c$, use one of the following methods.

Factoring by Grouping

Find m and n such that $mn = ac$ and $m + n = b$.

$$
\begin{array}{c}
mn = ac \\
\downarrow \qquad \downarrow \\
ax^2 + bx + c \\
\uparrow \\
m + n = b
\end{array}
$$

Factor $3x^2 + 14x - 5$ by grouping.

$$3x^2 + 14x - 5 \qquad \text{Here, } mn = -15 \text{ and } m + n = 14.$$

The required integers are $m = -1$ and $n = 15$.

$$3x^2 + 14x - 5$$
$$= 3x^2 - x + 15x - 5 \qquad 14x = -x + 15x$$
$$= (3x^2 - x) + (15x - 5) \qquad \text{Group the terms.}$$
$$= x(3x - 1) + 5(3x - 1) \qquad \text{Factor each group.}$$
$$= (3x - 1)(x + 5) \qquad \text{Factor out } 3x - 1.$$

CONCEPTS	EXAMPLES
Factoring by Trial and Error Use the FOIL method in reverse.	Factor $3x^2 + 14x - 5$ by trial and error. Because the only positive factors of 3 are 3 and 1, and -5 has possible factors of 1 and -5, or -1 and 5, the possible factored forms for this trinomial follow. $(3x - 5)(x + 1)$ Incorrect $(3x + 5)(x - 1)$ Incorrect $(3x + 1)(x - 5)$ Incorrect $(3x - 1)(x + 5)$ Correct Using grouping or trial and error, $$3x^2 + 14x - 5 \quad \text{factors as} \quad (3x - 1)(x + 5).$$

6.4   Special Factoring Techniques

CONCEPTS	EXAMPLES
Difference of Squares $$x^2 - y^2 = (x + y)(x - y)$$ **Perfect Square Trinomials** $$x^2 + 2xy + y^2 = (x + y)^2$$ $$x^2 - 2xy + y^2 = (x - y)^2$$ **Difference of Cubes** $$x^3 - y^3 = (x - y)(x^2 + xy + y^2)$$ **Sum of Cubes** $$x^3 + y^3 = (x + y)(x^2 - xy + y^2)$$	Factor. $$4x^2 - 9$$ $$= (2x + 3)(2x - 3)$$ $9x^2 + 6x + 1 \quad\mid\quad 4x^2 - 20x + 25$ $\quad = (3x + 1)^2 \quad\mid\quad = (2x - 5)^2$ $m^3 - 8 \quad\mid\quad z^3 + 27$ $= m^3 - 2^3 \quad\mid\quad = z^3 + 3^3$ $= (m - 2)(m^2 + 2m + 4) \quad\mid\quad = (z + 3)(z^2 - 3z + 9)$

6.5   Solving Quadratic Equations Using the Zero-Factor Property

CONCEPTS	EXAMPLES
Zero-Factor Property If a and b are real numbers and if $ab = 0$, then $a = 0$ or $b = 0$.	If $(x - 2)(x + 3) = 0$, then $x - 2 = 0$ or $x + 3 = 0$.
Solving a Quadratic Equation Using the Zero-Factor Property	Solve $2x^2 = 7x + 15$.
Step 1 Write the equation in standard form.	$2x^2 - 7x - 15 = 0$ Standard form
Step 2 Factor.	$(2x + 3)(x - 5) = 0$ Factor.
Step 3 Apply the zero-factor property.	$2x + 3 = 0 \quad$ or $\quad x - 5 = 0$ Zero-factor property
Step 4 Solve the resulting equations.	$2x = -3 \qquad\qquad x = 5$ Solve each equation. $$x = -\frac{3}{2}$$
Step 5 Check. Write the solution set.	**CHECK** $2x^2 = 7x + 15$ $2(5)^2 \overset{?}{=} 7(5) + 15$ Let $x = 5$. $50 \overset{?}{=} 35 + 15$ $50 = 50 \checkmark$ True The other value also checks. The solution set is $\left\{-\frac{3}{2}, 5\right\}$.

CONCEPTS

EXAMPLES

6.6 Applications of Quadratic Equations

Pythagorean Theorem
In a right triangle, the sum of the squares of the legs equals the square of the hypotenuse.

$$a^2 + b^2 = c^2$$

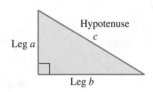

The longer leg of a right triangle is 2 ft longer than the shorter leg. The hypotenuse is 4 ft longer than the shorter leg. Find the lengths of the sides of the triangle.

Let x = the length of the shorter leg.
Then $x + 2$ = the length of the longer leg,
and $x + 4$ = the length of the hypotenuse.

$$x^2 + (x + 2)^2 = (x + 4)^2$$

$x^2 + x^2 + 4x + 4 = x^2 + 8x + 16$ Square each binomial.

$x^2 - 4x - 12 = 0$ Standard form

$(x - 6)(x + 2) = 0$ Factor.

$x - 6 = 0$ or $x + 2 = 0$ Zero-factor property

$x = 6$ or $x = -2$ Solve each equation.

Discard -2 as a solution. Check that the sides have lengths

6 ft, $6 + 2 = 8$ ft,   and $6 + 4 = 10$ ft.

Chapter 6   Review Exercises

1. $7(t + 2)$

2. $30z(2z^2 + 1)$

3. $-3x(x^2 - 2x - 1)$

4. $50m^2n^2(2n - mn^2 + 3)$

5. $(2y + 3)(x - 4)$

6. $(3y + 2x)(2y + 3)$

7. $(x + 3)(x + 2)$

8. $(y - 5)(y - 8)$

9. $(q + 9)(q - 3)$

10. $(r - 8)(r + 7)$

11. prime

12. $8p(p + 2)(p - 5)$

13. $3x^2(x + 2)(x + 8)$

14. $(r + 8s)(r - 12s)$

15. $(p + 12q)(p - 10q)$

16. $p^5(p - 2q)(p + q)$

17. $3r^3(r + 3s)(r - 5s)$

18. $2x^5(x - 2y)(x + 3y)$

19. r and $6r$, $2r$ and $3r$

20. Factor out z.

21. $(2k - 1)(k - 2)$

22. $(3r - 1)(r + 4)$

23. $(3r + 2)(2r - 3)$

24. $(5z + 1)(2z - 1)$

6.1 *Factor out the greatest common factor, or factor by grouping.*

1. $7t + 14$

2. $60z^3 + 30z$

3. $-3x^3 + 6x^2 + 3x$

4. $100m^2n^3 - 50m^3n^4 + 150m^2n^2$

5. $2xy - 8y + 3x - 12$

6. $6y^2 + 9y + 4xy + 6x$

6.2 *Factor completely.*

7. $x^2 + 5x + 6$ **8.** $y^2 - 13y + 40$ **9.** $q^2 + 6q - 27$

10. $r^2 - r - 56$ **11.** $x^2 + x + 1$ **12.** $8p^3 - 24p^2 - 80p$

13. $3x^4 + 30x^3 + 48x^2$ **14.** $r^2 - 4rs - 96s^2$ **15.** $p^2 + 2pq - 120q^2$

16. $p^7 - p^6q - 2p^5q^2$ **17.** $3r^5 - 6r^4s - 45r^3s^2$ **18.** $2x^7 + 2x^6y - 12x^5y^2$

6.3 *Answer each question.*

19. To begin factoring $6r^2 - 5r - 6$, what are the possible first terms of the two binomial factors if we consider only positive integer coefficients?

20. What is the first step to factor $2z^3 + 9z^2 - 5z$?

Factor completely.

21. $2k^2 - 5k + 2$ **22.** $3r^2 + 11r - 4$ **23.** $6r^2 - 5r - 6$

24. $10z^2 - 3z - 1$ **25.** $5t^2 - 11t + 12$ **26.** $24x^5 - 20x^4 + 4x^3$

27. $-6x^2 + 3x + 30$ **28.** $10r^3s + 17r^2s^2 + 6rs^3$ **29.** $48x^4y + 4x^3y^2 - 4x^2y^3$

25. prime

26. $4x^3(3x-1)(2x-1)$

27. $-3(2x-5)(x+2)$

28. $rs(5r+6s)(2r+s)$

29. $4x^2y(3x+y)(4x-y)$

30. The student stopped too soon. He needs to factor out the common factor $4x-1$ to obtain
$$(4x-1)(4x-5)$$
as the correct answer.

31. B   **32.** D

33. $(n+7)(n-7)$

34. $(5b+11)(5b-11)$

35. $(7y+5w)(7y-5w)$

36. $36(2p+q)(2p-q)$

37. prime   **38.** $(r-6)^2$

39. $(3t-7)^2$

40. $(m+10)(m^2-10m+100)$

41. $(5k+4x)(25k^2-20kx+16x^2)$

42. $(7x-4)(49x^2+28x+16)$

43. $(10-3x^2)(100+30x^2+9x^4)$

44. $(x-y)(x+y)(x^2+xy+y^2) \cdot$
(x^2-xy+y^2)

45. $\left\{-\dfrac{3}{4},1\right\}$ **46.** $\left\{0,\dfrac{5}{2}\right\}$

47. $\{-3,-1\}$ **48.** $\{1,4\}$

49. $\{3,5\}$ **50.** $\left\{-\dfrac{4}{3},5\right\}$

51. $\left\{-\dfrac{8}{9},\dfrac{8}{9}\right\}$ **52.** $\{0,8\}$

53. $\{-1,6\}$ **54.** $\{7\}$

55. $\{6\}$ **56.** $\{-3,3\}$

57. $\left\{-2,-1,-\dfrac{2}{5}\right\}$

58. $\left\{-\dfrac{3}{8},0,\dfrac{3}{8}\right\}$

59. length: 10 ft; width: 4 ft

60. 5 ft

61. 26 mi

62. length: 6 m; width: 4 m

---

30. On a quiz, a student factored $16x^2 - 24x + 5$ by grouping as follows.

$$16x^2 - 24x + 5$$
$$= 16x^2 - 4x - 20x + 5$$
$$= 4x(4x-1) - 5(4x-1) \qquad \text{His answer}$$

He thought his answer was correct because it checked by multiplication. *WHAT WENT WRONG?* Give the correct factored form.

6.4 *Answer each question.*

31. Which one of the following is a difference of squares?

  A. $32x^2 - 1$ **B.** $4x^2y^2 - 25z^2$ **C.** $x^2 + 36$ **D.** $25y^3 - 1$

32. Which one of the following is a perfect square trinomial?

  A. $x^2 + x + 1$ **B.** $y^2 - 4y + 9$ **C.** $4x^2 + 10x + 25$ **D.** $x^2 - 20x + 100$

Factor completely.

33. $n^2 - 49$

34. $25b^2 - 121$

35. $49y^2 - 25w^2$

36. $144p^2 - 36q^2$

37. $x^2 + 100$

38. $r^2 - 12r + 36$

39. $9t^2 - 42t + 49$

40. $m^3 + 1000$

41. $125k^3 + 64x^3$

42. $343x^3 - 64$

43. $1000 - 27x^6$

44. $x^6 - y^6$

6.5 *Solve each equation, and check the solutions.*

45. $(4t+3)(t-1) = 0$

46. $x(2x-5) = 0$

47. $z^2 + 4z + 3 = 0$

48. $m^2 - 5m + 4 = 0$

49. $x^2 = -15 + 8x$

50. $3z^2 - 11z - 20 = 0$

51. $81t^2 - 64 = 0$

52. $y^2 = 8y$

53. $n(n-5) = 6$

54. $t^2 - 14t + 49 = 0$

55. $t^2 = 12(t-3)$

56. $x^2 = 9$

57. $(5z+2)(z^2 + 3z + 2) = 0$

58. $64x^3 - 9x = 0$

6.6 *Solve each problem.*

59. The length of a rug is 6 ft more than the width. The area is 40 ft². Find the length and width of the rug.

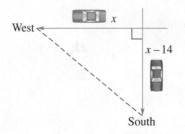

60. A treasure chest from a sunken galleon has the dimensions shown in the figure. Its surface area is 650 ft². Find its width.

61. Two cars left an intersection at the same time. One traveled west, and the other traveled 14 mi less, but to the south. How far apart were they at that time, if the distance between them was 16 mi more than the distance traveled south?

62. A pyramid has a rectangular base with a length that is 2 m more than its width. The height of the pyramid is 6 m, and its volume is 48 m³. Find the length and width of the base.

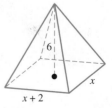

63. The product of the first and second of three consecutive integers is equal to 23 plus the third. Find the integers.

64. If an object is dropped, the distance d in feet it falls in t seconds (disregarding air resistance) is given by the quadratic equation

$$d = 16t^2.$$

Find the distance an object would fall in **(a)** 4 sec and **(b)** 8 sec.

65. The numbers of alternative-fueled vehicles in use in the United States are given in the table.

Year	Alternative-Fueled Vehicles (in thousands)
2001	425
2003	534
2005	592
2007	696
2009	826
2011	1192

Source: Energy Information Administration.

Using the data, we developed the quadratic equation

$$y = 7.02x^2 - 15.5x + 469,$$

which models the number of vehicles y (in thousands) in the year x, where $x = 1$ represents 2001, $x = 3$ represents 2003, and so on.

(a) Use the model to find the number of alternative-fueled vehicles in 2007 and 2011, to the nearest thousand.

(b) How do the results in part (a) compare with the actual data in the table?

Chapter 6 | Mixed Review Exercises

1. Which of the following is *not* factored completely?

A. $3(7t)$ **B.** $3x(7t + 4)$ **C.** $(3 + x)(7t + 4)$ **D.** $3(7t + 4) + x(7t + 4)$

2. A student did not receive full credit for factoring

$$6x^2 + 16x - 32 \quad \text{as} \quad (2x + 8)(3x - 4).$$

WHAT WENT WRONG? Give the completely factored form.

Factor completely.

3. $3k^2 + 11k + 10$

4. $z^2 - 11zx + 10x^2$

5. $y^4 - 625$

6. $6m^3 - 21m^2 - 45m$

7. $25a^2 + 15ab + 9b^2$

8. $2a^5 - 8a^4 - 24a^3$

9. $15m^2 + 20m - 12mp - 16p$

10. $24ab^3c^2 - 56a^2bc^3 + 72a^2b^2c$

11. $12x^2yz^3 + 12xy^2z - 30x^3y^2z^4$

12. $12r^2 + 18rq - 10r - 15q$

13. $49t^2 + 56t + 16$

14. $1000a^3 + 27$

Solve each equation.

15. $t(t - 7) = 0$

16. $x^2 + 3x = 10$

17. $25x^2 + 20x + 4 = 0$

Answers (left margin):

63. $-5, -4, -3$ or $5, 6, 7$

64. **(a)** 256 ft **(b)** 1024 ft

65. **(a)** 2007: 704 thousand;
2011: 1148 thousand

(b) 2007: The result is slightly higher than the actual number;
2011: The result is lower than the actual number.

1. D

2. The factor $(2x + 8)$ has a factor of 2. The completely factored form is

$$2(x + 4)(3x - 4).$$

3. $(3k + 5)(k + 2)$

4. $(z - x)(z - 10x)$

5. $(y^2 + 25)(y + 5)(y - 5)$

6. $3m(2m + 3)(m - 5)$

7. prime

8. $2a^3(a + 2)(a - 6)$

9. $(3m + 4)(5m - 4p)$

10. $8abc(3b^2c - 7ac^2 + 9ab)$

11. $6xyz(2xz^2 + 2y - 5x^2yz^3)$

12. $(2r + 3q)(6r - 5)$

13. $(7t + 4)^2$

14. $(10a + 3)(100a^2 - 30a + 9)$

15. $\{0, 7\}$ **16.** $\{-5, 2\}$

17. $\left\{-\dfrac{2}{5}\right\}$

18. 15 m, 36 m, 39 m

19. 6 m

20. width: 10 m; length: 17 m

Solve each problem.

18. A lot is in the shape of a right triangle. The hypotenuse is 3 m longer than the longer leg. The longer leg is 6 m longer than twice the length of the shorter leg. Find the lengths of the sides of the lot.

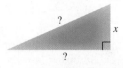

19. The triangular sail of a schooner has an area of 30 m². The height of the sail is 4 m more than the base. Find the base of the sail.

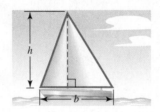

20. The floor plan for a house is a rectangle with length 7 m more than its width. The area is 170 m². Find the width and length of the house.

Chapter 6   **Test**   FOR EXTRA HELP   *Step-by-step test solutions are found on the Chapter Test Prep Videos available in* MyMathLab®, *or on* YouTube™.

▶ *View the complete solutions to all Chapter Test exercises in MyMathLab.*

[6.1–6.4]

1. D   **2.** $6x(2x - 5)$

3. $m^2n(2mn + 3m - 5n)$

4. $(2x + y)(a - b)$

5. $(x + 3)(x - 8)$

6. $(2x + 3)(x - 1)$

7. $(5z - 1)(2z - 3)$

8. $3(x + 1)(x - 5)$

9. prime   **10.** prime

11. $(2 - a)(6 + b)$

12. $(3y + 8)(3y - 8)$

13. $(9a + 11b)(9a - 11b)$

14. $(x + 8)^2$

15. $(2x - 7y)^2$

16. $3t^2(2t + 9)(t - 4)$

17. $(r - 5)(r^2 + 5r + 25)$

18. $8(k + 2)(k^2 - 2k + 4)$

19. $(x^2 + 9)(x + 3)(x - 3)$

20. $(3x + 2y)(3x - 2y)(9x^2 + 4y^2)$

[6.5]

21. $\{-3, 9\}$ **22.** $\left\{\dfrac{1}{2}, 6\right\}$

23. $\left\{-\dfrac{2}{5}, \dfrac{2}{5}\right\}$ **24.** $\{0, 9\}$

25. $\{10\}$

26. $\left\{-8, -\dfrac{5}{2}, \dfrac{1}{3}\right\}$

1. Which one of the following is the correct, completely factored form of $2x^2 - 2x - 24$?

A. $(2x + 6)(x - 4)$ **B.** $(x + 3)(2x - 8)$

C. $2(x + 4)(x - 3)$ **D.** $2(x + 3)(x - 4)$

Factor completely. If the polynomial is prime, say so.

2. $12x^2 - 30x$ **3.** $2m^3n^2 + 3m^3n - 5m^2n^2$ **4.** $2ax - 2bx + ay - by$

5. $x^2 - 5x - 24$ **6.** $2x^2 + x - 3$ **7.** $10z^2 - 17z + 3$

8. $3x^2 - 12x - 15$ **9.** $t^2 + 2t + 3$ **10.** $x^2 + 36$

11. $12 - 6a + 2b - ab$ **12.** $9y^2 - 64$ **13.** $81a^2 - 121b^2$

14. $x^2 + 16x + 64$ **15.** $4x^2 - 28xy + 49y^2$ **16.** $6t^4 + 3t^3 - 108t^2$

17. $r^3 - 125$ **18.** $8k^3 + 64$ **19.** $x^4 - 81$ **20.** $81x^4 - 16y^4$

Solve each equation.

21. $(x + 3)(x - 9) = 0$ **22.** $2r^2 - 13r + 6 = 0$

23. $25x^2 - 4 = 0$ **24.** $t^2 = 9t$

25. $x(x - 20) = -100$ **26.** $(s + 8)(6s^2 + 13s - 5) = 0$

Solve each problem.

27. The length of a rectangular flower bed is 3 ft less than twice its width. The area of the bed is 54 ft². Find the dimensions of the flower bed.

28. Find two consecutive integers such that the square of the sum of the two integers is 11 more than the first integer.

[6.6]

27. 6 ft by 9 ft

28. −2, −1

29. 17 ft

30. $13,672 billion

29. A carpenter needs to cut a brace to support a wall stud, as shown in the figure. The brace should be 7 ft less than three times the length of the stud. If the brace will be anchored on the floor 15 ft away from the stud, how long should the brace be?

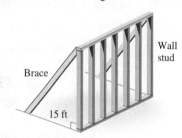

Wall stud

Brace

15 ft

30. The public debt y (in billions of dollars) of the United States from 2000 through 2014 can be approximated by the quadratic equation

$$y = 57.53x^2 - 72.93x + 3417,$$

where $x = 0$ represents 2000, $x = 1$ represents 2001, and so on. Use the model to estimate the public debt, to the nearest billion dollars, in the year 2014. (*Source:* Bureau of Public Debt.)

Chapters R–6	Cumulative Review Exercises

[2.1–2.3]

1. {0}

2. {0.05}

3. {6}

[2.5]

4. $P = \dfrac{A}{1 + rt}$

5. 110° and 70°

[2.4]

6. gold: 9; silver: 7; bronze: 12

[R.2, 2.6]

7. 230; 205; 38%; 12%

Solve each equation.

1. $3x + 2(x - 4) = 4(x - 2)$

2. $0.3x + 0.9x = 0.06$

3. $\dfrac{2}{3}m - \dfrac{1}{2}(m - 4) = 3$

4. Solve for P: $A = P + Prt$.

5. Find the measures of the marked angles.

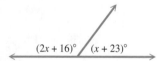

$(2x + 16)°$ $(x + 23)°$

Solve each problem.

6. At the 2014 Winter Olympics in Sochi, Russia, the United States won a total of 28 medals. The United States won 2 more gold medals than silver and 5 fewer silver medals than bronze. Find the number of each type of medal won. (*Source: The Gazette.*)

7. From a list of "technology-related items," 500 adults were surveyed as to those items they couldn't live without. Complete the results shown in the table.

Item	Percent That Couldn't Live Without	Number That Couldn't Live Without
Personal computer	46%	
Cell phone	41%	
High-speed Internet		190
MP3 player		60

(Other items included digital cable, HDTV, and electronic gaming console.)
Source: Ipsos for AP.

8. Fill in each blank with *positive* or *negative.* The point with coordinates (a, b) is in

  (a) quadrant II if a is _____ and b is _____ .

  (b) quadrant III if a is _____ and b is _____ .

9. Consider the equation $y = -2x - 4$. Find the following.

  (a) The x- and y-intercepts **(b)** The slope **(c)** The graph

10. The points on the graph show total retail sales of prescription drugs in the United States in the years 2004–2010, along with a graph of a linear equation that models the data.

  (a) Use the ordered pairs shown on the graph to find the slope of the line to the nearest whole number. Interpret the slope.

  (b) Use the graph to estimate sales in the year 2008. Write your answer as an ordered pair of the form (year, sales in billions of dollars).

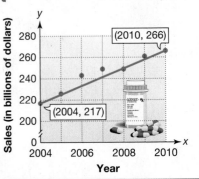

Retail Prescription Drug Sales

(2010, 266)

(2004, 217)

Source: National Association of Chain Drug Stores.

Solve each system of equations.

11. $4x - y = -6$
  $2x + 3y = 4$

12. $5x + 3y = 10$
  $2x + \dfrac{6}{5}y = 5$

Evaluate each expression.

13. $\left(\dfrac{3}{4}\right)^{-2}$

14. $\left(\dfrac{4^{-3} \cdot 4^4}{4^5}\right)^{-1}$

Simplify each expression, and write the answer using only positive exponents. Assume that no denominators are 0.

15. $\dfrac{(p^2)^3 p^{-4}}{(p^{-3})^{-1} p}$

16. $\dfrac{(m^{-2})^3 m}{m^5 m^{-4}}$

Perform each indicated operation.

17. $(2k^2 + 4k) - (5k^2 - 2) - (k^2 + 8k - 6)$ **18.** $(9x + 6)(5x - 3)$

19. $(3p + 2)^2$

20. $\dfrac{8x^4 + 12x^3 - 6x^2 + 20x}{2x}$

21. To make a pound of honey, bees may travel 55,000 mi and visit more than 2,000,000 flowers. Write the two given numbers in scientific notation. (*Source: Home & Garden.*)

Factor completely.

22. $2a^2 + 7a - 4$ **23.** $10m^2 + 19m + 6$ **24.** $8t^2 + 10tv + 3v^2$

25. $4p^2 - 12p + 9$ **26.** $25r^2 - 81t^2$ **27.** $2pq + 6p^3q + 8p^2q$

Solve.

28. $6m^2 + m - 2 = 0$ **29.** $8x^2 = 64x$

30. The length of the hypotenuse of a right triangle is twice the length of the shorter leg, plus 3 m. The longer leg is 7 m longer than the shorter leg. Find the lengths of the sides.

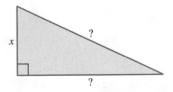

7 Rational Expressions and Applications

The formula that gives the rate of a speeding car in terms of its distance and its time traveled involves a *rational expression* (or *fraction*), the subject of this chapter.

7.1 The Fundamental Property of Rational Expressions

OBJECTIVES

1. Find the numerical value of a rational expression.
2. Find the values of the variable for which a rational expression is undefined.
3. Write rational expressions in lowest terms.
4. Recognize equivalent forms of rational expressions.

VOCABULARY

☐ rational expression
☐ lowest terms

The quotient of two integers (with denominator not 0), such as $\frac{2}{3}$ or $-\frac{3}{4}$, is a *rational number*. In the same way, the quotient of two polynomials with denominator not equal to 0 is a *rational expression*.

> **Rational Expression**
>
> A **rational expression** is an expression of the form $\frac{P}{Q}$, where P and Q are polynomials and $Q \neq 0$.
>
> *Examples:* $\dfrac{-6x}{x^3 + 8}$, $\dfrac{9x}{y + 3}$, and $\dfrac{2m^3}{8}$ Rational expressions

Our work with rational expressions requires much of what we learned in **Chapters 5 and 6** on polynomials and factoring, as well as the rules for fractions from **Section R.1.**

OBJECTIVE 1 Find the numerical value of a rational expression.

Remember that to *evaluate* an expression means to find its *value*. We use substitution to evaluate a rational expression for a given value of the variable.

EXAMPLE 1 Evaluating Rational Expressions

Find the numerical value of $\frac{3x + 6}{2x - 4}$ for each value of x.

(a) $x = 1$

$$\frac{3x + 6}{2x - 4}$$

$$= \frac{3(1) + 6}{2(1) - 4} \quad \text{Let } x = 1.$$

$$= \frac{9}{-2}$$

$$= -\frac{9}{2} \qquad \frac{a}{-b} = -\frac{a}{b}$$

(b) $x = 0$

$$\frac{3x + 6}{2x - 4}$$

$$= \frac{3(0) + 6}{2(0) - 4} \quad \text{Let } x = 0.$$

$$= \frac{6}{-4}$$

$$= -\frac{3}{2} \qquad \text{Lowest terms}$$

(c) $x = 2$

$$\frac{3x + 6}{2x - 4}$$

$$= \frac{3(2) + 6}{2(2) - 4} \quad \text{Let } x = 2.$$

$$= \frac{12}{0} \quad \boxed{\text{The expression is undefined for } x = 2.}$$

(d) $x = -2$

$$\frac{3x + 6}{2x - 4}$$

$$= \frac{3(-2) + 6}{2(-2) - 4} \quad \text{Let } x = -2.$$

$$= \frac{0}{-8}$$

$$= 0 \qquad \frac{0}{b} = 0$$

 NOW TRY EXERCISE 1

Find the numerical value of each expression for $x = -3$.

(a) $\dfrac{2x - 1}{x + 4}$ **(b)** $\dfrac{x + 3}{4}$

(c) $\dfrac{4}{x + 3}$

NOW TRY ANSWERS

1. **(a)** -7 **(b)** 0
   (c) The expression is undefined for $x = -3$.

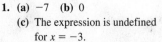 **NOW TRY**

NOTE *The numerator of a rational expression may be any real number.* If the numerator equals 0 and the denominator does not equal 0, then the rational expression equals 0. **See Example 1(d).**

OBJECTIVE 2 Find the values of the variable for which a rational expression is undefined.

In the definition of a rational expression $\frac{P}{Q}$, Q cannot equal 0. *The denominator of a rational expression cannot equal 0 because division by 0 is undefined.*

For instance, in the rational expression

$$\frac{8x^2}{x-3}, \leftarrow \text{Denominator cannot equal 0.}$$

the variable x can take on any real number value except 3. If x is 3, then the denominator becomes $3 - 3 = 0$, making the expression undefined. Thus, x cannot equal 3. We indicate this restriction by writing $x \neq 3$.

Determining When a Rational Expression Is Undefined

Step 1 Set the denominator of the rational expression equal to 0.

Step 2 Solve this equation.

Step 3 The solutions of the equation are the values that make the rational expression undefined. The variable *cannot* equal these values.

EXAMPLE 2 Finding Values That Make Rational Expressions Undefined

Find any values of the variable for which each rational expression is undefined.

(a) $\dfrac{x+5}{3x+2}$ We must find any value of x that makes the *denominator* equal to 0 because division by 0 is undefined.

Step 1 Set the denominator equal to 0.

$$3x + 2 = 0$$

Step 2 Solve. $3x = -2$ Subtract 2.

$$x = -\frac{2}{3}$$ Divide by 3.

Step 3 The given expression is undefined for $-\frac{2}{3}$, so $x \neq -\frac{2}{3}$.

(b) $\dfrac{8x^2+1}{x-3}$ The denominator $x - 3 = 0$ when x is 3. The given expression is undefined for 3, so $x \neq 3$.

(c) $\dfrac{9m^2}{m^2 - 5m + 6}$

$$m^2 - 5m + 6 = 0 \quad \text{Set the denominator equal to 0.}$$

$$(m-2)(m-3) = 0 \quad \text{Factor.}$$

$$m - 2 = 0 \quad \text{or} \quad m - 3 = 0 \quad \text{Zero-factor property}$$

$$m = 2 \quad \text{or} \qquad m = 3 \quad \text{Solve for } m.$$

The given expression is undefined for 2 and 3, so $m \neq 2$, $m \neq 3$.

**NOW TRY
EXERCISE 2**

Find any values of the variable for which each rational expression is undefined.

(a) $\dfrac{k-4}{2k-1}$

(b) $\dfrac{2x}{x^2+5x-14}$

(c) $\dfrac{y+10}{y^2+10}$

(d) $\dfrac{2r}{r^2+1}$ This denominator will not equal 0 for any value of r, because r^2 is always greater than or equal to 0, and adding 1 makes the sum greater than or equal to 1. There are no values for which this expression is undefined.

NOW TRY

OBJECTIVE 3   Write rational expressions in lowest terms.

A fraction such as $\frac{2}{3}$ is said to be in *lowest terms*.

Lowest Terms

A rational expression $\dfrac{P}{Q}$ (where $Q \neq 0$) is in **lowest terms** if the greatest common factor of its numerator and denominator is 1.

We use the **fundamental property of rational expressions** to write a rational expression in lowest terms.

Fundamental Property of Rational Expressions

If $\dfrac{P}{Q}$ (where $Q \neq 0$) is a rational expression and if K represents any polynomial (where $K \neq 0$), then the following holds.

$$\frac{PK}{QK} = \frac{P}{Q}$$

This property is based on the identity property of multiplication.

$$\frac{PK}{QK} = \frac{P}{Q} \cdot \frac{K}{K} = \frac{P}{Q} \cdot 1 = \frac{P}{Q}$$

**NOW TRY
EXERCISE 3**

Write each rational expression in lowest terms.

(a) $\dfrac{20}{48}$ **(b)** $\dfrac{21y^5}{7y^2}$

EXAMPLE 3   Writing in Lowest Terms

Write each rational expression in lowest terms.

(a) $\dfrac{30}{72}$

Begin by factoring.

$$\frac{30}{72} = \frac{2 \cdot 3 \cdot 5}{2 \cdot 2 \cdot 2 \cdot 3 \cdot 3}$$

(b) $\dfrac{14k^2}{2k^3}$

Write k^2 as $k \cdot k$ and k^3 as $k \cdot k \cdot k$.

$$\frac{14k^2}{2k^3} = \frac{2 \cdot 7 \cdot k \cdot k}{2 \cdot k \cdot k \cdot k}$$

Group any factors common to the numerator and denominator.

$$= \frac{5 \cdot (2 \cdot 3)}{2 \cdot 2 \cdot 3 \cdot (2 \cdot 3)}$$

$$= \frac{7(2 \cdot k \cdot k)}{k(2 \cdot k \cdot k)}$$

Use the fundamental property.

$$= \frac{5}{2 \cdot 2 \cdot 3}$$

$$= \frac{5}{12}$$

$$= \frac{7}{k}$$

NOW TRY ANSWERS

2. (a) $k \neq \frac{1}{2}$ **(b)** $x \neq -7, x \neq 2$
   (c) It is never undefined.

3. (a) $\frac{5}{12}$ **(b)** $3y^3$

NOW TRY

> **Writing a Rational Expression in Lowest Terms**
>
> **Step 1** **Factor** the numerator and denominator completely.
>
> **Step 2** **Use the fundamental property** to divide out any common factors.

**NOW TRY
EXERCISE 4**

Write each rational expression in lowest terms.

(a) $\dfrac{3x + 15}{5x + 25}$

(b) $\dfrac{k^2 - 36}{k^2 + 8k + 12}$

EXAMPLE 4 Writing in Lowest Terms

Write each rational expression in lowest terms.

(a) $\dfrac{3x - 12}{5x - 20}$

> $x \neq 4$ because the denominator is 0 for this value.

$= \dfrac{3(x - 4)}{5(x - 4)}$ Factor. (Step 1)

$= \dfrac{3}{5}$ Fundamental property (Step 2)

The given expression is equal to $\frac{3}{5}$ for all values of x, where $x \neq 4$ (because the denominator of the original rational expression is 0 when x is 4).

(b) $\dfrac{2y^2 - 8}{2y + 4}$

> $y \neq -2$ because the denominator is 0 for this value.

$= \dfrac{2(y^2 - 4)}{2(y + 2)}$ Factor. (Step 1)

$= \dfrac{2(y + 2)(y - 2)}{2(y + 2)}$ Factor the numerator completely.

$= y - 2$ Fundamental property (Step 2)

(c) $\dfrac{m^2 + 2m - 8}{2m^2 - m - 6}$

$= \dfrac{(m + 4)(m - 2)}{(2m + 3)(m - 2)}$ Factor. (Step 1)

> $m \neq -\frac{3}{2}, m \neq 2$

$= \dfrac{m + 4}{2m + 3}$ Fundamental property (Step 2)   **NOW TRY**

From now on, we write statements of equality of rational expressions with the understanding that they apply only to real numbers that make neither denominator equal to 0.

⚠ **CAUTION** *Rational expressions cannot be written in lowest terms until after the numerator and denominator have been factored. Only common* **factors** *(not terms) can be divided out.*

$$\dfrac{6x + 9}{4x + 6} = \dfrac{3(2x + 3)}{2(2x + 3)} = \dfrac{3}{2} \qquad \dfrac{6 + x}{4x} \leftarrow \text{Numerator cannot be factored.}$$

Divide out the common factor.   Already in lowest terms

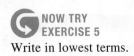
**NOW TRY
EXERCISE 5**

Write in lowest terms.

$$\frac{10 - a^2}{a^2 - 10}$$

EXAMPLE 5   Writing in Lowest Terms (Factors Are Opposites)

Write $\frac{x - y}{y - x}$ in lowest terms.

To find a common factor, the denominator $y - x$ can be factored as follows.

$$y - x \quad \boxed{\text{We are factoring out } -1, \text{ \textbf{NOT} multiplying by it.}}$$

$$= -1(-y + x) \qquad \text{Factor out } -1.$$

$$= -1(x - y) \qquad \text{Commutative property, } a + b = b + a.$$

With this result in mind, we simplify as follows.

$$\frac{x - y}{y - x}$$

$$= \frac{1(x - y)}{-1(x - y)} \qquad y - x = -1(x - y) \text{ from above.}$$

$$= \frac{1}{-1} \qquad \text{Fundamental property}$$

$$= -1 \qquad \text{Lowest terms} \qquad \text{NOW TRY} \; \circlearrowleft$$

NOTE   The numerator *or* the denominator could have been factored in the first step in **Example 5.** Factor -1 from the numerator, and confirm that the result is the same.

In **Example 5,** notice that $y - x$ is the **opposite** (or **additive inverse**) of $x - y$. A general rule for this situation follows.

Quotient of Opposites

If the numerator and the denominator of a rational expression are opposites, such as in $\frac{x - y}{y - x}$, then the rational expression is equal to -1.

Based on this result, the following are true.

Numerator and denominator are opposites. $\Rightarrow \dfrac{q - 7}{7 - q} = -1$ and $\dfrac{-5a + 2b}{5a - 2b} = -1$

However, the following expression cannot be simplified further.

$$\frac{x - 2}{x + 2} \quad \text{Numerator and denominator are *not* opposites.}$$

EXAMPLE 6   Writing in Lowest Terms (Factors Are Opposites)

Write each rational expression in lowest terms.

(a) $\dfrac{2 - m}{m - 2}$

NOW TRY ANSWER
5. -1

Because $2 - m$ and $m - 2$ are opposites, this expression equals -1.

NOW TRY EXERCISE 6

Write each rational expression in lowest terms.

(a) $\dfrac{p - 4}{4 - p}$ **(b)** $\dfrac{4m^2 - n^2}{2n - 4m}$

(c) $\dfrac{x + y}{x - y}$

(b) $\dfrac{4x^2 - 9}{6 - 4x}$

$= \dfrac{(2x + 3)(2x - 3)}{2(3 - 2x)}$ Factor the numerator and denominator.

$= \dfrac{(2x + 3)(2x - 3)}{2(-1)(2x - 3)}$ Write $3 - 2x$ in the denominator as $-1(2x - 3)$.

$= \dfrac{2x + 3}{2(-1)}$ Fundamental property

$= \dfrac{2x + 3}{-2}$ Multiply in the denominator.

$= -\dfrac{2x + 3}{2}$ $\dfrac{a}{-b} = -\dfrac{a}{b}$

(c) $\dfrac{3 + r}{3 - r}$ $3 - r$ is *not* the opposite of $3 + r$.

This rational expression is already in lowest terms.   **NOW TRY**

OBJECTIVE 4 Recognize equivalent forms of rational expressions.

It is important in algebra to recognize equivalent forms of expressions. For example,

$$0.5, \quad \frac{1}{2}, \quad 50\%, \quad \text{and} \quad \frac{50}{100}$$

all represent the *same* real number. On a number line, the exact same point would apply to all four of them.

A similar situation exists with negative common fractions. The common fraction $-\dfrac{5}{6}$ can also be written $\dfrac{-5}{6}$ and $\dfrac{5}{-6}$, with the negative sign appearing in any of three different positions. All represent the *same* rational number.

Consider the following rational expression.

$$-\frac{2x + 3}{2} \quad \text{Final result from \textbf{Example 6(b)}}$$

The $-$ sign representing the factor -1 is in front of the expression, aligned with the fraction bar. To obtain other equivalent forms of this rational expression, the factor -1 may instead be placed in the numerator or in the denominator.

Use parentheses.

$$\frac{-(2x + 3)}{2} \quad \text{and} \quad \frac{2x + 3}{-2}$$

In the first of these two expressions, the distributive property can be applied. Thus,

$$\frac{-(2x + 3)}{2} \quad \text{can also be written} \quad \frac{-2x - 3}{2}.$$

Multiply *each* term in the binomial by -1.

⚠ CAUTION $\dfrac{-2x + 3}{2}$ is *not* an equivalent form of $\dfrac{-(2x + 3)}{2}$. **Be careful to apply the distributive property correctly.**

NOW TRY
EXERCISE 7

Write four equivalent forms of the rational expression.

$$-\frac{4k-9}{k+3}$$

EXAMPLE 7   Writing Equivalent Forms of a Rational Expression

Write four equivalent forms of the rational expression.

$$-\frac{3x+2}{x-6}$$

If we apply the negative sign to the numerator, we obtain these equivalent forms.

① → $\dfrac{-(3x+2)}{x-6}$ and, by the distributive property, $\dfrac{-3x-2}{x-6}$ ← ②

If we apply the negative sign to the denominator, we obtain two more forms.

③ → $\dfrac{3x+2}{-(x-6)}$ and, by distributing once again, $\dfrac{3x+2}{-x+6}$ ← ④

NOW TRY ↻

NOW TRY ANSWER

7. $\dfrac{-(4k-9)}{k+3}$, $\dfrac{-4k+9}{k+3}$, $\dfrac{4k-9}{-(k+3)}$, $\dfrac{4k-9}{-k-3}$

⊘ **CAUTION**   Recall that $-\frac{5}{6} \neq \frac{-5}{-6}$. Thus, in **Example 7,** it would be incorrect to distribute the negative sign in $-\frac{3x+2}{x-6}$ to *both* the numerator *and* the denominator. (Doing this would actually lead to the *opposite* of the original expression.)

7.1 Exercises

FOR EXTRA HELP   ▶   MyMathLab®

▶ *Complete solution available in MyMathLab*

Concept Check   *Work each problem.*

1. Fill in each blank with the correct response: The rational expression $\frac{x+5}{x-3}$ is undefined when x is _____, so $x \neq$ _____. This rational expression is equal to 0 when $x =$ _____.

2. Which one of these rational expressions can be simplified?

A. $\dfrac{x^2+2}{x^2}$ **B.** $\dfrac{x^2+2}{2}$ **C.** $\dfrac{x^2+y^2}{y^2}$ **D.** $\dfrac{x^2-5x}{x}$

3. Which two of the following rational expressions equal -1?

A. $\dfrac{2x+3}{2x-3}$ **B.** $\dfrac{2x-3}{3-2x}$ **C.** $\dfrac{2x+3}{3+2x}$ **D.** $\dfrac{2x+3}{-2x-3}$

4. Make the correct choice: $\frac{4-r^2}{4+r^2}$ *(is / is not)* equal to -1.

5. Which one of these rational expressions is *not* equivalent to $\frac{x-3}{4-x}$?

A. $\dfrac{3-x}{x-4}$ **B.** $\dfrac{x+3}{4+x}$ **C.** $-\dfrac{3-x}{4-x}$ **D.** $-\dfrac{x-3}{x-4}$

6. Make the correct choice: $\frac{5+2x}{3-x}$ and $\frac{-5-2x}{x-3}$ *(are / are not)* equivalent rational expressions.

7. Find the numerical value of the rational expression for $x = -3$.

$$\frac{x}{2x+1}$$

$= \dfrac{\underline{\quad}}{2(\underline{\quad})+1}$ Let $x = -3$.

$= \dfrac{\underline{\quad}}{\underline{\quad}+1}$

$= \underline{\quad}$

8. Find any values of the variable for which the rational expression is undefined.

$$\frac{x+2}{x-5}$$

Step 1   _____ $= 0$

Step 2 $x =$ _____

Step 3   The given expression is undefined for _____. Thus,

$x \ (= / \neq)\ 5.$

Find the numerical value of each rational expression for **(a)** *x = 2 and* **(b)** *x = −3. See Example 1.*

▶ **9.** $\dfrac{3x + 1}{5x}$

10. $\dfrac{5x - 2}{4x}$

11. $\dfrac{x^2 - 4}{2x + 1}$

12. $\dfrac{2x^2 - 4x}{3x - 1}$

13. $\dfrac{(-2x)^3}{3x + 9}$

14. $\dfrac{(-3x)^2}{4x + 12}$

15. $\dfrac{7 - 3x}{3x^2 - 7x + 2}$

16. $\dfrac{5x + 2}{4x^2 - 5x - 6}$

17. $\dfrac{(x + 3)(x - 2)}{500x}$

18. $\dfrac{(x - 2)(x + 3)}{1000x}$

19. $\dfrac{x^2 - 4}{x^2 - 9}$

20. $\dfrac{x^2 - 9}{x^2 - 4}$

Find any values of the variable for which each rational expression is undefined. Write answers with the symbol ≠. See Example 2.

21. $-\dfrac{5}{x}$

22. $-\dfrac{2}{y}$

23. $\dfrac{12}{5y}$

24. $\dfrac{-7}{3z}$

25. $\dfrac{x + 1}{x - 6}$

26. $\dfrac{m - 2}{m - 5}$

▶ **27.** $\dfrac{4x^2}{3x + 5}$

28. $\dfrac{2x^3}{3x + 4}$

29. $\dfrac{5m + 2}{m^2 + m - 6}$

30. $\dfrac{2r - 5}{r^2 - 5r + 4}$

31. $\dfrac{x^2 + 3x}{4}$

32. $\dfrac{x^2 - 4x}{6}$

33. $\dfrac{3x - 1}{x^2 + 2}$

34. $\dfrac{4q + 2}{q^2 + 9}$

Concept Check   *Work each problem.*

35. Identify the two *terms* in the numerator and the two *terms* in the denominator of the rational expression $\dfrac{x^2 + 4x}{x + 4}$.

36. Describe the steps you would use to write the rational expression in **Exercise 35** in lowest terms. (*Hint:* It simplifies to *x*.)

Write each rational expression in lowest terms. See Examples 3 and 4.

▶ **37.** $\dfrac{18r^3}{6r}$

38. $\dfrac{27p^4}{3p}$

39. $\dfrac{4(y - 2)}{10(y - 2)}$

40. $\dfrac{15(m - 1)}{9(m - 1)}$

41. $\dfrac{(x + 1)(x - 1)}{(x + 1)^2}$

42. $\dfrac{(t + 5)(t - 3)}{(t + 5)^2}$

▶ **43.** $\dfrac{7m + 14}{5m + 10}$

44. $\dfrac{16x + 8}{14x + 7}$

45. $\dfrac{6m - 18}{7m - 21}$

46. $\dfrac{5r + 20}{3r + 12}$

47. $\dfrac{m^2 - n^2}{m + n}$

48. $\dfrac{a^2 - b^2}{a - b}$

49. $\dfrac{2t + 6}{t^2 - 9}$

50. $\dfrac{5s - 25}{s^2 - 25}$

51. $\dfrac{12m^2 - 3}{8m - 4}$

52. $\dfrac{20p^2 - 45}{6p - 9}$

53. $\dfrac{3m^2 - 3m}{5m - 5}$

54. $\dfrac{6t^2 - 6t}{5t - 5}$

55. $\dfrac{9r^2 - 4s^2}{9r + 6s}$

56. $\dfrac{16x^2 - 9y^2}{12x - 9y}$

57. $\dfrac{x - 6}{x^2 - 36}$

58. $\dfrac{x - 8}{x^2 - 64}$

59. $\dfrac{x^2 - 9}{x^2 - 6x + 9}$

60. $\dfrac{x^2 - 16}{x^2 - 8x + 16}$

61. $\dfrac{13x^2 - 39x^3}{7x - 21x^2}$

62. $\dfrac{30x^3 - 15x^5}{22x^2 - 11x^4}$

63. $\dfrac{5k^2 - 13k - 6}{5k + 2}$

64. $\dfrac{7t^2 - 31t - 20}{7t + 4}$

65. $\dfrac{x^2 + 2x - 15}{x^2 + 6x + 5}$

66. $\dfrac{y^2 - 5y - 14}{y^2 + y - 2}$

67. $\dfrac{2x^2 - 3x - 5}{2x^2 - 7x + 5}$

68. $\dfrac{3x^2 + 8x + 4}{3x^2 - 4x - 4}$

69. $\dfrac{3x^3 + 13x^2 + 14x}{3x^3 - 5x^2 - 28x}$

70. $\dfrac{2x^3 + 7x^2 - 30x}{2x^3 - 11x^2 + 15x}$ **71.** $\dfrac{-3t + 6t^2 - 3t^3}{7t^2 - 14t^3 + 7t^4}$ **72.** $\dfrac{-20r - 20r^2 - 5r^3}{24r^2 + 24r^3 + 6r^4}$

Extending Skills *Exercises 73–94 involve factoring by grouping (**Section 6.1**) and factoring sums and differences of cubes (**Section 6.4**). Write each rational expression in lowest terms.*

73. $\dfrac{zw + 4z - 3w - 12}{zw + 4z + 5w + 20}$ **74.** $\dfrac{km + 4k - 4m - 16}{km + 4k + 5m + 20}$ **75.** $\dfrac{pr + qr + ps + qs}{pr + qr - ps - qs}$

76. $\dfrac{wt + ws + xt + xs}{wt - xs - xt + ws}$ **77.** $\dfrac{ac - ad + bc - bd}{ac - ad - bc + bd}$ **78.** $\dfrac{ac - bc - ad + bd}{ac - ad - bd + bc}$

79. $\dfrac{m^2 - n^2 - 4m - 4n}{2m - 2n - 8}$ **80.** $\dfrac{x^2 - y^2 - 7y - 7x}{3x - 3y - 21}$ **81.** $\dfrac{x^2y + y + x^2z + z}{xy + xz}$

82. $\dfrac{y^2k + pk - y^2z - pz}{yk - yz}$ **83.** $\dfrac{1 + p^3}{1 + p}$ **84.** $\dfrac{8 + x^3}{2 + x}$

85. $\dfrac{x^3 - 27}{x - 3}$ **86.** $\dfrac{r^3 - 1000}{r - 10}$ **87.** $\dfrac{b^3 - a^3}{a^2 - b^2}$

88. $\dfrac{8y^3 - 27z^3}{9z^2 - 4y^2}$ **89.** $\dfrac{k^3 + 8}{k^2 - 4}$ **90.** $\dfrac{r^3 + 27}{r^2 - 9}$

91. $\dfrac{z^3 + 27}{z^3 - 3z^2 + 9z}$ **92.** $\dfrac{t^3 + 64}{t^3 - 4t^2 + 16t}$ **93.** $\dfrac{1 - 8r^3}{8r^2 + 4r + 2}$ **94.** $\dfrac{8 - 27x^3}{27x^2 + 18x + 12}$

Write each rational expression in lowest terms. **See Examples 5 and 6.**

▶ 95. $\dfrac{6 - t}{t - 6}$ **96.** $\dfrac{2 - k}{k - 2}$ **▶ 97.** $\dfrac{m^2 - 1}{1 - m}$ **98.** $\dfrac{a^2 - b^2}{b - a}$

99. $\dfrac{q^2 - 4q}{4q - q^2}$ **100.** $\dfrac{z^2 - 5z}{5z - z^2}$ **101.** $\dfrac{p + 6}{p - 6}$

102. $\dfrac{5 - x}{5 + x}$ **103.** $\dfrac{-2m + 2n}{m - n}$ **104.** $\dfrac{-5p + 5q}{p - q}$

Write four equivalent forms for each rational expression. **See Example 7.**

▶ 105. $-\dfrac{x + 4}{x - 3}$ **106.** $-\dfrac{x + 6}{x - 1}$ **107.** $-\dfrac{2x - 3}{x + 3}$

108. $-\dfrac{5x - 6}{x + 4}$ **109.** $-\dfrac{3x - 1}{5x - 6}$ **110.** $-\dfrac{2x - 9}{7x - 1}$

Solve each problem.

111. The area of the rectangle is represented by

$$x^4 + 10x^2 + 21.$$

What is the width? $\left(Hint:\ Use\ W = \dfrac{\mathcal{A}}{L}.\right)$

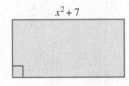

112. The volume of the box is represented by

$$(x^2 + 8x + 15)(x + 4).$$

Find the polynomial that represents the area of the bottom of the box.

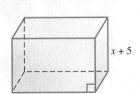

Solve each problem.

113. The average number of vehicles waiting in line to enter a sports arena parking area is approximated by the rational expression

$$\frac{x^2}{2(1-x)},$$

where x is a quantity between 0 and 1 known as the **traffic intensity.** (*Source:* Mannering, F., and W. Kilareski, *Principles of Highway Engineering and Traffic Control,* John Wiley and Sons.)

To the nearest tenth, find the average number of vehicles waiting if the traffic intensity is the given number.

(a) 0.1   **(b)** 0.8   **(c)** 0.9

(d) What happens to the number of vehicles waiting as traffic intensity increases?

114. The percent of deaths caused by smoking is modeled by the rational expression

$$\frac{x-1}{x},$$

where x is the number of times a smoker is more likely than a nonsmoker to die of lung cancer. This is called the **incidence rate.** (*Source:* Walker, A., *Observation and Inference: An Introduction to the Methods of Epidemiology,* Epidemiology Resources Inc.) For example, $x = 10$ means that a smoker is 10 times more likely than a nonsmoker to die of lung cancer.

Find the percent of deaths if the incidence rate is the given number.

(a) 5   **(b)** 10   **(c)** 20

(d) Can the incidence rate equal 0? Explain.

RELATING CONCEPTS For Individual or Group Work (Exercises 115–118)

In Section 5.7, we used long division to find a quotient of two polynomials. We obtain the same quotient by expressing a division problem as a rational expression (fraction) and writing this rational expression in lowest terms, as shown below.

$$
\begin{array}{r}
x + 4 \\
2x - 3 \overline{)2x^2 + 5x - 12} \\
\underline{2x^2 - 3x} \\
8x - 12 \\
\underline{8x - 12} \\
0
\end{array}
$$

$$\frac{2x^2 + 5x - 12}{2x - 3}$$

$$= \frac{(2x-3)(x+4)}{2x-3} \qquad \text{Factor.}$$

$$= x + 4 \qquad \text{Fundamental property}$$

Show that performing the long division and simplifying the rational expression yield the same result.

115. $4x + 7 \overline{)8x^2 + 26x + 21}$

and $\dfrac{8x^2 + 26x + 21}{4x + 7}$

116. $6x + 5 \overline{)12x^2 + 16x + 5}$

and $\dfrac{12x^2 + 16x + 5}{6x + 5}$

117. $x + 1 \overline{)x^3 + x^2 + x + 1}$

and $\dfrac{x^3 + x^2 + x + 1}{x + 1}$

118. $x + 1 \overline{)x^3 + x^2 + 2x + 2}$

and $\dfrac{x^3 + x^2 + 2x + 2}{x + 1}$

7.2 Multiplying and Dividing Rational Expressions

OBJECTIVES

1 Multiply rational expressions.

2 Divide rational expressions.

OBJECTIVE 1 Multiply rational expressions.

The product of two fractions is found by multiplying the numerators and multiplying the denominators. Rational expressions are multiplied in the same way.

> **Multiplying Rational Expressions**
>
> The product of the rational expressions $\frac{P}{Q}$ and $\frac{R}{S}$ is defined as follows.
>
> $$\frac{P}{Q} \cdot \frac{R}{S} = \frac{PR}{QS}$$
>
> That is, to multiply rational expressions, multiply the numerators and multiply the denominators.

NOW TRY EXERCISE 1

Multiply. Write each answer in lowest terms.

(a) $\dfrac{7}{18} \cdot \dfrac{9}{14}$ **(b)** $\dfrac{4k^2}{7} \cdot \dfrac{14}{11k}$

EXAMPLE 1 Multiplying Rational Expressions

Multiply. Write each answer in lowest terms.

(a) $\dfrac{3}{10} \cdot \dfrac{5}{9}$

(b) $\dfrac{6}{x} \cdot \dfrac{x^2}{12}$

Indicate the product of the numerators and the product of the denominators.

$= \dfrac{3 \cdot 5}{10 \cdot 9}$

$= \dfrac{6 \cdot x^2}{x \cdot 12}$

Leave the products in factored form. Factor the numerator and denominator to further identify any common factors. Then use the fundamental property to divide out any common factors and write each product in lowest terms.

$= \dfrac{3 \cdot 5}{2 \cdot 5 \cdot 3 \cdot 3}$

$= \dfrac{6 \cdot x \cdot x}{x \cdot 2 \cdot 6}$

$= \dfrac{1}{6}$ Remember to write 1 in the numerator.

$= \dfrac{x}{2}$

  NOW TRY

NOTE   It is also possible to divide out common factors in the numerator and denominator *before* multiplying the rational expressions. Consider the following.

$\dfrac{3}{10} \cdot \dfrac{5}{9}$ Example 1(a)

$= \dfrac{3}{5 \cdot 2} \cdot \dfrac{5}{3 \cdot 3}$ Identify the common factors.

$= \dfrac{1}{2 \cdot 3}$ Divide out the common factors. Insert a factor of 1 in the numerator.

$= \dfrac{1}{6}$ Multiply.

NOW TRY ANSWERS

1. (a) $\frac{1}{4}$ **(b)** $\frac{8k}{11}$

NOW TRY
EXERCISE 2

Multiply. Write the answer in lowest terms.

$$\frac{m - 3}{3m} \cdot \frac{9m^2}{8(m - 3)^2}$$

EXAMPLE 2 Multiplying Rational Expressions

Multiply. Write the answer in lowest terms.

$$\frac{x + y}{2x} \cdot \frac{x^2}{(x + y)^2}$$

> Use parentheses here around $x + y$.

$$= \frac{(x + y)x^2}{2x(x + y)^2}$$ Multiply numerators. Multiply denominators.

$$= \frac{(x + y)x \cdot x}{2x(x + y)(x + y)}$$ Factor. Identify the common factors.

$$= \frac{x}{2(x + y)}$$ $\frac{(x + y)x}{x(x + y)} = 1$; Lowest terms **NOW TRY**

NOW TRY
EXERCISE 3

Multiply. Write the answer in lowest terms.

$$\frac{y^2 - 3y - 28}{y^2 - 9y + 14} \cdot \frac{y^2 - 7y + 10}{y^2 + 4y}$$

EXAMPLE 3 Multiplying Rational Expressions

Multiply. Write the answer in lowest terms.

$$\frac{x^2 + 3x}{x^2 - 3x - 4} \cdot \frac{x^2 - 5x + 4}{x^2 + 2x - 3}$$

$$= \frac{(x^2 + 3x)(x^2 - 5x + 4)}{(x^2 - 3x - 4)(x^2 + 2x - 3)}$$ Definition of multiplication

$$= \frac{x(x + 3)(x - 4)(x - 1)}{(x - 4)(x + 1)(x + 3)(x - 1)}$$ Factor.

$$= \frac{x}{x + 1}$$ Divide out the common factors; lowest terms

The quotients $\frac{x + 3}{x + 3}, \frac{x - 4}{x - 4}$, and $\frac{x - 1}{x - 1}$ all equal 1, justifying the final product $\frac{x}{x + 1}$.

NOW TRY

OBJECTIVE 2 Divide rational expressions.

Suppose we have $\frac{7}{8}$ gal of milk and want to find how many quarts we have. Because 1 qt is $\frac{1}{4}$ gal, we ask, "How many $\frac{1}{4}$s are there in $\frac{7}{8}$?" This would be interpreted as follows.

$$\frac{7}{8} \div \frac{1}{4}, \quad \text{or} \quad \frac{\frac{7}{8}}{\frac{1}{4}} \leftarrow \text{The fraction bar means division.}$$

The fundamental property of rational expressions discussed earlier can be applied to rational number values of P, Q, and K.

$$\frac{P}{Q} = \frac{P \cdot K}{Q \cdot K} = \frac{\frac{7}{8} \cdot 4}{\frac{1}{4} \cdot 4} = \frac{\frac{7}{8} \cdot 4}{1} = \frac{7}{8} \cdot \frac{4}{1} \quad \begin{array}{l} \text{Let } P = \frac{7}{8}, Q = \frac{1}{4}, \text{ and } K = 4. \\ \left(K \text{ is the reciprocal of } Q = \frac{1}{4}. \right) \end{array}$$

NOW TRY ANSWERS

2. $\frac{3m}{8(m - 3)}$

3. $\frac{y - 5}{y}$

Therefore, to divide $\frac{7}{8}$ by $\frac{1}{4}$, we multiply $\frac{7}{8}$ by the reciprocal of $\frac{1}{4}$, namely 4. Because $\frac{7}{8}(4) = \frac{7}{2}$, there are $\frac{7}{2}$ qt, or $3\frac{1}{2}$ qt, in $\frac{7}{8}$ gal.

The preceding discussion illustrates dividing common fractions. Division of rational expressions is defined in the same way.

Dividing Rational Expressions

If $\frac{P}{Q}$ and $\frac{R}{S}$ are any two rational expressions where $\frac{R}{S} \neq 0$, then their quotient is defined as follows.

$$\frac{P}{Q} \div \frac{R}{S} = \frac{P}{Q} \cdot \frac{S}{R} = \frac{PS}{QR}$$

That is, to divide one rational expression by another rational expression, multiply the first rational expression (dividend) by the reciprocal of the second rational expression (divisor).

NOW TRY
EXERCISE 4

Divide. Write each answer in lowest terms.

(a) $\frac{3}{10} \div \frac{11}{20}$

(b) $\frac{2x - 5}{3x^2} \div \frac{2x - 5}{12x}$

EXAMPLE 4   Dividing Rational Expressions

Divide. Write each answer in lowest terms.

(a) $\frac{5}{8} \div \frac{7}{16}$ | **(b)** $\frac{y}{y + 3} \div \frac{4y}{y + 5}$

Multiply the dividend by the reciprocal of the divisor.

$= \frac{5}{8} \cdot \frac{16}{7}$ ← Reciprocal of $\frac{7}{16}$ | $= \frac{y}{y + 3} \cdot \frac{y + 5}{4y}$ ← Reciprocal of $\frac{4y}{y + 5}$

$= \frac{5 \cdot 16}{8 \cdot 7}$ Multiply. | $= \frac{y(y + 5)}{(y + 3)(4y)}$ Multiply.

$= \frac{5 \cdot 8 \cdot 2}{8 \cdot 7}$ Factor 16. | $= \frac{y + 5}{4(y + 3)}$ Lowest terms

$= \frac{10}{7}$ Lowest terms | **NOW TRY**

NOW TRY
EXERCISE 5

Divide. Write the answer in lowest terms.

$$\frac{(3k)^3}{2j^4} \div \frac{9k^2}{6j}$$

EXAMPLE 5   Dividing Rational Expressions

Divide. Write the answer in lowest terms.

$$\frac{(3m)^2}{(2p)^3} \div \frac{6m^3}{16p^2}$$

$= \frac{(3m)^2}{(2p)^3} \cdot \frac{16p^2}{6m^3}$ Multiply by the reciprocal of the divisor.

$(3m)^2 = 3^2 m^2;$
$(2p)^3 = 2^3 p^3$ $= \frac{9m^2}{8p^3} \cdot \frac{16p^2}{6m^3}$ Power rule for exponents

$= \frac{9 \cdot 16 m^2 p^2}{8 \cdot 6 p^3 m^3}$ Multiply numerators.
Multiply denominators.

$= \frac{3}{mp}$ Lowest terms   **NOW TRY**

NOW TRY ANSWERS

4. (a) $\frac{6}{11}$ (b) $\frac{4}{x}$ **5.** $\frac{9k}{j^3}$

NOW TRY
EXERCISE 6

Divide. Write the answer in lowest terms.

$$\frac{(t+2)(t-5)}{-4t} \div$$

$$\frac{t^2-25}{(t+5)(t+2)}$$

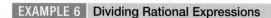

EXAMPLE 6 **Dividing Rational Expressions**

Divide. Write the answer in lowest terms.

$$\frac{x^2-4}{(x+3)(x-2)} \div \frac{(x+2)(x+3)}{-2x}$$

$$= \frac{x^2-4}{(x+3)(x-2)} \cdot \frac{-2x}{(x+2)(x+3)} \qquad \text{Multiply by the reciprocal of the divisor.}$$

$$= \frac{-2x(x^2-4)}{(x+3)(x-2)(x+2)(x+3)} \qquad \begin{array}{l}\text{Multiply numerators.}\\ \text{Multiply denominators.}\end{array}$$

$$= \frac{-2x(x+2)(x-2)}{(x+3)(x-2)(x+2)(x+3)} \qquad \text{Factor the numerator.}$$

$$= \frac{-2x}{(x+3)^2} \qquad \begin{array}{l}\text{Divide out the common factors;}\\ a \cdot a = a^2\end{array}$$

$$= -\frac{2x}{(x+3)^2} \qquad \frac{-a}{b} = -\frac{a}{b}; \text{ Lowest terms}$$

NOW TRY

NOW TRY
EXERCISE 7

Divide. Write the answer in lowest terms.

$$\frac{7-x}{2x+6} \div \frac{x^2-49}{x^2+6x+9}$$

EXAMPLE 7 **Dividing Rational Expressions (Factors Are Opposites)**

Divide. Write the answer in lowest terms.

$$\frac{m^2-4}{m^2-1} \div \frac{2m^2+4m}{1-m}$$

$$= \frac{m^2-4}{m^2-1} \cdot \frac{1-m}{2m^2+4m} \qquad \text{Multiply by the reciprocal of the divisor.}$$

$$= \frac{(m^2-4)(1-m)}{(m^2-1)(2m^2+4m)} \qquad \begin{array}{l}\text{Multiply numerators.}\\ \text{Multiply denominators.}\end{array}$$

$$= \frac{(m+2)(m-2)(1-m)}{(m+1)(m-1)(2m)(m+2)} \qquad \text{Factor. } 1-m \text{ and } m-1 \text{ are opposites.}$$

$$= \frac{-1(m-2)}{2m(m+1)} \qquad \begin{array}{l}\text{Divide out the common factors.}\\ \text{From Section 7.1, } \frac{1-m}{m-1} = -1.\end{array}$$

$$= \frac{-m+2}{2m(m+1)} \qquad \text{Distribute } -1 \text{ in the numerator.}$$

$$= \frac{2-m}{2m(m+1)} \qquad \text{Rewrite } -m+2 \text{ as } 2-m. \quad \text{NOW TRY}$$

Multiplying or Dividing Rational Expressions

Step 1 **Note the operation.** If the operation is division, use the definition of division to rewrite it as multiplication.

Step 2 **Multiply** numerators and multiply denominators.

Step 3 **Factor** all numerators and denominators completely.

Step 4 **Write in lowest terms** using the fundamental property.

Note: Steps 2 and 3 may be interchanged based on personal preference.

NOW TRY ANSWERS

6. $-\dfrac{(t+2)^2}{4t}$

7. $-\dfrac{x+3}{2(x+7)}$

7.2 Exercises

 FOR EXTRA HELP

 MyMathLab®

▶ *Complete solution available in MyMathLab*

1. *Concept Check* Match each multiplication problem in Column I with the correct product in Column II.

	I		II
(a)	$\dfrac{5x^3}{10x^4} \cdot \dfrac{10x^7}{4x}$	**A.**	$\dfrac{4}{5x^5}$
(b)	$\dfrac{10x^4}{5x^3} \cdot \dfrac{10x^7}{4x}$	**B.**	$\dfrac{5x^5}{4}$
(c)	$\dfrac{5x^3}{10x^4} \cdot \dfrac{4x}{10x^7}$	**C.**	$\dfrac{1}{5x^7}$
(d)	$\dfrac{10x^4}{5x^3} \cdot \dfrac{4x}{10x^7}$	**D.**	$5x^7$

2. *Concept Check* Match each division problem in Column I with the correct quotient in Column II.

	I		II
(a)	$\dfrac{5x^3}{10x^4} \div \dfrac{10x^7}{4x}$	**A.**	$\dfrac{5x^5}{4}$
(b)	$\dfrac{10x^4}{5x^3} \div \dfrac{10x^7}{4x}$	**B.**	$5x^7$
(c)	$\dfrac{5x^3}{10x^4} \div \dfrac{4x}{10x^7}$	**C.**	$\dfrac{4}{5x^5}$
(d)	$\dfrac{10x^4}{5x^3} \div \dfrac{4x}{10x^7}$	**D.**	$\dfrac{1}{5x^7}$

Multiply. Write each answer in lowest terms. **See Examples 1 and 2.**

▶ **3.** $\dfrac{15a^2}{14} \cdot \dfrac{7}{5a}$

4. $\dfrac{21b^6}{18} \cdot \dfrac{9}{7b^4}$

5. $\dfrac{12x^4}{18x^3} \cdot \dfrac{-8x^5}{4x^2}$

6. $\dfrac{12m^5}{-2m^2} \cdot \dfrac{6m^6}{28m^3}$

7. $\dfrac{2(c+d)}{3} \cdot \dfrac{18}{6(c+d)^2}$

8. $\dfrac{4(y-2)}{x} \cdot \dfrac{3x}{6(y-2)^2}$

▶ **9.** $\dfrac{(x-y)^2}{2} \cdot \dfrac{24}{3(x-y)}$

10. $\dfrac{(a+b)^2}{5} \cdot \dfrac{30}{2(a+b)}$

11. $\dfrac{t-4}{8} \cdot \dfrac{4t^2}{t-4}$

12. $\dfrac{z+9}{12} \cdot \dfrac{3z^2}{z+9}$

13. $\dfrac{3x}{x+3} \cdot \dfrac{(x+3)^2}{6x^2}$

14. $\dfrac{(t-2)^2}{4t^2} \cdot \dfrac{2t}{t-2}$

Concept Check *Multiply or divide. Write each answer in lowest terms.*

15. $\dfrac{5x-10}{6} \cdot \dfrac{9}{10x-20}$

$= \dfrac{5(\underline{})}{6} \cdot \dfrac{3 \cdot \underline{}}{10(\underline{})}$

$= \dfrac{5(x-2) \cdot 3 \cdot 3}{2 \cdot 3 \cdot 2 \cdot \underline{} \cdot (x-2)}$

$= \underline{}$

16. $\dfrac{6x-4}{3} \div \dfrac{15x-10}{9}$

$= \dfrac{6x-4}{3} \cdot \dfrac{\underline{}}{\underline{}}$

$= \dfrac{2(\underline{})}{3} \cdot \dfrac{9}{5(\underline{})}$

$= \dfrac{2(3x-2) \cdot 3 \cdot 3}{\underline{} \cdot 5(3x-2)}$

$= \underline{}$

Divide. Write each answer in lowest terms. **See Examples 4 and 5.**

17. $\dfrac{9z^4}{3z^5} \div \dfrac{3z^2}{5z^3}$

18. $\dfrac{35x^8}{7x^9} \div \dfrac{5x^5}{9x^6}$

▶ **19.** $\dfrac{4t^4}{2t^5} \div \dfrac{(2t)^3}{-6}$

20. $\dfrac{-12a^6}{3a^2} \div \dfrac{(2a)^3}{27a}$

▶ **21.** $\dfrac{3}{2y-6} \div \dfrac{6}{y-3}$

22. $\dfrac{4m+16}{10} \div \dfrac{3m+12}{18}$

23. $\dfrac{7t+7}{-6} \div \dfrac{4t+4}{15}$

24. $\dfrac{8z-16}{-20} \div \dfrac{3z-6}{40}$

25. $\dfrac{2x}{x-1} \div \dfrac{x^2}{x+2}$

26. $\dfrac{y^2}{y+1} \div \dfrac{3y}{y-3}$

27. $\dfrac{(x-3)^2}{6x} \div \dfrac{x-3}{x^2}$

28. $\dfrac{2a}{a+4} \div \dfrac{a^2}{(a+4)^2}$

29. $\dfrac{5x^3}{x^2 - 16} \div \dfrac{x^5}{(x-4)^2}$ **30.** $\dfrac{8x^4}{x^2 - 25} \div \dfrac{x^7}{(x-5)^2}$ **31.** $\dfrac{-4t^3}{t^2 - 1} \div \dfrac{t^2}{(t+1)^2}$

32. *Concept Check*   After factoring numerators and denominators in a multiplication problem, a student obtained the following.

$$\frac{(x+3)^2}{(x+3)} \cdot \frac{(x+5)}{(x+5)^3}$$

Is it permissible for the student to divide out common factors within the same fractions here?

Multiply or divide. Write each answer in lowest terms. **See Examples 3, 6, and 7.**

33. $\dfrac{5x - 15}{3x + 9} \cdot \dfrac{4x + 12}{6x - 18}$ **34.** $\dfrac{8r + 16}{24r - 24} \cdot \dfrac{6r - 6}{3r + 6}$ **35.** $\dfrac{2 - t}{8} \div \dfrac{t - 2}{6}$

36. $\dfrac{m - 2}{4} \div \dfrac{2 - m}{6}$ **37.** $\dfrac{27 - 3z}{4} \cdot \dfrac{12}{2z - 18}$ **38.** $\dfrac{35 - 5x}{6} \cdot \dfrac{12}{3x - 21}$

▶ **39.** $\dfrac{p^2 + 4p - 5}{p^2 + 7p + 10} \div \dfrac{p - 1}{p + 4}$ **40.** $\dfrac{z^2 - 3z + 2}{z^2 + 4z + 3} \div \dfrac{z - 1}{z + 1}$ ▶ **41.** $\dfrac{m^2 - 4}{16 - 8m} \div \dfrac{m + 2}{8}$

42. $\dfrac{r^2 - 36}{54 - 9r} \div \dfrac{r + 6}{9}$ **43.** $\dfrac{m^2 - 4}{16 - 8m} \div \dfrac{m^2 + 3m + 2}{8m + 16}$ **44.** $\dfrac{t^2 - 49}{42 - 6t} \div \dfrac{t^2 + 10t + 21}{6t + 42}$

45. $\dfrac{2x^2 - 7x + 3}{x - 3} \cdot \dfrac{x + 2}{x - 1}$ **46.** $\dfrac{3x^2 - 5x - 2}{x - 2} \cdot \dfrac{x - 3}{x + 1}$

47. $\dfrac{2k^2 - k - 1}{2k^2 + 5k + 3} \div \dfrac{4k^2 - 1}{2k^2 + k - 3}$ **48.** $\dfrac{3t^2 - 4t - 4}{3t^2 + 10t + 8} \div \dfrac{9t^2 + 21t + 10}{3t^2 - t - 10}$

▶ **49.** $\dfrac{2k^2 + 3k - 2}{6k^2 - 7k + 2} \cdot \dfrac{4k^2 - 5k + 1}{k^2 + k - 2}$ **50.** $\dfrac{2m^2 - 5m - 12}{m^2 - 10m + 24} \cdot \dfrac{m^2 - 9m + 18}{4m^2 - 9}$

51. $\dfrac{m^2 + 2mp - 3p^2}{m^2 - 3mp + 2p^2} \div \dfrac{m^2 + 4mp + 3p^2}{m^2 + 2mp - 8p^2}$ **52.** $\dfrac{x^2 - 2xy - 3y^2}{x^2 + xy - 30y^2} \div \dfrac{x^2 + xy - 12y^2}{x^2 - xy - 20y^2}$

53. $\dfrac{m^2 + 3m + 2}{m^2 + 5m + 4} \cdot \dfrac{m^2 + 10m + 24}{m^2 + 5m + 6}$ **54.** $\dfrac{z^2 - z - 6}{z^2 - 2z - 8} \cdot \dfrac{z^2 + 7z + 12}{z^2 - 9}$

55. $\dfrac{y^2 + y - 2}{y^2 + 3y - 4} \div \dfrac{y + 2}{y + 3}$ **56.** $\dfrac{r^2 + r - 6}{r^2 + 4r - 12} \div \dfrac{r + 3}{r - 1}$

57. $\dfrac{2m^2 + 7m + 3}{m^2 - 9} \cdot \dfrac{m^2 - 3m}{2m^2 + 11m + 5}$ **58.** $\dfrac{6s^2 + 17s + 10}{s^2 - 4} \cdot \dfrac{s^2 - 2s}{6s^2 + 29s + 20}$

59. $\dfrac{r^2 + rs - 12s^2}{r^2 - rs - 20s^2} \div \dfrac{r^2 - 2rs - 3s^2}{r^2 + rs - 30s^2}$ **60.** $\dfrac{m^2 + 8mn + 7n^2}{m^2 + mn - 42n^2} \div \dfrac{m^2 - 3mn - 4n^2}{m^2 - mn - 30n^2}$

61. $\dfrac{(q - 3)^4 (q + 2)}{q^2 + 3q + 2} \div \dfrac{q^2 - 6q + 9}{q^2 + 4q + 4}$ **62.** $\dfrac{(x + 4)^3 (x - 3)}{x^2 - 9} \div \dfrac{x^2 + 8x + 16}{x^2 + 6x + 9}$

Extending Skills   *Exercises 63–68 involve grouping symbols* (*Section 1.1*), *factoring by grouping* (*Section 6.1*), *and factoring sums and differences of cubes* (*Section 6.4*). *Multiply or divide as indicated. Write each answer in lowest terms.*

63. $\dfrac{3a - 3b - a^2 + b^2}{4a^2 - 4ab + b^2} \cdot \dfrac{4a^2 - b^2}{2a^2 - ab - b^2}$ **64.** $\dfrac{4r^2 - t^2 + 10r - 5t}{2r^2 + rt + 5r} \cdot \dfrac{4r^3 + 4r^2t + rt^2}{2r + t}$

65. $\dfrac{-x^3 - y^3}{x^2 - 2xy + y^2} \div \dfrac{3y^2 - 3xy}{x^2 - y^2}$

66. $\dfrac{b^3 - 8a^3}{4a^3 + 4a^2b + ab^2} \div \dfrac{4a^2 + 2ab + b^2}{-a^3 - ab^3}$

67. $\dfrac{x + 5}{x + 10} \div \left(\dfrac{x^2 + 10x + 25}{x^2 + 10x} \cdot \dfrac{10x}{x^2 + 15x + 50} \right)$

68. $\dfrac{m - 8}{m - 4} \div \left(\dfrac{m^2 - 12m + 32}{8m} \cdot \dfrac{m^2 - 8m}{m^2 - 8m + 16} \right)$

Answer each question.

69. If the rational expression $\dfrac{5x^2y^3}{2pq}$ represents the area of a rectangle and $\dfrac{2xy}{p}$ represents the length, what rational expression represents the width?

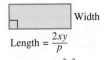

Width

Length $= \dfrac{2xy}{p}$

The area is $\dfrac{5x^2y^3}{2pq}$.

70. Given the following problem, what polynomial is represented by the red question mark?

$$\dfrac{4y + 12}{2y - 10} \div \dfrac{?}{y^2 - y - 20} = \dfrac{2(y + 4)}{y - 3}$$

7.3 Least Common Denominators

OBJECTIVES

1. Find the least common denominator for a group of fractions.
2. Write equivalent rational expressions.

VOCABULARY

☐ least common denominator (LCD)

OBJECTIVE 1 Find the least common denominator for a group of fractions.

Adding or subtracting rational expressions often requires finding the **least common denominator (LCD).** The LCD is the simplest expression that is divisible by all of the denominators in all of the expressions. For example, the fractions

$$\dfrac{2}{9} \quad \text{and} \quad \dfrac{5}{12} \quad \text{have LCD} \quad 36,$$

because 36 is the least positive number divisible by both 9 and 12.

We can often find least common denominators by inspection. In other cases, we find the LCD using the following procedure.

Finding the Least Common Denominator (LCD)

Step 1 **Factor** each denominator into prime factors.

Step 2 **List each different denominator factor** the *greatest* number of times it appears in any of the denominators.

Step 3 **Multiply** the denominator factors from Step 2 to find the LCD.

When each denominator is factored into prime factors, every prime factor must be a factor of the LCD.

⚠ CAUTION When finding the LCD, use each factor the *greatest* number of times it appears in any *single* denominator, not the *total* number of times it appears. For instance, the greatest number of times 2 appears as a factor in one denominator in **Example 1(b)** on the next page is 3, *not* 4.

**NOW TRY
EXERCISE 1**

Find the LCD for each pair of fractions.

(a) $\dfrac{5}{48}, \dfrac{1}{30}$ **(b)** $\dfrac{3}{10y}, \dfrac{1}{6y}$

EXAMPLE 1   Finding Least Common Denominators

Find the LCD for each pair of fractions.

(a) $\dfrac{1}{24}, \dfrac{7}{15}$ **(b)** $\dfrac{1}{8x}, \dfrac{3}{10x}$

Step 1   Write each denominator in factored form with numerical coefficients in prime factored form.

$$24 = 2 \cdot 2 \cdot 2 \cdot 3 = 2^3 \cdot 3 \qquad\qquad 8x = 2 \cdot 2 \cdot 2 \cdot x = 2^3 \cdot x$$
$$15 = 3 \cdot 5 \qquad\qquad\qquad\qquad 10x = 2 \cdot 5 \cdot x$$

Step 2   Find the LCD by taking each different factor the *greatest* number of times it appears as a factor in any of the denominators.

The factor 2 appears three times in one product and not at all in the other, so the greatest number of times 2 appears is three. The greatest number of times both 3 and 5 appear is one.

Here, 2 appears three times in one product and once in the other, so the greatest number of times 2 appears is three. The greatest number of times 5 appears is one. The greatest number of times x appears in either product is one.

Step 3   LCD $= 2 \cdot 2 \cdot 2 \cdot 3 \cdot 5$ LCD $= 2 \cdot 2 \cdot 2 \cdot 5 \cdot x$

$\qquad\qquad = 2^3 \cdot 3 \cdot 5 \qquad\qquad\qquad\qquad = 2^3 \cdot 5 \cdot x$

$\qquad\qquad = 120 \qquad\qquad\qquad\qquad\qquad = 40x$ NOW TRY

**NOW TRY
EXERCISE 2**

Find the LCD for the pair of fractions.

$\dfrac{5}{6x^4}$ and $\dfrac{7}{8x^3}$

EXAMPLE 2   Finding the LCD

Find the LCD for $\dfrac{5}{6r^2}$ and $\dfrac{3}{4r^3}$.

Step 1   Factor each denominator.

$$6r^2 = 2 \cdot 3 \cdot r^2$$
$$4r^3 = 2 \cdot 2 \cdot r^3 = 2^2 \cdot r^3$$

Step 2   The greatest number of times 2 appears is two, the greatest number of times 3 appears is one, and the greatest number of times r appears is three.

Step 3   LCD $= 2^2 \cdot 3 \cdot r^3 = 12r^3$ NOW TRY

EXAMPLE 3   Finding LCDs

Find the LCD for the fractions in each list.

(a) $\dfrac{6}{5m}, \dfrac{4}{m^2 - 3m}$

$$\left. \begin{array}{l} 5m = 5 \cdot m \\ m^2 - 3m = m(m - 3) \end{array} \right\} \text{Factor each denominator.}$$

Use each different factor the greatest number of times it appears.

$$\text{LCD} = 5 \cdot m \cdot (m - 3) = 5m(m - 3)$$

> Be sure to include m as a factor in the LCD.

Because m is not a *factor* of $m - 3$, **both** m and $m - 3$ must appear in the LCD.

**NOW TRY
EXERCISE 3**

Find the LCD for the fractions in each list.

(a) $\dfrac{3t}{2t^2 - 10t}, \dfrac{t+4}{t^2 - 25}$

(b) $\dfrac{1}{x^2 + 7x + 12},$
$\dfrac{2}{x^2 + 6x + 9}, \dfrac{5}{x^2 + 2x - 8}$

(c) $\dfrac{2}{a-4}, \dfrac{1}{4-a}$

(b) $\dfrac{1}{r^2 - 4r - 5}, \dfrac{3}{r^2 - r - 20}, \dfrac{1}{r^2 - 10r + 25}$

$$\left. \begin{array}{l} r^2 - 4r - 5 = (r-5)(r+1) \\ r^2 - r - 20 = (r-5)(r+4) \\ r^2 - 10r + 25 = (r-5)^2 \end{array} \right\} \text{ Factor each denominator.}$$

Use each different factor the greatest number of times it appears as a factor.

$$\text{LCD} = (r-5)^2(r+1)(r+4) \quad \boxed{\text{Be sure to include the exponent 2 on the factor } (r-5).}$$

(c) $\dfrac{1}{q-5}, \dfrac{3}{5-q}$

The expressions $q-5$ and $5-q$ are opposites of each other. This means that if we multiply $q-5$ by -1, we will obtain $5-q$.

$$-(q-5) = -q + 5 = 5 - q$$

Therefore, either $q-5$ or $5-q$ can be used as the LCD.   **NOW TRY**

OBJECTIVE 2 Write equivalent rational expressions.

Once we have the LCD, the next step in preparing to add or subtract two rational expressions is to use the fundamental property to write equivalent rational expressions.

> **Writing a Rational Expression with a Specified Denominator**
>
> **Step 1** **Factor** both denominators.
>
> **Step 2** **Decide what factor(s) the denominator must be multiplied by** in order to equal the specified denominator.
>
> **Step 3** **Multiply** the rational expression by that factor divided by itself. (That is, multiply by 1.)

**NOW TRY
EXERCISE 4**

Write each rational expression with the indicated denominator.

(a) $\dfrac{2}{9} = \dfrac{?}{27}$ **(b)** $\dfrac{4t}{11} = \dfrac{?}{33t}$

EXAMPLE 4 Writing Equivalent Rational Expressions

Write each rational expression with the indicated denominator.

(a) $\dfrac{3}{8} = \dfrac{?}{40}$ **(b)** $\dfrac{9k}{25} = \dfrac{?}{50k}$

Step 1 For each example, first factor the denominator on the right. Then compare the denominator on the left with the one on the right to decide what factors are missing. (It may sometimes be necessary to factor both denominators.)

$$\dfrac{3}{8} = \dfrac{?}{5 \cdot 8} \qquad\qquad \dfrac{9k}{25} = \dfrac{?}{25 \cdot 2k}$$

Step 2 A factor of 5 is missing.   Factors of 2 and k are missing.

Step 3 Multiply $\frac{3}{8}$ by $\frac{5}{5}$.   Multiply $\frac{9k}{25}$ by $\frac{2k}{2k}$.

$$\dfrac{3}{8} = \dfrac{3}{8} \cdot \dfrac{5}{5} = \dfrac{15}{40} \qquad \dfrac{9k}{25} = \dfrac{9k}{25} \cdot \dfrac{2k}{2k} = \dfrac{18k^2}{50k}$$

$$\underset{\uparrow}{\tfrac{5}{5} = 1} \qquad\qquad\qquad \underset{\uparrow}{\tfrac{2k}{2k} = 1}$$

NOW TRY

NOW TRY ANSWERS

3. (a) $2t(t-5)(t+5)$
  (b) $(x+3)^2(x+4)(x-2)$
  (c) either $a-4$ or $4-a$

4. (a) $\dfrac{6}{27}$ **(b)** $\dfrac{12t^2}{33t}$

**NOW TRY
EXERCISE 5**

Write each rational expression with the indicated denominator.

(a) $\dfrac{8k}{5k - 2} = \dfrac{?}{25k - 10}$

(b) $\dfrac{2t - 1}{t^2 + 4t} = \dfrac{?}{t^3 + 12t^2 + 32t}$

EXAMPLE 5 Writing Equivalent Rational Expressions

Write each rational expression with the indicated denominator.

(a) $\dfrac{8}{3x + 1} = \dfrac{?}{12x + 4}$

$\dfrac{8}{3x + 1} = \dfrac{?}{4(3x + 1)}$ Factor the denominator on the right.

The missing factor is 4, so multiply the fraction on the left by $\dfrac{4}{4}$.

$$\dfrac{8}{3x + 1} \cdot \dfrac{4}{4} = \dfrac{32}{12x + 4} \qquad \text{Fundamental property}$$

(b) $\dfrac{12p}{p^2 + 8p} = \dfrac{?}{p^3 + 4p^2 - 32p}$

Factor the denominator in each rational expression.

$$\dfrac{12p}{p(p + 8)} = \dfrac{?}{p(p + 8)(p - 4)}$$

$p^3 + 4p^2 - 32p$
$= p(p^2 + 4p - 32)$
$= p(p + 8)(p - 4)$

The factor $p - 4$ is missing, so multiply $\dfrac{12p}{p(p + 8)}$ by $\dfrac{p - 4}{p - 4}$.

$$\dfrac{12p}{p^2 + 8p} = \dfrac{12p}{p(p + 8)} \cdot \dfrac{p - 4}{p - 4} \qquad \text{Fundamental property}$$

$$= \dfrac{12p(p - 4)}{p(p + 8)(p - 4)} \qquad \begin{array}{l}\text{Multiply numerators.}\\ \text{Multiply denominators.}\end{array}$$

$$= \dfrac{12p^2 - 48p}{p^3 + 4p^2 - 32p} \qquad \text{Multiply the factors.}$$

NOW TRY

NOW TRY ANSWERS

5. (a) $\dfrac{40k}{25k - 10}$

(b) $\dfrac{2t^2 + 15t - 8}{t^3 + 12t^2 + 32t}$

NOTE   While it is beneficial to leave the denominator in factored form, we multiplied the factors in the denominator in **Example 5** to give the answer in the same form as the original problem.

7.3 Exercises

FOR EXTRA HELP   MyMathLab®

 Complete solution available in MyMathLab

Concept Check   Choose the correct response.

1. Suppose that the greatest common factor of x and y is 1. What is the least common denominator for $\dfrac{1}{x}$ and $\dfrac{1}{y}$?

A. x **B.** y **C.** xy **D.** 1

2. If x is a factor of y, what is the least common denominator for $\dfrac{1}{x}$ and $\dfrac{1}{y}$?

A. x **B.** y **C.** xy **D.** 1

3. What is the least common denominator for $\dfrac{9}{20}$ and $\dfrac{1}{2}$?

A. 40   **B.** 2   **C.** 20   **D.** None of these

4. Suppose that we wish to write the fraction $\frac{1}{(x-4)^2(y-3)}$ with denominator $(x-4)^3(y-3)^2$. By what must we multiply both the numerator and the denominator?

A. $(x-4)(y-3)$ **B.** $(x-4)^2$ **C.** $x-4$ **D.** $(x-4)^2(y-3)$

5. *Concept Check*   Find the LCD for the pair of fractions.

$$\frac{7}{10}, \frac{1}{25}$$

Step 1 $10 = 2 \cdot \underline{\quad\quad}$

$25 = \underline{\quad\quad} \cdot 5$

Step 2   The greatest number of times 2 appears is $\underline{\quad\quad}$. The greatest number of times $\underline{\quad\quad}$ appears is two.

Step 3   LCD $= \underline{\quad\quad} \cdot \underline{\quad\quad} \cdot 5$

LCD $= \underline{\quad\quad}$

6. *Concept Check*   Write the rational expression as an equivalent expression with the indicated denominator.

$$\frac{7k}{5} = \frac{?}{30p}$$

Step 1   Factor the denominator on the right.

$$\frac{7k}{5} = \frac{?}{5 \cdot \underline{\quad}}$$

Step 2   Factors of $\underline{\quad\quad}$ and $\underline{\quad\quad}$ are missing.

Step 3 $\dfrac{7k}{5} \cdot \dfrac{\overline{\underline{\quad\quad}}}{\underline{\quad\quad}} = \dfrac{\overline{\quad\quad}}{30p}$

Find the LCD for the fractions in each list. See Examples 1 and 2.

▶ 7. $\dfrac{7}{15}, \dfrac{21}{20}$

8. $\dfrac{9}{10}, \dfrac{13}{25}$

9. $\dfrac{17}{100}, \dfrac{23}{120}, \dfrac{43}{180}$

10. $\dfrac{17}{250}, \dfrac{21}{300}, \dfrac{1}{360}$

11. $\dfrac{9}{x^2}, \dfrac{8}{x^5}$

12. $\dfrac{12}{m^7}, \dfrac{14}{m^8}$

13. $\dfrac{-2}{5p}, \dfrac{13}{6p}$

14. $\dfrac{-14}{15k}, \dfrac{11}{4k}$

▶ 15. $\dfrac{17}{15y^2}, \dfrac{55}{36y^4}$

16. $\dfrac{4}{25m^3}, \dfrac{7}{10m^4}$

17. $\dfrac{5}{21r^3}, \dfrac{7}{12r^5}$

18. $\dfrac{6}{35t^2}, \dfrac{5}{49t^6}$

19. $\dfrac{13}{5a^2b^3}, \dfrac{29}{15a^5b}$

20. $\dfrac{7}{3r^4s^5}, \dfrac{23}{9r^6s^8}$

21. $\dfrac{1}{r^2t^3}, \dfrac{1}{r^5t}, \dfrac{1}{r^9t^2}$

22. $\dfrac{5}{x^8y^4}, \dfrac{5}{x^9y^3}, \dfrac{5}{xy^2}$

23. $\dfrac{7}{x+1}, \dfrac{9}{x-1}$

24. $\dfrac{3}{y+3}, \dfrac{2}{y-3}$

Find the LCD for the fractions in each list. See Example 3.

▶ 25. $\dfrac{7}{6p}, \dfrac{15}{4p-8}$

26. $\dfrac{7}{8k}, \dfrac{28}{12k-24}$

27. $\dfrac{9}{28m^2}, \dfrac{3}{12m-20}$

28. $\dfrac{14}{27a^3}, \dfrac{7}{9a-45}$

29. $\dfrac{7}{5b-10}, \dfrac{11}{6b-12}$

30. $\dfrac{3}{7x^2+21x}, \dfrac{2}{5x^2+15x}$

31. $\dfrac{37}{6r-12}, \dfrac{25}{9r-18}$

32. $\dfrac{14}{5p-30}, \dfrac{11}{6p-36}$

33. $\dfrac{5}{c-d}, \dfrac{8}{d-c}$

34. $\dfrac{4}{y-x}, \dfrac{8}{x-y}$

35. $\dfrac{12}{m-3}, \dfrac{-4}{3-m}$

36. $\dfrac{3}{a-8}, \dfrac{-17}{8-a}$

37. $\dfrac{29}{p-q}, \dfrac{18}{q-p}$

38. $\dfrac{16}{z-x}, \dfrac{9}{x-z}$

39. $\dfrac{13}{x^2-1}, \dfrac{-5}{2x+2}$

40. $\dfrac{9}{y^2-9}, \dfrac{-2}{2y+6}$

41. $\dfrac{4x^2}{(x-4)^2}, \dfrac{17x}{3x-12}$

42. $\dfrac{3y^2}{(y+6)^2}, \dfrac{5y}{2y+12}$

43. $\dfrac{5}{12p + 60}, \dfrac{-17}{p^2 + 5p}, \dfrac{-16}{p^2 + 10p + 25}$

44. $\dfrac{13}{r^2 + 7r}, \dfrac{-3}{5r + 35}, \dfrac{-4}{r^2 + 14r + 49}$

45. $\dfrac{-3}{8y + 16}, \dfrac{-22}{y^2 + 3y + 2}$

46. $\dfrac{-2}{9m - 18}, \dfrac{-6}{m^2 - 7m + 10}$

47. $\dfrac{3}{k^2 + 5k}, \dfrac{2}{k^2 + 3k - 10}$

48. $\dfrac{1}{z^2 - 4z}, \dfrac{9}{z^2 - 3z - 4}$

49. $\dfrac{6}{a^2 + 6a}, \dfrac{-5}{a^2 + 3a - 18}$

50. $\dfrac{8}{y^2 - 5y}, \dfrac{-5}{y^2 - 2y - 15}$

51. $\dfrac{5}{p^2 + 8p + 15}, \dfrac{3}{p^2 - 3p - 18}, \dfrac{12}{p^2 - p - 30}$

52. $\dfrac{10}{y^2 - 10y + 21}, \dfrac{2}{y^2 - 2y - 3}, \dfrac{15}{y^2 - 6y - 7}$

53. $\dfrac{-5}{k^2 + 2k - 35}, \dfrac{-8}{k^2 + 3k - 40}, \dfrac{19}{k^2 - 2k - 15}$

54. $\dfrac{-19}{z^2 + 4z - 12}, \dfrac{-16}{z^2 + z - 30}, \dfrac{16}{z^2 + 2z - 24}$

*Write each rational expression with the indicated denominator. **See Examples 4 and 5.***

▶ **55.** $\dfrac{4}{11} = \dfrac{?}{55}$

56. $\dfrac{8}{7} = \dfrac{?}{42}$

57. $\dfrac{-5}{k} = \dfrac{?}{9k}$

58. $\dfrac{-4}{q} = \dfrac{?}{6q}$

59. $\dfrac{15m^2}{8k} = \dfrac{?}{32k^4}$

60. $\dfrac{7t^2}{3y} = \dfrac{?}{9y^2}$

▶ **61.** $\dfrac{19z}{2z - 6} = \dfrac{?}{6z - 18}$

62. $\dfrac{3r}{5r - 5} = \dfrac{?}{15r - 15}$

63. $\dfrac{-2a}{9a - 18} = \dfrac{?}{18a - 36}$

64. $\dfrac{-7y}{6y + 18} = \dfrac{?}{24y + 72}$

65. $\dfrac{6}{k^2 - 4k} = \dfrac{?}{k(k - 4)(k + 1)}$

66. $\dfrac{25}{m^2 - 9m} = \dfrac{?}{m(m - 9)(m + 8)}$

67. $\dfrac{4r - t}{r^2 + rt + t^2} = \dfrac{?}{t^3 - r^3}$

68. $\dfrac{3x - 1}{x^2 + 2x + 4} = \dfrac{?}{x^3 - 8}$

69. $\dfrac{2(z - y)}{y^2 + yz + z^2} = \dfrac{?}{y^4 - z^3 y}$

70. $\dfrac{2p + 3q}{p^2 + 2pq + q^2} = \dfrac{?}{(p + q)(p^3 + q^3)}$

Extending Skills *Write each rational expression with the indicated denominator. **See Examples 4 and 5.***

71. $\dfrac{36r}{r^2 - r - 6} = \dfrac{?}{(r - 3)(r + 2)(r + 1)}$

72. $\dfrac{4m}{m^2 + m - 2} = \dfrac{?}{(m - 1)(m - 3)(m + 2)}$

73. $\dfrac{a + 2b}{2a^2 + ab - b^2} = \dfrac{?}{2a^3 b + a^2 b^2 - ab^3}$

74. $\dfrac{m - 4}{6m^2 + 7m - 3} = \dfrac{?}{12m^3 + 14m^2 - 6m}$

Work Exercises 75–80 in order.

75. Suppose that we want to write $\frac{3}{4}$ as an equivalent fraction with denominator 28. By what number must we multiply both the numerator and the denominator?

76. If we write $\frac{3}{4}$ as an equivalent fraction with denominator 28, by what number are we actually multiplying the fraction?

77. What property of multiplication is being used when we write a common fraction as an equivalent one with a larger denominator? (See **Section 1.6.**)

78. Suppose that we want to write $\frac{2x+5}{x-4}$ as an equivalent fraction with denominator $7x - 28$. By what number must we multiply both the numerator and the denominator?

79. If we write $\frac{2x+5}{x-4}$ as an equivalent fraction with denominator $7x - 28$, by what number are we actually multiplying the fraction?

80. Repeat **Exercise 77,** changing "a common" to "an algebraic."

7.4 Adding and Subtracting Rational Expressions

OBJECTIVES

1 Add rational expressions having the same denominator.

2 Add rational expressions having different denominators.

3 Subtract rational expressions.

OBJECTIVE 1 Add rational expressions having the same denominator.

We find the sum of two rational expressions with the same denominator using the same procedure that we used in **Section R.1** for adding two common fractions.

Adding Rational Expressions (Same Denominator)

The rational expressions $\frac{P}{Q}$ and $\frac{R}{Q}$ (where $Q \neq 0$) are added as follows.

$$\frac{P}{Q} + \frac{R}{Q} = \frac{P+R}{Q}$$

That is, to add rational expressions with the same denominator, add the numerators and keep the same denominator.

NOW TRY EXERCISE 1

Add. Write each answer in lowest terms.

(a) $\dfrac{2}{7k} + \dfrac{4}{7k}$

(b) $\dfrac{4y}{y+3} + \dfrac{12}{y+3}$

EXAMPLE 1   Adding Rational Expressions (Same Denominator)

Add. Write each answer in lowest terms.

(a) $\dfrac{4}{9} + \dfrac{2}{9}$ | **(b)** $\dfrac{3x}{x+1} + \dfrac{3}{x+1}$

The denominators are the same, so the sum is found by adding the two numerators and keeping the same (common) denominator.

$= \dfrac{4+2}{9}$ Add.

$= \dfrac{6}{9}$

$= \dfrac{2 \cdot 3}{3 \cdot 3}$ Factor.

$= \dfrac{2}{3}$ Lowest terms

$= \dfrac{3x+3}{x+1}$ Add.

$= \dfrac{3(x+1)}{x+1}$ Factor.

$= 3$ Lowest terms

NOW TRY ANSWERS

1. (a) $\frac{6}{7k}$ (b) 4

NOW TRY

OBJECTIVE 2 Add rational expressions having different denominators.

As in **Section R.1,** we use the following steps to add fractions having different denominators.

Adding Rational Expressions (Different Denominators)

Step 1 **Find the least common denominator (LCD).**

Step 2 **Write each rational expression** as an equivalent rational expression with the LCD as the denominator.

Step 3 **Add** the numerators to obtain the numerator of the sum. The LCD is the denominator of the sum.

Step 4 **Write in lowest terms** using the fundamental property.

NOW TRY EXERCISE 2

Add. Write each answer in lowest terms.

(a) $\dfrac{5}{12} + \dfrac{3}{20}$ **(b)** $\dfrac{3}{5x} + \dfrac{2}{7x}$

EXAMPLE 2 Adding Rational Expressions (Different Denominators)

Add. Write each answer in lowest terms.

(a) $\dfrac{1}{12} + \dfrac{7}{15}$

(b) $\dfrac{2}{3y} + \dfrac{1}{4y}$

Step 1 Find the LCD, using the methods of the previous section.

$$12 = 2 \cdot 2 \cdot 3 = 2^2 \cdot 3$$
$$15 = 3 \cdot 5$$
$$\text{LCD} = 2^2 \cdot 3 \cdot 5 = 60$$

$$3y = 3 \cdot y$$
$$4y = 2 \cdot 2 \cdot y = 2^2 \cdot y$$
$$\text{LCD} = 2^2 \cdot 3 \cdot y = 12y$$

Step 2 Write each rational expression as a fraction with the LCD (60 and 12y, respectively) as the denominator.

$$\dfrac{1}{12} + \dfrac{7}{15} \qquad \text{The LCD is 60.}$$

$$= \dfrac{1(5)}{12(5)} + \dfrac{7(4)}{15(4)}$$

$$= \dfrac{5}{60} + \dfrac{28}{60}$$

$$\dfrac{2}{3y} + \dfrac{1}{4y} \qquad \text{The LCD is 12}y.$$

$$= \dfrac{2(4)}{3y(4)} + \dfrac{1(3)}{4y(3)}$$

$$= \dfrac{8}{12y} + \dfrac{3}{12y}$$

Step 3 Add the numerators. The LCD is the denominator.

Step 4 Write in lowest terms if necessary.

$$= \dfrac{5 + 28}{60}$$

$$= \dfrac{33}{60}$$

$$= \dfrac{11}{20}$$

$$= \dfrac{8 + 3}{12y}$$

$$= \dfrac{11}{12y}$$

NOW TRY

**NOW TRY
EXERCISE 3**

Add. Write the answer in lowest terms.

$$\frac{6t}{t^2 - 9} + \frac{-3}{t + 3}$$

EXAMPLE 3   Adding Rational Expressions

Add. Write the answer in lowest terms.

$$\frac{2x}{x^2 - 1} + \frac{-1}{x + 1}$$

Step 1   The denominators are different, so find the LCD.

$$\left.\begin{array}{l} x^2 - 1 = (x + 1)(x - 1) \\ x + 1 \text{ is prime.} \end{array}\right\} \quad \text{The LCD is } (x + 1)(x - 1).$$

Step 2   Write each rational expression with the LCD as the denominator.

$$\frac{2x}{x^2 - 1} + \frac{-1}{x + 1} \qquad \text{LCD} = (x + 1)(x - 1)$$

$$= \frac{2x}{(x + 1)(x - 1)} + \frac{-1(x - 1)}{(x + 1)(x - 1)} \qquad \begin{array}{l}\text{Multiply the second} \\ \text{fraction by } \frac{x - 1}{x - 1}.\end{array}$$

$$= \frac{2x}{(x + 1)(x - 1)} + \frac{-x + 1}{(x + 1)(x - 1)} \qquad \text{Distributive property}$$

Step 3
$$= \frac{2x - x + 1}{(x + 1)(x - 1)} \qquad \begin{array}{l}\text{Add numerators.} \\ \text{Keep the same denominator.}\end{array}$$

$$= \frac{x + 1}{(x + 1)(x - 1)} \qquad \text{Combine like terms.}$$

Step 4
$$= \frac{1(x + 1)}{(x + 1)(x - 1)} \qquad \begin{array}{l}\text{Identity property of} \\ \text{multiplication}\end{array}$$

Remember to write 1 in the numerator.
$$= \frac{1}{x - 1} \qquad \text{Lowest terms} \qquad \textbf{NOW TRY} \circlearrowleft$$

**NOW TRY
EXERCISE 4**

Add. Write the answer in lowest terms.

$$\frac{x - 1}{x^2 + 6x + 8} + \frac{4x}{x^2 + x - 12}$$

EXAMPLE 4   Adding Rational Expressions

Add. Write the answer in lowest terms.

$$\frac{2x}{x^2 + 5x + 6} + \frac{x + 1}{x^2 + 2x - 3}$$

$$= \frac{2x}{(x + 2)(x + 3)} + \frac{x + 1}{(x + 3)(x - 1)} \qquad \text{Factor the denominators.}$$

$$= \frac{2x(x - 1)}{(x + 2)(x + 3)(x - 1)} + \frac{(x + 1)(x + 2)}{(x + 2)(x + 3)(x - 1)} \qquad \begin{array}{l}\text{The LCD is} \\ (x + 2)(x + 3)(x - 1).\end{array}$$

$$= \frac{2x(x - 1) + (x + 1)(x + 2)}{(x + 2)(x + 3)(x - 1)} \qquad \begin{array}{l}\text{Add numerators.} \\ \text{Keep the same denominator.}\end{array}$$

$$= \frac{2x^2 - 2x + x^2 + 3x + 2}{(x + 2)(x + 3)(x - 1)} \qquad \text{Multiply.}$$

$$= \frac{3x^2 + x + 2}{(x + 2)(x + 3)(x - 1)} \qquad \text{Combine like terms.}$$

The numerator cannot be factored here, so the expression is in lowest terms.

NOW TRY ↻

NOW TRY ANSWERS

3. $\dfrac{3}{t - 3}$

4. $\dfrac{5x^2 + 4x + 3}{(x + 4)(x + 2)(x - 3)}$

NOTE If the final expression in **Example 4** could be written in lower terms, the numerator would have a factor of $x + 2$, $x + 3$, or $x - 1$. Therefore, it is only necessary to check for possible factored forms of the numerator that would contain one of these binomials.

NOW TRY EXERCISE 5

Add. Write the answer in lowest terms.

$$\frac{2k}{k - 7} + \frac{5}{7 - k}$$

EXAMPLE 5 Adding Rational Expressions (Denominators Are Opposites)

Add. Write the answer in lowest terms.

$$\frac{y}{y - 2} + \frac{8}{2 - y}$$

The denominators are opposites. Use the process of multiplying one of the fractions by 1 in the form $\frac{-1}{-1}$ to obtain the same denominator for both fractions.

$$= \frac{y}{y - 2} + \frac{8(-1)}{(2 - y)(-1)} \qquad \text{Multiply } \frac{8}{2 - y} \text{ by } \frac{-1}{-1} \text{ to find a common denominator.}$$

$$= \frac{y}{y - 2} + \frac{-8}{-2 + y} \qquad \text{Distributive property}$$

$$= \frac{y}{y - 2} + \frac{-8}{y - 2} \qquad \text{Rewrite } -2 + y \text{ as } y - 2.$$

$$= \frac{y - 8}{y - 2} \qquad \text{Add numerators.} \\ \text{Keep the same denominator.}$$

NOW TRY

NOTE If we had chosen to use $2 - y$ as the common denominator, the final answer would be in the form $\frac{8 - y}{2 - y}$, which is equivalent to $\frac{y - 8}{y - 2}$.

$$\frac{y}{y - 2} + \frac{8}{2 - y} \qquad \text{See **Example 5**.}$$

$$= \frac{y(-1)}{(y - 2)(-1)} + \frac{8}{2 - y} \qquad \text{Multiply } \frac{y}{y - 2} \text{ by } \frac{-1}{-1}.$$

$$= \frac{-y}{2 - y} + \frac{8}{2 - y} \qquad \text{In the first denominator,} \\ (y - 2)(-1) = -y + 2 = 2 - y.$$

$$= \frac{-y + 8}{2 - y}, \quad \text{or} \quad \frac{8 - y}{2 - y} \qquad \text{Add numerators.} \\ \text{Keep the same denominator.}$$

OBJECTIVE 3 Subtract rational expressions.

Subtracting Rational Expressions (Same Denominator)

The rational expressions $\frac{P}{Q}$ and $\frac{R}{Q}$ (where $Q \neq 0$) are subtracted as follows.

$$\frac{P}{Q} - \frac{R}{Q} = \frac{P - R}{Q}$$

That is, to subtract rational expressions with the same denominator, subtract the numerators and keep the same denominator.

NOW TRY ANSWER

5. $\frac{2k - 5}{k - 7}$, or $\frac{5 - 2k}{7 - k}$

**NOW TRY
EXERCISE 6**

Subtract. Write the answer in lowest terms.

$$\frac{2x}{x+5} - \frac{x+1}{x+5}$$

EXAMPLE 6 Subtracting Rational Expressions (Same Denominator)

Subtract. Write the answer in lowest terms.

$$\frac{2m}{m-1} - \frac{m+3}{m-1}$$

> Use parentheses around the numerator of the subtrahend.

$$= \frac{2m - (m+3)}{m-1}$$ Subtract numerators.
Keep the same denominator.

> Be careful with signs.

$$= \frac{2m - m - 3}{m-1}$$ Distributive property

$$= \frac{m-3}{m-1}$$ Combine like terms.   **NOW TRY**

> **! CAUTION** Sign errors often occur in subtraction problems like the one in **Example 6.** The numerator of the fraction being subtracted must be treated as a single quantity. *Be sure to use parentheses after the subtraction symbol.*

**NOW TRY
EXERCISE 7**

Subtract. Write the answer in lowest terms.

$$\frac{6}{y-6} - \frac{2}{y}$$

EXAMPLE 7 Subtracting Rational Expressions (Different Denominators)

Subtract. Write the answer in lowest terms.

$$\frac{9}{x-2} - \frac{3}{x}$$ The LCD is $x(x-2)$.

$$= \frac{9x}{x(x-2)} - \frac{3(x-2)}{x(x-2)}$$ Write each expression with the LCD.

$$= \frac{9x - 3(x-2)}{x(x-2)}$$ Subtract numerators.
Keep the same denominator.

> Be careful with signs.

$$= \frac{9x - 3x + 6}{x(x-2)}$$ Distributive property

$$= \frac{6x + 6}{x(x-2)}$$ Combine like terms.

> These answers are equivalent.

$$= \frac{6(x+1)}{x(x-2)}$$ Factor the numerator.   **NOW TRY**

> **NOTE** We factored the final numerator in **Example 7** to obtain $\frac{6(x+1)}{x(x-2)}$. The fundamental property does not apply, however, because there are no common factors to divide out. The answer is in lowest terms.

EXAMPLE 8 Subtracting Rational Expressions (Denominators Are Opposites)

Subtract. Write the answer in lowest terms.

$$\frac{3x}{x-5} - \frac{2x-25}{5-x}$$ The denominators are opposites. We choose $x-5$ as the common denominator.

$$= \frac{3x}{x-5} - \frac{(2x-25)(-1)}{(5-x)(-1)}$$ Multiply $\frac{2x-25}{5-x}$ by $\frac{-1}{-1}$ to obtain a common denominator.

NOW TRY ANSWERS

6. $\frac{x-1}{x+5}$

7. $\frac{4(y+3)}{y(y-6)}$

NOW TRY
EXERCISE 8

Subtract. Write the answer in lowest terms.

$$\frac{2m}{m-4} - \frac{m-12}{4-m}$$

$$= \frac{3x}{x-5} - \frac{-2x+25}{x-5} \qquad \begin{array}{l}(2x-25)(-1) = -2x+25; \\ (5-x)(-1) = -5+x = x-5\end{array}$$

> Subtract the *entire* numerator. Use parentheses to show this.

$$= \frac{3x - (-2x+25)}{x-5}$$ Subtract numerators.

> Be careful with signs.

$$= \frac{3x + 2x - 25}{x-5}$$ Distributive property

$$= \frac{5x - 25}{x-5}$$ Combine like terms.

$$= \frac{5(x-5)}{x-5}$$ Factor.

$$= 5$$ Lowest terms **NOW TRY**

NOW TRY
EXERCISE 9

Subtract. Write the answer in lowest terms.

$$\frac{5}{t^2 - 6t + 9} - \frac{2t}{t^2 - 9}$$

EXAMPLE 9 Subtracting Rational Expressions

Subtract. Write the answer in lowest terms.

$$\frac{6x}{x^2 - 2x + 1} - \frac{1}{x^2 - 1}$$

$$= \frac{6x}{(x-1)^2} - \frac{1}{(x-1)(x+1)} \qquad \begin{array}{l}\text{Factor the denominators.} \\ \text{LCD} = (x-1)(x-1)(x+1), \\ \text{or } (x-1)^2(x+1)\end{array}$$

$$= \frac{6x(x+1)}{(x-1)^2(x+1)} - \frac{1(x-1)}{(x-1)(x-1)(x+1)}$$ Fundamental property

$$= \frac{6x(x+1) - 1(x-1)}{(x-1)^2(x+1)}$$ Subtract numerators.

$$= \frac{6x^2 + 6x - x + 1}{(x-1)^2(x+1)}$$ Distributive property

$$= \frac{6x^2 + 5x + 1}{(x-1)^2(x+1)}$$ Combine like terms.

$$= \frac{(2x+1)(3x+1)}{(x-1)^2(x+1)}$$ Factor the numerator.

NOW TRY

EXAMPLE 10 Subtracting Rational Expressions

Subtract. Write the answer in lowest terms.

$$\frac{q}{q^2 - 4q - 5} - \frac{3}{2q^2 - 13q + 15}$$

$$= \frac{q}{(q+1)(q-5)} - \frac{3}{(q-5)(2q-3)} \qquad \begin{array}{l}\text{Factor the denominators.} \\ \text{LCD} = (q+1)(q-5)(2q-3)\end{array}$$

$$= \frac{q(2q-3)}{(q+1)(q-5)(2q-3)} - \frac{3(q+1)}{(q+1)(q-5)(2q-3)}$$

Fundamental property

NOW TRY ANSWERS

8. 3

9. $\dfrac{-2t^2 + 11t + 15}{(t-3)^2(t+3)}$

NOW TRY
EXERCISE 10

Subtract. Write the answer in lowest terms.

$$\frac{q}{2q^2 + 5q - 3} - \frac{3q + 4}{3q^2 + 10q + 3}$$

$$= \frac{q(2q - 3) - 3(q + 1)}{(q + 1)(q - 5)(2q - 3)} \qquad \text{Subtract numerators.}$$

$$= \frac{2q^2 - 3q - 3q - 3}{(q + 1)(q - 5)(2q - 3)} \qquad \text{Distributive property}$$

$$= \frac{2q^2 - 6q - 3}{(q + 1)(q - 5)(2q - 3)} \qquad \text{Combine like terms.}$$

NOW TRY ANSWER

10. $\dfrac{-3q^2 - 4q + 4}{(2q - 1)(q + 3)(3q + 1)}$

The numerator cannot be factored, so the final answer is in lowest terms.

NOW TRY

7.4 Exercises

FOR EXTRA HELP   ▶   MyMathLab®

▶ *Complete solution available in MyMathLab*

Concept Check   *Match each expression in Column I with the correct sum or difference in Column II.*

I

1. $\dfrac{x}{x + 8} + \dfrac{8}{x + 8}$

2. $\dfrac{2x}{x - 8} - \dfrac{16}{x - 8}$

3. $\dfrac{8}{x - 8} - \dfrac{x}{x - 8}$

4. $\dfrac{8}{x + 8} - \dfrac{x}{x + 8}$

5. $\dfrac{x}{x + 8} - \dfrac{8}{x + 8}$

6. $\dfrac{1}{x} + \dfrac{1}{8}$

7. $\dfrac{1}{8} - \dfrac{1}{x}$

8. $\dfrac{1}{8x} - \dfrac{1}{8x}$

II

A. 2

B. $\dfrac{x - 8}{x + 8}$

C. -1

D. $\dfrac{8 + x}{8x}$

E. 1

F. 0

G. $\dfrac{x - 8}{8x}$

H. $\dfrac{8 - x}{x + 8}$

Concept Check   *Add or subtract. Write each answer in lowest terms.*

9. $\dfrac{6}{5x} + \dfrac{9}{2x}$

$$= \frac{6(\underline{\quad})}{5x(\underline{\quad})} + \frac{9(\underline{\quad})}{2x(\underline{\quad})}$$

$$= \frac{12 + 45}{\underline{\quad}}$$

$$= \underline{\quad}$$

10. $\dfrac{x}{2x + 3} - \dfrac{3x + 4}{2x + 3}$

$$= \frac{x - (\underline{\quad})}{2x + 3}$$

$$= \frac{x - \underline{\quad} - \underline{\quad}}{2x + 3}$$

$$= \frac{\underline{\quad}}{2x + 3}$$

If -2 is factored from the numerator, the answer can be written as _____.

Note: When adding and subtracting rational expressions, several different equivalent forms of the answer often exist. If your answer does not look exactly like the one given in the back of the book, check to see whether you have written an equivalent form.

*Add or subtract. Write each answer in lowest terms. **See Examples 1 and 6.***

▶ **11.** $\dfrac{4}{m} + \dfrac{7}{m}$

12. $\dfrac{5}{p} + \dfrac{12}{p}$

13. $\dfrac{5}{y + 4} - \dfrac{1}{y + 4}$

14. $\dfrac{6}{t + 3} - \dfrac{3}{t + 3}$

15. $\dfrac{x}{x + y} + \dfrac{y}{x + y}$

16. $\dfrac{a}{a + b} + \dfrac{b}{a + b}$

17. $\dfrac{5m}{m+1} - \dfrac{1+4m}{m+1}$

18. $\dfrac{4x}{x+2} - \dfrac{2+3x}{x+2}$

19. $\dfrac{a+b}{2} - \dfrac{a-b}{2}$

20. $\dfrac{x-y}{2} - \dfrac{x+y}{2}$

21. $\dfrac{x^2}{x+5} + \dfrac{5x}{x+5}$

22. $\dfrac{t^2}{t-3} + \dfrac{-3t}{t-3}$

23. $\dfrac{y^2-3y}{y+3} + \dfrac{-18}{y+3}$

24. $\dfrac{r^2-8r}{r-5} + \dfrac{15}{r-5}$

25. $\dfrac{x}{x^2-9} - \dfrac{-3}{x^2-9}$

26. $\dfrac{-4}{y^2-16} - \dfrac{-y}{y^2-16}$

27. $\dfrac{y^2+x^2}{x^2-y^2} - \dfrac{2x^2}{x^2-y^2}$

28. $\dfrac{3a^2+b^2}{a^2-b^2} - \dfrac{4a^2}{a^2-b^2}$

*Add or subtract. Write each answer in lowest terms. **See Examples 2, 3, 4, and 7.***

29. $\dfrac{z}{5} + \dfrac{1}{3}$

30. $\dfrac{p}{8} + \dfrac{4}{5}$

31. $\dfrac{5}{7} - \dfrac{r}{2}$

32. $\dfrac{20}{9} - \dfrac{z}{3}$

33. $-\dfrac{3}{4} - \dfrac{1}{2x}$

34. $-\dfrac{7}{8} - \dfrac{3}{2a}$

35. $\dfrac{7}{4t} + \dfrac{3}{7t}$

36. $\dfrac{8}{3r} + \dfrac{2}{5r}$

37. $\dfrac{x+1}{6} + \dfrac{3x+3}{9}$

38. $\dfrac{2x-6}{4} + \dfrac{x+5}{6}$

39. $\dfrac{x+3}{3x} + \dfrac{2x+2}{4x}$

40. $\dfrac{x+2}{5x} + \dfrac{6x+3}{3x}$

41. $\dfrac{7}{3p^2} - \dfrac{2}{p}$

42. $\dfrac{12}{5m^2} - \dfrac{5}{m}$

43. $\dfrac{1}{k+4} - \dfrac{2}{k}$

44. $\dfrac{3}{m+1} - \dfrac{4}{m}$

45. $\dfrac{x}{x-2} + \dfrac{-8}{x^2-4}$

46. $\dfrac{2x}{x-1} + \dfrac{-4}{x^2-1}$

47. $\dfrac{4m}{m^2+3m+2} + \dfrac{2m-1}{m^2+6m+5}$

48. $\dfrac{a}{a^2+3a-4} + \dfrac{4a}{a^2+7a+12}$

49. $\dfrac{4y}{y^2-1} - \dfrac{5}{y^2+2y+1}$

50. $\dfrac{2x}{x^2-16} - \dfrac{3}{x^2+8x+16}$

51. $\dfrac{t}{t+2} + \dfrac{5-t}{t} - \dfrac{4}{t^2+2t}$

52. $\dfrac{2p}{p-3} + \dfrac{2+p}{p} - \dfrac{-6}{p^2-3p}$

Concept Check   *Answer each question.*

53. What are the *two* possible LCDs that could be used for the sum $\dfrac{10}{m-2} + \dfrac{5}{2-m}$?

54. If one form of the correct answer to a sum or difference of rational expressions is $\dfrac{4}{k-3}$, what would an alternative form of the answer be if the denominator is $3-k$?

*Add or subtract. Write each answer in lowest terms. **See Examples 5 and 8.***

55. $\dfrac{4}{x-5} + \dfrac{6}{5-x}$

56. $\dfrac{10}{m-2} + \dfrac{5}{2-m}$

57. $\dfrac{-1}{1-y} - \dfrac{4y-3}{y-1}$

58. $\dfrac{-4}{p-3} - \dfrac{p+1}{3-p}$

59. $\dfrac{2}{x-y^2} + \dfrac{7}{y^2-x}$

60. $\dfrac{-8}{p-q^2} + \dfrac{3}{q^2-p}$

61. $\dfrac{x}{5x-3y} - \dfrac{y}{3y-5x}$

62. $\dfrac{t}{8t-9s} - \dfrac{s}{9s-8t}$

63. $\dfrac{3}{4p-5} + \dfrac{9}{5-4p}$

64. $\dfrac{8}{3-7y} - \dfrac{2}{7y-3}$

65. $\dfrac{15x}{5x-7} - \dfrac{-21}{7-5x}$

66. $\dfrac{24y}{6y-5} - \dfrac{-20}{5-6y}$

*In these subtraction problems, the rational expression that follows the subtraction sign has a numerator with more than one term. **Be careful with signs** and find each difference. **See Example 9.***

67. $\dfrac{2m}{m-n} - \dfrac{5m+n}{2m-2n}$

68. $\dfrac{5p}{p-q} - \dfrac{3p+1}{4p-4q}$

▶ 69. $\dfrac{5}{x^2-9} - \dfrac{x+2}{x^2+4x+3}$

70. $\dfrac{1}{a^2-1} - \dfrac{a-1}{a^2+3a-4}$

71. $\dfrac{2q+1}{3q^2+10q-8} - \dfrac{3q+5}{2q^2+5q-12}$

72. $\dfrac{4y-1}{2y^2+5y-3} - \dfrac{y+3}{6y^2+y-2}$

*Perform each indicated operation. **See Examples 1–10.***

73. $\dfrac{y^2}{y-2} - \dfrac{9y-14}{y-2}$

74. $\dfrac{y^2}{y-4} - \dfrac{y+12}{y-4}$

75. $\dfrac{3}{x+4} + 7$

76. $\dfrac{9}{x+7} + 2$

77. $\dfrac{-x+2}{x} - \dfrac{x-5}{4x}$

78. $\dfrac{-y+3}{y} - \dfrac{y+4}{3y}$

79. $\dfrac{5x}{x-7} - \dfrac{3x}{x-3}$

80. $\dfrac{6t}{t+4} - \dfrac{2t}{t+1}$

81. $\dfrac{5a}{3a-6} - \dfrac{a-7}{a-2}$

82. $\dfrac{4a}{5a-15} - \dfrac{a-1}{a-3}$

83. $\dfrac{4}{3-x} + \dfrac{x}{2x-6}$

84. $\dfrac{5}{4-x} + \dfrac{x}{2x-8}$

85. $\dfrac{5x+11}{x^2-11x+18} - \dfrac{4x+20}{x^2-11x+18}$

86. $\dfrac{4x+7}{x^2+2x-3} - \dfrac{3x+4}{x^2+2x-3}$

87. $\dfrac{4}{r^2-r} + \dfrac{6}{r^2+2r} - \dfrac{1}{r^2+r-2}$

88. $\dfrac{6}{k^2+3k} - \dfrac{1}{k^2-k} + \dfrac{2}{k^2+2k-3}$

89. $\dfrac{x+3y}{x^2+2xy+y^2} + \dfrac{x-y}{x^2+4xy+3y^2}$

90. $\dfrac{m}{m^2-1} + \dfrac{m-1}{m^2+2m+1}$

91. $\dfrac{r+y}{18r^2+9ry-2y^2} + \dfrac{3r-y}{36r^2-y^2}$

92. $\dfrac{2x-z}{2x^2+xz-10z^2} - \dfrac{x+z}{x^2-4z^2}$

Work each problem.

93. Refer to the rectangle in the figure.

   (a) Find an expression that represents its perimeter. Give the simplified form.

   (b) Find an expression that represents its area. Give the simplified form.

94. Refer to the triangle in the figure. Find an expression that represents its perimeter. Give the simplified form.

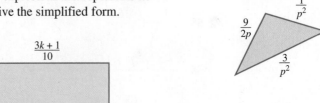

A ***concours d'elegance*** *is a competition in which a maximum of* 100 *points is awarded to a car based on its general attractiveness. The rational expression*

$$\frac{1010}{49(101 - x)} - \frac{10}{49}$$

approximates the cost, in thousands of dollars, of restoring a car so that it will win x points.
Use this information to work Exercises 95 and 96.

95. Simplify the given expression by performing the indicated subtraction.

96. Use the simplified expression from **Exercise 95** to determine how much it would cost to win 95 points.

7.5 Complex Fractions

VOCABULARY

☐ complex fraction

OBJECTIVE 1 Define and recognize a complex fraction.

The quotient of two mixed numbers in arithmetic, such as $2\frac{1}{2} \div 3\frac{1}{4}$, can be written as a fraction.

$$2\frac{1}{2} \div 3\frac{1}{4}$$

$$= \frac{2\frac{1}{2}}{3\frac{1}{4}}$$

$$= \frac{2 + \frac{1}{2}}{3 + \frac{1}{4}}$$

> We do this to illustrate a *complex fraction.*

Some rational expressions in algebra have fractions in the numerator, or denominator, or both.

> ### Complex Fraction
>
> A quotient with one or more fractions in the numerator, or denominator, or both, is a **complex fraction.**
>
> *Examples:* $\dfrac{2 + \dfrac{1}{2}}{3 + \dfrac{1}{4}}$, $\dfrac{\dfrac{3x^2 - 5x}{6x^2}}{2x - \dfrac{1}{x}}$, and $\dfrac{3 + x}{5 - \dfrac{2}{x}}$ Complex fractions

The parts of a complex fraction are named as follows.

$$\left.\frac{\dfrac{2}{p} - \dfrac{1}{q}}{\dfrac{3}{p} + \dfrac{5}{q}}\right.$$

$\left.\dfrac{2}{p} - \dfrac{1}{q}\right\}$ ← Numerator of complex fraction

← Main fraction bar

$\left.\dfrac{3}{p} + \dfrac{5}{q}\right\}$ ← Denominator of complex fraction

OBJECTIVE 2 Simplify a complex fraction by writing it as a division problem (Method 1).

Because the main fraction bar represents division in a complex fraction, one method of simplifying a complex fraction involves division.

Simplifying a Complex Fraction (Method 1)

Step 1   Write both the numerator and denominator as single fractions.

Step 2   Change the complex fraction to a division problem.

Step 3   Perform the indicated division.

EXAMPLE 1   Simplifying Complex Fractions (Method 1)

Simplify each complex fraction.

(a) $\dfrac{\dfrac{2}{3}+\dfrac{5}{9}}{\dfrac{1}{4}+\dfrac{1}{12}}$

(b) $\dfrac{6+\dfrac{3}{x}}{\dfrac{x}{4}+\dfrac{1}{8}}$

Step 1   First, write each numerator as a single fraction.

$$\frac{2}{3}+\frac{5}{9}$$

$$=\frac{2(3)}{3(3)}+\frac{5}{9}$$

$$=\frac{6}{9}+\frac{5}{9}$$

$$=\frac{11}{9}$$

$$6+\frac{3}{x}$$

$$=\frac{6}{1}+\frac{3}{x}$$

$$=\frac{6x}{x}+\frac{3}{x}$$

$$=\frac{6x+3}{x}$$

Repeat the process for each denominator.

$$\frac{1}{4}+\frac{1}{12}$$

$$=\frac{1(3)}{4(3)}+\frac{1}{12}$$

$$=\frac{3}{12}+\frac{1}{12}$$

$$=\frac{4}{12}$$

$$\frac{x}{4}+\frac{1}{8}$$

$$=\frac{x(2)}{4(2)}+\frac{1}{8}$$

$$=\frac{2x}{8}+\frac{1}{8}$$

$$=\frac{2x+1}{8}$$

Step 2   Write the equivalent complex fraction as a division problem.

$$\frac{\dfrac{11}{9}}{\dfrac{4}{12}}$$

$$\frac{\dfrac{6x+3}{x}}{\dfrac{2x+1}{8}}$$

$$=\frac{11}{9}\div\frac{4}{12}$$

$$=\frac{6x+3}{x}\div\frac{2x+1}{8}$$

**NOW TRY
EXERCISE 1**

Simplify each complex
fraction.

(a) $\dfrac{\dfrac{2}{5}+\dfrac{1}{4}}{\dfrac{1}{6}+\dfrac{3}{8}}$ **(b)** $\dfrac{2+\dfrac{4}{x}}{\dfrac{5}{6}+\dfrac{5x}{12}}$

Step 3 Now use the definition of division and multiply by the reciprocal. Then write in lowest terms using the fundamental property.

$$= \frac{11}{9} \cdot \frac{12}{4} \qquad\qquad = \frac{6x+3}{x} \cdot \frac{8}{2x+1}$$

$$= \frac{11 \cdot 3 \cdot 4}{3 \cdot 3 \cdot 4} \qquad\qquad = \frac{3(2x+1)}{x} \cdot \frac{8}{2x+1}$$

$$= \frac{11}{3} \qquad\qquad\qquad = \frac{24}{x}$$

NOW TRY

**NOW TRY
EXERCISE 2**

Simplify the complex fraction.

$$\frac{\dfrac{a^2b}{c}}{\dfrac{ab^2}{c^3}}$$

EXAMPLE 2 Simplifying a Complex Fraction (Method 1)

Simplify the complex fraction.

$\dfrac{\dfrac{xp}{q^3}}{\dfrac{p^2}{qx^2}}$ The numerator and denominator are single fractions, so use the definition of division and then the fundamental property.

$$\frac{xp}{q^3} \div \frac{p^2}{qx^2}$$

$$= \frac{xp}{q^3} \cdot \frac{qx^2}{p^2}$$

$$= \frac{x^3}{q^2p}$$

NOW TRY

**NOW TRY
EXERCISE 3**

Simplify the complex fraction.

$$\frac{5+\dfrac{2}{a-3}}{\dfrac{1}{a-3}-2}$$

EXAMPLE 3 Simplifying a Complex Fraction (Method 1)

Simplify the complex fraction.

$$\frac{\dfrac{3}{x+2}-4}{\dfrac{2}{x+2}+1}$$ *Find a common denominator before subtracting in the numerator and denominator.*

$$= \frac{\dfrac{3}{x+2}-\dfrac{4(x+2)}{x+2}}{\dfrac{2}{x+2}+\dfrac{1(x+2)}{x+2}}$$ Write both second terms with a denominator of $x+2$.

$$= \frac{\dfrac{3-4(x+2)}{x+2}}{\dfrac{2+1(x+2)}{x+2}}$$ Subtract in the numerator.

Add in the denominator.

Be careful with signs.
$$= \frac{\dfrac{3-4x-8}{x+2}}{\dfrac{2+x+2}{x+2}}$$ Distributive property

$$= \frac{\dfrac{-5-4x}{x+2}}{\dfrac{4+x}{x+2}}$$ Combine like terms.

$$= \frac{-5-4x}{x+2} \cdot \frac{x+2}{4+x}$$ Multiply by the reciprocal of the denominator (divisor).

$$= \frac{-5-4x}{4+x}$$ Divide out the common factor.

NOW TRY

NOW TRY ANSWERS

1. **(a)** $\dfrac{6}{5}$ **(b)** $\dfrac{24}{5x}$

2. $\dfrac{ac^2}{b}$

3. $\dfrac{5a-13}{7-2a}$

OBJECTIVE 3 Simplify a complex fraction by multiplying numerator and denominator by the LCD (Method 2).

If we multiply both the numerator and the denominator of a complex fraction by the LCD of all the fractions within the complex fraction, the result will no longer be complex. This is Method 2.

Simplifying a Complex Fraction (Method 2)

Step 1 Find the LCD of all fractions within the complex fraction.

Step 2 Multiply both the numerator and the denominator of the complex fraction by this LCD using the distributive property as necessary. Write in lowest terms.

NOW TRY
EXERCISE 4

Simplify each complex fraction.

(a) $\dfrac{\dfrac{3}{5} - \dfrac{1}{4}}{\dfrac{1}{8} + \dfrac{3}{20}}$ (b) $\dfrac{\dfrac{2}{x} - 3}{7 + \dfrac{x}{5}}$

EXAMPLE 4   **Simplifying Complex Fractions (Method 2)**

Simplify each complex fraction.

(a) $\dfrac{\dfrac{2}{3} + \dfrac{5}{9}}{\dfrac{1}{4} + \dfrac{1}{12}}$

(b) $\dfrac{6 + \dfrac{3}{x}}{\dfrac{x}{4} + \dfrac{1}{8}}$ (In **Example 1**, we simplified these same fractions using Method 1.)

Step 1   Find the LCD for all denominators in the complex fraction.

The LCD for 3, 9, 4, and 12 is 36. | The LCD for x, 4, and 8 is $8x$.

Step 2 $\dfrac{\dfrac{2}{3} + \dfrac{5}{9}}{\dfrac{1}{4} + \dfrac{1}{12}}$ $\dfrac{6 + \dfrac{3}{x}}{\dfrac{x}{4} + \dfrac{1}{8}}$ Multiply numerator and denominator of the complex fraction by the LCD.

$= \dfrac{36\left(\dfrac{2}{3} + \dfrac{5}{9}\right)}{36\left(\dfrac{1}{4} + \dfrac{1}{12}\right)}$ $= \dfrac{8x\left(6 + \dfrac{3}{x}\right)}{8x\left(\dfrac{x}{4} + \dfrac{1}{8}\right)}$

Multiply each term by 36. $= \dfrac{36\left(\dfrac{2}{3}\right) + 36\left(\dfrac{5}{9}\right)}{36\left(\dfrac{1}{4}\right) + 36\left(\dfrac{1}{12}\right)}$ Multiply each term by 8x. $= \dfrac{8x(6) + 8x\left(\dfrac{3}{x}\right)}{8x\left(\dfrac{x}{4}\right) + 8x\left(\dfrac{1}{8}\right)}$ Distributive property

$= \dfrac{24 + 20}{9 + 3}$ Multiply. $= \dfrac{48x + 24}{2x^2 + x}$ Multiply.

$= \dfrac{44}{12}$ Add. $= \dfrac{24(2x + 1)}{x(2x + 1)}$ Factor.

$= \dfrac{4 \cdot 11}{4 \cdot 3}$ Factor. $= \dfrac{24}{x}$ Lowest terms

$= \dfrac{11}{3}$ Lowest terms

NOW TRY ANSWERS

4. (a) $\dfrac{14}{11}$ (b) $\dfrac{10 - 15x}{x^2 + 35x}$

NOW TRY

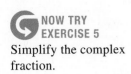 **NOW TRY EXERCISE 5**

Simplify the complex fraction.

$$\dfrac{\dfrac{1}{y} + \dfrac{2}{3y^2}}{\dfrac{5}{4y^2} - \dfrac{3}{2y^3}}$$

EXAMPLE 5 Simplifying a Complex Fraction (Method 2)

Simplify the complex fraction.

$$\dfrac{\dfrac{3}{5m} - \dfrac{2}{m^2}}{\dfrac{9}{2m} + \dfrac{3}{4m^2}}$$
The LCD for $5m$, m^2, $2m$, and $4m^2$ is $20m^2$.

$$= \dfrac{20m^2\left(\dfrac{3}{5m} - \dfrac{2}{m^2}\right)}{20m^2\left(\dfrac{9}{2m} + \dfrac{3}{4m^2}\right)}$$
Multiply numerator and denominator by $20m^2$.

$$= \dfrac{20m^2\left(\dfrac{3}{5m}\right) - 20m^2\left(\dfrac{2}{m^2}\right)}{20m^2\left(\dfrac{9}{2m}\right) + 20m^2\left(\dfrac{3}{4m^2}\right)}$$
Distributive property

$$= \dfrac{12m - 40}{90m + 15}$$
Multiply. Divide out the common factors.

$$= \dfrac{4(3m - 10)}{5(18m + 3)}$$
Factor.

NOW TRY

Some students prefer Method 1 for problems like **Example 2,** which is the quotient of two fractions, and Method 2 for problems like **Examples 1, 3, 4, and 5,** which have sums or differences in the numerators, or denominators, or both. (However, either method can be used for *any* complex fraction.)

EXAMPLE 6 Simplifying Complex Fractions

Simplify each complex fraction. Use either method.

(a) $\dfrac{\dfrac{1}{y} + \dfrac{2}{y + 2}}{\dfrac{4}{y} - \dfrac{3}{y + 2}}$
There are sums and differences in the numerator and denominator. Use Method 2.

$$= \dfrac{\left(\dfrac{1}{y} + \dfrac{2}{y + 2}\right) \cdot y(y + 2)}{\left(\dfrac{4}{y} - \dfrac{3}{y + 2}\right) \cdot y(y + 2)}$$
Multiply numerator and denominator by the LCD, $y(y + 2)$. Because y appears in two denominators, it must be a factor in the LCD.

$$= \dfrac{\left(\dfrac{1}{y}\right)y(y + 2) + \left(\dfrac{2}{y + 2}\right)y(y + 2)}{\left(\dfrac{4}{y}\right)y(y + 2) - \left(\dfrac{3}{y + 2}\right)y(y + 2)}$$
Distributive property

$$= \dfrac{1(y + 2) + 2y}{4(y + 2) - 3y}$$
Multiply. Divide out the common factors.

$$= \dfrac{y + 2 + 2y}{4y + 8 - 3y}$$
Distributive property

$$= \dfrac{3y + 2}{y + 8}$$
Combine like terms.

NOW TRY ANSWER

5. $\dfrac{12y^2 + 8y}{15y - 18}$, or $\dfrac{4y(3y + 2)}{3(5y - 6)}$

**NOW TRY
EXERCISE 6**

Simplify each complex fraction.

(a) $\dfrac{1 - \dfrac{2}{x} - \dfrac{15}{x^2}}{1 + \dfrac{5}{x} + \dfrac{6}{x^2}}$

(b) $\dfrac{\dfrac{9y^2 - 16}{y^2 - 100}}{\dfrac{3y - 4}{y + 10}}$

(b) $\dfrac{1 - \dfrac{2}{x} - \dfrac{3}{x^2}}{1 - \dfrac{5}{x} + \dfrac{6}{x^2}}$ There are sums and differences in the numerator and denominator. Use Method 2.

$= \dfrac{\left(1 - \dfrac{2}{x} - \dfrac{3}{x^2}\right)x^2}{\left(1 - \dfrac{5}{x} + \dfrac{6}{x^2}\right)x^2}$ Multiply numerator and denominator by the LCD, x^2.

$= \dfrac{x^2 - 2x - 3}{x^2 - 5x + 6}$ Distributive property

$= \dfrac{(x - 3)(x + 1)}{(x - 3)(x - 2)}$ Factor.

$= \dfrac{x + 1}{x - 2}$ Lowest terms

(c) $\dfrac{\dfrac{x + 2}{x - 3}}{\dfrac{x^2 - 4}{x^2 - 9}}$ This is a quotient of two rational expressions. Use Method 1.

$= \dfrac{x + 2}{x - 3} \div \dfrac{x^2 - 4}{x^2 - 9}$ Write as a division problem.

$= \dfrac{x + 2}{x - 3} \cdot \dfrac{x^2 - 9}{x^2 - 4}$ Multiply by the reciprocal of the divisor.

$= \dfrac{(x + 2)(x + 3)(x - 3)}{(x - 3)(x + 2)(x - 2)}$ Multiply, and then factor.

NOW TRY ANSWERS

6. **(a)** $\dfrac{x - 5}{x + 2}$ **(b)** $\dfrac{3y + 4}{y - 10}$

$= \dfrac{x + 3}{x - 2}$ Lowest terms   **NOW TRY**

7.5 Exercises

 FOR EXTRA HELP   ▶ **MyMathLab®**

● *Complete solution available in MyMathLab*

Concept Check   In Exercises 1 and 2, consider the following complex fraction.

$$\dfrac{\dfrac{1}{2} - \dfrac{1}{3}}{\dfrac{5}{6} - \dfrac{1}{12}}$$

1. Answer each part, outlining Method 1 for simplifying the complex fraction.

(a) To combine the terms in the numerator, we must find the LCD of $\frac{1}{2}$ and $\frac{1}{3}$. What is this LCD?

Determine the simplified form of the numerator of the complex fraction.

(b) To combine the terms in the denominator, we must find the LCD of $\frac{5}{6}$ and $\frac{1}{12}$. What is this LCD?

Determine the simplified form of the denominator of the complex fraction.

(c) Now use the results from parts (a) and (b) to write the complex fraction as a division problem using the symbol ÷.

(d) Perform the operation from part (c) to obtain the final simplification.

2. Answer each part, outlining Method 2 for simplifying the complex fraction.

(a) We must determine the LCD of all the fractions within the complex fraction.

What is this LCD?

(b) Multiply every term in the complex fraction by the LCD found in part (a), but at this time do not combine the terms in the numerator and the denominator.

(c) Now combine the terms from part (b) to obtain the simplified form of the complex fraction.

Concept Check   *Work each problem.*

3. Which complex fraction is equivalent to $\dfrac{2 - \frac{1}{4}}{3 - \frac{1}{2}}$? Answer this question without showing any work, and explain your reasoning.

   A. $\dfrac{2 + \frac{1}{4}}{3 + \frac{1}{2}}$ **B.** $\dfrac{2 - \frac{1}{4}}{-3 + \frac{1}{2}}$ **C.** $\dfrac{-2 - \frac{1}{4}}{-3 - \frac{1}{2}}$ **D.** $\dfrac{-2 + \frac{1}{4}}{-3 + \frac{1}{2}}$

4. Only one of these choices is equal to $\dfrac{\frac{1}{3} + \frac{1}{12}}{\frac{1}{2} + \frac{1}{4}}$. Which one is it? Answer this question without showing any work, and explain your reasoning.

   A. $\dfrac{5}{9}$ **B.** $-\dfrac{5}{9}$ **C.** $-\dfrac{9}{5}$ **D.** $-\dfrac{1}{12}$

5. Simplify the complex fraction using Method 1.

$$\frac{\frac{2}{5} + \frac{1}{4}}{\frac{1}{2} + \frac{1}{3}}$$

Step 1

Write the numerator as a single fraction. _____

Write the denominator as a single fraction. _____

Step 2

Write the equivalent fraction as a division problem.

_____

Step 3

Write the division problem as a multiplication problem.

_____

Multiply and write the answer in lowest terms. _____

6. Simplify the complex fraction using Method 2.

$$\frac{\frac{2}{3} - \frac{1}{4}}{\frac{4}{9} + \frac{1}{2}}$$

Step 1

The LCD for 3, _____, _____, and _____ is _____.

Step 2

Multiply each term in the complex fraction by _____. The result is as follows.

$$\frac{24 - \underline{}}{\underline{} + 18}$$

Simplify this expression.   _____

Simplify each complex fraction. Use either method. **See Examples 1–6.**

7. $\dfrac{-\dfrac{4}{3}}{\dfrac{2}{9}}$

8. $\dfrac{-\dfrac{5}{6}}{\dfrac{5}{4}}$

9. $\dfrac{\dfrac{x}{y^2}}{\dfrac{x^2}{y}}$

10. $\dfrac{\dfrac{p^4}{r}}{\dfrac{p^2}{r^2}}$

11. $\dfrac{\dfrac{4a^4b^3}{3a}}{\dfrac{2ab^4}{b^2}}$

12. $\dfrac{\dfrac{2r^4t^2}{3t}}{\dfrac{5r^2t^5}{3r}}$

13. $\dfrac{\dfrac{m+2}{3}}{\dfrac{m-4}{m}}$

14. $\dfrac{\dfrac{q-5}{q}}{\dfrac{q+5}{3}}$

15. $\dfrac{\dfrac{2}{x}-3}{\dfrac{2-3x}{2}}$

16. $\dfrac{6+\dfrac{2}{r}}{\dfrac{3r+1}{4}}$

17. $\dfrac{\dfrac{1}{x}+x}{\dfrac{x^2+1}{8}}$

18. $\dfrac{\dfrac{3}{m}-m}{\dfrac{3-m^2}{4}}$

19. $\dfrac{a-\dfrac{5}{a}}{a+\dfrac{1}{a}}$

20. $\dfrac{q+\dfrac{1}{q}}{q+\dfrac{4}{q}}$

21. $\dfrac{\dfrac{5}{8}+\dfrac{2}{3}}{\dfrac{7}{3}-\dfrac{1}{4}}$

22. $\dfrac{\dfrac{6}{5}-\dfrac{1}{9}}{\dfrac{2}{5}+\dfrac{5}{3}}$

23. $\dfrac{\dfrac{1}{x^2}+\dfrac{1}{y^2}}{\dfrac{1}{x}-\dfrac{1}{y}}$

24. $\dfrac{\dfrac{1}{a^2}-\dfrac{1}{b^2}}{\dfrac{1}{a}-\dfrac{1}{b}}$

25. $\dfrac{\dfrac{2}{p^2}-\dfrac{3}{5p}}{\dfrac{4}{p}+\dfrac{1}{4p}}$

26. $\dfrac{\dfrac{2}{m^2}-\dfrac{3}{m}}{\dfrac{2}{5m^2}+\dfrac{1}{3m}}$

27. $\dfrac{\dfrac{5}{x^2y}-\dfrac{2}{xy^2}}{\dfrac{3}{x^2y^2}+\dfrac{4}{xy}}$

28. $\dfrac{\dfrac{1}{m^3p}+\dfrac{2}{mp^2}}{\dfrac{4}{mp}+\dfrac{1}{m^2p}}$

29. $\dfrac{\dfrac{1}{4}-\dfrac{1}{a^2}}{\dfrac{1}{2}+\dfrac{1}{a}}$

30. $\dfrac{\dfrac{1}{9}-\dfrac{1}{m^2}}{\dfrac{1}{3}+\dfrac{1}{m}}$

31. $\dfrac{\dfrac{1}{z+5}}{\dfrac{4}{z^2-25}}$

32. $\dfrac{\dfrac{1}{a+1}}{\dfrac{2}{a^2-1}}$

33. $\dfrac{\dfrac{1}{m+1}-1}{\dfrac{1}{m+1}+1}$

34. $\dfrac{\dfrac{2}{x-1}+2}{\dfrac{2}{x-1}-2}$

35. $\dfrac{\dfrac{12}{x+2}+2}{\dfrac{18}{x+2}-2}$

36. $\dfrac{\dfrac{6}{x+3}+3}{\dfrac{9}{x+3}-3}$

37. $\dfrac{\dfrac{x}{y}+\dfrac{y}{x}}{\dfrac{x}{y}-\dfrac{y}{x}}$

38. $\dfrac{\dfrac{x}{y} - \dfrac{y}{x}}{\dfrac{x}{y} + \dfrac{y}{x}}$

39. $\dfrac{1}{\dfrac{1}{a} + \dfrac{1}{b}}$

40. $\dfrac{-1}{\dfrac{1}{a} - \dfrac{1}{b}}$

41. $\dfrac{\dfrac{1}{m-1} + \dfrac{2}{m+2}}{\dfrac{2}{m+2} - \dfrac{1}{m-3}}$

42. $\dfrac{\dfrac{5}{r+3} - \dfrac{1}{r-1}}{\dfrac{2}{r+2} + \dfrac{3}{r+3}}$

43. $\dfrac{2 + \dfrac{1}{x} - \dfrac{28}{x^2}}{3 + \dfrac{13}{x} + \dfrac{4}{x^2}}$

44. $\dfrac{4 - \dfrac{11}{x} - \dfrac{3}{x^2}}{2 - \dfrac{1}{x} - \dfrac{15}{x^2}}$

45. $\dfrac{\dfrac{y+8}{y-4}}{\dfrac{y^2-64}{y^2-16}}$

46. $\dfrac{\dfrac{t+5}{t-8}}{\dfrac{t^2-25}{t^2-64}}$

47. $\dfrac{\dfrac{15a^2 + 15b^2}{5}}{\dfrac{a^4 - b^4}{10}}$

48. $\dfrac{\dfrac{14x^2 + 14y^2}{21}}{\dfrac{x^4 - y^4}{27}}$

49. $\dfrac{\dfrac{1}{x^3 - y^3}}{\dfrac{1}{x^2 - y^2}}$

50. *Concept Check*   What property of real numbers justifies Method 2 of simplifying complex fractions?

Extending Skills   *Simplify each fraction.*

51. $\dfrac{1 + x^{-1} - 12x^{-2}}{1 - x^{-1} - 20x^{-2}}$

52. $\dfrac{1 + t^{-1} - 56t^{-2}}{1 - t^{-1} - 72t^{-2}}$

Extending Skills   *The fractions in Exercises 53–58 are* **continued fractions.** *Simplify by starting at "the bottom" and working upward.*

53. $1 + \dfrac{1}{1 + \dfrac{1}{1 + 1}}$

54. $5 + \dfrac{5}{5 + \dfrac{5}{5 + 5}}$

55. $7 - \dfrac{3}{5 + \dfrac{2}{4 - 2}}$

56. $3 - \dfrac{2}{4 + \dfrac{2}{4 - 2}}$

57. $r + \dfrac{r}{4 - \dfrac{2}{6 + 2}}$

58. $\dfrac{2q}{7} - \dfrac{q}{6 + \dfrac{8}{4 + 4}}$

RELATING CONCEPTS   For Individual or Group Work   (Exercises 59–62)

To find the average of two numbers, we add them and divide by 2. Suppose that we wish to find the average of $\frac{3}{8}$ and $\frac{5}{6}$. **Work Exercises 59–62 in order,** *to see how a complex fraction occurs in a problem like this.*

59. Write in symbols: The sum of $\frac{3}{8}$ and $\frac{5}{6}$, divided by 2. The result should be a complex fraction.

60. Use Method 1 to simplify the complex fraction from **Exercise 59.**

61. Use Method 2 to simplify the complex fraction from **Exercise 59.**

62. The answers in **Exercises 60 and 61** should be the same. Which method did you prefer? Why?

7.6   Solving Equations with Rational Expressions

**NOW TRY
EXERCISE 1**

Identify each of the following as an *expression* or an *equation*. Then simplify the expression or solve the equation.

(a) $\dfrac{3}{2}t - \dfrac{5}{7}t = \dfrac{11}{7}$

(b) $\dfrac{3}{2}t - \dfrac{5}{7}t$

OBJECTIVE 1   Distinguish between operations with rational expressions and equations with terms that are rational expressions.

Before solving equations with rational expressions, we emphasize the distinction between sums and differences of terms with rational coefficients, or rational *expressions,* and *equations* with terms that are rational expressions.

Sums and differences are expressions to simplify. Equations are solved.

EXAMPLE 1   Distinguishing between Expressions and Equations

Identify each of the following as an *expression* or an *equation*. Then simplify the expression or solve the equation.

(a) $\dfrac{3}{4}x - \dfrac{2}{3}x$ This is a difference of two terms. It represents an *expression* to simplify because there is no equality symbol.

$= \dfrac{3 \cdot 3}{3 \cdot 4}x - \dfrac{4 \cdot 2}{4 \cdot 3}x$ The LCD is 12. Write each coefficient with this LCD.

$= \dfrac{9}{12}x - \dfrac{8}{12}x$ Multiply.

$= \dfrac{1}{12}x$ Combine like terms, using the distributive property: $\dfrac{9}{12}x - \dfrac{8}{12}x = \left(\dfrac{9}{12} - \dfrac{8}{12}\right)x$.

(b) $\dfrac{3}{4}x - \dfrac{2}{3}x = \dfrac{1}{2}$ Because there is an equality symbol, this is an *equation* to be solved.

$12\left(\dfrac{3}{4}x - \dfrac{2}{3}x\right) = 12\left(\dfrac{1}{2}\right)$ Use the multiplication property of equality to clear fractions. Multiply by 12, the LCD.

Multiply *each* term by 12. $12\left(\dfrac{3}{4}x\right) - 12\left(\dfrac{2}{3}x\right) = 12\left(\dfrac{1}{2}\right)$ Distributive property

$9x - 8x = 6$ Multiply.

$x = 6$ Combine like terms.

CHECK $\dfrac{3}{4}x - \dfrac{2}{3}x = \dfrac{1}{2}$ Original equation

$\dfrac{3}{4}(6) - \dfrac{2}{3}(6) \overset{?}{=} \dfrac{1}{2}$ Let $x = 6$.

$\dfrac{9}{2} - 4 \overset{?}{=} \dfrac{1}{2}$ Multiply.

$\dfrac{1}{2} = \dfrac{1}{2}$ ✓   True

NOW TRY ANSWERS
1. (a) equation; $\{2\}$
    (b) expression; $\dfrac{11}{14}t$

Because a true statement results, $\{6\}$ is the solution set of the equation.

NOW TRY ↻

The ideas of **Example 1** can be summarized as follows.

Uses of the LCD

When adding or subtracting rational expressions, keep the LCD throughout the simplification. (See **Example 1(a).**)

When solving an equation with terms that are rational expressions, multiply each side by the LCD so that denominators are eliminated. (See **Example 1(b).**)

OBJECTIVE 2 Solve equations with rational expressions.

When an equation involves fractions, as in **Example 1(b),** we use the multiplication property of equality to clear the fractions. When we choose the LCD of all denominators as the multiplier, the resulting equation contains no fractions.

**NOW TRY
EXERCISE 2**

Solve, and check the solution.

$$\frac{x}{6} + \frac{x}{3} = 6 + x$$

EXAMPLE 2 Solving an Equation with Rational Expressions

Solve, and check the solution.

$$\frac{x}{3} + \frac{x}{4} = 10 + x$$

$$12\left(\frac{x}{3} + \frac{x}{4}\right) = 12(10 + x) \qquad \text{Multiply by the LCD, 12, to clear fractions.}$$

$$12\left(\frac{x}{3}\right) + 12\left(\frac{x}{4}\right) = 12(10) + 12x \qquad \text{Distributive property}$$

$$4x + 3x = 120 + 12x \qquad \text{Multiply.}$$

$$7x = 120 + 12x \qquad \text{Combine like terms.}$$

$$-5x = 120 \qquad \text{Subtract } 12x.$$

$$x = -24 \qquad \text{Divide by } -5.$$

CHECK

$$\frac{x}{3} + \frac{x}{4} = 10 + x \qquad \text{Original equation}$$

$$\frac{-24}{3} + \frac{-24}{4} \overset{?}{=} 10 - 24 \qquad \text{Let } x = -24.$$

$$-8 + (-6) \overset{?}{=} -14 \qquad \text{Divide. Subtract.}$$

$$-14 = -14 \checkmark \qquad \text{True}$$

The solution set is $\{-24\}$.

NOW TRY

⚠ **CAUTION** *Be careful not to confuse the following procedures.*

- In **Examples 2 and 3,** we use the multiplication property of equality to multiply each side of an *equation* by the LCD.

- In **Section 7.5,** we used the fundamental property to multiply a *fraction* (an expression) by another fraction that had the LCD as both its numerator and denominator.

NOW TRY ANSWER
2. $\{-12\}$

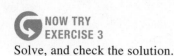

Solve, and check the solution.

$$\frac{x}{7} - \frac{x+5}{5} = -\frac{3}{7}$$

EXAMPLE 3   Solving an Equation with Rational Expressions

Solve, and check the solution.

$$\frac{p}{2} - \frac{p-1}{3} = 1$$

$$6\left(\frac{p}{2} - \frac{p-1}{3}\right) = 6(1) \qquad \text{Multiply each side by the LCD, 6.}$$

$$6\left(\frac{p}{2}\right) - 6\left(\frac{p-1}{3}\right) = 6(1) \qquad \text{Distributive property}$$

$$3p - 2(p-1) = 6 \qquad \text{Use parentheses around } p-1 \text{ to avoid errors.}$$

$$3p - 2(p) - 2(-1) = 6 \qquad \text{Distributive property}$$

Be careful with signs.

$$3p - 2p + 2 = 6 \qquad \text{Multiply.}$$

$$p + 2 = 6 \qquad \text{Combine like terms.}$$

$$p = 4 \qquad \text{Subtract 2.}$$

CHECK $\qquad \dfrac{p}{2} - \dfrac{p-1}{3} = 1 \qquad$ Original equation

$$\frac{4}{2} - \frac{4-1}{3} \stackrel{?}{=} 1 \qquad \text{Let } p = 4.$$

$$2 - 1 \stackrel{?}{=} 1 \qquad \text{Simplify.}$$

$$1 = 1 \checkmark \qquad \text{True}$$

The solution set is $\{4\}$.   NOW TRY

Recall that division by 0 is undefined. *When solving an equation with rational expressions that have variables in the denominator, the solution cannot be a number that makes the denominator equal 0.*

A value of the variable that appears to be a solution after both sides of a rational equation are multiplied by a variable expression is a **proposed solution.** *All proposed solutions must be checked in the original equation.*

EXAMPLE 4   Solving an Equation with Rational Expressions

Solve, and check the proposed solution.

$$\frac{x}{x-2} = \frac{2}{x-2} + 2 \qquad \begin{array}{l} x \text{ cannot equal 2, because 2} \\ \text{causes both denominators} \\ \text{to equal 0.} \end{array}$$

$$(x-2)\left(\frac{x}{x-2}\right) = (x-2)\left(\frac{2}{x-2} + 2\right) \qquad \begin{array}{l} \text{Multiply each side by the LCD,} \\ x - 2. \end{array}$$

$$(x-2)\left(\frac{x}{x-2}\right) = (x-2)\left(\frac{2}{x-2}\right) + (x-2)(2) \qquad \text{Distributive property}$$

$$x = 2 + 2x - 4 \qquad \text{Simplify.}$$

$$x = -2 + 2x \qquad \text{Combine like terms.}$$

$$-x = -2 \qquad \text{Subtract } 2x.$$

Proposed solution $\rightarrow \quad x = 2 \qquad \text{Multiply by } -1.$

NOW TRY ANSWER
3. $\{-10\}$

**NOW TRY
EXERCISE 4**

Solve, and check the proposed solution.

$$4 + \frac{6}{x-3} = \frac{2x}{x-3}$$

As noted, x cannot equal 2 because replacing x with 2 in the original equation causes the denominators to equal 0. We see this in the following check.

CHECK

$$\frac{x}{x-2} = \frac{2}{x-2} + 2 \qquad \text{Original equation}$$

$$\frac{2}{2-2} \overset{?}{=} \frac{2}{2-2} + 2 \qquad \text{Let } x = 2.$$

$$\boxed{\text{Division by 0 is undefined.}} \quad \frac{2}{0} \overset{?}{=} \frac{2}{0} + 2 \qquad \text{Subtract in the denominators.}$$

Thus, the proposed solution 2 must be rejected. The solution set is $\varnothing$. **NOW TRY**

A proposed solution that is not an actual solution of the original equation, such as 2 in **Example 4,** is an **extraneous solution,** or **extraneous value.** Some students like to determine which numbers cannot be solutions *before* solving the equation, as we did in **Example 4.**

Solving an Equation with Rational Expressions

Step 1 **Multiply each side of the equation by the LCD** to clear the equation of fractions. Be sure to distribute to *every* term on *both* sides.

Step 2 **Solve** the resulting equation for proposed solutions.

Step 3 **Check** each proposed solution by substituting it into the original equation. Reject any that cause a denominator to equal 0.

EXAMPLE 5 Solving an Equation with Rational Expressions

Solve, and check the proposed solution.

$$\frac{2}{x^2-x} = \frac{1}{x^2-1}$$

Step 1 Factor the denominators to find the LCD.

$$\frac{2}{x(x-1)} = \frac{1}{(x+1)(x-1)}$$

The LCD is $x(x+1)(x-1)$. Notice that 0, 1, and -1 cannot be solutions. Otherwise a denominator will equal 0. Multiply both sides of the equation by the LCD to clear the fractions.

$$x(x+1)(x-1)\frac{2}{x(x-1)} = x(x+1)(x-1)\frac{1}{(x+1)(x-1)} \qquad \text{Multiply by the LCD.}$$

Step 2
$$2(x+1) = x \qquad \text{Divide out the common factors.}$$
$$2x + 2 = x \qquad \text{Distributive property}$$
$$x + 2 = 0 \qquad \text{Subtract } x.$$
$$x = -2 \qquad \text{Subtract 2.}$$

NOW TRY ANSWER
4. $\varnothing$

Step 3 The proposed solution is -2, which does not make any denominator equal 0.

NOW TRY
EXERCISE 5
Solve, and check the proposed solution.

$$\frac{3}{2x^2 - 8x} = \frac{1}{x^2 - 16}$$

CHECK

$$\frac{2}{x^2 - x} = \frac{1}{x^2 - 1} \qquad \text{Original equation}$$

$$\frac{2}{(-2)^2 - (-2)} \overset{?}{=} \frac{1}{(-2)^2 - 1} \qquad \text{Let } x = -2.$$

$$\frac{2}{4 + 2} \overset{?}{=} \frac{1}{4 - 1} \qquad \begin{array}{l}\text{Apply the exponents;}\\ \text{definition of subtraction}\end{array}$$

$$\frac{1}{3} = \frac{1}{3} \checkmark \qquad \text{True}$$

The solution set is $\{-2\}$.   NOW TRY

NOW TRY
EXERCISE 6
Solve, and check the proposed solution.

$$\frac{2y}{y^2 - 25} = \frac{8}{y + 5} - \frac{1}{y - 5}$$

EXAMPLE 6   **Solving an Equation with Rational Expressions**

Solve, and check the proposed solution.

$$\frac{2m}{m^2 - 4} + \frac{1}{m - 2} = \frac{2}{m + 2}$$

$$\frac{2m}{(m + 2)(m - 2)} + \frac{1}{m - 2} = \frac{2}{m + 2} \qquad \begin{array}{l}\text{Factor the first denominator}\\ \text{on the left to find the LCD,}\\ (m + 2)(m - 2).\end{array}$$

Notice that -2 and 2 cannot be solutions of this equation.

$$(m + 2)(m - 2)\left(\frac{2m}{(m + 2)(m - 2)} + \frac{1}{m - 2}\right)$$

$$\text{Multiply by the LCD.}$$

$$= (m + 2)(m - 2)\frac{2}{m + 2}$$

$$(m + 2)(m - 2)\frac{2m}{(m + 2)(m - 2)} + (m + 2)(m - 2)\frac{1}{m - 2}$$

$$= (m + 2)(m - 2)\frac{2}{m + 2} \qquad \text{Distributive property}$$

$$2m + m + 2 = 2(m - 2) \qquad \text{Divide out the common factors.}$$

$$3m + 2 = 2m - 4 \qquad \text{Combine like terms; distributive property}$$

$$m + 2 = -4 \qquad \text{Subtract } 2m.$$

$$m = -6 \qquad \text{Subtract 2.}$$

CHECK $\dfrac{2m}{m^2 - 4} + \dfrac{1}{m - 2} = \dfrac{2}{m + 2}$ Original equation

$$\frac{2(-6)}{(-6)^2 - 4} + \frac{1}{-6 - 2} \overset{?}{=} \frac{2}{-6 + 2} \qquad \text{Let } m = -6.$$

$$\frac{-12}{32} + \frac{1}{-8} \overset{?}{=} \frac{2}{-4} \qquad \begin{array}{l}\text{Apply the exponent.}\\ \text{Subtract and add.}\end{array}$$

$$-\frac{1}{2} = -\frac{1}{2} \checkmark \qquad \text{True}$$

NOW TRY ANSWERS
5. $\{-12\}$ **6.** $\{9\}$

The solution set is $\{-6\}$.   NOW TRY

NOW TRY EXERCISE 7

Solve, and check the proposed solution(s).

$$\frac{3}{m^2 - 9} = \frac{1}{2(m - 3)} - \frac{1}{4}$$

EXAMPLE 7 Solving an Equation with Rational Expressions

Solve, and check the proposed solution(s).

$$\frac{1}{x - 1} + \frac{1}{2} = \frac{2}{x^2 - 1}$$

> $x \neq 1, -1$. Otherwise, a denominator is 0.

$$\frac{1}{x - 1} + \frac{1}{2} = \frac{2}{(x + 1)(x - 1)}$$

Factor the denominator on the right. The LCD is $2(x + 1)(x - 1)$.

$$2(x + 1)(x - 1)\left(\frac{1}{x - 1} + \frac{1}{2}\right) = 2(x + 1)(x - 1)\frac{2}{(x + 1)(x - 1)}$$

Multiply by the LCD.

$$2(x + 1)(x - 1)\frac{1}{x - 1} + 2(x + 1)(x - 1)\frac{1}{2} = 2(x + 1)(x - 1)\frac{2}{(x + 1)(x - 1)}$$

Distributive property

$$2(x + 1) + (x + 1)(x - 1) = 2(2)$$ Divide out the common factors.

$$2x + 2 + x^2 - 1 = 4$$ Multiply.

> Write in standard form.

$$x^2 + 2x - 3 = 0$$ Subtract 4. Combine like terms.

$$(x + 3)(x - 1) = 0$$ Factor.

$$x + 3 = 0 \quad \text{or} \quad x - 1 = 0$$ Zero-factor property

$$x = -3 \quad \text{or} \quad x = 1 \quad \leftarrow \text{Proposed solutions}$$

Because 1 makes a denominator equal 0, 1 is an extraneous value. Check that -3 is a solution.

CHECK

$$\frac{1}{x - 1} + \frac{1}{2} = \frac{2}{x^2 - 1}$$ Original equation

$$\frac{1}{-3 - 1} + \frac{1}{2} \overset{?}{=} \frac{2}{(-3)^2 - 1}$$ Let $x = -3$.

$$\frac{1}{-4} + \frac{1}{2} \overset{?}{=} \frac{2}{9 - 1}$$ Subtract. Apply the exponent.

$$\frac{1}{4} = \frac{1}{4} \checkmark$$ True

The solution set is $\{-3\}$.   **NOW TRY**

EXAMPLE 8 Solving an Equation with Rational Expressions

Solve, and check the proposed solution.

$$\frac{1}{k^2 + 4k + 3} + \frac{1}{2k + 2} = \frac{3}{4k + 12}$$

$$\frac{1}{(k + 1)(k + 3)} + \frac{1}{2(k + 1)} = \frac{3}{4(k + 3)}$$

Factor each denominator. The LCD is $4(k + 1)(k + 3)$.

> $k \neq -1, -3$

$$4(k + 1)(k + 3)\left(\frac{1}{(k + 1)(k + 3)} + \frac{1}{2(k + 1)}\right)$$

$$= 4(k + 1)(k + 3)\frac{3}{4(k + 3)}$$

Multiply by the LCD.

NOW TRY ANSWER
7. $\{-1\}$

**NOW TRY
EXERCISE 8**

Solve, and check the proposed solution.

$$\frac{5}{k^2 + k - 2} = \frac{1}{3k - 3} - \frac{1}{k + 2}$$

$$4(k + 1)(k + 3)\frac{1}{(k + 1)(k + 3)} + 2 \cdot 2(k + 1)(k + 3)\frac{1}{2(k + 1)}$$

$$= 4(k + 1)(k + 3)\frac{3}{4(k + 3)} \qquad \text{Distributive property}$$

> Do *not* add
> 4 + 2 here.

$$4 + 2(k + 3) = 3(k + 1) \qquad \text{Divide out the common factors.}$$

$$4 + 2k + 6 = 3k + 3 \qquad \text{Distributive property}$$

$$2k + 10 = 3k + 3 \qquad \text{Combine like terms.}$$

$$10 = k + 3 \qquad \text{Subtract } 2k.$$

$$7 = k \qquad \text{Subtract } 3.$$

The proposed solution, 7, does not make an original denominator equal 0. A check shows that the algebra is correct. (See **Exercise 82.**) The solution set is $\{7\}$.

NOW TRY

OBJECTIVE 3 Solve a formula for a specified variable.

When solving a formula for a specified variable, *remember to treat the variable for which you are solving as if it were the only variable, and all others as if they were constants.*

EXAMPLE 9 Solving for a Specified Variable

Solve each formula for the specified variable.

(a) $a = \dfrac{v - w}{t}$ for v

$$a = \frac{v - w}{t} \qquad \text{Our goal is to isolate } v.$$

$$at = \left(\frac{v - w}{t}\right)t \qquad \text{Multiply by } t \text{ to clear the fraction.}$$

$$at = v - w \qquad \text{Divide out the common factor.}$$

$$at + w = v, \quad \text{or} \quad v = at + w \qquad \text{Add } w. \text{ Rewrite.}$$

(b) $F = \dfrac{k}{d - D}$ for d

$$F = \frac{k}{d - D} \qquad \text{Given equation}$$

We must isolate d.

$$F(d - D) = \frac{k}{d - D}(d - D) \qquad \text{Multiply by } d - D \text{ to clear the fraction.}$$

$$F(d - D) = k \qquad \text{Divide out the common factor.}$$

$$Fd - FD = k \qquad \text{Distributive property}$$

$$Fd = k + FD \qquad \text{Add } FD.$$

$$d = \frac{k + FD}{F} \qquad \text{Divide by } F.$$

NOW TRY ANSWER
8. $\{-5\}$

**NOW TRY
EXERCISE 9**

Solve each formula for the specified variable.

(a) $p = \dfrac{x - y}{z}$ for x

(b) $a = \dfrac{b}{c + d}$ for d

We can write an equivalent form of this answer as follows.

$$d = \frac{k + FD}{F} \qquad \text{Answer on the preceding page}$$

$$d = \frac{k}{F} + \frac{FD}{F} \qquad \begin{array}{l}\text{Definition of addition of fractions,}\\ \frac{a + b}{c} = \frac{a}{c} + \frac{b}{c}\end{array}$$

$$d = \frac{k}{F} + D \qquad \text{Divide out the common factor from } \frac{FD}{F}.$$

Either answer is correct. **NOW TRY**

**NOW TRY
EXERCISE 10**

Solve the following formula for x.

$$\frac{2}{w} = \frac{1}{x} - \frac{3}{y}$$

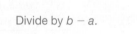

EXAMPLE 10 Solving for a Specified Variable

Solve the following formula for c.

$$\frac{1}{a} = \frac{1}{b} + \frac{1}{c} \quad \boxed{\begin{array}{l}\text{Goal: Isolate } c, \text{ the}\\ \text{specified variable.}\end{array}}$$

$$abc\left(\frac{1}{a}\right) = abc\left(\frac{1}{b} + \frac{1}{c}\right) \qquad \begin{array}{l}\text{Multiply by the LCD, } abc,\\ \text{to clear the fractions.}\end{array}$$

$$abc\left(\frac{1}{a}\right) = abc\left(\frac{1}{b}\right) + abc\left(\frac{1}{c}\right) \qquad \text{Distributive property}$$

$$bc = ac + ab \qquad \text{Divide out the common factors.}$$

$$bc - ac = ab \qquad \begin{array}{l}\text{Subtract } ac \text{ so that both terms}\\ \text{with } c \text{ are on the same side.}\end{array}$$

$\boxed{\begin{array}{l}\text{Pay careful}\\ \text{attention}\\ \text{here.}\end{array}}$ $c(b - a) = ab \qquad \text{Factor out } c.$

$$c = \frac{ab}{b - a} \qquad \text{Divide by } b - a. \qquad \textbf{NOW TRY} \;$$

⚠ CAUTION Students often have trouble in the step that involves factoring out the variable for which they are solving. In **Example 10,** we needed to transform so that both terms with c are on the same side of the equation. This allowed us to factor out c on the left, and then isolate it by dividing each side by $b - a$.

When solving an equation for a specified variable, be sure that the specified variable appears alone on only one side of the equality symbol in the final equation.

NOW TRY ANSWERS
9. (a) $x = pz + y$
    (b) $d = \dfrac{b - ac}{a}$
10. $x = \dfrac{wy}{2y + 3w}$

7.6 Exercises

$\boxed{\begin{array}{l}\text{FOR}\\ \text{EXTRA}\\ \text{HELP}\end{array}}$ ▶ **MyMathLab®**

▶ *Complete solution available in MyMathLab*

Concept Check   Provide a short answer to each of the following.

1. What is the least positive whole number by which we can multiply both sides of the equation

$$\frac{2}{3}x + \frac{1}{4}x = 6$$

to obtain an equation with only integer coefficients?

2. Before even beginning to solve the equation

$$\frac{1}{x - 3} + \frac{2}{3 - x} = 4,$$

what number do we know cannot be a solution? Why?

3. What is the simplest monomial by which we can multiply both sides of the equation

$$\frac{1}{x} - \frac{1}{y} = \frac{1}{z}$$

so that there are no variables in the denominators?

4. If we are solving an equation for the variable k, and our steps lead to the equation

$$kr - mr = km,$$

what would be the next step?

5. Suppose an equation includes the rational expression $\frac{1}{3-x}$. Is it acceptable to replace this expression by $\frac{-1}{x-3}$? Why or why not?

6. To combine the terms in $\frac{2}{3x} + \frac{7}{5y}$, is it acceptable to simply multiply through by $15xy$? Why or why not?

Identify each of the following as an expression *or an* equation. *Then simplify each expression or solve each equation.* **See Examples 1 and 2.**

▶ 7. $\frac{7}{8}x + \frac{1}{5}x$

8. $\frac{4}{7}x + \frac{4}{5}x$

9. $\frac{7x}{8} + \frac{x}{5} = 1$

10. $\frac{4x}{7} + \frac{4x}{5} = 1$

11. $\frac{3}{5}x - \frac{7}{10}x$

12. $\frac{2}{3}x - \frac{9}{4}x$

13. $\frac{3}{5}x - \frac{7}{10}x = 1$

14. $\frac{2}{3}x - \frac{9}{4}x = -19$

15. $\frac{3}{4}x - \frac{1}{2}x = 0$

16. *Concept Check* Why is the equation in **Exercise 15** easy to check?

When solving an equation with variables in denominators, we must determine the values that cause these denominators to equal 0, so that we can reject these extraneous values if they appear as potential solutions. Find all values for which at least one denominator is equal to 0. Write answers using the symbol $\neq$. Do not solve. **See Examples 4–8.**

17. $\frac{3}{x+2} - \frac{5}{x} = 1$

18. $\frac{7}{x} + \frac{9}{x-4} = 5$

19. $\frac{-1}{(x+3)(x-4)} = \frac{1}{2x+1}$

20. $\frac{8}{(x-7)(x+3)} = \frac{7}{3x-10}$

21. $\frac{4}{x^2+8x-9} + \frac{1}{x^2-4} = 0$

22. $\frac{-3}{x^2+9x-10} - \frac{12}{x^2-49} = 0$

Solve each equation, and check the solutions. **See Examples 1–4.**

23. $\frac{5}{m} - \frac{3}{m} = 8$

24. $\frac{4}{y} + \frac{1}{y} = 2$

25. $\frac{5}{y} + 4 = \frac{2}{y}$

26. $\frac{11}{q} - 3 = \frac{1}{q}$

27. $\frac{3x}{5} - 6 = x$

28. $\frac{5t}{4} + t = 9$

29. $\frac{4m}{7} + m = 11$

30. $x - \frac{3x}{2} = 1$

31. $\frac{z-1}{4} = \frac{z+3}{3}$

32. $\frac{r-5}{2} = \frac{r+2}{3}$

33. $\frac{3p+6}{8} = \frac{3p-3}{16}$

34. $\frac{2z+1}{5} = \frac{7z+5}{15}$

35. $\frac{2x+3}{x} = \frac{3}{2}$

36. $\frac{7-2x}{x} = \frac{-17}{5}$

▶ 37. $\dfrac{k}{k-4} - 5 = \dfrac{4}{k-4}$

38. $\dfrac{-5}{a+5} - 2 = \dfrac{a}{a+5}$

39. $\dfrac{q+2}{3} + \dfrac{q-5}{5} = \dfrac{7}{3}$

40. $\dfrac{x-6}{6} + \dfrac{x+2}{8} = \dfrac{11}{4}$

41. $\dfrac{x}{2} = \dfrac{5}{4} + \dfrac{x-1}{4}$

42. $\dfrac{8p}{5} = \dfrac{3p-4}{2} + \dfrac{5}{2}$

43. $x + \dfrac{17}{2} = \dfrac{x}{2} + x + 6$

44. $t + \dfrac{8}{3} = \dfrac{t}{3} + t + \dfrac{14}{3}$

45. $\dfrac{9}{3x+4} = \dfrac{36-27x}{16-9x^2}$

46. $\dfrac{25}{5x-6} = \dfrac{-150-125x}{36-25x^2}$

Solve each equation, and check the solutions. ***Be careful with signs. See Example 3.***

▶ 47. $\dfrac{a+7}{8} - \dfrac{a-2}{3} = \dfrac{4}{3}$

48. $\dfrac{x+3}{7} - \dfrac{x+2}{6} = \dfrac{1}{6}$

49. $\dfrac{p}{2} - \dfrac{p-1}{4} = \dfrac{5}{4}$

50. $\dfrac{r}{6} - \dfrac{r-2}{3} = -\dfrac{4}{3}$

51. $\dfrac{3x}{5} - \dfrac{x-5}{7} = 3$

52. $\dfrac{8k}{5} - \dfrac{3k-4}{2} = \dfrac{5}{2}$

Solve each equation, and check the solutions. ***See Examples 4–8.***

▶ 53. $\dfrac{4}{x^2-3x} = \dfrac{1}{x^2-9}$

54. $\dfrac{2}{t^2-4} = \dfrac{3}{t^2-2t}$

55. $\dfrac{2}{m} = \dfrac{m}{5m+12}$

56. $\dfrac{x}{4-x} = \dfrac{2}{x}$

57. $\dfrac{-2}{z+5} + \dfrac{3}{z-5} = \dfrac{20}{z^2-25}$

58. $\dfrac{3}{r+3} - \dfrac{2}{r-3} = \dfrac{-12}{r^2-9}$

59. $\dfrac{3}{x-1} + \dfrac{2}{4x-4} = \dfrac{7}{4}$

60. $\dfrac{2}{p+3} + \dfrac{3}{8} = \dfrac{5}{4p+12}$

61. $\dfrac{x}{3x+3} = \dfrac{2x-3}{x+1} - \dfrac{2x}{3x+3}$

62. $\dfrac{2k+3}{k+1} - \dfrac{3k}{2k+2} = \dfrac{-2k}{2k+2}$

▶ 63. $\dfrac{2p}{p^2-1} = \dfrac{2}{p+1} - \dfrac{1}{p-1}$

64. $\dfrac{2x}{x^2-16} - \dfrac{2}{x-4} = \dfrac{4}{x+4}$

65. $\dfrac{5x}{14x+3} = \dfrac{1}{x}$

66. $\dfrac{m}{8m+3} = \dfrac{1}{3m}$

67. $\dfrac{2}{x-1} - \dfrac{2}{3} = \dfrac{-1}{x+1}$

68. $\dfrac{5}{p-2} = 7 - \dfrac{10}{p+2}$

69. $\dfrac{x}{2x+2} = \dfrac{-2x}{4x+4} + \dfrac{2x-3}{x+1}$

70. $\dfrac{5t+1}{3t+3} = \dfrac{5t-5}{5t+5} + \dfrac{3t-1}{t+1}$

71. $\dfrac{8x+3}{x} = 3x$

72. $\dfrac{10x-24}{x} = x$

▶ 73. $\dfrac{1}{x+4} + \dfrac{x}{x-4} = \dfrac{-8}{x^2-16}$

74. $\dfrac{x}{x-3} + \dfrac{4}{x+3} = \dfrac{18}{x^2-9}$

▶ 75. $\dfrac{4}{3x+6} - \dfrac{3}{x+3} = \dfrac{8}{x^2+5x+6}$

76. $\dfrac{-13}{t^2+6t+8} + \dfrac{4}{t+2} = \dfrac{3}{2t+8}$

77. $\dfrac{3x}{x^2 + 5x + 6} = \dfrac{5x}{x^2 + 2x - 3} - \dfrac{2}{x^2 + x - 2}$

78. $\dfrac{m}{m^2 + m - 2} + \dfrac{m}{m^2 - 1} = \dfrac{m}{m^2 + 3m + 2}$

79. $\dfrac{x + 4}{x^2 - 3x + 2} - \dfrac{5}{x^2 - 4x + 3} = \dfrac{x - 4}{x^2 - 5x + 6}$

80. $\dfrac{3}{r^2 + r - 2} - \dfrac{1}{r^2 - 1} = \dfrac{7}{2(r^2 + 3r + 2)}$

81. $\dfrac{1}{x^2 - 1} = \dfrac{2}{x - 1} - \dfrac{1}{x - 1}$

82. Refer to **Example 8,** and show that 7 is a solution.

Solve each formula for the specified variable. **See Examples 9 and 10.**

83. $m = \dfrac{kF}{a}$ for F

84. $I = \dfrac{kE}{R}$ for E

85. $m = \dfrac{kF}{a}$ for a

86. $I = \dfrac{kE}{R}$ for R

▶ 87. $I = \dfrac{E}{R + r}$ for R

88. $I = \dfrac{E}{R + r}$ for r

89. $h = \dfrac{2\mathcal{A}}{B + b}$ for $\mathcal{A}$

90. $d = \dfrac{2S}{n(a + L)}$ for S

91. $d = \dfrac{2S}{n(a + L)}$ for a

92. $h = \dfrac{2\mathcal{A}}{B + b}$ for B

93. $\dfrac{1}{x} = \dfrac{1}{y} - \dfrac{1}{z}$ for y

94. $\dfrac{3}{k} = \dfrac{1}{p} + \dfrac{1}{q}$ for q

95. $\dfrac{2}{r} + \dfrac{3}{s} + \dfrac{1}{t} = 1$ for t

96. $\dfrac{5}{p} + \dfrac{2}{q} + \dfrac{3}{r} = 1$ for r

97. $9x + \dfrac{3}{z} = \dfrac{5}{y}$ for z

98. $-3t - \dfrac{4}{p} = \dfrac{6}{s}$ for p

99. $\dfrac{t}{x - 1} - \dfrac{2}{x + 1} = \dfrac{1}{x^2 - 1}$ for t

100. $\dfrac{5}{y + 2} - \dfrac{r}{y - 2} = \dfrac{3}{y^2 - 4}$ for r

RELATING CONCEPTS   For Individual or Group Work (Exercises 101–108)

In these exercises, we summarize various concepts involving rational expressions. **Work Exercises 101–108 in order.**

Let P, Q, and R be rational expressions defined as follows.

$$P = \dfrac{6}{x + 3}, \qquad Q = \dfrac{5}{x + 1}, \qquad R = \dfrac{4x}{x^2 + 4x + 3}$$

101. Find the values for which each expression is undefined.

    (a) P **(b)** Q **(c)** R

102. Find and express $(P \cdot Q) \div R$ in lowest terms.

103. Why is $(P \cdot Q) \div R$ not defined if $x = 0$?

104. Find the LCD for P, Q, and R.

105. Perform the operations and express $P + Q - R$ in lowest terms.

106. Simplify the complex fraction $\dfrac{P + Q}{R}$.

107. Solve the equation $P + Q = R$.

108. How does the answer to **Exercise 101** help when working **Exercise 107**?

Students often confuse *simplifying expressions* with *solving equations.* We review the four operations applied to the rational expressions $\frac{1}{x}$ and $\frac{1}{x-2}$ as follows.

Add: $\dfrac{1}{x} + \dfrac{1}{x-2}$

$$= \frac{1(x-2)}{x(x-2)} + \frac{x(1)}{x(x-2)} \qquad \text{Write with a common denominator.}$$

$$= \frac{x-2+x}{x(x-2)} \qquad \begin{array}{l}\text{Add numerators.}\\ \text{Keep the same denominator.}\end{array}$$

$$= \frac{2x-2}{x(x-2)} \qquad \text{Combine like terms.}$$

Subtract: $\dfrac{1}{x} - \dfrac{1}{x-2}$

$$= \frac{1(x-2)}{x(x-2)} - \frac{x(1)}{x(x-2)} \qquad \text{Write with a common denominator.}$$

$$= \frac{x-2-x}{x(x-2)} \qquad \begin{array}{l}\text{Subtract numerators.}\\ \text{Keep the same denominator.}\end{array}$$

$$= \frac{-2}{x(x-2)} \qquad \text{Combine like terms.}$$

Multiply: $\dfrac{1}{x} \cdot \dfrac{1}{x-2}$

$$= \frac{1}{x(x-2)} \qquad \begin{array}{l}\text{Multiply numerators.}\\ \text{Multiply denominators.}\end{array}$$

Divide: $\dfrac{1}{x} \div \dfrac{1}{x-2}$

$$= \frac{1}{x} \cdot \frac{x-2}{1} \qquad \begin{array}{l}\text{Multiply by the reciprocal of the}\\ \text{divisor.}\end{array}$$

$$= \frac{x-2}{x} \qquad \begin{array}{l}\text{Multiply numerators.}\\ \text{Multiply denominators.}\end{array}$$

By contrast, consider the following *equation.*

$$\frac{1}{x} + \frac{1}{x-2} = \frac{3}{4} \qquad \boxed{\begin{array}{l}x \neq 0, 2 \text{ because a}\\ \text{denominator is 0 for}\\ \text{these values.}\end{array}}$$

$$4x(x-2)\left(\frac{1}{x} + \frac{1}{x-2}\right) = 4x(x-2)\frac{3}{4} \qquad \begin{array}{l}\text{Multiply each side by the LCD,}\\ 4x(x-2), \text{ to clear fractions.}\end{array}$$

$$4x(x-2)\frac{1}{x} + 4x(x-2)\frac{1}{x-2} = 4x(x-2)\frac{3}{4} \qquad \text{Distributive property}$$

$$4(x-2) + 4x = 3x(x-2) \qquad \text{Divide out the common factors.}$$

$$4x - 8 + 4x = 3x^2 - 6x \qquad \text{Distributive property}$$

$$3x^2 - 14x + 8 = 0 \qquad \text{Standard form}$$
$$(3x - 2)(x - 4) = 0 \qquad \text{Factor.}$$
$$3x - 2 = 0 \quad \text{or} \quad x - 4 = 0 \qquad \text{Zero-factor property}$$
$$\text{Proposed solutions} \rightarrow x = \frac{2}{3} \quad \text{or} \qquad x = 4 \qquad \text{Solve for } x.$$

Neither $\frac{2}{3}$ nor 4 makes a denominator equal 0. Check to confirm that the solution set is $\left\{\frac{2}{3}, 4\right\}$.

Points to Remember

1. When simplifying rational expressions, the fundamental property is applied only after numerators and denominators have been *factored.*

2. When adding and subtracting rational expressions, the common denominator must be kept throughout the problem and in the final result.

3. When simplifying rational expressions, always check to see if the answer is in lowest terms. If it is not, use the fundamental property.

4. When solving equations with rational expressions, the LCD is used to clear the equation of fractions. Multiply each side by the LCD. (Notice how this use differs from that of the LCD in Point 2.)

5. When solving equations with rational expressions, reject any proposed solution that causes an original denominator to equal 0.

---

1. expression; $\dfrac{10}{p}$

2. expression; $\dfrac{y^3}{x^3}$

3. expression; $\dfrac{1}{2x^2(x + 2)}$

4. equation; $\{9\}$

5. equation; $\{39\}$

6. expression; $\dfrac{5k + 8}{k(k - 4)(k + 4)}$

7. expression; $\dfrac{y + 2}{y - 1}$

8. expression; $\dfrac{t - 5}{3(2t + 1)}$

9. expression; $\dfrac{13}{3(p + 2)}$

10. equation; $\left\{-1, \dfrac{12}{5}\right\}$

11. equation; $\left\{\dfrac{1}{7}, 2\right\}$

12. expression; $\dfrac{16}{3k}$

13. expression; $\dfrac{7}{12z}$

14. equation; $\{13\}$

15. expression;
$\dfrac{3m + 5}{(m + 3)(m + 2)(m + 1)}$

16. expression; $\dfrac{k + 3}{5(k - 1)}$

17. equation; $\varnothing$

18. equation; $\varnothing$

19. expression; $\dfrac{t + 2}{2(2t + 1)}$

20. equation; $\{-7\}$

For each exercise, indicate "expression" if an expression is to be simplified or "equation" if an equation is to be solved. Then simplify the expression or solve the equation.

1. $\dfrac{4}{p} + \dfrac{6}{p}$

2. $\dfrac{x^3 y^2}{x^2 y^4} \cdot \dfrac{y^5}{x^4}$

3. $\dfrac{1}{x^2 + x - 2} \div \dfrac{4x^2}{2x - 2}$

4. $\dfrac{8}{t - 5} = 2$

5. $\dfrac{x - 4}{5} = \dfrac{x + 3}{6}$

6. $\dfrac{2}{k^2 - 4k} + \dfrac{3}{k^2 - 16}$

7. $\dfrac{2y^2 + y - 6}{2y^2 - 9y + 9} \cdot \dfrac{y^2 - 2y - 3}{y^2 - 1}$

8. $\dfrac{3t^2 - t}{6t^2 + 15t} \div \dfrac{6t^2 + t - 1}{2t^2 - 5t - 25}$

9. $\dfrac{4}{p + 2} + \dfrac{1}{3p + 6}$

10. $\dfrac{1}{x} + \dfrac{1}{x - 3} = -\dfrac{5}{4}$

11. $\dfrac{3}{t - 1} + \dfrac{1}{t} = \dfrac{7}{2}$

12. $\dfrac{6}{k} - \dfrac{2}{3k}$

13. $\dfrac{5}{4z} - \dfrac{2}{3z}$

14. $\dfrac{x + 2}{3} = \dfrac{2x - 1}{5}$

15. $\dfrac{1}{m^2 + 5m + 6} + \dfrac{2}{m^2 + 4m + 3}$

16. $\dfrac{2k^2 - 3k}{20k^2 - 5k} \div \dfrac{2k^2 - 5k + 3}{4k^2 + 11k - 3}$

17. $\dfrac{2}{x + 1} + \dfrac{5}{x - 1} = \dfrac{10}{x^2 - 1}$

18. $\dfrac{3}{x + 3} + \dfrac{4}{x + 6} = \dfrac{9}{x^2 + 9x + 18}$

19. $\dfrac{4t^2 - t}{6t^2 + 10t} \div \dfrac{8t^2 + 2t - 1}{3t^2 + 11t + 10}$

20. $\dfrac{x}{x - 2} + \dfrac{3}{x + 2} = \dfrac{8}{x^2 - 4}$

7.7 Applications of Rational Expressions

**NOW TRY
EXERCISE 1**

In a certain fraction, the numerator is 4 less than the denominator. If 7 is added to both the numerator and denominator, the resulting fraction is equivalent to $\frac{7}{8}$. What is the original fraction?

For applications that lead to rational equations, the six-step problem-solving method of **Section 2.4** still applies.

OBJECTIVE 1 Solve problems about numbers.

EXAMPLE 1 Solving a Problem about an Unknown Number

If the same number is added to both the numerator and the denominator of the fraction $\frac{2}{5}$, the result is equivalent to $\frac{2}{3}$. Find the number.

Step 1 **Read** the problem carefully. We are trying to find a number.

Step 2 **Assign a variable.**

Let $x =$ the number added to the numerator and the denominator.

Step 3 **Write an equation.** The fraction

$$\frac{2 + x}{5 + x}$$

represents the result of adding the same number to both the numerator and the denominator. This result is equivalent to $\frac{2}{3}$, so the equation is written as follows.

$$\frac{2 + x}{5 + x} = \frac{2}{3}$$

Step 4 **Solve.** $3(5 + x)\dfrac{2 + x}{5 + x} = 3(5 + x)\dfrac{2}{3}$ Multiply by the LCD, $3(5 + x)$.

$$3(2 + x) = 2(5 + x) \qquad \text{Divide out the common factors.}$$

$$6 + 3x = 10 + 2x \qquad \text{Distributive property}$$

$$x = 4 \qquad \text{Subtract } 2x. \text{ Subtract } 6.$$

Step 5 **State the answer.** The number is 4.

Step 6 **Check** the solution in the words of the original problem. If 4 is added to both the numerator and the denominator of $\frac{2}{5}$, the result is $\frac{2 + 4}{5 + 4} = \frac{6}{9} = \frac{2}{3}$, as required.

NOW TRY ↺

OBJECTIVE 2 Solve problems about distance, rate, and time.

Recall the following formulas relating distance, rate, and time. Refer to **Example 6** in **Section 2.7** to review the basic use of these formulas.

Forms of the Distance Formula

$$d = rt \qquad r = \frac{d}{t} \qquad t = \frac{d}{r}$$

EXAMPLE 2 Solving a Problem about Distance, Rate, and Time

The Tickfaw River has a current of 3 mph. A motorboat takes as long to travel 12 mi downstream as to travel 8 mi upstream. What is the rate of the boat in still water?

Step 1 **Read** the problem again. We must find the rate (speed) of the boat in still water.

Step 2 **Assign a variable.**

Let x = the rate of the boat in still water.

Because the current pushes the boat when the boat is going downstream, the rate of the boat downstream will be the *sum* of the rate of the boat and the rate of the current, $(x + 3)$ mph.

Because the current slows the boat down when the boat is going upstream, the boat's rate going upstream will be the *difference* between the rate of the boat in still water and the rate of the current, $(x - 3)$ mph. See **FIGURE 1**.

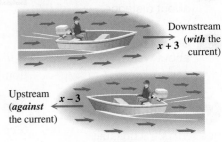

Downstream (*with* the current) $x + 3$

Upstream (*against* the current) $x - 3$

FIGURE 1

This information is summarized in the following table.

	d	r	t
Downstream	12	$x + 3$	
Upstream	8	$x - 3$	

Fill in the times using the formula $t = \frac{d}{r}$.

The time downstream is the distance divided by the rate.

$$t = \frac{d}{r} = \frac{12}{x + 3} \qquad \text{Time downstream}$$

The time upstream is that distance divided by that rate.

$$t = \frac{d}{r} = \frac{8}{x - 3} \qquad \text{Time upstream}$$

	d	r	t
Downstream	12	$x + 3$	$\dfrac{12}{x + 3}$
Upstream	8	$x - 3$	$\dfrac{8}{x - 3}$

Times are equal.

Step 3 **Write an equation.**

$$\frac{12}{x + 3} = \frac{8}{x - 3} \qquad \text{The time downstream equals the time upstream, so the two times from the table must be equal.}$$

Step 4 **Solve.**

$$(x + 3)(x - 3)\frac{12}{x + 3} = (x + 3)(x - 3)\frac{8}{x - 3} \qquad \text{Multiply by the LCD, } (x + 3)(x - 3).$$

$$12(x - 3) = 8(x + 3) \qquad \text{Divide out the common factors.}$$

$$12x - 36 = 8x + 24 \qquad \text{Distributive property}$$

$$4x = 60 \qquad \text{Subtract 8x. Add 36.}$$

$$x = 15 \qquad \text{Divide by 4.}$$

NOW TRY
EXERCISE 2
In her small boat, Jennifer can travel 12 mi downstream in the same amount of time that she can travel 4 mi upstream. The rate of the current is 2 mph. Find the rate of Jennifer's boat in still water.

Step 5   **State the answer.** The rate of the boat in still water is 15 mph.

Step 6   **Check.** The rate of the boat downstream is $15 + 3 = 18$ mph. Divide 12 mi by 18 mph to find the time.

$$t = \frac{d}{r} = \frac{12}{18} = \frac{2}{3} \text{ hr}$$

The rate of the boat upstream is $15 - 3 = 12$ mph. Divide 8 mi by 12 mph to find the time.

$$t = \frac{d}{r} = \frac{8}{12} = \frac{2}{3} \text{ hr}$$

The time upstream equals the time downstream, as required.

NOW TRY

OBJECTIVE 3   Solve problems about work.

Suppose that we can mow a lawn in 4 hr. Then after 1 hr, we will have mowed $\frac{1}{4}$ of the lawn. After 2 hr, we will have mowed $\frac{2}{4}$, or $\frac{1}{2}$, of the lawn, and so on. This idea is generalized as follows.

> ### Rate of Work
>
> If a job can be completed in t units of time, then the rate of work is
>
> $$\frac{1}{t} \text{ job per unit of time.}$$

PROBLEM-SOLVING HINT   Recall that the formula $d = rt$ says that distance traveled is equal to rate of travel multiplied by time traveled. Similarly, the fractional part of a job accomplished is equal to the rate of work multiplied by the time worked.

In the lawn-mowing example, after 3 hr, the fractional part of the job done is as follows.

$$\underbrace{\frac{1}{4}}_{\substack{\text{Rate of} \\ \text{work}}} \cdot \underbrace{3}_{\substack{\text{Time} \\ \text{worked}}} = \underbrace{\frac{3}{4}}_{\substack{\text{Fractional part} \\ \text{of job done}}}$$

After 4 hr, $\frac{1}{4}(4) = 1$ whole job has been done.

EXAMPLE 3   Solving a Problem about Work Rates

"If Joe can paint a house in 3 hr and Sam can paint the same house in 5 hr, how long does it take for them to do it together?" (*Source:* The movie *Little Big League.*)

Step 1   **Read** the problem again. We are looking for time working together.

Step 2   **Assign a variable.**

Let x = the number of hours it takes Joe and Sam to paint the house, working together.

NOW TRY ANSWER
2. 4 mph

NOW TRY
EXERCISE 3
Sarah can proofread a manuscript in 10 hr, while Joyce can proofread the same manuscript in 12 hr. How long will it take them to proofread the manuscript if they work together?

Certainly, x will be less than 3 because Joe alone can complete the job in 3 hr. We begin by making a table. Based on the preceding discussion, Joe's rate alone is $\frac{1}{3}$ job per hour, and Sam's rate is $\frac{1}{5}$ job per hour.

	Rate	Time Working Together	Fractional Part of the Job Done When Working Together
Joe	$\frac{1}{3}$	x	$\frac{1}{3}x$
Sam	$\frac{1}{5}$	x	$\frac{1}{5}x$

Sum is 1 whole job.

Step 3 **Write an equation.**

$$\underbrace{\text{Fractional part done by Joe}}_{} + \underbrace{\text{Fractional part done by Sam}}_{} = \underbrace{\text{1 whole job.}}_{}$$

$$\frac{1}{3}x \quad + \quad \frac{1}{5}x \quad = \quad 1$$

Together, Joe and Sam complete 1 whole job. Add their individual fractional parts and set the sum equal to 1.

Step 4 **Solve.** $15\left(\frac{1}{3}x + \frac{1}{5}x\right) = 15(1)$ Multiply by the LCD, 15.

$$15\left(\frac{1}{3}x\right) + 15\left(\frac{1}{5}x\right) = 15(1)$$ Distributive property

$$5x + 3x = 15$$ Simplify.

$$8x = 15$$ Combine like terms.

$$x = \frac{15}{8}$$ Divide by 8.

Step 5 **State the answer.** Working together, Joe and Sam can paint the house in $\frac{15}{8}$ hr, or $1\frac{7}{8}$ hr.

Step 6 **Check.** Substitute $\frac{15}{8}$ for x in the equation from Step 3.

$$\frac{1}{3}x + \frac{1}{5}x = 1$$ Equation from Step 3

$$\frac{1}{3}\left(\frac{15}{8}\right) + \frac{1}{5}\left(\frac{15}{8}\right) \stackrel{?}{=} 1$$ Let $x = \frac{15}{8}$.

$$\frac{5}{8} + \frac{3}{8} \stackrel{?}{=} 1$$ Multiply.

$$1 = 1 \ \checkmark$$ True

The answer $\frac{15}{8}$ hr, or $1\frac{7}{8}$ hr, is correct.

NOW TRY

NOW TRY ANSWER
3. $\frac{60}{11}$ hr, or $5\frac{5}{11}$ hr

NOTE An alternative approach in work problems is to consider the part of the job that can be done in 1 hr. For instance, in **Example 3** Joe can do the entire job in 3 hr and Sam can do it in 5 hr. Thus, their work rates, as we saw in **Example 3,** are $\frac{1}{3}$ and $\frac{1}{5}$, respectively. Because it takes them x hours to complete the job working together, in 1 hr they can paint $\frac{1}{x}$ of the house.

The amount painted by Joe in 1 hr plus the amount painted by Sam in 1 hr must equal the amount they can paint together in 1 hr. This relationship leads to the equation

$$\text{Amount by Joe} \rightarrow \frac{1}{3} + \overset{\overset{\textstyle\text{Amount by Sam}}{\textstyle\downarrow}}{\frac{1}{5}} = \frac{1}{x}. \leftarrow \text{Amount together}$$

Compare this equation with the one in **Example 3.** Multiplying each side by $15x$ leads to the same equation found in the third line of Step 4 in the example,

$$5x + 3x = 15.$$

The same solution results.

7.7 Exercises

FOR EXTRA HELP   ▶ MyMathLab®

▶ *Complete solution available in MyMathLab*

Concept Check Answer each question.

1. If a migrating hawk travels m mph in still air, what is its rate when it flies into a steady headwind of 5 mph? What is its rate with a tailwind of 5 mph?

2. Suppose Stephanie walks D miles at R mph in the same time that Wally walks d miles at r mph. What is an equation relating D, R, d, and r?

3. If it takes Katherine 10 hr to do a job, what is her rate?

4. If it takes Clayton 12 hr to do a job, how much of the job does he do in 8 hr?

*Use Steps 2 and 3 of the six-step problem-solving method to set up an equation to use in solving each problem. (Remember that Step 1 is to read the problem carefully.) Do not actually solve the equation. **See Example 1.***

5. The numerator of the fraction $\frac{5}{6}$ is increased by an amount so that the value of the resulting fraction is equivalent to $\frac{13}{3}$. By what amount was the numerator increased?

   (a) Let $x = $ _____ . (*Step 2*)

   (b) Write an expression for "the numerator of the fraction $\frac{5}{6}$ is increased by an amount."

   (c) Set up an equation to solve the problem. (*Step 3*)

6. If the same number is added to the numerator and subtracted from the denominator of the fraction $\frac{23}{12}$, the resulting fraction is equivalent to $\frac{3}{2}$. What is the number?

   (a) Let $x = $ _____ . (*Step 2*)

   (b) Write an expression for "a number is added to the numerator of $\frac{23}{12}$." Then write an expression for "the same number is subtracted from the denominator of $\frac{23}{12}$."

   (c) Set up an equation to solve the problem. (*Step 3*)

In each problem, state what x represents, write an equation, and answer the question. See Example 1.

7. In a certain fraction, the denominator is 4 less than the numerator. If 3 is added to both the numerator and the denominator, the resulting fraction is equivalent to $\frac{3}{2}$. What was the original fraction?

▶ **8.** In a certain fraction, the denominator is 6 more than the numerator. If 3 is added to both the numerator and the denominator, the resulting fraction is equivalent to $\frac{5}{7}$. What was the original fraction (*not* written in lowest terms)?

9. The denominator of a certain fraction is three times the numerator. If 2 is added to the numerator and subtracted from the denominator, the resulting fraction is equivalent to 1. What was the original fraction (*not* written in lowest terms)?

10. The numerator of a certain fraction is four times the denominator. If 6 is added to both the numerator and the denominator, the resulting fraction is equivalent to 2. What was the original fraction (*not* written in lowest terms)?

11. One-sixth of a number is 5 more than the same number. What is the number?

12. One-third of a number is 2 more than one-sixth of the same number. What is the number?

13. A quantity, its $\frac{3}{4}$, its $\frac{1}{2}$, and its $\frac{1}{3}$, added together, becomes 93. What is the quantity? (*Source: Rhind Mathematical Papyrus.*)

14. A quantity, its $\frac{2}{3}$, its $\frac{1}{2}$, and its $\frac{1}{7}$, added together, becomes 33. What is the quantity? (*Source: Rhind Mathematical Papyrus.*)

Solve each problem. See Example 6 in Section 2.7.

15. British explorer and endurance swimmer Lewis Gordon Pugh was the first person to swim at the North Pole. He swam 0.6 mi at 0.0319 mi per min in waters created by melted sea ice. What was his time (to three decimal places)? (*Source: The Gazette.*)

16. In the 2012 Summer Olympics, Ranomi Kromowidjojo of the Netherlands won the women's 100-m freestyle swimming event. Her rate was 1.8868 m per sec. What was her time (to two decimal places)? (*Source:* www.olympic.org)

17. Meseret Defar of Ethiopia won the women's 5000-m run in the 2012 Olympics with a time of 15.071 min. What was her rate (to three decimal places)? (*Source:* www.olympic.org)

18. Asli Cakir Alptekin of Turkey won the women's 1500-m run in the 2012 Olympics with a time of 4.1705 min. What was her rate (to three decimal places)? (*Source:* www.olympic.org)

19. In 2012, Matt Kenseth drove his Ford to victory in the Daytona 500 (mile) race with a rate of 140.256 mph. What was his time (to the nearest thousandth of an hour)? (*Source:* www.cbssports.com)

20. In 2011, Trevor Bayne drove his Ford to victory in the Daytona 500 (mile) race. His rate was 130.326 mph. What was his time (to the nearest thousandth of an hour)? (*Source:* www.cbssports.com)

Set up an equation to solve each problem. Do not actually solve the equation. ***See Example 2.***

21. Mitch flew his airplane 500 mi against the wind in the same time it took him to fly 600 mi with the wind. If the speed of the wind was 10 mph, what was the rate of his plane in still air? (Let x = rate of the plane in still air.)

	d	r	t
Against the Wind	500	$x - 10$	
With the Wind	600	$x + 10$	

22. Janet can row her boat 4 mph in still water. She takes as long to row 8 mi upstream as 24 mi downstream. What is the rate of the current? (Let x = rate of the current.)

	d	r	t
Upstream	8	$4 - x$	
Downstream	24	$4 + x$	

Solve each problem. ***See Example 2.***

▶ **23.** A boat can travel 20 mi against a current in the same time that it can travel 60 mi with the current. The rate of the current is 4 mph. Find the rate of the boat in still water.

24. Vince can fly his plane 200 mi against the wind in the same time it takes him to fly 300 mi with the wind. The wind blows at 30 mph. Find the rate of his plane in still air.

25. The sanderling is a small shorebird about 6.5 in. long, with a thin, dark bill and a wide, white wing stripe. If a sanderling can fly 30 mi with the wind in the same time it can fly 18 mi against the wind when the wind speed is 8 mph, what is the rate of the bird in still air?

26. Airplanes usually fly faster from west to east than from east to west because the prevailing winds go from west to east. The air distance between Chicago and London is about 4000 mi, while the air distance between New York and London is about 3500 mi. If a jet can fly eastbound from Chicago to London in the same time it can fly westbound from London to New York in a 35-mph wind, what is the rate of the plane in still air?

27. An airplane maintaining a constant airspeed takes as long to travel 450 mi with the wind as it does to travel 375 mi against the wind. If the wind is blowing at 15 mph, what is the rate of the plane in still air?

	d	r	t
Against the Wind			
With the Wind			

28. A river has a current of 4 km per hr. Find the rate of Jai's boat in still water if it travels 40 km downstream in the same time that it takes to travel 24 km upstream.

	d	r	t
Upstream			
Downstream			

29. Connie's boat travels at 12 mph. Find the rate of the current of the river if she can travel 6 mi upstream in the same amount of time she can travel 10 mi downstream.

30. Mohammed can travel 8 mi upstream in the same time it takes him to travel 12 mi downstream. His boat travels 15 mph in still water. What is the rate of the current?

31. The distance from Seattle, Washington, to Victoria, British Columbia, is about 148 mi by ferry. It takes about 4 hr less to travel by the same ferry from Victoria to Vancouver, British Columbia, a distance of about 74 mi. What is the average rate of the ferry?

32. Driving from Tulsa to Detroit, Dean averaged 50 mph. He figured that if he had averaged 60 mph, his driving time would have decreased 3 hr. How far is it from Tulsa to Detroit?

Set up an equation to solve each problem. Do not actually solve the equation. **See Example 3.**

33. Working alone, Edward can paint a room in 8 hr. Abdalla can paint the same room working alone in 6 hr. How long will it take them if they work together? (Let t represent the time they work together.)

	r	t	w
Edward		t	
Abdalla		t	

34. Donald can tune up his Chevy in 2 hr working alone. Jeff can do the same job in 3 hr working alone. How long would it take them if they worked together? (Let t represent the time they work together.)

	r	t	w
Donald		t	
Jeff		t	

Solve each problem. **See Example 3.**

▶ **35.** Heather, a high school mathematics teacher, gave a test on perimeter, area, and volume to her geometry classes. Working alone, it would take her 4 hr to grade the tests. Her student teacher, Courtney, would take 6 hr to grade the same tests. How long would it take them to grade these tests if they work together?

36. Zachary and Samuel are brothers who share a bedroom. By himself, Zachary can completely mess up their room in 20 min, while it would take Samuel only 12 min to do the same thing. How long would it take them to mess up the room together?

37. A pump can pump the water out of a flooded basement in 10 hr. A smaller pump takes 12 hr. How long would it take to pump the water from the basement with both pumps?

38. Lou's copier can do a printing job in 7 hr. Nora's copier can do the same job in 12 hr. How long would it take to do the job with both copiers?

39. An experienced employee can enter tax data into a computer twice as fast as a new employee. Working together, it takes the employees 2 hr. How long would it take the experienced employee working alone?

40. One roofer can put a roof on a house three times faster than another. Working together, they can roof a house in 4 days. How long would it take the faster roofer working alone?

41. One pipe can fill a swimming pool in 6 hr, and another pipe can do it in 9 hr. How long will it take the two pipes working together to fill the pool $\frac{3}{4}$ full?

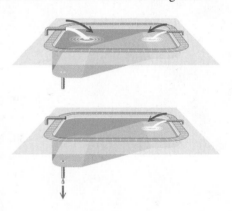

42. An inlet pipe can fill a swimming pool in 9 hr, and an outlet pipe can empty the pool in 12 hr. Through an error, both pipes are left open. How long will it take to fill the pool?

Extending Skills *Extend the concepts of* **Example 3** *to solve each problem.*

43. A cold-water faucet can fill a sink in 12 min, and a hot-water faucet can fill it in 15 min. The drain can empty the sink in 25 min. If both faucets are on and the drain is open, how long will it take to fill the sink?

44. Refer to **Exercise 42.** Assume that the error was discovered after both pipes had been running for 3 hr and the outlet pipe was then closed. How much more time would then be required to fill the pool? (*Hint:* Consider how much of the job had been done when the error was discovered.)

7.8 Variation

VOCABULARY
☐ direct variation
☐ constant of variation
☐ inverse variation

OBJECTIVE 1 Solve direct variation problems.

Suppose that gasoline costs $4.00 per gal. Then 1 gal costs $4.00, 2 gal costs $2($4.00) = \$8.00$, 3 gal costs $3($4.00) = \$12.00$, and so on. Each time, the total cost is obtained by multiplying the number of gallons by the price per gallon. In general, if k equals the price per gallon and x equals the number of gallons, then the total cost y is given by $y = kx$.

As the *number of gallons* **increases,**
the *total cost* **increases.**

The following is also true.

As the *number of gallons* **decreases,**
the *total cost* **decreases.**

The preceding discussion presents an example of *variation*. ***Two quantities vary directly if one is a constant multiple of the other.***

Direct Variation

y* varies directly as *x if there exists a constant k such that the following holds true.

$$y = kx$$

Also, y is said to be *proportional to x*. The constant k in the equation for direct variation is a numerical value, such as 4.00 in the gasoline price discussion. This value is the **constant of variation.**

Solving a Variation Problem

Step 1 Write the variation equation.

Step 2 Substitute the appropriate given values and solve for k.

Step 3 Write the variation equation with the value of k from Step 2.

Step 4 Substitute the remaining values, solve for the unknown, and find the required answer.

EXAMPLE 1 Using Direct Variation

Suppose y varies directly as x, and $y = 20$ when $x = 4$. Find y when $x = 9$.

Step 1 Because y varies directly as x, there is a constant k such that $y = kx$.

Step 2 We let $y = 20$ when $x = 4$, substitute these values into $y = kx$, and solve for k.

$$y = kx \qquad \text{Equation for direct variation}$$

$$20 = k \cdot 4 \qquad \text{Substitute the given values.}$$

$$k = 5 \quad \leftarrow \text{Constant of variation}$$

NOW TRY
EXERCISE 1

If W varies directly as r, and $W = 40$ when $r = 5$, find W when $r = 10$.

Step 3   Because $y = kx$ and $k = 5$, we have the following.

$$y = 5x \qquad \text{Let } k = 5.$$

Step 4   Now we can find the value of y when $x = 9$.

$$y = 5x \qquad \text{Variation equation}$$
$$y = 5 \cdot 9 \qquad \text{Let } x = 9.$$
$$y = 45 \qquad \text{Multiply.}$$

Thus, $y = 45$ when $x = 9$.

 NOW TRY

The direct variation equation $y = kx$ is a linear equation. However, other kinds of variation involve other types of equations.

Direct Variation as a Power

y **varies directly as the *n*th power of *x*** if there exists a real number k such that the following holds true.

$$y = kx^n$$

$\mathcal{A} = \pi r^2$

FIGURE 2

An example of direct variation as a power is the formula for the area of a circle, $\mathcal{A} = \pi r^2$. Here, π is the constant of variation, and the area varies directly as the square of the radius. See **FIGURE 2**.

NOW TRY
EXERCISE 2

If the height is constant, the volume of a right circular cylinder varies directly as the square of the radius of its base. If the volume is 80 ft³ when the radius is 4 ft, find the volume when the radius is 5 ft.

EXAMPLE 2   Solving a Direct Variation Problem

The distance a body falls from rest varies directly as the square of the time it falls (disregarding air resistance). If a sky diver falls 64 ft in 2 sec, how far will she fall in 8 sec?

Step 1   If d represents the distance the sky diver falls and t the time it takes to fall, then for some constant k, the following holds true.

$$d = kt^2$$

Step 2   To find the value of k, use the fact that the sky diver falls 64 ft in 2 sec.

$$d = kt^2 \qquad \text{Variation equation}$$
$$64 = k(2)^2 \qquad \text{Let } d = 64 \text{ and } t = 2.$$
$$64 = 4k \qquad \text{Apply the exponent.}$$
$$k = 16 \qquad \text{Divide by 4.}$$

Step 3   Substitute 16 for k to find the variation equation.

$$d = 16t^2$$

Step 4   Now let $t = 8$ to find the number of feet the sky diver will fall in 8 sec.

$$d = 16(8)^2 \qquad \text{Let } t = 8.$$
$$d = 16 \cdot 64 \qquad \text{Apply the exponent.}$$
$$d = 1024 \qquad \text{Multiply.}$$

NOW TRY ANSWERS
1. 80   2. 125 ft³

The sky diver will fall 1024 ft in 8 sec.

NOW TRY

OBJECTIVE 2 Solve inverse variation problems.

In direct variation, where $k > 0$, as x increases, y increases. Similarly, as x decreases, y decreases. Another type of variation is *inverse variation*. With inverse variation, where $k > 0$,

> As one variable ***increases,***
> the other variable ***decreases.***

As pressure increases, volume decreases.

For example, in a closed space, volume decreases as pressure increases, as illustrated by a trash compactor. See **FIGURE 3**. As the compactor presses down, the pressure on the trash increases. As a result, the trash occupies a smaller space.

FIGURE 3

Inverse Variation

y **varies inversely as** *x* if there exists a real number k such that the following holds true.

$$y = \frac{k}{x}$$

Also, *y* **varies inversely as the** *n***th power of** *x* if there exists a real number k such that the following holds true.

$$y = \frac{k}{x^n}$$

Another example of inverse variation comes from the distance formula.

$$d = rt \qquad \text{Distance formula}$$

$$t = \frac{d}{r} \qquad \text{Divide each side by } r.$$

In the form $t = \frac{d}{r}$, t (time) varies inversely as r (rate or speed), with d (distance) serving as the constant of variation. For example, if the distance between two cities is 300 mi, then

$$t = \frac{300}{r}$$

and the values of r and t might be any of the following.

$$
\left.\begin{array}{l}
r = 50, t = 6 \\
r = 60, t = 5 \\
r = 75, t = 4
\end{array}\right\} \begin{array}{l}\text{As } r \text{ increases,} \\ t \text{ decreases.}\end{array}
\qquad
\left.\begin{array}{l}
r = 30, t = 10 \\
r = 25, t = 12 \\
r = 20, t = 15
\end{array}\right\} \begin{array}{l}\text{As } r \text{ decreases,} \\ t \text{ increases.}\end{array}
$$

If we *increase* the rate (speed) at which we drive, the time *decreases.*

If we *decrease* the rate (speed) at which we drive, the time *increases.*

**NOW TRY
EXERCISE 3**

If t varies inversely as r, and $t = 12$ when $r = 3$, find t when $r = 6$.

EXAMPLE 3   Using Inverse Variation

Suppose y varies inversely as x, and $y = 3$ when $x = 8$. Find y when $x = 6$.

Because y varies inversely as x, there is a constant k such that $y = \frac{k}{x}$. We know that $y = 3$ when $x = 8$, so we can find k.

$$y = \frac{k}{x} \qquad \text{Equation for inverse variation}$$

$$3 = \frac{k}{8} \qquad \text{Substitute the given values.}$$

$$k = 24 \qquad \text{Multiply by 8. Rewrite } 24 = k \text{ as } k = 24.$$

We now have $y = \frac{24}{x}$, and so we let $x = 6$ to find y.

$$y = \frac{24}{x} = \frac{24}{6} = 4$$

Therefore, when $x = 6$, $y = 4$.   NOW TRY

 **NOW TRY
EXERCISE 4**

If the area is constant, the height of a triangle varies inversely as the base. If the height is 6 ft when the base is 4 ft, find the height when the base is 12 ft.

EXAMPLE 4   Using Inverse Variation

In the manufacturing of a phone-charging device, the cost of production varies inversely as the number of units produced. If 10,000 units are produced, the cost is $2 per unit. Find the cost per unit to produce 25,000 units.

$$\text{Let} \qquad x = \text{the number of units produced,}$$
$$\text{and} \qquad c = \text{the cost per unit.}$$

Because c varies inversely as x, there is a constant k such that the following holds true.

$$c = \frac{k}{x} \qquad \text{Equation for inverse variation}$$

$$2 = \frac{k}{10,000} \qquad \text{Substitute the given values.}$$

$$k = 20,000 \qquad \text{Multiply by 10,000. Rewrite.}$$

Now use $c = \frac{k}{x}$.

$$c = \frac{20,000}{25,000} = 0.80 \qquad \text{Let } k = 20,000 \text{ and } x = 25,000.$$

The cost per unit to make 25,000 devices is $0.80.   NOW TRY

NOW TRY ANSWERS
3. 6   4. 2 ft

7.8 Exercises

FOR EXTRA HELP

 MyMathLab®

 Complete solution available in MyMathLab

Concept Check   Use personal experience or intuition to determine whether the situation suggests direct *or* inverse *variation.**

1. The number of candy bars you buy and your total price for the candy

2. The rate and the distance traveled by a moving van in 3 hr

3. The amount of pressure put on the accelerator of a truck and the speed of the truck

*The authors thank Linda Kodama for suggesting the inclusion of exercises of this type.

4. The surface area of a beach ball and its diameter

5. For a triangle of constant area, the base and the height

6. For a rectangle of constant area, the length and the width

7. The intensity of a light source (such as a light bulb) and the distance from which a person views the light

8. The loudness of a sound source (such as a car horn) and the distance from which a person hears the sound

Concept Check *Determine whether each equation represents* direct *or* inverse *variation.*

9. $y = \dfrac{3}{x}$ **10.** $y = \dfrac{5}{x}$ **11.** $y = 10x^2$ **12.** $y = 3x^3$

13. $y = 50x$ **14.** $y = 200x$ **15.** $y = \dfrac{12}{x^2}$ **16.** $y = \dfrac{8}{x^3}$

17. *Concept Check* Complete each statement.

    (a) If the constant of variation is positive and y varies directly as x, then as x increases, y (*increases*/*decreases*).

    (b) If the constant of variation is positive and y varies inversely as x, then as x increases, y (*increases*/*decreases*).

18. Bill Veeck was the owner of several major league baseball teams in the 1950s and 1960s. He was known to often sit in the stands and enjoy games with his paying customers. Here is a quote attributed to him:

    I have discovered in 20 years of moving around a ballpark, that the knowledge of the game is usually in inverse proportion to the price of the seats.

    Explain in your own words the meaning of his statement. (To prove his point, Veeck once allowed the fans to vote on managerial decisions.)

Solve each problem involving direct or inverse variation. ***See Examples 1 and 3.***

▶ **19.** If x varies directly as y, and $x = 27$ when $y = 6$, find x when $y = 2$.

20. If z varies directly as x, and $z = 30$ when $x = 8$, find z when $x = 4$.

21. If d varies directly as t, and $d = 150$ when $t = 3$, find d when $t = 5$.

22. If d varies directly as r, and $d = 200$ when $r = 40$, find d when $r = 60$.

▶ **23.** If x varies inversely as y, and $x = 3$ when $y = 8$, find y when $x = 4$.

24. If z varies inversely as x, and $z = 50$ when $x = 2$, find z when $x = 25$.

25. If p varies inversely as q, and $p = 7$ when $q = 6$, find p when $q = 2$.

26. If m varies inversely as r, and $m = 12$ when $r = 8$, find m when $r = 16$.

27. If m varies inversely as p^2, and $m = 20$ when $p = 2$, find m when $p = 5$.

28. If a varies inversely as b^2, and $a = 48$ when $b = 4$, find a when $b = 7$.

29. If p varies inversely as q^2, and $p = 4$ when $q = \frac{1}{2}$, find p when $q = \frac{3}{2}$.

30. If z varies inversely as x^2, and $z = 9$ when $x = \frac{2}{3}$, find z when $x = \frac{5}{4}$.

Solve each problem. **See Examples 1–4.**

31. Simple interest on an investment varies directly as the rate of interest. If the interest is $48 when the interest rate is 3%, find the interest when the rate is 2.2%.

32. Simple interest on an investment varies directly as the rate of interest. If the interest is $60 when the interest rate is 4%, find the interest when the rate is 1.5%.

33. For a given base, the area of a triangle varies directly as its height. Find the area of a triangle with a height of 6 in., if the area is 10 in.^{2} when the height is 4 in.

34. For a given height, the area of a triangle varies directly as its base. Find the area of a triangle with a base of 8 in., if the area is 24 in.^{2} when the base is 12 in.

35. Hooke's law for an elastic spring states that the distance a spring stretches varies directly with the force applied. If a force of 75 lb stretches a certain spring 16 in., how much will a force of 200 lb stretch the spring?

36. The pressure exerted by water at a given point varies directly with the depth of the point beneath the surface of the water. Water exerts 4.34 lb per in.^{2} for every 10 ft traveled below the water's surface. What is the pressure exerted on a scuba diver at 20 ft?

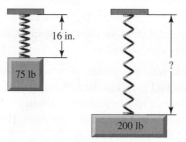

37. Over a specified distance, rate varies inversely with time. A Nissan 370Z on a test track goes a certain distance in one-half minute at 160 mph. Find the rate needed to go the same distance in three-fourths minute.

38. For a constant area, the length of a rectangle varies inversely as the width. The length of a rectangle is 27 ft when the width is 10 ft. Find the width of a rectangle with the same area if the length is 18 ft.

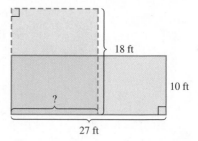

▶ 39. The current in a simple electrical circuit varies inversely as the resistance. If the current is 20 amps when the resistance is 5 ohms, what is the current when the resistance is 8 ohms?

40. If the temperature is constant, the pressure of a gas in a container varies inversely as the volume of the container. If the pressure is 10 lb per ft^2 in a container with volume 3 ft^3, what is the pressure in a container with volume 1.5 ft^3?

41. The force required to compress a spring varies directly as the change in the length of the spring. If a force of 12 lb is required to compress a certain spring 3 in., how much force is required to compress the spring 5 in.?

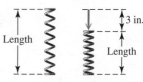

42. In the inversion of raw sugar, the rate of change of the amount of raw sugar varies directly as the amount of raw sugar remaining. The rate is 200 kg per hr when there are 800 kg left. What is the rate of change per hour when only 100 kg are left?

▶ **43.** The area of a circle varies directly as the square of its radius. A circle with radius 3 in. has area 28.278 in.^{2}. What is the area of a circle with radius 4.1 in. (to the nearest thousandth)?

44. For a body falling freely from rest (disregarding air resistance), the distance the body falls varies directly as the square of the time. If an object is dropped from the top of a tower 400 ft high and hits the ground in 5 sec, how far did it fall in the first 3 sec?

45. The amount of light (measured in *footcandles*) produced by a light source varies inversely as the square of the distance from the source. If the amount of light produced 4 ft from a light source is 75 footcandles, find the amount of light produced 9 ft from the same source.

46. The force with which Earth attracts an object above Earth's surface varies inversely as the square of the object's distance from the center of Earth. If an object 4000 mi from the center of Earth is attracted with a force of 160 lb, find the force of attraction on an object 6000 mi from the center of Earth.

Chapter 7	Summary

Key Terms

7.1

rational expression
lowest terms

7.3

least common
    denominator (LCD)

7.5

complex fraction

7.6

proposed solution
extraneous solution
    (extraneous value)

7.8

direct variation
constant of variation
inverse variation

Test Your Word Power

See how well you have learned the vocabulary in this chapter.

1. A **rational expression** is
  A. an algebraic expression made up of a term or the sum of a finite number of terms with real coefficients and whole number exponents
  B. a polynomial equation of degree 2
  C. an expression with one or more fractions in the numerator, or denominator, or both
  D. the quotient of two polynomials with denominator not 0.

2. A **complex fraction** is
  A. an algebraic expression made up of a term or the sum of a finite number of terms with real coefficients and whole number exponents
  B. a polynomial equation of degree 2
  C. a quotient with one or more fractions in the numerator, or denominator, or both
  D. the quotient of two polynomials with denominator not 0.

3. If two positive quantities x and y are in **direct variation,** and the constant of variation is positive, then
  A. as x increases, y decreases
  B. as x increases, y increases
  C. as x increases, y remains constant
  D. as x decreases, y remains constant.

4. If two positive quantities x and y are in **inverse variation,** and the constant of variation is positive, then
  A. as x increases, y decreases
  B. as x increases, y increases
  C. as x increases, y remains constant
  D. as x decreases, y remains constant.

ANSWERS

1. D; *Examples:* $-\dfrac{3}{4y}$, $\dfrac{5x^3}{x+2}$, $\dfrac{a+3}{a^2-4a-5}$ **2.** C; *Examples:* $\dfrac{\frac{2}{3}}{\frac{4}{7}}$, $\dfrac{x-\frac{1}{y}}{x+\frac{1}{y}}$, $\dfrac{\frac{2}{a+1}}{a^2-1}$

3. B; *Example:* The equation $y = 3x$ represents direct variation. When $x = 2$, $y = 6$. If x increases to 3, then y increases to $3(3) = 9$.

4. A; *Example:* The equation $y = \dfrac{3}{x}$ represents inverse variation. When $x = 1$, $y = 3$. If x increases to 2, then y decreases to $\dfrac{3}{2}$, or $1\frac{1}{2}$.

Quick Review

CONCEPTS

EXAMPLES

7.1   The Fundamental Property of Rational Expressions

To find the value(s) for which a rational expression is undefined, set the denominator equal to 0 and solve the equation.

Find the values for which the expression $\dfrac{x-4}{x^2-16}$ is undefined.

$$x^2 - 16 = 0$$

$$(x-4)(x+4) = 0 \qquad \text{Factor.}$$

$$x - 4 = 0 \quad \text{or} \quad x + 4 = 0 \qquad \text{Zero-factor property}$$

$$x = 4 \quad \text{or} \qquad x = -4 \qquad \text{Solve for } x.$$

The rational expression is undefined for 4 and -4, so $x \neq 4$ and $x \neq -4$.

Writing a Rational Expression in Lowest Terms

Step 1   Factor the numerator and denominator.

Step 2   Use the fundamental property to divide out common factors.

Write in lowest terms. $\dfrac{x^2-1}{(x-1)^2}$

$$= \dfrac{(x-1)(x+1)}{(x-1)(x-1)} \qquad \text{Factor.}$$

$$= \dfrac{x+1}{x-1} \qquad \text{Lowest terms}$$

There are often several different equivalent forms of a rational expression.

Give four equivalent forms of $-\dfrac{x-1}{x+2}$.

① → $\dfrac{-(x-1)}{x+2}$,   or $\dfrac{-x+1}{x+2}$ ← ②   Distribute the negative sign in the numerator.

③ → $\dfrac{x-1}{-(x+2)}$,   or $\dfrac{x-1}{-x-2}$ ← ④   Distribute the negative sign in the denominator.

7.2   Multiplying and Dividing Rational Expressions

Multiplying or Dividing Rational Expressions

Step 1   Note the operation. If the operation is division, use the definition of division to rewrite as multiplication.

Step 2   Multiply numerators and multiply denominators.

Step 3   Factor numerators and denominators completely.

Step 4   Write in lowest terms, using the fundamental property.

Note: Steps 2 and 3 may be interchanged based on personal preference.

Multiply. $\dfrac{3x+9}{x-5} \cdot \dfrac{x^2-3x-10}{x^2-9}$

$$= \dfrac{(3x+9)(x^2-3x-10)}{(x-5)(x^2-9)} \qquad \text{Multiply numerators and denominators.}$$

$$= \dfrac{3(x+3)(x-5)(x+2)}{(x-5)(x+3)(x-3)} \qquad \text{Factor.}$$

$$= \dfrac{3(x+2)}{x-3} \qquad \text{Lowest terms}$$

Divide. $\dfrac{2x+1}{x+5} \div \dfrac{6x^2-x-2}{x^2-25}$

$$= \dfrac{2x+1}{x+5} \cdot \dfrac{x^2-25}{6x^2-x-2} \qquad \text{Multiply by the reciprocal of the divisor.}$$

$$= \dfrac{(2x+1)(x^2-25)}{(x+5)(6x^2-x-2)} \qquad \text{Multiply numerators and denominators.}$$

$$= \dfrac{(2x+1)(x+5)(x-5)}{(x+5)(2x+1)(3x-2)} \qquad \text{Factor.}$$

$$= \dfrac{x-5}{3x-2} \qquad \text{Lowest terms}$$

CONCEPTS	EXAMPLES

7.3 Least Common Denominators

Finding the LCD

Step 1 Factor each denominator into prime factors.

Step 2 List each different factor the greatest number of times it appears.

Step 3 Multiply the factors from Step 2 to find the LCD.

Find the LCD for $\dfrac{3}{k^2 - 8k + 16}$ and $\dfrac{1}{4k^2 - 16k}$.

$$\left.\begin{array}{l} k^2 - 8k + 16 = (k - 4)^2 \\ 4k^2 - 16k = 4k(k - 4) \end{array}\right\} \begin{array}{l}\text{Factor each} \\ \text{denominator.}\end{array}$$

$$\text{LCD} = (k - 4)^2 \cdot 4 \cdot k$$
$$= 4k(k - 4)^2$$

Writing a Rational Expression with a Specified Denominator

Step 1 Factor both denominators.

Step 2 Decide what factor(s) the denominator must be multiplied by in order to equal the specified denominator.

Step 3 Multiply the rational expression by that factor divided by itself. (That is, multiply by 1.)

Find the numerator. $\dfrac{5}{2z^2 - 6z} = \dfrac{?}{4z^3 - 12z^2}$

$$\dfrac{5}{2z(z - 3)} = \dfrac{?}{4z^2(z - 3)}$$

$2z(z - 3)$ must be multiplied by $2z$ in order to obtain $4z^2(z - 3)$.

$$\dfrac{5}{2z(z - 3)} \cdot \dfrac{2z}{2z}$$

$$= \dfrac{10z}{4z^2(z - 3)}, \quad \text{or} \quad \dfrac{10z}{4z^3 - 12z^2}$$

7.4 Adding and Subtracting Rational Expressions

Adding Rational Expressions

Step 1 Find the LCD.

Step 2 Write each rational expression with the LCD as denominator.

Step 3 Add the numerators to obtain the numerator of the sum. The LCD is the denominator of the sum.

Step 4 Write in lowest terms.

Add. $\dfrac{2}{3m + 6} + \dfrac{m}{m^2 - 4}$

$$\left.\begin{array}{l} 3m + 6 = 3(m + 2) \\ m^2 - 4 = (m + 2)(m - 2) \end{array}\right\} \begin{array}{l}\text{The LCD is} \\ 3(m + 2)(m - 2).\end{array}$$

$$= \dfrac{2(m - 2)}{3(m + 2)(m - 2)} + \dfrac{3m}{3(m + 2)(m - 2)} \quad \begin{array}{l}\text{Write with} \\ \text{the LCD.}\end{array}$$

$$= \dfrac{2m - 4 + 3m}{3(m + 2)(m - 2)} \quad \begin{array}{l}\text{Add numerators.} \\ \text{Keep the same denominator.}\end{array}$$

$$= \dfrac{5m - 4}{3(m + 2)(m - 2)} \quad \text{Combine like terms.}$$

Subtracting Rational Expressions

Follow the same steps as for addition, but subtract in Step 3.

Subtract. $\dfrac{6}{k + 4} - \dfrac{2}{k}$ The LCD is $k(k + 4)$.

$$= \dfrac{6k}{(k + 4)k} - \dfrac{2(k + 4)}{k(k + 4)} \quad \text{Write with the LCD.}$$

$$= \dfrac{6k - 2(k + 4)}{k(k + 4)} \quad \begin{array}{l}\text{Subtract numerators.} \\ \text{Keep the same} \\ \text{denominator.}\end{array}$$

$$= \dfrac{6k - 2k - 8}{k(k + 4)} \quad \text{Distributive property}$$

$$= \dfrac{4k - 8}{k(k + 4)} \quad \text{Combine like terms.}$$

$$= \dfrac{4(k - 2)}{k(k + 4)} \quad \text{Factor.}$$

| **CONCEPTS** | **EXAMPLES** |

7.5 Complex Fractions

Simplifying Complex Fractions

Method 1 Simplify the numerator and denominator separately. Then divide the simplified numerator by the simplified denominator.

Method 2 Multiply the numerator and denominator of the complex fraction by the LCD of all the fractions in the numerator and denominator of the complex fraction. Write in lowest terms.

Simplify.

Method 1

$$\dfrac{\dfrac{1}{a} - a}{1 - a}$$

$$= \dfrac{\dfrac{1}{a} - \dfrac{a^2}{a}}{1 - a}$$

$$= \dfrac{\dfrac{1 - a^2}{a}}{1 - a}$$

$$= \dfrac{1 - a^2}{a} \div (1 - a)$$

$$= \dfrac{1 - a^2}{a} \cdot \dfrac{1}{1 - a}$$

$$= \dfrac{(1 - a)(1 + a)}{a(1 - a)}$$

$$= \dfrac{1 + a}{a}$$

Method 2

$$\dfrac{\dfrac{1}{a} - a}{1 - a}$$

$$= \dfrac{\left(\dfrac{1}{a} - a\right)a}{(1 - a)a}$$

$$= \dfrac{\dfrac{a}{a} - a^2}{(1 - a)a}$$

$$= \dfrac{1 - a^2}{(1 - a)a}$$

$$= \dfrac{(1 + a)(1 - a)}{(1 - a)a}$$

$$= \dfrac{1 + a}{a}$$

7.6 Solving Equations with Rational Expressions

Solving Equations with Rational Expressions

Step 1 Multiply each side of the equation by the LCD to clear the equation of fractions. Be sure to distribute to *every* term on *both* sides.

Solve. $\dfrac{x}{x - 3} + \dfrac{4}{x + 3} = \dfrac{18}{x^2 - 9}$

$\dfrac{x}{x - 3} + \dfrac{4}{x + 3} = \dfrac{18}{(x - 3)(x + 3)}$ Factor.

The LCD is $(x - 3)(x + 3)$. Note that 3 and -3 cannot be solutions, as they cause a denominator to equal 0.

$$(x - 3)(x + 3)\left(\dfrac{x}{x - 3} + \dfrac{4}{x + 3}\right)$$

$$= (x - 3)(x + 3)\dfrac{18}{(x - 3)(x + 3)}$$

Multiply by the LCD, $(x - 3)(x + 3)$.

Step 2 Solve the resulting equation.

$x(x + 3) + 4(x - 3) = 18$ Distributive property

$x^2 + 3x + 4x - 12 = 18$ Distributive property

$x^2 + 7x - 30 = 0$ Standard form

$(x - 3)(x + 10) = 0$ Factor.

$x - 3 = 0$ or $x + 10 = 0$ Zero-factor property

Step 3 Check each proposed solution by substituting it in the original equation. Reject any value that causes an original denominator to equal 0.

Reject ⟶ $x = 3$ or $x = -10$ Solve.

Because 3 causes denominators to equal 0, it is an extraneous value. Check that the only solution is -10. Thus, $\{-10\}$ is the solution set.

CONCEPTS	EXAMPLES

7.7   Applications of Rational Expressions

Solving Problems about Distance, Rate, and Time

Use the formulas relating d, r, and t.

$$d = rt, \quad r = \frac{d}{t}, \quad t = \frac{d}{r}$$

A small plane flew from Chicago to Kansas City averaging 145 mph. The trip took 3.5 hr. What is the distance between Chicago and Kansas City?

$$145 \;\cdot\; 3.5 \;=\; 507.5 \text{ mi}$$

Rate   Time   Distance

Solving Problems about Work

Step 1   Read the problem carefully.

Step 2   Assign a variable. State what the variable represents. Organize the information from the problem in a table. If a job is done in t units of time, then the rate is $\frac{1}{t}$.

It takes the regular mail carrier 6 hr to cover her route. A substitute takes 8 hr to cover the same route. How long would it take them to cover the route together?

Let $x = $ the number of hours required to cover the route together.

	Rate	Time	Part of the Job Done
Regular	$\frac{1}{6}$	x	$\frac{1}{6}x$
Substitute	$\frac{1}{8}$	x	$\frac{1}{8}x$

Multiply rate by time to find the fractional part done.

Step 3   Write an equation. The sum of the fractional parts should equal 1 (whole job).

$$\frac{1}{6}x + \frac{1}{8}x = 1 \qquad \text{The parts add to 1 whole job.}$$

Step 4   Solve the equation.

$$24\left(\frac{1}{6}x + \frac{1}{8}x\right) = 24(1) \qquad \text{The LCD is 24.}$$

$$4x + 3x = 24 \qquad \text{Multiply.}$$

$$7x = 24 \qquad \text{Combine like terms.}$$

$$x = \frac{24}{7} \qquad \text{Divide by 7.}$$

Step 5   State the answer.

It would take them $\frac{24}{7}$ hr, or $3\frac{3}{7}$ hr, to cover the route together.

Step 6   Check the solution.

The solution checks because $\frac{1}{6}\left(\frac{24}{7}\right) + \frac{1}{8}\left(\frac{24}{7}\right) = 1$.

7.8   Variation

Solving Variation Problems

Step 1   Write the variation equation.

$$y = kx \quad \text{or} \quad y = kx^n \qquad \text{Direct variation}$$

$$y = \frac{k}{x} \quad \text{or} \quad y = \frac{k}{x^n} \qquad \text{Inverse variation}$$

Step 2   Find k by substituting the appropriate given values of x and y into the equation.

Step 3   Write the variation equation with the value of k from Step 2.

Step 4   Substitute the remaining values, and solve for the unknown.

If y varies inversely as x, and $y = 4$ when $x = 9$, find y when $x = 6$.

$$y = \frac{k}{x} \qquad \text{Equation for } y \text{ varying inversely as } x$$

$$4 = \frac{k}{9} \qquad \text{Substitute given values.}$$

$$k = 36 \qquad \text{Solve for } k.$$

$$y = \frac{36}{x} \qquad \text{Let } k = 36.$$

$$y = \frac{36}{6} \qquad \text{Let } x = 6.$$

$$y = 6 \qquad \text{Divide.}$$

Chapter 7 Review Exercises

7.1 *Find the numerical value of each rational expression for (a) $x = -2$ and (b) $x = 4$.*

1. $\dfrac{4x - 3}{5x + 2}$

2. $\dfrac{3x}{x^2 - 4}$

3. Explain the process used to determine the values of the variable for which a rational expression is undefined.

Find any values of the variable for which each rational expression is undefined. Write answers with the symbol $\neq$.

4. $\dfrac{4}{x - 3}$

5. $\dfrac{y + 3}{2y}$

6. $\dfrac{2k + 1}{3k^2 + 17k + 10}$

Write each rational expression in lowest terms.

7. $\dfrac{5a^3b^3}{15a^4b^2}$

8. $\dfrac{m - 4}{4 - m}$

9. $\dfrac{4x^2 - 9}{6 - 4x}$

10. $\dfrac{4p^2 + 8pq - 5q^2}{10p^2 - 3pq - q^2}$

Write four equivalent forms for each rational expression.

11. $-\dfrac{4x - 9}{2x + 3}$

12. $-\dfrac{8 - 3x}{3 - 6x}$

7.2 *Multiply or divide. Write each answer in lowest terms.*

13. $\dfrac{18p^3}{6} \cdot \dfrac{24}{p^4}$

14. $\dfrac{8x^2}{12x^5} \cdot \dfrac{6x^4}{2x}$

15. $\dfrac{x - 3}{4} \cdot \dfrac{5}{2x - 6}$

16. $\dfrac{2r + 3}{r - 4} \cdot \dfrac{r^2 - 16}{6r + 9}$

17. $\dfrac{6a^2 + 7a - 3}{2a^2 - a - 6} \div \dfrac{a + 5}{a - 2}$

18. $\dfrac{y^2 - 6y + 8}{y^2 + 3y - 18} \div \dfrac{y - 4}{y + 6}$

19. $\dfrac{2p^2 + 13p + 20}{p^2 + p - 12} \cdot \dfrac{p^2 + 2p - 15}{2p^2 + 7p + 5}$

20. $\dfrac{3z^2 + 5z - 2}{9z^2 - 1} \cdot \dfrac{9z^2 + 6z + 1}{z^2 + 5z + 6}$

7.3 *Find the LCD for the fractions in each list.*

21. $\dfrac{4}{9y}, \dfrac{7}{12y^2}, \dfrac{5}{27y^4}$

22. $\dfrac{3}{x^2 + 4x + 3}, \dfrac{5}{x^2 + 5x + 4}$

Write each rational expression with the given denominator.

23. $\dfrac{3}{2a^3} = \dfrac{?}{10a^4}$

24. $\dfrac{9}{x - 3} = \dfrac{?}{18 - 6x}$

25. $\dfrac{-3y}{2y - 10} = \dfrac{?}{50 - 10y}$

26. $\dfrac{4b}{b^2 + 2b - 3} = \dfrac{?}{(b + 3)(b - 1)(b + 2)}$

7.4 *Add or subtract. Write each answer in lowest terms.*

27. $\dfrac{10}{x} + \dfrac{5}{x}$

28. $\dfrac{6}{3p} - \dfrac{12}{3p}$

29. $\dfrac{9}{k} - \dfrac{5}{k - 5}$

---

1. (a) $\dfrac{11}{8}$ (b) $\dfrac{13}{22}$

2. (a) undefined (b) 1

3. Set the denominator equal to 0 and solve the equation. Any solutions are values for which the rational expression is undefined.

4. $x \neq 3$

5. $y \neq 0$

6. $k \neq -5, -\dfrac{2}{3}$

7. $\dfrac{b}{3a}$

8. -1

9. $\dfrac{-(2x + 3)}{2}$

10. $\dfrac{2p + 5q}{5p + q}$

Answers may vary in Exercises 11 and 12.

11. $\dfrac{-(4x - 9)}{2x + 3}, \dfrac{-4x + 9}{2x + 3},$ $\dfrac{4x - 9}{-(2x + 3)}, \dfrac{4x - 9}{-2x - 3}$

12. $\dfrac{-(8 - 3x)}{3 - 6x}, \dfrac{-8 + 3x}{3 - 6x},$ $\dfrac{8 - 3x}{-(3 - 6x)}, \dfrac{8 - 3x}{-3 + 6x}$

13. $\dfrac{72}{p}$

14. 2

15. $\dfrac{5}{8}$

16. $\dfrac{r + 4}{3}$

17. $\dfrac{3a - 1}{a + 5}$

18. $\dfrac{y - 2}{y - 3}$

19. $\dfrac{p + 5}{p + 1}$

20. $\dfrac{3z + 1}{z + 3}$

21. $108y^4$

22. $(x + 3)(x + 1)(x + 4)$

23. $\dfrac{15a}{10a^4}$

24. $\dfrac{-54}{18 - 6x}$

25. $\dfrac{15y}{50 - 10y}$

26. $\dfrac{4b(b + 2)}{(b + 3)(b - 1)(b + 2)}$

27. $\dfrac{15}{x}$

28. $-\dfrac{2}{p}$

29. $\dfrac{4k - 45}{k(k - 5)}$

30. $\dfrac{28+11y}{y(7+y)}$

31. $\dfrac{-2-3m}{6}$ **32.** $\dfrac{3(16-x)}{4x^2}$

33. $\dfrac{7a+6b}{(a-2b)(a+2b)}$

34. $\dfrac{-k^2-6k+3}{3(k+3)(k-3)}$

35. $\dfrac{5z-16}{z(z+6)(z-2)}$

36. $\dfrac{-13p+33}{p(p-2)(p-3)}$

37. $\dfrac{4(y-3)}{y+3}$ **38.** $\dfrac{10}{13}$

39. $\dfrac{xw+1}{xw-1}$ **40.** $\dfrac{(q-p)^2}{pq}$

41. $(x-5)(x-3)$, or $x^2-8x+15$

42. $\dfrac{1-r-t}{1+r+t}$

43. $\varnothing$ **44.** $\{-16\}$

45. $\{0\}$ **46.** $\{3\}$

47. $t=\dfrac{Ry}{m}$ **48.** $y=\dfrac{4x+5}{3}$

49. $m=\dfrac{4+p^2q}{3p^2}$

50. $\dfrac{20}{15}$

51. $\dfrac{3}{18}$ **52.** 10 mph

53. $3\dfrac{1}{13}$ hr **54.** 2 hr

55. 4 cm **56.** $\dfrac{36}{5}$

---

30. $\dfrac{4}{y}+\dfrac{7}{7+y}$ **31.** $\dfrac{m}{3}-\dfrac{2+5m}{6}$ **32.** $\dfrac{12}{x^2}-\dfrac{3}{4x}$

33. $\dfrac{5}{a-2b}+\dfrac{2}{a+2b}$ **34.** $\dfrac{4}{k^2-9}-\dfrac{k+3}{3k-9}$

35. $\dfrac{8}{z^2+6z}-\dfrac{3}{z^2+4z-12}$ **36.** $\dfrac{11}{2p-p^2}-\dfrac{2}{p^2-5p+6}$

7.5 *Simplify each complex fraction.*

37. $\dfrac{\dfrac{y-3}{y}}{\dfrac{y+3}{4y}}$ **38.** $\dfrac{\dfrac{2}{3}-\dfrac{1}{6}}{\dfrac{1}{4}+\dfrac{2}{5}}$ **39.** $\dfrac{x+\dfrac{1}{w}}{x-\dfrac{1}{w}}$

40. $\dfrac{\dfrac{1}{p}-\dfrac{1}{q}}{\dfrac{1}{q-p}}$ **41.** $\dfrac{\dfrac{x^2-25}{x+3}}{\dfrac{x+5}{x^2-9}}$ **42.** $\dfrac{\dfrac{1}{r+t}-1}{\dfrac{1}{r+t}+1}$

7.6 *Solve each equation, and check the solutions.*

43. $\dfrac{3x-1}{x-2}=\dfrac{5}{x-2}+1$ **44.** $\dfrac{4-z}{z}+\dfrac{3}{2}=\dfrac{-4}{z}$

45. $\dfrac{3}{x+4}-\dfrac{2x}{5}=\dfrac{3}{x+4}$ **46.** $\dfrac{3}{m-2}+\dfrac{1}{m-1}=\dfrac{7}{m^2-3m+2}$

Solve each formula for the specified variable.

47. $m=\dfrac{Ry}{t}$ for t **48.** $x=\dfrac{3y-5}{4}$ for y **49.** $p^2=\dfrac{4}{3m-q}$ for m

7.7 *Solve each problem.*

50. In a certain fraction, the denominator is 5 less than the numerator. If 5 is added to both the numerator and the denominator, the resulting fraction is equivalent to $\dfrac{5}{4}$. Find the original fraction (*not* written in lowest terms).

51. The denominator of a certain fraction is six times the numerator. If 3 is added to the numerator and subtracted from the denominator, the resulting fraction is equivalent to $\dfrac{2}{5}$. Find the original fraction (*not* written in lowest terms).

52. A plane flies 350 mi with the wind in the same time that it can fly 310 mi against the wind. The plane has a speed of 165 mph in still air. Find the speed of the wind.

53. Sarita can plant her garden in 5 hr working alone. A friend can do the same job in 8 hr. How long would it take them if they worked together?

54. The head gardener can mow the lawns in the city park twice as fast as his assistant. Working together, they can complete the job in $1\dfrac{1}{3}$ hr. How long would it take the head gardener working alone?

7.8 *Solve each problem.*

55. If a parallelogram has a fixed area, the height varies inversely as the base. A parallelogram has a height of 8 cm and a base of 12 cm. Find the height if the base is changed to 24 cm.

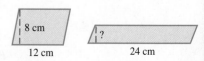

56. If y varies directly as x, and $x=12$ when $y=5$, find x when $y=3$.

Chapter 7 | Mixed Review Exercises

Perform each indicated operation.

1. $\dfrac{m+7}{(m-1)(m+1)}$

2. $8p^2$

3. $\dfrac{1}{6}$ **4.** 3

5. $\dfrac{z+7}{(z+1)(z-1)^2}$

6. $\dfrac{-t-1}{(t+2)(t-2)}$,

   or $\dfrac{t+1}{(2+t)(2-t)}$

7. $\{-2, 3\}$ **8.** $\{2\}$

9. $v = at + w$

10. 150 km per hr

11. $5\dfrac{1}{11}$ hr   **12.** 24

13. 4   **14.** -16

1. $\dfrac{4}{m-1} - \dfrac{3}{m+1}$

2. $\dfrac{8p^5}{5} \div \dfrac{2p^3}{10}$

3. $\dfrac{r-3}{8} \div \dfrac{3r-9}{4}$

4. $\dfrac{\dfrac{5}{x}-1}{\dfrac{5-x}{3x}}$

5. $\dfrac{4}{z^2-2z+1} - \dfrac{3}{z^2-1}$

6. $\dfrac{1}{t^2-4} + \dfrac{1}{2-t}$

Solve each equation.

7. $\dfrac{2}{z} - \dfrac{z}{z+3} = \dfrac{1}{z+3}$

8. $\dfrac{1}{x^2-1} = \dfrac{1}{x+1}$

9. $a = \dfrac{v-w}{t}$ for v

Solve each problem.

10. Rob flew his plane 400 km with the wind in the same time it took him to travel 200 km against the wind. The speed of the wind is 50 km per hr. Find the rate of the plane in still air.

11. With spraying equipment, Lizette can paint the woodwork in a small house in 8 hr. Seyed needs 14 hr to complete the same job painting by hand. If Lizette and Seyed work together, how long will it take them to paint the woodwork?

12. If w varies inversely as z, and $w = 16$ when $z = 3$, find w when $z = 2$.

13. In a rectangle of constant area, the length and the width vary inversely. When the length is 24, the width is 2. What is the width when the length is 12?

14. If x varies directly as the cube of y, and $x = 54$ when $y = 3$, find x when $y = -2$.

Chapter 7 | Test

FOR EXTRA HELP *Step-by-step test solutions are found on the Chapter Test Prep Videos available in* MyMathLab® *or on* You[Tube]™.

▶ *View the complete solutions to all Chapter Test exercises in MyMathLab.*

[7.1]

1. (a) $\dfrac{11}{6}$ (b) undefined

2. $x \neq -2, 4$

3. (Answers may vary.)

   $\dfrac{-(6x-5)}{2x+3}$, $\dfrac{-6x+5}{2x+3}$,

   $\dfrac{6x-5}{-(2x+3)}$, $\dfrac{6x-5}{-2x-3}$

4. $-3x^2y^3$ **5.** $\dfrac{3a+2}{a-1}$

[7.2]

6. $\dfrac{25}{27}$ **7.** $\dfrac{3k-2}{3k+2}$

8. $\dfrac{a-1}{a+4}$ **9.** $\dfrac{x-5}{3-x}$

[7.3]

10. $150p^5$

11. $(2r+3)(r+2)(r-5)$

1. Find the numerical value of $\dfrac{6r+1}{2r^2-3r-20}$ for **(a)** $r = -2$ and **(b)** $r = 4$.

2. Find any values for which $\dfrac{3x-1}{x^2-2x-8}$ is undefined. Write the answer with the symbol $\neq$.

3. Write four equivalent forms of the rational expression $-\dfrac{6x-5}{2x+3}$.

Write each rational expression in lowest terms.

4. $\dfrac{-15x^6y^4}{5x^4y}$

5. $\dfrac{6a^2+a-2}{2a^2-3a+1}$

Multiply or divide. Write each answer in lowest terms.

6. $\dfrac{5(d-2)}{9} \div \dfrac{3(d-2)}{5}$

7. $\dfrac{6k^2-k-2}{8k^2+10k+3} \cdot \dfrac{4k^2+7k+3}{3k^2+5k+2}$

8. $\dfrac{4a^2+9a+2}{3a^2+11a+10} \div \dfrac{4a^2+17a+4}{3a^2+2a-5}$

9. $\dfrac{x^2-10x+25}{9-6x+x^2} \cdot \dfrac{x-3}{5-x}$

Find the least common denominator for the fractions in each list.

10. $\dfrac{-3}{10p^2}$, $\dfrac{21}{25p^3}$, $\dfrac{-7}{30p^5}$

11. $\dfrac{r+1}{2r^2+7r+6}$, $\dfrac{-2r+1}{2r^2-7r-15}$

12. $\dfrac{240p^2}{64p^3}$

13. $\dfrac{21}{42m - 84}$

[7.4]

14. 2

15. $\dfrac{-14}{5(y+2)}$

16. $\dfrac{-x^2 + x + 1}{3 - x}$, or $\dfrac{x^2 - x - 1}{x - 3}$

17. $\dfrac{-m^2 + 7m + 2}{(2m+1)(m-5)(m-1)}$

[7.5]

18. $\dfrac{2k}{3p}$

19. $\dfrac{-2 - x}{4 + x}$

[7.6]

20. $\left\{ -\dfrac{1}{2}, 1 \right\}$

21. $\left\{ -\dfrac{1}{2} \right\}$

22. $D = \dfrac{dF - k}{F}$, or $D = d - \dfrac{k}{F}$

[7.7]

23. 3 mph   **24.** $2\dfrac{2}{9}$ hr

[7.8]

25. 27   **26.** 27 days

Write each rational expression with the given denominator.

12. $\dfrac{15}{4p} = \dfrac{?}{64p^3}$

13. $\dfrac{3}{6m - 12} = \dfrac{?}{42m - 84}$

Add or subtract. Write each answer in lowest terms.

14. $\dfrac{4x + 2}{x + 5} + \dfrac{-2x + 8}{x + 5}$

15. $\dfrac{-4}{y + 2} + \dfrac{6}{5y + 10}$

16. $\dfrac{x + 1}{3 - x} + \dfrac{x^2}{x - 3}$

17. $\dfrac{3}{2m^2 - 9m - 5} - \dfrac{m + 1}{2m^2 - m - 1}$

Simplify each complex fraction.

18. $\dfrac{\dfrac{2p}{k^2}}{\dfrac{3p^2}{k^3}}$

19. $\dfrac{\dfrac{1}{x + 3} - 1}{1 + \dfrac{1}{x + 3}}$

Solve each equation.

20. $\dfrac{3x}{x + 1} = \dfrac{3}{2x}$

21. $\dfrac{2x}{x - 3} + \dfrac{1}{x + 3} = \dfrac{-6}{x^2 - 9}$

22. $F = \dfrac{k}{d - D}$ for D

Solve each problem.

23. A boat travels 7 mph in still water. It takes as long to travel 20 mi upstream as 50 mi downstream. Find the rate of the current.

24. Sanford can paint a room in his house, working alone, in 5 hr. His neighbor can do the job in 4 hr. How long will it take them to paint the room if they work together?

25. If x varies directly as y, and $x = 12$ when $y = 4$, find x when $y = 9$.

26. Under certain conditions, the length of time that it takes for fruit to ripen during the growing season varies inversely as the average maximum temperature during the season. If it takes 25 days for fruit to ripen with an average maximum temperature of $80°$, find the number of days it would take at $75°$. Round the answer to the nearest whole number.

Chapters R–7   Cumulative Review Exercises

[R.1, 1.7] **1.** 2

[2.3] **2.** $\{17\}$

[2.5] **3.** $b = \dfrac{2\mathcal{A}}{h}$

[2.6] **4.** $\left\{ -\dfrac{2}{7} \right\}$

[2.8] **5.** $[-8, \infty)$

[3.1, 3.2]

6. **(a)** $(-3, 0)$ **(b)** $(0, -4)$

7.   **[5.4]** **8.**

[4.1–4.3]

9. $\{(-1, 3)\}$ **10.** $\varnothing$

1. Evaluate $3 + 4\left(\dfrac{1}{2} - \dfrac{3}{4} \right)$.

Solve.

2. $3(2y - 5) = 2 + 5y$

3. $\mathcal{A} = \dfrac{1}{2}bh$ for b

4. $\dfrac{2 + m}{3} = \dfrac{2 - m}{4}$

5. $5y \le 6y + 8$

6. Consider the graph of $4x + 3y = -12$.

   (a) What is the x-intercept?   **(b)** What is the y-intercept?

Graph each equation.

7. $y = -3x + 2$

8. $y = -x^2 + 1$

Solve each system.

9. $4x - y = -7$
    $5x + 2y = 1$

10. $5x + 2y = 7$
     $10x + 4y = 12$

[5.1, 5.2]

11. $\dfrac{1}{2^4 x^7}$ 12. $\dfrac{1}{m^6}$

[5.4]

13. $k^2 + 2k + 1$

[5.6]

14. $4a^2 - 4ab + b^2$

[5.5]

15. $3y^3 + 8y^2 + 12y - 5$

[5.7]

16. $6p^2 + 7p + 1 + \dfrac{3}{p-1}$

[6.3]

17. $(4t + 3v)(2t + v)$

18. prime

[6.4]

19. $(4x^2 + 1)(2x + 1)(2x - 1)$

[6.5]

20. $\{-3, 5\}$ 21. $\left\{5, -\dfrac{1}{2}, \dfrac{2}{3}\right\}$

[6.6]

22. -2 or -1 23. 6 m

[7.1]

24. A   25. D

[7.4]

26. $\dfrac{4}{q}$

27. $\dfrac{3r + 28}{7r}$ 28. $\dfrac{7}{15(q - 4)}$

29. $\dfrac{-k - 5}{k(k + 1)(k - 1)}$

[7.2]

30. $\dfrac{7(2z + 1)}{24}$

[7.5]

31. $\dfrac{195}{29}$

[7.6]

32. $\left\{\dfrac{21}{2}\right\}$ 33. $\{-2, 1\}$

[7.7]

34. $1\dfrac{1}{5}$ hr

[7.8]

35. 32.97 in.

Simplify each expression. Write answers with only positive exponents.

11. $\dfrac{(2x^3)^{-1} \cdot x}{2^3 x^5}$

12. $\dfrac{(m^{-2})^3 m}{m^5 m^{-4}}$

Perform each indicated operation.

13. $(2k^2 + 3k) - (k^2 + k - 1)$

14. $(2a - b)^2$

15. $(y^2 + 3y + 5)(3y - 1)$

16. $\dfrac{12p^3 + 2p^2 - 12p + 4}{2p - 2}$

Factor completely.

17. $8t^2 + 10tv + 3v^2$ 18. $8r^2 - 9rs + 12s^2$ 19. $16x^4 - 1$

Solve each equation.

20. $r^2 = 2r + 15$

21. $(r - 5)(2r + 1)(3r - 2) = 0$

Solve each problem.

22. One number is 4 greater than another. The product of the numbers is 2 less than the lesser number. Find the lesser number.

23. The length of a rectangle is 2 m less than twice the width. The area is 60 m². Find the width of the rectangle.

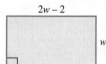

24. Which one of the following is equal to 1 for *all* real numbers?

   A. $\dfrac{k^2 + 2}{k^2 + 2}$ **B.** $\dfrac{4 - m}{4 - m}$ **C.** $\dfrac{2x + 9}{2x + 9}$ **D.** $\dfrac{x^2 - 1}{x^2 - 1}$

25. Which one of the following rational expressions is *not* equivalent to $\dfrac{4 - 3x}{7}$?

   A. $-\dfrac{-4 + 3x}{7}$ **B.** $-\dfrac{4 - 3x}{-7}$ **C.** $\dfrac{-4 + 3x}{-7}$ **D.** $\dfrac{-(3x + 4)}{7}$

Perform each operation, and write the answer in lowest terms.

26. $\dfrac{5}{q} - \dfrac{1}{q}$

27. $\dfrac{3}{7} + \dfrac{4}{r}$

28. $\dfrac{4}{5q - 20} - \dfrac{1}{3q - 12}$

29. $\dfrac{2}{k^2 + k} - \dfrac{3}{k^2 - k}$

30. $\dfrac{7z^2 + 49z + 70}{16z^2 + 72z - 40} \div \dfrac{3z + 6}{4z^2 - 1}$

31. $\dfrac{\dfrac{4}{a} + \dfrac{5}{2a}}{\dfrac{7}{6a} - \dfrac{1}{5a}}$

Solve each equation. Check the solutions.

32. $\dfrac{r + 2}{5} = \dfrac{r - 3}{3}$

33. $\dfrac{1}{x} = \dfrac{1}{x + 1} + \dfrac{1}{2}$

Solve each problem.

34. Jody can weed the yard in 3 hr. Pat can weed the same yard in 2 hr. How long will it take them if they work together?

35. The circumference of a circle varies directly as its radius. A circle with circumference 9.42 in. has radius approximately 1.5 in. Find the circumference of a circle with radius 5.25 in. Give the answer to the nearest hundredth.

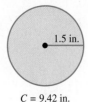

$C = 9.42$ in.

8

Roots and Radicals

Many formulas in mathematics and science, such as that for finding the time it takes for a pendulum to swing from one extreme to the other and back again, include *square roots,* the topic of this chapter.

8.1 Evaluating Roots

VOCABULARY

☐ square root
☐ positive (principal) square root
☐ negative square root
☐ radicand
☐ radical
☐ radical expression
☐ perfect square
☐ irrational number
☐ cube root
☐ fourth root
☐ index (order)
☐ perfect cube
☐ perfect fourth power

NOW TRY EXERCISE 1

Find all square roots of 81.

OBJECTIVE 1   Find square roots.

Recall that *squaring* a number means multiplying the number by itself.

$$7^2 \quad \text{means} \quad 7 \cdot 7, \quad \text{which equals} \quad 49.$$ The square of 7 is 49.

The opposite (inverse) of squaring a number is taking its *square root.* This is equivalent to asking

"What number when multiplied by itself equals 49?"

From the example above, one answer is 7 because $7 \cdot 7 = 49$.

> **Square Root**
>
> A number b is a **square root** of a if $b^2 = a$ (that is, $b \cdot b = a$).

EXAMPLE 1   Finding All Square Roots of a Number

Find all square roots of 49.

We ask, "What number when multiplied by itself equals 49?" As mentioned above, one square root is 7. Another square root of 49 is -7, because

$$(-7)(-7) = 49.$$

Thus, the number 49 has *two* square roots, 7 and -7. One square root is positive, and one is negative.

NOW TRY

The **positive** or **principal square root** of a number is written with the symbol $\sqrt{}$. For example, the positive square root of 121 is 11.

$$\sqrt{121} = 11 \quad 11^2 = 121$$

The symbol $-\sqrt{}$ is used for the **negative square root** of a number. For example, the negative square root of 121 is -11.

$$-\sqrt{121} = -11 \quad (-11)^2 = 121$$

The **radical symbol** $\sqrt{}$ always represents the positive square root $\big($except that $\sqrt{0} = 0\big)$. The number inside the radical symbol is the **radicand,** and the entire expression—radical symbol and radicand—is a **radical.**

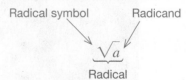

Radical symbol   Radicand

$$\sqrt{a}$$

Radical

An algebraic expression containing a radical is a **radical expression.**

The radical symbol $\sqrt{}$ has been used since 16th-century Germany and was likely derived from the letter R. The radical symbol at the right comes from the Latin word *radix,* for *root.* It was first used by Leonardo of Pisa (Fibonacci) in 1220.

Early radical symbol

NOW TRY ANSWER
1. 9, −9

We summarize our discussion of square roots as follows.

Square Roots of *a*

Let *a* be a positive real number.

$$\sqrt{a} \text{ is the positive or principal square root of } a.$$

$$-\sqrt{a} \text{ is the negative square root of } a.$$

For nonnegative *a*, the following hold.

$$\sqrt{a} \cdot \sqrt{a} = \left(\sqrt{a}\right)^2 = a \quad \text{and} \quad -\sqrt{a} \cdot \left(-\sqrt{a}\right) = \left(-\sqrt{a}\right)^2 = a$$

Also, $\sqrt{0} = 0$.

NOW TRY
EXERCISE 2

Find each square root.

(a) $\sqrt{400}$ **(b)** $-\sqrt{169}$

(c) $\sqrt{\dfrac{100}{121}}$

EXAMPLE 2   Finding Square Roots

Find each square root.

(a) $\sqrt{144}$

The radical $\sqrt{144}$ represents the positive or principal square root of 144. Think of a positive number whose square is 144.

$$12^2 = 144, \quad \text{so} \quad \sqrt{144} = 12.$$

(b) $-\sqrt{1024}$

This symbol represents the negative square root of 1024. A calculator with a square root key can be used to find $\sqrt{1024} = 32$. Therefore,

$$-\sqrt{1024} = -32.$$

> $(0.9)^2 = 0.9 \cdot 0.9$
> $= 0.81$

(c) $\sqrt{\dfrac{4}{9}} = \dfrac{2}{3}$ **(d)** $-\sqrt{\dfrac{16}{49}} = -\dfrac{4}{7}$ **(e)** $\sqrt{0.81} = 0.9$ **NOW TRY**

⚠️ **CAUTION** By definition, $\sqrt{4} = 2$ because $2^2 = 4$. *In general, however, the square root of a number is not half the number.*

As shown in the preceding definition, when the square root of a positive real number is squared, the result is that positive real number. $\left(\text{Also, } \left(\sqrt{0}\right)^2 = 0.\right)$

NOW TRY
EXERCISE 3

Find the square of each radical expression.

(a) $\sqrt{15}$ **(b)** $-\sqrt{23}$

(c) $\sqrt{2k^2 + 5}$

EXAMPLE 3   Squaring Radical Expressions

Find the square of each radical expression.

(a) $\sqrt{13}$

The square of $\sqrt{13}$ is $\left(\sqrt{13}\right)^2 = 13$.   Definition of square root

(b) $-\sqrt{29}$

$$\left(-\sqrt{29}\right)^2 = 29 \quad \text{The square of a } \textit{negative} \text{ number is positive.}$$

(c) $\sqrt{p^2 + 1}$

$$\left(\sqrt{p^2 + 1}\right)^2 = p^2 + 1$$

   NOW TRY

NOW TRY ANSWERS

2. **(a)** 20   **(b)** −13   **(c)** $\dfrac{10}{11}$

3. **(a)** 15   **(b)** 23   **(c)** $2k^2 + 5$

OBJECTIVE 2 Decide whether a given root is rational, irrational, or not a real number.

Numbers with square roots that are rational are **perfect squares.**

Perfect squares		Rational square roots
↓		↓
25		$\sqrt{25} = 5$
144	are perfect squares because	$\sqrt{144} = 12$
$\dfrac{4}{9}$		$\sqrt{\dfrac{4}{9}} = \dfrac{2}{3}$

A number that is not a perfect square has a square root that is not a rational number. For example, $\sqrt{5}$ is not a rational number because it cannot be written as the ratio of two integers. Its decimal equivalent neither terminates nor repeats. However, $\sqrt{5}$ is a real number and corresponds to a point on the number line.

A real number that is not rational is an **irrational number.** The number $\sqrt{5}$ is irrational. *Many square roots of integers are irrational.*

> If a is a *positive* real number that is *not* a perfect square, then $\sqrt{a}$ is irrational.

Not every number has a real number square root. For example, there is no real number that can be squared to obtain -36. (The square of a real number can never be negative.) Because of this, $\sqrt{-36}$ *is not a real number.*

> If a is a *negative* real number, then $\sqrt{a}$ is *not* a real number.

⚠ **CAUTION** Do not confuse $\sqrt{-36}$ and $-\sqrt{36}$. $\sqrt{-36}$ is not a real number because there is no real number that can be squared to obtain -36. However, $-\sqrt{36}$ is the negative square root of 36, which is -6.

NOW TRY EXERCISE 4

Tell whether each number is *rational, irrational,* or *not a real number.*

(a) $\sqrt{31}$ (b) $\sqrt{900}$
(c) $\sqrt{-16}$

EXAMPLE 4 Identifying Types of Square Roots

Tell whether each number is *rational, irrational,* or *not a real number.*

(a) $\sqrt{17}$ Because 17 is not a perfect square, $\sqrt{17}$ is irrational.

(b) $\sqrt{64}$ The number 64 is a perfect square, 8^2, so $\sqrt{64} = 8$, a rational number.

(c) $\sqrt{-25}$ There is no real number whose square is -25. Therefore, $\sqrt{-25}$ is not a real number.

NOW TRY ↻

NOW TRY ANSWERS
4. (a) irrational (b) rational
   (c) not a real number

NOTE Not all irrational numbers are square roots of integers. For example, the number π (approximately 3.14159) is an irrational number that is not a square root of any integer.

OBJECTIVE 3 Find decimal approximations for irrational square roots.

Even if a number is irrational, a decimal that *approximates* the number can be found using a calculator.

NOW TRY EXERCISE 5

Find a decimal approximation for each square root. Round answers to the nearest thousandth.

(a) $\sqrt{51}$ **(b)** $-\sqrt{360}$

EXAMPLE 5 Approximating Irrational Square Roots

Find a decimal approximation for each square root. Round answers to the nearest thousandth.

(a) $\sqrt{11}$

Using the square root key on a calculator gives $3.31662479 \approx 3.317$, where the symbol $\approx$ means "**is approximately equal to.**"

(b) $\sqrt{39} \approx 6.245$ Use a calculator.   **(c)** $-\sqrt{740} \approx -27.203$

NOW TRY

OBJECTIVE 4 Use the Pythagorean theorem.

Many applications of square roots require the use of the Pythagorean theorem. Recall from **Section 6.6** and **FIGURE 1** that if a and b are the lengths of the two legs of a right triangle, and c is the length of the hypotenuse, then

$$a^2 + b^2 = c^2.$$

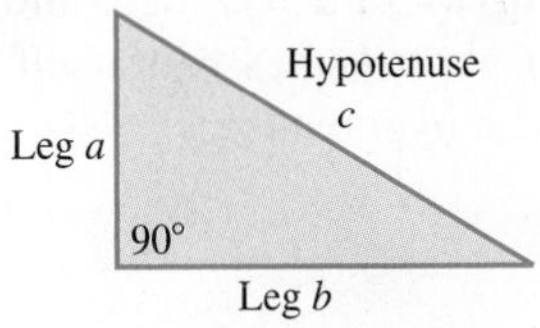

Hypotenuse
c
Leg a
$90°$
Leg b

FIGURE 1

In the next example, we use the fact that if $k > 0$, then the positive solution of the equation $x^2 = k$ is $\sqrt{k}$. (See **Section 9.1.**)

NOW TRY EXERCISE 6

Find the length of the unknown side in each right triangle with sides a, b, and c, where c is the hypotenuse. Give any decimal approximations to the nearest thousandth.

(a) $a = 5, b = 12$
(b) $a = 9, c = 14$

EXAMPLE 6 Using the Pythagorean Theorem

Find the length of the unknown side in each right triangle with sides a, b, and c, where c is the hypotenuse.

(a) $a = 3, b = 4$

$$a^2 + b^2 = c^2 \qquad \text{Use the Pythagorean theorem.}$$
$$3^2 + 4^2 = c^2 \qquad \text{Let } a = 3 \text{ and } b = 4.$$
$$9 + 16 = c^2 \qquad \text{Square.}$$
$$25 = c^2 \qquad \text{Add.}$$

Because the length of a side of a triangle must be a positive number, we find the positive square root of 25 to obtain c.

$$c = \sqrt{25} = 5$$

(b) $b = 5, c = 9$

$$a^2 + b^2 = c^2 \qquad \text{Use the Pythagorean theorem.}$$
Solve for a^2. $\quad a^2 + 5^2 = 9^2 \qquad \text{Let } b = 5 \text{ and } c = 9.$
$$a^2 + 25 = 81 \qquad \text{Square.}$$
$$a^2 = 56 \qquad \text{Subtract 25.}$$

NOW TRY ANSWERS
5. (a) 7.141   **(b)** -18.974
6. (a) $c = 13$ **(b)** $b \approx 10.724$

Use a calculator to find the positive square root of 56 and approximate a.

$$a = \sqrt{56} \approx 7.483$$

NOW TRY

! CAUTION Consider the following.

$$\sqrt{9 + 16} = \sqrt{25} = 5, \quad \text{but} \quad \sqrt{9} + \sqrt{16} = 3 + 4 = 7.$$

In general, $\sqrt{a^2 + b^2} \neq a + b$.

**NOW TRY
EXERCISE 7**

The length of a rectangle is 48 ft, and the width is 14 ft. Find the measure of the diagonal of the rectangle.

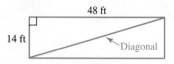

(Note that the diagonal divides the rectangle into two right triangles with itself as the hypotenuse.)

EXAMPLE 7 Using the Pythagorean Theorem to Solve an Application

A ladder 10 ft long leans against a wall. The foot of the ladder is 6 ft from the base of the wall. How high up the wall does the top of the ladder rest?

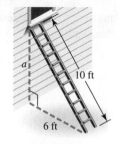

The symbol ⌐ indicates a 90° or right angle.

FIGURE 2

Step 1 **Read** the problem again.

Step 2 **Assign a variable.** As shown in **FIGURE 2**, a right triangle is formed with the ladder as the hypotenuse.

Let a = the height of the top of the ladder, measured straight down to the ground.

Step 3 **Write an equation** using the Pythagorean theorem.

$$a^2 + b^2 = c^2 \quad \boxed{\text{Substitute carefully.}}$$

$$a^2 + 6^2 = 10^2 \quad \text{Let } b = 6 \text{ and } c = 10.$$

Step 4 **Solve.** $\quad a^2 + 36 = 100 \quad$ Square.

$$a^2 = 64 \quad \text{Subtract 36.}$$

$$a = \sqrt{64} \quad \text{Solve for } a.$$

$$a = 8 \quad \sqrt{64} = 8$$

We choose the positive square root of 64 because a represents a length.

Step 5 **State the answer.** The top of the ladder rests 8 ft up the wall.

Step 6 **Check.** From **FIGURE 2**, we must have the following.

$$8^2 + 6^2 \overset{?}{=} 10^2 \quad a^2 + b^2 = c^2$$

$$100 = 100 \checkmark \quad \text{True; } 8^2 = 64 \text{ and } 6^2 = 36$$

The check confirms that the top of the ladder rests 8 ft up the wall.

NOW TRY

OBJECTIVE 5 Use the distance formula.

Consider **FIGURE 3**. The distance between the points (x_2, y_2) and (x_2, y_1) is $a = y_2 - y_1$, and the distance between the points (x_1, y_1) and (x_2, y_1) is $b = x_2 - x_1$. From the Pythagorean theorem,

$$d^2 = (x_2 - x_1)^2 + (y_2 - y_1)^2.$$

Taking the square root of each side gives the **distance formula.**

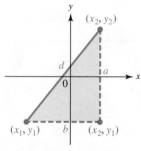

FIGURE 3

Distance Formula

The distance d between the points (x_1, y_1) and (x_2, y_2) is

$$d = \sqrt{(x_2 - x_1)^2 + (y_2 - y_1)^2}.$$

NOW TRY ANSWER
7. 50 ft

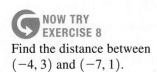

**NOW TRY
EXERCISE 8**

Find the distance between $(-4, 3)$ and $(-7, 1)$.

EXAMPLE 8 Using the Distance Formula

Find the distance between $(-3, 4)$ and $(2, 5)$.

$$d = \sqrt{(x_2 - x_1)^2 + (y_2 - y_1)^2}$$ Distance formula

$$= \sqrt{(2 - (-3))^2 + (5 - 4)^2}$$ Let $(x_1, y_1) = (-3, 4)$ and $(x_2, y_2) = (2, 5)$.

Start with the *x*-value and the *y*-value of the *same* point.

$$= \sqrt{5^2 + 1^2}$$ Subtract.

$$= \sqrt{26}$$ Apply the exponents. Add.

NOW TRY

OBJECTIVE 6 Find cube, fourth, and other roots.

Finding the square root of a number is the inverse (opposite) of squaring a number. In a similar way, there are inverses to finding the cube of a number and to finding the fourth or greater power of a number. These inverses are, respectively, the **cube root,** $\sqrt[3]{a}$, and the **fourth root,** $\sqrt[4]{a}$. Similar symbols are used for other roots.

$\sqrt[n]{a}$

The *n*th root of *a*, written $\sqrt[n]{a}$, is a number whose *n*th power equals *a*. That is,

$$\sqrt[n]{a} = b \quad \text{means} \quad b^n = a.$$

In $\sqrt[n]{a}$, the number *n* is the **index,** or **order,** of the radical.

Radical symbol   Index   Radicand

$$\sqrt[n]{a}$$

Radical

We could write $\sqrt[2]{a}$ instead of $\sqrt{a}$, but the simpler symbol $\sqrt{a}$ is customary because the square root is the most commonly used root.

NOTE When working with cube roots or fourth roots, it is helpful to learn the first few **perfect cubes** ($1^3 = 1$, $2^3 = 8$, $3^3 = 27$, and so on) and the first few **perfect fourth powers** ($1^4 = 1$, $2^4 = 16$, $3^4 = 81$, and so on). See **Exercises 99 and 100.**

 **NOW TRY
EXERCISE 9**

Find each cube root.

(a) $\sqrt[3]{343}$ **(b)** $\sqrt[3]{-1000}$

(c) $\sqrt[3]{27}$

NOW TRY ANSWERS

8. $\sqrt{13}$

9. (a) 7   **(b)** -10 **(c)** 3

EXAMPLE 9 Finding Cube Roots

Find each cube root. 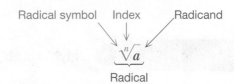 $2^3 = 2 \cdot 2 \cdot 2$

(a) $\sqrt[3]{8}$ What number can be cubed to give 8? Because $2^3 = 8$, $\sqrt[3]{8} = 2$.

(b) $\sqrt[3]{-8}$ Because $(-2)^3 = -8$, $\sqrt[3]{-8} = -2$.

(c) $\sqrt[3]{216}$ Because $6^3 = 216$, $\sqrt[3]{216} = 6$.   **NOW TRY**

In **Example 9(b),** $\sqrt[3]{-8} = -2$—that is, we can find the *cube root* of a negative number. (Contrast this with the *square root* of a negative number, which is not real.) In fact, the cube root of a positive number is positive, and the cube root of a negative number is negative. ***There is only one real number cube root for each real number.***

When a radical has an ***even index*** (square root, fourth root, and so on), ***the radicand must be nonnegative*** to yield a real number root. Also, for $a > 0$,

$$\sqrt{a}, \ \sqrt[4]{a}, \ \sqrt[6]{a}, \text{ and so on are positive (principal) roots.}$$

$$-\sqrt{a}, \ -\sqrt[4]{a}, \ -\sqrt[6]{a}, \text{ and so on are negative roots.}$$

 NOW TRY EXERCISE 10

Find each root.

(a) $\sqrt[4]{625}$ **(b)** $\sqrt[4]{-625}$

(c) $-\sqrt[4]{625}$ **(d)** $\sqrt[5]{3125}$

(e) $\sqrt[5]{-3125}$

NOW TRY ANSWERS

10. **(a)** 5   **(b)** not a real number
    (c) −5   **(d)** 5   **(e)** −5

EXAMPLE 10 Finding Other Roots

Find each root.

 $2^4 = 2 \cdot 2 \cdot 2 \cdot 2$

(a) $\sqrt[4]{16}$ Because 2 is positive and $2^4 = 16$, $\sqrt[4]{16} = 2$.

(b) $-\sqrt[4]{16}$ From part (a), $\sqrt[4]{16} = 2$, so the negative root is $-\sqrt[4]{16} = -2$.

(c) $\sqrt[4]{-16}$

For a fourth root to be a real number, the radicand must be nonnegative. There is no real number that equals $\sqrt[4]{-16}$.

(d) $-\sqrt[5]{32}$

First find $\sqrt[5]{32}$. Here 2 is the number whose fifth power is 32, so $\sqrt[5]{32} = 2$. Because $\sqrt[5]{32} = 2$, it follows that

$$-\sqrt[5]{32} = -2.$$

(e) $\sqrt[5]{-32}$ Because $(-2)^5 = -32$, $\sqrt[5]{-32} = -2$.   **NOW TRY**

8.1 Exercises

FOR EXTRA HELP   ▶ MyMathLab®

▶ *Complete solution available in MyMathLab*

Concept Check   *Decide whether each statement is* true *or* false. *If false, tell why.*

1. Every positive number has two real square roots.

2. A negative number has negative real square roots.

3. Every nonnegative number has two real square roots.

4. The positive square root of a positive number is its principal square root.

5. The cube root of every nonzero real number has the same sign as the number itself.

6. Every positive number has three real cube roots.

Concept Check   *What must be true about the variable a for each statement to be true?*

7. $\sqrt{a}$ represents a positive number.

8. $-\sqrt{a}$ represents a negative number.

9. $\sqrt{a}$ is not a real number.

10. $-\sqrt{a}$ is not a real number.

Find all square roots of each number. See Example 1.

▶ **11.** 9   **12.** 16   **13.** 64   **14.** 100   **15.** 169

16. 225   **17.** $\dfrac{25}{196}$ **18.** $\dfrac{81}{400}$ **19.** 900   **20.** 1600

Find each square root. See Examples 2 and 4(c).

21. $\sqrt{1}$ **22.** $\sqrt{4}$ ▶ **23.** $\sqrt{49}$ **24.** $\sqrt{81}$ **25.** $\sqrt{100}$

26. $\sqrt{400}$ **27.** $-\sqrt{16}$ **28.** $-\sqrt{64}$ **29.** $-\sqrt{256}$ **30.** $-\sqrt{196}$

▶ **31.** $-\sqrt{\dfrac{144}{121}}$ **32.** $-\sqrt{\dfrac{49}{36}}$ **33.** $\sqrt{0.64}$ **34.** $\sqrt{0.16}$

35. $\sqrt{-121}$ **36.** $\sqrt{-64}$ **37.** $-\sqrt{-49}$ **38.** $-\sqrt{-100}$

Find the square of each radical expression. See Example 3.

▶ **39.** $\sqrt{19}$ **40.** $\sqrt{59}$ **41.** $-\sqrt{19}$ **42.** $-\sqrt{59}$

43. $\sqrt{\dfrac{2}{3}}$ **44.** $\sqrt{\dfrac{5}{7}}$ ▶ **45.** $\sqrt{3x^2 + 4}$ **46.** $\sqrt{9y^2 + 3}$

Determine whether each number is rational, irrational, *or* not a real number. *If a number is rational, give its exact value. If a number is irrational, give a decimal approximation to the nearest thousandth. Use a calculator as necessary. See Examples 4 and 5.*

▶ **47.** $\sqrt{25}$ **48.** $\sqrt{169}$ ▶ **49.** $\sqrt{29}$ **50.** $\sqrt{33}$

51. $-\sqrt{64}$ **52.** $-\sqrt{81}$ ▶ **53.** $-\sqrt{300}$ **54.** $-\sqrt{500}$

▶ **55.** $\sqrt{-29}$ **56.** $\sqrt{-47}$ **57.** $\sqrt{1200}$ **58.** $\sqrt{1500}$

Concept Check   Without using a calculator, determine between which two consecutive integers each square root lies. For example,

$$\sqrt{75} \text{ is between 8 and 9, because } \sqrt{64} = 8,\ \sqrt{81} = 9, \text{ and } 64 < 75 < 81.$$

59. $\sqrt{94}$ **60.** $\sqrt{43}$ **61.** $\sqrt{51}$ **62.** $\sqrt{30}$

63. $-\sqrt{40}$ **64.** $-\sqrt{63}$ **65.** $\sqrt{23.2}$ **66.** $\sqrt{10.3}$

Work each problem without using a calculator.

67. Choose the best estimate for the length and width (in meters) of this rectangle.

    A. 11 by 6   **B.** 11 by 7

    C. 10 by 7   **D.** 10 by 6

68. Choose the best estimate for the base and height (in feet) of this triangle.

    A. $b = 8, h = 5$ **B.** $b = 8, h = 4$

    C. $b = 9, h = 5$ **D.** $b = 9, h = 4$

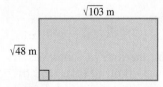

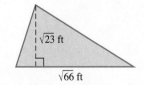

Find the length of the unknown side in each right triangle with sides a, b, and c, where c is the hypotenuse. Give any decimal approximations to the nearest thousandth. See **FIGURE 1** *and Example 6.*

▶ **69.** $a = 8, b = 15$ **70.** $a = 24, b = 10$ **71.** $a = 6, c = 10$

72. $a = 5, c = 13$ **73.** $a = 11, b = 4$ **74.** $a = 13, b = 9$

Solve each problem. Give any decimal approximations to the nearest tenth. ***See Example 7.***

75. The diagonal of a rectangle measures 25 cm. The width of the rectangle is 7 cm. Find the length of the rectangle.

76. The length of a rectangle is 40 m, and the width is 9 m. Find the measure of the diagonal of the rectangle.

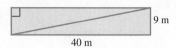

77. Tyler is flying a kite on 100 ft of string. How high is it above his hand (vertically) if the horizontal distance between Tyler and the kite is 60 ft?

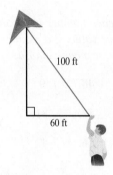

78. A guy wire is attached to the mast of a shortwave transmitting antenna at a point 96 ft above ground level. If the wire is staked to the ground 72 ft from the base of the mast, how long is the wire?

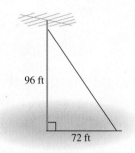

79. A surveyor measured the distances shown in the figure. Find the distance across the lake between points *R* and *S*.

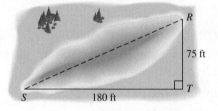

80. A boat is being pulled toward a dock with a rope attached at water level. When the boat is 24 ft from the dock, 30 ft of rope is extended. What is the height of the dock above the water?

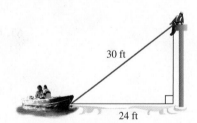

81. During Hurricane Katrina, thousands of pine trees snapped off to form right triangles. Suppose that for one such tree, the vertical distance from the base of the broken tree to the point of the break was 4.5 ft. The length of the broken part was 12.0 ft. How far along the ground was it from the base of the tree to the point where the broken part touched the ground?

82. A television set is "sized" according to the diagonal measurement of the viewing screen. A rectangular 46-in. TV measures 46 in. from one corner of the viewing screen diagonally to the other corner. The viewing screen is 40 in. wide. Find the height of the viewing screen.

83. A surveyor wants to find the height of a building. At a point 110.0 ft from the base of the building, he sights to the top of the building and finds the distance to be 193.0 ft. How tall is the building?

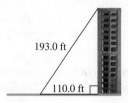

84. Two towns are separated by dense woods. To go from Town B to Town A, it is necessary to travel due west for 19.0 mi and then turn due north and travel for 14.0 mi. How far apart are the towns?

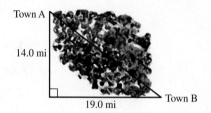

85. What is the value of x (to the nearest thousandth) in the figure?

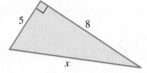

86. What is the value of y (to the nearest thousandth) in the figure?

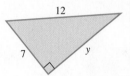

Extending Skills *Although Pythagoras may have written the first proof of the Pythagorean relationship, many other proofs have been given, as shown in Exercises 87–90.*

87. The Babylonians may have used a tile pattern like that shown here to illustrate the Pythagorean theorem.

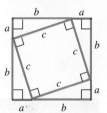

   (a) The side of the square along the hypotenuse measures 5 units. What are the measures of the sides along the legs?

   (b) Using the measures from part (a), show that the Pythagorean theorem is satisfied.

88. The diagram shown here can be used to verify the Pythagorean theorem.

   (a) Find the area of the large square.

   (b) Find the sum of the areas of the smaller square and the four right triangles.

   (c) Set the areas equal and write a simpler, equivalent equation.

89. Another proof of the Pythagorean theorem is attributed to the Hindu mathematician Bhaskara. The figure on the left is made up of the same square and triangles as the figure on the right. Use this information to derive the Pythagorean theorem.

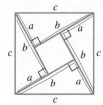
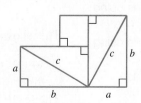

90. James A. Garfield, the twentieth president of the United States, provided a proof of the Pythagorean theorem using the given figure.

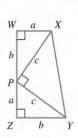

(a) Find the area of the trapezoid *WXYZ* using the formula for the area of a trapezoid.

(b) Find the area of each of the right triangles *PWX*, *PZY*, and *PXY*.

(c) Because the sum of the areas of the three right triangles must equal the area of the trapezoid, set the expression from part (a) equal to the sum of the three expressions from part (b). Simplify as much as possible.

Find the distance between each pair of points. Express the answer as a whole number or as a square root. Do not use a calculator. See Example 8.

91. $(5, 7)$ and $(1, 4)$ 　　　**92.** $(8, 13)$ and $(3, 1)$ 　　　**93.** $(2, 9)$ and $(-3, -3)$

94. $(4, 6)$ and $(-4, -9)$ 　▶ **95.** $(-1, -2)$ and $(-3, 1)$ 　　**96.** $(-3, -6)$ and $(-4, 0)$

97. $\left(-\dfrac{1}{4}, \dfrac{2}{3}\right)$ and $\left(\dfrac{3}{4}, -\dfrac{1}{3}\right)$ 　　　　**98.** $\left(\dfrac{2}{5}, \dfrac{3}{2}\right)$ and $\left(-\dfrac{8}{5}, \dfrac{1}{2}\right)$

99. *Concept Check* 　To help find cube roots, complete this list of perfect cubes.

$1^3 =$ _____ 　　　$2^3 =$ _____ 　　　$3^3 =$ _____ 　　　$4^3 =$ _____ 　　　$5^3 =$ _____

$6^3 =$ _____ 　　　$7^3 =$ _____ 　　　$8^3 =$ _____ 　　　$9^3 =$ _____ 　　　$10^3 =$ _____

100. *Concept Check* 　To help find fourth roots, complete this list of perfect fourth powers.

$1^4 =$ _____ 　　　$2^4 =$ _____ 　　　$3^4 =$ _____ 　　　$4^4 =$ _____ 　　　$5^4 =$ _____

$6^4 =$ _____ 　　　$7^4 =$ _____ 　　　$8^4 =$ _____ 　　　$9^4 =$ _____ 　　　$10^4 =$ _____

Find each root. See Examples 9 and 10.

101. $\sqrt[3]{1}$ 　　　**102.** $\sqrt[3]{27}$ 　▶ **103.** $\sqrt[3]{125}$ 　　　**104.** $\sqrt[3]{64}$ 　　　**105.** $\sqrt[3]{729}$

106. $\sqrt[3]{1000}$ 　▶ **107.** $\sqrt[3]{-27}$ 　　**108.** $\sqrt[3]{-64}$ 　　**109.** $\sqrt[3]{-216}$ 　**110.** $\sqrt[3]{-343}$

111. $-\sqrt[3]{-8}$ 　　　**112.** $-\sqrt[3]{-343}$ 　　　**113.** $\sqrt[4]{81}$ 　　　　**114.** $\sqrt[4]{16}$

▶ **115.** $\sqrt[4]{625}$ 　　　**116.** $\sqrt[4]{256}$ 　　　**117.** $\sqrt[4]{1296}$ 　　　**118.** $\sqrt[4]{10{,}000}$

119. $\sqrt[4]{-1}$ 　　　**120.** $\sqrt[4]{-625}$ 　　　**121.** $-\sqrt[4]{81}$ 　　　**122.** $-\sqrt[4]{256}$

123. $\sqrt[5]{32}$ 　　　**124.** $\sqrt[5]{1}$ 　　▶ **125.** $\sqrt[5]{-1024}$ 　　**126.** $\sqrt[5]{-100{,}000}$

8.2 　Multiplying, Dividing, and Simplifying Radicals

OBJECTIVES

1. Multiply square root radicals.
2. Simplify radicals using the product rule.
3. Simplify radicals using the quotient rule.
4. Simplify radicals involving variables.
5. Simplify other roots.

OBJECTIVE 1 　Multiply square root radicals.

Consider the following.

$$\sqrt{4} \cdot \sqrt{9} = 2 \cdot 3 = 6 \quad \text{and} \quad \sqrt{4 \cdot 9} = \sqrt{36} = 6$$

This shows that

$$\sqrt{4} \cdot \sqrt{9} = \sqrt{4 \cdot 9}.$$

The result here is a particular case of the **product rule for radicals.**

Product Rule for Radicals

If a and b are nonnegative real numbers, then the following hold.

$$\sqrt{a} \cdot \sqrt{b} = \sqrt{a \cdot b} \quad \text{and} \quad \sqrt{a \cdot b} = \sqrt{a} \cdot \sqrt{b}$$

That is, the product of two square roots is the square root of the product, and the square root of a product is the product of the two square roots.

⊘ **CAUTION** In general, $\sqrt{x + y} \neq \sqrt{x} + \sqrt{y}$. To see why this is so, let $x = 16$ and $y = 9$.

$$\sqrt{16 + 9} = \sqrt{25} = 5, \quad \text{but} \quad \sqrt{16} + \sqrt{9} = 4 + 3 = 7.$$

NOW TRY EXERCISE 1

Find each product.

(a) $\sqrt{5} \cdot \sqrt{11}$

(b) $\sqrt{11} \cdot \sqrt{11}$

(c) $\sqrt{7} \cdot \sqrt{k}$ $(k \geq 0)$

(d) $\sqrt{5} \cdot \sqrt{5}$

EXAMPLE 1 Using the Product Rule to Multiply Radicals

Use the product rule to find each product.

(a) $\sqrt{2} \cdot \sqrt{3}$ **(b)** $\sqrt{7} \cdot \sqrt{5}$ **(c)** $\sqrt{11} \cdot \sqrt{a}$ **(d)** $\sqrt{31} \cdot \sqrt{31}$

$ = \sqrt{2 \cdot 3}$ $ = \sqrt{35}$ $ = \sqrt{11a}$ $ = 31$

$ = \sqrt{6}$ Assume $a \geq 0$.   **NOW TRY** ⟳

OBJECTIVE 2 Simplify radicals using the product rule.

A square root radical is simplified when no perfect square factor other than 1 remains under the radical symbol.

EXAMPLE 2 Using the Product Rule to Simplify Radicals

Simplify each radical.

(a) $\sqrt{20}$ ⟵ 20 has a perfect square factor of 4.

$ = \sqrt{4 \cdot 5}$ Factor; 4 is a perfect square.

$ = \sqrt{4} \cdot \sqrt{5}$ Product rule in the form $\sqrt{a \cdot b} = \sqrt{a} \cdot \sqrt{b}$

$ = 2\sqrt{5}$ $\sqrt{4} = 2$

Thus, $\sqrt{20} = 2\sqrt{5}$. Because 5 has no perfect square factor (other than 1), $2\sqrt{5}$ is the **simplified form** of $\sqrt{20}$. Note that $2\sqrt{5}$ represents a product whose factors are 2 and $\sqrt{5}$.

We could also factor 20 into prime factors and look for pairs of like factors.

$$\sqrt{20}$$

$$= \sqrt{2 \cdot 2 \cdot 5} \qquad \text{Each pair of like factors produces one factor outside the radical.}$$

$$= 2\sqrt{5} \qquad \text{The result is the same.}$$

NOW TRY ANSWERS

1. (a) $\sqrt{55}$ **(b)** 11 **(c)** $\sqrt{7k}$
   (d) 5

NOW TRY
EXERCISE 2

Simplify each radical.

(a) $\sqrt{28}$ **(b)** $\sqrt{99}$

(c) $\sqrt{85}$

(b) $\sqrt{72}$ ◁ Look for the *greatest* perfect square factor of 72.

$= \sqrt{36 \cdot 2}$ Factor; 36 is a perfect square.

$= \sqrt{36} \cdot \sqrt{2}$ Product rule

$= 6\sqrt{2}$ $\sqrt{36} = 6$

We could also factor 72 into its prime factors and look for pairs of like factors.

$\sqrt{72}$

$= \sqrt{2 \cdot 2 \cdot 2 \cdot 3 \cdot 3}$ Factor into primes.

$= 2 \cdot 3 \cdot \sqrt{2}$ $\sqrt{2 \cdot 2} = 2; \sqrt{3 \cdot 3} = 3$

$= 6\sqrt{2}$ The result is the same.

(c) $\sqrt{300}$

$= \sqrt{100 \cdot 3}$ Factor; 100 is a perfect square.

$= \sqrt{100} \cdot \sqrt{3}$ Product rule

$= 10\sqrt{3}$ $\sqrt{100} = 10$

(d) $\sqrt{15}$

Because 15 has no perfect square factors (except 1), $\sqrt{15}$ cannot be simplified.

NOW TRY

EXAMPLE 3   **Multiplying and Simplifying Radicals**

Find each product and simplify.

(a) $\sqrt{9} \cdot \sqrt{75}$

$= 3\sqrt{75}$ $\sqrt{9} = 3$

$= 3\sqrt{25 \cdot 3}$ Factor; 25 is a perfect square.

$= 3\sqrt{25} \cdot \sqrt{3}$ Product rule

$= 3 \cdot 5\sqrt{3}$ $\sqrt{25} = 5$

$= 15\sqrt{3}$ Multiply.

We could have used the product rule to obtain $\sqrt{9} \cdot \sqrt{75} = \sqrt{675}$ and then simplified. The method above uses smaller numbers.

(b) $\sqrt{8} \cdot \sqrt{12}$

$= \sqrt{8 \cdot 12}$ Product rule

$= \sqrt{4 \cdot 2 \cdot 4 \cdot 3}$ Factor; 4 is a perfect square.

$= \sqrt{4} \cdot \sqrt{4} \cdot \sqrt{2 \cdot 3}$ Commutative property; product rule

$= 2 \cdot 2 \cdot \sqrt{6}$ $\sqrt{4} = 2$; Multiply.

$= 4\sqrt{6}$ Multiply.

NOW TRY ANSWERS

2. (a) $2\sqrt{7}$ **(b)** $3\sqrt{11}$
   (c) It cannot be simplified.

**NOW TRY
EXERCISE 3**

Find each product and simplify.

(a) $\sqrt{16} \cdot \sqrt{50}$

(b) $\sqrt{6} \cdot \sqrt{30}$

(c) $4\sqrt{3} \cdot 5\sqrt{15}$

(c) $2\sqrt{3} \cdot 3\sqrt{6}$

$= 2 \cdot 3 \cdot \sqrt{3 \cdot 6}$ Commutative property; product rule

$= 6\sqrt{18}$ Multiply.

$= 6\sqrt{9 \cdot 2}$ Factor; 9 is a perfect square.

$= 6\sqrt{9} \cdot \sqrt{2}$ Product rule

$= 6 \cdot 3 \cdot \sqrt{2}$ $\sqrt{9} = 3$

$= 18\sqrt{2}$ Multiply.   NOW TRY

NOTE   There is often more than one way to find a product of radicals.

$\sqrt{8} \cdot \sqrt{12}$ See **Example 3(b)**.

$= \sqrt{4 \cdot 2} \cdot \sqrt{4 \cdot 3}$ Factor.

$= \sqrt{4} \cdot \sqrt{2} \cdot \sqrt{4} \cdot \sqrt{3}$ Product rule

$= 2\sqrt{2} \cdot 2\sqrt{3}$ $\sqrt{4} = 2$

$= 2 \cdot 2 \cdot \sqrt{2} \cdot \sqrt{3}$ Commutative property

Same result $\longrightarrow$ $= 4\sqrt{6}$ Multiply; product rule

---

OBJECTIVE 3 Simplify radicals using the quotient rule.

The **quotient rule for radicals** is similar to the product rule.

Quotient Rule for Radicals

If a and b are nonnegative real numbers and $b \neq 0$, then the following hold.

$$\sqrt{\frac{a}{b}} = \frac{\sqrt{a}}{\sqrt{b}} \quad \text{and} \quad \frac{\sqrt{a}}{\sqrt{b}} = \sqrt{\frac{a}{b}}$$

That is, the square root of a quotient is the quotient of the two square roots, and the quotient of two square roots is the square root of the quotient.

**NOW TRY
EXERCISE 4**

Simplify each radical.

(a) $\sqrt{\dfrac{81}{100}}$ **(b)** $\dfrac{\sqrt{245}}{\sqrt{5}}$

(c) $\sqrt{\dfrac{11}{49}}$

EXAMPLE 4 Using the Quotient Rule to Simplify Radicals

Use the quotient rule to simplify each radical.

(a) $\sqrt{\dfrac{25}{9}}$

$= \dfrac{\sqrt{25}}{\sqrt{9}}$

$= \dfrac{5}{3}$

(b) $\dfrac{\sqrt{288}}{\sqrt{2}}$

$= \sqrt{\dfrac{288}{2}}$

$= \sqrt{144}$

$= 12$

(c) $\sqrt{\dfrac{3}{4}}$

$= \dfrac{\sqrt{3}}{\sqrt{4}}$

$= \dfrac{\sqrt{3}}{2}$ NOW TRY

NOW TRY ANSWERS

3. **(a)** $20\sqrt{2}$ **(b)** $6\sqrt{5}$ **(c)** $60\sqrt{5}$

4. **(a)** $\dfrac{9}{10}$ **(b)** 7   **(c)** $\dfrac{\sqrt{11}}{7}$

NOW TRY
EXERCISE 5

Simplify.

$$\frac{24\sqrt{39}}{4\sqrt{13}}$$

NOW TRY
EXERCISE 6

Simplify.

$$\sqrt{\frac{1}{2}} \cdot \sqrt{\frac{5}{18}}$$

EXAMPLE 5   **Using the Quotient Rule to Divide Radicals**

Simplify.

$$\frac{27\sqrt{15}}{9\sqrt{3}}$$

$$= \frac{27}{9} \cdot \frac{\sqrt{15}}{\sqrt{3}} \qquad \text{Multiplication of fractions}$$

$$= \frac{27}{9} \cdot \sqrt{\frac{15}{3}} \qquad \text{Quotient rule}$$

$$= 3\sqrt{5} \qquad \text{Divide.} \qquad \text{NOW TRY}$$

EXAMPLE 6   **Using Both the Product and Quotient Rules**

Simplify.

$$\sqrt{\frac{3}{5}} \cdot \sqrt{\frac{1}{5}}$$

$$= \sqrt{\frac{3}{5} \cdot \frac{1}{5}} \qquad \text{Product rule}$$

$$= \sqrt{\frac{3}{25}} \qquad \text{Multiply fractions.}$$

$$= \frac{\sqrt{3}}{\sqrt{25}} \qquad \text{Quotient rule}$$

$$= \frac{\sqrt{3}}{5} \qquad \sqrt{25} = 5 \qquad \text{NOW TRY}$$

OBJECTIVE 4 Simplify radicals involving variables.

Consider a radical with a variable radicand, such as $\sqrt{x^2}$.

> If x represents a nonnegative number, then $\sqrt{x^2} = x$.
>
> If x represents a negative number, then $\sqrt{x^2} = -x$, the *opposite* of x (which is positive).

For example, $\sqrt{5^2} = 5$,   but $\sqrt{(-5)^2} = \sqrt{25} = 5$,   the *opposite* of -5.

This means that the square root of a squared number is always nonnegative. We can use absolute value to express this.

$\sqrt{a^2}$

For any real number a, the following holds.

$$\sqrt{a^2} = |a|$$

NOW TRY ANSWERS
5. $6\sqrt{3}$ 6. $\frac{\sqrt{5}}{6}$

The product and quotient rules apply when variables appear under radical symbols, as long as the variables represent *nonnegative* real numbers. **To avoid negative radicands, we assume variables under radical symbols are nonnegative.** In such cases, absolute value bars are not necessary because, for all $x \geq 0$, $|x| = x$.

**NOW TRY
EXERCISE 7**

Simplify each radical. Assume that all variables represent nonnegative real numbers.

(a) $\sqrt{16y^8}$ **(b)** $\sqrt{x^5}$

(c) $\sqrt{\dfrac{13}{t^2}}$ $(t \neq 0)$

EXAMPLE 7   Simplifying Radicals Involving Variables

Simplify each radical. Assume that all variables represent nonnegative real numbers.

(a) $\sqrt{x^4}$

$= x^2$

This is true because $(x^2)^2 = x^4$.

(b) $\sqrt{25m^6}$

$= \sqrt{25} \cdot \sqrt{m^6}$ Product rule

$= 5m^3$ $(m^3)^2 = m^6$

(c) $\sqrt{8p^{10}}$

$= \sqrt{4 \cdot 2 \cdot p^{10}}$ Factor; 4 is a perfect square.

$= \sqrt{4} \cdot \sqrt{2} \cdot \sqrt{p^{10}}$ Product rule

$= 2 \cdot \sqrt{2} \cdot p^5$ $\sqrt{4} = 2; (p^5)^2 = p^{10}$

$= 2p^5\sqrt{2}$ Commutative property

(d) $\sqrt{r^9}$

$= \sqrt{r^8 \cdot r}$

$= \sqrt{r^8} \cdot \sqrt{r}$ Product rule

$= r^4\sqrt{r}$ $(r^4)^2 = r^8$

(e) $\sqrt{\dfrac{5}{x^2}}$ $(x \neq 0)$

$= \dfrac{\sqrt{5}}{\sqrt{x^2}}$ Quotient rule

$= \dfrac{\sqrt{5}}{x}$ **NOW TRY**

NOTE   A quick way to find the square root of a variable raised to an even power is to divide the exponent by the index, 2.

Examples: $\sqrt{x^6} = x^3$ and $\sqrt{x^{10}} = x^5$

$6 \div 2 = 3$ $10 \div 2 = 5$

OBJECTIVE 5   Simplify other roots.

The product and quotient rules for radicals also apply to other roots.

Properties of Radicals

For all real numbers a and b for which the indicated roots exist, the following hold.

$$\sqrt[n]{a} \cdot \sqrt[n]{b} = \sqrt[n]{ab} \quad \text{and} \quad \frac{\sqrt[n]{a}}{\sqrt[n]{b}} = \sqrt[n]{\frac{a}{b}} \quad (\text{where } b \neq 0)$$

NOW TRY ANSWERS

7. (a) $4y^4$ **(b)** $x^2\sqrt{x}$

(c) $\dfrac{\sqrt{13}}{t}$

NOW TRY
EXERCISE 8

Simplify each radical.

(a) $\sqrt[3]{250}$ (b) $\sqrt[4]{48}$

(c) $\sqrt[3]{\dfrac{1}{125}}$

EXAMPLE 8   Simplifying Other Roots

Simplify each radical.

(a)

$$\sqrt[3]{32}$$

> Because the index is 3, look for factors that are perfect cubes.

$$= \sqrt[3]{8 \cdot 4} \qquad \text{Factor; 8 is a perfect cube.}$$

> Remember to write the root index 3 in each radical.

$$= \sqrt[3]{8} \cdot \sqrt[3]{4} \qquad \text{Product rule}$$

$$= 2\sqrt[3]{4} \qquad \text{Take the cube root.}$$

(b)

$$\sqrt[4]{32}$$

$$= \sqrt[4]{16 \cdot 2} \qquad \text{Factor; 16 is a perfect fourth power.}$$

> Remember to write the root index 4 in each radical.

$$= \sqrt[4]{16} \cdot \sqrt[4]{2} \qquad \text{Product rule}$$

$$= 2\sqrt[4]{2} \qquad \text{Take the fourth root.}$$

(c) $\sqrt[3]{\dfrac{27}{125}}$

$$= \dfrac{\sqrt[3]{27}}{\sqrt[3]{125}} \qquad \text{Quotient rule}$$

$$= \dfrac{3}{5} \qquad \text{Take cube roots.}$$

NOW TRY

Other roots of radicals involving variables can also be simplified. To simplify cube roots with variables, use the fact that for any real number a,

$$\sqrt[3]{a^3} = a.$$

This is true whether a is positive, negative, or 0.

NOW TRY
EXERCISE 9

Simplify each radical.

(a) $\sqrt[3]{x^{12}}$ (b) $\sqrt[3]{64t^3}$

(c) $\sqrt[3]{40a^7}$ (d) $\sqrt[3]{\dfrac{x^{15}}{1000}}$

EXAMPLE 9   Simplifying Cube Roots Involving Variables

Simplify each radical.

(a) $\sqrt[3]{m^6}$

$$= m^2 \qquad (m^2)^3 = m^6$$

(b) $\sqrt[3]{27x^{12}}$

$$= \sqrt[3]{27} \cdot \sqrt[3]{x^{12}} \qquad \text{Product rule}$$

$$= 3x^4 \qquad \begin{array}{l} 3^3 = 27; \\ (x^4)^3 = x^{12} \end{array}$$

(c) $\sqrt[3]{32a^4}$

$$= \sqrt[3]{8a^3 \cdot 4a} \qquad \begin{array}{l} \text{Factor; } 8a^3 \text{ is} \\ \text{a perfect cube.} \end{array}$$

$$= \sqrt[3]{8a^3} \cdot \sqrt[3]{4a} \qquad \text{Product rule}$$

$$= 2a\sqrt[3]{4a} \qquad (2a)^3 = 8a^3$$

(d) $\sqrt[3]{\dfrac{y^3}{125}}$

$$= \dfrac{\sqrt[3]{y^3}}{\sqrt[3]{125}} \qquad \text{Quotient rule}$$

$$= \dfrac{y}{5} \qquad \text{Take cube roots.}$$

NOW TRY

NOW TRY ANSWERS

8. (a) $5\sqrt[3]{2}$ (b) $2\sqrt[4]{3}$ (c) $\dfrac{1}{5}$

9. (a) x^4 (b) $4t$

(c) $2a^2\sqrt[3]{5a}$ (d) $\dfrac{x^5}{10}$

8.2 Exercises

   FOR EXTRA HELP   ▶ MyMathLab®

▶ *Complete solution available in MyMathLab*

Concept Check Decide whether each statement is true or false. *If false, tell why.*

1. $\sqrt{(-6)^2} = -6$

2. $\sqrt[3]{(-6)^3} = -6$

3. The radical $2\sqrt{7}$ represents a sum.

4. In $3\sqrt{11}$, the numbers 3 and $\sqrt{11}$ are factors.

5. *Concept Check* Which one of the following radicals is simplified?

  A. $\sqrt{47}$ **B.** $\sqrt{45}$ **C.** $\sqrt{48}$ **D.** $\sqrt{44}$

6. *Concept Check* If p is a prime number, is $\sqrt{p}$ in simplified form?

*Find each product. **See Example 1.***

▶ **7.** $\sqrt{3} \cdot \sqrt{5}$

8. $\sqrt{3} \cdot \sqrt{7}$

9. $\sqrt{2} \cdot \sqrt{11}$

10. $\sqrt{2} \cdot \sqrt{15}$

11. $\sqrt{6} \cdot \sqrt{7}$

12. $\sqrt{5} \cdot \sqrt{6}$

13. $\sqrt{3} \cdot \sqrt{27}$

14. $\sqrt{2} \cdot \sqrt{8}$

15. $\sqrt{13} \cdot \sqrt{13}$

16. $\sqrt{17} \cdot \sqrt{17}$

17. $\sqrt{13} \cdot \sqrt{r}$ $(r \geq 0)$

18. $\sqrt{19} \cdot \sqrt{k}$ $(k \geq 0)$

*Simplify each radical. **See Example 2.***

19. $\sqrt{45}$ **20.** $\sqrt{27}$ **21.** $\sqrt{24}$ **22.** $\sqrt{44}$

23. $\sqrt{90}$ **24.** $\sqrt{56}$ **25.** $\sqrt{75}$ **26.** $\sqrt{18}$

27. $\sqrt{125}$ **28.** $\sqrt{80}$ ▶ **29.** $\sqrt{145}$ **30.** $\sqrt{110}$

31. $\sqrt{160}$ **32.** $\sqrt{128}$ **33.** $-\sqrt{700}$ **34.** $-\sqrt{600}$

35. $3\sqrt{27}$ **36.** $9\sqrt{8}$ **37.** $5\sqrt{50}$ **38.** $6\sqrt{40}$

*Use the Pythagorean theorem to find the length of the unknown side of each right triangle. Express answers as simplified radicals. **See Section 8.1.***

39.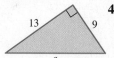

40.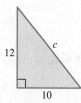

41.

42.

*Use the distance formula to find the length of each line segment. Express answers as simplified radicals. **See Section 8.1.***

43.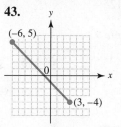

44.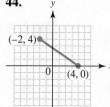

45.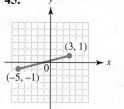

46.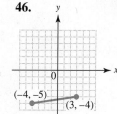

Find each product and simplify. ***See Example 3.***

▶ 47. $\sqrt{9} \cdot \sqrt{32}$ **48.** $\sqrt{9} \cdot \sqrt{50}$ **49.** $\sqrt{3} \cdot \sqrt{18}$

50. $\sqrt{3} \cdot \sqrt{21}$ **51.** $\sqrt{12} \cdot \sqrt{48}$ **52.** $\sqrt{50} \cdot \sqrt{72}$

53. $\sqrt{12} \cdot \sqrt{30}$ **54.** $\sqrt{30} \cdot \sqrt{24}$ **55.** $2\sqrt{10} \cdot 3\sqrt{2}$

56. $5\sqrt{6} \cdot 2\sqrt{10}$ **57.** $5\sqrt{3} \cdot 2\sqrt{15}$ **58.** $4\sqrt{6} \cdot 3\sqrt{2}$

Simplify each radical expression. ***See Examples 4–6.***

▶ 59. $\sqrt{\dfrac{16}{225}}$ **60.** $\sqrt{\dfrac{9}{100}}$ **61.** $\sqrt{\dfrac{7}{16}}$ **62.** $\sqrt{\dfrac{13}{25}}$

63. $\sqrt{\dfrac{4}{50}}$ **64.** $\sqrt{\dfrac{14}{72}}$ **65.** $\dfrac{\sqrt{75}}{\sqrt{3}}$ **66.** $\dfrac{\sqrt{200}}{\sqrt{2}}$

▶ 67. $\dfrac{30\sqrt{10}}{5\sqrt{2}}$ **68.** $\dfrac{50\sqrt{20}}{2\sqrt{10}}$ **▶ 69.** $\sqrt{\dfrac{5}{2}} \cdot \sqrt{\dfrac{125}{8}}$ **70.** $\sqrt{\dfrac{8}{3}} \cdot \sqrt{\dfrac{512}{27}}$

Simplify each radical. Assume that all variables represent nonnegative real numbers. ***See Example 7.***

▶ 71. $\sqrt{m^2}$ **72.** $\sqrt{k^2}$ **73.** $\sqrt{y^4}$ **74.** $\sqrt{s^4}$

75. $\sqrt{36z^2}$ **76.** $\sqrt{49n^2}$ **77.** $\sqrt{400x^6}$ **78.** $\sqrt{900y^8}$

79. $\sqrt{18x^8}$ **80.** $\sqrt{20r^{10}}$ **81.** $\sqrt{45c^{14}}$ **82.** $\sqrt{50d^{20}}$

83. $\sqrt{z^5}$ **84.** $\sqrt{y^3}$ **85.** $\sqrt{a^{13}}$ **86.** $\sqrt{p^{17}}$

87. $\sqrt{64x^7}$ **88.** $\sqrt{25t^{11}}$ **89.** $\sqrt{x^6y^{12}}$ **90.** $\sqrt{a^8b^{10}}$

91. $\sqrt{81m^4n^2}$ **92.** $\sqrt{100c^4d^6}$ **93.** $\sqrt{\dfrac{7}{x^{10}}}$ $(x \neq 0)$ **94.** $\sqrt{\dfrac{14}{z^{12}}}$ $(z \neq 0)$

95. $\sqrt{\dfrac{y^4}{100}}$ **96.** $\sqrt{\dfrac{w^8}{144}}$ **97.** $\sqrt{\dfrac{x^4y^6}{169}}$ **98.** $\sqrt{\dfrac{w^8z^{10}}{400}}$

Simplify each radical. ***See Example 8.***

▶ 99. $\sqrt[3]{40}$ **100.** $\sqrt[3]{48}$ **101.** $\sqrt[3]{54}$ **102.** $\sqrt[3]{135}$

103. $\sqrt[3]{128}$ **104.** $\sqrt[3]{192}$ **105.** $\sqrt[4]{80}$ **106.** $\sqrt[4]{243}$

107. $\sqrt[3]{\dfrac{8}{27}}$ **108.** $\sqrt[3]{\dfrac{64}{125}}$ **109.** $\sqrt[3]{-\dfrac{216}{125}}$ **110.** $\sqrt[3]{-\dfrac{1}{64}}$

Simplify each radical. ***See Example 9.***

111. $\sqrt[3]{p^3}$ **112.** $\sqrt[3]{w^3}$ **▶ 113.** $\sqrt[3]{x^9}$ **114.** $\sqrt[3]{y^{18}}$

115. $\sqrt[3]{64z^6}$ **116.** $\sqrt[3]{125a^{15}}$ **117.** $\sqrt[3]{343a^9b^3}$ **118.** $\sqrt[3]{216m^3n^6}$

119. $\sqrt[3]{16t^5}$ **120.** $\sqrt[3]{24x^4}$ **121.** $\sqrt[3]{\dfrac{m^{12}}{8}}$ **122.** $\sqrt[3]{\dfrac{n^9}{27}}$

The volume V of a cube is found with the formula V = s³, where s is the length of an edge of the cube. Use this information in Exercises 123 and 124.

123. A container in the shape of a cube has a volume of 216 cm³. What is the length of each side of the container?

124. A cube-shaped box must be constructed to contain 128 ft³. What should the dimensions (height, width, and length) of the box be?

The volume V of a sphere is found with the formula $V = \frac{4}{3}\pi r^3$, where r is the length of the radius of the sphere. Use this information in Exercises 125 and 126.

125. A ball in the shape of a sphere has a volume of 288π in.³. What is the radius of the ball?

126. Suppose that the volume of the ball described in **Exercise 125** is multiplied by 8. How is the radius affected?

Work each problem without using a calculator.

127. Choose the best estimate for the area (in square inches) of this rectangle.

A. 45   **B.** 72   **C.** 80   **D.** 90

2√26 in.

√83 in.

128. Choose the best estimate for the area (in square feet) of this triangle.

A. 20   **B.** 40   **C.** 60   **D.** 80

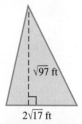

√97 ft

2√17 ft

8.3   Adding and Subtracting Radicals

OBJECTIVES

1 Add and subtract radicals.

2 Simplify radical sums and differences.

3 Simplify more complicated radical expressions.

VOCABULARY

☐ like radicals
☐ unlike radicals

OBJECTIVE 1 Add and subtract radicals.

We add or subtract radicals using the distributive property, $ac + bc = (a + b)c$.

$$8\sqrt{3} + 6\sqrt{3}$$ Distributive property
$$= (8 + 6)\sqrt{3}$$
$$= 14\sqrt{3}$$ Add.

$$2\sqrt{11} - 7\sqrt{11}$$ Distributive property
$$= (2 - 7)\sqrt{11}$$
$$= -5\sqrt{11}$$ Subtract.

Only **like radicals**—those that are *multiples of the same root of the same number*—can be combined in this way. By contrast, the following are **unlike radicals.**

$$2\sqrt{5} \quad \text{and} \quad 2\sqrt{3},$$ Radicands are different.

$$2\sqrt{3} \quad \text{and} \quad 2\sqrt[3]{3}$$ Indexes are different.

EXAMPLE 1 Adding and Subtracting Like Radicals

Add or subtract, as indicated.

(a) $3\sqrt{6} + 5\sqrt{6}$

$$= (3 + 5)\sqrt{6}$$ We are factoring out √6 here.
$$= 8\sqrt{6}$$

(b) $5\sqrt{10} - 7\sqrt{10}$

$$= (5 - 7)\sqrt{10}$$
$$= -2\sqrt{10}$$

**NOW TRY
EXERCISE 1**

Add or subtract, as indicated.

(a) $4\sqrt{3} + \sqrt{3}$

(b) $2\sqrt{11} - 6\sqrt{11}$

(c) $\sqrt{5} + \sqrt{14}$

(c) $\sqrt{7} + 2\sqrt{7}$

$= 1\sqrt{7} + 2\sqrt{7}$

$= (1 + 2)\sqrt{7}$

$= 3\sqrt{7}$

(d) $\sqrt{5} + \sqrt{5}$

$= 1\sqrt{5} + 1\sqrt{5}$

$= (1 + 1)\sqrt{5}$

$= 2\sqrt{5}$

(e) $\sqrt{3} + \sqrt{7}$

These unlike radicals cannot be added using the distributive property.

NOW TRY

OBJECTIVE 2 Simplify radical sums and differences.

**NOW TRY
EXERCISE 2**

Add or subtract, as indicated.

(a) $\sqrt{2} + \sqrt{18}$

(b) $3\sqrt{48} - 2\sqrt{75}$

(c) $8\sqrt[3]{5} + 10\sqrt[3]{40}$

EXAMPLE 2   Adding and Subtracting Radicals

Add or subtract, as indicated.

(a) $3\sqrt{2} + \sqrt{8}$

$= 3\sqrt{2} + \sqrt{4 \cdot 2}$ Factor; 4 is a perfect square.

$= 3\sqrt{2} + \sqrt{4} \cdot \sqrt{2}$ Product rule

$= 3\sqrt{2} + 2\sqrt{2}$ $\sqrt{4} = 2$

$= (3 + 2)\sqrt{2}$ Distributive property

$= 5\sqrt{2}$ Add.

(b) $\sqrt{18} - \sqrt{27}$

$= \sqrt{9 \cdot 2} - \sqrt{9 \cdot 3}$ Factor; 9 is a perfect square.

$= \sqrt{9} \cdot \sqrt{2} - \sqrt{9} \cdot \sqrt{3}$ Product rule

Stop here. **These are unlike radicals.** They cannot be combined. $= 3\sqrt{2} - 3\sqrt{3}$ $\sqrt{9} = 3$

(c) $2\sqrt{12} + 3\sqrt{75}$

$= 2(\sqrt{4} \cdot \sqrt{3}) + 3(\sqrt{25} \cdot \sqrt{3})$ Product rule

$= 2(2\sqrt{3}) + 3(5\sqrt{3})$ $\sqrt{4} = 2; \sqrt{25} = 5$

$= 4\sqrt{3} + 15\sqrt{3}$ Multiply.

Think: $(4 + 15)\sqrt{3}$

$= 19\sqrt{3}$ Add like radicals.

(d) $3\sqrt[3]{16} + 5\sqrt[3]{2}$

$= 3(\sqrt[3]{8} \cdot \sqrt[3]{2}) + 5\sqrt[3]{2}$ Product rule

$= 3(2\sqrt[3]{2}) + 5\sqrt[3]{2}$ $\sqrt[3]{8} = 2$

$= 6\sqrt[3]{2} + 5\sqrt[3]{2}$ Multiply.

$= 11\sqrt[3]{2}$ Add like radicals.   NOW TRY

NOW TRY ANSWERS

1. (a) $5\sqrt{3}$ **(b)** $-4\sqrt{11}$
   (c) These unlike radicals cannot be added using the distributive property.

2. (a) $4\sqrt{2}$ **(b)** $2\sqrt{3}$
   (c) $28\sqrt[3]{5}$

OBJECTIVE 3 Simplify more complicated radical expressions.

EXAMPLE 3 Simplifying Radical Expressions

Simplify. Assume that all variables represent nonnegative real numbers.

(a) $\sqrt{5} \cdot \sqrt{15} + 4\sqrt{3}$ *We simplify this expression by performing the operations.*

$$= \sqrt{5 \cdot 15} + 4\sqrt{3} \qquad \text{Product rule}$$

$$= \sqrt{75} + 4\sqrt{3} \qquad \text{Multiply.}$$

$$= \sqrt{25 \cdot 3} + 4\sqrt{3} \qquad \text{Factor; 25 is a perfect square.}$$

$$= \sqrt{25} \cdot \sqrt{3} + 4\sqrt{3} \qquad \text{Product rule}$$

$$= 5\sqrt{3} + 4\sqrt{3} \qquad \sqrt{25} = 5$$

$$= 9\sqrt{3} \qquad \text{Add like radicals.}$$

(b) $\sqrt{12k} + \sqrt{27k}$

$$= \sqrt{4 \cdot 3k} + \sqrt{9 \cdot 3k} \qquad \text{Factor.}$$

$$= \sqrt{4} \cdot \sqrt{3k} + \sqrt{9} \cdot \sqrt{3k} \qquad \text{Product rule}$$

$$= 2\sqrt{3k} + 3\sqrt{3k} \qquad \sqrt{4} = 2; \sqrt{9} = 3$$

$$= 5\sqrt{3k} \qquad \text{Add like radicals.}$$

(c) $3x\sqrt{50} + \sqrt{2x^2}$

$$= 3x\sqrt{25 \cdot 2} + \sqrt{x^2 \cdot 2} \qquad \text{Factor.}$$

$$= 3x\sqrt{25} \cdot \sqrt{2} + \sqrt{x^2} \cdot \sqrt{2} \qquad \text{Product rule}$$

$$= 3x \cdot 5\sqrt{2} + x\sqrt{2} \qquad \sqrt{25} = 5; \sqrt{x^2} = x$$

$$= 15x\sqrt{2} + x\sqrt{2} \qquad \text{Multiply.}$$

$$= 16x\sqrt{2} \qquad \text{Add like radicals.} \qquad \textit{Think:} \; (15x + 1x)\sqrt{2}$$

(d) $2\sqrt[3]{32m^3} - \sqrt[3]{108m^3}$

$$= 2\sqrt[3]{8m^3 \cdot 4} - \sqrt[3]{27m^3 \cdot 4} \qquad \text{Factor.}$$

$$= 2 \cdot 2m\sqrt[3]{4} - 3m\sqrt[3]{4} \qquad \sqrt[3]{8m^3} = 2m; \sqrt[3]{27m^3} = 3m$$

$$= 4m\sqrt[3]{4} - 3m\sqrt[3]{4} \qquad \text{Multiply.}$$

$$= m\sqrt[3]{4} \qquad \text{Subtract like radicals.} \qquad \text{NOW TRY}$$

! CAUTION *Only like radicals can be combined.*

$$\left. \begin{array}{l} \sqrt{5} + 3\sqrt{5} \\ = 4\sqrt{5} \end{array} \right\} \quad \text{Add like radicals.} \qquad \left. \begin{array}{l} \sqrt{5} + 5\sqrt{3} \\ 2\sqrt{3} + 5\sqrt[3]{3} \end{array} \right\} \quad \text{Unlike radicals cannot be combined.}$$

NOW TRY EXERCISE 3

Simplify. Assume that all variables represent nonnegative real numbers.

(a) $\sqrt{7} \cdot \sqrt{14} + 5\sqrt{2}$

(b) $\sqrt{150x} + 2\sqrt{24x}$

(c) $5k^2\sqrt{12} - 4\sqrt{27k^4}$

(d) $\sqrt[3]{128y^5} + 5y\sqrt[3]{16y^2}$

NOW TRY ANSWERS

3. (a) $12\sqrt{2}$ **(b)** $9\sqrt{6x}$

   (c) $-2k^2\sqrt{3}$ **(d)** $14y\sqrt[3]{2y^2}$

8.3 Exercises

 Complete solution available in MyMathLab

Concept Check   *Work each problem.*

1. Like radicals have the same _____ and the same _____, or order. For example, $5\sqrt{2}$ and $-3\sqrt{2}$ are (*like / unlike*) radicals, as are $\sqrt{7}$, $-\sqrt{\underline{}}$, and $8\sqrt{\underline{}}$.

2. The radicals $\sqrt[4]{3xy^3}$ and $-6\sqrt[4]{3xy^3}$ are (*like / unlike*) radicals because both have the same root index, _____, and the same radicand, _____.

3. Are the radicals in each pair *like* or *unlike*? Tell why.

    (a) $5\sqrt{6}$ and $4\sqrt{6}$ (b) $2\sqrt{3}$ and $3\sqrt{2}$ (c) $\sqrt{10}$ and $\sqrt[3]{10}$

    (d) $7\sqrt{2x}$ and $8\sqrt{2x}$ (e) $\sqrt{3y}$ and $\sqrt{6y}$ (f) $\sqrt[3]{2ab}$ and $4\sqrt[3]{2ba}$

4. The radicals $\sqrt{5x}$ and $4\sqrt{5x}$ are (*like / unlike*) radicals. To add them, we use the identity property of multiplication to write $\sqrt{5x}$ as _____ $\cdot \sqrt{5x}$. Then we write $1\sqrt{5x} + 4\sqrt{5x}$ as $(1 + 4)$_____, which is an application of the _____ property, and add to obtain _____ .

Perform the indicated operations. **See Examples 1, 2, and 3(a).**

▶ 5. $2\sqrt{3} + 5\sqrt{3}$ 6. $6\sqrt{5} + 8\sqrt{5}$ 7. $4\sqrt{7} - 9\sqrt{7}$

8. $6\sqrt{2} - 8\sqrt{2}$ 9. $\sqrt{6} + \sqrt{6}$ 10. $\sqrt{11} + \sqrt{11}$

11. $\sqrt{17} + 2\sqrt{17}$ 12. $\sqrt{19} + 3\sqrt{19}$ 13. $5\sqrt{3} - \sqrt{3}$

14. $6\sqrt{7} - \sqrt{7}$ 15. $\sqrt{6} + \sqrt{7}$ 16. $\sqrt{14} + \sqrt{17}$

▶ 17. $5\sqrt{3} + \sqrt{12}$ 18. $3\sqrt{2} + \sqrt{50}$ 19. $2\sqrt{75} - \sqrt{12}$

20. $2\sqrt{27} - \sqrt{300}$ 21. $2\sqrt{50} - 5\sqrt{72}$ 22. $6\sqrt{18} - 4\sqrt{32}$

23. $\frac{1}{4}\sqrt{288} + \frac{1}{6}\sqrt{72}$ 24. $\frac{2}{3}\sqrt{27} + \frac{3}{4}\sqrt{48}$ 25. $\frac{3}{5}\sqrt{75} - \frac{2}{3}\sqrt{45}$

26. $\frac{5}{8}\sqrt{128} - \frac{3}{4}\sqrt{160}$ 27. $4\sqrt[3]{16} - 3\sqrt[3]{54}$ 28. $3\sqrt[3]{250} - 4\sqrt[3]{128}$

29. $3\sqrt[3]{24} + 6\sqrt[3]{81}$ 30. $2\sqrt[4]{48} - \sqrt[4]{243}$ ▶ 31. $\sqrt{3} \cdot \sqrt{7} + 2\sqrt{21}$

32. $\sqrt{13} \cdot \sqrt{2} + 3\sqrt{26}$ 33. $\sqrt{6} \cdot \sqrt{2} + 3\sqrt{3}$ 34. $\sqrt{15} \cdot \sqrt{3} + 2\sqrt{5}$

35. $5\sqrt{7} - 2\sqrt{28} + 6\sqrt{63}$ 36. $3\sqrt{11} - 3\sqrt{99} + 5\sqrt{44}$

37. $9\sqrt{24} - 2\sqrt{54} + 3\sqrt{20}$ 38. $2\sqrt{8} - 5\sqrt{32} + 2\sqrt{48}$

39. $5\sqrt{72} - 3\sqrt{48} - 4\sqrt{128}$ 40. $4\sqrt{50} - 3\sqrt{12} - 5\sqrt{200}$

41. $5\sqrt[4]{32} + 2\sqrt[4]{32} \cdot \sqrt[4]{4}$ 42. $8\sqrt[3]{48} + 10\sqrt[3]{3} \cdot \sqrt[3]{18}$

Find the perimeter of each figure.

43.
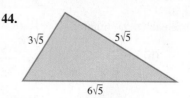
$7\sqrt{2}$

$4\sqrt{2}$

44.

$3\sqrt{5}$ $5\sqrt{5}$

$6\sqrt{5}$

Simplify. Assume that all variables represent nonnegative real numbers. **See Example 3.**

45. $\sqrt{32x} - \sqrt{18x}$ **46.** $\sqrt{125t} - \sqrt{80t}$ **47.** $\sqrt{27r} + \sqrt{48r}$

48. $\sqrt{24x} + \sqrt{54x}$ **49.** $\sqrt{75x^2} + x\sqrt{300}$ **50.** $\sqrt{20y^2} + y\sqrt{125}$

51. $3\sqrt{8x^2} - 4x\sqrt{2}$ **52.** $2\sqrt{18b^2} - 3b\sqrt{2}$ **53.** $5\sqrt{75p^2} - 4\sqrt{27p^2}$

54. $4\sqrt{32k^2} - 6\sqrt{8k^2}$ **55.** $2\sqrt{125x^2z} + 8x\sqrt{80z}$ **56.** $4p\sqrt{63m} + 6\sqrt{28mp^2}$

57. $3k\sqrt{24k^2h^2} + 9h\sqrt{54k^3}$ **58.** $6r\sqrt{27r^2s^2} + 3r\sqrt{12s^3}$

59. $6\sqrt[3]{8p^2} - 2\sqrt[3]{27p^2}$ **60.** $5\sqrt[3]{27x^2} + 8\sqrt[3]{8x^2}$

61. $5\sqrt[4]{m^3} + 8\sqrt[4]{16m^3}$ **62.** $5\sqrt[4]{x^3} + 3\sqrt[4]{81x^3}$

63. $2\sqrt[4]{p^5} - 5p\sqrt[4]{16p}$ **64.** $8k\sqrt[3]{54k} + 6\sqrt[3]{16k^4}$

65. $-5\sqrt[3]{256z^4} - 2z\sqrt[3]{32z}$ **66.** $-10\sqrt[3]{4m^4} - 3m\sqrt[3]{32m}$

67. $2\sqrt[4]{6k^7} - k\sqrt[4]{96k^3}$ **68.** $\dfrac{3}{2}\sqrt[3]{16a^4b^5} - ab\sqrt[3]{54ab^2}$

8.4 Rationalizing the Denominator

OBJECTIVES

1 Rationalize denominators with square roots.

2 Write radicals in simplified form.

3 Rationalize denominators with cube roots.

OBJECTIVE 1 Rationalize denominators with square roots.

Although calculators make it fairly easy to divide by a radical in an expression such as $\dfrac{1}{\sqrt{2}}$, it is sometimes easier to work with radical expressions if the denominators do not contain radicals.

For example, the radical in the denominator of $\dfrac{1}{\sqrt{2}}$ can be eliminated by multiplying the numerator and denominator by $\sqrt{2}$ because $\sqrt{2} \cdot \sqrt{2} = \sqrt{4} = 2$.

$$\frac{1}{\sqrt{2}} = \frac{1 \cdot \sqrt{2}}{\sqrt{2} \cdot \sqrt{2}} = \frac{\sqrt{2}}{2} \quad \text{Multiply by } \tfrac{\sqrt{2}}{\sqrt{2}} = 1.$$

Rationalizing a Denominator

The process of changing a denominator from one with a radical to one without a radical is called **rationalizing the denominator.**

The value of the radical expression is not changed. Only the form is changed, because the expression has been multiplied by a form of 1.

**NOW TRY
EXERCISE 1**

Rationalize each denominator.

(a) $\dfrac{15}{\sqrt{5}}$ **(b)** $\dfrac{3}{\sqrt{24}}$

EXAMPLE 1   **Rationalizing Denominators**

Rationalize each denominator.

(a) $\dfrac{9}{\sqrt{6}}$

$= \dfrac{9 \cdot \sqrt{6}}{\sqrt{6} \cdot \sqrt{6}}$ Multiply by $\dfrac{\sqrt{6}}{\sqrt{6}} = 1$.

$= \dfrac{9\sqrt{6}}{6}$ In the denominator, $\sqrt{6} \cdot \sqrt{6} = \sqrt{36} = 6$.

$= \dfrac{3\sqrt{6}}{2}$ Write in lowest terms.

(b) $\dfrac{12}{\sqrt{8}}$ While the denominator could be rationalized by multiplying by $\sqrt{8}$, simplifying the denominator first is more direct.

$= \dfrac{12}{2\sqrt{2}}$ $\sqrt{8} = \sqrt{4} \cdot \sqrt{2} = 2\sqrt{2}$

$= \dfrac{12 \cdot \sqrt{2}}{2\sqrt{2} \cdot \sqrt{2}}$ Multiply by $\dfrac{\sqrt{2}}{\sqrt{2}} = 1$.

$= \dfrac{12 \cdot \sqrt{2}}{2 \cdot 2}$ $\sqrt{2} \cdot \sqrt{2} = \sqrt{4} = 2$

$= \dfrac{12\sqrt{2}}{4}$ Multiply.

$= 3\sqrt{2}$ Write in lowest terms; $\frac{12}{4} = 3$.   **NOW TRY**

NOTE   In **Example 1(b),** we could also have rationalized the original denominator, $\sqrt{8}$, by multiplying by $\sqrt{2}$ because $\sqrt{8} \cdot \sqrt{2} = \sqrt{16} = 4$.

$$\dfrac{12}{\sqrt{8}} = \dfrac{12 \cdot \sqrt{2}}{\sqrt{8} \cdot \sqrt{2}} = \dfrac{12\sqrt{2}}{\sqrt{16}} = \dfrac{12\sqrt{2}}{4} = 3\sqrt{2}$$ Either approach yields the same correct answer.

OBJECTIVE 2   Write radicals in simplified form.

Conditions for Simplified Form of a Radical

1. The radicand contains no factor (except 1) that is a perfect square (when dealing with square roots), a perfect cube (when dealing with cube roots), and so on.

2. The radicand has no fractions.

3. No denominator contains a radical.

**NOW TRY
EXERCISE 2**

Simplify $\sqrt{\dfrac{27}{7}}$.

EXAMPLE 2 **Simplifying a Radical**

Simplify.

$$\sqrt{\frac{27}{5}}$$

$$= \frac{\sqrt{27}}{\sqrt{5}} \qquad \text{Quotient rule}$$

$$= \frac{\sqrt{27} \cdot \sqrt{5}}{\sqrt{5} \cdot \sqrt{5}} \qquad \text{Rationalize the denominator.}$$

$$= \frac{\sqrt{9 \cdot 3} \cdot \sqrt{5}}{5} \qquad \begin{array}{l}\text{Factor in the numerator.} \\ \sqrt{5} \cdot \sqrt{5} = 5 \text{ in the denominator.}\end{array}$$

$$= \frac{\sqrt{9} \cdot \sqrt{3} \cdot \sqrt{5}}{5} \qquad \text{Product rule}$$

$$= \frac{3 \cdot \sqrt{3} \cdot \sqrt{5}}{5} \qquad \sqrt{9} = 3$$

> The three conditions for a simplified radical are met.

$$= \frac{3\sqrt{15}}{5} \qquad \text{Product rule} \qquad \text{NOW TRY}$$

**NOW TRY
EXERCISE 3**

Simplify.

$$\sqrt{\frac{1}{6}} \cdot \sqrt{\frac{3}{10}}$$

EXAMPLE 3 **Simplifying a Product of Radicals**

Simplify.

$$\sqrt{\frac{5}{8}} \cdot \sqrt{\frac{1}{6}}$$

$$= \sqrt{\frac{5}{8} \cdot \frac{1}{6}} \qquad \text{Product rule}$$

$$= \sqrt{\frac{5}{48}} \qquad \text{Multiply fractions.}$$

$$= \frac{\sqrt{5}}{\sqrt{48}} \qquad \text{Quotient rule}$$

$$= \frac{\sqrt{5}}{\sqrt{16} \cdot \sqrt{3}} \qquad \text{Product rule}$$

$$= \frac{\sqrt{5}}{4\sqrt{3}} \qquad \sqrt{16} = 4$$

$$= \frac{\sqrt{5} \cdot \sqrt{3}}{4\sqrt{3} \cdot \sqrt{3}} \qquad \text{Rationalize the denominator.}$$

$$= \frac{\sqrt{15}}{4 \cdot 3} \qquad \text{Product rule; } \sqrt{3} \cdot \sqrt{3} = 3$$

$$= \frac{\sqrt{15}}{12} \qquad \text{Multiply.} \qquad \text{NOW TRY}$$

NOW TRY ANSWERS

2. $\dfrac{3\sqrt{21}}{7}$ **3.** $\dfrac{\sqrt{5}}{10}$

NOW TRY EXERCISE 4

Simplify. Assume that m and n represent positive real numbers.

(a) $\dfrac{\sqrt{9m}}{\sqrt{n}}$ **(b)** $\sqrt{\dfrac{16m^2n}{5}}$

EXAMPLE 4 Simplifying Quotients Involving Radicals

Simplify. Assume that x and y represent positive real numbers.

(a) $\dfrac{\sqrt{4x}}{\sqrt{y}}$

$= \dfrac{\sqrt{4x} \cdot \sqrt{y}}{\sqrt{y} \cdot \sqrt{y}}$ Rationalize the denominator.

$= \dfrac{\sqrt{4xy}}{y}$ Product rule; $\sqrt{y} \cdot \sqrt{y} = y$

$= \dfrac{\sqrt{4} \cdot \sqrt{xy}}{y}$ Product rule

$= \dfrac{2\sqrt{xy}}{y}$ $\sqrt{4} = 2$

(b) $\sqrt{\dfrac{2x^2y}{3}}$

$= \dfrac{\sqrt{2x^2y}}{\sqrt{3}}$ Quotient rule

$= \dfrac{\sqrt{2x^2y} \cdot \sqrt{3}}{\sqrt{3} \cdot \sqrt{3}}$ Rationalize the denominator.

$= \dfrac{\sqrt{6x^2y}}{3}$ Product rule; $\sqrt{3} \cdot \sqrt{3} = 3$

$= \dfrac{\sqrt{x^2}\sqrt{6y}}{3}$ Product rule

$= \dfrac{x\sqrt{6y}}{3}$ $\sqrt{x^2} = x$ because $x > 0$.

NOW TRY

OBJECTIVE 3 Rationalize denominators with cube roots.

To rationalize a denominator with a cube root, we change the radicand in the denominator to a perfect cube.

EXAMPLE 5 Rationalizing Denominators with Cube Roots

Rationalize each denominator.

(a) $\sqrt[3]{\dfrac{3}{2}}$

First write the expression as a quotient of radicals. Then multiply the numerator and denominator by a sufficient number of factors of 2 to make the radicand in the denominator a perfect cube. This will eliminate the radical in the denominator. Here, multiply by $\sqrt[3]{2 \cdot 2}$, or $\sqrt[3]{2^2}$.

$$\sqrt[3]{\dfrac{3}{2}} = \dfrac{\sqrt[3]{3}}{\sqrt[3]{2}} = \dfrac{\sqrt[3]{3} \cdot \sqrt[3]{2 \cdot 2}}{\sqrt[3]{2} \cdot \sqrt[3]{2 \cdot 2}} = \dfrac{\sqrt[3]{3 \cdot 2 \cdot 2}}{\sqrt[3]{2 \cdot 2 \cdot 2}} = \dfrac{\sqrt[3]{12}}{2}$$

We need 3 factors of 2 in the radicand in the denominator.

$\sqrt[3]{2 \cdot 2 \cdot 2} = \sqrt[3]{2^3} = 2$
Denominator radicand is a perfect cube.

(b) $\dfrac{\sqrt[3]{3}}{\sqrt[3]{4}}$

Because $\sqrt[3]{4} = \sqrt[3]{2 \cdot 2}$, multiply the numerator and denominator by a sufficient number of factors of 2 to obtain a perfect cube in the radicand in the denominator.

$$\dfrac{\sqrt[3]{3}}{\sqrt[3]{4}} = \dfrac{\sqrt[3]{3} \cdot \sqrt[3]{2}}{\sqrt[3]{2 \cdot 2} \cdot \sqrt[3]{2}} = \dfrac{\sqrt[3]{6}}{\sqrt[3]{2 \cdot 2 \cdot 2}} = \dfrac{\sqrt[3]{6}}{2}$$

NOW TRY ANSWERS

4. (a) $\dfrac{3\sqrt{mn}}{n}$ **(b)** $\dfrac{4m\sqrt{5n}}{5}$

NOW TRY EXERCISE 5

Rationalize each denominator.

(a) $\sqrt[3]{\dfrac{2}{7}}$ **(b)** $\dfrac{\sqrt[3]{2}}{\sqrt[3]{5}}$

(c) $\dfrac{\sqrt[3]{4}}{\sqrt[3]{9t}}$ $(t \neq 0)$

(c) $\dfrac{\sqrt[3]{2}}{\sqrt[3]{3x^2}}$ $(x \neq 0)$

Multiply the numerator and denominator by a sufficient number of factors of 3 and of x to obtain a perfect cube in the radicand in the denominator.

$$\frac{\sqrt[3]{2}}{\sqrt[3]{3x^2}} = \frac{\sqrt[3]{2} \cdot \sqrt[3]{3 \cdot 3 \cdot x}}{\sqrt[3]{3 \cdot x \cdot x} \cdot \sqrt[3]{3 \cdot 3 \cdot x}} = \frac{\sqrt[3]{18x}}{\sqrt[3]{(3x)^3}} = \frac{\sqrt[3]{18x}}{3x}$$

We need 3 factors of 3 and 3 factors of x in the radicand in the denominator.

Denominator radicand is a perfect cube.

NOW TRY

NOW TRY ANSWERS

5. **(a)** $\dfrac{\sqrt[3]{98}}{7}$ **(b)** $\dfrac{\sqrt[3]{50}}{5}$

(c) $\dfrac{\sqrt[3]{12t^2}}{3t}$

⚠ **CAUTION** A common error in a problem like the one in **Example 5(a)** is to multiply both the numerator and denominator of

$$\frac{\sqrt[3]{3}}{\sqrt[3]{2}}$$

by $\sqrt[3]{2}$ instead of $\sqrt[3]{2^2}$. Doing this would give a denominator of

$$\sqrt[3]{2} \cdot \sqrt[3]{2} = \sqrt[3]{4}.$$

Because 4 is not a perfect cube, the denominator is still not rationalized.

8.4 Exercises

FOR EXTRA HELP ▶ MyMathLab®

 Complete solution available in MyMathLab

1. *Concept Check* To rationalize the denominator of a radical expression such as $\dfrac{4}{\sqrt{3}}$, we multiply both the numerator and denominator by $\sqrt{3}$. By what number are we actually multiplying the given expression, and what property of real numbers justifies the fact that our result is equal to the given expression?

2. In **Example 1(a)**, we showed algebraically that $\dfrac{9}{\sqrt{6}} = \dfrac{3\sqrt{6}}{2}$. Support this result numerically by finding the decimal approximation of $\dfrac{9}{\sqrt{6}}$ on a calculator and then finding the decimal approximation of $\dfrac{3\sqrt{6}}{2}$. What do you notice?

Rationalize each denominator. ***See Examples 1 and 2.***

3. $\dfrac{6}{\sqrt{5}}$ **4.** $\dfrac{3}{\sqrt{2}}$ **5.** $\dfrac{5}{\sqrt{5}}$ **6.** $\dfrac{15}{\sqrt{15}}$

 7. $\dfrac{4}{\sqrt{6}}$ **8.** $\dfrac{15}{\sqrt{10}}$ **9.** $\dfrac{8\sqrt{3}}{\sqrt{5}}$ **10.** $\dfrac{9\sqrt{6}}{\sqrt{5}}$

11. $\dfrac{12\sqrt{10}}{8\sqrt{3}}$ **12.** $\dfrac{9\sqrt{15}}{6\sqrt{2}}$ **13.** $\dfrac{8}{\sqrt{27}}$ **14.** $\dfrac{12}{\sqrt{18}}$

15. $\dfrac{6}{\sqrt{200}}$ **16.** $\dfrac{10}{\sqrt{300}}$ **17.** $\dfrac{12}{\sqrt{72}}$ **18.** $\dfrac{21}{\sqrt{45}}$

19. $\dfrac{\sqrt{10}}{\sqrt{5}}$ **20.** $\dfrac{\sqrt{6}}{\sqrt{3}}$ ▶ **21.** $\sqrt{\dfrac{40}{3}}$ **22.** $\sqrt{\dfrac{5}{8}}$

23. $\sqrt{\dfrac{1}{32}}$ **24.** $\sqrt{\dfrac{1}{8}}$ **25.** $\sqrt{\dfrac{9}{5}}$ **26.** $\sqrt{\dfrac{16}{7}}$

27. $\dfrac{-3}{\sqrt{50}}$ **28.** $\dfrac{-5}{\sqrt{75}}$ **29.** $\dfrac{63}{\sqrt{45}}$ **30.** $\dfrac{27}{\sqrt{32}}$ **31.** $\dfrac{\sqrt{8}}{\sqrt{24}}$

32. $\dfrac{\sqrt{5}}{\sqrt{10}}$ **33.** $-\sqrt{\dfrac{1}{5}}$ **34.** $-\sqrt{\dfrac{1}{6}}$ **35.** $\sqrt{\dfrac{13}{5}}$ **36.** $\sqrt{\dfrac{17}{11}}$

Simplify. See Example 3.

37. $\sqrt{\dfrac{7}{13}} \cdot \sqrt{\dfrac{13}{3}}$ **38.** $\sqrt{\dfrac{19}{20}} \cdot \sqrt{\dfrac{20}{3}}$ **39.** $\sqrt{\dfrac{21}{7}} \cdot \sqrt{\dfrac{21}{8}}$ **40.** $\sqrt{\dfrac{5}{8}} \cdot \sqrt{\dfrac{5}{6}}$

41. $\sqrt{\dfrac{1}{12}} \cdot \sqrt{\dfrac{1}{3}}$ **42.** $\sqrt{\dfrac{1}{8}} \cdot \sqrt{\dfrac{1}{2}}$ **43.** $\sqrt{\dfrac{2}{9}} \cdot \sqrt{\dfrac{9}{2}}$ **44.** $\sqrt{\dfrac{4}{3}} \cdot \sqrt{\dfrac{3}{4}}$

▶ **45.** $\sqrt{\dfrac{3}{4}} \cdot \sqrt{\dfrac{1}{5}}$ **46.** $\sqrt{\dfrac{1}{10}} \cdot \sqrt{\dfrac{10}{3}}$ **47.** $\sqrt{\dfrac{17}{3}} \cdot \sqrt{\dfrac{17}{6}}$ **48.** $\sqrt{\dfrac{1}{11}} \cdot \sqrt{\dfrac{33}{16}}$

49. $\sqrt{\dfrac{2}{5}} \cdot \sqrt{\dfrac{3}{10}}$ **50.** $\sqrt{\dfrac{9}{8}} \cdot \sqrt{\dfrac{7}{16}}$ **51.** $\sqrt{\dfrac{16}{27}} \cdot \sqrt{\dfrac{1}{9}}$ **52.** $\sqrt{\dfrac{256}{125}} \cdot \sqrt{\dfrac{1}{16}}$

Simplify. Assume that all variables represent positive real numbers. See Example 4.

53. $\sqrt{\dfrac{6}{p}}$ **54.** $\sqrt{\dfrac{5}{x}}$ **55.** $\sqrt{\dfrac{3}{y}}$ **56.** $\sqrt{\dfrac{9}{k}}$

57. $\sqrt{\dfrac{16}{m}}$ **58.** $\sqrt{\dfrac{36}{x}}$ ▶ **59.** $\dfrac{\sqrt{3p^2}}{\sqrt{q}}$ **60.** $\dfrac{\sqrt{5a^2}}{\sqrt{b}}$

61. $\dfrac{\sqrt{7x^3}}{\sqrt{y}}$ **62.** $\dfrac{\sqrt{4r^3}}{\sqrt{s}}$ **63.** $\sqrt{\dfrac{6p^3}{3m}}$ **64.** $\sqrt{\dfrac{12x^2}{4y}}$

65. $\sqrt{\dfrac{a^3b}{6}}$ **66.** $\sqrt{\dfrac{m^2n}{2}}$ ▶ **67.** $\sqrt{\dfrac{9a^2r}{5}}$ **68.** $\sqrt{\dfrac{2x^2z}{3}}$

69. *Concept Check* Which one of the following would be an appropriate choice for multiplying the numerator and denominator of $\dfrac{\sqrt[3]{2}}{\sqrt[3]{5}}$ in order to rationalize the denominator?

A. $\sqrt[3]{5}$ **B.** $\sqrt[3]{25}$ **C.** $\sqrt[3]{2}$ **D.** $\sqrt[3]{3}$

70. *Concept Check* In **Example 5(b),** we multiplied the numerator and denominator of $\dfrac{\sqrt[3]{3}}{\sqrt[3]{4}}$ by $\sqrt[3]{2}$ to rationalize the denominator. Suppose we had chosen to multiply by $\sqrt[3]{16}$ instead. Would we have obtained the correct answer after all simplifications were done?

Rationalize each denominator. Assume that variables in denominators represent nonzero real numbers. **See Example 5.**

71. $\sqrt[3]{\dfrac{1}{2}}$ **72.** $\sqrt[3]{\dfrac{1}{4}}$ **73.** $\sqrt[3]{\dfrac{1}{32}}$ **74.** $\sqrt[3]{\dfrac{1}{5}}$

75. $\sqrt[3]{\dfrac{1}{11}}$ **76.** $\sqrt[3]{\dfrac{3}{2}}$ ▶ **77.** $\sqrt[3]{\dfrac{2}{5}}$ **78.** $\sqrt[3]{\dfrac{4}{9}}$

79. $\dfrac{\sqrt[3]{4}}{\sqrt[3]{7}}$ **80.** $\dfrac{\sqrt[3]{5}}{\sqrt[3]{10}}$ **81.** $\sqrt[3]{\dfrac{3}{4y^2}}$ **82.** $\sqrt[3]{\dfrac{3}{25x^2}}$

83. $\dfrac{\sqrt[3]{7m}}{\sqrt[3]{36n}}$ **84.** $\dfrac{\sqrt[3]{11p}}{\sqrt[3]{49q}}$ **85.** $\sqrt[4]{\dfrac{1}{8}}$ **86.** $\sqrt[4]{\dfrac{1}{27}}$

In each problem, **(a)** *give the answer as a simplified radical and* **(b)** *use a calculator to give the answer correct to the nearest thousandth.*

87. The period p of a pendulum is the time it takes for it to swing from one extreme to the other and back again. The value of p in seconds is given by

$$p = k \cdot \sqrt{\dfrac{L}{g}},$$

where L is the length of the pendulum, g is the acceleration due to gravity, and k is a constant. Find the period when $k = 6$, $L = 9$ ft, and $g = 32$ ft per sec^2.

88. The velocity v in kilometers per second of a meteor approaching Earth is given by

$$v = \dfrac{k}{\sqrt{d}},$$

where d is the distance of the meteor from the center of Earth and k is a constant. What is the velocity of a meteor that is 6000 km away from the center of Earth if $k = 450$?

8.5 More Simplifying and Operations with Radicals

OBJECTIVES

1 Simplify products of radical expressions.

2 Use conjugates to rationalize denominators of radical expressions.

3 Write radical expressions with quotients in lowest terms.

VOCABULARY

☐ conjugates

Apply these guidelines when simplifying radical expressions.

Guidelines for Simplifying Radical Expressions

1. If a radical represents a rational number, use that rational number in place of the radical.

Examples: $\sqrt{49} = 7, \quad \sqrt{\dfrac{169}{9}} = \dfrac{13}{3}$

2. If a radical expression contains products of radicals, use the product rule for radicals, $\sqrt[n]{a} \cdot \sqrt[n]{b} = \sqrt[n]{ab},$ to obtain a single radical.

Examples: $\sqrt{5} \cdot \sqrt{x} = \sqrt{5x}, \quad \sqrt[3]{3} \cdot \sqrt[3]{2} = \sqrt[3]{6}$

3. If a radicand of a square root radical has a factor that is a perfect square, express the radical as the product of the positive square root of the perfect square and the remaining radical factor. Higher roots are treated similarly.

Examples: $\sqrt{20} = \sqrt{4 \cdot 5} = \sqrt{4} \cdot \sqrt{5} = 2\sqrt{5}$

$\sqrt[3]{16} = \sqrt[3]{8 \cdot 2} = \sqrt[3]{8} \cdot \sqrt[3]{2} = 2\sqrt[3]{2}$

4. If a radical expression contains sums or differences of radicals, use the distributive property to combine like radicals.

Examples: $3\sqrt{2} + 4\sqrt{2}$ can be combined to obtain $7\sqrt{2}.$

$3\sqrt{2} + 4\sqrt{3}$ cannot be combined in this way.

5. Rationalize any denominator containing a radical.

Examples: $\dfrac{5}{\sqrt{3}} = \dfrac{5 \cdot \sqrt{3}}{\sqrt{3} \cdot \sqrt{3}} = \dfrac{5\sqrt{3}}{3}$

$\sqrt[3]{\dfrac{1}{4}} = \dfrac{\sqrt[3]{1}}{\sqrt[3]{4}} = \dfrac{\sqrt[3]{1} \cdot \sqrt[3]{2}}{\sqrt[3]{4} \cdot \sqrt[3]{2}} = \dfrac{\sqrt[3]{2}}{\sqrt[3]{8}} = \dfrac{\sqrt[3]{2}}{2}$

OBJECTIVE 1 Simplify products of radical expressions.

EXAMPLE 1 Multiplying Radical Expressions

Find each product and simplify.

(a) $\sqrt{5}(\sqrt{8} - \sqrt{32})$ ⟵ Simplify inside the parentheses.

$= \sqrt{5}(2\sqrt{2} - 4\sqrt{2})$ $\sqrt{8} = 2\sqrt{2}; \sqrt{32} = 4\sqrt{2}$

$= \sqrt{5}(-2\sqrt{2})$ Subtract like radicals.

$= -2\sqrt{5 \cdot 2}$ Product rule

$= -2\sqrt{10}$ Multiply.

NOW TRY
EXERCISE 1
Find each product and simplify.

(a) $\sqrt{3}(\sqrt{45} - \sqrt{20})$

(b) $(2\sqrt{3} + \sqrt{7})(\sqrt{3} + 3\sqrt{7})$

(c) $(\sqrt{10} - 8)(2\sqrt{10} + 3\sqrt{2})$

(b) $(\sqrt{3} + 2\sqrt{5})(\sqrt{3} - 4\sqrt{5})$ ⟵ Use the FOIL method to multiply.

$= \underbrace{\sqrt{3}(\sqrt{3})}_{\text{First}} + \underbrace{\sqrt{3}(-4\sqrt{5})}_{\text{Outer}} + \underbrace{2\sqrt{5}(\sqrt{3})}_{\text{Inner}} + \underbrace{2\sqrt{5}(-4\sqrt{5})}_{\text{Last}}$

$= 3 - 4\sqrt{15} + 2\sqrt{15} - 8 \cdot 5$ Product rule

$= 3 - 2\sqrt{15} - 40$ Add like radicals. Multiply.

This does *not* equal $-39\sqrt{15}$.

$= -37 - 2\sqrt{15}$ Combine like terms.

(c) $(\sqrt{3} + \sqrt{21})(\sqrt{3} - \sqrt{7})$

$= \sqrt{3}(\sqrt{3}) + \sqrt{3}(-\sqrt{7}) + \sqrt{21}(\sqrt{3}) + \sqrt{21}(-\sqrt{7})$ FOIL method

$= 3 - \sqrt{21} + \sqrt{63} - \sqrt{147}$ Product rule

$= 3 - \sqrt{21} + \sqrt{9} \cdot \sqrt{7} - \sqrt{49} \cdot \sqrt{3}$ Factor; 9 and 49 are perfect squares.

$= 3 - \sqrt{21} + 3\sqrt{7} - 7\sqrt{3}$ $\sqrt{9} = 3$; $\sqrt{49} = 7$

Because there are no like radicals, no terms can be combined.   NOW TRY

Example 2 uses the rules for squaring binomials from **Section 5.6**.

$$(x + y)^2 = x^2 + 2xy + y^2 \quad \text{and} \quad (x - y)^2 = x^2 - 2xy + y^2$$

NOW TRY
EXERCISE 2
Find each product. In part (c), assume that $y \geq 0$.

(a) $(\sqrt{7} - 4)^2$

(b) $(3\sqrt{2} - 5)^2$

(c) $(3 + \sqrt{y})^2$

EXAMPLE 2 Using Special Products with Radicals

Find each product. In part (c), assume that $x \geq 0$.

(a) $(\sqrt{10} - 7)^2$

$= (\sqrt{10})^2 - 2(\sqrt{10})(7) + 7^2$ $(x - y)^2 = x^2 - 2xy + y^2$
Let $x = \sqrt{10}$ and $y = 7$.

Do *not* try to combine further here. $= 10 - 14\sqrt{10} + 49$ $(\sqrt{10})^2 = 10$; $7^2 = 49$

$= 59 - 14\sqrt{10}$ Add 10 and 49.

(b) $(2\sqrt{3} + 4)^2$

$= (2\sqrt{3})^2 + 2(2\sqrt{3})(4) + 4^2$ $(x + y)^2 = x^2 + 2xy + y^2$
Let $x = 2\sqrt{3}$ and $y = 4$.

Do *not* try to combine further here. $= 12 + 16\sqrt{3} + 16$ $(2\sqrt{3})^2 = 4 \cdot 3 = 12$

$= 28 + 16\sqrt{3}$ Add 12 and 16.

NOW TRY ANSWERS
1. (a) $\sqrt{15}$ (b) $27 + 7\sqrt{21}$
   (c) $20 + 6\sqrt{5} - 16\sqrt{10} - 24\sqrt{2}$
2. (a) $23 - 8\sqrt{7}$ (b) $43 - 30\sqrt{2}$
   (c) $9 + 6\sqrt{y} + y$

(c) $(5 - \sqrt{x})^2$

$= 5^2 - 2(5)(\sqrt{x}) + (\sqrt{x})^2$ Square the binomial.

$= 25 - 10\sqrt{x} + x$ Apply the exponents. Multiply.   NOW TRY

> **!** **CAUTION**   *Only like radicals can be combined.* In Examples 2(a) and (b),
>
> $$59 - 14\sqrt{10} \ne 45\sqrt{10} \quad \text{and} \quad 28 + 16\sqrt{3} \ne 44\sqrt{3}.$$

Example 3 uses the rule for a product of a sum and difference of two terms.

$$(x + y)(x - y) = x^2 - y^2$$

**NOW TRY
EXERCISE 3**

Find each product. In part (b), assume that $x \ge 0$.

(a) $\left(8 + \sqrt{10}\right)\left(8 - \sqrt{10}\right)$

(b) $\left(\sqrt{x} + 2\sqrt{3}\right)\left(\sqrt{x} - 2\sqrt{3}\right)$

EXAMPLE 3   Using a Special Product with Radicals

Find each product. In part (b), assume that $x \ge 0$.

(a) $\left(4 + \sqrt{3}\right)\left(4 - \sqrt{3}\right)$

$= 4^2 - \left(\sqrt{3}\right)^2$ $(x + y)(x - y) = x^2 - y^2$
Let $x = 4$ and $y = \sqrt{3}$.

$= 16 - 3$ $4^2 = 16; \left(\sqrt{3}\right)^2 = 3$

$= 13$ Subtract.

(b) $\left(\sqrt{x} - \sqrt{6}\right)\left(\sqrt{x} + \sqrt{6}\right)$

$= \left(\sqrt{x}\right)^2 - \left(\sqrt{6}\right)^2$ Product of the sum and
difference of two terms

$= x - 6$ $\left(\sqrt{x}\right)^2 = x; \left(\sqrt{6}\right)^2 = 6$ **NOW TRY**

In **Example 3,** the expressions $4 + \sqrt{3}$ and $4 - \sqrt{3}$ are **conjugates** of each other, as are $\sqrt{x} - \sqrt{6}$ and $\sqrt{x} + \sqrt{6}$. Recall from **Section 5.6** that the expressions $x + y$ and $x - y$ are conjugates.

OBJECTIVE 2   Use conjugates to rationalize denominators of radical expressions.

To rationalize the denominator in a quotient such as

$$\frac{2}{4 - \sqrt{3}},$$

we multiply the numerator and denominator by the conjugate of the denominator, here $4 + \sqrt{3}$.

See Example 3(a). $\dfrac{2\left(4 + \sqrt{3}\right)}{\left(4 - \sqrt{3}\right)\left(4 + \sqrt{3}\right)}$ gives $\dfrac{2\left(4 + \sqrt{3}\right)}{13}.$

Rationalizing a Binomial Denominator

To rationalize a binomial denominator, where at least one of those terms is a square root radical, multiply the numerator and denominator by the conjugate of the denominator.

NOW TRY ANSWERS
3. (a) 540   **(b)** $x - 12$

**NOW TRY
EXERCISE 4**

Rationalize each denominator. In part (c), assume that $k \geq 0$.

(a) $\dfrac{6}{4 + \sqrt{3}}$ **(b)** $\dfrac{5 + \sqrt{7}}{\sqrt{7} - 2}$

(c) $\dfrac{9}{\sqrt{k} - 6}$ $(k \neq 36)$

EXAMPLE 4 Using Conjugates to Rationalize Denominators

Rationalize each denominator. In part (c), assume that $x \geq 0$.

(a) $\dfrac{5}{3 + \sqrt{5}}$

$= \dfrac{5(3 - \sqrt{5})}{(3 + \sqrt{5})(3 - \sqrt{5})}$ Multiply the numerator and denominator by the conjugate of the denominator.

$= \dfrac{5(3 - \sqrt{5})}{3^2 - (\sqrt{5})^2}$ $(x + y)(x - y) = x^2 - y^2$

$= \dfrac{5(3 - \sqrt{5})}{9 - 5}$ $3^2 = 9;\ (\sqrt{5})^2 = 5$

$= \dfrac{5(3 - \sqrt{5})}{4}$ Subtract in the denominator.

(b) $\dfrac{6 + \sqrt{2}}{\sqrt{2} - 5}$

$= \dfrac{(6 + \sqrt{2})(\sqrt{2} + 5)}{(\sqrt{2} - 5)(\sqrt{2} + 5)}$ Multiply the numerator and denominator by the conjugate of the denominator.

$= \dfrac{6\sqrt{2} + 30 + 2 + 5\sqrt{2}}{2 - 25}$ FOIL method; $(x + y)(x - y) = x^2 - y^2$

$= \dfrac{11\sqrt{2} + 32}{-23}$ Combine like terms.

$= \dfrac{-(11\sqrt{2} + 32)}{23}$ $\dfrac{x}{-y} = \dfrac{-x}{y}$

> Be careful. Distribute the $-$ sign to *both* terms in the numerator.

$= \dfrac{-11\sqrt{2} - 32}{23}$ Distributive property

The last three lines above give three equivalent forms of the answer.

(c) $\dfrac{4}{3 + \sqrt{x}}$

$= \dfrac{4(3 - \sqrt{x})}{(3 + \sqrt{x})(3 - \sqrt{x})}$ Multiply by $\dfrac{3 - \sqrt{x}}{3 - \sqrt{x}} = 1$.

$= \dfrac{4(3 - \sqrt{x})}{9 - x}$ $3^2 = 9;\ (\sqrt{x})^2 = x$ **NOW TRY**

> We assume here that $x \neq 9$.

NOW TRY ANSWERS

4. (a) $\dfrac{6(4 - \sqrt{3})}{13}$

    (b) $\dfrac{17 + 7\sqrt{7}}{3}$

    (c) $\dfrac{9(\sqrt{k} + 6)}{k - 36}$

NOW TRY
EXERCISE 5

Write the quotient in lowest terms.

$$\frac{12\sqrt{6} + 28}{20}$$

OBJECTIVE 3 Write radical expressions with quotients in lowest terms.

EXAMPLE 5 Writing a Radical Quotient in Lowest Terms

Write the quotient in lowest terms.

$$\frac{3\sqrt{3} + 9}{12}$$ Each term in the numerator and denominator has a common factor of 3.

$$= \frac{3(\sqrt{3} + 3)}{3(4)}$$ Factor first.

$$= 1 \cdot \frac{\sqrt{3} + 3}{4}$$ Now divide out the common factor; $\frac{3}{3} = 1$

$$= \frac{\sqrt{3} + 3}{4}$$ Identity property; lowest terms   NOW TRY

⚠ **CAUTION** An expression like the one in **Example 5** can be simplified only by factoring a common factor from the denominator and *each* term of the numerator. ***First factor, and then divide out the common factor.***

NOW TRY ANSWER

5. $\dfrac{3\sqrt{6} + 7}{5}$

Factor $\dfrac{4 + 8\sqrt{5}}{4}$ as $\dfrac{4(1 + 2\sqrt{5})}{4}$ to obtain $1 + 2\sqrt{5}$.

8.5 Exercises

FOR EXTRA HELP   ▶ MyMathLab®

▶ *Complete solution available in MyMathLab*

In this exercise set, we assume that variables are such that no negative numbers appear as radicals in square roots and no denominators are 0.

Concept Check   *Perform the operations mentally, and write the answers without doing intermediate steps.*

1. $\sqrt{25} + \sqrt{64}$ **2.** $\sqrt{100} - \sqrt{49}$ **3.** $\sqrt{8} \cdot \sqrt{2}$ **4.** $\sqrt{6} \cdot \sqrt{6}$

5. *Concept Check*   In **Example 1(b),** a student simplified $-37 - 2\sqrt{15}$ by combining the -37 and the -2 to obtain $-39\sqrt{15}$, which is incorrect. *WHAT WENT WRONG?*

6. *Concept Check*   Find each product mentally.

(a) $(\sqrt{x} + \sqrt{y})(\sqrt{x} - \sqrt{y})$ (b) $(\sqrt{28} - \sqrt{14})(\sqrt{28} + \sqrt{14})$

Find each product. Refer to the five guidelines given in this section to be sure answers are simplified. See Examples 1–3.

▶ **7.** $\sqrt{5}(\sqrt{3} - \sqrt{7})$ **8.** $\sqrt{7}(\sqrt{10} + \sqrt{3})$ **9.** $2\sqrt{5}(3\sqrt{5} + \sqrt{2})$

10. $3\sqrt{7}(2\sqrt{7} + 4\sqrt{5})$ **11.** $3\sqrt{14} \cdot \sqrt{2} - \sqrt{28}$ **12.** $7\sqrt{6} \cdot \sqrt{3} - 2\sqrt{18}$

13. $(2\sqrt{6} + 3)(3\sqrt{6} + 7)$ **14.** $(4\sqrt{5} - 2)(2\sqrt{5} - 4)$ ▶ **15.** $(8 - \sqrt{7})^2$

16. $(6 - \sqrt{11})^2$ **17.** $(2\sqrt{7} + 3)^2$ **18.** $(4\sqrt{5} + 5)^2$

19. $(\sqrt{6} + 1)^2$ **20.** $(\sqrt{7} + 2)^2$ **21.** $(\sqrt{a} + 1)^2$

22. $\left(\sqrt{y}+4\right)^2$

23. $\left(7+\sqrt{x}\right)^2$

24. $\left(12-\sqrt{r}\right)^2$

25. $\left(5\sqrt{7}-2\sqrt{3}\right)^2$

26. $\left(8\sqrt{2}-3\sqrt{3}\right)^2$

▶ **27.** $\left(5-\sqrt{2}\right)\left(5+\sqrt{2}\right)$

28. $\left(3-\sqrt{5}\right)\left(3+\sqrt{5}\right)$

29. $\left(\sqrt{8}-\sqrt{7}\right)\left(\sqrt{8}+\sqrt{7}\right)$

30. $\left(\sqrt{12}-\sqrt{11}\right)\left(\sqrt{12}+\sqrt{11}\right)$

31. $\left(\sqrt{78}-\sqrt{76}\right)\left(\sqrt{78}+\sqrt{76}\right)$

32. $\left(\sqrt{85}-\sqrt{82}\right)\left(\sqrt{85}+\sqrt{82}\right)$

33. $\left(\sqrt{y}-\sqrt{10}\right)\left(\sqrt{y}+\sqrt{10}\right)$

34. $\left(\sqrt{t}-\sqrt{13}\right)\left(\sqrt{t}+\sqrt{13}\right)$

35. $\left(\sqrt{2}+\sqrt{3}\right)\left(\sqrt{6}-\sqrt{2}\right)$

36. $\left(\sqrt{3}+\sqrt{5}\right)\left(\sqrt{15}-\sqrt{5}\right)$

37. $\left(\sqrt{10}-\sqrt{5}\right)\left(\sqrt{5}+\sqrt{20}\right)$

38. $\left(\sqrt{6}-\sqrt{3}\right)\left(\sqrt{3}+\sqrt{18}\right)$

39. $\left(\sqrt{5}+\sqrt{30}\right)\left(\sqrt{6}+\sqrt{3}\right)$

40. $\left(\sqrt{10}-\sqrt{20}\right)\left(\sqrt{2}-\sqrt{5}\right)$

41. $\left(5\sqrt{7}-2\sqrt{3}\right)\left(3\sqrt{7}+4\sqrt{3}\right)$

42. $\left(2\sqrt{10}+5\sqrt{2}\right)\left(3\sqrt{10}-3\sqrt{2}\right)$

43. $\left(3\sqrt{t}+\sqrt{7}\right)\left(2\sqrt{t}-\sqrt{14}\right)$

44. $\left(2\sqrt{z}-\sqrt{3}\right)\left(\sqrt{z}-\sqrt{5}\right)$

45. $\left(\sqrt{3m}+\sqrt{2n}\right)\left(\sqrt{3m}-\sqrt{2n}\right)$

46. $\left(\sqrt{4p}-\sqrt{3k}\right)\left(\sqrt{4p}+\sqrt{3k}\right)$

47. *Concept Check*   Determine the expression by which we should multiply the numerator and denominator to rationalize each denominator.

(a) $\dfrac{1}{\sqrt{5}+\sqrt{3}}$ (b) $\dfrac{3}{\sqrt{6}-\sqrt{5}}$

48. *Concept Check*   A student tried to rationalize the denominator of $\dfrac{2}{4+\sqrt{3}}$ by multiplying by $\dfrac{4+\sqrt{3}}{4+\sqrt{3}}$. *WHAT WENT WRONG?* By what should he multiply?

Rationalize each denominator. Write quotients in lowest terms. ***See Example 4.***

49. $\dfrac{1}{2+\sqrt{5}}$

50. $\dfrac{1}{4+\sqrt{15}}$

▶ **51.** $\dfrac{7}{2-\sqrt{11}}$

52. $\dfrac{38}{5-\sqrt{6}}$

53. $\dfrac{\sqrt{12}}{\sqrt{3}+1}$

54. $\dfrac{\sqrt{18}}{\sqrt{2}-1}$

55. $\dfrac{2\sqrt{3}}{\sqrt{3}+5}$

56. $\dfrac{2\sqrt{3}}{2-\sqrt{10}}$

57. $\dfrac{\sqrt{2}+3}{\sqrt{3}-1}$

58. $\dfrac{\sqrt{5}+2}{2-\sqrt{3}}$

59. $\dfrac{6-\sqrt{5}}{\sqrt{2}+2}$

60. $\dfrac{3+\sqrt{2}}{\sqrt{2}+1}$

61. $\dfrac{2\sqrt{6}+1}{\sqrt{2}+5}$

62. $\dfrac{3\sqrt{2}-4}{\sqrt{3}+2}$

63. $\dfrac{\sqrt{7}+\sqrt{2}}{\sqrt{3}-\sqrt{2}}$

64. $\dfrac{\sqrt{6}+\sqrt{5}}{\sqrt{3}+\sqrt{5}}$

65. $\dfrac{\sqrt{5}}{\sqrt{2}+\sqrt{3}}$

66. $\dfrac{\sqrt{3}}{\sqrt{2}+\sqrt{3}}$

67. $\dfrac{\sqrt{108}}{3+3\sqrt{3}}$

68. $\dfrac{9\sqrt{8}}{6\sqrt{2}-6}$

69. $\dfrac{8}{4-\sqrt{x}}$

70. $\dfrac{12}{6+\sqrt{y}}$

71. $\dfrac{1}{\sqrt{x}+\sqrt{y}}$

72. $\dfrac{2}{\sqrt{x}-\sqrt{y}}$

Write each quotient in lowest terms. **See Example 5.**

▶ **73.** $\dfrac{5\sqrt{7} - 10}{5}$

74. $\dfrac{6\sqrt{5} - 9}{3}$

75. $\dfrac{2\sqrt{3} + 10}{8}$

76. $\dfrac{4\sqrt{6} + 6}{10}$

77. $\dfrac{12 - 2\sqrt{10}}{4}$

78. $\dfrac{9 - 6\sqrt{2}}{12}$

79. $\dfrac{16 + \sqrt{128}}{24}$

80. $\dfrac{25 + \sqrt{75}}{10}$

Extending Skills   *Perform each operation and express the answer in simplest form.*

81. $\sqrt[3]{4}\left(\sqrt[3]{2} - 3\right)$

82. $\sqrt[3]{5}\left(4\sqrt[3]{5} - \sqrt[3]{25}\right)$

83. $2\sqrt[4]{2}\left(3\sqrt[4]{8} + 5\sqrt[4]{4}\right)$

84. $6\sqrt[4]{9}\left(2\sqrt[4]{9} - \sqrt[4]{27}\right)$

85. $\left(\sqrt[3]{2} - 1\right)\left(\sqrt[3]{4} + 3\right)$

86. $\left(\sqrt[3]{9} + 5\right)\left(\sqrt[3]{3} - 4\right)$

87. $\left(\sqrt[3]{5} - \sqrt[3]{4}\right)\left(\sqrt[3]{25} + \sqrt[3]{20} + \sqrt[3]{16}\right)$

88. $\left(\sqrt[3]{4} + \sqrt[3]{2}\right)\left(\sqrt[3]{16} - \sqrt[3]{8} + \sqrt[3]{4}\right)$

Solve each problem.

89. The radius of the circular top or bottom of a tin can with surface area S and height h is given by

$$r = \frac{-h + \sqrt{h^2 + 0.64S}}{2}.$$

What radius should be used to make a can with height 12 in. and surface area 400 in.²?

90. If an investment of P dollars grows to A dollars in 2 yr, the annual rate of return on the investment is given by

$$r = \frac{\sqrt{A} - \sqrt{P}}{\sqrt{P}}.$$

First rationalize the denominator, and then find the annual rate of return (as a percent) if $50,000 increases to $54,080.

RELATING CONCEPTS For Individual or Group Work (Exercises 91–96)

Work Exercises 91–96 in order, *to see why a common student error is indeed an error.*

91. Use the distributive property to write $6(5 + 3x)$ as a sum.

92. The answer in **Exercise 91** should be $30 + 18x$. Why can we not combine these two terms to obtain $48x$?

93. Repeat **Exercise 42** from earlier in this exercise set, and find the product.

$$\left(2\sqrt{10} + 5\sqrt{2}\right)\left(3\sqrt{10} - 3\sqrt{2}\right)$$

94. The answer in **Exercise 93** should be $30 + 18\sqrt{5}$. Many students will, in error, try to combine these terms to obtain $48\sqrt{5}$. Why is this wrong?

95. Write the expression similar to $30 + 18x$ that simplifies to $48x$. Then write the expression similar to $30 + 18\sqrt{5}$ that simplifies to $48\sqrt{5}$.

96. Write a short explanation of the similarities between combining like terms and combining like radicals.

SUMMARY EXERCISES   Applying Operations with Radicals

Perform all indicated operations, and express each answer in simplest form. Assume that all variables represent positive real numbers.

1. $-3\sqrt{10}$ **2.** $5 - \sqrt{15}$

3. $2 - \sqrt{6} + 2\sqrt{3} - 3\sqrt{2}$

4. $6\sqrt{2}$

5. $73 - 12\sqrt{35}$

6. $\dfrac{\sqrt{6}}{2}$

7. $-3 - 2\sqrt{2}$

8. $4\sqrt{7} + 4\sqrt{5}$

9. -33 **10.** $\dfrac{\sqrt{t} - \sqrt{3}}{t - 3}$

11. $2xyz^2\sqrt[3]{y^2}$ **12.** $4\sqrt[3]{3}$

13. $\sqrt{6} + 1$ **14.** $\dfrac{\sqrt{6x}}{3x}$

15. $\dfrac{3}{5}$ **16.** $4\sqrt{2}$

17. $-2\sqrt[3]{2}$ **18.** $11 - 2\sqrt{30}$

19. $3\sqrt{3x}$ **20.** $52 + 30\sqrt{3}$

21. $\dfrac{2\sqrt[3]{18}}{9}$ **22.** 1

23. $-x^2\sqrt[4]{x}$ **24.** $3\sqrt[3]{2t^2}$

25. These unlike radicals cannot be combined.

26. (a) 6   (b) $\{-6, 6\}$

27. (a) 9   (b) $\{-9, 9\}$

28. (a) $\{-2, 2\}$ (b) -2

29. (a) $\{-3, 3\}$ (b) -3

30. $x^2 = 25$

$x^2 - 25 = 0$

$(x + 5)(x - 5) = 0$

$x = -5$ or $x = 5$

Solution set: $\{-5, 5\}$

1. $5\sqrt{10} - 8\sqrt{10}$ **2.** $\sqrt{5}(\sqrt{5} - \sqrt{3})$ **3.** $(1 + \sqrt{3})(2 - \sqrt{6})$

4. $\sqrt{98} - \sqrt{72} + \sqrt{50}$ **5.** $(3\sqrt{5} - 2\sqrt{7})^2$ **6.** $\dfrac{3}{\sqrt{6}}$

7. $\dfrac{1 + \sqrt{2}}{1 - \sqrt{2}}$ **8.** $\dfrac{8}{\sqrt{7} - \sqrt{5}}$ **9.** $(\sqrt{3} + 6)(\sqrt{3} - 6)$

10. $\dfrac{1}{\sqrt{t} + \sqrt{3}}$ **11.** $\sqrt[3]{8x^3y^5z^6}$ **12.** $\dfrac{12}{\sqrt[3]{9}}$

13. $\dfrac{5}{\sqrt{6} - 1}$ **14.** $\sqrt{\dfrac{2}{3x}}$ **15.** $\dfrac{6\sqrt{3}}{5\sqrt{12}}$

16. $\dfrac{8\sqrt{50}}{2\sqrt{25}}$ **17.** $\dfrac{-4}{\sqrt[3]{4}}$ **18.** $\dfrac{\sqrt{6} - \sqrt{5}}{\sqrt{6} + \sqrt{5}}$

19. $\sqrt{75x} - \sqrt{12x}$ **20.** $(5 + 3\sqrt{3})^2$ **21.** $\sqrt[3]{\dfrac{16}{81}}$

22. $(\sqrt{107} - \sqrt{106})(\sqrt{107} + \sqrt{106})$ **23.** $x\sqrt[4]{x^5} - 3\sqrt[4]{x^9} + x^2\sqrt[4]{x}$

24. $\sqrt[3]{16t^2} - \sqrt[3]{54t^2} + \sqrt[3]{128t^2}$ **25.** $\sqrt{14} + \sqrt{5}$

Students often have trouble distinguishing between the following two types of problems.

Simplifying a Radical Involving a Square Root	**Solving an Equation Using Square Roots**
Exercise: Simplify $\sqrt{25}$.	*Exercise:* Solve $x^2 = 25$.
Answer: 5	*Answer:* $\{-5, 5\}$
In this situation, $\sqrt{25}$ represents the positive square root of 25, namely, 5.	In this situation, $x^2 = 25$ has either of two solutions: the negative square root of 25 or the positive square root of 25—that is, -5 or 5. (See **Exercise 30.**)

Use the preceding information to work Exercises 26–29.

26. (a) Simplify $\sqrt{36}$.

  (b) Solve $x^2 = 36$.

27. (a) Simplify $\sqrt{81}$.

  (b) Solve $x^2 = 81$.

28. (a) Solve $x^2 = 4$.

  (b) Simplify $-\sqrt{4}$.

29. (a) Solve $x^2 = 9$.

  (b) Simplify $-\sqrt{9}$.

30. Use the zero-factor property (**Section 6.5**) to show that the solution set of $x^2 = 25$ is $\{-5, 5\}$.

8.6 Solving Equations with Radicals

OBJECTIVES

1. Solve radical equations having square root radicals.
2. Identify equations with no solutions.
3. Solve equations by squaring a binomial.
4. Solve radical equations having cube root radicals.

VOCABULARY

☐ radical equation
☐ extraneous solution

OBJECTIVE 1 Solve radical equations having square root radicals.

A **radical equation** is an equation having a variable in the radicand.

$$\sqrt{x} = 6, \quad \sqrt{x+1} = 3, \quad \text{and} \quad 3\sqrt{x} = \sqrt{8x+9} \qquad \text{Radical equations}$$

To solve such equations, we use the **squaring property of equality.**

> **Squaring Property of Equality**
>
> If each side of a given equation is squared, then all solutions of the original equation are *among* the solutions of the squared equation.

⚠ **CAUTION** Using the squaring property can give a new equation with *more* solutions than the original equation. For example, starting with $x = 4$ and squaring each side gives

$$x^2 = 4^2, \quad \text{or} \quad x^2 = 16.$$

This last equation, $x^2 = 16$, has *two* solutions, 4 or -4, while the original equation, $x = 4$, has only *one* solution, 4.

Because of this possibility, checking is more than just a guard against algebraic errors when solving an equation with radicals. It is an essential part of the solution process. *All proposed solutions from the squared equation must be checked in the original equation.*

The squaring property allows us to eliminate the radicals in an equation. Then we can solve the resulting equation, which will be either linear or quadratic, using the methods of **Section 2.3 or 6.5.**

NOW TRY EXERCISE 1

Solve $\sqrt{x-5} = 6$.

EXAMPLE 1 Using the Squaring Property of Equality

Solve $\sqrt{x+1} = 3$.

$$\sqrt{x+1} = 3$$

$$\left(\sqrt{x+1}\right)^2 = 3^2 \qquad \text{To eliminate the radical, use the squaring property and square each side.}$$

This equation is linear. $\longrightarrow$ $x + 1 = 9 \qquad$ On the left, $\left(\sqrt{a}\right)^2 = a$.

Proposed solution $\longrightarrow$ $x = 8 \qquad$ Subtract 1.

CHECK $\sqrt{x+1} = 3 \qquad$ Original equation

A check is essential. $\sqrt{8+1} \overset{?}{=} 3 \qquad$ Let $x = 8$.

$$\sqrt{9} \overset{?}{=} 3 \qquad \text{Add.}$$

$$3 = 3 \checkmark \quad \text{True}$$

Because this statement is true, $\{8\}$ is the solution set of $\sqrt{x+1} = 3$. In this case, the equation obtained by squaring had just one solution, which also satisfied the original equation.

NOW TRY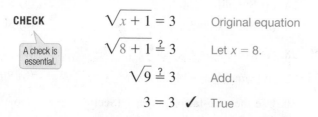

NOW TRY ANSWER
1. $\{41\}$

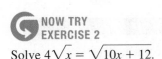

NOW TRY
EXERCISE 2
Solve $4\sqrt{x} = \sqrt{10x + 12}$.

EXAMPLE 2   Using the Squaring Property with a Radical on Each Side

Solve $3\sqrt{x} = \sqrt{x + 8}$.

$$3\sqrt{x} = \sqrt{x + 8}$$ We need to eliminate *both* radicals.

$$\left(3\sqrt{x}\right)^2 = \left(\sqrt{x + 8}\right)^2$$ Squaring property

$$3^2\left(\sqrt{x}\right)^2 = \left(\sqrt{x + 8}\right)^2$$ On the left, $(ab)^2 = a^2b^2$.

Be careful here.

$$9x = x + 8$$ $\left(\sqrt{x}\right)^2 = x;\ \left(\sqrt{x+8}\right)^2 = x + 8$

$$8x = 8$$ Subtract x.

Proposed solution $\longrightarrow x = 1$ Divide by 8.

CHECK $$3\sqrt{x} = \sqrt{x + 8}$$ Original equation

$$3\sqrt{1} \stackrel{?}{=} \sqrt{1 + 8}$$ Let $x = 1$.

$$3(1) \stackrel{?}{=} \sqrt{9}$$

This is *not* the solution.

$$3 = 3\ \checkmark$$ True

Because a true statement results, the solution set is $\{1\}$.   NOW TRY

⚠ **CAUTION**   Do not write the final result obtained in the check in the solution set. In **Example 2**, the solution set is $\{1\}$, **not** $\{3\}$.

OBJECTIVE 2   Identify equations with no solutions.

Not all radical equations have real number solutions.

NOW TRY
EXERCISE 3
Solve $\sqrt{x} = -6$.

EXAMPLE 3   Using the Squaring Property When One Side Is Negative

Solve $\sqrt{x} = -3$.

$$\sqrt{x} = -3$$

$$\left(\sqrt{x}\right)^2 = (-3)^2$$ Squaring property

Proposed solution $\longrightarrow x = 9$ Apply the exponents.

CHECK $$\sqrt{x} = -3$$ Original equation

$$\sqrt{9} \stackrel{?}{=} -3$$ Let $x = 9$.

$$3 = -3$$ False

Because the statement $3 = -3$ is false, the number 9 is *not* a solution of the given equation. It is an **extraneous solution** and must be rejected. In fact, $\sqrt{x} = -3$ has no real number solution. The solution set is $\varnothing$.

Do *not* write $\{\varnothing\}$ to represent the empty set.   NOW TRY

NOTE   Because $\sqrt{x}$ represents the *principal* or *nonnegative* square root of x, we might have seen immediately in **Example 3** that there is no real number solution.

NOW TRY ANSWERS
2. $\{2\}$ **3.** $\varnothing$

Solving a Radical Equation

Step 1   **Isolate a radical.** Arrange the terms so that a radical is isolated on one side of the equation.

Step 2   **Square each side.**

Step 3   **Combine like terms.**

Step 4   **Repeat Steps 1–3,** if there is still a term with a radical.

Step 5   **Solve the equation.** Find all proposed solutions.

Step 6   **Check all proposed solutions** in the original equation. Write the solution set.

NOW TRY
EXERCISE 4

Solve $t = \sqrt{t^2 + 3t + 9}$.

EXAMPLE 4   **Using the Squaring Property with a Quadratic Expression**

Solve $x = \sqrt{x^2 + 5x + 10}$.

Step 1   The radical is already isolated on the right side of the equation.

Step 2   Square each side.

$$x^2 = \left(\sqrt{x^2 + 5x + 10}\right)^2 \qquad \text{Squaring property}$$

$$x^2 = x^2 + 5x + 10 \qquad \text{On the right, } (\sqrt{a})^2 = a.$$

Step 3 $$0 = 5x + 10 \qquad \text{Subtract } x^2.$$

Step 4   This step is not needed. No remaining terms contain radicals.

Step 5 $$-10 = 5x \qquad \text{Subtract 10.}$$

Proposed solution $\longrightarrow$ $-2 = x$ Divide by 5.

Step 6   **CHECK** $$x = \sqrt{x^2 + 5x + 10} \qquad \text{Original equation}$$

$$-2 \overset{?}{=} \sqrt{(-2)^2 + 5(-2) + 10} \qquad \text{Let } x = -2.$$

The principal square root of a quantity *cannot* be negative.

$$-2 \overset{?}{=} \sqrt{4 - 10 + 10} \qquad \text{Multiply.}$$

$$-2 = 2 \qquad \text{False}$$

Because substituting -2 for x leads to a false result, the equation has no real number solution. The solution set is $\varnothing$.

NOW TRY

OBJECTIVE 3   Solve equations by squaring a binomial.

Recall the rules for squaring binomials from **Section 5.6.**

$$(x + y)^2 = x^2 + 2xy + y^2 \quad \text{and} \quad (x - y)^2 = x^2 - 2xy + y^2$$

We apply the second rule in **Example 5** when finding $(x - 3)^2$.

$$(x - 3)^2$$

Remember the middle term when squaring.

$$= x^2 - 2x(3) + 3^2$$

$$= x^2 - 6x + 9$$

NOW TRY ANSWER
4. $\varnothing$

Verify the result by multiplying $(x - 3)(x - 3)$ using the FOIL method.

NOW TRY
EXERCISE 5

Solve $\sqrt{4x + 1} = x - 5$.

EXAMPLE 5 Using the Squaring Property When One Side Has Two Terms

Solve $\sqrt{2x - 3} = x - 3$.

$$\sqrt{2x - 3} = x - 3$$

$$\left(\sqrt{2x - 3}\right)^2 = (x - 3)^2 \qquad \text{Square each side.}$$

> Be careful squaring.

$$2x - 3 = x^2 - 6x + 9 \qquad (\sqrt{a})^2 = a;\ (x - y)^2 = x^2 - 2xy + y^2$$

The equation $2x - 3 = x^2 - 6x + 9$ is quadratic because of the x^2-term. To solve it as shown in **Section 6.5,** we must write the equation in standard form.

$$\text{Standard form} \rightarrow \quad x^2 - 8x + 12 = 0 \qquad \begin{array}{l}\text{Subtract } 2x, \text{ add } 3, \text{ and}\\ \text{interchange sides.}\end{array}$$

$$(x - 6)(x - 2) = 0 \qquad \text{Factor.}$$

$$x - 6 = 0 \quad \text{or} \quad x - 2 = 0 \qquad \text{Zero-factor property}$$

$$\text{Proposed solutions} \rightarrow \quad x = 6 \quad \text{or} \qquad x = 2 \qquad \text{Solve each equation.}$$

CHECK

$$\sqrt{2x - 3} = x - 3$$
$$\sqrt{2(6) - 3} \overset{?}{=} 6 - 3 \qquad \text{Let } x = 6.$$
$$\sqrt{12 - 3} \overset{?}{=} 3$$
$$\sqrt{9} \overset{?}{=} 3$$
$$3 = 3 \ \checkmark \qquad \text{True}$$

$$\sqrt{2x - 3} = x - 3$$
$$\sqrt{2(2) - 3} \overset{?}{=} 2 - 3 \qquad \text{Let } x = 2.$$
$$\sqrt{4 - 3} \overset{?}{=} -1$$
$$\sqrt{1} \overset{?}{=} -1$$
$$1 = -1 \qquad \text{False}$$

Only 6 is a valid solution. (2 is extraneous.) The solution set is $\{6\}$.   **NOW TRY**

EXAMPLE 6 Rewriting an Equation before Using the Squaring Property

Solve $\sqrt{9x} - 1 = 2x$.

If we begin by squaring each side, we obtain the following.

$$\left(\sqrt{9x} - 1\right)^2 = (2x)^2$$

$$9x - 2\sqrt{9x} + 1 = 4x^2 \qquad \text{This equation still contains a radical.}$$

We must apply Step 1 here and isolate the radical *before* squaring each side.

$$\sqrt{9x} - 1 = 2x$$

> This is a key step.

$$\sqrt{9x} = 2x + 1 \qquad \text{Add 1 to isolate the radical. (Step 1)}$$

$$\left(\sqrt{9x}\right)^2 = (2x + 1)^2 \qquad \text{Square each side. (Step 2)}$$

> No terms contain radicals.

$$9x = 4x^2 + 4x + 1 \qquad (\sqrt{a})^2 = a;\ (x + y)^2 = x^2 + 2xy + y^2$$

$$4x^2 - 5x + 1 = 0 \qquad \text{Standard form (Step 3)}$$

$$(4x - 1)(x - 1) = 0 \qquad \text{Factor. (Step 5; Step 4 is not needed.)}$$

$$4x - 1 = 0 \quad \text{or} \quad x - 1 = 0 \qquad \text{Zero-factor property}$$

$$\text{Proposed solutions} \rightarrow x = \frac{1}{4} \quad \text{or} \qquad x = 1 \qquad \text{Solve each equation.}$$

NOW TRY ANSWER
5. $\{12\}$

NOW TRY
EXERCISE 6

Solve $\sqrt{27x} - 3 = 2x$.

CHECK $\sqrt{9x} - 1 = 2x$ (Step 6) $\sqrt{9x} - 1 = 2x$ (Step 6)

$$\sqrt{9\left(\frac{1}{4}\right)} - 1 \stackrel{?}{=} 2\left(\frac{1}{4}\right) \quad \text{Let } x = \tfrac{1}{4}.$$ $$\sqrt{9(1)} - 1 \stackrel{?}{=} 2(1) \quad \text{Let } x = 1.$$

$$\frac{3}{2} - 1 \stackrel{?}{=} \frac{1}{2}$$ $$3 - 1 \stackrel{?}{=} 2$$

$$\frac{1}{2} = \frac{1}{2} \ \checkmark \quad \text{True}$$ $$2 = 2 \ \checkmark \quad \text{True}$$

Both proposed solutions check, so the solution set is $\left\{\frac{1}{4}, 1\right\}$.   **NOW TRY**

⚠ **CAUTION**   When squaring each side of the equation

$$\sqrt{9x} = 2x + 1, \quad \text{See Example 6.}$$

the *entire* binomial $2x + 1$ must be squared to obtain $4x^2 + 4x + 1$. It is incorrect to square the $2x$ and the 1 separately and write $4x^2 + 1$.

In an equation like the one in **Example 7,** squaring the binomial on the right results in a trinomial that still has a radical expression as the middle term. This requires squaring both sides of the equation a *second* time to eliminate the radical term.

NOW TRY
EXERCISE 7

Solve $\sqrt{x} + 2 = \sqrt{x + 8}$.

EXAMPLE 7   **Using the Squaring Property Twice**

Solve $\sqrt{21 + x} = 3 + \sqrt{x}$.

$$\sqrt{21 + x} = 3 + \sqrt{x} \qquad \text{A radical is isolated on the left. (Step 1)}$$

$$\left(\sqrt{21 + x}\right)^2 = \left(3 + \sqrt{x}\right)^2 \qquad \text{Square each side. (Step 2)}$$

$$21 + x = 9 + 6\sqrt{x} + x \quad \boxed{\text{Be careful here.}}$$

$$12 = 6\sqrt{x} \qquad \text{Subtract 9. Subtract } x. \text{ (Step 3)}$$

$$2 = \sqrt{x} \qquad \text{Divide by 6.}$$

$$2^2 = \left(\sqrt{x}\right)^2 \qquad \text{Square each side again. (Step 4)}$$

Proposed solution $\longrightarrow \ 4 = x \qquad \text{Apply the exponents. (Step 5)}$

CHECK $\sqrt{21 + x} = 3 + \sqrt{x} \qquad \text{Original equation (Step 6)}$

$$\sqrt{21 + 4} \stackrel{?}{=} 3 + \sqrt{4} \qquad \text{Let } x = 4.$$

$$\sqrt{25} \stackrel{?}{=} 3 + 2 \qquad \text{Simplify.}$$

$$5 = 5 \ \checkmark \qquad \text{True}$$

NOW TRY ANSWERS

6. $\left\{\frac{3}{4}, 3\right\}$ **7.** $\{1\}$ The solution set is $\{4\}$.   **NOW TRY**

OBJECTIVE 4 Solve radical equations having cube root radicals.

We do this by extending the concept of raising both sides of an equation to a power. Instead of squaring each side, we *cube* each side, using the fact that

$$\left(\sqrt[3]{a}\right)^3 = a, \quad \text{for any real number } a.$$

⟳ NOW TRY
EXERCISE 8

Solve each equation.

(a) $\sqrt[3]{8x - 3} = \sqrt[3]{4x}$

(b) $\sqrt[3]{2x^2} = \sqrt[3]{10x - 12}$

EXAMPLE 8 Solving Equations with Cube Root Radicals

Solve each equation.

(a)
$$\sqrt[3]{5x} = \sqrt[3]{3x + 1}$$

$\left(\sqrt[3]{5x}\right)^3 = \left(\sqrt[3]{3x + 1}\right)^3$	Cube each side.
$5x = 3x + 1$	Apply the exponents.
$2x = 1$	Subtract $3x$.
$x = \dfrac{1}{2}$	Divide by 2.

CHECK	$\sqrt[3]{5x} = \sqrt[3]{3x + 1}$	Original equation
	$\sqrt[3]{5\left(\dfrac{1}{2}\right)} \stackrel{?}{=} \sqrt[3]{3\left(\dfrac{1}{2}\right) + 1}$	Let $x = \frac{1}{2}$.
	$\sqrt[3]{\dfrac{5}{2}} \stackrel{?}{=} \sqrt[3]{\dfrac{3}{2} + \dfrac{2}{2}}$	Multiply; $1 = \frac{2}{2}$
	$\sqrt[3]{\dfrac{5}{2}} = \sqrt[3]{\dfrac{5}{2}}$ ✓	True

The solution set is $\left\{\dfrac{1}{2}\right\}$.

(b)
$$\sqrt[3]{x^2} = \sqrt[3]{26x + 27}$$

$\left(\sqrt[3]{x^2}\right)^3 = \left(\sqrt[3]{26x + 27}\right)^3$	Cube each side.
$x^2 = 26x + 27$	Apply the exponents.
$x^2 - 26x - 27 = 0$	Standard form
$(x + 1)(x - 27) = 0$	Factor.
$x + 1 = 0 \quad \text{or} \quad x - 27 = 0$	Zero-factor property
$x = -1 \quad \text{or} \quad x = 27$	Solve each equation.

CHECK

$\sqrt[3]{x^2} = \sqrt[3]{26x + 27}$	$\sqrt[3]{x^2} = \sqrt[3]{26x + 27}$
$\sqrt[3]{(-1)^2} \stackrel{?}{=} \sqrt[3]{26(-1) + 27}$	$\sqrt[3]{(27)^2} \stackrel{?}{=} \sqrt[3]{26(27) + 27}$
Let $x = -1$.	Let $x = 27$.
$\sqrt[3]{1} \stackrel{?}{=} \sqrt[3]{1}$	$\sqrt[3]{729} \stackrel{?}{=} \sqrt[3]{729}$
$1 = 1$ ✓ True	$9 = 9$ ✓ True

NOW TRY ANSWERS
8. (a) $\left\{\frac{3}{4}\right\}$ **(b)** $\{2, 3\}$

Both proposed solutions check, so the solution set is $\{-1, 27\}$.

NOW TRY ⟳

8.6 Exercises

    MyMathLab®

▶ *Complete solution available in MyMathLab*

Concept Check   *Fill in the blanks to complete the following.*

1. To solve an equation involving a radical, such as $\sqrt{2x-1} = 5$, use the _____ property of equality. This property says that if each side of an equation is _____, all solutions of the _____ equation are among the solutions of the squared equation.

2. Solving some radical equations involves squaring a binomial using the following rules.
$$(x+y)^2 = \text{_____} \qquad (x-y)^2 = \text{_____}$$

Solve each equation. **See Examples 1–4.**

3. $\sqrt{x} = 7$ 4. $\sqrt{x} = 10$ ▶ 5. $\sqrt{x+2} = 3$ 6. $\sqrt{x+7} = 5$

7. $\sqrt{r-4} = 9$ 8. $\sqrt{x-12} = 3$ 9. $\sqrt{4-t} = 7$ 10. $\sqrt{9-s} = 5$

11. $\sqrt{2t+3} = 0$ 12. $\sqrt{5x-4} = 0$ ▶ 13. $\sqrt{t} = -5$ 14. $\sqrt{p} = -8$

15. $\sqrt{w-4} = 7$ 16. $\sqrt{t+3} = 10$ ▶ 17. $\sqrt{10x-8} = 3\sqrt{x}$

18. $\sqrt{17t-4} = 4\sqrt{t}$ 19. $5\sqrt{x} = \sqrt{10x+15}$ 20. $4\sqrt{x} = \sqrt{20x-16}$

21. $\sqrt{3x-5} = \sqrt{2x+1}$ 22. $\sqrt{5x+2} = \sqrt{3x+8}$ ▶ 23. $k = \sqrt{k^2-5k-15}$

24. $x = \sqrt{x^2-2x-6}$ 25. $7x = \sqrt{49x^2+2x-10}$ 26. $6x = \sqrt{36x^2+5x-5}$

27. $\sqrt{2x+2} = \sqrt{3x-5}$ 28. $\sqrt{x+2} = \sqrt{2x-5}$ 29. $\sqrt{5x-5} = \sqrt{4x+1}$

30. $\sqrt{3m+3} = \sqrt{5m-1}$ 31. $\sqrt{3x-8} = -2$ 32. $\sqrt{6t+4} = -3$

33. *Concept Check*   Consider the following "solution." **WHAT WENT WRONG?**

$-\sqrt{x-1} = -4$

$-(x-1) = 16$ Square each side.

$-x+1 = 16$ Distributive property

$-x = 15$ Subtract 1.

$x = -15$ Multiply by -1.

Solution set: $\{-15\}$

34. *Concept Check*   The first step in solving the equation
$$\sqrt{2x+1} = x-7$$
is to square each side of the equation. When a student did this, he obtained
$$2x+1 = x^2+49.$$

WHAT WENT WRONG? Square each side correctly.

Solve each equation. **See Examples 5 and 6.**

35. $\sqrt{5x+11} = x+3$ 36. $\sqrt{5x+1} = x+1$ ▶ 37. $\sqrt{2x+1} = x-7$

38. $\sqrt{3x+10} = 2x-5$ 39. $\sqrt{x+2}-2 = x$ 40. $\sqrt{x+1}-1 = x$

41. $\sqrt{12x+12}+10 = 2x$ 42. $\sqrt{4x+5}+5 = 2x$ 43. $\sqrt{6x+7}-1 = x+1$

44. $\sqrt{8x+8}-1 = 2x+1$ 45. $2\sqrt{x+7} = x-1$ 46. $3\sqrt{x+13} = x+9$

47. $\sqrt{2x+4} = x$ 48. $\sqrt{3x+6} = x$

49. $\sqrt{x+9} = x+3$ 50. $\sqrt{x+3} = x-9$

▶ 51. $3\sqrt{x-2} = x-2$ 52. $2\sqrt{x+4} = x+1$

Solve each equation. See Example 7.

53. $\sqrt{3x+3} + \sqrt{x+2} = 5$

54. $\sqrt{2x+1} + \sqrt{x+4} = 3$

▶ **55.** $\sqrt{x+6} = \sqrt{x+72}$

56. $\sqrt{x-4} = \sqrt{x-32}$

57. $\sqrt{3x+4} - \sqrt{2x-4} = 2$

58. $\sqrt{1-x} + \sqrt{x+9} = 4$

59. $\sqrt{2x+11} + \sqrt{x+6} = 2$

60. $\sqrt{x+9} + \sqrt{x+16} = 7$

Solve each equation. (Hint: In Exercises 67 and 68, extend the concepts to fourth root radicals.) See Example 8.

▶ **61.** $\sqrt[3]{2x} = \sqrt[3]{5x+2}$

62. $\sqrt[3]{4x+3} = \sqrt[3]{2x-1}$

63. $\sqrt[3]{x^2} = \sqrt[3]{8+7x}$

64. $\sqrt[3]{x^2} = \sqrt[3]{8-7x}$

65. $\sqrt[3]{3x^2-9x+8} = \sqrt[3]{x}$

66. $\sqrt[3]{5x^2-6x+2} = \sqrt[3]{x}$

67. $\sqrt[4]{x^2+24x} = 3$

68. $\sqrt[4]{x^2+6x} = 2$

Solve each problem.

69. The square root of the sum of a number and 4 is 5. Find the number.

70. A certain number is the same as the square root of the product of 8 and the number. Find the number.

71. Three times the square root of 2 equals the square root of the sum of some number and 10. Find the number.

72. The negative square root of a number equals that number decreased by 2. Find the number.

Solve each problem. Give answers to the nearest tenth.

73. To estimate the speed at which a car was traveling at the time of an accident, a police officer drives the car under conditions similar to those during which the accident took place and then skids to a stop. If the car is driven at 30 mph, then the speed *s* at the time of the accident is given by

$$s = 30\sqrt{\frac{a}{p}},$$

where *a* is the length of the skid marks left at the time of the accident and *p* is the length of the skid marks in the police test. Find *s* for the following values of *a* and *p*.

(a) $a = 862$ ft; $p = 156$ ft   **(b)** $a = 382$ ft; $p = 96$ ft   **(c)** $a = 84$ ft; $p = 26$ ft

74. A formula for calculating the distance *d* in miles one can see from an airplane to the horizon on a clear day is

$$d = 1.22\sqrt{x},$$

where *x* is the altitude of the plane in feet. How far can one see to the horizon in a plane flying at the following altitudes?

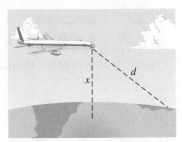

(a) 15,000 ft   **(b)** 18,000 ft   **(c)** 24,000 ft

On a clear day, the maximum distance in kilometers that can be seen from a tall building is given by the formula

$$\text{sight distance} = 111.7\sqrt{\text{height of building in kilometers}}.$$

(*Source: A Sourcebook of Applications of School Mathematics*, NCTM, 1980.)

Use the conversion equations 1 ft ≈ 0.3048 m *and* 1 km ≈ 0.621371 mi *as necessary to solve each problem. Round answers to the nearest mile.*

75. The London Eye is a unique structure that features 32 observation capsules and has a diameter of 135 m. Does the formula justify the claim that on a clear day passengers on the London Eye can see Windsor Castle, 25 mi away? (*Source:* www.londoneye.com)

76. The Empire State Building in New York City is 1250 ft high. The observation deck, located on the 102nd floor, is at a height of 1050 ft. How far could we see on a clear day from the observation deck? (*Source:* www.esbnyc.com)

Chapter 8   Summary

Key Terms

8.1

square root
positive (principal)
  square root
negative square
  root
radicand

radical
radical expression
perfect square
irrational number
cube root
fourth root
index (order)

perfect cube
perfect fourth
  power

8.3

like radicals
unlike radicals

8.5

conjugates

8.6

radical equation
extraneous solution

New Symbols

$\sqrt{}$ radical symbol

≈   is approximately
  equal to

$\sqrt[3]{a}$ cube root of a

$\sqrt[n]{a}$ nth root of a

Test Your Word Power

See how well you have learned the vocabulary in this chapter.

1. The **square root** of a number is
  A. the number raised to the second power
  B. the number under a radical symbol
  C. a number that, when multiplied by itself, gives the original number
  D. the inverse of the number.

2. A **radical** is
  A. a symbol that indicates the nth root
  B. an algebraic expression containing a square root
  C. the positive nth root of a number
  D. a radical symbol and the number or expression under it.

3. The **principal root** of a positive number with even index n is
  A. the positive nth root of the number
  B. the negative nth root of the number
  C. the square root of the number
  D. the cube root of the number.

4. An **irrational number** is
  A. the quotient of two integers, with denominator not 0
  B. a decimal number that neither terminates nor repeats
  C. the principal square root of a number
  D. a nonreal number.

5. **Like radicals** are
  A. radicals in simplest form
  B. algebraic expressions containing radicals
  C. multiples of the same root of the same number
  D. radicals with the same index.

6. The **conjugate** of $a + b$ is
  A. $a - b$
  B. $a \cdot b$
  C. $a \div b$
  D. $(a + b)^2$.

ANSWERS

1. C; *Examples:* 6 is a square root of 36 because $6^2 = 6 \cdot 6 = 36$. -6 is also a square root of 36.   **2.** D; *Examples:* $\sqrt{144}$, $\sqrt{4xy^2}$, $\sqrt{4 + t^2}$
3. A; *Examples:* $\sqrt{36} = 6$, $\sqrt[4]{81} = 3$, $\sqrt[6]{64} = 2$ **4.** B; *Examples:* π, $\sqrt{2}$, $-\sqrt{5}$ **5.** C; *Examples:* $\sqrt{7}$ and $3\sqrt{7}$ are like radicals, as are $2\sqrt[3]{6k}$ and $5\sqrt[3]{6k}$.   **6.** A; *Example:* The conjugate of $\sqrt{3} + 1$ is $\sqrt{3} - 1$.

Quick Review

CONCEPTS	**EXAMPLES**

8.1 Evaluating Roots

Let a be a positive real number.

$\sqrt{a}$ is the positive or principal square root of a.

$-\sqrt{a}$ is the negative square root of a. Also, $\sqrt{0} = 0$.

If a is a negative real number, then $\sqrt{a}$ is not a real number.

Let a be a positive rational number.

$\sqrt{a}$ is rational if a is a perfect square.

$\sqrt{a}$ is irrational if a is not a perfect square.

Each real number has exactly one real cube root.

$$\sqrt{49} = 7$$

$$-\sqrt{81} = -9$$

$$\sqrt{-25} \text{ is not a real number.}$$

$\sqrt{\dfrac{4}{9}}, \sqrt{16}$ are rational. $\sqrt{\dfrac{2}{3}}, \sqrt{21}$ are irrational.

$$\sqrt[3]{27} = 3 \qquad \sqrt[3]{-8} = -2$$

Distance Formula

The distance d between the points (x_1, y_1) and (x_2, y_2) is

$$d = \sqrt{(x_2 - x_1)^2 + (y_2 - y_1)^2}.$$

The distance between $(0, -2)$ and $(-1, 1)$ is

$\sqrt{(-1 - 0)^2 + [1 - (-2)]^2}$ Substitute.

$= \sqrt{(-1)^2 + 3^2}$ Subtract.

$= \sqrt{1 + 9}$ Apply the exponents.

$= \sqrt{10}.$ Add.

8.2 Multiplying, Dividing, and Simplifying Radicals

Product Rule for Radicals

If a and b are nonnegative real numbers, then the following hold.

$$\sqrt{a} \cdot \sqrt{b} = \sqrt{ab} \quad \text{and} \quad \sqrt{a \cdot b} = \sqrt{a} \cdot \sqrt{b}$$

Quotient Rule for Radicals

If a and b are nonnegative real numbers and $b \neq 0$, then the following hold.

$$\sqrt{\dfrac{a}{b}} = \dfrac{\sqrt{a}}{\sqrt{b}} \quad \text{and} \quad \dfrac{\sqrt{a}}{\sqrt{b}} = \sqrt{\dfrac{a}{b}}$$

For all real numbers a and b for which the indicated roots exist, the following hold.

$$\sqrt[n]{a} \cdot \sqrt[n]{b} = \sqrt[n]{ab}$$

and $\dfrac{\sqrt[n]{a}}{\sqrt[n]{b}} = \sqrt[n]{\dfrac{a}{b}}$ **(where $b \neq 0$)**

$$\sqrt{5} \cdot \sqrt{7} = \sqrt{35}$$

$$\sqrt{48} = \sqrt{16 \cdot 3} = \sqrt{16} \cdot \sqrt{3} = 4\sqrt{3}$$

$\sqrt{\dfrac{25}{64}} = \dfrac{\sqrt{25}}{\sqrt{64}} = \dfrac{5}{8} \qquad \dfrac{\sqrt{8}}{\sqrt{2}} = \sqrt{\dfrac{8}{2}} = \sqrt{4} = 2$

$$\sqrt[3]{5} \cdot \sqrt[3]{3} = \sqrt[3]{15}$$

$$\dfrac{\sqrt[4]{12}}{\sqrt[4]{4}} = \sqrt[4]{\dfrac{12}{4}} = \sqrt[4]{3}$$

8.3 Adding and Subtracting Radicals

Add and subtract like radicals using the distributive property. *Only like radicals can be combined in this way.*

$2\sqrt{5} + 4\sqrt{5}$

$= (2 + 4)\sqrt{5}$

$= 6\sqrt{5}$

$\sqrt{8} - \sqrt{32}$

$= 2\sqrt{2} - 4\sqrt{2}$

$= -2\sqrt{2}$

CONCEPTS	EXAMPLES

8.4 Rationalizing the Denominator

To rationalize the denominator of a radical expression, multiply both the numerator and denominator by a number that will eliminate the radical from the denominator.

$$\frac{2}{\sqrt{3}} = \frac{2 \cdot \sqrt{3}}{\sqrt{3} \cdot \sqrt{3}} = \frac{2\sqrt{3}}{3}$$

$$\sqrt[3]{\frac{5}{6}} = \frac{\sqrt[3]{5} \cdot \sqrt[3]{6^2}}{\sqrt[3]{6} \cdot \sqrt[3]{6^2}} = \frac{\sqrt[3]{180}}{6}$$

8.5 More Simplifying and Operations with Radicals

When appropriate, use the rules for adding and multiplying polynomials to simplify radical expressions.

$$\sqrt{6}(\sqrt{5} - \sqrt{7}) = \sqrt{30} - \sqrt{42} \qquad \text{Distributive property}$$

$$(\sqrt{3} + 1)(\sqrt{3} - 2)$$

$$= 3 - 2\sqrt{3} + \sqrt{3} - 2 \qquad \text{FOIL method}$$

$$= 1 - \sqrt{3} \qquad \text{Combine like terms.}$$

When simplifying radical expressions, the following rules are useful.

$$(x + y)^2 = x^2 + 2xy + y^2$$
$$(x - y)^2 = x^2 - 2xy + y^2$$
$$(x + y)(x - y) = x^2 - y^2$$

$$(\sqrt{13} - \sqrt{2})^2$$

$$= (\sqrt{13})^2 - 2(\sqrt{13})(\sqrt{2}) + (\sqrt{2})^2$$

$$= 13 - 2\sqrt{26} + 2$$

$$= 15 - 2\sqrt{26}$$

$$(\sqrt{5} + \sqrt{3})(\sqrt{5} - \sqrt{3})$$

$$= 5 - 3$$

$$= 2$$

Any denominators with radicals should be rationalized. If a radical expression contains two terms in the denominator and at least one of those terms is a square root radical, multiply both the numerator and denominator by the conjugate of the denominator.

$$\frac{6}{\sqrt{7} - \sqrt{2}}$$

$$= \frac{6(\sqrt{7} + \sqrt{2})}{(\sqrt{7} - \sqrt{2})(\sqrt{7} + \sqrt{2})} \qquad \begin{array}{l}\text{Multiply by the} \\ \text{conjugate of the} \\ \text{denominator.}\end{array}$$

$$= \frac{6(\sqrt{7} + \sqrt{2})}{7 - 2} \qquad \text{Multiply.}$$

$$= \frac{6(\sqrt{7} + \sqrt{2})}{5} \qquad \text{Subtract.}$$

8.6 Solving Equations with Radicals

Solving a Radical Equation

Step 1 Isolate a radical.

Step 2 Square each side. (By the squaring property of equality, all solutions of the original equation are *among* the solutions of the squared equation.)

Solve.

$$\sqrt{2x - 3} + x = 3$$

$$\sqrt{2x - 3} = 3 - x \qquad \text{Isolate the radical.}$$

$$(\sqrt{2x - 3})^2 = (3 - x)^2 \qquad \text{Square each side.}$$

$$2x - 3 = 9 - 6x + x^2 \qquad \begin{array}{l}\text{Remember the middle} \\ \text{term when squaring.}\end{array}$$

CONCEPTS	EXAMPLES
Step 3 Combine like terms.	$x^2 - 8x + 12 = 0$ Standard form
Step 4 If there is still a term with a radical, repeat Steps 1–3.	$(x - 2)(x - 6) = 0$ Factor.
Step 5 Solve the equation for all proposed solutions.	$x - 2 = 0$ or $x - 6 = 0$ Zero-factor property
Step 6 Check all proposed solutions in the original equation. Write the solution set.	$x = 2$ or $x = 6 \leftarrow$ Proposed solutions

A check is essential here. As shown below, 6 is extraneous.

CHECK $\sqrt{2x - 3} + x = 3$ Original equation

$\sqrt{2(6) - 3} + 6 \overset{?}{=} 3$ Let $x = 6$.

$9 = 3$ False

2 is the only solution because it leads to a true statement, $3 = 3$. The solution set is $\{2\}$.

Chapter 8   Review Exercises

8.1 *Find all square roots of each number.*

1. 49   **2.** 81   **3.** 196   **4.** 121   **5.** 225   **6.** 729

Find each indicated root. If the root is not a real number, say so.

7. $\sqrt{16}$ **8.** $-\sqrt{36}$ **9.** $\sqrt[3]{1000}$ **10.** $\sqrt[4]{81}$

11. $\sqrt{-8100}$ **12.** $-\sqrt{4225}$ **13.** $\sqrt{\dfrac{49}{36}}$ **14.** $\sqrt{0.25}$

15. Find the distance between $(-3, -5)$ and $(4, -3)$.

16. Find the value of x in the figure.

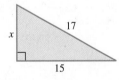

17. An HP f1905 computer monitor has viewing screen dimensions as shown in the figure. Find the diagonal measure of the viewing screen to the nearest tenth.

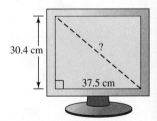

Determine whether each number is rational, irrational, *or* not a real number. *If a number is rational, give its exact value. If a number is irrational, give a decimal approximation rounded to the nearest thousandth. Use a calculator as necessary.*

18. $\sqrt{111}$ **19.** $-\sqrt{25}$ **20.** $\sqrt{-4}$

8.2 *Simplify each radical expression.*

21. $\sqrt{48}$ **22.** $-\sqrt{27}$ **23.** $\sqrt{160}$ **24.** $\sqrt[3]{16}$

25. $\sqrt[3]{375}$ **26.** $\sqrt{5} \cdot \sqrt{15}$ **27.** $\sqrt{12} \cdot \sqrt{27}$ **28.** $\sqrt{32} \cdot \sqrt{48}$

---

1. $-7, 7$ **2.** $-9, 9$

3. $-14, 14$ **4.** $-11, 11$

5. $-15, 15$ **6.** $-27, 27$

7. 4   **8.** -6

9. 10   **10.** 3

11. It is not a real number.

12. -65

13. $\dfrac{7}{6}$ **14.** 0.5

15. $\sqrt{53}$ **16.** 8

17. 48.3 cm

18. irrational; 10.536

19. rational; -5

20. It is not a real number.

21. $4\sqrt{3}$ **22.** $-3\sqrt{3}$

23. $4\sqrt{10}$ **24.** $2\sqrt[3]{2}$

25. $5\sqrt[3]{3}$ **26.** $5\sqrt{3}$

27. 18   **28.** $16\sqrt{6}$

29. $\dfrac{\sqrt{3}}{7}$ **30.** $\dfrac{\sqrt{3}}{5}$

31. 4 **32.** $\dfrac{\sqrt{5}}{6}$

33. $\dfrac{2}{15}$ **34.** $3\sqrt{2}$

35. 8 **36.** $2\sqrt{2}$

37. r^9 **38.** x^5y^8

39. $a^7b^{10}\sqrt{ab}$ **40.** $11x^3y^5$

41. y^2 **42.** $6x^5$

43. $8\sqrt{11}$ **44.** $9\sqrt{2}$

45. $21\sqrt{3}$ **46.** $12\sqrt{3}$

47. 0 **48.** $3\sqrt{7}$

49. $2\sqrt{3} + 3\sqrt{10}$

50. $2\sqrt{2}$

51. $6\sqrt{30}$ **52.** $5\sqrt{x}$

53. $-m\sqrt{5}$ **54.** $11k^2\sqrt{2n}$

55. $\dfrac{8\sqrt{10}}{5}$ **56.** $\sqrt{5}$

57. $\sqrt{6}$ **58.** $\dfrac{\sqrt{30}}{15}$

59. $\dfrac{\sqrt{10}}{5}$ **60.** $\dfrac{1}{8}$

61. $\dfrac{\sqrt{42}}{21}$ **62.** $\dfrac{r\sqrt{x}}{4x}$

63. $\dfrac{\sqrt[3]{9}}{3}$ **64.** $\dfrac{\sqrt[3]{98}}{7}$

65. $-\sqrt{15} - 9$

66. $3\sqrt{6} + 12$

67. $22 - 16\sqrt{3}$

68. $179 + 20\sqrt{7}$

69. -2 **70.** -13

71. $-1 + \sqrt{2}$ **72.** $\dfrac{-2 + 6\sqrt{2}}{17}$

73. $\dfrac{2\sqrt{3} + 2 + 3\sqrt{2} + \sqrt{6}}{2}$

74. $\dfrac{3 + 2\sqrt{6}}{3}$

75. $\dfrac{1 - 3\sqrt{7}}{4}$ **76.** $3 + 4\sqrt{3}$

77. $\{25\}$ **78.** $\varnothing$

79. $\{48\}$ **80.** $\{1\}$

81. $\{2\}$ **82.** $\{-2\}$

83. $\{-3, -1\}$ **84.** $\{-2\}$

85. $\{4\}$ **86.** $\{-1\}$

---

29. $\sqrt{\dfrac{3}{49}}$ **30.** $\sqrt{\dfrac{6}{50}}$ **31.** $\dfrac{\sqrt{48}}{\sqrt{3}}$ **32.** $\sqrt{\dfrac{1}{6}} \cdot \sqrt{\dfrac{5}{6}}$

33. $\sqrt{\dfrac{2}{5}} \cdot \sqrt{\dfrac{2}{45}}$ **34.** $\dfrac{3\sqrt{10}}{\sqrt{5}}$ **35.** $\dfrac{24\sqrt{12}}{6\sqrt{3}}$ **36.** $\dfrac{8\sqrt{150}}{4\sqrt{75}}$

Simplify each expression. Assume that all variables represent nonnegative real numbers.

37. $\sqrt{r^{18}}$ **38.** $\sqrt{x^{10}y^{16}}$ **39.** $\sqrt{a^{15}b^{21}}$

40. $\sqrt{121x^6y^{10}}$ **41.** $\sqrt[3]{y^6}$ **42.** $\sqrt[3]{216x^{15}}$

8.3 *Perform the indicated operations.*

43. $7\sqrt{11} + \sqrt{11}$ **44.** $3\sqrt{2} + 6\sqrt{2}$ **45.** $3\sqrt{75} + 2\sqrt{27}$

46. $4\sqrt{12} + \sqrt{48}$ **47.** $4\sqrt{24} - 3\sqrt{54} + \sqrt{6}$ **48.** $2\sqrt{7} - 4\sqrt{28} + 3\sqrt{63}$

49. $\dfrac{2}{5}\sqrt{75} + \dfrac{3}{4}\sqrt{160}$ **50.** $\dfrac{1}{3}\sqrt{18} + \dfrac{1}{4}\sqrt{32}$ **51.** $\sqrt{15} \cdot \sqrt{2} + 5\sqrt{30}$

Simplify. Assume that all variables represent nonnegative real numbers.

52. $\sqrt{4x} + \sqrt{36x} - \sqrt{9x}$ **53.** $\sqrt{20m^2} - m\sqrt{45}$ **54.** $3k\sqrt{8k^2n} + 5k^2\sqrt{2n}$

8.4 *Simplify. Assume that all variables represent positive real numbers.*

55. $\dfrac{8\sqrt{2}}{\sqrt{5}}$ **56.** $\dfrac{5}{\sqrt{5}}$ **57.** $\dfrac{12}{\sqrt{24}}$ **58.** $\dfrac{\sqrt{2}}{\sqrt{15}}$ **59.** $\sqrt{\dfrac{2}{5}}$

60. $\sqrt{\dfrac{1}{2}} \cdot \sqrt{\dfrac{1}{32}}$ **61.** $\sqrt{\dfrac{2}{7}} \cdot \sqrt{\dfrac{1}{3}}$ **62.** $\sqrt{\dfrac{r^2}{16x}}$ **63.** $\sqrt[3]{\dfrac{1}{3}}$ **64.** $\sqrt[3]{\dfrac{2}{7}}$

8.5 *Find each product.*

65. $-\sqrt{3}(\sqrt{5} + \sqrt{27})$ **66.** $3\sqrt{2}(\sqrt{3} + 2\sqrt{2})$

67. $(2\sqrt{3} - 4)(5\sqrt{3} + 2)$ **68.** $(5\sqrt{7} + 2)^2$

69. $(\sqrt{5} - \sqrt{7})(\sqrt{5} + \sqrt{7})$ **70.** $(2\sqrt{3} + 5)(2\sqrt{3} - 5)$

Rationalize each denominator.

71. $\dfrac{1}{1 + \sqrt{2}}$ **72.** $\dfrac{\sqrt{8}}{\sqrt{2} + 6}$ **73.** $\dfrac{2 + \sqrt{6}}{\sqrt{3} - 1}$

Write each quotient in lowest terms.

74. $\dfrac{15 + 10\sqrt{6}}{15}$ **75.** $\dfrac{3 - 9\sqrt{7}}{12}$ **76.** $\dfrac{6 + \sqrt{192}}{2}$

8.6 *Solve each equation.*

77. $\sqrt{m} - 5 = 0$ **78.** $\sqrt{p} + 4 = 0$ **79.** $\sqrt{x + 1} = 7$

80. $\sqrt{5m + 4} = 3\sqrt{m}$ **81.** $\sqrt{2p + 3} = \sqrt{5p - 3}$ **82.** $\sqrt{-2t - 4} = t + 2$

83. $\sqrt{13 + 4t} = t + 4$ **84.** $\sqrt{2 - x} + 3 = x + 7$

85. $\sqrt[3]{x + 4} = \sqrt[3]{16 - 2x}$ **86.** $\sqrt{5x + 6} + \sqrt{3x + 4} = 2$

Chapter 8 | Mixed Review Exercises

Simplify. Assume that all variables represent positive real numbers.

1. $\sqrt{\dfrac{1}{3}} \cdot \sqrt{\dfrac{24}{5}}$

2. $\dfrac{1}{5 + \sqrt{2}}$

3. $\sqrt[3]{-125}$

4. $\sqrt{50y^2}$

5. $\sqrt{\dfrac{16r^3}{3s}}$

6. $\dfrac{12 + 6\sqrt{13}}{12}$

7. $-\sqrt{121}$

8. $\left(\sqrt{5} - \sqrt{2}\right)^2$

9. $-\sqrt{5}\left(\sqrt{2} + \sqrt{75}\right)$

10. $\left(6\sqrt{7} + 2\right)\left(4\sqrt{7} - 1\right)$

11. $2\sqrt{27} + 3\sqrt{75} - \sqrt{300}$

Solve.

12. $\sqrt{x + 2} = x - 4$

13. $\sqrt{x} + 3 = 0$

14. $\sqrt{1 + 3t} - t = -3$

15. A biologist has shown that the number of different plant species S on a Galápagos island is related to the area $\mathcal{A}$ (in square miles) of the island by

$$S = 28.6\sqrt[3]{\mathcal{A}}.$$

How many plant species would exist on such an island with the following areas?

(a) 8 mi²   **(b)** 1790 mi²

Answers (left margin):

1. $\dfrac{2\sqrt{10}}{5}$ **2.** $\dfrac{5 - \sqrt{2}}{23}$

3. -5 **4.** $5y\sqrt{2}$

5. $\dfrac{4r\sqrt{3rs}}{3s}$ **6.** $\dfrac{2 + \sqrt{13}}{2}$

7. -11 **8.** $7 - 2\sqrt{10}$

9. $-\sqrt{10} - 5\sqrt{15}$

10. $166 + 2\sqrt{7}$

11. $11\sqrt{3}$

12. $\{7\}$

13. $\varnothing$ **14.** $\{8\}$

15. (a) 57 species
    (b) 347 species

Chapter 8 | Test

FOR EXTRA HELP

Step-by-step test solutions are found on the Chapter Test Prep Videos available in MyMathLab® *or on* YouTube.

▶ *View the complete solutions to all Chapter Test exercises in MyMathLab.*

[8.1]
1. $-14, 14$
2. (a) irrational **(b)** 11.916
3. (a) B **(b)** F **(c)** D
    (d) A **(e)** C **(f)** A

[8.2]
4. $\dfrac{8\sqrt{2}}{5}$

5. $2\sqrt[3]{4}$ **6.** $4\sqrt{6}$

7. $2y\sqrt[3]{4x^2}$

[8.3–8.5]
8. 31
9. $9\sqrt{7}$ **10.** $11 + 2\sqrt{30}$

11. $-6x\sqrt[3]{2x}$ **12.** $\dfrac{\sqrt[3]{18}}{3}$

13. $6\sqrt{2} + 2 - 3\sqrt{14} - \sqrt{7}$

14. $-5\sqrt{3x}$

1. Find all square roots of 196.

2. Consider $\sqrt{142}$.

    (a) Determine whether it is rational or irrational.

    (b) Find a decimal approximation to the nearest thousandth.

3. Match each radical in Column I with the equivalent choice in Column II. Choices may be used once, more than once, or not at all.

I		II	
(a) $\sqrt{64}$	**(b)** $-\sqrt{64}$	**A.** 4	**B.** 8
(c) $\sqrt{-64}$	**(d)** $\sqrt[3]{64}$	**C.** -4	**D.** Not a real number
(e) $\sqrt[3]{-64}$	**(f)** $-\sqrt[3]{-64}$	**E.** 16	**F.** -8

Simplify. Assume that all variables represent positive real numbers.

4. $\sqrt{\dfrac{128}{25}}$

5. $\sqrt[3]{32}$

6. $\dfrac{20\sqrt{18}}{5\sqrt{3}}$

7. $\sqrt[3]{32x^2y^3}$

8. $\left(6 - \sqrt{5}\right)\left(6 + \sqrt{5}\right)$

9. $3\sqrt{28} + \sqrt{63}$

10. $\left(\sqrt{5} + \sqrt{6}\right)^2$

11. $\sqrt[3]{16x^4} - 2\sqrt[3]{128x^4}$

12. $\sqrt[3]{\dfrac{2}{3}}$

13. $\left(2 - \sqrt{7}\right)\left(3\sqrt{2} + 1\right)$

14. $3\sqrt{27x} - 4\sqrt{48x} + 2\sqrt{3x}$

[8.1]
15. (a) $6\sqrt{2}$ in.   (b) 8.485 in.
16. 50 ohms

[8.4]
17. $\dfrac{5\sqrt{14}}{7}$ **18.** $\dfrac{\sqrt{6x}}{3x}$

19. $-\sqrt[3]{2}$

[8.5]
20. $\dfrac{-12 - 3\sqrt{3}}{13}$

21. $\dfrac{1 + \sqrt{2}}{2}$

[8.6]
22. $\varnothing$

23. $\{3\}$ **24.** $\left\{\dfrac{1}{4}, 1\right\}$

25. $\{-4\}$

26. 12 is not a solution. A check shows that it does not satisfy the original equation. The solution set is $\varnothing$.

Solve each problem.

15. Find the measure of the unknown leg of this right triangle.

    (**a**) Give its length in simplified radical form.

    (**b**) Give its length to the nearest thousandth.

16. In electronics, the impedance Z of an alternating series circuit is given by the formula

$$Z = \sqrt{R^2 + X^2},$$

where R is the resistance and X is the reactance, both in ohms. Find the value of the impedance Z if $R = 40$ ohms and $X = 30$ ohms. (*Source:* Cooke, Nelson M., and Orleans, Joseph B., *Mathematics Essential to Electricity and Radio,* McGraw-Hill.)

Rationalize each denominator.

17. $\dfrac{5\sqrt{2}}{\sqrt{7}}$ **18.** $\sqrt{\dfrac{2}{3x}}$ **19.** $\dfrac{-2}{\sqrt[3]{4}}$ **20.** $\dfrac{-3}{4 - \sqrt{3}}$

21. Write $\dfrac{2 + \sqrt{8}}{4}$ in lowest terms.

Solve each equation.

22. $\sqrt{2x + 6} + 4 = 2$ **23.** $\sqrt{x + 1} = 5 - x$

24. $3\sqrt{x - 1} = 2x$ **25.** $\sqrt{2x + 9} + \sqrt{x + 5} = 2$

26. Consider the following "solution." ***WHAT WENT WRONG?***

$$\sqrt{2x + 1} + 5 = 0$$

$$\sqrt{2x + 1} = -5 \qquad \text{Subtract 5.}$$

$$2x + 1 = 25 \qquad \text{Square each side.}$$

$$2x = 24 \qquad \text{Subtract 1.}$$

$$x = 12 \qquad \text{Divide by 2.}$$

The solution set is $\{12\}$.

Chapters R–8   Cumulative Review Exercises

[1.1]
1. 54   **2.** 6

[1.3]
3. 3

[2.3]
4. $\{3\}$

[2.8]
5. $[-16, \infty)$ **6.** $(5, \infty)$

[2.4]
7. 2011: \$298,626;
2012: \$337,601

Simplify each expression.

1. $3(6 + 7) + 6 \cdot 4 - 3^2$ **2.** $\dfrac{3(6 + 7) + 3}{2(4) - 1}$ **3.** $|-6| - |-3|$

Solve each equation or inequality.

4. $5(k - 4) - k = k - 11$ **5.** $-\dfrac{3}{4}x \le 12$ **6.** $5z + 3 - 4 > 2z + 9 + z$

7. Trevor Brazile won the ProRodeo All-Around Championship in both 2011 and 2012. He won \$38,975 more in 2012 than in 2011. He won a total of \$636,227 in these two years. How much did he win each year? (*Source: World Almanac and Book of Facts.*)

[3.2]

8.   9.

[3.6] 10.

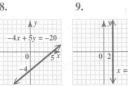

[3.3–3.5]

11. (a) 18.1; The number of
    subscribers increased by an
    average of 18.1 million per
    year.
    (b) $y = 18.1x + 109.5$
    (c) 362.9 million ($x = 14$)

[4.1–4.3]

12. $\{(3, -7)\}$

13. $\{(x, y) \mid 2x - y = 6\}$

[4.4]

14. from Chicago: 61 mph;
     from Des Moines: 54 mph

[5.1]

15. $12x^{10}y^2$

[5.2]

16. $\dfrac{y^{15}}{2^3 \cdot 3^6}$, or $\dfrac{y^{15}}{5832}$

[5.4]

17. $3x^3 + 11x^2 - 13$

[5.7]

18. $4t^2 - 8t + 5$

[6.2–6.4]

19. $(m + 8)(m + 4)$

20. $(6a + 5b)(2a - b)$

21. $(9z + 4)^2$

Graph.

8. $-4x + 5y = -20$ **9.** $x = 2$ **10.** $2x - 5y > 10$

11. The graph shows a line that models the number of cell phone subscribers in millions in
the United States from 2000 to 2012.

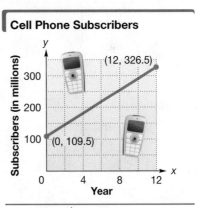

Cell Phone Subscribers

(12, 326.5)

(0, 109.5)

Source: CITA.

(a) Use the ordered pairs shown on the graph to find the slope of the line to the nearest
tenth. Interpret the slope.

(b) Use the slope from part (a) and the ordered pair (0, 109.5) to find an equation of
the line that models the data, where $x = 0$ represents 2000.

(c) Use the equation from part (b) to estimate the number of cell phone subscribers in 2014.
(*Hint:* What value of x corresponds to 2014?)

Solve each system of equations.

12. $4x - \ y = 19$ **13.** $2x - y = 6$
       $3x + 2y = -5$ $3y = 6x - 18$

14. Des Moines and Chicago are 345 mi apart. Two cars start from these cities, traveling toward
each other. They meet after 3 hr. The car from Chicago has an average rate 7 mph faster than
the other car. Find the average rate of each car. (*Source: State Farm Road Atlas.*)

Des Moines   345 mi   Chicago

Simplify and write each expression without negative exponents. Assume that variables represent positive real numbers.

15. $(3x^6)(2x^2y)^2$ **16.** $\left(\dfrac{3^2 y^{-2}}{2^{-1} y^3}\right)^{-3}$

17. Subtract $7x^3 - 8x^2 + 4$ from $10x^3 + 3x^2 - 9$.

18. Divide $\dfrac{8t^3 - 4t^2 - 14t + 15}{2t + 3}$.

Factor each polynomial completely.

19. $m^2 + 12m + 32$ **20.** $12a^2 + 4ab - 5b^2$ **21.** $81z^2 + 72z + 16$

Perform each indicated operation. Express answers in lowest terms.

22. $\dfrac{x^2 - 3x - 4}{x^2 + 3x} \cdot \dfrac{x^2 + 2x - 3}{x^2 - 5x + 4}$

23. $\dfrac{t^2 + 4t - 5}{t + 5} \div \dfrac{t - 1}{t^2 + 8t + 15}$

24. $\dfrac{y}{y^2 - 1} + \dfrac{y}{y + 1}$

25. $\dfrac{2}{x + 3} - \dfrac{4}{x - 1}$

26. $\dfrac{\frac{2}{3} + \frac{1}{2}}{\frac{1}{9} - \frac{1}{6}}$

Solve each equation.

27. $x^2 - 7x = -12$

28. $(x + 4)(x - 1) = -6$

29. $z^2 + 144 = 24z$

30. $\dfrac{x}{x + 8} - \dfrac{3}{x - 8} = \dfrac{128}{x^2 - 64}$

31. $A = \dfrac{B + CD}{BC + D}$ for B

32. $\sqrt{x} + 2 = x - 10$

Simplify each expression. Assume that all variables represent nonnegative real numbers.

33. $\sqrt{27} - 2\sqrt{12} + 6\sqrt{75}$

34. $\left(3\sqrt{2} + 1\right)\left(4\sqrt{2} - 3\right)$

35. $\dfrac{2}{\sqrt{3} + \sqrt{5}}$

9

Quadratic Equations

The trajectory of a water spout follows a *parabolic* path, which can be described by a *quadratic* (second-degree) *equation,* the topic of this chapter.

9.1 Solving Quadratic Equations by the Square Root Property

OBJECTIVES

1 Review the zero-factor property.

2 Solve equations of the form $x^2 = k$, where $k > 0$.

3 Solve equations of the form $(ax + b)^2 = k$, where $k > 0$.

4 Use formulas involving second-degree variables.

VOCABULARY

☐ quadratic equation

OBJECTIVE 1 Review the zero-factor property.

Recall from **Section 6.5** that a *quadratic equation* is defined as follows.

Quadratic Equation

A **quadratic equation** (in x here) can be written in the form

$$ax^2 + bx + c = 0,$$

where a, b, and c are real numbers and $a \neq 0$. The given form is called **standard form.**

Examples: $4x^2 + 4x - 5 = 0$ and $3x^2 = 4x - 8$

Quadratic equations
(The first equation is in standard form.)

A quadratic equation is a *second-degree equation,* that is, an equation with a squared variable term and no terms of greater degree.

In **Section 6.5** we used the zero-factor property to solve quadratic equations.

Zero-Factor Property

If a and b are real numbers and if $ab = 0$, then $a = 0$ or $b = 0$.

That is, if the product of two numbers is 0, then at least one of the numbers must be 0. One number must be 0, but both *may* be 0.

NOW TRY
EXERCISE 1

Solve each equation using the zero-factor property.

(a) $x^2 - x - 20 = 0$

(b) $x^2 = 36$

EXAMPLE 1   Solving Quadratic Equations Using the Zero-Factor Property

Solve each equation using the zero-factor property.

(a)
$$x^2 + 4x + 3 = 0$$
$$(x + 3)(x + 1) = 0 \qquad \text{Factor.}$$
$$x + 3 = 0 \quad \text{or} \quad x + 1 = 0 \qquad \text{Zero-factor property}$$
$$x = -3 \quad \text{or} \quad x = -1 \qquad \text{Solve each equation.}$$

The solution set is $\{-3, -1\}$.

(b)
$$x^2 = 9$$
$$x^2 - 9 = 0 \qquad \text{Subtract 9.}$$
$$(x + 3)(x - 3) = 0 \qquad \text{Factor.}$$
$$x + 3 = 0 \quad \text{or} \quad x - 3 = 0 \qquad \text{Zero-factor property}$$
$$x = -3 \quad \text{or} \quad x = 3 \qquad \text{Solve each equation.}$$

The solution set is $\{-3, 3\}$.

 NOW TRY

OBJECTIVE 2 Solve equations of the form $x^2 = k$, where $k > 0$.

In **Example 1(b),** we might also have solved $x^2 = 9$ by noticing that x must be a number whose square is 9. Thus, $x = \sqrt{9} = 3$ or $x = -\sqrt{9} = -3$.

This is generalized as the **square root property.**

NOW TRY ANSWERS
1. (a) $\{-4, 5\}$ **(b)** $\{-6, 6\}$

Square Root Property

If k is a positive number and if $x^2 = k$, then

$$x = \sqrt{k} \quad \text{or} \quad x = -\sqrt{k}.$$

The solution set is $\left\{-\sqrt{k}, \sqrt{k}\right\}$, which can be written $\left\{\pm\sqrt{k}\right\}$. (The symbol $\pm$ is read "positive or negative" or "plus or minus.")

NOTE When we solve an equation, we must find *all* values of the variable that satisfy the equation. Therefore, we want both the positive and negative square roots of k.

EXAMPLE 2 Solving Quadratic Equations of the Form $x^2 = k$

Solve each equation. Write radicals in simplified form.

(a) $x^2 = 16$

By the square root property, if $x^2 = 16$, then

$$x = \sqrt{16} = 4 \quad \text{or} \quad x = -\sqrt{16} = -4.$$

Check each solution by substituting it for x in the original equation. The solution set is

$$\{4, -4\}, \quad \text{or} \quad \{\pm 4\}.$$

> This notation indicates *two* solutions, one **positive** and one **negative**.

(b) $x^2 = 5$

By the square root property, if $x^2 = 5$, then

$$x = \sqrt{5} \quad \text{or} \quad x = -\sqrt{5}.$$

> Don't forget the negative solution.

The solutions set is $\left\{\sqrt{5}, -\sqrt{5}\right\}$, or $\left\{\pm\sqrt{5}\right\}$.

(c)

$$5m^2 - 40 = 0$$

$$5m^2 = 40 \qquad \text{Add 40.}$$

$$m^2 = 8 \qquad \text{Divide by 5.}$$

> Don't stop here. Simplify the radicals.

$$m = \sqrt{8} \quad \text{or} \quad m = -\sqrt{8} \qquad \text{Square root property}$$

$$m = 2\sqrt{2} \quad \text{or} \quad m = -2\sqrt{2} \qquad \sqrt{8} = \sqrt{4} \cdot \sqrt{2} = 2\sqrt{2}$$

CHECK Substitute each value in the original equation.

$$5m^2 - 40 = 0$$
$$5\left(2\sqrt{2}\right)^2 - 40 \stackrel{?}{=} 0 \qquad \text{Let } m = 2\sqrt{2}.$$
$$5(8) - 40 \stackrel{?}{=} 0$$
$$40 - 40 \stackrel{?}{=} 0$$
$$0 = 0 \quad \checkmark \quad \text{True}$$

> $\left(2\sqrt{2}\right)^2$
> $= 2^2 \cdot \left(\sqrt{2}\right)^2$
> $= 4 \cdot 2$
> $= 8$

$$5m^2 - 40 = 0$$
$$5\left(-2\sqrt{2}\right)^2 - 40 \stackrel{?}{=} 0 \qquad \text{Let } m = -2\sqrt{2}.$$
$$5(8) - 40 \stackrel{?}{=} 0$$
$$40 - 40 \stackrel{?}{=} 0$$
$$0 = 0 \quad \checkmark \quad \text{True}$$

The solution set is $\left\{2\sqrt{2}, -2\sqrt{2}\right\}$, or $\left\{\pm 2\sqrt{2}\right\}$.

**NOW TRY
EXERCISE 2**

Solve each equation. Write radicals in simplified form.

(a) $t^2 = 25$

(b) $x^2 = 13$

(c) $x^2 = -144$

(d) $2x^2 - 5 = 35$

(d)

$$3x^2 + 5 = 11$$
$$3x^2 = 6 \qquad \text{Subtract 5.}$$
$$x^2 = 2 \qquad \text{Divide by 3.}$$
$$x = \sqrt{2} \quad \text{or} \quad x = -\sqrt{2} \qquad \text{Square root property}$$

The solution set is $\left\{ \sqrt{2}, -\sqrt{2} \right\}$, or $\left\{ \pm \sqrt{2} \right\}$.

(e) $x^2 = -4$

Because -4 is a negative number and because the square of a real number cannot be negative, **there is no real solution** of this equation.* In later courses, there are solutions in the *complex number system.*

NOW TRY

OBJECTIVE 3   Solve equations of the form $(ax + b)^2 = k$, where $k > 0$.

In each equation so far, the exponent 2 appeared with a single variable as its base. We can extend the square root property to solve equations in which the base is a binomial.

**NOW TRY
EXERCISE 3**

Solve $(x - 2)^2 = 32$, and check the solutions.

EXAMPLE 3   Solving Quadratic Equations of the Form $(x + b)^2 = k$

Solve each equation, and check the solutions.

(a) [Use $(x - 3)$ as the base.] $(x - 3)^2 = 16$

$$x - 3 = \sqrt{16} \quad \text{or} \quad x - 3 = -\sqrt{16} \qquad \text{Square root property}$$
$$x - 3 = 4 \quad \text{or} \quad x - 3 = -4 \qquad \sqrt{16} = 4$$
$$x = 7 \quad \text{or} \quad x = -1 \qquad \text{Add 3.}$$

CHECK Substitute each value in the original equation.

$(x - 3)^2 = 16$	$(x - 3)^2 = 16$
$(7 - 3)^2 \stackrel{?}{=} 16 \qquad$ Let $x = 7$.	$(-1 - 3)^2 \stackrel{?}{=} 16 \qquad$ Let $x = -1$.
$4^2 \stackrel{?}{=} 16 \qquad$ Subtract.	$(-4)^2 \stackrel{?}{=} 16 \qquad$ Subtract.
$16 = 16 \checkmark \quad$ True	$16 = 16 \checkmark \quad$ True

The solution set is $\{-1, 7\}$.

(b)

$$(x - 1)^2 = 6$$
$$x - 1 = \sqrt{6} \qquad \text{or} \qquad x - 1 = -\sqrt{6} \qquad \text{Square root property}$$
$$x = 1 + \sqrt{6} \qquad \text{or} \qquad x = 1 - \sqrt{6} \qquad \text{Add 1.}$$

CHECK Substitute each value in the original equation.

$(x - 1)^2 = 6$	$(x - 1)^2 = 6$
$\left(1 + \sqrt{6} - 1\right)^2 \stackrel{?}{=} 6 \qquad$ Let $x = 1 + \sqrt{6}$.	$\left(1 - \sqrt{6} - 1\right)^2 \stackrel{?}{=} 6 \qquad$ Let $x = 1 - \sqrt{6}$.
$\left(\sqrt{6}\right)^2 \stackrel{?}{=} 6 \qquad$ Simplify.	$\left(-\sqrt{6}\right)^2 \stackrel{?}{=} 6 \qquad$ Simplify.
$6 = 6 \checkmark \quad$ True	$6 = 6 \checkmark \quad$ True

NOW TRY ANSWERS

2. (a) $\{\pm 5\}$ **(b)** $\left\{ \pm \sqrt{13} \right\}$
   (c) no real solution
   (d) $\left\{ \pm 2\sqrt{5} \right\}$

3. $\left\{ 2 \pm 4\sqrt{2} \right\}$

The solution set is $\left\{ 1 + \sqrt{6}, 1 - \sqrt{6} \right\}$, or $\left\{ 1 \pm \sqrt{6} \right\}$.   NOW TRY

---

*The equation in **Example 2(e)** has no solution over the *real number system.* In the **complex number system,** however, this equation does have solutions. The complex numbers include numbers whose squares are negative. These numbers are discussed in intermediate and college algebra courses.

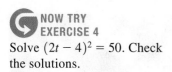

NOW TRY
EXERCISE 4

Solve $(2t - 4)^2 = 50$. Check the solutions.

EXAMPLE 4 Solving a Quadratic Equation of the Form $(ax + b)^2 = k$

Solve $(3r - 2)^2 = 27$. Check the solutions.

$$(3r - 2)^2 = 27$$

$3r - 2 = \sqrt{27}$ or $3r - 2 = -\sqrt{27}$ Square root property

$3r - 2 = 3\sqrt{3}$ or $3r - 2 = -3\sqrt{3}$ $\sqrt{27} = \sqrt{9} \cdot \sqrt{3} = 3\sqrt{3}$

$3r = 2 + 3\sqrt{3}$ or $3r = 2 - 3\sqrt{3}$ Add 2.

$r = \dfrac{2 + 3\sqrt{3}}{3}$ or $r = \dfrac{2 - 3\sqrt{3}}{3}$ Divide by 3.

CHECK
$$(3r - 2)^2 = 27$$

$$\left(3 \cdot \frac{2 + 3\sqrt{3}}{3} - 2\right)^2 \overset{?}{=} 27 \qquad \text{Let } r = \frac{2 + 3\sqrt{3}}{3}.$$

$$\left(2 + 3\sqrt{3} - 2\right)^2 \overset{?}{=} 27 \qquad \text{Multiply.}$$

$$\left(3\sqrt{3}\right)^2 \overset{?}{=} 27 \qquad \text{Subtract.}$$

$(ab)^2 = a^2 b^2$

$$27 = 27 \ \checkmark \quad \text{True}$$

The check of the other value is similar. The solution set is $\left\{\dfrac{2 \pm 3\sqrt{3}}{3}\right\}$.   **NOW TRY**

⚠ **CAUTION** The solutions in **Example 4** are fractions that cannot be simplified because 3 is *not* a common factor in the numerator.

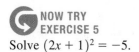

NOW TRY
EXERCISE 5

Solve $(2x + 1)^2 = -5$.

EXAMPLE 5 Recognizing a Quadratic Equation with No Real Solutions

Solve $(x + 3)^2 = -9$.

Because the square root of -9 is not a real number, there is no real solution.

NOW TRY

OBJECTIVE 4 Use formulas involving second-degree variables.

EXAMPLE 6 Finding the Length of a Bass

The weight w in pounds, length L in inches, and girth g in inches of a bass are related by the formula

$$w = \frac{L^2 g}{1200}.$$

NOW TRY ANSWERS

4. $\left\{\dfrac{4 \pm 5\sqrt{2}}{2}\right\}$

5. no real solution

Approximate the length, to the nearest inch, of a bass weighing 2.20 lb and having girth 10 in. (*Source: Sacramento Bee.*)

NOW TRY EXERCISE 6
Use the formula in **Example 6** to approximate, to the nearest inch, the length of a bass weighing 2.10 lb and having girth 9 in.

Substitute the given values 2.20 for the weight w and 10 for the girth g in the formula. Then solve for the length L.

$$w = \frac{L^2 g}{1200} \quad \text{Given formula}$$

$$2.20 = \frac{L^2 \cdot 10}{1200} \quad w = 2.20,\ g = 10$$

$$2640 = 10L^2 \quad \text{Multiply by 1200.}$$

$$L^2 = 264 \quad \text{Divide by 10. Interchange sides.}$$

$$L = \pm\sqrt{264} \quad \text{Square root property}$$

A calculator shows that $\sqrt{264} \approx 16.25$, so the length of the bass, to the nearest inch, is 16 in. (We must reject the negative solution $-\sqrt{264} \approx -16.25$ here, because L represents length.)

NOW TRY

NOW TRY ANSWER
6. 17 in.

9.1 Exercises

FOR EXTRA HELP MyMathLab®

Complete solution available in MyMathLab

1. *Concept Check* Which of the following are quadratic equations?
A. $x + 2y = 0$ **B.** $x^2 - 8x + 16 = 0$ **C.** $2x^2 - 5x = 3$ **D.** $x^3 + x^2 + 4 = 0$

2. *Concept Check* Which quadratic equation identified in **Exercise 1** is in standard form?

Concept Check Match each equation in Column I with the correct description of its solution in Column II.

I	II
3. $x^2 = 12$	**A.** No real solutions
4. $x^2 = -9$	**B.** Two integer solutions
5. $x^2 = \frac{25}{36}$	**C.** Two irrational solutions
6. $x^2 = 16$	**D.** Two rational solutions that are not integers

Solve each equation using the zero-factor property. ***See Example 1.***

7. $x^2 - x - 56 = 0$ **8.** $x^2 - 2x - 99 = 0$ **9.** $x^2 = 121$

10. $x^2 = 144$ **11.** $3x^2 - 13x = 30$ **12.** $5x^2 - 14x = 3$

Solve each equation using the square root property. Simplify all radicals. ***See Example 2.***

13. $x^2 = 81$ **14.** $z^2 = 169$ **15.** $x^2 = 14$ **16.** $m^2 = 22$

17. $t^2 = 48$ **18.** $x^2 = 54$ **19.** $x^2 = -100$ **20.** $m^2 = -64$

21. $x^2 = \frac{25}{4}$ **22.** $m^2 = \frac{36}{121}$ **23.** $x^2 = 2.25$ **24.** $w^2 = 56.25$

25. $r^2 - 3 = 0$ **26.** $x^2 - 13 = 0$ **27.** $7x^2 = 4$

28. $2x^2 = 9$ **29.** $3n^2 - 72 = 0$ **30.** $2x^2 - 80 = 0$

31. $5x^2 + 4 = 8$ **32.** $7p^2 - 5 = 11$ **33.** $2t^2 + 7 = 61$

34. $3x^2 + 8 = 80$ **35.** $-8x^2 = -64$ **36.** $-12x^2 = -144$

37. *Concept Check*   When a student was asked to solve $x^2 = 81$, she wrote $\{9\}$ as her answer. Her teacher did not give her full credit. The student argued that because $9^2 = 81$, her answer had to be correct. *WHAT WENT WRONG?* Give the correct solution set.

38. *Concept Check*   Illustrate the square root property for solving quadratic equations with an example.

*Solve each equation using the square root property. Simplify all radicals. **See Examples 3–5.***

▶ **39.** $(x - 3)^2 = 25$ **40.** $(x - 7)^2 = 16$ **41.** $(x - 8)^2 = 27$

42. $(p - 5)^2 = 40$ ▶ **43.** $(z + 5)^2 = -13$ **44.** $(m + 2)^2 = -17$

45. $(3x + 2)^2 = 49$ **46.** $(5t + 3)^2 = 36$ **47.** $(4x - 3)^2 = 9$

48. $(7z - 5)^2 = 25$ **49.** $(5 - 2x)^2 = 30$ **50.** $(3 - 2x)^2 = 70$

▶ **51.** $(3k + 1)^2 = 18$ **52.** $(5z + 6)^2 = 75$ **53.** $\left(\dfrac{1}{2}x + 5\right)^2 = 12$

54. $\left(\dfrac{1}{3}m + 4\right)^2 = 27$ **55.** $\left(x - \dfrac{1}{8}\right)^2 = \dfrac{1}{64}$ **56.** $\left(x - \dfrac{1}{9}\right)^2 = \dfrac{1}{81}$

57. $\left(x - \dfrac{1}{3}\right)^2 = \dfrac{4}{9}$ **58.** $\left(x - \dfrac{1}{5}\right)^2 = \dfrac{16}{25}$ **59.** $\left(x + \dfrac{1}{4}\right)^2 = \dfrac{3}{16}$

60. $\left(x + \dfrac{1}{7}\right)^2 = \dfrac{11}{49}$ **61.** $(4x - 1)^2 - 48 = 0$ **62.** $(2x - 5)^2 - 180 = 0$

63. *Concept Check*   Jeff solved the equation in **Exercise 49** and wrote his answer as $\left\{\dfrac{5 \pm \sqrt{30}}{2}\right\}$. Linda solved the same equation and wrote her answer as $\left\{\dfrac{-5 \pm \sqrt{30}}{-2}\right\}$. The teacher gave them both full credit. Explain why both students were correct.

64. *Concept Check*   In the solutions $\dfrac{2 \pm 3\sqrt{3}}{3}$ found in **Example 4** of this section, why is it not valid to simplify the answers by dividing out the 3's in the numerator and denominator?

Use a calculator to solve each equation. Round answers to the nearest hundredth.

65. $(k + 2.14)^2 = 5.46$ **66.** $(r - 3.91)^2 = 9.28$

67. $(2.11p + 3.42)^2 = 9.58$ **68.** $(1.71m - 6.20)^2 = 5.41$

*Solve each problem. **See Example 6.***

69. An expert marksman can hold a silver dollar at forehead level, drop it, draw his gun, and shoot the coin as it passes waist level. The distance traveled by a falling object is given by

$$d = 16t^2,$$

where d is the distance (in feet) the object falls in t seconds. If the coin falls 4 ft, use the formula to find the time that elapses between the dropping of the coin and the shot.

70. The illumination produced by a light source depends on the distance from the source. For a particular light source, this relationship can be expressed as

$$I = \frac{4050}{d^2},$$

where I is the amount of illumination in foot-candles and d is the distance from the light source (in feet). How far from the source is the illumination equal to 50 footcandles?

▶ **71.** The area $\mathcal{A}$ of a circle with radius r is given by the formula

$$\mathcal{A} = \pi r^2.$$

If a circle has area 81π in.², what is its radius?

72. The surface area S of a sphere with radius r is given by the formula

$$S = 4\pi r^2.$$

If a sphere has surface area 36π ft², what is its radius?

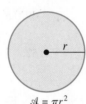

$\mathcal{A} = \pi r^2$

$S = 4\pi r^2$

The amount A that P dollars invested at an annual rate of interest r will grow to in 2 yr is

$$A = P(1 + r)^2.$$

73. At what interest rate will $100 grow to $104.04 in 2 yr?

74. At what interest rate will $500 grow to $530.45 in 2 yr?

9.2 Solving Quadratic Equations by Completing the Square

OBJECTIVES

1 Solve quadratic equations by completing the square when the coefficient of the second-degree term is 1.

2 Solve quadratic equations by completing the square when the coefficient of the second-degree term is not 1.

3 Simplify the terms of an equation before solving.

4 Solve applied problems that require quadratic equations.

OBJECTIVE 1 Solve quadratic equations by completing the square when the coefficient of the second-degree term is 1.

The methods we have studied so far are not enough to solve an equation such as

$$x^2 + 6x + 7 = 0.$$

If we could write the equation in the form $(x + 3)^2$ equals a constant, we could solve it with the square root property discussed in **Section 9.1.** To do that, we need to have a perfect square trinomial on one side of the equation.

Recall from **Section 6.4** that a perfect square trinomial has the form

$$x^2 + 2kx + k^2 \quad \text{or} \quad x^2 - 2kx + k^2, \quad \text{where } k \text{ represents a real number.}$$

EXAMPLE 1 Creating Perfect Square Trinomials

Complete each trinomial so that it is a perfect square. Then factor the trinomial.

(a) $x^2 + 8x + $ _____

The perfect square trinomial will have the form $x^2 + 2kx + k^2$. Thus, the middle term, $8x$, must equal $2kx$.

$$8x = 2kx \leftarrow \text{Solve this equation for } k.$$

$$4 = k \qquad \text{Divide each side by } 2x.$$

**NOW TRY
EXERCISE 1**

Complete each trinomial so that it is a perfect square. Then factor the trinomial.

(a) $x^2 + 4x + $ _____

(b) $x^2 - 22x + $ _____

Therefore, $k = 4$ and $k^2 = 4^2 = 16$. The required perfect square trinomial is

$$x^2 + 8x + 16, \quad \text{which factors as} \quad (x + 4)^2.$$

(b) $x^2 - 18x + $ _____

Here the perfect square trinomial will have the form $x^2 - 2kx + k^2$. The middle term, $-18x$, must equal $-2kx$.

$$-18x = -2kx \leftarrow \text{Solve this equation for } k.$$

$$9 = k \qquad \text{Divide each side by } -2x.$$

Thus, $k = 9$ and $k^2 = 9^2 = 81$. The required perfect square trinomial is

$$x^2 - 18x + 81, \quad \text{which factors as} \quad (x - 9)^2. \qquad \text{NOW TRY} $$

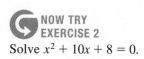

**NOW TRY
EXERCISE 2**

Solve $x^2 + 10x + 8 = 0$.

EXAMPLE 2 Rewriting an Equation to Use the Square Root Property

Solve $x^2 + 6x + 7 = 0$.

$$x^2 + 6x = -7 \qquad \text{Subtract 7 from each side.}$$

To solve this equation with the square root property, the quantity on the left side, $x^2 + 6x$, must be written as a perfect square trinomial in the form $x^2 + 2kx + k^2$.

$$x^2 + 6x + \rule{2cm}{0.4pt} \quad \boxed{\text{A square must go here.}}$$

Here, $2kx = 6x$, so $k = 3$ and $k^2 = 9$. The required perfect square trinomial is

$$x^2 + 6x + 9, \quad \text{which factors as} \quad (x + 3)^2.$$

Therefore, if we add 9 to each side of $x^2 + 6x = -7$, the equation will have a perfect square trinomial on the left side, as needed.

$$x^2 + 6x = -7 \qquad \text{Transformed equation}$$

$$\boxed{\text{This is a key step.}} \quad x^2 + 6x + 9 = -7 + 9 \qquad \text{Add 9.}$$

$$(x + 3)^2 = 2 \qquad \text{Factor on the left. Add on the right.}$$

Now use the square root property to complete the solution.

$$x + 3 = \sqrt{2} \qquad \text{or} \quad x + 3 = -\sqrt{2}$$

$$x = -3 + \sqrt{2} \quad \text{or} \qquad x = -3 - \sqrt{2}$$

Check by substituting $-3 + \sqrt{2}$ and $-3 - \sqrt{2}$ for x in the original equation. The solution set is $\left\{ -3 \pm \sqrt{2} \right\}$.

NOW TRY

Completing the square is the process of changing the form of the equation in **Example 2** from

$$x^2 + 6x + 7 = 0 \quad \text{to} \quad (x + 3)^2 = 2.$$

Completing the square changes only the form of the equation. To see this, multiply out the left side of $(x + 3)^2 = 2$ and combine like terms. Then subtract 2 from each side to see that the result is $x^2 + 6x + 7 = 0$.

NOW TRY ANSWERS

1. (a) 4; $(x + 2)^2$

 (b) 121; $(x - 11)^2$

2. $\left\{ -5 \pm \sqrt{17} \right\}$

Look again at the original equation in **Example 2,** $x^2 + 6x + 7 = 0$. If we take half the coefficient of x, which is 6 here, and square it, we obtain 9.

$$\frac{1}{2} \cdot 6 = 3 \quad \text{and} \quad 3^2 = 9$$

Coefficient of x \quad Quantity added to each side

To complete the square in **Example 2,** we added 9 to each side.

NOW TRY EXERCISE 3

Solve $x^2 - 6x = 9$.

EXAMPLE 3 Completing the Square to Solve a Quadratic Equation

Solve $x^2 - 8x = 5$.

To complete the square on $x^2 - 8x$, take half the coefficient of x and square it.

$$\frac{1}{2}(-8) = -4 \quad \text{and} \quad (-4)^2 = 16$$

Coefficient of x

Add the result, 16, to each side of the equation.

$$x^2 - 8x = 5 \qquad \text{Given equation}$$
$$x^2 - 8x + 16 = 5 + 16 \qquad \text{Add 16.}$$
$$(x - 4)^2 = 21 \qquad \text{Factor on the left. Add on the right.}$$
$$x - 4 = \sqrt{21} \quad \text{or} \quad x - 4 = -\sqrt{21} \qquad \text{Square root property}$$
$$x = 4 + \sqrt{21} \quad \text{or} \quad x = 4 - \sqrt{21} \qquad \text{Add 4.}$$

A check confirms that the solution set is $\{4 \pm \sqrt{21}\}$. NOW TRY

NOW TRY EXERCISE 4

Solve $x(x + 5) = 3$

EXAMPLE 4 Solving a Quadratic Equation by Completing the Square ($a = 1$)

Solve $x(x + 7) = 2$.

$$x(x + 7) = 2$$
$$x^2 + 7x = 2 \qquad \text{Multiply.}$$
$$x^2 + 7x + \frac{49}{4} = 2 + \frac{49}{4} \qquad \text{To complete the square, add } \left(\frac{1}{2} \cdot 7\right)^2 = \frac{49}{4} \text{ to each side.}$$
$$\left(x + \frac{7}{2}\right)^2 = \frac{57}{4} \qquad \text{Factor; } 2 + \frac{49}{4} = \frac{8}{4} + \frac{49}{4} = \frac{57}{4}.$$
$$x + \frac{7}{2} = \sqrt{\frac{57}{4}} \quad \text{or} \quad x + \frac{7}{2} = -\sqrt{\frac{57}{4}} \qquad \text{Square root property}$$
$$x = -\frac{7}{2} + \frac{\sqrt{57}}{2} \quad \text{or} \quad x = -\frac{7}{2} - \frac{\sqrt{57}}{2} \qquad \text{Subtract } \frac{7}{2}; \text{ quotient rule for radicals}$$
$$x = \frac{-7 + \sqrt{57}}{2} \quad \text{or} \quad x = \frac{-7 - \sqrt{57}}{2} \qquad \text{Add and subtract the fractions.}$$

NOW TRY ANSWERS

3. $\{3 \pm 3\sqrt{2}\}$

4. $\left\{\dfrac{-5 \pm \sqrt{37}}{2}\right\}$

A check confirms that the solution set is $\left\{\dfrac{-7 \pm \sqrt{57}}{2}\right\}$. NOW TRY

OBJECTIVE 2 Solve quadratic equations by completing the square when the coefficient of the second-degree term is not 1.

If a quadratic equation has the form

$$ax^2 + bx + c = 0, \quad \text{where} \quad a \neq 1,$$

we obtain 1 as the coefficient of x^2 by dividing each side of the equation by a.

Completing the Square

To solve $ax^2 + bx + c = 0$ (where $a \neq 0$) by completing the square, follow these steps.

Step 1 **Be sure the second-degree term has coefficient 1.**
  • If the coefficient of the second-degree term is 1, go to Step 2.
  • If it is not 1, but some other nonzero number a, divide each side of the equation by a.

Step 2 **Write the equation in correct form.** Make sure that all variable terms are on one side of the equality symbol and the constant term is on the other side.

Step 3 **Complete the square.**
  • Take half the coefficient of the first-degree term, and square it.
  • Add the square to each side of the equation.
  • Factor the variable side, which should be a perfect square trinomial, as the square of a binomial. Combine terms on the other side.

Step 4 **Solve** the equation using the square root property.

EXAMPLE 5 Solving a Quadratic Equation by Completing the Square ($a \neq 1$)

Solve $4x^2 + 16x - 9 = 0$.

Step 1 *Before completing the square, the coefficient of x^2 must be 1,* not 4. We obtain 1 as the coefficient of x^2 here by dividing each side by 4.

$$4x^2 + 16x - 9 = 0 \qquad \text{Given equation}$$

The coefficient of x^2 must be 1. $\longrightarrow$
$$x^2 + 4x - \frac{9}{4} = 0 \qquad \text{Divide by 4.}$$

Step 2 Write the equation so that all variable terms are on one side of the equation and all constant terms are on the other side.

$$x^2 + 4x = \frac{9}{4} \qquad \text{Add } \tfrac{9}{4}.$$

Step 3 Complete the square by taking half the coefficient of x, and squaring it.

$$\frac{1}{2}(4) = 2 \quad \text{and} \quad 2^2 = 4$$

We add the result, 4, to each side of the equation.

$$x^2 + 4x + 4 = \frac{9}{4} + 4 \qquad \text{Add 4.}$$

$$(x + 2)^2 = \frac{25}{4} \qquad \text{Factor; } \tfrac{9}{4} + 4 = \tfrac{9}{4} + \tfrac{16}{4} = \tfrac{25}{4}.$$

**NOW TRY
EXERCISE 5**
Solve $4t^2 - 4t - 3 = 0$.

Step 4 Solve the equation using the square root property.

$$x + 2 = \sqrt{\frac{25}{4}} \qquad \text{or} \qquad x + 2 = -\sqrt{\frac{25}{4}} \qquad \text{Square root property}$$

$$x + 2 = \frac{5}{2} \qquad \text{or} \qquad x + 2 = -\frac{5}{2} \qquad \text{Take square roots.}$$

$$x = -2 + \frac{5}{2} \qquad \text{or} \qquad x = -2 - \frac{5}{2} \qquad \text{Subtract 2.}$$

$$x = \frac{1}{2} \qquad \text{or} \qquad x = -\frac{9}{2} \qquad -2 = -\frac{4}{2}$$

CHECK

$$4x^2 + 16x - 9 = 0 \qquad\qquad\qquad 4x^2 + 16x - 9 = 0$$

$$4\left(\frac{1}{2}\right)^2 + 16\left(\frac{1}{2}\right) - 9 \overset{?}{=} 0 \quad \text{Let } x = \tfrac{1}{2}. \quad 4\left(-\frac{9}{2}\right)^2 + 16\left(-\frac{9}{2}\right) - 9 \overset{?}{=} 0 \quad \text{Let } x = -\tfrac{9}{2}.$$

$$4\left(\frac{1}{4}\right) + 8 - 9 \overset{?}{=} 0 \qquad\qquad\qquad 4\left(\frac{81}{4}\right) - 72 - 9 \overset{?}{=} 0$$

$$1 + 8 - 9 \overset{?}{=} 0 \qquad\qquad\qquad\qquad 81 - 72 - 9 \overset{?}{=} 0$$

$$0 = 0 \checkmark \text{ True} \qquad\qquad\qquad\qquad 0 = 0 \checkmark \quad \text{True}$$

The two values, $\frac{1}{2}$ and $-\frac{9}{2}$, check, so the solution set is $\left\{-\frac{9}{2}, \frac{1}{2}\right\}$.   **NOW TRY** 🔄

EXAMPLE 6   **Solving a Quadratic Equation by Completing the Square ($a \neq 1$)**

Solve $2x^2 - 7x - 9 = 0$.

Step 1 Transform the equation so that 1 is the coefficient of the x^2-term.

$$2x^2 - 7x - 9 = 0 \qquad \text{Given equation}$$

$$x^2 - \frac{7}{2}x - \frac{9}{2} = 0 \qquad \text{Divide by 2.}$$

Step 2 Add $\frac{9}{2}$ to each side so that the variable terms are on the left and the constant is on the right.

$$x^2 - \frac{7}{2}x = \frac{9}{2} \qquad \text{Add } \tfrac{9}{2}.$$

Step 3 To complete the square, take half the coefficient of x and square it.

$$\left[\frac{1}{2}\left(-\frac{7}{2}\right)\right]^2 = \left(-\frac{7}{4}\right)^2 = \frac{49}{16}$$

Add the result, $\frac{49}{16}$, to each side of the equation.

$$x^2 - \frac{7}{2}x + \frac{49}{16} = \frac{9}{2} + \frac{49}{16} \quad \boxed{\text{Be sure to add } \tfrac{49}{16} \text{ to } each \text{ side.}}$$

NOW TRY ANSWER
5. $\left\{-\frac{1}{2}, \frac{3}{2}\right\}$

$$\left(x - \frac{7}{4}\right)^2 = \frac{121}{16} \qquad \text{Factor; } \tfrac{9}{2} + \tfrac{49}{16} = \tfrac{72}{16} + \tfrac{49}{16} = \tfrac{121}{16}.$$

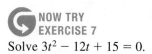

Solve $4x^2 + 9x - 9 = 0$.

Step 4 Solve the equation using the square root property.

$$x - \frac{7}{4} = \sqrt{\frac{121}{16}} \quad \text{or} \quad x - \frac{7}{4} = -\sqrt{\frac{121}{16}} \qquad \text{Square root property}$$

$$x = \frac{7}{4} + \frac{11}{4} \quad \text{or} \quad x = \frac{7}{4} - \frac{11}{4} \qquad \text{Add } \tfrac{7}{4}; \sqrt{\tfrac{121}{16}} = \tfrac{11}{4}.$$

$$x = \frac{18}{4} \quad \text{or} \quad x = -\frac{4}{4} \qquad \begin{array}{l}\text{Add and subtract}\\ \text{the fractions.}\end{array}$$

$$x = \frac{9}{2} \quad \text{or} \quad x = -1 \qquad \text{Lowest terms}$$

A check confirms that the solution set is $\left\{-1, \frac{9}{2}\right\}$. **NOW TRY**

NOW TRY EXERCISE 7

Solve $3t^2 - 12t + 15 = 0$.

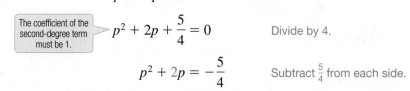

Solve $4p^2 + 8p + 5 = 0$.

$$4p^2 + 8p + 5 = 0$$

> The coefficient of the second-degree term must be 1.

$$p^2 + 2p + \frac{5}{4} = 0 \qquad \text{Divide by 4.}$$

$$p^2 + 2p = -\frac{5}{4} \qquad \text{Subtract } \tfrac{5}{4} \text{ from each side.}$$

The coefficient of p is 2. Take half of 2, square the result, and add it to each side.

$$p^2 + 2p + 1 = -\frac{5}{4} + 1 \qquad \left[\tfrac{1}{2}(2)\right]^2 = 1^2 = 1; \text{ Add } 1.$$

$$(p + 1)^2 = -\frac{1}{4} \qquad \begin{array}{l}\text{Factor on the left.}\\ \text{Add on the right.}\end{array}$$

If we apply the square root property to solve this equation, we obtain the square root of $-\frac{1}{4}$, which is not a real number. There is no real solution. **NOW TRY**

OBJECTIVE 3 Simplify the terms of an equation before solving.

NOW TRY EXERCISE 8

Solve $(x - 5)(x + 1) = 2$.

EXAMPLE 8 Simplifying the Terms of an Equation before Solving

Solve $(x + 3)(x - 1) = 2$.

$$(x + 3)(x - 1) = 2$$

$$x^2 + 2x - 3 = 2 \qquad \text{Multiply using the FOIL method.}$$

$$x^2 + 2x = 5 \qquad \text{Add 3.}$$

$$x^2 + 2x + 1 = 5 + 1 \qquad \text{Add } \left[\tfrac{1}{2}(2)\right]^2 = 1^2 = 1.$$

$$(x + 1)^2 = 6 \qquad \text{Factor on the left. Add on the right.}$$

$$x + 1 = \sqrt{6} \quad \text{or} \quad x + 1 = -\sqrt{6} \qquad \text{Square root property}$$

$$x = -1 + \sqrt{6} \quad \text{or} \quad x = -1 - \sqrt{6} \qquad \text{Subtract 1.}$$

A check confirms that the solution set is $\left\{-1 \pm \sqrt{6}\right\}$. **NOW TRY**

NOW TRY ANSWERS
6. $\left\{-3, \frac{3}{4}\right\}$ **7.** no real solution
8. $\left\{2 \pm \sqrt{11}\right\}$

NOTE The solutions $-1 \pm \sqrt{6}$ given in **Example 8** are *exact*. In applications, decimal solutions are often required. Using the square root key of a calculator yields $\sqrt{6} \approx 2.449$. Approximating the two solutions gives

$$x \approx 1.449 \quad \text{and} \quad x \approx -3.449.$$

OBJECTIVE 4 Solve applied problems that require quadratic equations.

EXAMPLE 9 Solving a Velocity Problem

If a ball is projected into the air from ground level with an initial velocity of 64 ft per sec, its altitude (height) s in feet in t seconds is given by the formula

$$s = -16t^2 + 64t.$$

At what times will the ball be 48 ft above the ground?

   Because s represents the height, we let $s = 48$ in the formula and solve this equation for the time t by completing the square.

$48 = -16t^2 + 64t$	Let $s = 48$.
$-3 = t^2 - 4t$	Divide by -16.
$t^2 - 4t = -3$	Interchange sides.
$t^2 - 4t + 4 = -3 + 4$	Add $\left[\frac{1}{2}(-4)\right]^2 = (-2)^2 = 4$.
$(t - 2)^2 = 1$	Factor. Add.
$t - 2 = 1 \quad$ or $\quad t - 2 = -1$	Square root property
$t = 3 \quad$ or $\quad t = 1$	Add 2.

The ball reaches a height of 48 ft twice, once on the way up and again on the way down. It takes 1 sec to reach 48 ft on the way up, and then after 3 sec, the ball reaches 48 ft again on the way down.   **NOW TRY**

NOW TRY
EXERCISE 9
At what times will the ball in **Example 9** be 28 ft above the ground? Give answers to the nearest tenth.

NOW TRY ANSWER
9. 0.5 sec and 3.5 sec

9.2 Exercises

FOR EXTRA HELP   ▶ MyMathLab®

1. *Concept Check* Which step is an appropriate way to begin solving the quadratic equation $2x^2 - 4x = 9$ by completing the square?

A. Add 4 to each side of the equation.   **B.** Factor the left side as $2x(x - 2)$.

C. Factor the left side as $x(2x - 4)$.   **D.** Divide each side by 2.

2. *Concept Check* In **Example 3** of **Section 6.5**, we solved the quadratic equation $4x^2 + 40 = 26x$ by factoring. If we were to solve by completing the square, would we obtain the same solution set, $\left\{\frac{5}{2}, 4\right\}$?

Complete each trinomial so that it is a perfect square. Then factor the trinomial. **See Example 1.**

▶ 3. $x^2 + 10x +$ _____   **4.** $x^2 + 16x +$ _____   **▶ 5.** $z^2 - 20z +$ _____

6. $a^2 - 32a +$ _____   **7.** $x^2 + 2x +$ _____   **8.** $m^2 - 2m +$ _____

9. $p^2 - 5p +$ _____   **10.** $x^2 + 3x +$ _____   **11.** $x^2 + \frac{1}{2}x +$ _____

12. $x^2 + \frac{1}{3}x +$ _____   **13.** $x^2 - 0.4x +$ _____   **14.** $x^2 - 0.8x +$ _____

Concept Check   Solve each equation by completing the square.

15. $x^2 + 4x = 1$

Take half the coefficient of x and square it.

$$\frac{1}{2} \cdot \underline{\hspace{1cm}} = 2, \quad \text{and} \quad \underline{\hspace{1cm}}^2 = 4.$$

Add $\underline{\hspace{1cm}}$ to each side of the equation.

$$x^2 + 4x + \underline{\hspace{1cm}} = 1 + 4$$

Factor and add.

$$\underline{\hspace{3cm}}$$

Complete the solution.

16. $3x^2 + 5x - 2 = 0$

Divide each side by $\underline{\hspace{1cm}}$ to obtain $\underline{\hspace{3cm}}$. Add $\frac{2}{3}$ to each side to obtain $\underline{\hspace{3cm}}$.

Take half the coefficient of x and square it. Add $\underline{\hspace{1cm}}$ to each side of the equation.

$$x^2 + \frac{5}{3}x + \underline{\hspace{1cm}} = \frac{2}{3} + \frac{25}{36}$$

Factor and add.

$$(\underline{\hspace{2cm}})^2 = \underline{\hspace{1cm}}$$

Complete the solution.

Solve each equation by completing the square. See Examples 2–4.

▶ **17.** $x^2 - 4x = -3$

18. $p^2 - 2p = 8$

▶ **19.** $x^2 + 2x - 5 = 0$

20. $r^2 + 4r + 1 = 0$

21. $x^2 - 8x = -4$

22. $m^2 - 4m = 14$

23. $x^2 + 6x + 9 = 0$ **24.** $x^2 - 8x + 16 = 0$ **25.** $x^2 - 16x = 0$ **26.** $x^2 - 20x = 0$

27. $x(x - 3) = 1$ **28.** $x(x - 5) = 2$ **29.** $x(x + 3) = -1$ **30.** $x(x + 7) = -2$

Solve each equation by completing the square. See Examples 5–8.

▶ **31.** $4x^2 + 4x = 3$

32. $9x^2 + 3x = 2$

▶ **33.** $2p^2 - 2p + 3 = 0$

34. $3q^2 - 3q + 4 = 0$

35. $3x^2 - 9x + 5 = 0$

36. $6x^2 - 8x - 3 = 0$

37. $3x^2 + 7x = 4$

38. $2x^2 + 5x = 1$

▶ **39.** $(x + 3)(x - 1) = 5$

40. $(x - 8)(x + 2) = 24$

41. $(r - 3)(r - 5) = 2$

42. $(x - 1)(x - 7) = 1$

43. $-x^2 + 2x = -5$

44. $-x^2 + 4x = 1$

45. $5x^2 + 6x - 11 = 0$

46. $7x^2 - 9x - 10 = 0$

47. $-3x^2 + 11x + 42 = 0$

48. $-9x^2 - 20x + 21 = 0$

Solve each equation by completing the square. Give (a) exact solutions and (b) solutions rounded to the nearest thousandth. See the Note following Example 8.

49. $3r^2 - 2 = 6r + 3$

50. $4p + 3 = 2p^2 + 2p$

51. $(x + 1)(x + 3) = 2$

52. $(x - 3)(x + 1) = 1$

Solve each problem. See Example 9.

▶ **53.** If an object is projected upward on the surface of Mars from ground level with an initial velocity of 104 ft per sec, its altitude (height) s in feet in t seconds is given by the formula

$$s = -13t^2 + 104t.$$

At what times will the object be 195 ft above the ground?

54. After how many seconds will the object in **Exercise 53** return to the surface? (*Hint:* When it returns to the surface, $s = 0$.)

55. If an object is projected upward from ground level on Earth with an initial velocity of 96 ft per sec, its altitude (height) s in feet in t seconds is given by the formula

$$s = -16t^2 + 96t.$$

At what times will the object be at a height of 80 ft? (*Hint:* Let $s = 80$.)

56. At what times will the object described in **Exercise 55** be at a height of 100 ft? Give answers to the nearest tenth.

57. A farmer has a rectangular cattle pen with perimeter 350 ft and area 7500 ft². What are the dimensions of the pen? (*Hint:* Use the figure to set up the equation.)

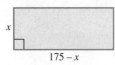

58. The base of a triangle measures 1 m more than three times the height of the triangle. The area of the triangle is 15 m². Find the lengths of the base and the height.

59. Two cars travel at right angles to each other from an intersection until they are 17 mi apart. At that point, one car has gone 7 mi farther than the other. How far did the slower car travel? (*Hint:* Use the Pythagorean theorem.)

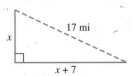

60. Two painters are painting a house. One painter takes 2 hr longer to paint the house working alone than the other painter. When they do the job together, they can complete it in 4.8 hr. How long would it take the faster painter alone to paint the house? Give the answer to the nearest tenth.

RELATING CONCEPTS For Individual or Group Work (Exercises 61–64)

We have discussed "completing the square" in an algebraic sense. This procedure can literally be applied to a geometric figure so that it becomes a square.

   For example, to complete the square for $x^2 + 8x$, begin with a square having a side of length x. Add four rectangles of width 1 to the right side and to the bottom, as shown in the top figure. To "complete the square," fill in the bottom right corner with 16 squares of area 1, as shown in the bottom figure.

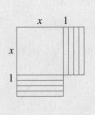

Work Exercises 61–64 in order.

61. What is the area of the original square?

62. What is the area of the figure after the 8 rectangles are added?

63. What is the area of the figure after the 16 small squares are added?

64. At what point did we "complete the square"?

9.3 Solving Quadratic Equations by the Quadratic Formula

OBJECTIVES

1 Identify the values of *a*, *b*, and *c* in a quadratic equation.

2 Use the quadratic formula to solve quadratic equations.

3 Solve quadratic equations with a double solution.

4 Solve quadratic equations with fractions as coefficients.

Although we can solve any quadratic equation by completing the square, the method can be tedious. By completing the square on the general quadratic equation

$$ax^2 + bx + c = 0, \quad \text{where } a \neq 0, \qquad \text{Standard form}$$

we obtain the *quadratic formula,* which gives the solution(s) of *any* quadratic equation.

NOTE In $ax^2 + bx + c = 0$, there is a restriction that *a* is not zero. If it were, the equation would be linear, not quadratic.

OBJECTIVE 1 Identify the values of *a*, *b*, and *c* in a quadratic equation.

To solve a quadratic equation with the quadratic formula, we must first identify the values of *a*, *b*, and *c* in the standard form.

VOCABULARY

☐ quadratic formula
☐ discriminant
☐ double solution

 NOW TRY
EXERCISE 1

Write each equation in standard form, if necessary, with 0 on the right side. Then identify the values of a, b, and c.

(a) $3x^2 - 7x + 4 = 0$

(b) $x^2 - 3 = -2x$

(c) $2x^2 - 4x = 0$

(d) $2(2x + 1)(x - 5) = -3$

EXAMPLE 1 Determining Values of a, b, and c in Quadratic Equations

Write each equation in standard form, if necessary, with 0 on the right side. Then identify the values of a, b, and c.

$$\overset{a}{\downarrow} \quad \overset{b}{\downarrow} \quad \overset{c}{\downarrow}$$

(a) $2x^2 + 3x - 5 = 0$ [This equation is in standard form.]

    Here, $a = 2$, $b = 3$, and $c = -5$.

(b) $-x^2 + 2 = 6x$

    First write the equation in standard form $ax^2 + bx + c = 0$.

$$-x^2 + 2 = 6x$$

[$-x^2$ means $-1x^2$.] $-x^2 - 6x + 2 = 0$ Subtract 6x.

Here, $a = -1$, $b = -6$, and $c = 2$.

(c) $5x^2 - 12 = 0$

    No x-term appears. We can write it as $0x$.

$$5x^2 + 0x - 12 = 0 \qquad \text{Standard form of the equation}$$

Thus, $a = 5$, $b = 0$, and $c = -12$.

(d) [The equation is not in standard form.] $(2x - 7)(x + 4) = -23$

$$2x^2 + x - 28 = -23 \qquad \text{Multiply using the FOIL method.}$$

$$2x^2 + x - 5 = 0 \qquad \begin{array}{l}\text{Add 23. The equation is}\\\text{now in standard form.}\end{array}$$

Here, $a = 2$, $b = 1$, and $c = -5$.   NOW TRY

OBJECTIVE 2 Use the quadratic formula to solve quadratic equations.

To develop the quadratic formula, we follow the steps given in **Section 9.2** for completing the square on $ax^2 + bx + c = 0$ (where $a > 0$). For comparison, we also show the corresponding steps for solving $2x^2 + x - 5 = 0$ (from **Example 1(d)**).

Step 1   Transform so that the coefficient of the second-degree term is equal to 1.

$2x^2 + x - 5 = 0$ Standard form	$ax^2 + bx + c = 0$ Standard form
$x^2 + \dfrac{1}{2}x - \dfrac{5}{2} = 0$ Divide by 2.	$x^2 + \dfrac{b}{a}x + \dfrac{c}{a} = 0$ Divide by a.

Step 2   Write so that the variable terms with x are alone on the left side.

$x^2 + \dfrac{1}{2}x = \dfrac{5}{2}$ Add $\frac{5}{2}$.	$x^2 + \dfrac{b}{a}x = -\dfrac{c}{a}$ Subtract $\frac{c}{a}$.

Step 3   Add the square of half the coefficient of x to each side, factor the left side, and combine like terms on the right.

$x^2 + \dfrac{1}{2}x + \dfrac{1}{16} = \dfrac{5}{2} + \dfrac{1}{16}$ Add $\frac{1}{16}$.	$x^2 + \dfrac{b}{a}x + \dfrac{b^2}{4a^2} = -\dfrac{c}{a} + \dfrac{b^2}{4a^2}$ Add $\frac{b^2}{4a^2}$.
$\left(x + \dfrac{1}{4}\right)^2 = \dfrac{41}{16}$ Factor. Add on right.	$\left(x + \dfrac{b}{2a}\right)^2 = \dfrac{b^2 - 4ac}{4a^2}$ Factor. Add on right.

NOW TRY ANSWERS
1. (a) $a = 3$, $b = -7$, $c = 4$
   (b) $a = 1$, $b = 2$, $c = -3$
   (c) $a = 2$, $b = -4$, $c = 0$
   (d) $a = 4$, $b = -18$, $c = -7$

Step 4   Use the square root property to complete the solution.

$$x + \frac{1}{4} = \pm\sqrt{\frac{41}{16}} \qquad\qquad x + \frac{b}{2a} = \pm\sqrt{\frac{b^2 - 4ac}{4a^2}}$$

$$x + \frac{1}{4} = \pm\frac{\sqrt{41}}{4} \qquad\qquad x + \frac{b}{2a} = \pm\frac{\sqrt{b^2 - 4ac}}{2a}$$

$$x = -\frac{1}{4} \pm \frac{\sqrt{41}}{4} \qquad\qquad x = -\frac{b}{2a} \pm \frac{\sqrt{b^2 - 4ac}}{2a}$$

$$x = \frac{-1 \pm \sqrt{41}}{4} \qquad\qquad x = \frac{-b \pm \sqrt{b^2 - 4ac}}{2a}$$

The final result on the right (which is also valid for $a < 0$) is the **quadratic formula.** *It gives two values: one for the $+$ sign and one for the $-$ sign.*

Quadratic Formula

The quadratic equation $ax^2 + bx + c = 0$ (where $a \neq 0$) has solutions

$$x = \frac{-b + \sqrt{b^2 - 4ac}}{2a} \quad \text{and} \quad x = \frac{-b - \sqrt{b^2 - 4ac}}{2a},$$

or, in compact form, $x = \dfrac{-b \pm \sqrt{b^2 - 4ac}}{2a}.$

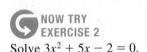

NOW TRY EXERCISE 2

Solve $3x^2 + 5x - 2 = 0$.

EXAMPLE 2   Solving a Quadratic Equation by the Quadratic Formula

Solve $2x^2 - 7x - 9 = 0$.

$$x = \frac{-b \pm \sqrt{b^2 - 4ac}}{2a} \qquad \text{Quadratic formula}$$

> Be sure to write $-b$ in the numerator and extend the fraction bar below it.

$$x = \frac{-(-7) \pm \sqrt{(-7)^2 - 4(2)(-9)}}{2(2)} \qquad \begin{array}{l}\text{Substitute } a = 2,\\ b = -7, \text{ and } c = -9.\end{array}$$

$$x = \frac{7 \pm \sqrt{49 + 72}}{4} \qquad \text{Simplify.}$$

$$x = \frac{7 \pm \sqrt{121}}{4} \qquad \text{Add.}$$

> This represents **two** solutions.

$$x = \frac{7 \pm 11}{4} \qquad \sqrt{121} = 11$$

Find the two solutions by first using the plus sign and then using the minus sign.

$$x = \frac{7 + 11}{4} = \frac{18}{4} = \frac{9}{2} \quad \text{or} \quad x = \frac{7 - 11}{4} = \frac{-4}{4} = -1$$

Check each value. The solution set is $\left\{-1, \frac{9}{2}\right\}$.   NOW TRY

NOW TRY ANSWER

2. $\left\{-2, \frac{1}{3}\right\}$

⚠ CAUTION   In the quadratic formula, the fraction bar is under $-b$ as well as the radical. *Find the values of $-b \pm \sqrt{b^2 - 4ac}$ first. Then divide by the value of 2a.*

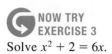 **NOW TRY EXERCISE 3**

Solve $x^2 + 2 = 6x$.

| EXAMPLE 3 | Rewriting a Quadratic Equation before Solving |

Solve $x^2 = 2x + 1$.

Write the given equation in standard form as $x^2 - 2x - 1 = 0$.

$$x = \frac{-b \pm \sqrt{b^2 - 4ac}}{2a}$$ Quadratic formula

$$x = \frac{-(-2) \pm \sqrt{(-2)^2 - 4(1)(-1)}}{2(1)}$$ Substitute $a = 1$, $b = -2$, and $c = -1$.

Be careful substituting the negative values.

$$x = \frac{2 \pm \sqrt{8}}{2}$$ Simplify.

$$x = \frac{2 \pm 2\sqrt{2}}{2}$$ $\sqrt{8} = \sqrt{4} \cdot \sqrt{2} = 2\sqrt{2}$

Factor first. Then divide out the common factor.

$$x = \frac{2(1 \pm \sqrt{2})}{2}$$ Factor.

$$x = 1 \pm \sqrt{2}$$ Divide out the common factor to write in lowest terms.

A check confirms that the solution set is $\left\{1 \pm \sqrt{2}\right\}$. **NOW TRY**

OBJECTIVE 3 Solve quadratic equations with a double solution.

In the quadratic formula, the quantity under the radical, $b^2 - 4ac$, is the **discriminant**. When the discriminant for $ax^2 + bx + c = 0$ equals 0 and a, b, and c are integers, the equation has just one *distinct* rational number solution, called a **double solution**. Furthermore, the trinomial $ax^2 + bx + c$ is a perfect square.

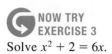 **NOW TRY EXERCISE 4**

Solve $16x^2 = 8x - 1$.

| EXAMPLE 4 | Solving a Quadratic Equation with One Distinct Solution |

Solve $4x^2 + 25 = 20x$.

Write the given equation in standard form as $4x^2 - 20x + 25 = 0$.

$$x = \frac{-b \pm \sqrt{b^2 - 4ac}}{2a}$$ Quadratic formula

$$x = \frac{-(-20) \pm \sqrt{(-20)^2 - 4(4)(25)}}{2(4)}$$ Substitute $a = 4$, $b = -20$, and $c = 25$.

$$x = \frac{20 \pm \sqrt{400 - 400}}{8}$$ ◄ $400 - 400 = 0$ Simplify.

$$x = \frac{20 \pm 0}{8}$$ $\sqrt{0} = 0$

$$x = \frac{5}{2}$$ Lowest terms

The discriminant $b^2 - 4ac$ is 0, and the trinomial $4x^2 - 20x + 25$ is a perfect square, $(2x - 5)^2$. This would lead to two solutions, each of which is $\frac{5}{2}$. Thus, there is just one distinct solution, $\frac{5}{2}$. A check confirms that the solution set is $\left\{\frac{5}{2}\right\}$. **NOW TRY**

NOW TRY ANSWERS

3. $\left\{3 \pm \sqrt{7}\right\}$ **4.** $\left\{\frac{1}{4}\right\}$

NOTE The single solution of the equation in **Example 4** is a rational number. If all solutions of a quadratic equation are rational, the equation can be solved by the zero-factor property.

$$4x^2 - 20x + 25 = 0 \qquad \text{See \textbf{Example 4}.}$$

$$(2x - 5)^2 = 0 \qquad \text{Factor.}$$

$$2x - 5 = 0 \qquad \text{Zero-factor property}$$

The same solution results. $\longrightarrow x = \dfrac{5}{2} \qquad$ Solve for x.

OBJECTIVE 4 Solve quadratic equations with fractions as coefficients.

In general, it is easier to apply the quadratic formula if a, b, and c are integers. We can always multiply both sides of an equation by a constant (for example, the LCD) to clear any fractions.

NOW TRY EXERCISE 5

Solve $\dfrac{1}{12}x^2 = \dfrac{1}{2}x - \dfrac{1}{3}$.

EXAMPLE 5   Solving a Quadratic Equation with Fractions

Solve $\dfrac{1}{10}t^2 = \dfrac{2}{5}t + \dfrac{1}{5}$.

$$\frac{1}{10}t^2 = \frac{2}{5}t + \frac{1}{5}$$

$$10\left(\frac{1}{10}t^2\right) = 10\left(\frac{2}{5}t + \frac{1}{5}\right) \qquad \text{To clear fractions, multiply by the LCD, 10.}$$

$$10\left(\frac{1}{10}t^2\right) = 10\left(\frac{2}{5}t\right) + 10\left(\frac{1}{5}\right) \qquad \text{Distributive property}$$

$$t^2 = 4t + 2 \qquad \text{Multiply.}$$

$$t^2 - 4t - 2 = 0 \qquad \text{Subtract } 4t \text{ and 2 to write in standard form.}$$

The values for the quadratic formula are $a = 1$, $b = -4$, and $c = -2$.

$$t = \frac{-b \pm \sqrt{b^2 - 4ac}}{2a} \qquad \text{Quadratic formula}$$

$$t = \frac{-(-4) \pm \sqrt{(-4)^2 - 4(1)(-2)}}{2(1)} \qquad \begin{array}{l}\text{Substitute into the}\\ \text{quadratic formula.}\end{array}$$

$$t = \frac{4 \pm \sqrt{16 + 8}}{2} \qquad \text{Simplify.}$$

$$t = \frac{4 \pm \sqrt{24}}{2} \qquad \text{Add.}$$

$$t = \frac{4 \pm 2\sqrt{6}}{2} \qquad \sqrt{24} = \sqrt{4 \cdot 6} = 2\sqrt{6}$$

$$t = \frac{2\left(2 \pm \sqrt{6}\right)}{2} \qquad \text{Factor.}$$

Be careful here.

$$t = 2 \pm \sqrt{6} \qquad \begin{array}{l}\text{Divide out the common factor}\\ \text{to write in lowest terms.}\end{array}$$

NOW TRY ANSWER

5. $\left\{3 \pm \sqrt{5}\right\}$

A check confirms that the solution set is $\left\{2 \pm \sqrt{6}\right\}$.

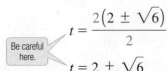 NOW TRY

9.3 Exercises

FOR EXTRA HELP MyMathLab®

▶ *Complete solution available in MyMathLab*

Concept Check The equations in Column I are in standard form $ax^2 + bx + c = 0$, or can be transformed easily. Match each equation with the correct choice in Column II. Do not actually solve the equation.

I	**II**
1. $3x^2 + 7x + 4 = 0$	**A.** $a = 4, b = 3, c = 7$
2. $7x^2 + 3x + 4 = 0$	**B.** $a = 7, b = 3, c = 4$
3. $7 + 3x + 4x^2 = 0$	**C.** $a = 4, b = 7, c = 3$
4. $7 + 4x + 3x^2 = 0$	**D.** $a = 7, b = 4, c = 3$
5. $7x^2 + 3 + 4x = 0$	**E.** $a = 3, b = 7, c = 4$
6. $4x^2 + 3 + 7x = 0$	**F.** $a = 3, b = 4, c = 7$

If necessary, write each equation in standard form $ax^2 + bx + c = 0$. Then identify the values of a, b, and c. Do not actually solve the equation. **See Example 1.**

▶ **7.** $3x^2 + 4x = 8$ **8.** $9x^2 + 2x = 3$ **9.** $-8x^2 = 2x + 3$

10. $-2x^2 = -3x + 8$ **11.** $3x^2 = 4x + 2$ **12.** $5x^2 = 3x - 6$

13. $3x^2 = -7x$ **14.** $9x^2 = 8x$ **15.** $(x - 3)(x + 4) = 0$

16. $(x + 7)(x - 2) = 0$ **17.** $9(x - 1)(x + 2) = 8$ **18.** $2(3x - 1)(2x + 5) = 5$

19. *Concept Check* A student writes the quadratic formula as

$$x = -b \pm \frac{\sqrt{b^2 - 4ac}}{2a}. \quad \text{Incorrect}$$

WHAT WENT WRONG? Explain the error, and give the correct formula.

20. *Concept Check* To solve the quadratic equation $-2x^2 - 4x + 3 = 0$, we might choose to use $a = -2, b = -4,$ and $c = 3$. Or, we might decide to first multiply both sides by -1, obtaining the equation $2x^2 + 4x - 3 = 0$, and then use $a = 2, b = 4,$ and $c = -3$. Show that in either case we obtain the same solution set.

Use the quadratic formula to solve each equation. Simplify all radicals, and write all answers in lowest terms. **See Examples 2–4.**

▶ **21.** $x^2 + 12x - 13 = 0$ **22.** $x^2 - 8x - 9 = 0$ ▶ **23.** $2x^2 + 12x = -5$

24. $5m^2 + m = 1$ ▶ **25.** $p^2 - 4p + 4 = 0$ **26.** $x^2 - 10x + 25 = 0$

27. $2x^2 = 5 + 3x$ **28.** $2x^2 = 30 + 7x$ **29.** $6x^2 + 6x = 0$

30. $4n^2 - 12n = 0$ **31.** $7x^2 = 12x$ **32.** $9r^2 = 11r$

33. $x^2 - 24 = 0$ **34.** $x^2 - 96 = 0$ **35.** $25x^2 - 4 = 0$

36. $16x^2 - 9 = 0$ **37.** $3x^2 - 2x + 5 = 10x + 1$ **38.** $4x^2 - x + 4 = x + 7$

39. $-2x^2 = -3x + 2$ **40.** $-2x^2 = -5x + 20$ **41.** $2x^2 + x + 5 = 0$

42. $3x^2 + 2x + 8 = 0$ **43.** $(x + 3)(x + 2) = 15$ **44.** $(2x + 1)(x + 1) = 7$

*Use the quadratic formula to solve each equation. Give (**a**) exact solutions and (**b**) solutions rounded to the nearest thousandth.*

45. $2x^2 = 5 - 2x$ **46.** $5x^2 = 3 - x$ **47.** $x^2 = 1 + x$ **48.** $x^2 = 2 + 4x$

Use the quadratic formula and a calculator to solve each equation. Give all solutions to the nearest tenth.

49. $0.1x^2 + 0.3x - 0.2 = 0$ **50.** $0.2x^2 + 0.1x - 0.4 = 0$

51. $-0.2x^2 - 0.3x + 0.1 = 0$ **52.** $-0.3x^2 - 0.5x + 0.3 = 0$

53. $5.1x^2 + 2.3x = 1.2$ **54.** $4.4x^2 + 3.3x = 2.2$

Use the quadratic formula to solve each equation. **See Example 5.**

55. $\dfrac{3}{2}x^2 - x - \dfrac{4}{3} = 0$ **56.** $\dfrac{2}{5}x^2 - \dfrac{3}{5}x - 1 = 0$ **57.** $\dfrac{1}{2}x^2 + \dfrac{1}{6}x = 1$

58. $\dfrac{2}{3}x^2 - \dfrac{4}{9}x = \dfrac{1}{3}$ ▶ **59.** $\dfrac{3}{8}x^2 - x + \dfrac{17}{24} = 0$ **60.** $\dfrac{1}{3}x^2 + \dfrac{8}{9}x + \dfrac{7}{9} = 0$

61. $0.5x^2 = x + 0.5$ **62.** $0.25x^2 = -1.5x - 1$ **63.** $0.6x - 0.4x^2 = -1$

64. $0.25x + 0.5x^2 = 1.5$ **65.** $0.25x^2 = 1.25 - 0.75x$ **66.** $0.75x^2 = 8.25 - 3.75x$

Solve each problem.

67. Solve the formula $S = 2\pi rh + \pi r^2$ for r by writing it in the form $ar^2 + br + c = 0$ and then using the quadratic formula. (Leave $\pm$ in the answer.)

68. Solve the formula $V = \pi r^2 h + \pi R^2 h$ for r, using the method described in **Exercise 67.** (Leave $\pm$ in the answer.)

69. A frog is sitting on a stump 3 ft above the ground. He hops off the stump and lands on the ground 4 ft away. During his leap, his height h in feet with respect to the ground is given by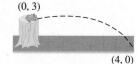

$$h = -0.5x^2 + 1.25x + 3,$$

where x is the distance in feet from the base of the stump. How far was the frog from the base of the stump when he was 1.25 ft above the ground?

70. An astronaut on the moon throws a baseball upward. The altitude (height) h of the ball, in feet, x seconds after he throws it, is given by the equation

$$h = -2.7x^2 + 30x + 6.5.$$

At what times to the nearest tenth is the ball 12 ft above the moon's surface?

71. A rule for estimating the number of board feet of lumber that can be cut from a log depends on the diameter of the log. To find the diameter d required to get 9 board feet of lumber, we use the equation

$$\left(\frac{d-4}{4}\right)^2 = 9.$$

Solve this equation for d. Are both answers reasonable?

72. A Babylonian problem asks for the length of the side of a square, where the area of the square minus the length of a side is 870. Find the length of the side. (*Source:* Eves, Howard, *An Introduction to the History of Mathematics,* Sixth Edition, Saunders College Publishing.)

SUMMARY EXERCISES · Applying Methods for Solving Quadratic Equations

The table summarizes four methods for solving a quadratic equation $ax^2 + bx + c = 0$.

Method	Advantages	Disadvantages
1. Zero-factor property	It is usually the fastest method.	Not all equations can be solved using this property. Some factorable polynomials are difficult to factor.
2. Square root property	It is the simplest method for solving equations of the form $(ax + b)^2 = $ a number.	Few equations are given in this form.
3. Completing the square	It can always be used. (Also, the procedure is useful in other areas of mathematics.)	It requires more steps than other methods.
4. Quadratic formula	It can always be used.	Sign errors are common due to the presence of the expression $\sqrt{b^2 - 4ac}$.

Solve each quadratic equation using the method of your choice.

1. $x^2 = 36$ **2.** $x^2 + 3x = -1$ **3.** $(x + 2)(x - 4) = 16$

4. $81t^2 = 49$ **5.** $x^2 - 4x + 3 = 0$ **6.** $w^2 + 3w + 2 = 0$

7. $x(x - 9) = -20$ **8.** $x^2 + 3x - 2 = 0$ **9.** $(3x - 2)^2 = 9$

10. $(2x - 1)^2 = 10$ **11.** $(x + 6)^2 = 121$ **12.** $(5x + 1)^2 = 36$

13. $(3r - 7)^2 = 24$ **14.** $(7p - 1)^2 = 32$ **15.** $(5x - 8)^2 = -6$

16. $2t^2 + 1 = t$ **17.** $-2x^2 = -3x - 2$ **18.** $-2x^2 + x = -1$

19. $8x^2 = 15 + 2x$ **20.** $3x^2 = 3 - 8x$ **21.** $0.1x^2 - 0.2x = 0.1$

22. $0.3x^2 + 0.5x = -0.1$ **23.** $5x^2 - 22x = -8$ **24.** $x(x + 6) + 4 = 0$

25. $(x + 2)(x + 1) = 10$ **26.** $16x^2 + 40x + 25 = 0$ **27.** $4x^2 = -1 + 5x$

28. $2p^2 = 2p + 1$ **29.** $3m(3m + 4) = 7$ **30.** $5x - 1 + 4x^2 = 0$

31. $\dfrac{r^2}{2} + \dfrac{7r}{4} + \dfrac{11}{8} = 0$ **32.** $\dfrac{1}{5}x^2 + x + 1 = 0$ **33.** $9x^2 = 16(3x + 4)$

34. $t(15t + 58) = -48$ **35.** $x^2 - x + 3 = 0$ **36.** $x^2 - \dfrac{100}{81} = 0$

37. $-3x^2 + 4x = -4$ **38.** $x^2 - \dfrac{5}{12}x = \dfrac{1}{6}$ **39.** $5x^2 + 19x = 2x + 12$

40. $\dfrac{1}{2}x^2 - x = \dfrac{15}{2}$ **41.** $x^2 - \dfrac{4}{15} = -\dfrac{4}{15}x$ **42.** $4m^2 - 11m + 8 = -2$

1. $\{\pm 6\}$ **2.** $\left\{\dfrac{-3 \pm \sqrt{5}}{2}\right\}$

3. $\{-4, 6\}$ **4.** $\left\{\pm\dfrac{7}{9}\right\}$

5. $\{1, 3\}$ **6.** $\{-2, -1\}$

7. $\{4, 5\}$ **8.** $\left\{\dfrac{-3 \pm \sqrt{17}}{2}\right\}$

9. $\left\{-\dfrac{1}{3}, \dfrac{5}{3}\right\}$ **10.** $\left\{\dfrac{1 \pm \sqrt{10}}{2}\right\}$

11. $\{-17, 5\}$ **12.** $\left\{-\dfrac{7}{5}, 1\right\}$

13. $\left\{\dfrac{7 \pm 2\sqrt{6}}{3}\right\}$

14. $\left\{\dfrac{1 \pm 4\sqrt{2}}{7}\right\}$

15. no real solution

16. no real solution

17. $\left\{-\dfrac{1}{2}, 2\right\}$ **18.** $\left\{-\dfrac{1}{2}, 1\right\}$

19. $\left\{-\dfrac{5}{4}, \dfrac{3}{2}\right\}$ **20.** $\left\{-3, \dfrac{1}{3}\right\}$

21. $\{1 \pm \sqrt{2}\}$ **22.** $\left\{\dfrac{-5 \pm \sqrt{13}}{6}\right\}$

23. $\left\{\dfrac{2}{5}, 4\right\}$ **24.** $\{-3 \pm \sqrt{5}\}$

25. $\left\{\dfrac{-3 \pm \sqrt{41}}{2}\right\}$

26. $\left\{-\dfrac{5}{4}\right\}$

27. $\left\{\dfrac{1}{4}, 1\right\}$ **28.** $\left\{\dfrac{1 \pm \sqrt{3}}{2}\right\}$

29. $\left\{\dfrac{-2 \pm \sqrt{11}}{3}\right\}$

30. $\left\{\dfrac{-5 \pm \sqrt{41}}{8}\right\}$

31. $\left\{\dfrac{-7 \pm \sqrt{5}}{4}\right\}$

32. $\left\{\dfrac{-5 \pm \sqrt{5}}{2}\right\}$

33. $\left\{\dfrac{8 \pm 8\sqrt{2}}{3}\right\}$

34. $\left\{-\dfrac{8}{3}, -\dfrac{6}{5}\right\}$

35. no real solution

36. $\left\{\pm\dfrac{10}{9}\right\}$

37. $\left\{-\dfrac{2}{3}, 2\right\}$ **38.** $\left\{-\dfrac{1}{4}, \dfrac{2}{3}\right\}$

39. $\left\{-4, \dfrac{3}{5}\right\}$ **40.** $\{-3, 5\}$

41. $\left\{-\dfrac{2}{3}, \dfrac{2}{5}\right\}$ **42.** no real solution

9.4 Graphing Quadratic Equations

OBJECTIVE

1 Graph quadratic equations of the form $y = ax^2 + bx + c$ $(a \neq 0)$.

VOCABULARY

☐ parabola
☐ vertex
☐ axis of symmetry (axis)

In **Section 5.4,** we graphed the quadratic equation $y = x^2$. By plotting points, we obtained the graph of a **parabola,** shown here in **FIGURE 1**.

Recall that the lowest point on this graph is the **vertex** of the parabola. (If the parabola opens downward, the vertex is the highest point.) The vertical line through the vertex is the **axis of symmetry,** or just **axis.** The two halves of the parabola are mirror images of each other across this axis.

x	y
3	9
2	4
1	1
0	0
−1	1
−2	4
−3	9

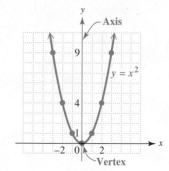

FIGURE 1

OBJECTIVE 1 Graph quadratic equations of the form $y = ax^2 + bx + c$ $(a \neq 0)$.

Every equation of the form

$$y = ax^2 + bx + c,$$

where $a \neq 0$, has a graph that is a parabola. The vertex is an important point to locate when graphing a quadratic equation.

EXAMPLE 1 Graphing a Parabola by Finding the Vertex and Intercepts

Graph $y = x^2 - 2x - 3$.

Because of its symmetry, if a parabola has two x-intercepts, the x-value of the vertex is exactly halfway between them. Therefore, we begin by finding the x-intercepts. We let $y = 0$ in the equation, and solve for x.

$$0 = x^2 - 2x - 3 \qquad \text{Let } y = 0.$$
$$0 = (x + 1)(x - 3) \qquad \text{Factor.}$$
$$x + 1 = 0 \quad \text{or} \quad x - 3 = 0 \qquad \text{Zero-factor property}$$
$$x = -1 \quad \text{or} \qquad x = 3 \qquad \text{Solve each equation.}$$

There are two x-intercepts, $(-1, 0)$ and $(3, 0)$. Because the x-value of the vertex is halfway between the x-values of the two x-intercepts, it is half their sum.

$$x = \frac{1}{2}(-1 + 3)$$

$$x = 1 \longleftarrow x\text{-value of the vertex}$$

We find the corresponding y-value by substituting 1 for x in the given equation.

$$y = x^2 - 2x - 3$$
$$y = 1^2 - 2(1) - 3 \qquad \text{Let } x = 1.$$
$$y = -4 \longleftarrow y\text{-value of the vertex}$$

The vertex is $(1, -4)$. The axis of symmetry is the vertical line $x = 1$.

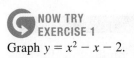
**NOW TRY
EXERCISE 1**
Graph $y = x^2 - x - 2$.

To find the y-intercept, we substitute 0 for x in the equation $y = x^2 - 2x - 3$.

$$y = 0^2 - 2(0) - 3 \quad \text{Let } x = 0.$$
$$y = -3 \quad \text{Simplify.}$$

The y-intercept is $(0, -3)$.

We plot the three intercepts and the vertex, and find additional ordered pairs as needed. For example, we let $x = 2$.

$$y = 2^2 - 2(2) - 3 \quad \text{Let } x = 2.$$
$$y = -3 \quad \text{Simplify.}$$

This leads to the ordered pair $(2, -3)$. A table that includes the ordered pairs we found is shown with the graph in **FIGURE 2**.

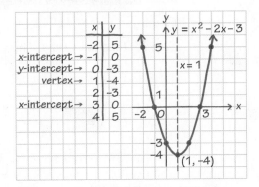

FIGURE 2

 NOW TRY

We can generalize from **Example 1.** The x-coordinates of the x-intercepts for the equation $y = ax^2 + bx + c$, by the quadratic formula, are

$$x = \frac{-b + \sqrt{b^2 - 4ac}}{2a} \quad \text{and} \quad x = \frac{-b - \sqrt{b^2 - 4ac}}{2a}.$$

Thus, the x-value of the vertex is half their sum.

$$x = \frac{1}{2}\left(\frac{-b + \sqrt{b^2 - 4ac}}{2a} + \frac{-b - \sqrt{b^2 - 4ac}}{2a}\right)$$

$$x = \frac{1}{2}\left(\frac{-b + \sqrt{b^2 - 4ac} - b - \sqrt{b^2 - 4ac}}{2a}\right)$$

$$x = \frac{1}{2}\left(\frac{-2b}{2a}\right) \quad \text{Combine like terms.}$$

$$x = -\frac{b}{2a} \quad \text{Multiply; lowest terms}$$

For the equation in **Example 1**, $y = x^2 - 2x - 3$, we have $a = 1$ and $b = -2$. Thus, the x-value of the vertex is

$$x = -\frac{b}{2a} = -\frac{-2}{2(1)} = 1,$$

which is the same value we found in **Example 1.** (It can be shown that the x-value of the vertex is always $x = -\frac{b}{2a}$, even if the graph has no x-intercepts.)

NOW TRY ANSWER
1.

Graphing a Quadratic Equation $y = ax^2 + bx + c$

Step 1   **Find the vertex.** Let $x = -\dfrac{b}{2a}$, and find the corresponding y-value by substituting for x in the equation.

Step 2   **Find the y-intercept.** Let $x = 0$ and solve for y.

Step 3   **Find the x-intercepts** (if they exist). Let $y = 0$ and solve for x.

Step 4   **Plot** the intercepts and the vertex.

Step 5   **Find and plot additional ordered pairs** near the vertex and intercepts as needed, using symmetry about the axis of the parabola.

EXAMPLE 2   Graphing a Parabola

Graph $y = -x^2 + 4x - 1$.

Step 1   Find the vertex. The x-value of the vertex is

$$x = -\frac{b}{2a} = -\frac{4}{2(-1)} = 2. \qquad a = -1, b = 4$$

Let $x = 2$ in $y = -x^2 + 4x - 1$ to obtain the y-value of the vertex.

$$y = -2^2 + 4(2) - 1 \qquad \text{Let } x = 2.$$
$$y = 3 \qquad \text{Simplify.}$$

The vertex is $(2, 3)$. The axis is the line $x = 2$.

Step 2   Now find the y-intercept. Let $x = 0$ in $y = -x^2 + 4x - 1$.

$$y = -0^2 + 4(0) - 1 \qquad \text{Let } x = 0.$$
$$y = -1 \qquad \text{Simplify.}$$

The y-intercept is $(0, -1)$.

Step 3   Let $y = 0$ to determine the x-intercepts. The equation is $0 = -x^2 + 4x - 1$, which cannot be solved using the zero-factor property, so we use the quadratic formula.

$$x = \frac{-4 \pm \sqrt{4^2 - 4(-1)(-1)}}{2(-1)} \qquad \begin{array}{l} \text{Let } a = -1, b = 4, c = -1 \\ \text{in the quadratic formula.} \end{array}$$

$$x = \frac{-4 \pm \sqrt{12}}{-2} \qquad \text{Simplify.}$$

$$x = \frac{-4 \pm 2\sqrt{3}}{-2} \qquad \sqrt{12} = \sqrt{4} \cdot \sqrt{3} = 2\sqrt{3}$$

Factor first. Then divide out the common factor -2. →
$$x = \frac{-2\left(2 \mp \sqrt{3}\right)}{-2} \qquad \text{Factor.}$$

$$x = 2 \pm \sqrt{3} \qquad \begin{array}{l} \text{Divide out } -2 \text{ to write} \\ \text{in lowest terms.} \end{array}$$

Using a calculator, we find that the x-intercepts are $(3.7, 0)$ and $(0.3, 0)$ to the nearest tenth.

Graph $y = -x^2 + 4x + 2$.

Steps 4 Plot the vertex, intercepts, and the additional points shown in the table.
and 5 Connect these points with a smooth curve. See **FIGURE 3**.

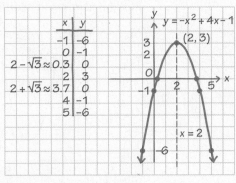

x	y
-1	-6
0	-1
$2 - \sqrt{3} \approx 0.3$	0
2	3
$2 + \sqrt{3} \approx 3.7$	0
4	-1
5	-6

FIGURE 3

NOW TRY

Notice in **Example 1** that the graph of $y = x^2 - 2x - 3$ opens *upward,* while in **Example 2** the graph of $y = -x^2 + 4x - 1$ opens *downward*. The sign of a in $ax^2 + bx + c$ determines the direction in which the graph opens. ***The graph of***

$$y = ax^2 + bx + c$$

opens upward if $a > 0$ or downward if $a < 0$.

EXAMPLE 3 Graphing a Quadratic Equation

Graph each equation. Give the vertex and intercepts.

(a) $y = x^2 - 3$

Step 1 For this equation, $a = 1$ and $b = 0$. The vertex has x-value $-\frac{b}{2a} = -\frac{0}{2(1)} = 0$ and y-value $0^2 - 3 = -3$. The vertex is $(0, -3)$.

Step 2 To find the y-intercept, let $x = 0$ in $y = x^2 - 3$ and find y. The y-intercept is the same as the vertex, $(0, -3)$.

Step 3 To find the x-intercepts, let $y = 0$ in $y = x^2 - 3$ and solve for x.

$$0 = x^2 - 3 \qquad \text{Let } y = 0.$$
$$3 = x^2 \qquad \text{Add 3.}$$
$$x = \pm\sqrt{3} \qquad \text{Square root property}$$

The x-intercepts are $(-1.7, 0)$ and $(1.7, 0)$ to the nearest tenth.

Steps 4 Plot the vertex, intercepts, and additional points. See **FIGURE 4**.
and 5

2.
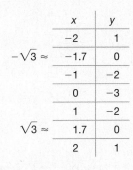

	x	y
	-2	1
$-\sqrt{3} \approx$	-1.7	0
	-1	-2
	0	-3
	1	-2
$\sqrt{3} \approx$	1.7	0
	2	1

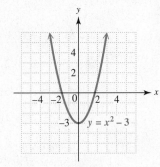

FIGURE 4

NOW TRY
EXERCISE 3
Graph $y = x^2 - 4$. Give the vertex and intercepts.

(b) $y = x^2 + 2$

Follow the same steps as in part (a) to find that the vertex and the y-intercept are both $(0, 2)$. There are no x-intercepts because $x^2 + 2 = 0$ has no real solutions. Plot the vertex and additional points shown in the table. See **FIGURE 5**.

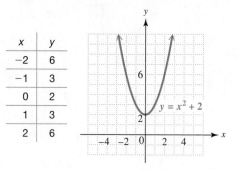

x	y
-2	6
-1	3
0	2
1	3
2	6

x	y
-1	9
0	4
1	1
2	0
3	1
4	4
5	9

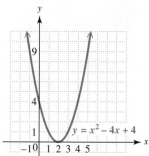

FIGURE 5

FIGURE 6

(c) $y = x^2 - 4x + 4$

For this equation, $a = 1$, $b = -4$, and $c = 4$. The vertex of the graph has the following coordinates.

$$x = \frac{-b}{2a} = \frac{-(-4)}{2(1)} = 2$$

$$y = 2^2 - 4(2) + 4 = 0$$

The graph opens upward because $a = 1$ and $1 > 0$. It has vertex $(2, 0)$. This is the only x-intercept because the equation $x^2 - 4x + 4 = 0$ has a double solution, -2. Plot the vertex, the y-intercept $(0, 4)$, and additional points. See **FIGURE 6**.

NOW TRY

NOW TRY ANSWER

3. $(0, -4)$; y-intercept $(0, -4)$;
x-intercepts $(-2, 0)$
and $(2, 0)$

9.4 Exercises

FOR EXTRA HELP

▶ MyMathLab®

● *Complete solution available in MyMathLab*

Concept Check *Answer each question.*

1. In what direction does the graph of $y = ax^2 + bx + c$ open if $a > 0$?

2. If the vertex of the graph of $y = ax^2 + bx + c$ is below the x-axis and $a > 0$, how many x-intercepts does the graph have?

3. In what direction does the graph of $y = ax^2 + bx + c$ open if $a < 0$?

4. If the vertex of the graph of $y = ax^2 + bx + c$ is above the x-axis and $a > 0$, how many x-intercepts does the graph have?

Identify the vertex and sketch the graph of each equation. ***See Examples 1 and 2.***

5. $y = x^2 - 6$ **6.** $y = x^2 - 5$ **7.** $y = -x^2 + 2$

8. $y = -x^2 + 4$ **9.** $y = (x + 3)^2$ **10.** $y = (x - 4)^2$

● **11.** $y = x^2 + 2x + 3$ **12.** $y = x^2 - 4x + 5$ ● **13.** $y = x^2 - 8x + 16$

14. $y = x^2 + 6x + 9$ **15.** $y = -x^2 + 6x - 5$ **16.** $y = -x^2 - 4x - 3$

17. $y = x^2 + 4x$ **18.** $y = x^2 - 2x$

Graph each equation. Give the vertex, the y-intercept, and the x-intercepts, if any (rounded to the nearest tenth, if necessary). ***See Example 3.***

19. $y = x^2 + 1$ **20.** $y = x^2 + 3$ **21.** $y = -x^2 + 2$

22. $y = -x^2 + 5$ **23.** $y = x^2 + 4x + 4$ **24.** $y = x^2 + 8x + 16$

25. $y = x^2 - 4x + 3$ **26.** $y = x^2 - 6x + 5$

Extending Skills *Solve each problem.*

27. Find two numbers whose sum is 80 and whose product is a maximum. (*Hint:* Let x represent one of the numbers. Then $80 - x$ represents the other.)

28. Find two numbers whose sum is 300 and whose product is a maximum.

▶ **29.** The U.S. Naval Research Laboratory designed a giant radio telescope that had a diameter of 300 ft and a maximum depth of 44 ft. The graph here depicts a cross section of that telescope. Find the equation of this parabola. (*Source:* Mar, J., and H. Liebowitz, *Structure Technology for Large Radio and Radar Telescope Systems,* The MIT Press.)

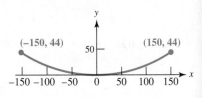

30. Suppose the telescope in **Exercise 29** had a diameter of 400 ft and a maximum depth of 50 ft. Find the equation of this parabola.

| Chapter 9 | Summary |

Key Terms

9.1

quadratic equation

9.3

quadratic formula
discriminant
double solution

9.4

parabola
vertex
axis of symmetry (axis)

New Symbols

$\pm$ positive or negative
       (plus or minus)

Test Your Word Power

See how well you have learned the vocabulary in this chapter.

1. A **quadratic equation** is an equation that can be written in the form
  A. $Ax + By = C$
  B. $ax^2 + bx + c = 0$
  C. $Ax + B = 0$
  D. $y = mx + b$.

2. A **parabola** is the graph of
  A. any equation in two variables
  B. a linear equation
  C. an equation of degree three
  D. a quadratic equation in two variables.

3. The **vertex** of a parabola is
  A. the point where the graph intersects the y-axis
  B. the point where the graph intersects the x-axis
  C. the lowest point on a parabola that opens up or the highest point on a parabola that opens down
  D. the origin.

4. The **axis of symmetry** of a vertical parabola is
  A. either the x-axis or the y-axis
  B. the vertical line through the vertex
  C. the horizontal line through the vertex
  D. the x-axis.

ANSWERS

1. B; *Examples:* $x^2 + 6x + 9 = 0$, $y^2 - 2y = 8$, $(x + 3)(x - 1) = 5$ **2.** D; *Examples:* See **FIGURES 1–6** in **Section 9.4.**
3. C; *Example:* The graph of $y = (x + 3)^2$ has vertex $(-3, 0)$, which is the lowest point on the graph.
4. B; *Example:* The axis of symmetry of the graph of $y = (x + 3)^2$ is the line $x = -3$.

Quick Review

CONCEPTS	EXAMPLES

9.1 **Solving Quadratic Equations by the Square Root Property**

Square Root Property

If k is a positive number and if $x^2 = k$, then

$$x = \sqrt{k} \quad \text{or} \quad x = -\sqrt{k}.$$

The solution set $\{-\sqrt{k}, \sqrt{k}\}$ can be written $\{\pm\sqrt{k}\}$.

Solve.

$$(2x + 1)^2 = 5$$

$$2x + 1 = \sqrt{5} \qquad \text{or} \quad 2x + 1 = -\sqrt{5}$$

$$2x = -1 + \sqrt{5} \quad \text{or} \qquad 2x = -1 - \sqrt{5}$$

$$x = \frac{-1 + \sqrt{5}}{2} \quad \text{or} \qquad x = \frac{-1 - \sqrt{5}}{2}$$

The solution set is $\left\{\dfrac{-1 \pm \sqrt{5}}{2}\right\}$.

9.2 **Solving Quadratic Equations by Completing the Square**

Solving a Quadratic Equation by Completing the Square

Step 1 If the coefficient of the second-degree term is 1, go to Step 2. If it is not 1, divide each side of the equation by this coefficient.

Step 2 Make sure that all variable terms are on one side of the equation and the constant term is on the other.

Step 3 Take half the coefficient of x, square it, and add the square to each side of the equation. Factor the variable side and combine terms on the other side.

Step 4 Use the square root property to solve the equation.

Solve.

$$2x^2 + 4x - 1 = 0$$

$$x^2 + 2x - \frac{1}{2} = 0 \qquad \text{Divide by 2.}$$

$$x^2 + 2x = \frac{1}{2} \qquad \text{Add } \frac{1}{2}.$$

$$x^2 + 2x + 1 = \frac{1}{2} + 1 \qquad \text{Add } \left[\frac{1}{2}(2)\right]^2 = 1^2 = 1.$$

$$(x + 1)^2 = \frac{3}{2} \qquad \text{Factor. Add.}$$

$$x + 1 = \sqrt{\frac{3}{2}} \qquad \text{or} \quad x + 1 = -\sqrt{\frac{3}{2}}$$

$$x + 1 = \frac{\sqrt{3} \cdot \sqrt{2}}{\sqrt{2} \cdot \sqrt{2}} \quad \text{or} \quad x + 1 = -\frac{\sqrt{3} \cdot \sqrt{2}}{\sqrt{2} \cdot \sqrt{2}}$$

$$x + 1 = \frac{\sqrt{6}}{2} \qquad \text{or} \quad x + 1 = -\frac{\sqrt{6}}{2}$$

$$x = -1 + \frac{\sqrt{6}}{2} \quad \text{or} \qquad x = -1 - \frac{\sqrt{6}}{2}$$

$$x = \frac{-2 + \sqrt{6}}{2} \quad \text{or} \qquad x = \frac{-2 - \sqrt{6}}{2} \qquad {\scriptstyle -1 = \frac{-2}{2}}$$

The solution set is $\left\{\dfrac{-2 \pm \sqrt{6}}{2}\right\}$.

CONCEPTS

EXAMPLES

9.3 Solving Quadratic Equations by the Quadratic Formula

Quadratic Formula

The solutions of $ax^2 + bx + c = 0$, where $a \neq 0$, are

$$x = \frac{-b \pm \sqrt{b^2 - 4ac}}{2a}.$$

The discriminant of this quadratic equation is

$$b^2 - 4ac.$$

Solve $3x^2 - 4x - 2 = 0$.

$$x = \frac{-(-4) \pm \sqrt{(-4)^2 - 4(3)(-2)}}{2(3)}$$ 　Let $a = 3$, $b = -4$, $c = -2$.

$$x = \frac{4 \pm \sqrt{40}}{6}$$ 　Simplify.

$$x = \frac{4 \pm 2\sqrt{10}}{6}$$ 　$\sqrt{40} = \sqrt{4 \cdot 10}$ $= 2\sqrt{10}$

$$x = \frac{2(2 \pm \sqrt{10})}{2 \cdot 3}$$ 　Factor out 2.

$$x = \frac{2 \pm \sqrt{10}}{3}$$ 　Divide out 2 to write in lowest terms.

The solution set is $\left\{ \dfrac{2 \pm \sqrt{10}}{3} \right\}$.

9.4 Graphing Quadratic Equations

Graphing $y = ax^2 + bx + c$

Step 1 　Find the vertex. Let $x = -\frac{b}{2a}$ and find y by substituting this value for x in the equation.

Graph $y = 2x^2 - 5x - 3$.

$$x = -\frac{b}{2a} = -\frac{-5}{2(2)} = \frac{5}{4}$$ 　$a = 2$, $b = -5$

$$y = 2\left(\frac{5}{4}\right)^2 - 5\left(\frac{5}{4}\right) - 3$$ 　Let $x = \frac{5}{4}$.

$$y = 2\left(\frac{25}{16}\right) - \frac{25}{4} - 3$$ 　Apply the exponent. Multiply.

$$y = \frac{25}{8} - \frac{50}{8} - \frac{24}{8}$$ 　Multiply. Find a common denominator.

$$y = -\frac{49}{8}$$ 　Subtract.

The vertex is $\left(\frac{5}{4}, -\frac{49}{8}\right)$.

Step 2 　Let $x = 0$ to find the y-intercept.

$$y = 2(0)^2 - 5(0) - 3$$ 　Let $x = 0$.

$$y = -3$$ 　Simplify.

The y-intercept is $(0, -3)$.

Step 3 　Let $y = 0$ and solve to find the x-intercepts (if they exist).

$$0 = 2x^2 - 5x - 3$$ 　Let $y = 0$.

$$0 = (2x + 1)(x - 3)$$ 　Factor.

$$2x + 1 = 0 \quad \text{or} \quad x - 3 = 0$$ 　Zero-factor property

$$x = -\frac{1}{2} \quad \text{or} \quad x = 3$$ 　Solve each equation.

The x-intercepts are $\left(-\frac{1}{2}, 0\right)$ and $(3, 0)$.

CONCEPTS	EXAMPLES

Step 4 Plot the intercepts and the vertex.

Step 5 Find and plot additional points near the vertex and intercepts as needed.

x	y
$-\dfrac{1}{2}$	0
0	-3
$\dfrac{5}{4}$	$-\dfrac{49}{8}$
2	-5
3	0

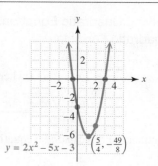

$y = 2x^2 - 5x - 3$ $\left(\dfrac{5}{4}, -\dfrac{49}{8}\right)$

Chapter 9   Review Exercises

1. $\{\pm 12\}$ **2.** $\{\pm\sqrt{37}\}$

3. $\{\pm 8\sqrt{2}\}$ **4.** $\{-7, 3\}$

5. $\{3 \pm \sqrt{10}\}$ **6.** $\left\{\dfrac{-1 \pm \sqrt{14}}{2}\right\}$

7. no real solution

8. $\left\{\dfrac{3 \pm 2\sqrt{2}}{5}\right\}$

9. $\{-5, -1\}$ **10.** $\{-2 \pm \sqrt{11}\}$

11. $\{-1 \pm \sqrt{6}\}$

12. $\left\{\dfrac{-4 \pm \sqrt{22}}{2}\right\}$

13. $\left\{-\dfrac{2}{5}, 1\right\}$

14. no real solution

15. 2.5 sec   **16.** 6, 8, 10

17. (a) $\{\pm 3\}$ (b) $\{\pm 3\}$ (c) $\{\pm 3\}$
(d) Because there is only one solution set, we will always get the same results, no matter which method of solution is used.

18. There are no real solutions.

19. $\{1 \pm \sqrt{5}\}$

20. no real solution

21. $\left\{\dfrac{2 \pm \sqrt{10}}{2}\right\}$

22. $\left\{\dfrac{-1 \pm \sqrt{29}}{4}\right\}$

23. $\left\{\dfrac{-3 \pm \sqrt{41}}{2}\right\}$

24. $\left\{-\dfrac{2}{3}, 1\right\}$

9.1 *Solve each equation using the square root property. Simplify all radicals.*

1. $x^2 = 144$ **2.** $x^2 = 37$ **3.** $m^2 = 128$ **4.** $(x + 2)^2 = 25$

5. $(r - 3)^2 = 10$ **6.** $(2p + 1)^2 = 14$ **7.** $(3x + 2)^2 = -3$ **8.** $(3 - 5x)^2 = 8$

9.2 *Solve each equation by completing the square.*

9. $m^2 + 6m + 5 = 0$ **10.** $p^2 + 4p = 7$

11. $-x^2 + 5 = 2x$ **12.** $2x^2 - 3 = -8x$

13. $5x^2 - 3x - 2 = 0$ **14.** $(4x + 1)(x - 1) = -7$

Solve each problem.

15. If an object is projected upward on Earth from a height of 50 ft, with an initial velocity of 32 ft per sec, then its altitude (height) after t seconds is given by

$$h = -16t^2 + 32t + 50,$$

where h is in feet. At what times will the object be at a height of 30 ft?

16. Find the lengths of the three sides of the right triangle shown.

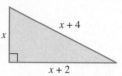

x $x + 4$ $x + 2$

9.3 *Work each problem.*

17. Consider the equation $x^2 - 9 = 0$.

(a) Solve the equation using the zero-factor property.

(b) Solve the equation using the square root property.

(c) Solve the equation using the quadratic formula.

(d) Compare the answers. If a quadratic equation can be solved using the zero-factor property and the quadratic formula, should we always get the same results? Explain.

18. How many real solutions are there for a quadratic equation that has a negative number as its radicand in the quadratic formula?

25. $(0, 5)$ **26.** $(-4, 0)$

27. $(1, 4)$ **28.** $(-2, -2)$

Solve each equation using the quadratic formula.

19. $x^2 - 2x - 4 = 0$ **20.** $3k^2 + 2k = -3$ **21.** $2p^2 + 8 = 4p + 11$

22. $-4x^2 + 7 = 2x$ **23.** $\frac{1}{4}p^2 = 2 - \frac{3}{4}p$ **24.** $3x^2 - x - 2 = 0$

9.4 *Identify the vertex and sketch the graph of each equation.*

25. $y = -x^2 + 5$ **26.** $y = (x + 4)^2$

27. $y = -x^2 + 2x + 3$ **28.** $y = x^2 + 4x + 2$

Chapter 9   Mixed Review Exercises

1. $\left\{-\dfrac{11}{2}, 5\right\}$ **2.** $\left\{-\dfrac{11}{2}, \dfrac{9}{2}\right\}$

3. $\left\{\dfrac{-1 \pm \sqrt{21}}{2}\right\}$

4. $\left\{-\dfrac{3}{2}, \dfrac{1}{3}\right\}$

5. $\left\{\dfrac{-5 \pm \sqrt{17}}{2}\right\}$

6. $\{-1 \pm \sqrt{3}\}$ **7.** no real solution

8. $\left\{\dfrac{9 \pm \sqrt{41}}{2}\right\}$ **9.** $\left\{-\dfrac{5}{3}\right\}$

10. $\{-1 \pm 2\sqrt{2}\}$

11. $\{-2 \pm \sqrt{5}\}$ **12.** $\{\pm 2\sqrt{2}\}$

13. 0.7 sec and 10.4 sec

14.

Solve using the method of your choice.

1. $(2t - 1)(t + 1) = 54$ **2.** $(2p + 1)^2 = 100$

3. $(x + 2)(x - 1) = 3$ **4.** $6t^2 + 7t - 3 = 0$

5. $2x^2 + 3x + 2 = x^2 - 2x$ **6.** $x^2 + 2x + 5 = 7$

7. $m^2 - 4m + 10 = 0$ **8.** $k^2 - 9k + 10 = 0$

9. $(3x + 5)^2 = 0$ **10.** $0.5r^2 = 3.5 - r$

11. $x^2 + 4x = 1$ **12.** $7x^2 - 8 = 5x^2 + 8$

13. If a toy rocket is launched from ground level on the moon, the altitude (height) h in feet x seconds after it is launched is given by

$$h = -2.7x^2 + 30x.$$

After how many seconds to the nearest tenth is the rocket 20 ft above ground level?

14. Graph $y = x^2 - 6x + 8$.

Chapter 9   Test   FOR EXTRA HELP *Step-by-step test solutions are found on the Chapter Test Prep Videos available in* MyMathLab® *or on* YouTube.

▶ *View the complete solutions to all Chapter Test exercises in MyMathLab.*

[9.1]
1. $\{\pm \sqrt{39}\}$ **2.** $\{-11, 5\}$

3. $\left\{\dfrac{-3 \pm 2\sqrt{6}}{4}\right\}$

[9.2]
4. $\{2 \pm \sqrt{10}\}$ **5.** $\left\{\dfrac{-6 \pm \sqrt{42}}{2}\right\}$

Solve using the square root property.

1. $x^2 = 39$ **2.** $(x + 3)^2 = 64$ **3.** $(4x + 3)^2 = 24$

Solve by completing the square.

4. $x^2 - 4x = 6$ **5.** $2x^2 + 12x - 3 = 0$

Solve using the quadratic formula.

6. $5x^2 + 2x = 0$ **7.** $2x^2 + 5x - 3 = 0$ **8.** $3w^2 + 2 = 6w$

9. $4x^2 + 8x + 11 = 0$ **10.** $t^2 - \dfrac{5}{3}t + \dfrac{1}{3} = 0$

[9.3]

6. $\left\{0, -\dfrac{2}{5}\right\}$　　**7.** $\left\{-3, \dfrac{1}{2}\right\}$

8. $\left\{\dfrac{3 \pm \sqrt{3}}{3}\right\}$　**9.** no real solution

10. $\left\{\dfrac{5 \pm \sqrt{13}}{6}\right\}$

[9.1–9.3]

11. $\{1 \pm \sqrt{2}\}$　**12.** $\left\{\dfrac{-1 \pm 3\sqrt{2}}{2}\right\}$

13. $\left\{\dfrac{11 \pm \sqrt{89}}{4}\right\}$

14. $\{5\}$
15. 2 sec　　**16.** 12, 16, 20

[9.4]
17. $(0, 6)$　　**18.** $(-3, -2)$

Solve using the method of your choice.

11. $p^2 - 2p - 1 = 0$　　**12.** $(2x + 1)^2 = 18$

13. $(x - 5)(2x - 1) = 1$　　**14.** $t^2 + 25 = 10t$

Solve each problem.

15. If an object is projected vertically into the air from ground level on Earth with an initial velocity of 64 ft per sec, its altitude (height) s in feet after t seconds is given by the formula

$$s = -16t^2 + 64t.$$

At what time will the object be at a height of 64 ft?

16. Find the lengths of the three sides of the right triangle.

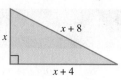

Identify the vertex and sketch the graph of each equation.

17. $y = -x^2 + 6$　　　　**18.** $y = x^2 + 6x + 7$

Chapters R–9　　Cumulative Review Exercises

[1.5]
1. 15　　　　**2.** 5

[1.7]
3. $-r + 7$　　**4.** $19m - 17$

[2.1–2.3]
5. $\{18\}$　　**6.** $\{5\}$

7. $\left\{\dfrac{8}{3}\right\}$　　**8.** $\{2\}$

[2.5]
9. 100°, 80°
10. width: 50 ft; length: 94 ft
11. $L = \dfrac{P - 2W}{2}$, or $L = \dfrac{P}{2} - W$

[2.8]
12. $(-2, \infty)$

13. $(-\infty, 4]$

[3.2]

14.　　　**15.**

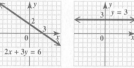

Perform each indicated operation.

1. $\dfrac{-4 \cdot 3^2 + 2 \cdot 3}{2 - 4 \cdot 1}$　　**2.** $-9 - (-8)(2) + 6 - (6 + 2)$

3. $-4r + 14 + 3r - 7$　　**4.** $5(4m - 2) - (m + 7)$

Solve each equation.

5. $x - 5 = 13$　　**6.** $3k - 9k - 8k + 6 = -64$

7. $\dfrac{3}{5}t - \dfrac{1}{10} = \dfrac{3}{2}$　　**8.** $2(m - 1) - 6(3 - m) = -4$

Solve each problem.

9. Find the measures of the marked angles.

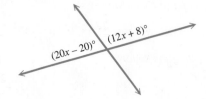

10. The perimeter of a basketball court is 288 ft. The width of the court is 44 ft less than the length. What are the dimensions of the court?

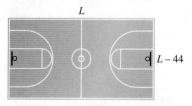

11. Solve the formula $P = 2L + 2W$ for L.

[3.3]

16. $-\dfrac{1}{3}$

[3.4, 3.5]

17. $2x - y = -3$

[4.1–4.3]

18. $\{(-3, 2)\}$ 19. $\varnothing$

[4.4]

20. Motorola Theory: $79.99;
Motorola i412: $69.99

[4.5]

21.

[5.1, 5.2]

22. $\dfrac{x^4}{3^2}$, or $\dfrac{x^4}{9}$ 23. $\dfrac{b^{16}}{c^2}$

[5.4]

24. $8x^5 - 17x^4 - x^2$

[5.5]

25. $2x^4 + x^3 - 19x^2 + 2x + 20$

[5.7]

26. $3x^2 - 2x + 1$

27. $2x + 2 + \dfrac{7}{x - 3}$

[5.3]

28. (a) 6.35×10^9 (b) 0.00023

[6.1]

29. $16x^2(x - 3y)$

[6.3]

30. $(2a + 1)(a - 3)$

[6.4]

31. $(4x^2 + 1)(2x + 1)(2x - 1)$

32. $(5m - 2)^2$

[6.5]   [6.6]

33. $\{-9, 6\}$ 34. 50 m

[7.2]

35. $\dfrac{4}{5}$

[7.4]

36. $\dfrac{-k - 1}{k(k - 1)}$

37. $\dfrac{5a + 2}{(a - 2)^2(a + 2)}$

[7.5]

38. $\dfrac{b + a}{b - a}$

Solve each inequality, and graph the solution set.

12. $-8m < 16$

13. $-9p + 2(8 - p) - 6 \geq 4p - 50$

Graph each equation.

14. $2x + 3y = 6$

15. $y = 3$

16. Find the slope of the line passing through $(-1, 4)$ and $(5, 2)$.

17. Write an equation of a line with slope 2 and y-intercept $(0, 3)$. Give the equation in the form $Ax + By = C$.

Solve each system of equations.

18. $2x + \ \ y = -4$
     $-3x + 2y = 13$

19. $3x - \ \ 5y = 8$
     $-6x + 10y = 16$

20. Three Motorola Theory cell phones and two Motorola i412 cell phones cost $379.95. Two Theory cell phones and three i412 phones cost $369.95. Find the cost for a single phone of each model. (*Source:* www.boostmobile.com)

21. Graph the solution set of the system of inequalities.

$$2x + y \leq 4$$
$$x - y > 2$$

Simplify each expression. Write answers with positive exponents.

22. $(3^2 \cdot x^{-4})^{-1}$

23. $\left(\dfrac{b^{-3}c^4}{b^5c^3}\right)^{-2}$

Perform each indicated operation.

24. $(5x^5 - 9x^4 + 8x^2) - (9x^2 + 8x^4 - 3x^5)$ 25. $(2x - 5)(x^3 + 3x^2 - 2x - 4)$

26. $\dfrac{3x^3 + 10x^2 - 7x + 4}{x + 4}$

27. $x - 3\overline{)2x^2 - 4x + 1}$

28. Answer each question.

(a) The number of possible hands in contract bridge is about 6,350,000,000. What is this number in scientific notation?

(b) The body of a 150-lb person contains about 2.3×10^{-4} lb of copper. What is this number in standard notation?

Factor.

29. $16x^3 - 48x^2y$

30. $2a^2 - 5a - 3$

31. $16x^4 - 1$

32. $25m^2 - 20m + 4$

Solve.

33. $x^2 + 3x - 54 = 0$

34. The length of a rectangle is 2.5 times its width. The area is 1000 m². Find the length.

Simplify each expression.

35. $\dfrac{2}{a - 3} \div \dfrac{5}{2a - 6}$

36. $\dfrac{1}{k} - \dfrac{2}{k - 1}$

37. $\dfrac{2}{a^2 - 4} + \dfrac{3}{a^2 - 4a + 4}$

38. $\dfrac{\dfrac{1}{a} + \dfrac{1}{b}}{\dfrac{1}{a} - \dfrac{1}{b}}$

Solve each equation.

39. $\dfrac{1}{x+3} + \dfrac{1}{x} = \dfrac{7}{10}$ **40.** $\dfrac{2}{t^2-1} - \dfrac{1}{t-1} = \dfrac{1}{2}$

Simplify each expression.

41. $\sqrt{100}$ **42.** $\dfrac{6\sqrt{6}}{\sqrt{5}}$

43. $\sqrt[3]{\dfrac{7}{16}}$ **44.** $3\sqrt{5} - 2\sqrt{20} + \sqrt{125}$

45. Solve $\sqrt{x+2} = x - 4$.

Solve each quadratic equation, using the method indicated.

46. $7 - x^2 = 0$ (square root property)   **47.** $(3x+2)^2 = 12$ (square root property)

48. $-x^2 + 5 = 2x$ (completing the square)   **49.** $2x(x-2) - 3 = 0$ (quadratic formula)

50. Give the coordinates of the vertex and the intercepts, and graph the equation $y = x^2 - 4x$.

ANSWERS TO SELECTED EXERCISES

In this section we provide the answers that we think most students will obtain when they work the exercises using the methods explained in the text. If your answer does not look exactly like the one given here, it is not necessarily wrong. In many cases, there are equivalent forms of the answer that are correct. For example, if the answer section shows $\frac{3}{4}$ and your answer is 0.75, you have obtained the right answer, but written it in a different (yet equivalent) form. Unless the directions specify otherwise, 0.75 is just as valid an answer as $\frac{3}{4}$.

In general, if your answer does not agree with the one given in the text, see whether it can be transformed into the other form. If it can, then it is the correct answer. If you still have doubts, talk with your instructor.

R PREALGEBRA REVIEW

Section R.1 (pages 12–16)

1. true **3.** false; This is an improper fraction. Its value is 1.
5. false; The fraction $\frac{13}{39}$ is written in lowest terms as $\frac{1}{3}$.
7. false; *Product* refers to multiplication, so the product of 10 and 2 is 20. **9.** C **11.** A **13.** prime **15.** composite; $2 \cdot 3 \cdot 5$
17. composite; $2 \cdot 2 \cdot 2 \cdot 2 \cdot 2 \cdot 2$ **19.** neither **21.** composite; $3 \cdot 19$
23. prime **25.** composite; $2 \cdot 2 \cdot 31$ **27.** composite; $2 \cdot 2 \cdot 5 \cdot 5 \cdot 5$
29. composite; $2 \cdot 7 \cdot 13 \cdot 19$ **31.** $\frac{1}{2}$ **33.** $\frac{5}{6}$ **35.** $\frac{16}{25}$ **37.** $\frac{1}{5}$ **39.** $\frac{6}{5}$
41. $1\frac{5}{7}$ **43.** $6\frac{5}{12}$ **45.** $7\frac{6}{11}$ **47.** $\frac{13}{5}$ **49.** $\frac{83}{8}$ **51.** $\frac{51}{5}$ **53.** $\frac{24}{35}$
55. $\frac{5}{8}$ **57.** $\frac{6}{25}$ **59.** $\frac{6}{5}$, or $1\frac{1}{5}$ **61.** 9 **63.** $\frac{65}{12}$, or $5\frac{5}{12}$ **65.** $\frac{38}{5}$, or $7\frac{3}{5}$
67. $\frac{10}{3}$, or $3\frac{1}{3}$ **69.** 12 **71.** $\frac{1}{16}$ **73.** 10 **75.** 18 **77.** $\frac{35}{24}$, or $1\frac{11}{24}$
79. $\frac{84}{47}$, or $1\frac{37}{47}$ **81.** $\frac{11}{15}$ **83.** $\frac{2}{3}$ **85.** $\frac{8}{9}$ **87.** $\frac{29}{24}$, or $1\frac{5}{24}$ **89.** $\frac{43}{8}$, or $5\frac{3}{8}$
91. $\frac{101}{20}$, or $5\frac{1}{20}$ **93.** $\frac{5}{9}$ **95.** $\frac{2}{3}$ **97.** $\frac{1}{4}$ **99.** $\frac{17}{36}$ **101.** $\frac{67}{20}$, or $3\frac{7}{20}$
103. $\frac{11}{12}$ **105.** (a) $\frac{1}{2}$ (b) $\frac{1}{4}$ (c) $\frac{1}{3}$ (d) $\frac{1}{6}$ **107.** 6 cups **109.** $1\frac{1}{8}$ in.
111. $\frac{9}{16}$ in. **113.** $618\frac{3}{4}$ ft **115.** $5\frac{5}{24}$ in. **117.** 8 cakes (There will be some sugar left over.) **119.** $16\frac{5}{8}$ yd **121.** $3\frac{3}{8}$ in. **123.** $\frac{3}{50}$
125. $4\frac{4}{5}$ million, or 4,800,000 **127.** C

Section R.2 (pages 23–25)

1. (a) 6 (b) 9 (c) 1 (d) 7 (e) 4 **3.** (a) 46.25 (b) 46.2
(c) 46 (d) 50 **5.** $\frac{4}{10}$ **7.** $\frac{64}{100}$ **9.** $\frac{138}{1000}$ **11.** $\frac{43}{1000}$ **13.** $\frac{3805}{1000}$
15. 143.094 **17.** 25.61 **19.** 15.33 **21.** 21.77 **23.** 81.716
25. 15.211 **27.** 116.48 **29.** 0.006 **31.** 7.15 **33.** 2.05
35. 5711.6 **37.** 94 **39.** 0.162 **41.** 1.2403 **43.** 1% **45.** $\frac{1}{20}$
47. $12\frac{1}{2}\%$, or 12.5% **49.** 0.25; 25% **51.** $\frac{1}{2}$; 0.5 **53.** $\frac{3}{4}$; 75%
55. 0.375 **57.** 1.25 **59.** $0.\overline{5}$; 0.556 **61.** $0.1\overline{6}$; 0.167 **63.** 0.54
65. 0.07 **67.** 1.17 **69.** 0.024 **71.** 0.0625 **73.** 0.008 **75.** 79%
77. 2% **79.** 0.4% **81.** 128% **83.** 40% **85.** 600% **87.** $\frac{51}{100}$

89. $\frac{3}{20}$ **91.** $\frac{1}{50}$ **93.** $\frac{7}{5}$, or $1\frac{2}{5}$ **95.** $\frac{3}{40}$ **97.** 80% **99.** 14%
101. $18.\overline{18}\%$ **103.** 225% **105.** $216.\overline{6}\%$ **107.** 160 **109.** 4.8
111. 109.2 **113.** $17.80; $106.80 **115.** $119.25; $675.75
117. 19.8 million **119.** 13%

1 THE REAL NUMBER SYSTEM

Section 1.1 (pages 33–35)

1. false; $3^2 = 3 \cdot 3 = 9$ **3.** false; A number raised to the first power is that number, so $3^1 = 3$. **5.** false; $4 + 3(8 - 2)$ means $4 + 3(6)$, which simplifies to $4 + 18$, or 22. The common error leading to 42 is adding 4 to 3 and then multiplying by 6. One must follow the rules for order of operations. **7.** ①, ② **9.** ①, ③, ② **11.** ②, ④, ③, ① **13.** 9
15. 49 **17.** 144 **19.** 64 **21.** 1000 **23.** 81 **25.** 1024 **27.** $\frac{1}{36}$
29. $\frac{16}{81}$ **31.** 0.064 **33.** 32 **35.** 58 **37.** 22.2 **39.** $\frac{49}{30}$, or $1\frac{19}{30}$
41. 12 **43.** 13 45. 26 **47.** 4 **49.** 42 **51.** 5 **53.** 41
55. 95 **57.** 90 **59.** 14 **61.** 9 **63.** $3 \cdot (6 + 4) \cdot 2$
65. $10 - (7 - 3)$ **67.** $(8 + 2)^2$ **69.** $16 \le 16$; true
71. $61 \le 60$; false **73.** $0 \ge 0$; true **75.** $45 \ge 46$; false
77. $66 > 72$; false **79.** $2 \ge 3$; false **81.** $3 \ge 3$; true
83. Five is less than seventeen; true **85.** Five is not equal to eight; true
87. Seven is greater than or equal to fourteen; false **89.** Fifteen is less than or equal to fifteen; true **91.** One-third is equal to three-tenths; false
93. Two and five-tenths is greater than two and fifty-hundredths; false
95. $15 = 5 + 10$ **97.** $9 > 5 - 4$ **99.** $16 \ne 19$ **101.** $\frac{1}{2} \le \frac{2}{4}$
103. $20 > 5$ **105.** $\frac{3}{4} < \frac{4}{5}$ **107.** $1.3 \le 2.5$ **109.** (a) $14.7 - 40 \cdot 0.13$
(b) 9.5 (c) 8.075; walking (5 mph) (d) $14.7 - 55 \cdot 0.11$;
8.65; 7.3525, swimming **111.** Alaska, Texas, California, Idaho
113. Alaska, Texas, California, Idaho, Missouri

Section 1.2 (pages 40–42)

1. B **3.** A **5.** $2x^3 = 2 \cdot x \cdot x \cdot x$, while $2x \cdot 2x \cdot 2x = (2x)^3$.
7. The exponent 2 applies only to its base, which is x. **9.** (a) 11
(b) 13 **11.** (a) 16 (b) 24 **13.** (a) 64 (b) 144 **15.** (a) $\frac{5}{3}$
(b) $\frac{7}{3}$ **17.** (a) $\frac{7}{8}$ (b) $\frac{13}{12}$ **19.** (a) 52 (b) 114 **21.** (a) 25.836
(b) 38.754 **23.** (a) 24 (b) 28 **25.** (a) 12 (b) 33 **27.** (a) 6
(b) $\frac{9}{5}$ **29.** (a) $\frac{4}{3}$ (b) $\frac{13}{6}$ **31.** (a) $\frac{2}{7}$ (b) $\frac{16}{27}$ **33.** (a) 12 (b) 55
35. (a) 1 (b) $\frac{28}{17}$ **37.** (a) 3.684 (b) 8.841 **39.** $12x$ **41.** $x + 9$
43. $x - 4$ **45.** $7 - x$ **47.** $x - 8$ **49.** $\frac{18}{x}$ **51.** $6(x - 4)$
53. yes **55.** no **57.** yes **59.** yes **61.** yes **63.** no
65. $x + 8 = 18$; 10 **67.** $2x + 1 = 5$; 2 **69.** $16 - \frac{3}{4}x = 13$; 4
71. $3x = 2x + 8$; 8 **73.** expression **75.** equation **77.** equation
79. 70 yr **81.** 76 yr

Section 1.3 (pages 49–51)

1. 0   **3.** positive   **5.** quotient; denominator   **7. (a)** A   **(b)** A   **(c)** B   **(d)** B   **9.** This is not true. The absolute value of 0 is 0, and 0 is not positive. We could say that *absolute value is never negative,* or *absolute value is always nonnegative.*   **11.** 4   **13.** 0   **15.** One example is $\sqrt{13}$. There are others.   **17.** true   **19.** true   **21.** false

In Exercises 23–27, answers will vary.

23. $\frac{1}{2}, \frac{5}{8}, 1\frac{3}{4}$ **25.** $-3\frac{1}{2}, -\frac{2}{3}, \frac{3}{7}$ **27.** $\sqrt{5}, \pi, -\sqrt{3}$
29. 2,259,105   **31.** -3424 **33.** 46.77

35. ◄──●●─┼┼┼┼●─┼●─┼●─┼─►
        -6 -4 -2 0 2

37. ◄──┼●─┼●─┼┼┼┼┼●─┼●─►
         -6 -4 -2 0 2 4

39. ◄──●─┼●─┼●─┼●─┼●─┼●─►
        -3.8 $-1\frac{5}{8}$ $\frac{1}{4}$ $2\frac{1}{2}$
         -4 -2 0 2 4

41. (a) 3, 7   **(b)** 0, 3, 7
(c) $-9, 0, 3, 7$ **(d)** $-9, -1\frac{1}{4}, -\frac{3}{5}, 0, 0.\overline{1}, 3, 5.9, 7$ **(e)** $-\sqrt{7}, \sqrt{5}$
(f) All are real numbers.   **43. (a)** 7   **(b)** 7   **45. (a)** -8 **(b)** 8
47. (a) $\frac{3}{4}$ **(b)** $\frac{3}{4}$ **49. (a)** -5.6 **(b)** 5.6   **51.** 6   **53.** -12
55. $-\frac{2}{3}$ **57.** 3   **59.** -3 **61.** -11 **63.** -7 **65.** 4
67. $|-3.5|$, or 3.5   **69.** $-|-6|$, or -6 **71.** $|5-3|$, or 2   **73.** true
75. true   **77.** true   **79.** false   **81.** true   **83.** false
85. Public transportation, 2010 to 2011
87. Communication, 2009 to 2010

Section 1.4 (pages 60–65)

1. negative

3. negative

5. $-8; -6; 2$ **7.** positive   **9.** negative   **11.** -8 **13.** -12
15. 2   **17.** -2 **19.** -9 **21.** 0   **23.** $-\frac{3}{5}$ **25.** $\frac{1}{2}$ **27.** $-\frac{19}{24}$
29. $-\frac{3}{4}$ **31.** 8.9   **33.** -6.01 **35.** 12   **37.** 5   **39.** 2   **41.** -9
43. 0   **45.** -7.7 **47.** -8 **49.** 0   **51.** -20 **53.** -3 **55.** -4
57. -8 **59.** -14 **61.** 9   **63.** -4 **65.** 4   **67.** $\frac{3}{4}$ **69.** $-\frac{11}{8}$, or $-1\frac{3}{8}$
71. $\frac{15}{8}$, or $1\frac{7}{8}$ **73.** 11.6   **75.** -9.9 **77.** 10   **79.** -5 **81.** 11
83. -10 **85.** 22   **87.** -2 **89.** $-\frac{17}{8}$, or $-2\frac{1}{8}$ **91.** $-\frac{1}{4}$, or -0.25
93. -6 **95.** -12 **97.** -5.891 **99.** $-5 + 12 + 6; 13$
101. $[-19 + (-4)] + 14; -9$ **103.** $[-4 + (-10)] + 12; -2$
105. $\left[\frac{5}{7} + \left(-\frac{9}{7}\right)\right] + \frac{2}{7}; -\frac{2}{7}$ **107.** $4 - (-8); 12$
109. $-2 - 8; -10$ **111.** $[9 + (-4)] - 7; -2$
113. $[8 - (-5)] - 12; 1$ **115.** -10 **117.** $+10$ **119.** -12
121. -184 m   **123.** 120°F   **125.** $-69°$F   **127.** 17
129. (a) 4.9%   **(b)** Americans spent more money than they earned, which means they had to dip into savings or increase borrowing.
131. $5540   **133.** $1045.55   **135.** $323.83   **137.** -11.03%
139. 13.8%   **141.** 50,395 ft   **143.** 1345 ft   **145.** 136 ft

Section 1.5 (pages 74–77)

1. greater than 0   **3.** less than 0   **5.** greater than 0   **7.** equal to 0
9. undefined; 0; Examples include $\frac{1}{0}$, which is undefined, and $\frac{0}{1}$, which equals 0.   **11.** -30 **13.** 30   **15.** 120   **17.** -33 **19.** 0   **21.** -2.38

23. $\frac{5}{12}$ **25.** $-\frac{1}{6}$ **27.** 6   **29.** $-32, -16, -8, -4, -2, -1, 1, 2, 4, 8, 16, 32$
31. $-40, -20, -10, -8, -5, -4, -2, -1, 1, 2, 4, 5, 8, 10, 20, 40$
33. $-31, -1, 1, 31$ **35.** 3   **37.** -7 **39.** 8   **41.** -6
43. -4 **45.** $\frac{32}{3}$, or $10\frac{2}{3}$ **47.** $-\frac{15}{16}$ **49.** 0   **51.** undefined
53. -11 **55.** -2 **57.** 35   **59.** 13   **61.** -22 **63.** 6   **65.** -18
67. 67   **69.** -8 **71.** 3   **73.** 7   **75.** 4   **77.** -1 **79.** 4   **81.** -3
83. 29   **85.** 47   **87.** 72   **89.** $-\frac{78}{25}$ **91.** 0   **93.** -23
95. undefined   **97.** $9 + (-9)(2); -9$ **99.** $-4 - 2(-1)(6); 8$
101. $(1.5)(-3.2) - 9; -13.8$ **103.** $12[9 - (-8)]; 204$
105. $\frac{-12}{-5 + (-1)}; 2$ **107.** $\frac{15 + (-3)}{4(-3)}; -1$ **109.** $\frac{2}{3}[8 - (-1)]; 6$
111. $0.20(-5 \cdot 6); -6$ **113.** $\left(\frac{1}{2} + \frac{5}{8}\right)\left(\frac{3}{5} - \frac{1}{3}\right); \frac{3}{10}$ **115.** $\frac{-\frac{1}{2}\left(\frac{3}{4}\right)}{-\frac{2}{3}}; \frac{9}{16}$
117. $\frac{x}{3} = -3; -9$ **119.** $x - 6 = 4; 10$ **121.** $x + 5 = -5; -10$
123. $8\frac{2}{5}$ **125.** 4   **127.** 2   **129. (a)** 6 is divisible by 2.
(b) 9 is not divisible by 2.   **131. (a)** 64 is divisible by 4.
(b) 35 is not divisible by 4.   **133. (a)** 2 is divisible by 2 and $1 + 5 + 2 + 4 + 8 + 2 + 2 = 24$ is divisible by 3.   **(b)** Although 0 is divisible by 2, $2 + 8 + 7 + 3 + 5 + 9 + 0 = 34$ is not divisible by 3.
135. (a) $4 + 1 + 1 + 4 + 1 + 0 + 7 = 18$ is divisible by 9.
(b) $2 + 2 + 8 + 7 + 3 + 2 + 1 = 25$ is not divisible by 9.

Section 1.6 (pages 85–87)

1. (a) B   **(b)** F   **(c)** C   **(d)** I   **(e)** B   **(f)** D, F   **(g)** B   **(h)** A
(i) G   **(j)** H   **3.** yes   **5.** no   **7.** no   **9.** (foreign sales) clerk; foreign (sales clerk)   **11.** Subtraction is not associative.
13. row 1: $-5, \frac{1}{5}$; row 2: $10, -\frac{1}{10}$; row 3: $\frac{1}{2}, -2$; row 4: $-\frac{3}{8}, \frac{8}{3}$; row 5: $-x$, $\frac{1}{x}$; row 6: $y, -\frac{1}{y}$; opposite; the same   **15.** -15; commutative property
17. 3; commutative property   **19.** 6; associative property
21. 7; associative property   **23.** commutative property
25. associative property   **27.** associative property
29. inverse property   **31.** inverse property   **33.** identity property
35. commutative property   **37.** distributive property
39. identity property   **41.** distributive property   **43.** 150   **45.** 2010
47. 400   **49.** 1400   **51.** 470   **53.** -9300 **55.** 11   **57.** 0
59. -0.38 **61.** 1   **63.** The student made a sign error. The expression following the first equality symbol should be $-3(4) - 3(-6)$. $-3(4 - 6)$ means $-3(4) - 3(-6)$, which simplifies to $-12 + 18$, or 6.
65. We must multiply $\frac{3}{4}$ by 1 in the form of a fraction, $\frac{3}{3}$; $\frac{3}{4} \cdot \frac{3}{3} = \frac{9}{12}$.
67. 85   **69.** $4t + 12$ **71.** $7z - 56$ **73.** $-8r - 24$ **75.** $-2x - \frac{3}{4}$
77. $-5y + 20$ **79.** $12x + 10$ **81.** $-6x + 15$ **83.** $-48x - 6$
85. $-16y - 20z$ **87.** $24r + 32s - 40y$ **89.** $-24x - 9y - 12z$
91. $-4t - 3m$ **93.** $5c + 4d$ **95.** $q - 5r + 8s$

Section 1.7 (pages 91–93)

1. B   **3.** C   **5.** The student made a sign error when applying the distributive property: $7x - 2(3 - 2x)$ means $7x - 2(3) - 2(-2x)$, which simplifies to $7x - 6 + 4x$, or $11x - 6$.   **7.** $4r + 11$ **9.** $21x - 28y$
11. $5 + 2x - 6y$ **13.** $-7 + 3p$ **15.** $2 - 3x$ **17.** -12 **19.** 3

21. 1 **23.** -1 **25.** $\frac{1}{2}$ **27.** $\frac{2}{5}$ **29.** -0.5 **31.** 10 **33.** like

35. unlike **37.** like **39.** unlike **41.** $13y$ **43.** $-9x$ **45.** $13b$

47. $7k + 15$ **49.** $-4y$ **51.** $2x + 6$ **53.** $14 - 7m$ **55.** $-17 + x$

57. $23x$ **59.** $-\frac{28}{3} - \frac{1}{3}t$ **61.** $9y^2$ **63.** $-14p^3 + 5p^2$ **65.** $8x + 15$

67. $22 - 4y$ **69.** $-19p + 16$ **71.** $-t + 3$ **73.** $5x + 15$

75. $15 - 9x$ **77.** $-16y + 63$ **79.** $4r + 15$ **81.** $12k - 5$

83. $-\frac{3}{2}y + 16$ **85.** $-2x + 4$ **87.** $-\frac{14}{3}x - \frac{22}{3}$ **89.** $-23.7y - 12.6$

91. $-2k - 3$ **93.** $4x - 7$ **95.** $(4x + 8) + (3x - 2); 7x + 6$

97. $(5x + 1) - (x - 7); 4x + 8$ **99.** $(x + 3) + 5x; 6x + 3$

101. $(13 + 6x) - (-7x); 13 + 13x$ **103.** $2(3x + 4) - (-4 + 6x); 12$

105. $1000 + 5x$ (dollars) **106.** $750 + 3y$ (dollars)

107. $1000 + 5x + 750 + 3y$ (dollars) **108.** $1750 + 5x + 3y$ (dollars)

2 LINEAR EQUATIONS AND INEQUALITIES IN ONE VARIABLE

Section 2.1 (pages 109–110)

1. equation; expression **3.** equivalent equations **5. (a)** expression;
$x + 15$ **(b)** expression; $y + 7$ **(c)** equation; $\{-1\}$ **(d)** equation;
$\{-17\}$ **7.** A, B **9.** $\{12\}$ **11.** $\{31\}$ **13.** $\{-3\}$ **15.** $\{4\}$

17. $\{-9\}$ **19.** $\left\{-\frac{3}{4}\right\}$ **21.** $\{-10\}$ **23.** $\{-13\}$ **25.** $\{10\}$

27. $\left\{\frac{4}{15}\right\}$ **29.** $\{6.3\}$ **31.** $\{-16.9\}$ **33.** $\{7\}$ **35.** $\{-4\}$

37. $\{-3\}$ **39.** $\{2\}$ **41.** $\{-6\}$ **43.** $\{-5\}$ **45.** $\{-2\}$ **47.** $\{3\}$

49. $\{0\}$ **51.** $\{-7\}$ **53.** $\{-3\}$ **55.** $\{0\}$ **57.** $\{2\}$ **59.** $\{-16\}$

61. $\{0\}$ **63.** $\{2\}$ **65.** $\{13\}$ **67.** $\{-4\}$ **69.** $\{0\}$ **71.** $\left\{\frac{7}{15}\right\}$

73. $\{7\}$ **75.** $\{-4\}$ **77.** $\{13\}$ **79.** $\{29\}$ **81.** $\{18\}$ **83.** $\{12\}$

85. Answers will vary. One example is $x - 6 = -8$. **87.** $3x = 2x + 17$;
$\{17\}$ **89.** $7x - 6x = -9; \{-9\}$

Section 2.2 (pages 115–117)

1. (a) multiplication property of equality **(b)** addition property of
equality **(c)** multiplication property of equality **(d)** addition property
of equality **3.** B **5.** $\frac{5}{4}$ **7.** 10 **9.** $-\frac{2}{9}$ **11.** -1 **13.** 6 **15.** -4

17. 0.12 **19.** -1 **21.** $\{6\}$ **23.** $\left\{\frac{15}{2}\right\}$ **25.** $\{-5\}$ **27.** $\{-4\}$

29. $\left\{-\frac{18}{5}\right\}$, or $\{-3.6\}$ **31.** $\{12\}$ **33.** $\{0\}$ **35.** $\{-12\}$ **37.** $\left\{\frac{3}{4}\right\}$

39. $\{40\}$ **41.** $\{-30\}$ **43.** $\{-2.4\}$ **45.** $\{3.5\}$ **47.** $\{-12.2\}$

49. $\{-48\}$ **51.** $\{72\}$ **53.** $\{-35\}$ **55.** $\{14\}$ **57.** $\{18\}$

59. $\left\{-\frac{27}{35}\right\}$ **61.** $\{3\}$ **63.** $\{-5\}$ **65.** $\{20\}$ **67.** $\{7\}$ **69.** $\{0\}$

71. $\left\{-\frac{3}{5}\right\}$ **73.** $\{18\}$ **75.** $\{-6\}$ **77.** Answers will vary. One
example is $\frac{3}{2}x = -6$. **79.** $4x = 6; \left\{\frac{3}{2}\right\}$ **81.** $\frac{x}{-5} = 2; \{-10\}$

Section 2.3 (pages 125–126)

1. Use the addition property of equality to subtract 8 from each side.
3. Clear parentheses by using the distributive property. **5.** Clear
fractions by multiplying by the LCD, 6. **7. (a)** identity; B
(b) conditional; A **(c)** contradiction; C **9.** $\{4\}$ **11.** $\{-5\}$ **13.** $\left\{\frac{5}{2}\right\}$

15. $\{-1\}$ **17.** $\left\{-\frac{1}{2}\right\}$ **19.** $\{-3\}$ **21.** $\{5\}$ **23.** $\{0\}$ **25.** $\left\{\frac{4}{3}\right\}$

27. $\left\{-\frac{5}{3}\right\}$ **29.** $\varnothing$ **31.** $\{5\}$ **33.** $\{0\}$ **35.** $\{$all real numbers$\}$

37. $\varnothing$ **39.** $\{5\}$ **41.** $\{12\}$ **43.** $\{11\}$ **45.** $\{0\}$ **47.** $\{18\}$

49. $\{3\}$ **51.** $\left\{\frac{5}{4}\right\}$ **53.** $\{120\}$ **55.** $\{6\}$ **57.** $\{15,000\}$ **59.** $\{8\}$

61. $\{0\}$ **63.** $\{$all real numbers$\}$ **65.** $\{4\}$ **67.** $\{20\}$ **69.** $\varnothing$

71. $11 - q$ **73.** $\frac{9}{x}$ **75.** $65 - h$ **77.** $x + 15; x - 5$ **79.** $25r$

81. $\frac{t}{5}$ **83.** $3x + 2y$

Section 2.4 (pages 137–141)

1. D; There cannot be a fractional number of cars. **3.** A; Distance
cannot be negative. **5.** $x + 9 = -26; -35$ **7.** $8(x + 6) = 104; 7$

9. $5x + 2 = 4x + 5; 3$ **11.** $3(x - 2) = x + 6; 6$ **13.** $\frac{3}{4}x + 6 = x - 4; 40$

15. $3x + (x + 7) = -11 - 2x; -3$ **17.** *Step 1:* Republicans;
Step 2: $x - 4$; Democrats; *Step 3:* $(x - 4) + x = 150$; Democrats: 73,
Republicans: 77 **19.** New York: 29 screens; Ohio: 27 screens

21. Democrats: 52; Republicans: 46 **23.** Madonna: $228 million;
Springsteen: $199 million **25.** wins: 66; losses: 16 **27.** orange: 97 mg;
pineapple: 25 mg **29.** 112 DVDs **31.** onions: 81.3 kg; grilled steak:
536.3 kg **33.** 1950 Denver nickel: $16.00; 1945 Philadelphia nickel: $8.00

35. whole wheat: 25.6 oz; rye: 6.4 oz **37.** American: 18; United: 11;
Southwest: 26 **39.** shortest piece: 15 in.; middle piece: 20 in.; longest
piece: 24 in. **41.** gold: 46; silver: 29; bronze: 29 **43.** 36 million mi

45. A and B: $40°$; C: $100°$ **47.** 68, 69 **49.** 101, 102 **51.** 10, 12

53. 17, 19 **55.** 10, 11 **57.** 18 **59.** 15, 17, 19 **61.** $18°$ **63.** $20°$

65. $39°$ **67.** $50°$

Section 2.5 (pages 148–152)

1. area **3.** perimeter **5.** area **7.** area **9.** $P = 26$ **11.** $\mathcal{A} = 64$

13. $b = 4$ **15.** $t = 5.6$ **17.** $B = 14$ **19.** $r = 2.6$ **21.** $r = 10$

23. $\mathcal{A} = 50.24$ **25.** $r = 6$ **27.** $V = 150$ **29.** $V = 52$

31. $V = 7234.56$ **33.** $I = \$600$ **35.** $p = \$550$ **37.** $t = 1.5$ yr

39. length: 18 in.; width: 9 in. **41.** length: 14 m; width: 4 m

43. shortest: 5 in.; medium: 7 in.; longest: 8 in. **45.** two equal sides: 7 m;
third side: 10 m **47.** perimeter: 5.4 m; area: 1.8 m^2 **49.** 10 ft

51. 154,000 ft^2 **53.** 194.48 ft^2; 49.42 ft **55.** 23,800.10 ft^2

57. length: 36 in.; volume: 11,664 in.^{3} **59.** $48°, 132°$ **61.** $55°, 35°$

63. $51°, 51°$ **65.** $105°, 105°$

We give one answer for Exercises 67–103. There are other correct forms.

67. $t = \frac{d}{r}$ **69.** $b = \frac{\mathcal{A}}{h}$ **71.** $d = \frac{C}{\pi}$ **73.** $H = \frac{V}{LW}$ **75.** $r = \frac{I}{pt}$

77. $h = \frac{2\mathcal{A}}{b}$ **79.** $h = \frac{3V}{\pi r^2}$ **81.** $b = P - a - c$ **83.** $W = \frac{P - 2L}{2}$

85. $m = \frac{y - b}{x}$ **87.** $y = \frac{C - Ax}{B}$ **89.** $r = \frac{M - C}{C}$ **91.** $a = \frac{P - 2b}{2}$

93. $b = 2S - a - c$ **95.** $F = \frac{9C + 160}{5}$ **97.** $y = -6x + 4$

99. $y = 5x - 2$ **101.** $y = \frac{3}{5}x - 3$ **103.** $y = \frac{1}{3}x - 4$

Section 2.6 (pages 157–162)

1. (a) C **(b)** D **(c)** B **(d)** A **3.** $\frac{4}{3}$ **5.** $\frac{4}{3}$ **7.** $\frac{15}{2}$ **9.** $\frac{1}{5}$ **11.** $\frac{5}{6}$
13. 10 lb; $0.749 **15.** 64 oz; $0.047 **17.** 32 oz; $0.531
19. 32 oz; $0.056 **21.** 263 oz; $0.076 **23.** true **25.** false
27. true **29.** $\{35\}$ **31.** $\{7\}$ **33.** $\left\{\frac{45}{2}\right\}$ **35.** $\{2\}$ **37.** $\{-1\}$
39. $\{5\}$ **41.** $\left\{-\frac{31}{5}\right\}$ **43.** $\{-2\}$ **45.** $30.00 **47.** $8.75
49. $67.50 **51.** $56.40 **53.** 50,000 fish **55.** 4 ft **57.** 2.7 in.
59. 2.0 in. **61.** $2\frac{5}{8}$ cups **63.** $404.76 **65.** $x = 4$ **67.** $x = 8$
69. $x = 22.5$; $y = 25.5$

71. (a)
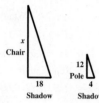
(b) 54 ft **73.** $239 **75.** $280
77. (a) 2625 mg
(b) $\dfrac{125 \text{ mg}}{5 \text{ mL}} = \dfrac{2625 \text{ mg}}{x \text{ mL}}$
(c) 105 mL

79. 140.4 **81.** 700 **83.** 425 **85.** 8% **87.** 120% **89.** 80%
91. 28% **93.** 32% **95.** $3000

Section 2.7 (pages 169–173)

1. A **3.** C **5.** D **7.** A **9.** 45 L **11.** $750 **13.** $17.50
15. 160 L **17.** $13\frac{1}{3}$ L **19.** 4 L **21.** 20 mL **23.** 4 L **25.** $2100
at 5%; $900 at 4% **27.** $2500 at 6%; $13,500 at 5% **29.** $1700 at 8%;
$800 at 2% **31.** 10 nickels **33.** 46-cent stamps: 25; 20-cent stamps: 20
35. Arabian Mocha: 7 lb; Colombian Decaf: 3.5 lb **37.** 530 mi
39. 2.668 hr **41.** 9.14 m per sec **43.** 8.51 m per sec **45.** $7\frac{1}{2}$ hr
47. 5 hr **49.** $1\frac{3}{4}$ hr **51.** eastbound: 300 mph; westbound: 450 mph
53. slower car: 40 mph; faster car: 60 mph **55.** Bob: 7 yr old; Kevin:
21 yr old **57.** width: 3 ft; length: 9 ft **59.** $650

Section 2.8 (pages 184–187)

1. >, < (or <, >); ≥, ≤ (or ≤, ≥) **3.** $(0, \infty)$ **5.** $x > -4$ **7.** $x \le 4$
9. $(-\infty, 4]$

11. $(-\infty, -3)$

13. $(4, \infty)$

15. $(-\infty, 0]$

17. $\left[-\frac{1}{2}, \infty\right)$

19. $[1, \infty)$

21. $[5, \infty)$

23. $(-\infty, -11)$

25. It must be reversed when multiplying or dividing by a negative
number.

27. $(-\infty, 6)$

29. $[-10, \infty)$

31. $(-\infty, -3)$

33. $(-\infty, 0]$

35. $(20, \infty)$

37. $[-3, \infty)$

39. $(-\infty, -3]$

41. $(-1, \infty)$

43. $[-5, \infty)$

45. $(-\infty, 2]$

47. $\left(-\infty, \frac{3}{2}\right)$

49. $(-\infty, 1)$

51. $(-\infty, 0]$

53. $\left(-\frac{1}{2}, \infty\right)$

55. $[4, \infty)$

57. $[2, \infty)$

59. $(-\infty, 32)$

61. $(-\infty, 6]$

63. $\left[\frac{5}{12}, \infty\right)$

65. $(-21, \infty)$

67. $x \ge 18$ **69.** $x > 5$ **71.** $x \le 20$ **73.** 83 or greater **75.** more
than 3.8 in. **77.** all numbers greater than 16 **79.** It is never less
than $-13°$F. **81.** 32 or greater **83.** 12 min **85.** $5x - 100$
87. $(5x - 100) - (125 + 4x)$, which simplifies to $x - 225$; $x > 225$
89. $-1 < x < 2$ **91.** $-1 < x \le 2$

93. $[8, 10]$

95. $(0, 10]$

97. $(-3, 4)$

99. $(-2, 1]$

101. $[-1, 6]$

103. $(1, 3)$

105. $(-6, 2)$

107. $\left(-\frac{11}{6}, -\frac{2}{3}\right)$

109. $[3, 7)$

111. $[-26, 6]$

113. $[-3, 6]$

115. $\left[-\frac{24}{5}, 0\right]$

117. $\{4\}$

118. $(4, \infty)$

119. $(-\infty, 4)$

120. The graph would be the set of all real numbers.

(c)

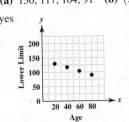

(d) The points lie approximately in a linear pattern. Rates at which 2-year college students complete a degree within 3 years were decreasing.

83. (a) 130; 117; 104; 91 **(b)** $(20, 130), (40, 117), (60, 104), (80, 91)$

(c) yes

85. between 130 and 170 beats per minute; between 117 and 153 beats per minute

3 LINEAR EQUATIONS AND INEQUALITIES IN TWO VARIABLES; FUNCTIONS

Section 3.1 (pages 207–210)

1. does; do not **3.** II **5.** 3 **7.** negative; negative **9.** positive; negative **11.** If $xy < 0$, then either $x < 0$ and $y > 0$ or $x > 0$ and $y < 0$. If $x < 0$ and $y > 0$, then the point lies in quadrant II. If $x > 0$ and $y < 0$, then the point lies in quadrant IV. **13.** between 2010 and 2011 and 2011 and 2012 **15.** 2011: 9%; 2012: 8%; decline: 1%
17. yes **19.** yes **21.** no **23.** yes **25.** yes **27.** no **29.** 17
31. -5 **33.** -1 **35.** -7 **37.** 8; 6; 3; $(0, 8)$; $(6, 0)$; $(3, 4)$
39. -9; 4; 9; $(-9, 0)$; $(0, 4)$; $(9, 8)$ **41.** 12; 12; 12; $(12, 3)$; $(12, 8)$; $(12, 0)$ **43.** -10; -10; -10; $(4, -10)$; $(0, -10)$; $(-4, -10)$
45. -2; -2; -2; $(9, -2)$; $(2, -2)$; $(0, -2)$ **47.** 4; 4; 4; $(4, 4)$; $(4, 0)$; $(4, -4)$ **49.** No, the ordered pair $(3, 4)$ represents the point 3 units to the right of the origin and 4 units up from the x-axis. The ordered pair $(4, 3)$ represents the point 4 units to the right of the origin and 3 units up from the x-axis. **51.** $(2, 4)$; I **53.** $(-5, 4)$; II
55. $(3, 0)$; no quadrant **57.** $(4; -4)$; IV

59.–69.

71. -3; 6; -2; 4 **73.** -3; 4; -6; $-\frac{4}{3}$

75. -4; -4; -4; -4 **77.** The points in each graph appear to lie on a straight line.

79. (a) $(5, 45)$ **(b)** $(6, 50)$

81. (a) $(2008, 29.3), (2009, 28.3), (2010, 28.0), (2011, 26.9), (2012, 25.4), (2013, 22.5)$ **(b)** $(2013, 22.5)$ means that 22.5 percent of 2-year college students in 2013 received a degree within 3 years.

Section 3.2 (pages 220–223)

1. By; C; 0 **3. (a)** A **(b)** C **(c)** D **(d)** B **5.** x-intercept: $(4, 0)$; y-intercept: $(0, -4)$ **7.** x-intercept: $(-2, 0)$; y-intercept: $(0, -3)$
9. (a) D **(b)** C **(c)** B **(d)** A

11. 5; 5; 3 **13.** 1; 3; -1 **15.** -6; -2; -5

17. $(8, 0)$; $(0, -8)$ **19.** $(4, 0)$; $(0, -10)$ **21.** $(0, 0)$; $(0, 0)$
23. $(2, 0)$; $(0, 4)$ **25.** $(6, 0)$; $(0, -2)$ **27.** $(0, 0)$; $(0, 0)$ **29.** $(4, 0)$; no y-intercept **31.** no x-intercept; $(0, 2.5)$

33. **35.** **37.**

39. **41.** **43.**

45. **47.** 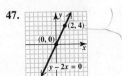 **49.**

51. **53.** 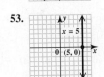 **55.**

57. **59.**

47. **41.** **43.**

In Exercises 61–67, descriptions may vary.

61. The graph is a line with *x*-intercept $(-3, 0)$ and *y*-intercept $(0, 9)$.

63. The graph is a vertical line with *x*-intercept $(11, 0)$.

65. The graph is a horizontal line with *y*-intercept $(0, -2)$.

67. The graph has *x*- and *y*-intercepts $(0, 0)$. It passes through the points $(2, 1)$ and $(4, 2)$.

69. 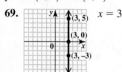   **71.**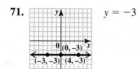

73. (a) 121 lb, 143 lb, 176 lb   **75. (a)** $62.50; $100   **(b)** 200
(b) $(62, 121)$, $(66, 143)$,   **(c)** $(50, 62.50)$, $(100, 100)$,
$(72, 176)$ $(200, 175)$

(c) 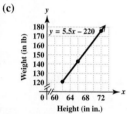   **(d)**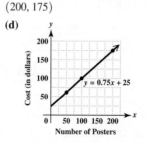

(d) 68 in.; 68 in.

77. (a) $30,000 **(b)** $15,000 **(c)** $5000 **(d)** After 5 yr, the SUV has a value of $5000. **79. (a)** 2000: 30.1 lb; 2008: 32.6 lb; 2012: 33.8 lb
(b) 2000: 29.8 lb; 2008: 32.8 lb; 2012: 33.5 lb **(c)** The values are quite close. **(d)** 36.9 lb; It is very close to the USDA projection.

Section 3.3 (pages 232–236)

1. steepness; vertical; horizontal **3. (a)** 6 **(b)** 4 **(c)** $\frac{6}{4}$, or $\frac{3}{2}$; slope of the line **5. (a)** C **(b)** A **(c)** D **(d)** B

In Exercises 7 and 9, sketches will vary.

7. The line must fall from left to right. **9.** The line must be vertical.
11. (a) negative **(b)** zero **13. (a)** positive **(b)** negative
15. (a) zero **(b)** negative **17.** Because he found the difference $3 - 5 = -2$ in the numerator, he should have subtracted in the same order in the denominator to obtain $-1 - 2 = -3$. The correct slope is $\frac{-2}{-3} = \frac{2}{3}$.
19. $\frac{8}{27}$ **21.** $-\frac{2}{3}$ **23.** 4 **25.** $-\frac{1}{2}$ **27.** 0 **29.** $\frac{5}{4}$ **31.** $\frac{3}{2}$ **33.** 0
35. -3 **37.** undefined **39.** $\frac{1}{4}$ **41.** $-\frac{1}{2}$ **43.** 5 **45.** $\frac{1}{4}$ **47.** $\frac{3}{2}$ **49.** $\frac{3}{2}$
51. -1 **53.** 0 **55.** undefined

In part (a) of Exercises 57 and 59, we used the intercepts. Other points can be used.

57. (a) $(5, 0)$ and $(0, 10)$; -2 **(b)** $y = -2x + 10$; -2
59. (a) $(3, 0)$ and $(0, -5)$; $\frac{5}{3}$ **(b)** $y = \frac{5}{3}x - 5$; $\frac{5}{3}$

61. (a) 1 **(b)** $(-4, 0)$; $(0, 4)$ **63. (a)** $-\frac{1}{3}$ **(b)** $(-6, 0)$; $(0, -2)$
(c)   **(c)**

65. $-3; \frac{1}{3}$ **67.** A **69.** $-\frac{2}{5}$; $-\frac{2}{5}$; parallel **71.** $\frac{8}{9}$; $-\frac{4}{3}$; neither
73. $\frac{3}{2}$; $-\frac{2}{3}$; perpendicular **75.** 5; $\frac{1}{5}$; neither
77. (a) $(2004, 817)$, $(2012, 1661)$ **(b)** 105.5 **(c)** Music purchases increased by 844 million units in 8 yr, or 105.5 million units per year.
79. 0.5 **80.** positive; increased **81.** 0.5% **82.** -0.2 **83.** negative; decreased **84.** 0.2%

Section 3.4 (pages 242–245)

1. m; $(0, b)$ **3. (a)** C **(b)** B **(c)** A **(d)** D **5.** slope: $\frac{5}{2}$; *y*-intercept: $(0, -4)$ **7.** slope: -1; *y*-intercept: $(0, 9)$ **9.** slope: $\frac{1}{5}$; *y*-intercept: $\left(0, -\frac{3}{10}\right)$

11.   **13.**   **15.**

17.   **19.**   **21.**

23.   **25.**   **27.**

29.   **31.**

33. *y*-axis **35.** $y = 3x - 3$ **37.** $y = -x + 3$ **39.** $y = -\frac{1}{2}x + 2$
41. $y = 4x - 3$ **43.** $y = -x - 7$ **45.** $y = 2x - 7$ **47.** $y = -4x - 1$
49. $y = x - 6$ **51.** $y = \frac{3}{4}x + 4$ **53.** $y = 3$ **55.** $x = 2$
57. $x = 0$ **59.** $y = -6$
61. (a) 2 **(b)** $(0, -1)$ **63. (a)** $-\frac{1}{3}$ **(b)** $(0, -2)$
(c) $y = 2x - 1$ **(c)** $y = -\frac{1}{3}x - 2$
(d)   **(d)**

65. (a) 0.05; commission rate **(b)** $(0; 2000)$; base salary per month
(c) $2500 **(d)** $30,000 **67. (a)** $400 **(b)** $0.25
(c) $y = 0.25x + 400$ **(d)** $425 **(e)** 1500 **69.** $y = -\frac{A}{B}x + \frac{C}{B}$
70. (a) $-\frac{2}{3}$ **(b)** 2 **(c)** $\frac{3}{7}$ **71.** $\left(0, \frac{C}{B}\right)$ **72. (a)** $(0, 6)$ **(b)** $\left(0, \frac{1}{2}\right)$
(c) $(0, -3)$

Section 3.5 (pages 249–252)

1. **(a)** D **(b)** C **(c)** B **(d)** E **(e)** A **3.** A, B, D **5.** $y = 5x + 2$ **7.** $y = x - 9$ **9.** $y = -3x - 4$ **11.** $y = -x + 1$ **13.** $y = \frac{2}{3}x + \frac{19}{3}$ **15.** $y = -\frac{4}{5}x + \frac{9}{5}$ **17.** **(a)** $y = x + 6$ **(b)** $x - y = -6$ **19.** **(a)** $y = \frac{1}{2}x + 2$ **(b)** $x - 2y = -4$ **21.** **(a)** $y = -\frac{3}{5}x - \frac{11}{5}$ **(b)** $3x + 5y = -11$ **23.** **(a)** $y = -\frac{1}{3}x + \frac{22}{9}$ **(b)** $3x + 9y = 22$ **25.** **(a)** $y = 3x - 9$ **(b)** $3x - y = 9$ **27.** **(a)** $y = -\frac{2}{3}x + \frac{4}{3}$ **(b)** $2x + 3y = 4$ **29.** $y = -2x - 3$ **31.** $y = 4x - 5$ **33.** $y = \frac{3}{4}x - \frac{9}{2}$

35. **(a)** (1, 2530), (2, 2790), (3, 2940), (4, 3070), (5, 3220)

(b) yes

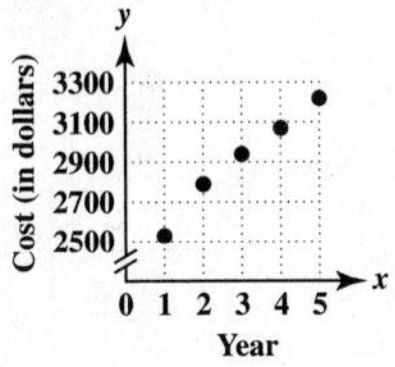

(c) $y = 180x + 2350$

(d) \$3430 ($x = 6$)

37. $y = 15x + 37$ **39.** $y = 0.2375x + 59.7$

Section 3.6 (pages 258–259)

1. $>, >$ **3.** $\leq$ **5.** $<$ **7.** false; The point (4, 0) lies on the boundary line $3x - 4y = 12$, which is *not* part of the graph because the symbol $<$ does not involve equality. **9.** false; Because (0, 0) is on the boundary line $x + 4y = 0$, it cannot be used as a test point. Use a test point *off* the line. **11.** true

13.
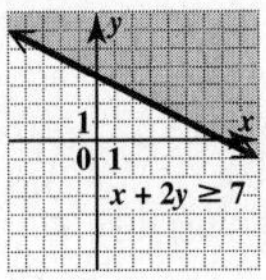

15.
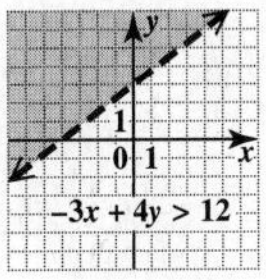

17.
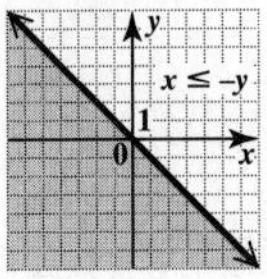

19.
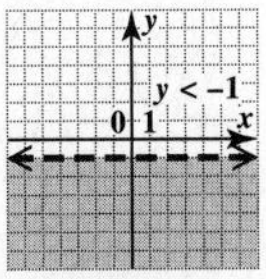

21. Use a dashed line if the symbol is $<$ or $>$. Use a solid line if the symbol is $\leq$ or $\geq$.

23.
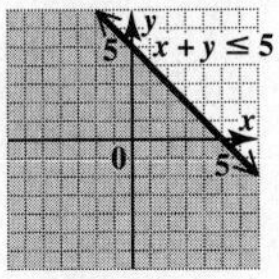

25.
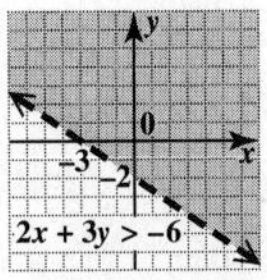

27.
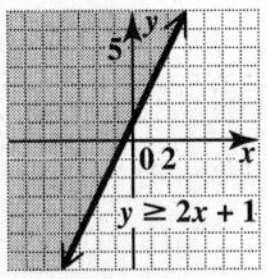

29.
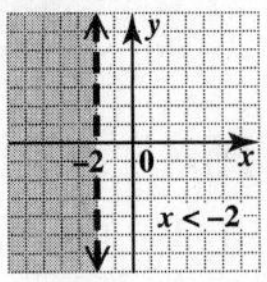

31.
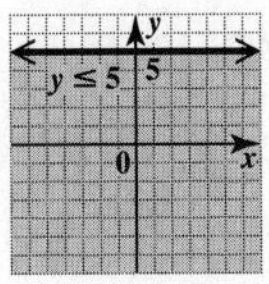

33.
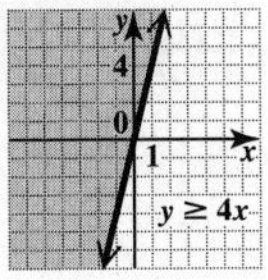

35.
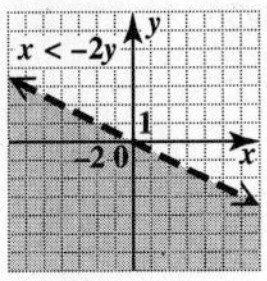

37. **(a)**
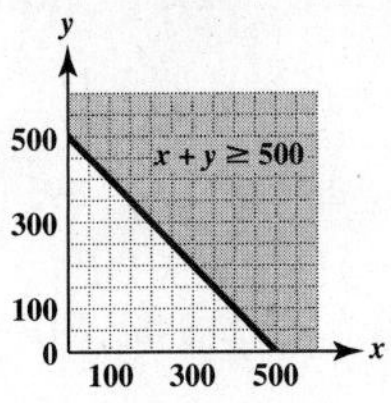

(b) (500, 0), (200, 400) (Other answers are possible.)

Section 3.7 (pages 265–267)

1. relation; domain; range **3.** 3; 3; (1, 3) **5.** 5; 5; (3, 5) **7.** The graph consists of the five points (0, 2), (1, 3), (2, 4), (3, 5), and (4, 6). **9.** function; domain: $\{3, 1, 0, -1\}$; range: $\{7, 4, -2, -1\}$ **11.** function; domain: $\{1, 2, 3\}$; range: $\{-1, -2\}$ **13.** not a function; domain: $\{-4, -2, 0\}$; range: $\{3, 1, 5, -8\}$ **15.** function; domain: $\{3, 1, 5, 7, 9\}$; range: $\{2, 3, 6, 4\}$ **17.** not a function; domain: $\{-4, -2, 0, 2, 3\}$; range: $\{-2, 0, 1, 2, 3\}$ **19.** function; domain: $(-\infty, \infty)$; range: $[-1, \infty)$ **21.** not a function; domain: $[0, \infty)$; range: $(-\infty, \infty)$ **23.** function **25.** not a function **27.** function **29.** function **31.** not a function **33.** domain: $(-\infty, \infty)$; range: $(-\infty, \infty)$ **35.** domain: $(-\infty, \infty)$; range: $(-\infty, \infty)$ **37.** domain: $(-\infty, \infty)$; range: $[2, \infty)$ **39.** domain: $(-\infty, \infty)$; range: $(-\infty, 0]$ **41.** **(a)** 11 **(b)** 3 **(c)** -9 **(d)** 5 **(e)** $\frac{5}{3}$ **43.** **(a)** -4 **(b)** -2 **(c)** 1 **(d)** $-\frac{5}{2}$ **(e)** $-\frac{5}{3}$ **45.** **(a)** 4 **(b)** 2 **(c)** 14 **(d)** $\frac{7}{4}$ **(e)** $\frac{22}{9}$ **47.** **(a)** 2 **(b)** 0 **(c)** 3 **(d)** $\frac{1}{2}$ **(e)** $\frac{1}{3}$ **49.** **(a)** 9 **(b)** 7 **(c)** 4 **(d)** $\frac{15}{2}$ **(e)** $\frac{20}{3}$ **51.** $\{(1970, 9.6), (1980, 14.1), (1990, 19.8), (2000, 28.4), (2010, 40.0)\}$; yes **53.** $g(1980) = 14.1$; $g(2000) = 28.4$ **55.** 2010 **57.** **(a)** 4 **(b)** 2 **59.** (2, 4) **60.** $(-1, -4)$ **61.** $\frac{8}{3}$ **62.** $f(x) = \frac{8}{3}x - \frac{4}{3}$

4 SYSTEMS OF LINEAR EQUATIONS AND INEQUALITIES

Section 4.1 (pages 283–286)

1. system of linear equations; same **3.** inconsistent; no; independent **5.** dependent; consistent; infinitely many **7.** It is not a solution of the system because it is not a solution of the second equation, $2x + y = 4$. **9.** A; The ordered-pair solution must be in quadrant II, and $(-4, -4)$ is in quadrant III. **11.** **(a)** B **(b)** C **(c)** D **(d)** A **13.** no **15.** yes **17.** yes **19.** no **21.** yes

23. $\{(4, 2)\}$
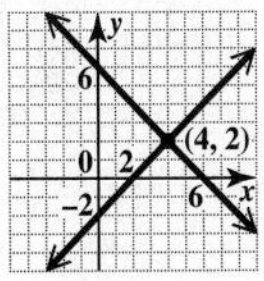

25. $\{(0, 4)\}$
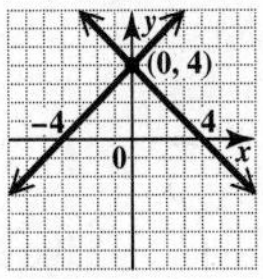

27. $\{(4, -1)\}$
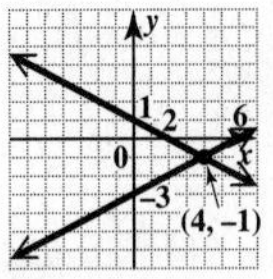

In Exercises 29–41, we do not show the graphs.

29. $\{(1, 3)\}$ **31.** $\{(x, y) \mid 3x + y = 5\}$ (dependent equations) **33.** $\{(0, 2)\}$ **35.** $\emptyset$ (inconsistent system) **37.** $\{(x, y) \mid 3x - y = -6\}$ (dependent equations) **39.** $\{(4, -3)\}$ **41.** $\emptyset$ (inconsistent system) **43.** **(a)** neither **(b)** intersecting lines **(c)** one solution **45.** **(a)** dependent **(b)** one line **(c)** infinite number of solutions **47.** **(a)** neither **(b)** intersecting lines **(c)** one solution **49.** **(a)** inconsistent **(b)** parallel lines **(c)** no solution **51.** **(a)** 1980–2000 **(b)** 2001; about 750 newspapers **(c)** (2001, 750) **53.** 40; 30 **55.** Supply exceeds demand.

Section 4.2 (pages 292–293)

1. The student must find the value of y and write the solution as an ordered pair. The solution set is $\{(3, 0)\}$.   **3.** A false statement, such as $0 = 3$, occurs.   **5.** $\{(3, 9)\}$ **7.** $\{(7, 3)\}$ **9.** $\{(-4, 8)\}$
11. $\{(3, -2)\}$ **13.** $\{(0, 5)\}$ **15.** $\{(x, y) \mid 3x - y = 5\}$
17. $\left\{\left(\frac{1}{4}, -\frac{1}{2}\right)\right\}$ **19.** $\varnothing$ **21.** $\{(x, y) \mid 2x - y = -12\}$ **23.** $\{(1, 5)\}$
25. $\varnothing$ **27.** $\{(0, 0)\}$ **29.** $\{(2, 6)\}$ **31.** $\{(2, -4)\}$ **33.** $\{(-2, 1)\}$
35. $\{(x, y) \mid x + 2y = 48\}$ **37.** $\{(10, 4)\}$ **39.** $\{(4, -9)\}$
41. To find the total cost, multiply the number of bicycles (x) by the cost per bicycle ($400), and add the fixed cost ($5000). Thus, $y_1 = 400x + 5000$ gives this total cost (in dollars).
42. $y_2 = 600x$ **43.** $y_1 = 400x + 5000$, $y_2 = 600x$; solution set: $\{(25, 15{,}000)\}$ **44.** 25; 15,000; 15,000

Section 4.3 (pages 298–299)

1. false; The solution set is $\varnothing$.   **3.** $\{(4, 6)\}$ **5.** $\{(-1, -3)\}$
7. $\{(-2, 3)\}$ **9.** $\left\{\left(\frac{1}{2}, 4\right)\right\}$ **11.** $\{(3, -6)\}$ **13.** $\{(0, 4)\}$
15. $\{(0, 0)\}$ **17.** $\{(7, 4)\}$ **19.** $\{(0, 3)\}$ **21.** $\{(3, 0)\}$
23. $\{(x, y) \mid x - 3y = -4\}$ **25.** $\varnothing$ **27.** $\{(-3, 2)\}$ **29.** $\{(11, 15)\}$
31. $\left\{\left(13, -\frac{7}{5}\right)\right\}$ **33.** $\{(6, -4)\}$ **35.** $\{(x, y) \mid x + 3y = 6\}$
37. $\varnothing$ **39.** $\left\{\left(-\frac{5}{7}, -\frac{2}{7}\right)\right\}$ **41.** $\left\{\left(\frac{1}{8}, -\frac{5}{6}\right)\right\}$ **43.** $6.21 = 2004a + b$
44. $7.96 = 2012a + b$ **45.** $2004a + b = 6.21$, $2012a + b = 7.96$; solution set: $\{(0.21875, -432.165)\}$ **46.** (a) $y = 0.21875x - 432.165$
(b) 7.74 ($7.74); This is less ($0.19) than the actual figure.

Section 4.4 (pages 307–312)

1. D   **3.** B   **5.** D   **7.** C   **9.** B   **11.** *Step 2:* the second number; *Step 3:* $x - y = 48$; The two numbers are 73 and 25.   **13.** *The Phantom of the Opera:* 10,703; *Cats:* 7485   **15.** *Marvel's The Avengers:* $623 million; *The Dark Knight Rises:* $448 million   **17.** Terminal Tower: 708 ft; Key Tower: 947 ft   **19.** *Steps 1 and 2:* variables, width; *Step 3:* $38 + y$, length, twice, $2y$, 188; length: 66 yd; width: 28 yd
21. (a) 45 units   (b) Do not produce—the product will lead to a loss.
23. quarters: 24; dimes: 15   **25.** 2 DVDs of *Iron Man 3;* 5 Blu-ray discs of *The Hunger Games: Catching Fire*   **27.** $2500 at 4%; $5000 at 5%
29. Madonna: $140; Bruce Springsteen and the E Street Band: $92
31. 40% solution: 80 L; 70% solution: 40 L   **33.** 30 lb at $6 per lb; 60 lb at $3 per lb   **35.** nuts: 40 lb; raisins: 20 lb   **37.** 60 mph; 50 mph
39. 35 mph; 65 mph   **41.** bicycle: 13.75 mph; car: 53.75 mph   **43.** plane: 470 mph; wind: 30 mph   **45.** boat: 10 mph; current: 2 mph
47. Yady: 17.5 mph; Dane: 12.5 mph

Section 4.5 (pages 315–316)

1. C   **3.** B

5. (a) no   (b) no   (c) yes   **7.** (a) no   (b) yes   (c) no

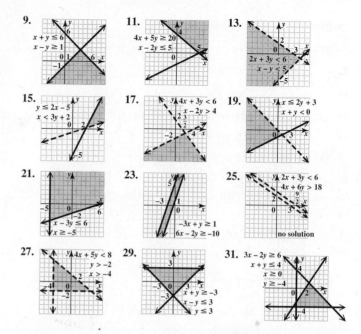

9. $x + y \leq 6$, $x - y \geq 1$ **11.** $4x + 5y \geq 20$, $x - 2y \leq 5$ **13.** $2x + 3y < 6$, $x - y < 5$ **15.** $y \leq 2x - 5$, $x < 3y + 2$ **17.** $4x + 3y < 6$, $x - 2y > 4$ **19.** $x \leq 2y + 3$, $x + y < 0$ **21.** $x - 3y \leq 6$, $x \geq -5$ **23.** $-3x + y \geq 1$, $6x - 2y \geq -10$ **25.** $2x + 3y < 6$, $4x + 6y > 18$; no solution   **27.** $4x + 5y < 8$, $y > -2$, $x > -4$ **29.** $x + y \geq -3$, $x - y \leq 3$, $y \leq 3$ **31.** $3x - 2y \geq 6$, $x + y \leq 4$, $x \geq 0$, $y \geq -4$

5   **EXPONENTS AND POLYNOMIALS**

Section 5.1 (pages 332–334)

1. false; $3^3 = 3 \cdot 3 \cdot 3 = 27$ **3.** true   **5.** w^6 **7.** $\left(\frac{1}{2}\right)^6$ **9.** $(-4)^4$
11. $(-7y)^4$ **13.** In $(-3)^4$, -3 is the base. In -3^4, 3 is the base. $(-3)^4 = 81$; $-3^4 = -81$ **15.** base: 3; exponent: 5; 243
17. base: -3; exponent: 5; -243 **19.** base: $-6x$; exponent: 4
21. base: x; exponent: 4   **23.** 2; 5; 8^7 **25.** 5^8 **27.** 4^{12} **29.** $(-7)^9$
31. t^{24} **33.** $-56r^7$ **35.** $42p^{10}$ **37.** $-30x^9$ **39.** The product rule does not apply.   **41.** The product rule does not apply.   **43.** 4^6 **45.** t^{20}
47. 7^3r^3 **49.** $5^5x^5y^5$ **51.** 5^{12} **53.** -8^{15} **55.** $8q^3r^3$ **57.** $\frac{9^8}{5^8}$
59. $\frac{1}{2^3}$ **61.** $\frac{a^3}{b^3}$ **63.** $\frac{x^3}{2^3}$ **65.** $-\frac{2^5x^5}{y^5}$ **67.** $\frac{5^5}{2^5}$ **69.** $\frac{9^5}{8^3}$ **71.** $2^{12}x^{12}$
73. -6^5p^5 **75.** $6^5x^{10}y^{15}$ **77.** x^{21} **79.** $4w^4x^{26}y^7$ **81.** $-r^{18}s^{17}$
83. $\frac{5^3a^6b^{15}}{c^{18}}$, or $\frac{125a^6b^{15}}{c^{18}}$ **85.** Using the product rule, the expression should be simplified as follows: $(10^2)^3 = 10^{2 \cdot 3} = 10^6$.   **87.** $12x^5$
89. $6p^7$ **91.** $125x^6$ **93.** $304.16   **95.** $1640.16

Section 5.2 (pages 340–342)

1. negative   **3.** negative   **5.** positive   **7.** 0   **9.** 1   **11.** 1   **13.** -1
15. -1 **17.** 0   **19.** 0   **21.** 0   **23.** 0   **25.** (a) B   (b) C   (c) D
(d) B   (e) E   (f) B   **27.** 2   **29.** $\frac{1}{64}$ **31.** 16   **33.** $\frac{49}{36}$ **35.** $\frac{1}{81}$
37. $\frac{8}{15}$ **39.** $-\frac{7}{18}$ **41.** $\frac{7}{2}$ **43.** 11; 8; 3; 125   **45.** 125   **47.** $\frac{125}{9}$ **49.** 25
51. x^{15} **53.** 216   **55.** $2r^4$ **57.** $\frac{25}{64}$ **59.** $\frac{p^5}{q^8}$ **61.** r^9 **63.** $\frac{yz^2}{4x^3}$
65. $a + b$ **67.** $(x + 2y)^2$ **69.** 343   **71.** $\frac{1}{x^2}$ **73.** $\frac{64x}{9}$ **75.** $\frac{x^2z^4}{y^2}$
77. $6x$ **79.** $\frac{1}{m^{10}n^5}$ **81.** $\frac{1}{xyz}$ **83.** x^3y^9 **85.** $\frac{1}{2r}$ **87.** $\frac{1}{-4y}$

89. The student attempted to use the quotient rule with unequal bases. The correct way to simplify is $\dfrac{16^3}{2^2} = \dfrac{(2^4)^3}{2^2} = \dfrac{2^{12}}{2^2} = 2^{10} = 1024$.

91. $\dfrac{a^{11}}{2b^5}$ **93.** $\dfrac{108}{y^5 z^3}$ **95.** $\dfrac{9z^2}{400x^3}$

Section 5.3 (pages 347–351)

1. (a) C **(b)** A **(c)** B **(d)** D **3.** in scientific notation **5.** not in scientific notation; 5.6×10^6 **7.** not in scientific notation; 8×10^1
9. not in scientific notation; 4×10^{-3} **11. (a)** 6; 4; 6.3; 4 **(b)** 5; 2; 5.71; -2 **13.** 5.876×10^9 **15.** 8.235×10^4 **17.** 7×10^{-6}
19. 2.03×10^{-3} **21.** -1.3×10^7 **23.** -6×10^{-3} **25.** 750,000
27. 5,677,000,000,000 **29.** 1,000,000,000,000 **31.** 6.21 **33.** 0.00078
35. 0.000000005134 **37.** -0.004 **39.** $-810,000$ **41. (a)** 6×10^{11}
(b) 600,000,000,000 **43. (a)** 1.5×10^7 **(b)** 15,000,000
45. (a) -6×10^4 **(b)** $-60,000$ **47. (a)** 2.4×10^2 **(b)** 240
49. (a) 6.3×10^{-2} **(b)** 0.063 **51. (a)** 3×10^{-4} **(b)** 0.0003
53. (a) -4×10 **(b)** -40 **55. (a)** 1.3×10^{-5} **(b)** 0.000013
57. (a) 5×10^2 **(b)** 500 **59. (a)** -3×10^6 **(b)** $-3,000,000$
61. (a) 2×10^{-7} **(b)** 0.0000002 **63.** 4.7E-7 **65.** 2E7 **67.** 1E1
69. 1.04×10^8 **71.** 9.2×10^{-3} **73.** 6×10^9 **75.** 0.000002
77. 4.2×10^{42} **79.** 1.5×10^{17} mi **81.** $3186 **83.** $53,185
85. 3.59×10^2, or 359 sec **87.** $98.28 **89.** 6×10^{17}, or
600,000,000,000,000,000; 3.6×10^{19}, or 36,000,000,000,000,000,000
91. The Chile earthquake was 10 times as intense as the Southern Sumatra earthquake. **92.** The Obihoro earthquake was 10 times as intense as the Hindu Kush earthquake. **93.** The Kamchatka earthquake was about 3.16 times as intense as the Southern Sumatra earthquake.
94. The Chile earthquake would be 100 times as intense.

Section 5.4 (pages 358–361)

1. 4; 6 **3.** 9 **5.** 19 **7.** 0 **9.** one; 6 **11.** one; 1 **13.** two; $-19, -1$
15. three; 1, 8, 5 **17.** $2m^5$ **19.** $-r^5$ **21.** It cannot be simplified.
23. $-5x^5$ **25.** $5p^9 + 4p^7$ **27.** 0 **29.** $-2xy^2$ **31.** $-\frac{22}{15}tu^7$
33. already simplified; 4; binomial **35.** $11m^4 - 7m^3 - 3m^2$; 4; trinomial
37. x^4; 4; monomial **39.** 7; 0; monomial **41.** $-13ab$; 2; monomial
43. (a) -3 **(b)** 0 **45. (a)** 14 **(b)** -19 **47. (a)** 36 **(b)** -12
49. $5x^2 - 2x$ **51.** $5m^2 + 3m + 2$ **53.** $\frac{7}{6}x^2 - \frac{2}{15}x + \frac{5}{6}$
55. $6m^3 + m^2 + 4m - 14$ **57.** $3y^3 - 11y^2$ **59.** $4x^4 - 4x^2 + 4x$
61. $15m^3 - 13m^2 + 8m + 11$ **63.** $5m^2 - 14m + 6$ **65.** $4x^3 + 2x^2 + 5x$
67. $-11y^4 + 8y^2 + y$ **69.** $a^4 - a^2 + 1$ **71.** $5m^2 + 8m - 10$
73. $-6x^2 - 12x + 12$ **75.** -10 **77.** $4b - 5c$ **79.** $6x - xy - 7$
81. $-3x^2y - 15xy - 3xy^2$ **83.** $8x^2 + 8x + 6$ **85.** $2x^2 + 8x$
87. $8t^2 + 8t + 13$ **89. (a)** $23y + 5t$ **(b)** $25°, 67°, 88°$ **91.** $-7x - 1$
93. $0, -3, -4, -3, 0$ **95.** $7, 1, -1, 1, 7$ **97.** $0, 3, 4, 3, 0$

99. 4, 1, 0, 1, 4

101. 63; If a dog is 9 in dog years, then it is 63 in human years. **102. (a)** 37 **(b)** 68 **(c)** 80
103. 2.5; 130 **104.** 6; $27

Section 5.5 (pages 366–369)

1. (a) B **(b)** D **(c)** A **(d)** C **3.** distributive **5.** one
7. $15y^{11}$ **9.** $30a^9$ **11.** $15pq^2$ **13.** $-18m^3n^2$ **15.** $9y^{10}$
17. $-8x^{10}$ **19.** $6m^2 + 4m$ **21.** $-6p^4 + 12p^3$
23. $-16z^2 - 24z^3 - 24z^4$ **25.** $6y^3 + 4y^4 + 10y^7$
27. $28r^5 - 32r^4 + 36r^3$ **29.** $6a^4 - 12a^3b + 15a^2b^2$
31. $3m^2$; $2mn$; $-n^3$; $21m^5n^2 + 14m^4n^3 - 7m^3n^5$
33. $12x^3 + 26x^2 + 10x + 1$ **35.** $72y^3 - 70y^2 + 21y - 2$
37. $20m^4 - m^3 - 8m^2 - 17m - 15$
39. $6x^6 - 3x^5 - 4x^4 + 4x^3 - 5x^2 + 8x - 3$ **41.** $5x^4 - 13x^3 +$
$20x^2 + 7x + 5$ **43.** $3x^5 + 18x^4 - 2x^3 - 8x^2 + 24x$ **45.** first row:
$x^2, 4x$; second row: $3x, 12$; Product: $x^2 + 7x + 12$ **47.** first row:
$2x^3, 6x^2, 4x$; second row: $x^2, 3x, 2$; Product: $2x^3 + 7x^2 + 7x + 2$
49. (a) $2p$; $3p$; $6p^2$ **(b)** $2p$; 7; $14p$ **(c)** -5; $3p$; $-15p$ **(d)** -5; 7;
-35 **(e)** $6p^2 - p - 35$ **51.** $m^2 + 12m + 35$ **53.** $n^2 + 3n - 4$
55. $12x^2 + 10x - 12$ **57.** $81 - t^2$ **59.** $9x^2 - 12x + 4$
61. $10a^2 + 37a + 7$ **63.** $12 + 8m - 15m^2$ **65.** $20 - 7x - 3x^2$
67. $3t^2 + 5st - 12s^2$ **69.** $8xy - 4x + 6y - 3$ **71.** $15x^2 + xy - 6y^2$
73. $6y^5 - 21y^4 - 45y^3$ **75.** $-200r^7 + 32r^3$ **77. (a)** $3y^2 + 10y + 7$
(b) $8y + 16$ **79.** $x^2 + 14x + 49$ **81.** $a^2 - 16$ **83.** $4p^2 - 20p + 25$
85. $25k^2 + 30kq + 9q^2$ **87.** $m^3 - 15m^2 + 75m - 125$
89. $8a^3 + 12a^2 + 6a + 1$ **91.** $-9a^3 + 33a^2 + 12a$
93. $56m^2 - 14m - 21$ **95.** $81r^4 - 216r^3s + 216r^2s^2 - 96rs^3 + 16s^4$
97. $6p^8 + 15p^7 + 12p^6 + 36p^5 + 15p^4$ **99.** $-24x^8 - 28x^7 +$
$32x^6 + 20x^5$ **101.** $6p^4 - \frac{5}{2}p^2q - \frac{25}{12}q^2$ **103.** $14x + 49$
105. $\pi x^2 - 9$ **107.** $30x + 60$ **108.** $30x + 60 = 600$; $\{18\}$
109. 10 ft by 60 ft **110.** 140 ft **111.** $450 **112.** $2870

Section 5.6 (pages 372–374)

1. (a) $4x$; $16x^2$ **(b)** $4x$; 3; $24x$ **(c)** 3; 9 **(d)** $16x^2 + 24x + 9$
3. $m^2 + 4m + 4$ **5.** $r^2 - 6r + 9$ **7.** $x^2 + 4xy + 4y^2$
9. $25p^2 + 20pq + 4q^2$ **11.** $16x^2 - 24x + 9$ **13.** $16a^2 + 40ab + 25b^2$
15. $36m^2 - \frac{48}{5}mn + \frac{16}{25}n^2$ **17.** $\frac{1}{4}x^2 + \frac{1}{3}x + \frac{1}{9}$ **19.** $2x^2 + 24x + 72$
21. $9t^3 - 6t^2 + t$ **23.** $48t^3 + 24t^2 + 3t$ **25.** $-16r^2 + 16r - 4$
27. $k^2 - 25$ **29.** $16 - 9t^2$ **31.** $25x^2 - 4$ **33.** $25y^2 - 9x^2$
35. $100x^2 - 9y^2$ **37.** $4x^4 - 25$ **39.** $\frac{9}{16} - x^2$ **41.** $81y^2 - \frac{4}{9}$
43. $25q^3 - q$ **45.** $-5a^2 + 5b^6$ **47.** $2k^2 - \frac{1}{2}$ **49.** $-x^2 + 1$
51. $x^3 + 3x^2 + 3x + 1$ **53.** $t^3 - 9t^2 + 27t - 27$ **55.** $r^3 + 15r^2 +$
$75r + 125$ **57.** $8a^3 + 12a^2 + 6a + 1$ **59.** $256x^4 - 256x^3 + 96x^2 -$
$16x + 1$ **61.** $81r^4 - 216r^3t + 216r^2t^2 - 96rt^3 + 16t^4$
63. $2x^4 + 6x^3 + 6x^2 + 2x$ **65.** $-4t^4 - 36t^3 - 108t^2 - 108t$
67. $x^4 - 2x^2y^2 + y^4$ **69.** no **71.** 9999 **73.** 39,999 **75.** $399\frac{3}{4}$

77. $\frac{1}{2}m^2 - 2n^2$ **79.** $9a^2 - 4$ **81.** $\pi x^2 + 4\pi x + 4\pi$
83. $x^3 + 6x^2 + 12x + 8$ **85.** $(a+b)^2$ **86.** a^2 **87.** $2ab$ **88.** b^2
89. $a^2 + 2ab + b^2$ **90.** They both represent the area of the entire large
square.   **91.** 1225   **92.** $30^2 + 2(30)(5) + 5^2$ **93.** 1225
94. They are equal.

Section 5.7 (pages 380–383)

1. $10x^2 + 8$; 2; $5x^2 + 4$ **3.** $5x^2 + 4$; 2 (These may be reversed.);
$10x^2 + 8$ **5.** $6p^4$; $18p^7$; $2p^2 + 6p^5$ **7.** $30x^3 - 10x + 5$
9. $4m^3 - 2m^2 + 1$ **11.** $4t^4 - 2t^2 + 2t$ **13.** $a^4 - a + \dfrac{2}{a}$
15. $-3p^2 - 2 + \dfrac{1}{p}$ **17.** $7r^2 - 6 + \dfrac{1}{r}$ **19.** $4x^3 - 3x^2 + 2x$
21. $-9x^2 + 5x + 1$ **23.** $2x + 8 + \dfrac{12}{x}$ **25.** $\dfrac{4x^2}{3} + x + \dfrac{2}{3x}$
27. $-27x^3 + 10x^2 + 4$ **29.** $9r^3 - 12r^2 + 2r + \dfrac{26}{3} - \dfrac{2}{3r}$
31. $-m^2 + 3m - \dfrac{4}{m}$ **33.** $-3a + 4 + \dfrac{5}{a}$ **35.** $\dfrac{12}{x} - \dfrac{6}{x^2} + \dfrac{14}{x^3} - \dfrac{10}{x^4}$
37. $6x^4y^2 - 4xy + 2xy^2 - x^4y$ **39.** $x + 2$ **41.** $2y - 5$
43. $p - 4 + \dfrac{44}{p+6}$ **45.** $6m - 1$ **47.** $2a - 14 + \dfrac{74}{2a+3}$
49. $4x^2 - 7x + 3$ **51.** $4k^3 - k + 2$ **53.** $5y^3 + 2y - 3 + \dfrac{-5}{y+1}$
55. $3k^2 + 2k - 2 + \dfrac{6}{k-2}$ **57.** $2p^3 - 6p^2 + 7p - 4 + \dfrac{14}{3p+1}$
59. $x^2 + 3x + 3$ **61.** $2x^2 - 6x + 19 + \dfrac{-55}{x+3}$ **63.** $r^2 + 2 + \dfrac{13}{r^2-4}$
65. $3x^2 + 3x - 1 + \dfrac{1}{x-1}$ **67.** $y^2 - 3y + 9$ **69.** $a^2 + 5$
71. $x^2 - 4x + 2 + \dfrac{9x-4}{x^2+3}$ **73.** $x^3 + 3x^2 - x + 5$
75. $\dfrac{3}{2}a - 10 + \dfrac{77}{2a+6}$ **77.** $x^2 + \dfrac{8}{3}x - \dfrac{1}{3} + \dfrac{4}{3x-3}$ **79.** $x^2 + x - 3$
81. $48m^2 + 96m + 24$ **83.** $5x^2 - 11x + 14$

6   FACTORING AND APPLICATIONS

Section 6.1 (pages 400–402)

1. product; mutliplying   **3.** 4   **5.** 4   **7.** 6   **9.** 1   **11.** 8   **13.** $10x^3$
15. xy^2 **17.** 6   **19.** $6m^3n^2$ **21.** factored   **23.** not factored
25. The correct factored form is $18x^3y^2 + 9xy = 9xy(2x^2y + 1)$. If a
polynomial has two terms, then the product of the factors must have two
terms. $9xy(2x^2y) = 18x^3y^2$ is just one term.   **27.** $3m^2$ **29.** $2z^4$
31. $2mn^4$ **33.** $y + 2$ **35.** $a - 2$ **37.** $2 + 3xy$ **39.** $x(x-4)$
41. $3t(2t + 5)$ **43.** $9m(3m^2 - 1)$ **45.** $m^2(m-1)$ **47.** $8z^2(2z^2 + 3)$
49. $-6x^2(2x + 1)$ **51.** $5y^6(13y^4 + 7)$ **53.** no common factor
(except 1)   **55.** $8mn^3(1 + 3m)$ **57.** $13y^2(y^6 + 2y^2 - 3)$
59. $-2x(2x^2 - 5x + 3)$ **61.** $9p^3q(4p^3 + 5p^2q^3 + 9q)$
63. $a^3(a^2 + 2b^2 - 3a^2b^2 + 4ab^3)$ **65.** $(x + 2)(c - d)$

67. $(m + 2n)(m + n)$ **69.** $(p - 4)(q^2 + 1)$ **71.** not in factored
form; $(7t + 4)(8 + x)$ **73.** in factored form   **75.** not in factored form
77. The student should factor out -2, instead of 2, in the second step to
obtain $x^2(x + 4) - 2(x + 4)$, which can be factored as $(x + 4)(x^2 - 2)$.
79. $(p + 4)(p + q)$ **81.** $(a - 2)(a + b)$ **83.** $(z + 2)(7z - a)$
85. $(3r + 2y)(6r - x)$ **87.** $(a^2 + b^2)(3a + 2b)$ **89.** $(3 - a)(4 - b)$
91. $(4m - p^2)(4m^2 - p)$ **93.** $(y + 3)(y + x)$ **95.** $(5 - 2p)(m + 3)$
97. $(3r + 2y)(6r - t)$ **99.** $(1 + 2b)(a^5 - 3)$
101. $8(2a + 5b^2)(a + b)$ **103.** $2p(p + q^2)(p - q)$

Section 6.2 (pages 406–408)

1. a and b must have different signs, one positive and one negative.
3. C   **5.** $a^2 + 13a + 36$ **7.** 1 and 48, -1 and -48, 2 and 24, -2 and
-24, 3 and 16, -3 and -16, 4 and 12, -4 and -12, 6 and 8, -6 and -8;
The pair with a sum of -19 is -3 and -16.   **9.** 1 and -24, -1 and 24,
2 and -12, -2 and 12, 3 and -8, -3 and 8, 4 and -6, -4 and 6; The pair
with a sum of -5 is 3 and -8.   **11.** 20; 12; table entries: 2, 2, 12, 4, 4,
9; 10 and 2; $(y + 10)(y + 2)$ **13.** $p + 6$ **15.** $x + 11$ **17.** $x - 8$
19. $y - 5$ **21.** $x + 11$ **23.** $y - 9$ **25.** $(y + 8)(y + 1)$
27. $(b + 3)(b + 5)$ **29.** $(m + 5)(m - 4)$ **31.** $(y - 5)(y - 3)$
33. prime   **35.** $(z - 7)(z - 8)$ **37.** $(r - 6)(r + 5)$
39. $(a + 4)(a - 12)$ **41.** prime   **43.** $(x + 16)(x - 2)$
45. $(r + 2a)(r + a)$ **47.** $(x + y)(x + 3y)$ **49.** $(t + 2z)(t - 3z)$
51. $(v - 5w)(v - 6w)$ **53.** $(m + 6n)(m - 2n)$ **55.** $(a - 6b)(a - 3b)$
57. $4(x + 5)(x - 2)$ **59.** $2t(t + 1)(t + 3)$ **61.** $2x^4(x - 3)(x + 7)$
63. $6z^2(z - 3)(z - 1)$ **65.** $5m^2(m^3 - 5m^2 + 8)$
67. $x(x - 4y)(x - 3y)$ **69.** $a^3(a + 4b)(a - b)$
71. $z^8(z - 7y)(z + 3y)$ **73.** $mn(m - 6n)(m - 4n)$
75. $yz(y + 3z)(y - 2z)$ **77.** $(a + b)(x + 4)(x - 3)$
79. $(2p + q)(r - 9)(r - 3)$

Section 6.3 (pages 414–416)

1. $(2t + 1)(5t + 2)$ **3.** $(3z - 2)(5z - 3)$ **5.** $(2s - t)(4s + 3t)$
7. (a) 2, 12, 24, 11   (b) 3, 8 (Order is irrelevant.)   (c) $3m$, $8m$
(d) $2m^2 + 3m + 8m + 12$ (e) $(2m + 3)(m + 4)$
(f) $(2m + 3)(m + 4) = 2m^2 + 8m + 3m + 12$, and combining like terms
gives the original trinomial $2m^2 + 11m + 12$.   **9.** B   **11.** B   **13.** A
15. $(4x + 4)$ cannot be a factor because its terms have a common factor
of 4, but those of the polynomial do not. The correct factored form is
$(4x - 3)(3x + 4)$.   **17.** $2a + 5b$ **19.** $x^2 + 3x - 4$; $x + 4$, $x - 1$, or
$x - 1$, $x + 4$ **21.** $2z^2 - 5z - 3$; $2z + 1$, $z - 3$, or $z - 3$, $2z + 1$
23. $(3a + 7)(a + 1)$ **25.** $(2y + 3)(y + 2)$ **27.** $(3m - 1)(5m + 2)$
29. $(3s - 1)(4s + 5)$ **31.** $(5m - 4)(2m - 3)$ **33.** $(4w - 1)(2w - 3)$
35. $(4y + 1)(5y - 11)$ **37.** prime   **39.** $2(5x + 3)(2x + 1)$
41. $3(4x - 1)(2x - 3)$ **43.** $-q(5m + 2)(8m - 3)$
45. $3n^2(5n - 3)(n - 2)$ **47.** $y^2(5x - 4)(3x + 1)$
49. $(5a + 3b)(a - 2b)$ **51.** $(4s + 5t)(3s - t)$
53. $m^4n(3m + 2n)(2m + n)$ **55.** prime   **57.** $(3x + 4)(x + 4)$

59. $-5x(2x + 7)(x - 4)$ **61.** prime   **63.** $(24y + 7x)(y - 2x)$
65. $(18x^2 - 5y)(2x^2 - 3y)$ **67.** $2(24a + b)(a - 2b)$
69. $x^2y^5(10x - 1)(x + 4)$ **71.** $4ab^2(9a + 1)(a - 3)$
73. $(12x - 5)(2x - 3)$ **75.** $(8x^2 - 3)(3x^2 + 8)$
77. $(4x + 3y)(6x + 5y)$ **79.** $-1(x + 7)(x - 3)$
81. $-1(3x + 4)(x - 1)$ **83.** $-1(a + 2b)(2a + b)$
85. $(m + 1)^3(5q - 2)(5q + 1)$ **87.** $(r + 3)^3(3x + 2y)^2$
89. $-4, 4$ **91.** $-11, -7, 7, 11$

Section 6.4 (pages 423–426)

1. $1; 4; 9; 16; 25; 36; 49; 64; 81; 100; 121; 144; 169; 196; 225; 256; 289;$
$324; 361; 400$ **3.** A, D   **5.** The binomial $4x^2 + 16$ can be factored
as $4(x^2 + 4)$. *After* any common factor is removed, a sum of squares
(like $x^2 + 4$ here) *cannot* be factored.   **7.** $(y + 5)(y - 5)$
9. $(x + 12)(x - 12)$ **11.** prime   **13.** prime   **15.** $4(m^2 + 4)$
17. $(3r + 2)(3r - 2)$ **19.** $4(3x + 2)(3x - 2)$
21. $(14p + 15)(14p - 15)$ **23.** $(4r + 5a)(4r - 5a)$
25. $(9x + 7y)(9x - 7y)$ **27.** $6(3x + y)(3x - y)$ **29.** prime
31. $(2 + x)(2 - x)$ **33.** $(6 + 5t)(6 - 5t)$ **35.** $x(x^2 + 4)$
37. $x^2(x + 1)(x - 1)$ **39.** $(p^2 + 7)(p^2 - 7)$
41. $(x^2 + 1)(x + 1)(x - 1)$ **43.** $(p^2 + 16)(p + 4)(p - 4)$
45. B, C   **47.** 10   **49.** 9   **51.** $(w + 1)^2$ **53.** $(x - 4)^2$ **55.** prime
57. $2(x + 6)^2$ **59.** $(2x + 3)^2$ **61.** $(4x - 5)^2$ **63.** $(7x - 2y)^2$
65. $(8x + 3y)^2$ **67.** $2(5h - 2y)^2$ **69.** $k(4k^2 - 4k + 9)$
71. $z^2(25z^2 + 5z + 1)$ **73.** $1; 8; 27; 64; 125; 216; 343; 512; 729; 1000$
75. C, D   **77.** **(a)** neither of these   **(b)** perfect cube   **(c)** perfect
square   **(d)** perfect square   **(e)** both of these   **(f)** perfect cube
79. $(a - 1)(a^2 + a + 1)$ **81.** $(m + 2)(m^2 - 2m + 4)$
83. $(y - 6)(y^2 + 6y + 36)$ **85.** $(k + 10)(k^2 - 10k + 100)$
87. $(3x - 4)(9x^2 + 12x + 16)$ **89.** $6(p + 1)(p^2 - p + 1)$
91. $5(x + 2)(x^2 - 2x + 4)$ **93.** $(y - 2x)(y^2 + 2yx + 4x^2)$
95. $2(x - 2y)(x^2 + 2xy + 4y^2)$ **97.** $(2p + 9q)(4p^2 - 18pq + 81q^2)$
99. $(3a + 4b)(9a^2 - 12ab + 16b^2)$
101. $(5t + 2s)(25t^2 - 10ts + 4s^2)$
103. $(2x - 5y^2)(4x^2 + 10xy^2 + 25y^4)$
105. $(3m^2 + 2n)(9m^4 - 6m^2n + 4n^2)$
107. $(x + y)(x^2 - xy + y^2)(x^6 - x^3y^3 + y^6)$ **109.** $\left(p + \frac{1}{3}\right)\left(p - \frac{1}{3}\right)$
111. $\left(6m + \frac{4}{5}\right)\left(6m - \frac{4}{5}\right)$ **113.** $(x + 0.8)(x - 0.8)$
115. $\left(t + \frac{1}{2}\right)^2$ **117.** $(x - 0.9)^2$ **119.** $\left(x + \frac{1}{2}\right)\left(x^2 - \frac{1}{2}x + \frac{1}{4}\right)$
121. $4mn$ **123.** $(m - p + 2)(m + p)$

Section 6.5 (pages 435–437)

1. $ax^2 + bx + c$ **3.** 0; zero-factor   **5.** the term with greatest degree is
greater than two (It is *cubic*.)   **7.** **(a)** linear   **(b)** quadratic
(c) quadratic   **(d)** linear   **9.** The variable x is another factor to set
equal to 0, so the solution set is $\left\{0, \frac{1}{7}\right\}$.   **11.** $\{-5, 2\}$ **13.** $\left\{3, \frac{7}{2}\right\}$

15. $\left\{-\frac{1}{2}, \frac{1}{6}\right\}$ **17.** $\left\{-\frac{5}{6}, 0\right\}$ **19.** $\left\{0, \frac{4}{3}\right\}$ **21.** $\{6\}$ **23.** $\{-2, -1\}$
25. $\{1, 2\}$ **27.** $\{-8, 3\}$ **29.** $\{-1, 3\}$ **31.** $\{-2, -1\}$ **33.** $\{-4\}$
35. $\left\{-2, \frac{1}{3}\right\}$ **37.** $\left\{-\frac{4}{3}, \frac{1}{2}\right\}$ **39.** $\left\{-\frac{2}{3}\right\}$ **41.** $\{-3, 3\}$ **43.** $\left\{-\frac{7}{4}, \frac{7}{4}\right\}$
45. $\{-11, 11\}$ **47.** $\{-6, 0\}$ **49.** $\{0, 7\}$ **51.** $\left\{0, \frac{1}{2}\right\}$ **53.** $\{2, 5\}$
55. $\left\{-4, \frac{1}{2}\right\}$ **57.** $\{-17, 4\}$ **59.** $\{-4, 12\}$ **61.** $\{-9, -2\}$
63. $\left\{-\frac{7}{3}, 0, \frac{7}{3}\right\}$ **65.** $\{-2, 0, 4\}$ **67.** $\{-5, 0, 4\}$ **69.** $\left\{0, \frac{1}{2}, 4\right\}$
71. $\{-3, 0, 5\}$ **73.** $\left\{-\frac{5}{2}, \frac{1}{3}, 5\right\}$ **75.** $\left\{-\frac{7}{2}, -3, 1\right\}$ **77.** $\{-1, 3\}$
79. $\{-1, 3\}$ **81.** $\{3\}$ **83.** $\left\{-\frac{2}{3}, 4\right\}$ **85.** $\left\{-\frac{4}{3}, -1, \frac{1}{2}\right\}$
87. **(a)** $64; 144; 4; 6$ **(b)** No time has elapsed, so the object hasn't
fallen (been released) yet.

Section 6.6 (pages 442–448)

1. Read; variable; equation; Solve; answer; Check, original
3. $\mathscr{A} = bh$; *Step 3:* $45 = (2x + 1)(x + 1)$; *Step 4:* $x = 4$ or $x = -\frac{11}{2}$;
Step 5: base: 9 units; height: 5 units; *Step 6:* $9 \cdot 5 = 45$
5. $\mathscr{A} = LW$; *Step 3:* $80 = (x + 8)(x - 8)$; *Step 4:* $x = 12$ or $x = -12$;
Step 5: length: 20 units; width: 4 units; *Step 6:* $20 \cdot 4 = 80$
7. length: 14 cm; width: 12 cm   **9.** base: 12 in.; height: 5 in.
11. height: 13 in.; width: 10 in.   **13.** length: 15 in.; width: 12 in.
15. mirror: 7 ft; painting: 9 ft   **17.** 20, 21   **19.** 0, 1, 2 or 7, 8, 9
21. $-3, -1$ or 7, 9   **23.** 7, 9, 11   **25.** $-2, 0, 2$ or 6, 8, 10   **27.** 12 cm
29. length: 20 in.; width: 15 in.; diagonal: 25 in.   **31.** 12 mi   **33.** 8 ft
35. 112 ft   **37.** 256 ft   **39.** **(a)** 1 sec   **(b)** $\frac{1}{2}$ sec and $1\frac{1}{2}$ sec   **(c)** 3 sec
(d) The negative solution, -1, does not make sense because t represents
time, which cannot be negative.   **41.** **(a)** 10   **(b)** 111 million; The
result obtained from the model is more than 109 million, the actual
number for 2000.   **(c)** 22   **(d)** 343 million; The result is more than
326 million, the actual number.   **(e)** 24   **(f)** 391 million

7   RATIONAL EXPRESSIONS AND APPLICATIONS

Section 7.1 (pages 464–467)

1. $3; 3; -5$ **3.** B, D   **5.** B   **7.** $-3; -3; -3; -6; \frac{3}{5}$
9. **(a)** $\frac{7}{10}$ **(b)** $\frac{8}{15}$ **11.** **(a)** 0   **(b)** -1 **13.** **(a)** $-\frac{64}{15}$ **(b)** undefined
15. **(a)** undefined   **(b)** $\frac{8}{25}$ **17.** **(a)** 0   **(b)** 0   **19.** **(a)** 0
(b) undefined   **21.** $x \neq 0$ **23.** $y \neq 0$ **25.** $x \neq 6$ **27.** $x \neq -\frac{5}{3}$
29. $m \neq -3, m \neq 2$ **31.** It is never undefined.   **33.** It is never
undefined.   **35.** numerator: $x^2, 4x$; denominator: $x, 4$ **37.** $3r^2$ **39.** $\frac{2}{5}$
41. $\frac{x - 1}{x + 1}$ **43.** $\frac{7}{5}$ **45.** $\frac{6}{7}$ **47.** $m - n$ **49.** $\frac{2}{t - 3}$ **51.** $\frac{3(2m + 1)}{4}$
53. $\frac{3m}{5}$ **55.** $\frac{3r - 2s}{3}$ **57.** $\frac{1}{x + 6}$ **59.** $\frac{x + 3}{x - 3}$ **61.** $\frac{13x}{7}$ **63.** $k - 3$
65. $\frac{x - 3}{x + 1}$ **67.** $\frac{x + 1}{x - 1}$ **69.** $\frac{x + 2}{x - 4}$ **71.** $-\frac{3}{7t}$ **73.** $\frac{z - 3}{z + 5}$

75. $\dfrac{r+s}{r-s}$ **77.** $\dfrac{a+b}{a-b}$ **79.** $\dfrac{m+n}{2}$ **81.** $\dfrac{x^2+1}{x}$ **83.** $1-p+p^2$

85. x^2+3x+9 **87.** $-\dfrac{b^2+ba+a^2}{a+b}$ **89.** $\dfrac{k^2-2k+4}{k-2}$ **91.** $\dfrac{z+3}{z}$

93. $\dfrac{1-2r}{2}$ **95.** -1 **97.** $-(m+1)$ **99.** -1 **101.** It is already in

lowest terms. **103.** -2

Answers may vary in Exercises 105, 107, and 109.

105. $\dfrac{-(x+4)}{x-3}, \dfrac{-x-4}{x-3}, \dfrac{x+4}{-(x-3)}, \dfrac{x+4}{-x+3}$ **107.** $\dfrac{-(2x-3)}{x+3},$

$\dfrac{-2x+3}{x+3}, \dfrac{2x-3}{-(x+3)}, \dfrac{2x-3}{-x-3}$ **109.** $\dfrac{-(3x-1)}{5x-6}, \dfrac{-3x+1}{5x-6}, \dfrac{3x-1}{-(5x-6)},$

$\dfrac{3x-1}{-5x+6}$ **111.** x^2+3 **113.** **(a)** 0 **(b)** 1.6 **(c)** 4.1

(d) The number of vehicles waiting also increases. **115.** Both yield

$2x+3$. **116.** Both yield $2x+1$. **117.** Both yield x^2+1.

118. Both yield x^2+2.

Section 7.2 (pages 472–474)

1. **(a)** B **(b)** D **(c)** C **(d)** A **3.** $\dfrac{3a}{2}$ **5.** $-\dfrac{4x^4}{3}$ **7.** $\dfrac{2}{c+d}$

9. $4(x-y)$ **11.** $\dfrac{t^2}{2}$ **13.** $\dfrac{x+3}{2x}$ **15.** $x-2$; 3; $x-2$; 5; $\dfrac{3}{4}$ **17.** 5

19. $-\dfrac{3}{2t^4}$ **21.** $\dfrac{1}{4}$ **23.** $-\dfrac{35}{8}$ **25.** $\dfrac{2(x+2)}{x(x-1)}$ **27.** $\dfrac{x(x-3)}{6}$

29. $\dfrac{5(x-4)}{x^2(x+4)}$ **31.** $\dfrac{-4t(t+1)}{t-1}$ **33.** $\dfrac{10}{9}$ **35.** $-\dfrac{3}{4}$ **37.** $-\dfrac{9}{2}$

39. $\dfrac{p+4}{p+2}$ **41.** -1 **43.** $-\dfrac{m+2}{m+1}$ **45.** $\dfrac{(2x-1)(x+2)}{x-1}$

47. $\dfrac{(k-1)^2}{(k+1)(2k-1)}$ **49.** $\dfrac{4k-1}{3k-2}$ **51.** $\dfrac{m+4p}{m+p}$ **53.** $\dfrac{m+6}{m+3}$

55. $\dfrac{y+3}{y+4}$ **57.** $\dfrac{m}{m+5}$ **59.** $\dfrac{r+6s}{r+s}$ **61.** $\dfrac{(q-3)^2(q+2)^2}{q+1}$

63. $\dfrac{3-a-b}{2a-b}$ **65.** $-\dfrac{(x+y)^2(x^2-xy+y^2)}{3y(y-x)(x-y)}$, or

$\dfrac{(x+y)^2(x^2-xy+y^2)}{3y(x-y)^2}$ **67.** $\dfrac{x+10}{10}$ **69.** $\dfrac{5xy^2}{4q}$

Section 7.3 (pages 477–480)

1. C **3.** C **5.** 5; 5; one; 5; 2; 5; 50 **7.** 60 **9.** 1800 **11.** x^5

13. $30p$ **15.** $180y^4$ **17.** $84r^5$ **19.** $15a^5b^3$ **21.** r^9t^3

23. $(x+1)(x-1)$ **25.** $12p(p-2)$ **27.** $28m^2(3m-5)$

29. $30(b-2)$ **31.** $18(r-2)$ **33.** $c-d$ or $d-c$ **35.** $m-3$ or $3-m$

37. $p-q$ or $q-p$ **39.** $2(x+1)(x-1)$ **41.** $3(x-4)^2$

43. $12p(p+5)^2$ **45.** $8(y+2)(y+1)$ **47.** $k(k+5)(k-2)$

49. $a(a+6)(a-3)$ **51.** $(p+3)(p+5)(p-6)$

53. $(k+3)(k-5)(k+7)(k+8)$ **55.** $\dfrac{20}{55}$ **57.** $\dfrac{-45}{9k}$

59. $\dfrac{60m^2k^3}{32k^4}$ **61.** $\dfrac{57z}{6z-18}$ **63.** $\dfrac{-4a}{18a-36}$ **65.** $\dfrac{6(k+1)}{k(k-4)(k+1)}$

67. $\dfrac{(t-r)(4r-t)}{t^3-r^3}$ **69.** $\dfrac{2y(z-y)(y-z)}{y^4-z^3y}$, or $\dfrac{-2y(y-z)^2}{y^4-z^3y}$

71. $\dfrac{36r(r+1)}{(r-3)(r+2)(r+1)}$ **73.** $\dfrac{ab(a+2b)}{2a^3b+a^2b^2-ab^3}$ **75.** 7

76. 1 **77.** identity property of multiplication **78.** 7 **79.** 1

80. identity property of multiplication

Section 7.4 (pages 486–489)

1. E **3.** C **5.** B **7.** G **9.** 2; 2; 5; 5; 10x; $\dfrac{57}{10x}$ **11.** $\dfrac{11}{m}$

13. $\dfrac{4}{y+4}$ **15.** 1 **17.** $\dfrac{m-1}{m+1}$ **19.** b **21.** x **23.** $y-6$ **25.** $\dfrac{1}{x-3}$

27. -1 **29.** $\dfrac{3z+5}{15}$ **31.** $\dfrac{10-7r}{14}$ **33.** $\dfrac{-3x-2}{4x}$ **35.** $\dfrac{61}{28t}$

37. $\dfrac{x+1}{2}$ **39.** $\dfrac{5x+9}{6x}$ **41.** $\dfrac{7-6p}{3p^2}$ **43.** $\dfrac{-k-8}{k(k+4)}$ **45.** $\dfrac{x+4}{x+2}$

47. $\dfrac{6m^2+23m-2}{(m+2)(m+1)(m+5)}$ **49.** $\dfrac{4y^2-y+5}{(y+1)^2(y-1)}$ **51.** $\dfrac{3}{t}$

53. $m-2$; $2-m$ **55.** $\dfrac{-2}{x-5}$, or $\dfrac{2}{5-x}$ **57.** -4 **59.** $\dfrac{-5}{x-y^2}$, or

$\dfrac{5}{y^2-x}$ **61.** $\dfrac{x+y}{5x-3y}$, or $\dfrac{-x-y}{3y-5x}$ **63.** $\dfrac{-6}{4p-5}$, or $\dfrac{6}{5-4p}$

65. 3 **67.** $\dfrac{-m-n}{2(m-n)}$ **69.** $\dfrac{-x^2+6x+11}{(x+3)(x-3)(x+1)}$

71. $\dfrac{-5q^2-13q+7}{(3q-2)(q+4)(2q-3)}$ **73.** $y-7$ **75.** $\dfrac{7x+31}{x+4}$

77. $\dfrac{-5x+13}{4x}$ **79.** $\dfrac{2x^2+6x}{(x-7)(x-3)}$, or $\dfrac{2x(x+3)}{(x-7)(x-3)}$

81. $\dfrac{2a+21}{3(a-2)}$ **83.** $\dfrac{x-8}{2(x-3)}$, or $\dfrac{8-x}{2(3-x)}$ **85.** $\dfrac{1}{x-2}$

87. $\dfrac{9r+2}{r(r+2)(r-1)}$ **89.** $\dfrac{2(x^2+3xy+4y^2)}{(x+y)(x+y)(x+3y)}$, or

$\dfrac{2(x^2+3xy+4y^2)}{(x+y)^2(x+3y)}$ **91.** $\dfrac{15r^2+10ry-y^2}{(3r+2y)(6r-y)(6r+y)}$

93. **(a)** $\dfrac{9k^2+6k+26}{5(3k+1)}$ **(b)** $\dfrac{1}{4}$ **95.** $\dfrac{10x}{49(101-x)}$

Section 7.5 (pages 494–497)

1. **(a)** 6; $\dfrac{1}{6}$ **(b)** 12; $\dfrac{3}{4}$ **(c)** $\dfrac{1}{6} \div \dfrac{3}{4}$ **(d)** $\dfrac{2}{9}$ **3.** Choice D is correct,

because every sign has been changed in the fraction. This means it

was multiplied by $\dfrac{-1}{-1}=1$. **5.** *Step 1:* $\dfrac{13}{20}, \dfrac{5}{6}$; *Step 2:* $\dfrac{13}{20} \div \dfrac{5}{6}$;

Step 3: $\dfrac{13}{20} \cdot \dfrac{6}{5}, \dfrac{39}{50}$ **7.** -6 **9.** $\dfrac{1}{xy}$ **11.** $\dfrac{2a^2b}{3}$ **13.** $\dfrac{m(m+2)}{3(m-4)}$

15. $\dfrac{2}{x}$ **17.** $\dfrac{8}{x}$ **19.** $\dfrac{a^2-5}{a^2+1}$ **21.** $\dfrac{31}{50}$ **23.** $\dfrac{y^2+x^2}{xy(y-x)}$

25. $\dfrac{40-12p}{85p}$, or $\dfrac{4(10-3p)}{85p}$ **27.** $\dfrac{5y-2x}{3+4xy}$ **29.** $\dfrac{a-2}{2a}$ **31.** $\dfrac{z-5}{4}$

33. $\dfrac{-m}{m+2}$ **35.** $\dfrac{x+8}{-x+7}$ **37.** $\dfrac{x^2+y^2}{x^2-y^2}$, or $\dfrac{x^2+y^2}{(x+y)(x-y)}$

39. $\dfrac{ab}{a+b}$ **41.** $\dfrac{3m(m-3)}{(m-1)(m-8)}$ **43.** $\dfrac{2x-7}{3x+1}$ **45.** $\dfrac{y+4}{y-8}$

47. $\dfrac{30}{(a+b)(a-b)}$ **49.** $\dfrac{x+y}{x^2+xy+y^2}$ **51.** $\dfrac{x-3}{x-5}$ **53.** $\dfrac{5}{3}$ **55.** $\dfrac{13}{2}$

57. $\dfrac{19r}{15}$ **59.** $\dfrac{\frac{3}{8}+\frac{5}{6}}{2}$ **60.** $\dfrac{29}{48}$ **61.** $\dfrac{29}{48}$ **62.** Answers will vary.

Section 7.6 (pages 505–508)

1. 12 **3.** xyz **5.** Yes, it is acceptable because $\dfrac{1}{3-x}$ is equivalent to

$\dfrac{-1}{x-3}$. **7.** expression; $\dfrac{43}{40}x$ **9.** equation; $\left\{\dfrac{40}{43}\right\}$ **11.** expression; $-\dfrac{1}{10}x$

13. equation; $\{-10\}$ **15.** equation; $\{0\}$ **17.** $x \neq -2, 0$

19. $x \neq -3, 4, -\dfrac{1}{2}$ **21.** $x \neq -9, 1, -2, 2$ **23.** $\left\{\dfrac{1}{4}\right\}$ **25.** $\left\{-\dfrac{3}{4}\right\}$

27. $\{-15\}$ **29.** $\{7\}$ **31.** $\{-15\}$ **33.** $\{-5\}$ **35.** $\{-6\}$ **37.** $\varnothing$

39. $\{5\}$ **41.** $\{4\}$ **43.** $\{5\}$ **45.** $\left\{x \,|\, x \neq \pm\dfrac{4}{3}\right\}$ **47.** $\{1\}$ **49.** $\{4\}$

51. $\{5\}$ **53.** $\{-4\}$ **55.** $\{-2, 12\}$ **57.** $\varnothing$ **59.** $\{3\}$ **61.** $\{3\}$

63. $\{-3\}$ **65.** $\left\{-\dfrac{1}{5}, 3\right\}$ **67.** $\left\{-\dfrac{1}{2}, 5\right\}$ **69.** $\{3\}$ **71.** $\left\{-\dfrac{1}{3}, 3\right\}$

73. $\{-1\}$ **75.** $\{-6\}$ **77.** $\left\{-6, \dfrac{1}{2}\right\}$ **79.** $\{6\}$ **81.** $\{0\}$

83. $F = \dfrac{ma}{k}$ **85.** $a = \dfrac{kF}{m}$ **87.** $R = \dfrac{E-Ir}{I}$, or $R = \dfrac{E}{I} - r$

89. $\mathcal{A} = \dfrac{h(B+b)}{2}$ **91.** $a = \dfrac{2S - ndL}{nd}$, or $a = \dfrac{2S}{nd} - L$ **93.** $y = \dfrac{xz}{x+z}$

95. $t = \dfrac{rs}{rs - 2s - 3r}$, or $t = \dfrac{-rs}{-rs + 2s + 3r}$ **97.** $z = \dfrac{3y}{5 - 9xy}$, or

$z = \dfrac{-3y}{9xy - 5}$ **99.** $t = \dfrac{2x-1}{x+1}$, or $t = \dfrac{-2x+1}{-x-1}$ **101. (a)** -3 **(b)** -1

(c) $-3, -1$ **102.** $\dfrac{15}{2x}$ **103.** If $x = 0$, the divisor R is equal to 0, and

division by 0 is undefined. **104.** $(x+3)(x+1)$

105. $\dfrac{7}{x+1}$ **106.** $\dfrac{11x+21}{4x}$ **107.** $\varnothing$ **108.** We know that -3 is not

allowed, because P and R are undefined for $x = -3$.

Section 7.7 (pages 515–518)

1. into a headwind: $(m-5)$ mph; with a tailwind: $(m+5)$ mph

3. $\dfrac{1}{10}$ job per hr **5. (a)** the amount **(b)** $5+x$ **(c)** $\dfrac{5+x}{6} = \dfrac{13}{3}$

7. x represents the original numerator; $\dfrac{x+3}{(x-4)+3} = \dfrac{3}{2}$; $\dfrac{9}{5}$

9. x represents the original numerator; $\dfrac{x+2}{3x-2} = 1$; $\dfrac{2}{6}$

11. x represents the number; $\dfrac{1}{6}x = x + 5$; -6

13. x represents the quantity; $x + \dfrac{3}{4}x + \dfrac{1}{2}x + \dfrac{1}{3}x = 93$; 36

15. 18.809 min **17.** 331.763 m per min **19.** 3.565 hr

21. $\dfrac{500}{x-10} = \dfrac{600}{x+10}$ **23.** 8 mph **25.** 32 mph **27.** 165 mph

29. 3 mph **31.** 18.5 mph **33.** $\dfrac{1}{8}t + \dfrac{1}{6}t = 1$, or $\dfrac{1}{8} + \dfrac{1}{6} = \dfrac{1}{t}$

35. $2\dfrac{2}{5}$ hr **37.** $5\dfrac{5}{11}$ hr **39.** 3 hr **41.** $2\dfrac{7}{10}$ hr **43.** $9\dfrac{1}{11}$ min

Section 7.8 (pages 522–525)

1. direct **3.** direct **5.** inverse **7.** inverse **9.** inverse **11.** direct
13. direct **15.** inverse **17. (a)** increases **(b)** decreases **19.** 9
21. 250 **23.** 6 **25.** 21 **27.** $\dfrac{16}{5}$ **29.** $\dfrac{4}{9}$ **31.** $35.20 **33.** 15 in.^{2}
35. $42\dfrac{2}{3}$ in. **37.** $106\dfrac{2}{3}$ mph **39.** $12\dfrac{1}{2}$ amps **41.** 20 lb
43. 52.817 in.^{2} **45.** $14\dfrac{22}{27}$ footcandles

8 ROOTS AND RADICALS

Section 8.1 (pages 542–546)

1. true **3.** false; Zero has only one square root. **5.** true **7.** a must be
positive. **9.** a must be negative. **11.** $-3, 3$ **13.** $-8, 8$ **15.** $-13, 13$
17. $-\dfrac{5}{14}, \dfrac{5}{14}$ **19.** $-30, 30$ **21.** 1 **23.** 7 **25.** 10 **27.** -4
29. -16 **31.** $-\dfrac{12}{11}$ **33.** 0.8 **35.** It is not a real number. **37.** It is not
a real number. **39.** 19 **41.** 19 **43.** $\dfrac{2}{3}$ **45.** $3x^2 + 4$ **47.** rational; 5
49. irrational; 5.385 **51.** rational; -8 **53.** irrational; -17.321
55. It is not a real number. **57.** irrational; 34.641 **59.** 9 and 10
61. 7 and 8 **63.** -7 and -6 **65.** 4 and 5 **67.** C **69.** $c = 17$
71. $b = 8$ **73.** $c \approx 11.705$ **75.** 24 cm **77.** 80 ft **79.** 195 ft
81. 11.1 ft **83.** 158.6 ft **85.** 9.434 **87. (a)** 3 units, 4 units
(b) If we let $a = 3$, $b = 4$, and $c = 5$, then the Pythagorean theorem
is satisfied because $3^2 + 4^2 = 5^2$ is a true statement. **89.** The area of
the square on the left is c^2. The small square inside that figure has area
$(b-a)^2 = b^2 - 2ba + a^2$. The sum of the areas of the two rectangles
that measure a by b in the figure on the right is $2ab$. Because the areas of
the two figures are the same, we have $c^2 = 2ab + b^2 - 2ba + a^2$, which
is equivalent to $c^2 = a^2 + b^2$. **91.** 5 **93.** 13 **95.** $\sqrt{13}$ **97.** $\sqrt{2}$
99. 1; 8; 27; 64; 125; 216; 343; 512; 729; 1000 **101.** 1 **103.** 5
105. 9 **107.** -3 **109.** -6 **111.** 2 **113.** 3 **115.** 5 **117.** 6
119. It is not a real number. **121.** -3 **123.** 2 **125.** -4

Section 8.2 (pages 553–555)

1. false; $\sqrt{(-6)^2} = \sqrt{36} = 6$ **3.** false; $2\sqrt{7}$ represents the product of
2 and $\sqrt{7}$. **5.** A **7.** $\sqrt{15}$ **9.** $\sqrt{22}$ **11.** $\sqrt{42}$ **13.** $\sqrt{81}$, or 9
15. 13 **17.** $\sqrt{13r}$ **19.** $3\sqrt{5}$ **21.** $2\sqrt{6}$ **23.** $3\sqrt{10}$ **25.** $5\sqrt{3}$
27. $5\sqrt{5}$ **29.** It cannot be simplified. **31.** $4\sqrt{10}$ **33.** $-10\sqrt{7}$
35. $9\sqrt{3}$ **37.** $25\sqrt{2}$ **39.** $5\sqrt{10}$ **41.** $6\sqrt{2}$ **43.** $9\sqrt{2}$
45. $2\sqrt{17}$ **47.** $12\sqrt{2}$ **49.** $3\sqrt{6}$ **51.** 24 **53.** $6\sqrt{10}$ **55.** $12\sqrt{5}$
57. $30\sqrt{5}$ **59.** $\dfrac{4}{15}$ **61.** $\dfrac{\sqrt{7}}{4}$ **63.** $\dfrac{\sqrt{2}}{5}$ **65.** 5 **67.** $6\sqrt{5}$ **69.** $\dfrac{25}{4}$
71. m **73.** y^2 **75.** $6z$ **77.** $20x^3$ **79.** $3x^4\sqrt{2}$ **81.** $3c^7\sqrt{5}$

83. $z^2\sqrt{z}$ **85.** $a^6\sqrt{a}$ **87.** $8x^3\sqrt{x}$ **89.** x^3y^6 **91.** $9m^2n$ **93.** $\dfrac{\sqrt{7}}{x^5}$

95. $\dfrac{y^2}{10}$ **97.** $\dfrac{x^2y^3}{13}$ **99.** $2\sqrt[3]{5}$ **101.** $3\sqrt[3]{2}$ **103.** $4\sqrt[3]{2}$ **105.** $2\sqrt[4]{5}$

107. $\dfrac{2}{3}$ **109.** $-\dfrac{6}{5}$ **111.** p **113.** x^3 **115.** $4z^2$ **117.** $7a^3b$

119. $2t\sqrt[3]{2t^2}$ **121.** $\dfrac{m^4}{2}$ **123.** 6 cm **125.** 6 in. **127.** D

Section 8.3 (pages 558–559)

1. radicand; index; like; 7; 7 **3. (a)** like; Both are square roots and have the same radicand, 6. **(b)** unlike; The radicands are different—one is 3 and one is 2. **(c)** unlike; The indexes are different—one is a square root and one is a cube root. **(d)** like; Both are square roots and have the same radicand, $2x$. **(e)** unlike; The radicands are different—one is $3y$ and one is $6y$. **(f)** like; $\sqrt[3]{2ba} = \sqrt[3]{2ab}$, so both are cube roots of the same radicand, $2ab$. **5.** $7\sqrt{3}$ **7.** $-5\sqrt{7}$ **9.** $2\sqrt{6}$ **11.** $3\sqrt{17}$ **13.** $4\sqrt{3}$ **15.** These unlike radicals cannot be added using the distributive property. **17.** $7\sqrt{3}$ **19.** $8\sqrt{3}$ **21.** $-20\sqrt{2}$ **23.** $4\sqrt{2}$ **25.** $3\sqrt{3} - 2\sqrt{5}$ **27.** $-\sqrt[3]{2}$ **29.** $24\sqrt[3]{3}$ **31.** $3\sqrt{21}$ **33.** $5\sqrt{3}$ **35.** $19\sqrt{7}$ **37.** $12\sqrt{6} + 6\sqrt{5}$ **39.** $-2\sqrt{2} - 12\sqrt{3}$ **41.** $10\sqrt[4]{2} + 4\sqrt[4]{8}$ **43.** $22\sqrt{2}$ **45.** $\sqrt{2x}$ **47.** $7\sqrt{3r}$ **49.** $15x\sqrt{3}$ **51.** $2x\sqrt{2}$ **53.** $13p\sqrt{3}$ **55.** $42x\sqrt{5z}$ **57.** $6k^2h\sqrt{6} + 27hk\sqrt{6k}$ **59.** $6\sqrt[3]{p^2}$ **61.** $21\sqrt[4]{m^3}$ **63.** $-8p\sqrt[4]{p}$ **65.** $-24z\sqrt[3]{4z}$ **67.** 0

Section 8.4 (pages 563–565)

1. 1; identity property of multiplication **3.** $\dfrac{6\sqrt{5}}{5}$ **5.** $\sqrt{5}$ **7.** $\dfrac{2\sqrt{6}}{3}$ **9.** $\dfrac{8\sqrt{15}}{5}$ **11.** $\dfrac{\sqrt{30}}{2}$ **13.** $\dfrac{8\sqrt{3}}{9}$ **15.** $\dfrac{3\sqrt{2}}{10}$ **17.** $\sqrt{2}$ **19.** $\sqrt{2}$ **21.** $\dfrac{2\sqrt{30}}{3}$ **23.** $\dfrac{\sqrt{2}}{8}$ **25.** $\dfrac{3\sqrt{5}}{5}$ **27.** $\dfrac{-3\sqrt{2}}{10}$ **29.** $\dfrac{21\sqrt{5}}{5}$ **31.** $\dfrac{\sqrt{3}}{3}$ **33.** $-\dfrac{\sqrt{5}}{5}$ **35.** $\dfrac{\sqrt{65}}{5}$ **37.** $\dfrac{\sqrt{21}}{3}$ **39.** $\dfrac{3\sqrt{14}}{4}$ **41.** $\dfrac{1}{6}$ **43.** 1 **45.** $\dfrac{\sqrt{15}}{10}$ **47.** $\dfrac{17\sqrt{2}}{6}$ **49.** $\dfrac{\sqrt{3}}{5}$ **51.** $\dfrac{4\sqrt{3}}{27}$ **53.** $\dfrac{\sqrt{6p}}{p}$ **55.** $\dfrac{\sqrt{3y}}{y}$ **57.** $\dfrac{4\sqrt{m}}{m}$ **59.** $\dfrac{p\sqrt{3q}}{q}$ **61.** $\dfrac{x\sqrt{7xy}}{y}$ **63.** $\dfrac{p\sqrt{2pm}}{m}$ **65.** $\dfrac{a\sqrt{6ab}}{6}$ **67.** $\dfrac{3a\sqrt{5r}}{5}$ **69.** B **71.** $\dfrac{\sqrt[3]{4}}{2}$ **73.** $\dfrac{\sqrt[3]{2}}{4}$ **75.** $\dfrac{\sqrt[3]{121}}{11}$ **77.** $\dfrac{\sqrt[3]{50}}{5}$ **79.** $\dfrac{\sqrt[3]{196}}{7}$ **81.** $\dfrac{\sqrt[3]{6y}}{2y}$ **83.** $\dfrac{\sqrt[3]{42mn^2}}{6n}$ **85.** $\dfrac{\sqrt[4]{2}}{2}$ **87. (a)** $\dfrac{9\sqrt{2}}{4}$ sec **(b)** 3.182 sec

Section 8.5 (pages 570–572)

1. 13 **3.** 4 **5.** Multiplication must be performed before addition, so it is incorrect to add -37 and -2. Because $-2\sqrt{15}$ cannot be simplified further, the expression cannot be written in a simpler form. The final answer is $-37 - 2\sqrt{15}$. **7.** $\sqrt{15} - \sqrt{35}$ **9.** $30 + 2\sqrt{10}$ **11.** $4\sqrt{7}$

13. $57 + 23\sqrt{6}$ **15.** $71 - 16\sqrt{7}$ **17.** $37 + 12\sqrt{7}$ **19.** $7 + 2\sqrt{6}$ **21.** $a + 2\sqrt{a} + 1$ **23.** $49 + 14\sqrt{x} + x$ **25.** $187 - 20\sqrt{21}$ **27.** 23 **29.** 1 **31.** 2 **33.** $y - 10$ **35.** $2\sqrt{3} - 2 + 3\sqrt{2} - \sqrt{6}$ **37.** $15\sqrt{2} - 15$ **39.** $\sqrt{30} + \sqrt{15} + 6\sqrt{5} + 3\sqrt{10}$ **41.** $81 + 14\sqrt{21}$ **43.** $6t - 3\sqrt{14t} + 2\sqrt{7t} - 7\sqrt{2}$ **45.** $3m - 2n$ **47. (a)** $\sqrt{5} - \sqrt{3}$ **(b)** $\sqrt{6} + \sqrt{5}$ **49.** $-2 + \sqrt{5}$ **51.** $-2 - \sqrt{11}$ **53.** $3 - \sqrt{3}$ **55.** $\dfrac{-3 + 5\sqrt{3}}{11}$ **57.** $\dfrac{\sqrt{6} + \sqrt{2} + 3\sqrt{3} + 3}{2}$ **59.** $\dfrac{-6\sqrt{2} + 12 + \sqrt{10} - 2\sqrt{5}}{2}$ **61.** $\dfrac{-4\sqrt{3} - \sqrt{2} + 10\sqrt{6} + 5}{23}$ **63.** $\sqrt{21} + \sqrt{14} + \sqrt{6} + 2$ **65.** $-\sqrt{10} + \sqrt{15}$ **67.** $3 - \sqrt{3}$ **69.** $\dfrac{8(4 + \sqrt{x})}{16 - x}$ **71.** $\dfrac{\sqrt{x} - \sqrt{y}}{x - y}$ **73.** $\sqrt{7} - 2$ **75.** $\dfrac{\sqrt{3} + 5}{4}$ **77.** $\dfrac{6 - \sqrt{10}}{2}$ **79.** $\dfrac{2 + \sqrt{2}}{3}$ **81.** $2 - 3\sqrt[3]{4}$ **83.** $12 + 10\sqrt[4]{8}$ **85.** $-1 + 3\sqrt[3]{2} - \sqrt[3]{4}$ **87.** 1 **89.** 4 in. **91.** $30 + 18x$ **92.** They are not like terms. **93.** $30 + 18\sqrt{5}$ **94.** They are not like radicals. **95.** Make the first term $30x$, so that $30x + 18x = 48x$. Make the first term $30\sqrt{5}$, so that $30\sqrt{5} + 18\sqrt{5} = 48\sqrt{5}$. **96.** Both like terms and like radicals are combined by adding their numerical coefficients. The variables in like terms are replaced by radicals in like radicals.

Section 8.6 (pages 579–582)

1. squaring; squared; original **3.** $\{49\}$ **5.** $\{7\}$ **7.** $\{85\}$ **9.** $\{-45\}$ **11.** $\left\{-\dfrac{3}{2}\right\}$ **13.** $\varnothing$ **15.** $\{121\}$ **17.** $\{8\}$ **19.** $\{1\}$ **21.** $\{6\}$ **23.** $\varnothing$ **25.** $\{5\}$ **27.** $\{7\}$ **29.** $\{6\}$ **31.** $\varnothing$ **33.** When the left side is squared, the result should be $x - 1$, not $-(x - 1)$. The correct solution set is $\{17\}$. **35.** $\{-2, 1\}$ **37.** $\{12\}$ **39.** $\{-2, -1\}$ **41.** $\{11\}$ **43.** $\{-1, 3\}$ **45.** $\{9\}$ **47.** $\{8\}$ **49.** $\{9\}$ **51.** $\{2, 11\}$ **53.** $\{2\}$ **55.** $\{9\}$ **57.** $\{4, 20\}$ **59.** $\{-5\}$ **61.** $\left\{-\dfrac{2}{3}\right\}$ **63.** $\{-1, 8\}$ **65.** $\left\{\dfrac{4}{3}, 2\right\}$ **67.** $\{-27, 3\}$ **69.** 21 **71.** 8 **73. (a)** 70.5 mph **(b)** 59.8 mph **(c)** 53.9 mph **75.** yes; 26 mi

9 **QUADRATIC EQUATIONS**

Section 9.1 (pages 596–598)

1. B, C **3.** C **5.** D **7.** $\{-7, 8\}$ **9.** $\{\pm 11\}$ **11.** $\left\{-\dfrac{5}{3}, 6\right\}$ **13.** $\{\pm 9\}$ **15.** $\{\pm\sqrt{14}\}$ **17.** $\{\pm 4\sqrt{3}\}$ **19.** no real solution **21.** $\left\{\pm\dfrac{5}{2}\right\}$ **23.** $\{\pm 1.5\}$ **25.** $\{\pm\sqrt{3}\}$ **27.** $\left\{\pm\dfrac{2\sqrt{7}}{7}\right\}$ **29.** $\{\pm 2\sqrt{6}\}$ **31.** $\left\{\pm\dfrac{2\sqrt{5}}{5}\right\}$ **33.** $\{\pm 3\sqrt{3}\}$ **35.** $\{\pm 2\sqrt{2}\}$ **37.** According to the square root property, -9 is also a solution, so her answer was not completely correct. The solution set is $\{\pm 9\}$.

39. $\{-2, 8\}$ **41.** $\{8 \pm 3\sqrt{3}\}$ **43.** no real solution **45.** $\{-3, \frac{5}{3}\}$

47. $\{0, \frac{3}{2}\}$ **49.** $\left\{\frac{5 \pm \sqrt{30}}{2}\right\}$ **51.** $\left\{\frac{-1 \pm 3\sqrt{2}}{3}\right\}$

53. $\{-10 \pm 4\sqrt{3}\}$ **55.** $\{0, \frac{1}{4}\}$ **57.** $\{-\frac{1}{3}, 1\}$ **59.** $\left\{\frac{-1 \pm \sqrt{3}}{4}\right\}$

61. $\left\{\frac{1 \pm 4\sqrt{3}}{4}\right\}$ **63.** If Jeff's solutions are multiplied by $\frac{-1}{-1}$, which equals 1, then they will have the same forms as Linda's. Both versions are correct. **65.** $\{-4.48, 0.20\}$ **67.** $\{-3.09, -0.15\}$ **69.** $\frac{1}{2}$ sec

71. 9 in. **73.** 2%

Section 9.2 (pages 604–606)

1. D **3.** 25; $(x + 5)^2$ **5.** 100; $(z - 10)^2$ **7.** 1; $(x + 1)^2$

9. $\frac{25}{4}$; $\left(p - \frac{5}{2}\right)^2$ **11.** $\frac{1}{16}$; $\left(x + \frac{1}{4}\right)^2$ **13.** 0.04; $(x - 0.2)^2$

15. 4; 2; 4; 4; $(x + 2)^2 = 5$; $\{-2 \pm \sqrt{5}\}$ **17.** $\{1, 3\}$

19. $\{-1 \pm \sqrt{6}\}$ **21.** $\{4 \pm 2\sqrt{3}\}$ **23.** $\{-3\}$ **25.** $\{0, 16\}$

27. $\left\{\frac{3 \pm \sqrt{13}}{2}\right\}$ **29.** $\left\{\frac{-3 \pm \sqrt{5}}{2}\right\}$ **31.** $\{-\frac{3}{2}, \frac{1}{2}\}$

33. no real solution **35.** $\left\{\frac{9 \pm \sqrt{21}}{6}\right\}$ **37.** $\left\{\frac{-7 \pm \sqrt{97}}{6}\right\}$

39. $\{-4, 2\}$ **41.** $\{4 \pm \sqrt{3}\}$ **43.** $\{1 \pm \sqrt{6}\}$ **45.** $\{-\frac{11}{5}, 1\}$

47. $\{-\frac{7}{3}, 6\}$ **49. (a)** $\left\{\frac{3 \pm 2\sqrt{6}}{3}\right\}$ **(b)** $\{-0.633, 2.633\}$

51. (a) $\{-2 \pm \sqrt{3}\}$ **(b)** $\{-3.732, -0.268\}$

53. 3 sec and 5 sec **55.** 1 sec and 5 sec **57.** 75 ft by 100 ft

59. 8 mi **61.** x^2 **62.** $x^2 + 8x$ **63.** $x^2 + 8x + 16$

64. It occurred when we added the 16 squares.

Section 9.3 (pages 611–612)

1. E **3.** A **5.** D **7.** $a = 3, b = 4, c = -8$ **9.** $a = -8, b = -2,$
$c = -3$ **11.** $a = 3, b = -4, c = -2$ **13.** $a = 3, b = 7, c = 0$

15. $a = 1, b = 1, c = -12$ **17.** $a = 9, b = 9, c = -26$

19. $2a$ should be the denominator for $-b$ as well. The correct formula is

$$x = \frac{-b \pm \sqrt{b^2 - 4ac}}{2a}.$$ **21.** $\{-13, 1\}$ **23.** $\left\{\frac{-6 \pm \sqrt{26}}{2}\right\}$

25. $\{2\}$ **27.** $\{-1, \frac{5}{2}\}$ **29.** $\{-1, 0\}$ **31.** $\{0, \frac{12}{7}\}$

33. $\{\pm 2\sqrt{6}\}$ **35.** $\{\pm\frac{2}{5}\}$ **37.** $\left\{\frac{6 \pm 2\sqrt{6}}{3}\right\}$

39. no real solution **41.** no real solution **43.** $\left\{\frac{-5 \pm \sqrt{61}}{2}\right\}$

45. (a) $\left\{\frac{-1 \pm \sqrt{11}}{2}\right\}$ **(b)** $\{-2.158, 1.158\}$

47. (a) $\left\{\frac{1 \pm \sqrt{5}}{2}\right\}$ **(b)** $\{-0.618, 1.618\}$ **49.** $\{-3.6, 0.6\}$

51. $\{-1.8, 0.3\}$ **53.** $\{-0.8, 0.3\}$ **55.** $\{-\frac{2}{3}, \frac{4}{3}\}$ **57.** $\left\{\frac{-1 \pm \sqrt{73}}{6}\right\}$

59. no real solution **61.** $\{1 \pm \sqrt{2}\}$ **63.** $\{-1, \frac{5}{2}\}$

65. $\left\{\frac{-3 \pm \sqrt{29}}{2}\right\}$ **67.** $r = \frac{-\pi h \pm \sqrt{\pi^2 h^2 + \pi S}}{\pi}$ **69.** 3.5 ft

71. $\{16, -8\}$; Only 16 ft is a reasonable answer.

Section 9.4 (pages 618–619)

1. upward **3.** downward

5. $(0, -6)$ **7.** $(0, 2)$ **9.** $(-3, 0)$

11. $(-1, 2)$ **13.** $(4, 0)$ **15.** $(3, 4)$

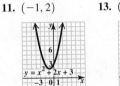

17. $(-2, -4)$ **19.** $(0, 1)$; $(0, 1)$; no x-intercepts **21.** $(0, 2)$; $(0, 2)$; $(-1.4, 0)$ and $(1.4, 0)$

23. $(-2, 0)$; $(0, 4)$; $(-2, 0)$ **25.** $(2, -1)$; $(0, 3)$; $(1, 0)$ and $(3, 0)$ **27.** 40 and 40 **29.** $y = \frac{11}{5625}x^2$

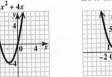

GLOSSARY

For a more complete discussion, see the section(s) in parentheses.

A

absolute value   The absolute value of a number is the distance between 0 and the number on a number line. (Section 1.3)

addends   In the addition $x + y$, the addends are x and y. (Section 1.4)

additive inverse (opposite)   The additive inverse of a number x, symbolized $-x$, is the number that is the same distance from 0 on the number line as x, but on the opposite side of 0. The number 0 is its own additive inverse. For all real numbers x, $x + (-x) = (-x) + x = 0$. (Section 1.3)

algebraic expression   An algebraic expression is a sequence of constants, variables, operation symbols, and/or grouping symbols (such as parentheses) formed according to the rules of algebra. (Section 1.2)

area   The area of a plane (two-dimensional) geometric figure is the measure (in square units) of the surface covered by the figure. (Section 2.5)

axis of symmetry (axis)   The axis of symmetry of a parabola is the vertical line or horizontal line (depending on the orientation of the graph) through the vertex of the parabola. (Sections 5.4, 9.4)

B

base (in an exponential expression)   The base in an exponential expression is the expression that is the repeated factor. In b^x, b is the base. (Sections 1.1, 5.1)

base (in percents)   The base in a percent equation is the whole amount being considered. (Section 2.6)

binomial   A binomial is a polynomial consisting of exactly two terms. (Section 5.4)

boundary line   In the graph of a linear inequality, the boundary line separates the region that satisfies the inequality from the region that does not satisfy the inequality. (Section 3.6)

C

circle graph (pie chart)   A circle graph (or pie chart) is a circle divided into sectors, or wedges, whose sizes show the relative magnitudes of the categories of data being represented. (Section R.1)

coefficient   (See **numerical coefficient.**)

common factor   An integer that is a factor of two or more integers is a common factor of those integers. (Section 6.1)

complementary angles (complements)   Complementary angles are two angles whose measures have a sum of 90°. (Section 2.4)

complex fraction   A complex fraction is a quotient with one or more fractions in the numerator, denominator, or both. (Section 7.5)

components   In an ordered pair (x, y), x and y are the components. (Section 3.7)

composite number   A natural number greater than 1 that is not prime is a composite number. It is composed of prime factors represented in one and only one way. (Section R.1)

conditional equation   A conditional equation is only true under certain conditions. (Section 2.3)

conjugate   The conjugate of $a + b$ is $a - b$. (Sections 5.6, 8.5)

consecutive even (or odd) integers   Two consecutive even integers, such as 4 and 6, differ by 2. Two consecutive odd integers, such as 3 and 5, also differ by 2. (Sections 2.4, 6.6)

consecutive integers   Two integers that differ by 1 are consecutive integers. (Sections 2.4, 6.6)

consistent system   A system of equations with a solution is a consistent system. (Section 4.1)

constant   A fixed, unchanging number is a constant. (Section 1.2)

constant of variation   In the equation $y = kx$ or $y = \frac{k}{x}$, the nonzero real number k is the constant of variation. (Section 7.8)

contradiction   A contradiction is an equation that has no solution. (Section 2.3)

coordinate on a number line   Every point on a number line is associated with a unique real number, which is the coordinate of the point. (Section 1.3)

coordinates of a point in a plane   The numbers in an ordered pair are called the coordinates of the corresponding point in the plane. (Section 3.1)

cross products   The cross products in the proportion $\frac{a}{b} = \frac{c}{d}$ are ad and bc. (Section 2.6)

cube root   A number b is a cube root of a if $b^3 = a$ is true. (Section 8.1)

D

decimal   A number written with place values as powers of ten, often using a decimal point, is a decimal number (Section R.2)

decimal places   In a decimal number, each digit occupies a decimal place that is a power of ten. (Section R.2)

degree   A degree is a basic unit of measure for angles in which one degree (1°) is $\frac{1}{360}$ of a complete revolution. (Section 2.4)

degree of a polynomial   The degree of a polynomial is the greatest degree of any of the terms in the polynomial. (Section 5.4)

degree of a term   The degree of a term is the sum of the exponents on the variables in the term. (Section 5.4)

denominator   The number below the fraction bar in a fraction is the denominator. It indicates the number of equal parts in a whole. (Section R.1)

dependent equations   Equations of a system that have the same graph (because they are different forms of the same equation) are dependent equations. (Section 4.1)

descending powers   A polynomial in one variable is written in descending powers of the variable if the exponents on the variables of the terms of the polynomial decrease from left to right. (Section 5.4)

difference   The result of subtracting two numbers is their difference. (Sections R.1, 1.4)

direct variation y varies directly as x if there exists a nonzero real number (constant) k such that $y = kx$. (Section 7.8)

discriminant   The discriminant of the quadratic equation $ax^2 + bx + c = 0$ is the quantity $b^2 - 4ac$ under the radical in the quadratic formula. (Section 9.3)

dividend   In the quotient $\frac{a}{b}$, the dividend is a. (Sections R.1, R.2, 1.5)

divisor   In the quotient $\frac{a}{b}$, the divisor is b. (Sections R.1, R.2, 1.5)

domain   The set of all first components (x-values) in the ordered pairs of a relation is the domain. (Section 3.7)

double solution   A double solution of a quadratic equation is a solution that appears twice in the solution process but represents only one distinct value. (Sections 6.5, 9.3)

E

elements   The elements of a set are the objects that belong to the set. (Section 1.2)

empty set (null set)   The empty set, denoted by $\{\ \}$ or $\varnothing$, is the set containing no elements. (Section 2.3)

equation   An equation is a statement that two algebraic expressions are equal. (Sections 1.2, 2.1)

equivalent equations   Equivalent equations are equations that have the same solution set. (Section 2.1)

exponent (power)   An exponent, or power, is a number that indicates how many times its base is used as a factor. In b^x, x is the exponent (power). (Sections 1.1, 5.1)

exponential expression   A number or letter (variable) written with an exponent is an exponential expression. (Sections 1.1, 5.1)

extraneous solution   A proposed solution to an equation that does not satisfy the original equation is an extraneous solution. (Sections 7.6, 8.6)

extremes of a proportion   In the proportion $\frac{a}{b} = \frac{c}{d}$, the a- and d-terms are the extremes. (Section 2.6)

F

factor   If a, b, and c represent numbers and $a \cdot b = c$, then a and b are factors of c. (Sections R.1, 1.5, 6.1)

factored form   An expression is in factored form when it is written as a product. (Section 6.1)

FOIL method   FOIL is a mnemonic device that represents a method for multiplying two binomials $(a + b)(c + d)$. Multiply **F**irst terms ac, **O**uter terms ad, **I**nner terms bc, and **L**ast terms bd. Then combine like terms. (Section 5.5)

formula   A formula is an equation in which variables are used to describe a relationship among several quantities. (Section 2.5)

fourth root   A number b is a fourth root of a if $b^4 = a$ is true. (Section 8.1)

fraction   A number expressed as a quotient of integers in the form $\frac{a}{b}$ is a fraction. (Section R.1)

function   A function is a set of ordered pairs (x, y) in which each value of the first component x corresponds to exactly one value of the second component y. (Section 3.7)

G

graph of a number   The point on a number line that corresponds to a number is its graph. (Section 1.3)

graph of an equation   The graph of an equation in two variables is the set of all points that correspond to all of the ordered pairs that satisfy the equation. (Section 3.2)

greatest common factor (GCF)   The greatest common factor of a list of integers is the largest factor of all those integers. The greatest common factor of the terms of a polynomial is the largest factor of all the terms in the polynomial. (Sections R.1, 6.1)

H

horizontal line   In a plane, a horizontal line is parallel to the y-axis and has no x-intercept. (Section 3.2)

hypotenuse   The side opposite the right angle in a right triangle is the longest side and is the hypotenuse. (Section 6.6)

I

identity   An identity is an equation that is true for all valid replacements of the variable. It has an infinite number of solutions. (Section 2.3)

identity element for addition   For all real numbers a, $a + 0 = 0 + a = a$. The number 0 is the identity element for addition. (Section 1.6)

identity element for multiplication   For all real numbers a, $a \cdot 1 = 1 \cdot a = a$. The number 1 is the identity element for multiplication. (Section 1.6)

improper fraction   A fraction with numerator greater than or equal to denominator is an improper fraction. Its value is greater than or equal to 1. (Section R.1)

inconsistent system   An inconsistent system of equations is a system with no solution. (Section 4.1)

independent equations   Equations of a system that have different graphs are independent equations. (Section 4.1)

index (order)   In a radical of the form $\sqrt[n]{a}$, n is the index or order. (Section 8.1)

inequality   An inequality is a statement that two expressions may not be equal. (Sections 1.1, 2.8)

inner product   When using the FOIL method to multiply two binomials $(a + b)(c + d)$, the inner product is bc. (Section 5.5)

integers $\{\ldots, -2, -1, 0, 1, 2, \ldots\}$ is the set of integers. (Section 1.3)

interval   An interval is a portion of a number line. (Section 2.8)

inverse variation y varies inversely as x if there exists a nonzero real number (constant) k such that $y = \frac{k}{x}$. (Section 7.8)

irrational number   An irrational number cannot be written as the quotient of two integers, but can be represented by a point on a number line. (Sections 1.3, 8.1)

L

least common denominator (LCD)   Given several denominators, the least multiple that is divisible by all the denominators is the least common denominator. (Sections R.1, 7.3)

legs (of a right triangle)   The two shorter perpendicular sides of a right triangle are the legs. (Section 6.6)

like radicals   Like radicals are multiples of the same root of the same number or expression. (Section 8.3)

like terms   Terms with exactly the same variables raised to exactly the same powers are like terms. (Sections 1.7, 5.4)

line graph   A line graph is a series of line segments in two dimensions that connect points representing data. (Section 3.1)

linear equation in one variable   A linear equation in one variable (here x) can be written in the form $Ax + B = C$, where A, B, and C are real numbers and $A \neq 0$. (Section 2.1)

linear equation in two variables   A linear equation in two variables (here x and y) is an equation that can be written in the form $Ax + By = C$, where A, B, and C are real numbers and A and B are not both 0. (Section 3.1)

linear inequality in one variable   A linear inequality in one variable (here x) can be written in the form $Ax + B < C$, $Ax + B \leq C$, $Ax + B > C$, or $Ax + B \geq C$, where A, B, and C are real numbers and $A \neq 0$. (Section 2.8)

linear inequality in two variables   A linear inequality in two variables (here x and y) can be written in the form $Ax + By < C$, $Ax + By \leq C$, $Ax + By > C$, or $Ax + By \geq C$, where A, B, and C are real numbers and A and B are not both 0. (Section 3.6)

lowest terms A fraction is in lowest terms if the greatest common factor of the numerator and denominator is 1. (Sections R.1, 7.1)

M

means of a proportion In the proportion $\frac{a}{b} = \frac{c}{d}$, the b- and c-terms are the means. (Section 2.6)

minuend In the subtraction $x - y$, the minuend is x. (Section 1.4)

mixed number A mixed number includes a whole number and a fraction written together and is understood to be the sum of the whole number and the fraction. (Section R.1)

monomial A monomial is a polynomial consisting of exactly one term. (Section 5.4)

multiplicative inverse (reciprocal) The multiplicative inverse (reciprocal) of a nonzero number x, symbolized $\frac{1}{x}$, is the real number which has the property that the product of the two numbers is 1. For all nonzero real numbers x, $\frac{1}{x} \cdot x = x \cdot \frac{1}{x} = 1$. (Sections R.1, 1.5)

N

natural (counting) numbers The set of natural numbers is the set of numbers $\{1, 2, 3, 4, \ldots\}$. (Sections R.1, 1.3)

negative number A negative number is located to the left of 0 on a number line. (Section 1.3)

negative square root For even indexes, the symbols $-\sqrt{\ }$ $-\sqrt[4]{\ }$, $-\sqrt[6]{\ }, \ldots, -\sqrt[n]{\ }$ are used for negative roots. (Section 8.1)

number line A line that has a point designated to correspond to the real number 0, and a standard unit chosen to represent the distance between 0 and 1, is a number line. All real numbers correspond to one and only one number on such a line. (Section 1.3)

numerator The number above the fraction bar in a fraction is the numerator. It shows how many of the equivalent parts are being considered. (Section R.1)

numerical coefficient (coefficient) The numerical factor in a term is the numerical coefficient, or simply the coefficient. (Sections 1.7, 5.4)

O

ordered pair An ordered pair is a pair of numbers written within parentheses in the form (x, y). (Section 3.1)

origin The point at which the x-axis and y-axis of a rectangular coordinate system intersect is the origin. (Section 3.1)

outer product When using the FOIL method to multiply two binomials $(a + b)(c + d)$, the outer product is ad. (Section 5.5)

P

parabola The graph of a second-degree (quadratic) equation in two variables is a parabola. (Sections 5.4, 9.4)

parallel lines Parallel lines are two lines in the same plane that never intersect. (Section 3.3)

percent Percent, written with the symbol %, means per one hundred. (Sections R.2, 2.6)

percentage A percentage is a part of a whole. (Section 2.6)

perfect cube A perfect cube is a number with a rational cube root. (Sections 6.4, 8.1)

perfect fourth power A perfect fourth power is a number with a rational fourth root. (Section 8.1)

perfect square A perfect square is a number with a rational square root. (Sections 6.4, 8.1)

perfect square trinomial A perfect square trinomial is a trinomial that can be factored as the square of a binomial. (Section 6.4)

perimeter The perimeter of a plane (two-dimensional) geometric figure is the measure of the outer boundary of the figure. For a polygon (e.g., a rectangle, square, or triangle), it is the sum of the lengths of the sides. (Section 2.5)

perpendicular lines Perpendicular lines are two lines that intersect to form a right (90°) angle. (Section 3.3)

plane In a rectangular coordinate system, the x- and y-axes form a plane—a flat surface illustrated by a sheet of paper. (Section 3.1)

plot To plot an ordered pair is to locate it on a rectangular coordinate system. (Section 3.1)

polynomial A polynomial is a term or a finite sum of terms in which all coefficients are real, all variables have whole number exponents, and no variables appear in denominators. (Section 5.4)

positive number A positive number is located to the right of 0 on a number line. (Section 1.3)

positive (principal) square root For even indexes, the symbols $\sqrt{\ }$, $\sqrt[4]{\ }$, $\sqrt[6]{\ }, \ldots, \sqrt[n]{\ }$ are used for positive roots, which are the principal roots. (Section 8.1)

prime number A natural number greater than 1 is prime if it has only 1 and itself as factors. (Section R.1)

prime polynomial A prime polynomial is a polynomial that cannot be factored into factors having only integer coefficients. (Section 6.2)

product The result of multiplying two numbers is their product. (Sections R.1, 1.5)

product of the sum and difference of two terms The product of the sum and difference of two terms is the difference of the squares of the terms, or $(x + y)(x - y) = x^2 - y^2$. (Section 5.6)

proper fraction A fraction with numerator less than denominator is a proper fraction. Its value is less than 1. (Section R.1)

proportion A proportion is a statement that two ratios are equal. (Section 2.6)

proposed solution A value that appears as an apparent solution after a rational or radical equation has been solved according to standard methods is a proposed solution for the original equation. It may or may not be an actual solution and must be checked. (Sections 7.6, 8.6)

Q

quadrant A quadrant is one of the four regions in the plane determined by the axes in a rectangular coordinate system. (Section 3.1)

quadratic equation A quadratic equation (in x here) is an equation that can be written in the form $ax^2 + bx + c = 0$, where a, b, and c are real numbers, with $a \neq 0$. (Sections 6.5, 9.1)

quadratic formula The quadratic formula is a general formula used to solve a quadratic equation of the form $ax^2 + bx + c = 0$, where $a \neq 0$. It is $x = \dfrac{-b \pm \sqrt{b^2 - 4ac}}{2a}$. (Section 9.3)

quotient The result of dividing two numbers is their quotient. (Sections R.1, 1.5)

R

radical An expression consisting of a radical symbol, root index, and radicand is a radical. (Section 8.1)

radical equation A radical equation is an equation with a variable in at least one radicand. (Section 8.6)

radical expression A radical expression is an algebraic expression that contains a radical. (Section 8.1)

radicand   The number or expression under a radical symbol is the radicand. (Section 8.1)

range   The set of all second components (*y*-values) in the ordered pairs of a relation is the range. (Section 3.7)

ratio   A ratio is a comparison of two quantities using a quotient. (Section 2.6)

rational expression   The quotient of two polynomials with denominator not 0 is a rational expression. (Section 7.1)

rational numbers   Rational numbers can be written as the quotient of two integers, with denominator not 0. (Section 1.3)

real numbers   Real numbers include all numbers that can be represented by points on a number line—that is, all rational and irrational numbers. (Section 1.3)

reciprocal (See **multiplicative inverse.**)

rectangular (Cartesian) coordinate system   The *x*-axis and *y*-axis placed at a right angle at their zero points form a rectangular coordinate system. It is also called the Cartesian coordinate system. (Section 3.1)

relation   A relation is a set of ordered pairs. (Section 3.7)

repeating decimal   A decimal number that does not have a final digit (such as 0.333 . . .) is a repeating decimal. (Section R.2)

right angle   A right angle measures 90°. (Section 2.4)

rise   Rise refers to the vertical change between two points on a line—that is, the change in *y*-values. (Section 3.3)

run   Run refers to the horizontal change between two points on a line—that is, the change in *x*-values. (Section 3.3)

S

scatter diagram   A scatter diagram is a graph of ordered pairs of data. (Section 3.1)

scientific notation   A number is written in scientific notation when it is expressed in the form $a \times 10^n$, where $1 \le |a| < 10$ and *n* is an integer. (Section 5.3)

set   A set is a collection of objects. (Section 1.2)

signed numbers   Signed numbers are numbers that can be written with a positive or negative sign. (Section 1.3)

slope   The ratio of the vertical change in *y* to the horizontal change in *x* for any two points on a line is the slope of the line. (Section 3.3)

solution of an equation   A solution of an equation is any replacement for the variable that makes the equation true. (Sections 1.2, 2.1)

solution of a system   A solution of a system of equations is an ordered pair (x, y) that makes all equations true at the same time. (Section 4.1)

solution set   The set of all solutions of an equation is the solution set. (Section 2.1)

solution set of a system of linear equations   The set of all ordered pairs that satisfy all equations of a linear system at the same time is the solution set. (Section 4.1)

solution set of a system of linear inequalities   The set of all ordered pairs that satisfy all inequalities of a linear system true at the same time is the solution set. (Section 4.5)

square root   The inverse of squaring a number is taking its square root. That is, a number *a* is a square root of *k* if $a^2 = k$ is true. (Section 8.1)

standard notation   If a decimal number is written in its usual expanded form showing all decimal places (in contrast to *scientific notation*), it is in standard form. (Section 5.3)

straight angle   A straight angle measures 180°. (Section 2.4)

subtrahend   In the subtraction $x - y$, the subtrahend is *y*. (Section 1.4)

sum   The result of adding two numbers is their sum. (Sections R.1, 1.4)

supplementary angles (supplements)   Supplementary angles are two angles whose measures have a sum of 180°. (Section 2.4)

system of linear equations (linear system)   A system of linear equations consists of two or more linear equations to be solved at the same time. (Section 4.1)

system of linear inequalities   A system of linear inequalities consists of two or more linear inequalities to be solved at the same time. (Section 4.5)

T

table of values   A table of values is an organized way of displaying ordered pairs. (Section 3.1)

term   A term is a number, a variable, or the product or quotient of a number and one or more variables raised to powers. (Sections 1.7, 5.4)

terminating decimal   A decimal number that has a final digit (such as 5 in 0.25) is a terminating decimal. (Section R.2)

terms of a proportion   The terms of the proportion $\frac{a}{b} = \frac{c}{d}$ are *a*, *b*, *c*, and *d*. (Section 2.6)

three-part inequality   An inequality that says that one number is between two other numbers is a three-part inequality. (Section 2.8)

trinomial   A trinomial is a polynomial consisting of exactly three terms. (Section 5.4)

U

unlike radicals   Radicals that have different root indexes and/or different radicands are unlike radicals. (Section 8.3)

unlike terms   Unlike terms are terms that do not have the same variable, or terms with the same variables but whose variables are not raised to the same powers. (Sections 1.7, 5.4)

V

variable   A variable is a symbol, usually a letter, used to represent an unknown number. (Section 1.2)

vertex   The point on a parabola that has the least *y*-value (if the parabola opens up) or the greatest *y*-value (if the parabola opens down) is the vertex of the parabola. (Sections 5.4, 9.4)

vertical angles   When two intersecting lines are drawn, the angles that lie opposite each other have the same measure and are vertical angles. (Section 2.5)

vertical line   In a plane, a vertical line is parallel to the *x*-axis and has no *y*-intercept. (Section 3.2)

volume   The volume (in cubic units) of a three-dimensional figure is a measure of the space occupied by the figure. (Section 2.5)

W

whole numbers   The set of whole numbers is {0, 1, 2, 3, 4, . . .}. (Sections R.1, 1.3)

X

x-axis   The horizontal number line in a rectangular coordinate system is the *x*-axis. (Section 3.1)

x-intercept   A point where a graph intersects the *x*-axis is an *x*-intercept. (Section 3.2)

Y

y-axis   The vertical number line in a rectangular coordinate system is the *y*-axis. (Section 3.1)

y-intercept   A point where a graph intersects the *y*-axis is a *y*-intercept. (Section 3.2)

PHOTO CREDITS

FRONT MATTER
p. **vT** Margaret L. Lial; p. **vB** John Hornsby

CHAPTER R
p. **11** Cristovao/Shutterstock; p. **15** Ryan McVay/Stockbyte/Getty Images; p. **26** WavebreakMediaMicro/Fotolia

CHAPTER 1
p. **27** Michael Flippo/Fotolia; p. **35** Monkey Business/Fotolia; p. **42** Monkey Business/Fotolia; p. **50** Kenishirotie/Fotolia; p. **51** Gstockstudio/Fotolia; p. **62** Orhan Cam/Shutterstock; p. **63** Shefkate/Fotolia; p. **65** GaryLHampton/iStockPhoto; p. **75** Anjelagr/Fotolia; p. **80** Terry McGinnis; p. **94** James Thew/Fotolia; p. **102** Sascha Burkard/Fotolia

CHAPTER 2
p. **103** Huaxiadragon/Fotolia; p. **111** Alex_SK/Fotolia; p. **130** PCN Photography/Alamy; p. **138** Emkaplin/Fotolia; p. **139T** DenisNata/Fotolia; p. **139B** Les Cunliffe/Fotolia; p. **153** Jaimie Duplass/Fotolia; **159T** Xtr2007/Fotolia; p. **159BL** Chokkicx/iStock/360/Getty Images; p. **159BR** Petro Feketa/Fotolia; p. **161** Sumire8/Fotolia; p. **163** Stuart Jenner/Shutterstock; p. **170** Wavebreakmedia/Shutterstock; p. **172** Christopher Nolan/Fotolia; p. **181** Kali9/Getty Images; p. **185** Micromonkey/Fotolia; p. **186** Rob/Fotolia; p. **188** Studio DMM Photography/Designs & Art/Shutterstock

CHAPTER 3
p. **199** Eppic/Fotolia; p. **200** ArenaCreative/Fotolia; p. **204** KingPhoto/Fotolia; p. **206** Felix Mizioznikov/Fotolia; p. **210** Andresr/Shutterstock; p. **219** Lofoto/Shutterstock; p. **248** Karen Roach/Fotolia; p. **251** Armadillo Stock/Shutterstock

CHAPTER 4
p. **277** Geewhiz/Fotolia; p. **293** Vadim Ponomarenko/Fotolia; p. **299** Sashkin/Fotolia; p. **301** Julijah/Fotolia; p. **302** ZanyZeus/Shutterstock; p. **308TL** Scott Rothstein/Shutterstock; p. **308TR** Hemera Technologies/Thinkstock; p. **308B** James Steidl/Shutterstock; p. **310** Dusan Jankovic/Shutterstock; p. **320** Boleslaw Kubica/Shutterstock

CHAPTER 5
p. **325** Sara Piaseczynski; p. **334** Goodluz/Fotolia; p. **350T** Romario Ien/Fotolia; p. **350BL** Mystique/Fotolia; p. **350BR** Rafael Ramirez Lee/Shutterstock; p. **361** Brodetskaya Elena/Fotolia; p. **390** James Thew/Fotolia

CHAPTER 6
p. **393** Image Asset Management Ltd./Alamy; p. **429** Nickolae/Fotolia; p. **437** 4745052183/Shutterstock; p. **442** Gino Santa Maria/Fotolia; p. **447** Philipp Nemenz/Getty Images

CHAPTER 7
p. **457** SMI/Newscom; p. **489** Denis Tabler/Fotolia; p. **513** Rafael Ben-Ari/Fotolia; p. **514** Johnny Habell/Shutterstock; p. **516** Alma_sacra/Fotolia; p. **519** John Hornsby; p. **520** Joggie Botma/Shutterstock; p. **523** Richard Coencas/Shutterstock; p. **524T** Specta/Shutterstock; p. **524B** Harvey Schwartz/ZUMA Press/Alamy

CHAPTER 8
p. **535** Danny E Hooks/Shutterstock; p. **565** NASA; p. **572** Cristovao31/Fotolia; p. **581** Tim Large/Shutterstock; p. **587** James Harrison/Shutterstock; p. **588** Thinkstock

CHAPTER 9
p. **591** Badmanproduction/Fotolia; p. **596** Sablin/Fotolia

INDEX

Triangles and Angles

Right Triangle
Triangle has one 90° (right) angle.

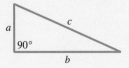

Pythagorean Theorem
(for right triangles)
$a^2 + b^2 = c^2$

Right Angle
Measure is 90°.

Isosceles Triangle
Two sides are equal.

$AB = BC$

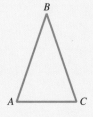

Straight Angle
Measure is 180°.

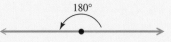

Equilateral Triangle
All sides are equal.

$AB = BC = CA$

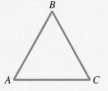

Complementary Angles
The sum of the measures of two complementary angles is 90°.

Angles ① and ② are complementary.

Sum of the Angles of Any Triangle
$A + B + C = 180°$

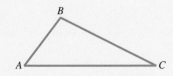

Supplementary Angles
The sum of the measures of two supplementary angles is 180°.

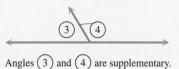

Angles ③ and ④ are supplementary.

Similar Triangles
Corresponding angles are equal. Corresponding sides are proportional.

$A = D, B = E, C = F$

$\dfrac{AB}{DE} = \dfrac{AC}{DF} = \dfrac{BC}{EF}$

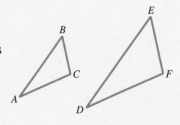

Vertical Angles
Vertical angles have equal measures.

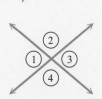

Angle ① = Angle ③

Angle ② = Angle ④

Formulas

Figure	*Formulas*	*Illustration*
Square	Perimeter: $P = 4s$ Area: $\mathcal{A} = s^2$	
Rectangle	Perimeter: $P = 2L + 2W$ Area: $\mathcal{A} = LW$	
Triangle	Perimeter: $P = a + b + c$ Area: $\mathcal{A} = \dfrac{1}{2}bh$	
Parallelogram	Perimeter: $P = 2a + 2b$ Area: $\mathcal{A} = bh$	
Trapezoid	Perimeter: $P = a + b + c + B$ Area: $\mathcal{A} = \dfrac{1}{2}h(b + B)$	
Circle	Diameter: $d = 2r$ Circumference: $C = 2\pi r$ $C = \pi d$ Area: $\mathcal{A} = \pi r^2$	